The TTL Data Book
Volume 3

Department of Physics
Randolph-Macon College
Ashland, Virginia

| General Information | 1 |

| ALS and AS Circuits | 2 |

| Mechanical Data | 3 |

Department of Physics
Randolph-Macon College
Ashland, Virginia

The TTL Data Book

Volume 3

Texas Instruments

IMPORTANT NOTICE

Texas Instruments reserves the right to make changes at any time in order to improve design and to supply the best product possible.

TI cannot assume any responsibility for any circuits shown or represent that they are free from patent infringement.

Information contained herein supersedes data published in the *ALS/AS Logic Circuits Data Book 1983*, SDAD001.

ISBN 0-89512-153-0
Library of Congress No. 82-074480

Copyright © 1984 Texas Instruments Incorporated

INTRODUCTION

In this volume, Texas Instruments presents technical information on the most advanced families of TTL integrated circuits, Advanced Low-Power Schottky[†] (ALS), and Advanced Schottky[†] (AS). The choice provided by selecting from either ALS or AS functions provides this systems design engineer with a management tools to achieve performance budgeting. Efficient design parameters can be met easily by taking advantage of this low power of ALS in noncritical paths and using high performance AS in speed critical paths. The efficiencies realized by ALS and AS are being offered to system designers in the following forms:

1. Pin-to-pin compatible, plug-in versions of most popular LSTTL and STTL functions.
2. Higher density MSI and LSI functions.

The ease of use of pin-to-pin compatible functions enables designers to upgrade existing TTL based systems with the following benefits:

- Reduce system power requirements
- Enhance critical system performance
- Improved overall system reliability

New system designs can can capitalize on both the improved efficiency of the pin-compatible devices and the higher densities of the MSI/LSI series of devices, such as the 'ALS632 32-bit parallel error detection and connection circuitry, with the following system benefits:

- Reduced system component count
- Expanded functional capabilities
- Improved levels of cost effectiveness
- DC and AC specifications across the full rated supply voltage and temperature ranges with AC specifications at 50 pF loading.

ALS and AS devices utilize an advanced wafer fabrication process that includes ion-implanted transistors, oxide isolations, and composed mask sets. This process is coupled with circuit design techniques to implement the following:

- Improve input threshold and noise margins
- Improve line driving and receiving
- Maintain or increase drive capability
- Increase density of LSI functions
- Implement universal logic solutions
- Take advantage of new packaging
  — 24-pin 300-mil DIP
  — Ceramic and plastic chip carriers

The ALS/AS family will grow to well over 300 devices through the end of 1984. Included among the new functions are:

- 16-bit by 16-bit universal multiplier
- 20-MHz 8-bit-slice universal processor element with on-board register file
- 20-MHz 14-bit controller with 9-word stack
- 20-MHz 16-bit barrel shifter
- Many additional pin-compatible ALS and AS devices

Included in this volume is a Functional Index to all bipolar digital devices types available or under development showing the available technologies for each type (Standard TTL, Schottky, Low-Power Schottky, Advanced Low-Power Schottky, Advanced Schottky, etc.). Logic symbols prepared in accordance with IEEE activity and pin assignments for all bipolar devices are shown in the Product Guide section of Volume 1 with typical performance data and chip carrier information. Package dimensions given in the Mechanical Data section of this book are in metric measurement (and parenthetically in inches), which should simplify board layout for designers involved in metric conversion and new designs.

While this volume offers design and specification data for bipolar Advanced Low-Power Schottky (ALS) and Advanced Schottky (AS) components, complete technical data for any TI semiconductor product is available from your nearest TI field sales office, local authorized TI distributor, or by writing direct to: Marketing and Information Services, Texas Instruments Incorporated, P.O. Box 225012, MS 308, Dallas, Texas 75265.

[†]Integrated Schottky-Barrier diode-clamped transistor is patented by Texas Instruments, U.S. Patent Number 3,463,975.

The TTL Data Book
Volume 3

General Information | **1**

ALS and AS Circuits | **2**

Mechanical Data | **3**

GENERAL INFORMATION

NUMERICAL INDEX

SN54ALS00A	SN74ALS00A 2-3	SN54ALS153	SN74ALS153 2-125
SN54AS00	SN74AS00 2-3	SN54AS153	SN74AS153 2-125
SN54ALS01	SN74ALS01 2-7	SN54ALS157	SN74ALS157 2-129
SN54ALS02	SN74ALS02 2-9	SN54AS157	SN74AS157 2-129
SN54AS02	SN74AS02 2-9	SN54ALS158	SN74ALS158 2-129
SN54ALS03A	SN74ALS03A 2-13	SN54AS158	SN74AS158 2-129
SN54ALS04A	SN74ALS04A 2-15	SN54ALS160A	SN74ALS160A 2-137
SN54AS04	SN74AS04 2-15	SN54AS160	SN74AS160 2-137
SN54ALS05A	SN74ALS05A 2-19	SN54ALS161A	SN74ALS161A 2-137
SN54ALS08	SN74ALS08 2-21	SN54AS161	SN74AS161 2-137
SN54AS08	SN74AS08 2-21	SN54ALS162A	SN74ALS162A 2-137
SN54ALS09	SN74ALS09 2-25	SN54AS162	SN74AS162 2-137
SN54ALS10	SN74ALS10 2-27	SN54ALS163A	SN74ALS163A 2-137
SN54AS10	SN74AS10 2-27	SN54AS163	SN74AS163 2-137
SN54ALS11	SN74ALS11 2-31	SN54ALS164	SN74ALS164 2-147
SN54AS11	SN74AS11 2-31	SN54ALS165	SN74ALS165 2-151
SN54ALS12	SN74ALS12 2-35	SN54ALS166	SN74ALS166 2-153
SN54ALS15	SN74ALS15 2-37	SN54ALS168A	SN74ALS168A 2-157
SN54ALS20A	SN74ALS20A 2-39	SN54AS168	SN74AS168 2-157
SN54AS20	SN74AS20 2-39	SN54ALS169A	SN74ALS169A 2-157
SN54ALS21	SN74ALS21 2-43	SN54AS169	SN74AS169 2-157
SN54AS21	SN74AS21 2-43	SN54ALS174	SN74ALS174 2-167
SN54ALS22A	SN74ALS22A 2-47	SN54AS174	SN74AS174 2-167
SN54ALS27	SN74ALS27 2-49	SN54ALS175	SN74ALS175 2-167
SN54AS27	SN74AS27 2-49	SN54AS175	SN74AS175 2-167
SN54ALS28A	SN74ALS28A 2-53	SN54ALS181A	SN74ALS181A 2-173
SN54ALS30	SN74ALS30 2-55	SN54AS182	SN74AS182 2-185
SN54AS30	SN74AS30 2-55	SN54ALS190	SN74ALS190 2-189
SN54ALS32	SN74ALS32 2-59	SN54ALS191	SN74ALS191 2-189
SN54AS32	SN74AS32 2-59	SN54ALS192	SN74ALS192 2-197
SN54ALS33A	SN74ALS33A 2-63	SN54ALS193	SN74ALS193 2-197
SN54ALS34	SN74ALS34 2-65	SN54ALS194	SN74ALS194 2-205
SN54AS34	SN74AS34 2-65	SN54ALS195	SN74ALS195 2-211
SN54ALS35	SN74ALS35 2-69	SN54AS230	SN74AS230 2-213
SN54ALS37A	SN74ALS37A 2-71	SN54AS231	SN74AS231 2-213
SN54ALS38A	SN74ALS38A 2-73	SN54ALS240A	SN74ALS240A 2-217
SN54ALS40A	SN74ALS40A 2-75	SN54AS240	SN74AS240 2-217
SN54ALS74	SN74ALS74 2-77	SN54ALS241A	SN74ALS241A 2-217
SN54AS74	SN74AS74 2-77	SN54AS241	SN74AS241 2-217
SN54ALS86	SN74ALS86 2-81	SN54ALS242A	SN74ALS242A 2-223
SN54AS95	SN74AS94 2-83	SN54AS242	SN74AS242 2-223
SN54ALS109	SN74ALS109 2-87	SN54ALS243A	SN74ALS243A 2-223
SN54AS109	SN74AS109 2-87	SN54AS243	SN74AS243 2-223
SN54ALS112A	SN74ALS112A 2-91	SN54ALS244A	SN74ALS244A 2-229
SN54ALS112	SN74ALS112 2-91	SN54AS244	SN74AS244 2-229
SN54ALS113A	SN74ALS113A 2-95	SN54ALS245A	SN74ALS245A 2-235
SN54AS113	SN74AS113 2-95	SN54AS245	SN74AS245 2-235
SN54ALS114A	SN74ALS114A 2-99	SN54AS250	SN74AS250 2-241
SN54AS114	SN74AS114 2-99	SN54ALS251	SN74ALS251 2-245
SN54ALS131	SN74ALS131 2-103	SN54AS251	SN74AS251 2-245
SN54AS131	SN74AS131 2-103	SN54ALS253	SN74ALS253 2-251
SN54ALS133	SN74ALS133 2-107	SN54AS253	SN74AS253 2-251
SN54ALS137	SN74ALS137 2-109	SN54ALS257	SN74ALS257 2-255
SN54AS137	SN74AS137 2-109	SN54AS257	SN74AS257 2-255
SN54ALS138	SN74ALS138 2-113	SN54ALS258	SN74ALS258 2-255
SN54AS138	SN74AS138 2-113	SN54AS258	SN74AS258 2-255
SN54ALS139	SN74ALS139 2-117	SN54ALS259	SN74ALS259 2-261
SN54AS139	SN74AS139 2-117	SN54AS264	SN74AS264 2-263
SN54ALS151	SN74ALS151 2-121	SN54ALS273	SN74ALS273 2-269
SN54AS151	SN74AS151 2-121	SN54AS280	SN74AS280 2-273

NUMERICAL INDEX

NUMERICAL INDEX

SN54AS282	SN74AS282	2-277	SN54ALS620A	SN74ALS620A	2-415
SN54AS286	SN74AS286	2-281	SN54AS620	SN74AS620	2-415
SN54AS298	SN74AS298	2-287	SN54ALS621A	SN74ALS621A	2-415
SN54ALS299	SN74ALS299	2-291	SN54ALS621	SN74ALS621	2-415
SN54AS299	SN74AS299	2-291	SN54ALS622A	SN74ALS622A	2-415
SN54ALS323	SN74ALS323	2-291	SN54AS622	SN74AS622	2-415
SN54AS323	SN74AS323	2-291	SN54ALS623A	SN74ALS623A	2-415
SN54ALS352	SN74ALS352	2-301	SN54AS623	SN74AS623	2-415
SN54AS352	SN74AS352	2-301	SN54ALS632	SN74ALS632	2-425
SN54ALS353	SN74ALS353	2-305	SN54ALS633	SN74ALS633	2-425
SN54AS353	SN74AS353	2-305	SN54ALS634	SN74ALS634	2-425
SN54ALS365	SN74ALS365	2-309	SN54ALS635	SN74ALS635	2-425
SN54ALS366	SN74ALS366	2-309	SN54ALS638A	SN74ALS638A	2-439
SN54ALS367	SN74ALS367	2-309	SN54AS638	SN74AS638	2-439
SN54ALS368	SN74ALS368	2-309	SN54ALS639A	SN74ALS639A	2-439
SN54ALS373	SN74ALS373	2-313	SN54AS639	SN74AS639	2-439
SN54AS373	SN74AS373	2-313	SN54ALS640A	SN74ALS640A	2-445
SN54ALS374	SN74ALS374	2-319	SN54AS640	SN74AS640	2-445
SN54AS374	SN74AS374	2-319	SN54ALS641A	SN74ALS641A	2-445
SN54AS395	SN74AS395	2-327	SN54AS641	SN74AS641	2-445
SN54ALS465A	SN74ALS465A	2-327	SN54ALS642A	SN74ALS642A	2-445
SN54ALS466A	SN74ALS466A	2-327	SN54AS642	SN74AS642	2-445
SN54ALS467A	SN74ALS467A	2-327	SN54ALS643A	SN74ALS643A	2-445
SN54ALS468A	SN74ALS468A	2-327	SN54AS643	SN74AS643	2-445
SN54ALS518	SN74ALS518	2-333	SN54ALS644A	SN74ALS644A	2-445
SN54ALS519	SN74ALS519	2-333	SN54AS644	SN74AS644	2-445
SN54ALS520	SN74ALS520	2-333	SN54ALS645A	SN74ALS645A	2-445
SN54ALS521	SN74ALS521	2-333	SN54AS645	SN74AS645	2-445
SN54ALS522	SN74ALS522	2-333	SN54ALS646	SN74ALS646	2-455
SN54ALS526	SN74ALS526	2-339	SN54AS646	SN74AS646	2-455
SN54ALS527	SN74ALS527	2-339	SN54ALS647	SN74ALS647	2-455
SN54ALS528	SN74ALS528	2-339	SN54ALS648	SN74ALS648	2-455
SN54ALS533	SN74ALS533	2-345	SN54AS648	SN74AS648	2-455
SN54AS533	SN74AS533	2-345	SN54ALS649	SN74ALS649	2-455
SN54ALS534	SN74ALS534	2-351	SN54ALS651	SN74ALS651	2-465
SN54AS534	SN74AS534	2-351	SN54AS651	SN74AS651	2-465
SN54ALS538	SN74ALS538	2-357	SN54ALS652	SN74ALS652	2-465
SN54ALS539	SN74ALS539	2-361	SN54AS652	SN74AS652	2-465
SN54ALS540	SN74ALS540	2-365	SN54ALS653	SN74ALS653	2-465
SN54ALS541	SN74ALS541	2-365	SN54ALS654	SN74ALS654	2-465
SN54ALS560A	SN74ALS560A	2-369	SN54ALS677	SN74ALS677	2-475
SN54ALS561A	SN74ALS561A	2-369	SN54ALS678	SN74ALS678	2-475
SN54ALS563	SN74ALS563	2-379	SN54ALS679	SN74ALS679	2-481
SN54ALS564	SN74ALS564	2-383	SN54ALS680	SN74ALS680	2-481
SN54ALS568A	SN74ALS568A	2-387	SN54ALS688	SN74ALS688	2-487
SN54ALS569A	SN74ALS569A	2-387	SN54ALS689	SN74ALS689	2-487
SN54ALS573	SN74ALS573	2-397	SN54AS756	SN74AS756	2-491
SN54AS573	SN74AS573	2-397	SN54AS757	SN74AS757	2-491
SN54ALS574	SN74ALS574	2-403	SN54AS758	SN74AS758	2-495
SN54AS574	SN74AS574	2-403	SN54AS759	SN74AS759	2-495
SN54ALS575	SN74ALS575	2-403	SN54AS760	SN74AS760	2-499
SN54AS575	SN74AS575	2-403	SN54AS762	SN74AS762	2-501
SN54ALS576	SN74ALS576	2-409	SN54AS763	SN74AS763	2-501
SN54AS576	SN74AS576	2-409	SN54AS800	SN74AS800	2-505
SN54ALS577	SN74ALS577	2-409	SN54AS802	SN74AS802	2-509
SN54AS577	SN74AS577	2-409	SN54ALS804	SN74ALS804	2-513
SN54ALS580	SN74ALS580	2-397	SN54AS804A	SN74AS804A	2-513
SN54AS580	SN74AS580	2-397	SN54ALS805	SN74ALS805	2-517
			SN54AS805A	SN74AS805A	2-517

NUMERICAL INDEX

SN54ALS808	SN74ALS808	2-521
SN54AS808A	SN74AS808A	2-521
SN54AS821	SN74AS821	2-525
SN54AS822	SN74AS822	2-525
SN54AS823	SN74AS823	2-531
SN54AS824	SN74AS824	2-531
SN54AS825	SN74AS825	2-537
SN54AS826	SN74AS826	2-537
SN54ALS832	SN74ALS832	2-543
SN54ALS832A	SN74ALS832A	2-543
SN54ALS841	SN74ALS841	2-547
SN54AS841	SN74AS841	2-547
SN54ALS842	SN74ALS842	2-547
SN54AS842	SN74AS842	2-547
SN54ALS843	SN74ALS843	2-555
SN54AS843	SN74AS843	2-555
SN54ALS844	SN74ALS844	2-555
SN54AS844	SN74AS844	2-555
SN54ALS845	SN74ALS845	2-563
SN54AS845	SN74AS845	2-563
SN54ALS846	SN74ALS846	2-563
SN54AS846	SN74AS846	2-563
SN54AS850	SN74AS850	2-571
SN54AS851	SN74AS851	2-571
SN54AS852	SN74AS852	2-581
SN54AS856	SN74AS856	2-587
SN54AS857	SN74ALS857	2-593
SN54AS857	SN74AS857	2-593
SN54AS866	SN74AS866	2-601
SN54AS867	SN74AS867	2-607
SN54AS869	SN74AS869	2-607
SN54AS870	SN74AS870	2-613
SN54AS871	SN74AS871	2-613
SN54ALS873	SN74ALS873	2-619
SN54AS873	SN74AS873	2-619
SN54ALS874	SN74ALS874	2-625
SN54AS874	SN74AS874	2-625
SN54ALS876	SN74ALS876	2-625
SN54AS876	SN74AS876	2-625
SN54AS877	SN74AS877	2-631
SN54ALS878	SN74ALS878	2-637
SN54AS878	SN74AS878	2-637
SN54ALS879	SN74ALS879	2-637
SN54AS879	SN74AS879	2-637
SN54ALS880	SN74ALS880	2-643
SN54AS880	SN74AS880	2-643
SN54AS881A	SN74AS881A	2-649
SN54AS882	SN74AS882	2-651
SN54AS885	SN74AS885	2-657
SN54ALS1000A	SN74ALS1000A	2-663
SN54AS1000	SN74AS1000	2-663
SN54ALS1002A	SN74ALS1002A	2-667
SN54ALS1003A	SN74ALS1003A	2-669
SN54ALS1004	SN74ALS1004	2-671
SN54AS1004	SN74AS1004	2-671
SN54ALS1005	SN74ALS1005	2-675
SN54ALS1008A	SN74ALS1008A	2-677
SN54AS1008	SN74AS1008	2-677
SN54ALS1010A	SN74ALS1010A	2-681
SN54ALS1011A	SN74ALS1011A	2-683
SN54ALS1020A	SN74ALS1020A	2-685
SN54ALS1032A	SN74ALS1032A	2-687
SN54AS1032	SN74AS1032	2-687
SN54ALS1034	SN74ALS1034	2-691
SN54AS1034	SN74AS1034	2-691
SN54ALS1035	SN74ALS1035	2-695
SN54AS1036	SN74AS1036	2-697
SN54ALS1240	SN74ALS1240	2-699
SN54ALS1241	SN74ALS1241	2-699
SN54ALS1242	SN74ALS1242	2-703
SN54ALS1243	SN74ALS1243	2-703
SN54ALS1244A	SN74ALS1244A	2-707
SN54ALS1245	SN74ALS1245	2-711
SN54ALS1620	SN74ALS1620	2-715
SN54ALS1621	SN74ALS1621	2-715
SN54ALS1622	SN74ALS1622	2-715
SN54ALS1623	SN74ALS1623	2-715
SN54ALS1638	SN74ALS1638	2-721
SN54ALS1639	SN74ALS1639	2-721
SN54ALS1640A	SN74ALS1640A	2-725
SN54ALS1641	SN74ALS1641	2-725
SN54ALS1642	SN74ALS1642	2-725
SN54ALS1643	SN74ALS1643	2-725
SN54ALS1644	SN74ALS1644	2-725
SN54ALS1645A	SN74ALS1645A	2-725
SN54AS2620	SN74AS2620	2-731
SN54AS2623	SN74AS2623	2-731
SN54AS2640	SN74AS2640	2-735
SN54AS2645	SN74AS2645	2-735
SN54ALS8003	SN74ALS8003	2-739

GLOSSARY
ALS/AS TTL SYMBOLS, TERMS, AND DEFINITIONS

INTRODUCTION

These symbols, terms, and definitions are in accordance with those currently agreed upon by the JEDEC Council of the Electronic Industries Association (EIA) for use in the USA and by the International Electrotechnical Commission (IEC) for international use.

PART I — OPERATING CONDITIONS AND CHARACTERISTICS (IN SEQUENCE BY LETTER SYMBOLS)

f_{max} **Maximum clock frequency**
The highest rate at which the clock input of a bistable circuit can be driven through its required sequence while maintaining stable transitions of logic level at the output with input conditions established that should cause changes of output logic level in accordance with the specification.

I_{CC} **Supply current**
The current into* the V_{CC} supply terminal of an integrated circuit.

I_{CCH} **Supply current, outputs high**
The current into* the V_{CC} supply terminal of an integrated circuit when all (or a specified number) of the outputs are at the high level.

I_{CCL} **Supply current, outputs low**
The current into* the V_{CC} supply terminal of an integrated circuit when all (or a specified number) of the outputs are at the low level.

I_{IH} **High-level input current**
The current into* an input when a high-level voltage is applied to that input.

I_{IL} **Low-level input current**
The current into* an input when a low-level voltage is applied to that input.

I_{OH} **High-level output current**
The current into* an output with input conditions applied that, according to the product specification, will establish a high level at the output.

I_{OL} **Low-level output current**
The current into* an output with input conditions applied that, according to the product specification, will establish a low level at the output.

I_{OS} **Short-circuit output current**
The current into* an output when that output is short-circuited to ground (or other specified potential) with input conditions applied to establish the output logic level farthest from ground potential (or other specified potential).

I_{OZH} **Off-state (high-impedance-state) output current (of a three-state output) with high-level voltage applied**
The current flowing into* an output having three-state capability with input conditions established that, according to the product specification, will establish the high-impedance state at the output and with a high-level voltage applied to the output.
NOTE: This parameter is measured with other input conditions established that would cause the output to be at a low level if it were enabled.

*Current out of a terminal is given as a negative value.

GLOSSARY
ALS/AS TTL SYMBOLS, TERMS, AND DEFINITIONS

I_{OZL} **Off-state (high-impedance-state) output current (of a three-state output) with low-level voltage applied**
The current flowing into* an output having three-state capability with input conditions established that, according to the product specification, will establish the high-impedance state at the output and with a low-level voltage applied to the output.
NOTE: This parameter is measured with other input conditions established that would cause the output to be at a high level if it were enabled.

V_{IH} **High-level input voltage**
An input voltage within the more positive (less negative) of the two ranges of values used to represent the binary variables.
NOTE: A minimum is specified that is the least-positive value of high-level input voltage for which operation of the logic element within specification limits is guaranteed.

V_{IK} **Input clamp voltage**
An input voltage in a region of relatively low differential resistance that serves to limit the input voltage swing.

V_{IL} **Low-level input voltage**
An input voltage level within the less positive (more negative) of the two ranges of values used to represent the binary variables.
NOTE: A maximum is specified that is the most-positive value of low-level input voltage for which operation of the logic element within specification limits is guaranteed.

V_{OH} **High-level output voltage**
The voltage at an output terminal with input conditions applied that, according to the product specification, will establish a high level at the output.

V_{OL} **Low-level output voltage**
The voltage at an output terminal with input conditions applied that, according to the product specification, will establish a low level at the output.

t_a **Access time**
The time interval between the application of a specific input pulse and the availability of valid signals at an output.

t_{dis} **Disable time (of a three-state output)**
The time interval between the specified reference points on the input and output voltage waveforms, with the three-state output changing from either of the defined active levels (high or low) to a high-impedance (off) state. (t_{dis} = t_{PHZ} or t_{PLZ}).

t_{en} **Enable time (of a three-state output)**
The time interval between the specified reference points on the input and output voltage waveforms, with the three-state output changing from a high-impedance (off) state to either of the defined active levels (high or low). (t_{en} = t_{PZH} or t_{PZL}).

*Current out of a terminal is given as a negative value.

GLOSSARY
ALS/AS TTL SYMBOLS, TERMS, AND DEFINITIONS

t_h **Hold time**
The time interval during which a signal is retained at a specified input terminal after an active transition occurs at another specified input terminal.
NOTES: 1. The hold time is the actual time interval between two signal events and is determined by the system in which the digital circuit operates. A minimum value is specified that is the shortest interval for which correct operation of the digital circuit is guaranteed.
2. The hold time may have a negative value in which case the minimum limit defines the longest interval (between the release of the signal and the active transition) for which correct operation of the digital circuit is guaranteed.

t_{pd} **Propagation delay time**
The time between the specified reference points on the input and output voltage waveforms with the output changing from one defined level (high or low) to the other defined level. (t_{pd} = t_{PHL} or t_{PLH}).

t_{PHL} **Propagation delay time, high-to-low-level output**
The time between the specified reference points on the input and output voltage waveforms with the output changing from the defined high level to the defined low level.

t_{PHZ} **Disable time (of a three-state output) from high level**
The time interval between the specified reference points on the input and output voltage waveforms with the three-state output changing from the defined high level to a high-impedance (off) state.

t_{PLH} **Propagation delay time, low-to-high-level output**
The time between the specified reference points on the input and output voltage waveforms with the output changing from the defined low level to the defined high level.

t_{PLZ} **Disable time (of a three-state output) from low level**
The time interval between the specified reference points on the input and output voltage waveforms with the three-state output changing from the defined low level to a high-impedance (off) state.

t_{PZH} **Enable time (of a three-state output) to high level**
The time interval between the specified reference points on the input and output voltage waveforms with the three-state output changing from a high-impedance (off) state to the defined high level.

t_{PZL} **Enable time (of a three-state output) to low level**
The time interval between the specified reference points on the input and output voltage waveforms with the three-state output changing from a high-impedance (off) state to the defined low level.

t_{sr} **Sense recovery time**
The time interval needed to switch a memory from a write mode to a read mode and to obtain valid data signals at the output.

t_{su} **Setup time**
The time interval between the application of a signal at a specified input terminal and a subsequent active transition at another specified input terminal.
NOTES: 1. The setup time is the actual time interval between two signal events and is determined by the system in which the digital circuit operates. A minimum value is specified that is the shortest interval for which correct operation of the digital circuit is guaranteed.
2. The setup time may have a negative value in which case the minimum limit defines the longest interval (between the active transition and the application of the other signal) for which correct operation of the digital circuit is guaranteed.

t_w **Pulse duration (width)**
The time interval between specified reference points on the leading and trailing edges of the pulse waveform.

GLOSSARY
ALS/AS TTL SYMBOLS, TERMS, AND DEFINITIONS

PART II — CLASSIFICATION OF CIRCUIT COMPLEXITY

Gate Equivalent Circuit

A basic unit-of-measure of relative digital-circuit complexity. The number of gate equivalent circuits is that number of individual logic gates that would have to be interconnected to perform the same function.

Large-Scale Integration, LSI

A concept whereby a complete major subsystem or system function is fabricated as a single microcircuit. In this context a major subsystem or system, whether digital or linear, is considered to be one that contains 100 or more equivalent gates or circuitry of similar complexity.

Medium-Scale Integration, MSI

A concept whereby a complete subsystem or system function is fabricated as a single microcircuit. The subsystem or system is smaller than for LSI, but whether digital or linear, is considered to be one that contains 12 or more equivalent gates or circuitry of similar complexity.

Small-Scale Integration, SSI

Integrated circuits of less complexity than medium-scale integration (MSI).

Very-Large-Scale Integration, VLSI

The description of any IC technology that is much more complex than large-scale integration (LSI), and involves a much higher equivalent gate count. At this time an exact definition including a minimum gate count has not been standardized by JEDEC or the IEEE.

EXPLANATION OF FUNCTION TABLES

The following symbols are used in function tables on TI data sheets:

H	=	high level (steady state)
L	=	low level (steady state)
↑	=	transition from low to high level
↓	=	transition from high to low level
→	=	value/level or resulting value/level is routed to indicated destination
↷	=	value/level is re-entered
X	=	irrelevant (any input, including transitions)
Z	=	off (high-impedance) state of a 3-state-output
a..h	=	the level of steady-state inputs at inputs A through H respectively
Q_0	=	level of Q before the indicated steady-state input conditions were established
$\bar{Q}_0$	=	complement of Q_0 or level of $\bar{Q}$ before the indicated steady-state input conditions were established
Q_n	=	level of Q before the most recent active transition indicated by ↓ or ↑
⊓	=	one high-level pulse
⊔	=	one low-level pulse
TOGGLE	=	each output changes to the complement of its previous level on each active transition indicated by ↓ or ↑.

If, in the input columns, a row contains only the symbols H, L, and/or X, this means the indicated output is valid whenever the input configuration is achieved and regardless of the sequence in which it is achieved. The output persists so long as the input configuration is maintained.

If, in the input columns, a row contains H, L, and/or X together with ↑ and/or ↓, this means the output is valid whenever the input configuration is achieved but the transition(s) must occur following the achievement of the steady-state levels. If the output is shown as a level (H, L, Q_0, or $\bar{Q}_0$), it persists so long as the steady-state input levels and the levels that terminate indicated transitions are maintained. Unless otherwise indicated, input transitions in the opposite direction to those shown have no effect at the output. (If the output is shown as a pulse, ⊓ or ⊔, the pulse follows the indicated input transition and persists for an interval dependent on the circuit.)

EXPLANATION OF FUNCTION TABLES

Among the most complex function tables in this book are those of the shift registers. These embody most of the symbols used in any of the function tables, plus more. Below is the function table of a 4-bit bidirectional universal shift register, e.g., type SN74194.

FUNCTION TABLE

INPUTS										OUTPUTS			
CLEAR	MODE		CLOCK	SERIAL		PARALLEL				Q_A	Q_B	Q_C	Q_D
	S1	S0		LEFT	RIGHT	A	B	C	D				
L	X	X	X	X	X	X	X	X	X	L	L	L	L
H	X	X	L	X	X	X	X	X	X	Q_{A0}	Q_{B0}	Q_{C0}	Q_{D0}
H	H	H	↑	X	X	a	b	c	d	a	b	c	d
H	L	H	↑	X	H	X	X	X	X	H	Q_{An}	Q_{Bn}	Q_{Cn}
H	L	H	↑	X	L	X	X	X	X	L	Q_{An}	Q_{Bn}	Q_{Cn}
H	H	L	↑	H	X	X	X	X	X	Q_{Bn}	Q_{Cn}	Q_{Dn}	H
H	H	L	↑	L	X	X	X	X	X	Q_{Bn}	Q_{Cn}	Q_{Dn}	L
H	L	L	X	X	X	X	X	X	X	Q_{A0}	Q_{B0}	Q_{C0}	Q_{D0}

The first line of the table represents a synchronous clearing of the register and says that if clear is low, all four outputs will be reset low regardless of the other inputs. In the following lines, clear is inactive (high) and so has no effect.

The second line shows that so long as the clock input remains low (while clear is high), no other input has any effect and the outputs maintain the levels they assumed before the steady-state combination of clear high and clock low was established. Since on other lines of the table only the rising transition of the clock is shown to be active, the second line implicitly shows that no further change in the outputs will occur while the clock remains high or on the high-to-low transition of the clock.

The third line of the table represents synchronous parallel loading of the register and says that if S1 and S0 are both high then, without regard to the serial input, the data entered at A will be at output Q_A, data entered at B will be at Q_B, and so forth, following a low-to-high clock transition.

The fourth and fifth lines represent the loading of high- and low-level data, respectively, from the shift-right serial input and the shifting of previously entered data one bit; data previously at Q_A is now at Q_B, the previous levels of Q_B and Q_C are now at Q_C and Q_D respectively, and the data previously at Q_D is no longer in the register. This entry of serial data and shift takes place on the low-to-high transition of the clock when S1 is low and S0 is high and the levels at inputs A through D have no effect.

The sixth and seventh lines represent the loading of high- and low-level data, respectively, from the shift-left serial input and the shifting of previously entered data one bit; data previously at Q_B is now at Q_A, the previous levels of Q_C and Q_D are now at Q_B and Q_C, respectively, and the data previously at Q_A is no longer in the register. This entry of serial data and shift takes place on the low-to-high transition of the clock when S1 is high and S0 is low and the levels at inputs A through D have no effect.

The last line shows that as long as both mode inputs are low, no other input has any effect and, as in the second line, the outputs maintain the levels they assumed before the steady-state combination of clear high and both mode inputs low was established.

SERIES 54ALS/74ALS AND 54AS/74AS DEVICES

PARAMETER MEASUREMENT INFORMATION

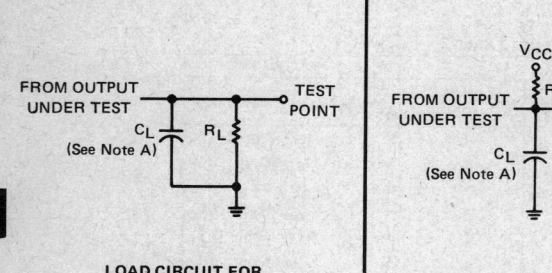
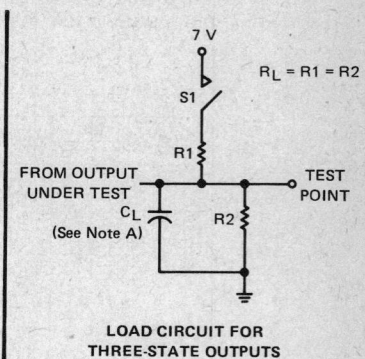

LOAD CIRCUIT FOR BI-STATE TOTEM-POLE OUTPUTS

LOAD CIRCUIT FOR OPEN-COLLECTOR OUTPUTS

LOAD CIRCUIT FOR THREE-STATE OUTPUTS

NOTE A. C_L includes probe and jig capacitance.

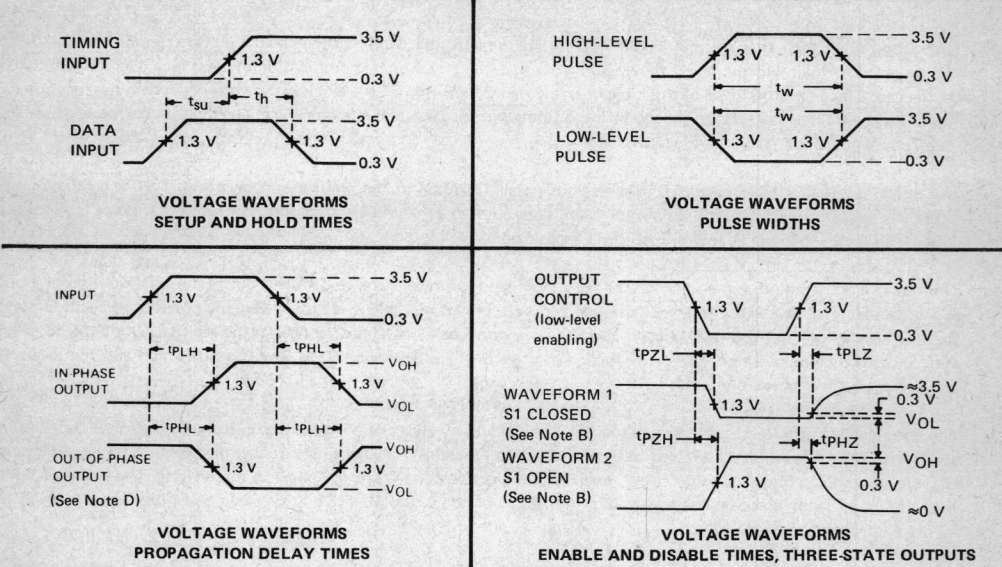

VOLTAGE WAVEFORMS SETUP AND HOLD TIMES

VOLTAGE WAVEFORMS PULSE WIDTHS

VOLTAGE WAVEFORMS PROPAGATION DELAY TIMES

VOLTAGE WAVEFORMS ENABLE AND DISABLE TIMES, THREE-STATE OUTPUTS

NOTES: B. Waveform 1 is for an output with internal conditions such that the output is low except when disabled by the output control. Waveform 2 is for an output with internal conditions such that the output is high except when disabled by the output control.
C. All input pulses have the following characteristics: PRR ⩽ 1 MHz, $t_r = t_f = 2$ ns, duty cyle = 50%.
D. When measuring propagation delay times of 3-state outputs, switch S1 is open.

SERIES 54ALS/74ALS
ADVANCED LOW-POWER SCHOTTKY TRANSISTOR-TRANSISTOR LOGIC

TYPICAL CHARACTERISTICS†

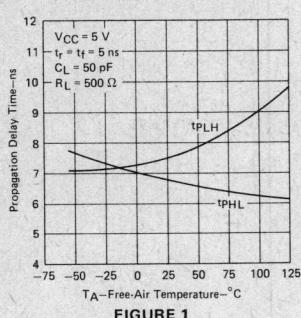

FIGURE 1 — 'ALS00A PROPAGATION DELAY TIMES vs FREE-AIR TEMPERATURE

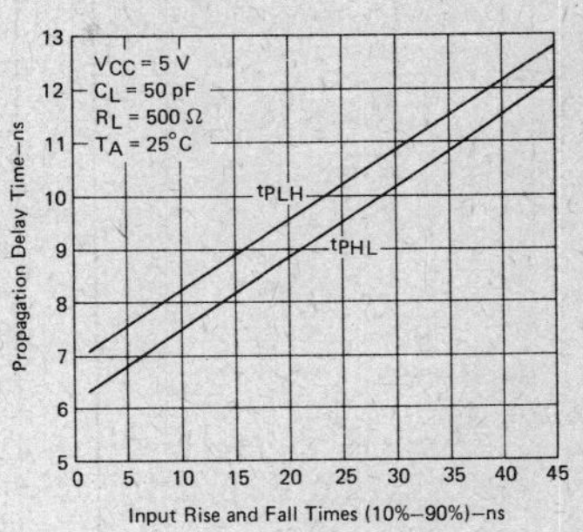

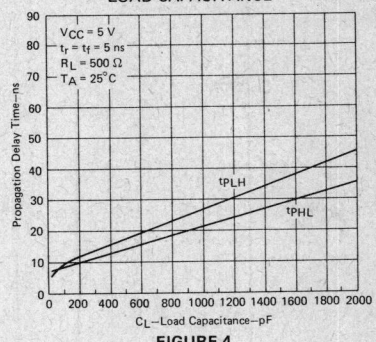

FIGURE 2 — 'ALS00A PROPAGATION DELAY TIMES vs INPUT RISE & FALL TIMES

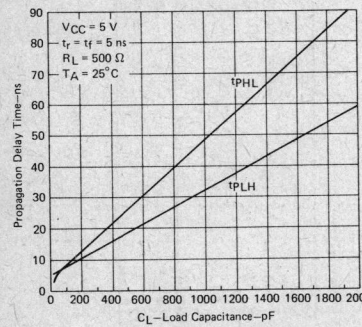

FIGURE 3 — 'ALS00A PROPAGATION DELAY TIMES vs LOAD CAPACITANCE

FIGURE 4 — 'ALS244A PROPAGATION DELAY TIMES vs LOAD CAPACITANCE

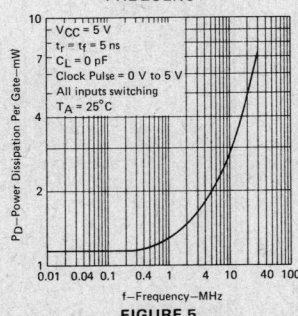

FIGURE 5 — 'ALS00A POWER DISSIPATION PER GATE vs FREQUENCY

†Data for temperatures below 0°C and above 70°C are applicable for Series 54ALS circuits only.

Texas Instruments
POST OFFICE BOX 225012 • DALLAS, TEXAS 75265

SERIES 54AS/74AS
ADVANCED SCHOTTKY TRANSISTOR-TRANSISTOR LOGIC

TYPICAL CHARACTERISTICS†

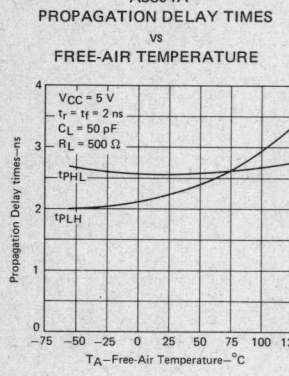

FIGURE 1

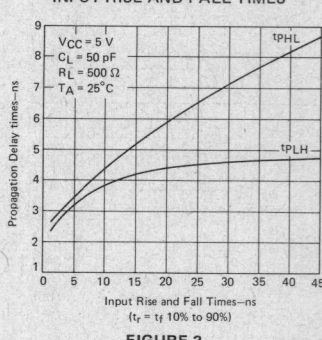

FIGURE 2

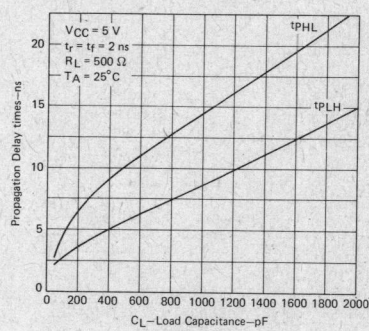

FIGURE 3

†Data for temperatures below 0°C and above 70°C are applicable for Series 54AS Circuits only.

FUNCTIONAL INDEX

GATES AND INVERTERS

POSITIVE-NAND GATES AND INVERTERS

DESCRIPTION	TYPE	STD TTL	ALS	AS	H	L	LS	S	VOLUME
Hex 2-Input Gates	'804		●	A					3
Hex Inverters	'04	●			●	●	●	●	2
	'1004		A	●					3
Quadruple 2-Input Gates	'00	●			●	●	●	●	2
	'1000		A	●					3
Triple 3-Input Gates	'10	●			●	●	●	●	2
	'1010		A						3
Dual 4-Input Gates	'20	●			●	●	●	●	2
	'1020		A	●					3
8-Input Gates	'30	●			●	●	●	●	2
			A	●					3
13-Input Gates	'133							●	2
Dual 2-Input Gates	'8003		●						3

POSITIVE-NAND GATES AND INVERTERS WITH OPEN-COLLECTOR OUTPUTS

DESCRIPTION	TYPE	STD TTL	ALS	AS	H	L	LS	S	VOLUME
Hex Inverters	'05	●			●		●	●	2
	'1005		A						3
			●						
Quadruple 2-Input Gates	'01	●			●		●	●	2
			●						3
	'03	●				●	●	●	2
	'1003		A						3
			A						
Triple 3-Input Gates	'12	●					●		2
			●						3
Dual 4-Input Gates	'22	●			●		●	●	2
			A						3

POSITIVE-AND GATES

DESCRIPTION	TYPE	STD TTL	ALS	AS	H	L	LS	S	VOLUME
Hex 2-Input Gates	'808		●	A					3
Quadruple 2-Input Gates	'08	●					●	●	2
	'1008		A	●					3
Triple 3-Input Gates	'11	●				●	●	●	2
	'1011		●						3
Dual 4-Input Gates	'21	●					●		2
			●	●					3
Triple 4-Input AND/NAND	'800			▲					

POSITIVE-AND GATES WITH OPEN-COLLECTOR OUTPUTS

DESCRIPTION	TYPE	STD TTL	ALS	AS	H	L	LS	S	VOLUME
Quadruple 2-Input Gates	'09	●					●	●	2
			●						3
Triple 3-Input Gates	'15						●	●	2
			●						3

POSITIVE-OR GATES

DESCRIPTION	TYP	STD TTL	ALS	AS	LS	S	VOLUME
Hex 2-Input Gates	'832		●	A			3
Quadruple 2-Input Gates	'32	●			●	●	2
	'1032		●	●			3
			A	●			
Triple 4-Input OR/NOR	'802		▲				

POSITIVE-NOR GATES

DESCRIPTION	TYPE	STD TTL	ALS	AS	L	LS	S	VOLUME
Hex 2-Input Gates	'805		●	A				3
Quadruple 2-Input Gates	'02	●			●	●	●	2
	'1002		A	●				3
			A					
Triple 3-Input Gates	'27	●				●		2
			●					3
Dual 4-Input Gates with Strobe	'25	●						2
Dual 5-Input Gates	'260						●	

SCHMITT-TRIGGER POSITIVE-NAND GATES AND INVERTERS

DESCRIPTION	TYPE	STD TTL	ALS	AS	LS	S	VOLUME
Hex Inverters	'14	●			●		
	'19				●		
Octal Inverters	'619				●		
Dual 4-Input Positive-NAND	'13	●			●		2
	'18				●		
Triple 4-Input Positive-NAND	'618				●		
Quadruple 2-Input Positive-NAND	'24				●		
	'132	●			●	●	

CURRENT-SENSING GATES

DESCRIPTION	TYPE	ALS	AS	LS	VOLUME
Hex	'63			●	2

DELAY ELEMENTS

DESCRIPTION	TYP	ALS	AS	LS	VOLUME
Inverting & Non-Inverting Elements, 2-INPUT NAND Buffers	'31			●	2

● Denotes available technology.
▲ Denotes planned new products.
A Denotes "A" suffix version available in the technology indicated.

FUNCTIONAL INDEX

GATES, EXPANDERS, BUFFERS, DRIVERS, AND TRANSCEIVERS

AND-OR-INVERT GATES

DESCRIPTION	TYPE	TECHNOLOGY							VOLUME
		STD TTL	ALS	AS	H	L	LS	S	
2-Wide 4-Input	'55				•	•	•		
4-Wide 4-2-3-2 Input	'64							•	
4-Wide 2-2-3-2 input	'54			•					2
4-Wide 2-Input	'54	•							
4-Wide 2-3-3-2 input	'54						•	•	
Dual 2-Wide 2-Input	'51	•			•	•	•	•	

AND-OR-INVERT GATES WITH OPEN-COLLECTOR OUTPUTS

DESCRIPTION	TYPE	TECHNOLOGY				VOLUME
		STD TTL	ALS	AS	S	
4-Wide 4-2-3-2-Input	'65				•	2

EXPANDABLE GATES

DESCRIPTION	TYPE	TECHNOLOGY						VOLUME
		STD TTL	ALS	AS	H	L	LS	
Dual 4-Input Positive-NOR With Strobe	'23	•						
4-Wide AND-OR	'52				•			
4-Wide AND-OR-INVERT	'53	•			•			2
2-Wide AND-OR-INVERT	'55					•	•	
Dual 2-Wide AND-OR-INVERT	'50	•			•			

EXPANDERS

DESCRIPTION	TYPE	TECHNOLOGY				VOLUME
		STD TTL	ALS	AS	H	
Dual 4-Input	'60	•			•	
Triple 3-Input	'61				•	2
3-2-2-3-Input AND-OR	'62				•	

BUFFER AND INTERFACE GATES WITH OPEN-COLLECTOR OUTPUTS

DESCRIPTION	TYPE	TECHNOLOGY					VOLUME
		STD TTL	ALS	AS	LS	S	
Hex	'07	•					2
	'17	•					
	'35	▲					3
	'1035			•			
Hex Inverter	'06	•					2
	'16	•					
	'1005			•			3
Quad 2-Input Positive-NAND	'26	•					2
	'38	•		•	•		
			A				3
	'39	•					2
	'1003		A				3
Quad 2-Input Positive-NOR	'33	•			•		2
			A				3

BUFFERS, DRIVERS, AND BUS TRANSCEIVERS WITH OPEN-COLLECTOR OUTPUTS

DESCRIPTION	TYPE	TECHNOLOGY					VOLUME
		STD TTL	ALS	AS	LS	S	
Non-inverting Octal Buffers/Drivers	'757		▲				
	'760		▲				
Inverting Octal Buffers/Drivers	'756		▲				
	'763		▲				3
Inverting and Non-Inverting Octal Buffers/Drivers	'762		▲				
Non-Inverting Quad Transceivers	'759		▲				
Inverting Quad Transceivers	'758		▲				

GATES, BUFFERS, DRIVERS, AND BUS TRANSCEIVERS WITH 3-STATE OUTPUTS

DESCRIPTION	TYPE	TECHNOLOGY				VOLUME
		STD TTL	ALS	AS	LS	S	
Non-Inverting Octal Buffers/Drivers	'241			•	•		2
			A	•			3
	'244			•	•		2
			A	•			3
	'465				•		2
			A				3
	'467				•		2
			A				3
	'541				•		2
		▲					
	'1241¶		▲				3
	'1244¶		A				
	'231		•				
Inverting Octal Buffers/Drivers	'240			•	•		2
			A	•			3
	'466				•		2
			A				3
	'468				•		2
			A				3
	'540				•		2
		▲					
	'1240¶		•				3
Inverting and Non-Inverting Octal Buffers/Drivers	'230			•			
Octal Transceivers	'245				•		2
			A	▲			
	'1245		•				3
Non-inverting Hex Buffers/Drivers	'365	A			A		2
			▲				3
	'367	A			A		2
			▲				3
Inverting Hex Buffers/Drivers	'366	A			A		2
			▲				3
	'368	A			A		2
			▲				3
Quad Buffers/Drivers with Independent Output Controls	'125	•			A		
	'126	•			A		2
	'425	•					
	'426	•					
Non-Inverting Quad Transceivers	'243			•			3
	'1243¶		▲				
Inverting Quad Transceivers	'242			•			2
			A	•			3
	'1242¶		▲				
Quad Transceivers with Storage	'226					•	
12-Input NAND Gate	'134					•	2
Controller and Bus Driver for 8080A System	'428					•	

50-OHM/75-OHM LINE DRIVERS

DESCRIPTION	TYPE	TECHNOLOGY				VOLUME
		STD TTL	ALS	AS	S	
Hex 2-Input Positive-NAND	'804		•	A		
Hex 2-Input Positive-NOR	'805		•	A		3
Hex 2-Input Positive-AND	'808		•	A		
Hex 2-Input Positive-OR	'832		•	A		
Quad 2-Input Positive-NOR	'128	•				2
Dual 4-Input Positive-NAND	'140				•	

• Denotes available technology.
▲ Denotes planned new products.
¶ Denotes very low power.
A Denotes "A" suffix version available in the technology indicated.

Texas Instruments
POST OFFICE BOX 225012 • DALLAS, TEXAS 75265

FUNCTIONAL INDEX

BUFFERS, DRIVERS, TRANSCEIVERS, AND CLOCK GENERATORS

BUFFERS, CLOCK/MEMORY DRIVERS

DESCRIPTION	TYPE	STD TTL	ALS	AS	H	LS	S	VOLUME
Hex 2-Input Positive-NAND	'804		●	A				
Hex 2-Input Positive-NOR	'805		●	A				
Hex 2-Input Positive-AND	'808		●	A				3
Hex 2-Input Positive-OR	'832		●	A				
Hex Inverter	'1004		●	●				
Hex Buffer	'34		▲	●				
	'1034		●	●				
Quad 2-Input Positive-NAND	'37	●				●	●	2
	'1000		A					3
Quad 2-Input Positive-NOR	'28	●	A			●		2
	'1002		A					
	'1036			●				
Quad 2-Input Positive-AND	'1008		A	●				
Quad 2-Input Positive-OR	'1032		A	●				
Triple 3-Input Positive-NAND	'1010		A					3
Triple 3-Input Positive-AND	'1011		A					
Triple 4-Input AND-NAND	'800			▲				
Triple 4-Input OR-NOR	'802			▲				
Dual 4-Input Positive-NAND	'40	●			●	●	●	2
	'1020		A					3
Line Driver/Memory Driver with Series Damping Resistor	'436					●		2
Line Driver/Memory Driver	'437					●		

BI-/TRI-DIRECTIONAL BUS TRANSCEIVERS AND DRIVERS

DESCRIPTION	TYPE OF OUTPUT	TYPE	ALS	AS	LS	S	VOLUME
Quad with Bit Direction Controls	3-State	'446			●		
	3-State	'449			●		
Quad Tridirection	OC	'440			●		
	OC	'441			●		2
	3-State	'442			●		
	3-State	'443			●		
	3-State	'444			●		
	OC	'448			●		
4-Bit with Storage	3-State	'226				●	
Controller and Bus Driver for 8080A Systems		'428				●	4

OCTAL BUS TRANSCEIVERS/MOS DRIVERS

DESCRIPTION	TYPE	STD TTL	ALS	AS	LS	S	VOLUME
Inverting Outputs, 3-State	'2620			▲			
	'2640			▲			3
True Outputs, 3-State	'2623			▲			
	'2645			▲			

OCTAL BI-/TRI-DIRECTIONAL BUS TRANSCEIVERS

DESCRIPTION	TYPE OF OUTPUT	TYPE	ALS	AS	LS	VOLUME
	3-State	'245	A	▲		3
					●	2
	OC	'621	A	●		3
					●	2
12 mA/24 mA/48 mA/64 mA Sink, True Outputs	Low Power	3-State	'623	A	▲	3
					●	2
	OC, 3-State	'639	A	●		3
					●	2
	3-State	'652	▲	●		3
					●	2
	OC, 3-State	'654	▲			3
					●	2
	Very Low Power	OC	'1621	▲		3
		3-State	'1623	▲		
		OC, 3-State	'1639	▲		
	3-State	'620	A	●		3
					●	2
	OC	'622	A	●		3
					●	2
12 mA/24 mA/48 mA/64 mA Sink, Inverting Outputs	Low Power	OC, 3-State	'638	A	●	3
					●	2
	3-State	'651	▲	●		3
					●	2
	OC, 3-State	'653	▲			3
					●	2
	Very Low Power	3-State	'1620	▲		
		OC	'1622	▲		3
		OC, 3-State	'1638	▲		
12 mA/24 mA/48 mA/64 mA Sink, True Outputs	Low Power	OC	'641	A	●	2
		3-State	'645	A	●	3
					●	2
	Very Low Power	OC	'1641	▲		3
		3-State	'1645	▲		
		3-State	'640	A	●	3
					●	2
12 mA/24 mA/48 mA/64 mA Sink, Inverting Outputs	Low Power	OC	'642	A	●	3
					●	2
	Very Low Power	3-State	'1640	A		
		3-State	'1642	▲		3
		3-State	'643	A	●	2
12 mA/24 mA/48 mA/64 mA Sink, True and Inverting Outputs	Low Power	OC	'644	A	●	3
					●	2
	Very Low Power	3-State	'1643	▲		3
		OC	'1644	▲		
Registered with Multiplex 12 mA/24 mA/48 mA/64 mA True Outputs		3-State	'646		●	2
		OC	'647		●	2
Registered with Multiplexed 12 mA/24 mA/48 mA/64 mA Inverting Outputs		3-State	'648	A	●	3
					●	2
		OC	'649	▲		3
					●	2
Universal Transceiver/ Port Controllers		3-State	'877	▲		
			'852	▲		3
			'856	▲		

● Denotes available technology.
▲ Denotes planned new products.
A Denotes "A" suffix version available in the technology indicated.

FUNCTIONAL INDEX

FLIP-FLOPS

DUAL AND SINGLE FLIP-FLOPS

DESCRIPTION	TYPE	STD TTL	ALS	AS	H	L	LS	S	VOLUME
Dual J-K Edge-Triggered	'73						A		2
	'76						A		
	'78						A		
	'103			●					
	'106			●					
	'107						A		
	'108			●					
	'109	●				A			
			●	●					3
	'112					A	●		2
			A	▲					3
	'113					A	●		2
			A	▲					3
	'114					A	●		2
			A	▲					3
Single J-K Edge-Triggered	'70	●							
	'101			●					
	'102			●					
Dual Pulse-Triggered	'73	●			●	●			2
	'76	●			●	●			
	'78	●			●	●			
	'107	●							
Single Pulse-Triggered	'71				●	●			
	'72	●			●	●			
	'104	●							
	'105	●							
Dual J-K with Data Lockout	'111	●							
Single J-K with Data Lockout	'110	●							
Dual D-Type	'74	●			●	●	A	●	
			●	●					3

QUAD AND HEX FLIP-FLOPS

DESCRIPTION	NO. OF FFs	OUTPUTS	TYPE	STD TTL	ALS	AS	LS	S	VOLUME
D Type	6	Q	'174	●			●	●	2
			'378		●	●	●		3
	4	Q, Q̄	'171				●	●	2
			'175	●	●	●	●	●	3
			'379				●		
J-K	4	Q	'276	●					2
			'376	●					

OCTAL, 9-BIT, AND 10-BIT D-TYPE FLIP-FLOPS

DESCRIPTION	NO. OF BITS	OUTPUT	TYPE	STD TTL	ALS	AS	LS	S	VOLUME
True Data	Octal	3-State	'374	●	●		●		3
		3-State	'574		●	●			2
True Data with Clear	Octal	2-State	'273	●	●		●		3
		3-State	'575		●	●			2
		3-State	'874		●	●			
		3-State	'878		●	●			3
True with Enable	Octal	3-State	'377	●		●			2
Inverting	Octal	3-State	'534	●	●				
		3-State	'564	●					
		3-State	'576		●	●			
Inverting with Clear	Octal	3-State	'577		●	●			3
		3-State	'879		●	●			
Inverting with Preset	Octal	3-State	'876		●	●			
True	Octal	3-State	'825		▲				
Inverting	Octal	3-State	'826		▲				
True	9-Bit	3-State	'823		▲				
Inverting	9-Bit	3-State	'824		▲				
True	10-Bit	3-State	'821		●				
Inverting	10-Bit	3-State	'822		●				

● Denotes available technology.
▲ Denotes planned new products.
A Denotes "A" suffix version available in the technology indicated.

FUNCTIONAL INDEX

LATCHES AND MULTIVIBRATORS

QUAD LATCHES

DESCRIPTION	OUTPUT	TYPE	STD TTL	ALS	AS	L	LS	VOLUME
Dual 2-Bit Transparent	2-State	'75				●	●	2
	2-State	'77	●			●	●	
	2-State	'375					●	
S-R	2-State	'279	●				A	

RETRIGGERABLE MONOSTABLE MULTIVIBRATORS

DESCRIPTION	TYPE	STD TTL	ALS	AS	LS	L	VOLUME
Single	'122	●			●	●	
	'130	●					2
	'422				●		
Dual	'123	●			●	●	
	'423				●		

OCTAL, 9-BIT, AND 10-BIT LATCHES

DESCRIPTION	NO. OF BITS	OUTPUT	TYPE	STD TTL	ALS	AS	LS	S	VOLUME
Transparent	Octal	3-State	'268					●	2
			'373			●	●		3
		3-State	'573		●	●			
Dual 4-Bit Transparent	Octal	2-State	'100	●					2
		2-State	'116	●					
		3-State	'873		●	●			
Inverting Transparent	Octal	3-State	'533				●		
		3-State	'563		●				3
		3-State	'580		●	●			
Dual 4-Bit Inverting Transparent	Octal	3-State	'880		●	●			
2-Input Multiplexed	Octal	3-State	'604				●		
		OC	'605				●		
		3-State	'606				●		2
		OC	'607				●		
Addressable	Octal	2-State	'259	●	▲				3
Multi-Mode Buffered	Octal	3-State	'412					●	2
True	Octal	3-State	'845		▲	▲			
Inverting	Octal	3-State	'846		▲	▲			
True	9-Bit	3-State	'843		▲	▲			3
Inverting	9-Bit	3-State	'844		▲	▲			
True	10-Bit	3-State	'841		▲	▲			
Inverting	10-Bit	3-State	'842		▲	▲			

MONOSTABLE MULTIVIBRATORS WITH SCHMITT-TRIGGER INPUTS

DESCRIPTION	TYPE	STD TTL	ALS	AS	LS	S	VOLUME
Single	'121	●					2
Dual	'221	●			●		

● Denotes available technology.
▲ Denotes planned new products.
A Denotes "A" suffix version available in the technology indicated.

TEXAS INSTRUMENTS
POST OFFICE BOX 225012 • DALLAS, TEXAS 75265

FUNCTIONAL INDEX

REGISTERS AND PROGRAMMABLE LOGIC ARRAYS

SHIFT REGISTERS

DESCRIPTION	NO. OF BITS	MODES S-R	MODES LOAD	MODES HOLD	TYPE	STD TTL	ALS	AS	L	LS	S	VOLUME
Sign-Protected		X	X	X	'322					A		
		X	X	X	X	'198	●					2
Parallel-In, Parallel-Out, Bidirectional	8	X	X	X	X	'299				●	●	3
		X	X	X	X	'323		●	▲		●	2
												3
	4	X	X	X	X	'194	●			A	●	2
							▲					3
Parallel-In, Parallel-Out, Registered Outputs	4	X	X	X	X	'671					●	
		X	X	X	X	'672					●	2
	8	X		X	X	'199	●					
	5	X		X		'96	●			●	●	
		X		X		'95	A		B			2
								▲				3
		X		X		'99			●			
Parallel-In, Parallel-Out	4	X		X	X	'178	●					2
		X		X	X	'179	●					
		X		X		'195	●			A		2
							▲					3
		X		X		'295			B			2
		X		X		'395			A			2
							▲					3
Serial-In Parallel-Out	16	X		X	X	'673				●		2
	8	X				'164	●	▲	●	●		3
	16	X		X	X	'674			A			2
Parallel-In, Serial-Out	8	X		X	X	'165	●	▲	A			2
		X		X	X	'166	●	▲	A			3
Serial-In, Serial-Out	8	X				'91	A		●	●		2
	4	X		X		'94	A					

SHIFT REGISTERS WITH LATCHES

DESCRIPTION	NO. OF BITS	OUTPUTS	TYPE	ALS	AS	LS	VOLUME
Parallel-In, Parallel-Out with Output Latches	4	3-State	'671			●	
		3-State	'672			●	
Serial-In, Parallel-Out with Output Latches	16	2-State	'673			●	
	8	Buffered	'594			●	
		3-State	'595			●	
		OC	'596			●	2
		OC	'599			●	
Parallel-In, Serial-Out, with Input Latches	8	2-State	'597			●	
		3-State	'589			●	
Parallel I/O Ports with Input Latches, Multiplexed Serial Inputs	8	3-State	'598			●	

SIGN-PROTECTED REGISTERS

DESCRIPTION	NO. OF BITS	MODES S-R	MODES LOAD	MODES HOLD	TYPE	ALS	AS	LS	VOLUME
Sign-Protected Register	8	X	X	X	'322			A	2

REGISTER FILES

DESCRIPTION	OUTPUT	TYPE	STD TTL	ALS	AS	LS	VOLUME
8 Words × 2 Bits	3-State	'172	●				
4 Words × 4 Bits	OC	'170	●			●	2
	3-State	'670				●	
Dual 16 Words × 4 Bits	3-State	'870		▲			3
	3-State	'871		▲			

OTHER REGISTERS

DESCRIPTION	TYPE	STD TTL	ALS	AS	L	LS	S	VOLUME
Quadruple Multiplexers with Storage	'98				●			
	'298	●				●		2
	'398/9					●		
8-Bit Universal Shift Registers	'299		●	▲		●	●	3
Quadruple Bus-Buffer Registers	'173	●			A			2
Octal Storage Register	'396					●		

PROGRAMMABLE LOGIC ARRAYS

DESCRIPTION	INPUTS	OUTPUTS NO.	OUTPUTS TYPE	TYPE NO	NO. OF PINS	VOLUME
	16	8	Active-Low	'PAL16L8	20	
		8		'PAL16R8		
		6	Registered	'PAL16R6		
		4		'PAL16R4		
	19 Registered ▲	8	Active-Low	'PLR19L8	24	
		8		'PLR19R8		
		6	Registered	'PLR19R6		
Fixed-OR Arrays		4		'PLR19R4		4
	19 Latched ▲	8	Active-Low	'PLT19L8	24	
		8		'PLT19R8		
		6	Registered	'PLT19R6		
		4		'PLT19R4		
	20 ▲	8	Active-Low	'PL20L8	24	
		8		'PL20R8		
		6	Registered	'PL20R6		
		4		'PL20R4		
Field-Programmable 14 × 32 × 6 Logic Arrays	14	6	3-State	'PL839	24	
			OC	'PL840		

● Denotes available technology.
▲ Denotes planned new products.
A Denotes "A" suffix version available in the technology indicated.
B Denotes "B" suffix version available in the technology indicated.

TEXAS INSTRUMENTS
POST OFFICE BOX 225012 • DALLAS, TEXAS 75265

FUNCTIONAL INDEX

COUNTERS

SYNCHRONOUS COUNTERS — POSITIVE-EDGE TRIGGERED

DESCRIPTION	PARALLEL LOAD	TYPE	STD TTL	ALS	AS	L	LS	S	VOLUME
Decade	Sync	'160	●				A		2
				A	▲				3
	Sync	'162	●				A	●	2
				A	▲				3
	Sync	'560	A						
	Sync	'668					●		
	Sync	'690					●		2
	Sync	'692					B	●	
Decade Up/Down	Sync	'168		A	▲				3
	Async	'190	●				●		2
				●					3
	Async	'192	●			●	●		2
				●					3
	Sync	'568	A						
	Sync	'696					●		
	Sync	'698					●		2
Decade Rate Multipler, 1/N10	Async Set-to-9	'167	●						
4-Bit Binary	Sync	'161	●				A		3
				A	▲				
	Sync	'163	●				A	●	2
				A	▲				3
	Sync	'561	A						
	Sync	'669					●		
	Sync	'691					●		2
	Sync	'693					B	●	
4-Bit Binary Up/Down	Sync	'169		A	▲				3
	Async	'191	●				●		2
				●					3
	Async	'193	●			●	●		2
				●					3
	Sync	'569	A						
	Sync	'697					●		
	Sync	'699					●		2
6-Bit Binary Rate Multiplier, 1/N2		'97	●						
8-Bit Up/Down	Async CLR	'867					●		3
	Sync CLR	'869					●		

ASYNCHRONOUS COUNTERS (RIPPLE CLOCK) — NEGATIVE-EDGE TRIGGERED

DESCRIPTION	PARALLEL LOAD	TYPE	STD TTL	ALS	AS	L	LS	S	VOLUME
Decade	Set-to-9	'90	A			●	●		
		'68					●		
	Yes	'176	●						
	Yes	'196	●				●	●	
	Set-to-9	'290	●				●		
4-Bit Binary	None	'93	A			●	●		2
		'69					●		
	Yes	'177	●						
	Yes	'197	●				●	●	
	None	'293	●				●		
Divide-by-12	None	'92	A				●		
Dual Decade	None	'390	●				●		
	Set-to-9	'490	●				●		
Dual 4-Bit Binary	None	'393	●				●		

8-BIT BINARY COUNTERS WITH REGISTERS

DESCRIPTION	TYPE OF OUTPUT	TYPE	ALS	AS	LS	VOLUME
Parallel Register Outputs	3-State	'590			●	2
	OC	'591			●	
Parallel Register Inputs	2-State	'592			●	
Parallel I/O	3-State	'593			●	

FREQUENCY DIVIDERS, RATE MULTIPLIERS

DESCRIPTION	TYPE	STD TTL	ALS	AS	LS	VOLUME
50-to-1 Frequency Divider	'56				●	2
60-to-1 Frequency Divider	'57				●	
60-Bit Binary Rate Multiplier,	'97	●				
Decade Rate Multiplier,	'167	●				

● Denotes available technology.
▲ Denotes planned new products.
A Denotes "A" suffix version available in the technology indicated.
B Denotes "B" suffix version available in the technology indicated.

GENERAL INFORMATION

TEXAS INSTRUMENTS
POST OFFICE BOX 225012 • DALLAS, TEXAS 75265

FUNCTIONAL INDEX

DECODERS, ENCODERS, DATA SELECTORS/MULTIPLEXERS AND SHIFTERS

DATA SELECTORS/MULTIPLEXERS

DESCRIPTION	TYPE OF OUTPUT	TYPE	STD TTL	ALS	AS	L	LS	S	VOLUME
16-To-1	2-State	'150	●						2
	3-State	'250		▲					
	3-State	'850			▲				3
	3-State	'851			▲				
Dual 8-To-1	3-State	'351	●						2
8-To-1	2-State	'151	A				●	●	3
				●	●				
	2-State	'152	A				●		2
			●				●	●	
	3-State	'251		●	▲		●	●	3
	3-State	'354					●		
	2-State	'355					●		
	3-State	'356					●		2
	OC	'357					●		
Dual 4-To-1	2-State	'153	●			●	●	●	3
				●	●			●	2
	3-State	'253		●	●		●		3
	2-State	'352					●		2
				●			●		3
	3-State	'353					●		2
				●			●		3
Octal 2-To-1 with Storage	3-State	'604					●		
	OC	'605					●		2
	3-State	'606					●		
	OC	'607					●		
Quad 2-To-1 with Storage	2-State	'98				●			
	2-State	'298	●				●		2
				▲					3
	2-State	'398					●		
	2-State	'399					●		2
Quad 2-To-1	2-State	'157	●			●	●	●	3
				●	●				2
	2-State	'158				●	●	●	3
	3-State	'257				B	●		2
				●	●				3
	3-State	'258				B	●		2
6-to-1 Universal Multiplexer	3-State	'857					●	●	3

DECODERS/DEMULTIPLEXERS

DESCRIPTION	TYPE OF OUTPUT	TYPE	STD TTL	ALS	AS	L	LS	S	VOLUME
4-To-16	3-State	'154	●			●			2
	OC	'159	●						
4-To-10 BCD-To-Decimal	2-State	'42	A			●	●		2
4-To-10 Excess 3-To-Decimal	2-State	'43	A			●			
4-To-10 Excess 3-Gray-To-Decimal	2-State	'44	A			●			
3-To-8 with Address Latches		'131		●	▲				3
	2-State	'137		●	▲		●		2
3-To-8	2-State	'138		●	▲		●		3
							●	●	2
	3-State	'538		▲			●		3
Dual 2-To-4	2-State	'139		▲	●	A	●		3
	2-State	'155	●			A	●		2
	OC	'156	●				●		
Dual 1-To-4 Decoders	3-State	'539		▲					3

CODE CONVERTERS

DESCRIPTION	TYPE	STD TTL	S	VOLUME
6-Line-BCD to 6-Line Binary, Or 4-Line to 4-Line BCD 9's/BCD 10's Converters	'184	●		2
6-Bit-Binary to 6-Bit BCD Converters	'185	A		
BCD-to-Binary Converters	'484		A	4
Binary-to-BCD Converters	'485		A	

PRIORITY ENCODERS/REGISTERS

DESCRIPTION	TYPE	STD TTL	ALS	AS	LS	VOLUME
Full BCD	'147	●			●	2
Cascadable Octal	'148	●			●	
Cascadable Octal with 3-State Outputs	'348	●			●	
4-Bit Cascadable with Registers	'278	●				

SHIFTERS

DESCRIPTION	OUTPUT	TYPE	STD TTL	ALS	AS	L	LS	S	VOLUME
4-Bit Shifter	3-State	'350					●		2
Parallel 16-Bit Multi-Mode Barrel Shifter	3-State	'897			▲				4

● Denotes available technology.
▲ Denotes planned new products.
A Denotes "A" suffix version available in the technology indicated.
B Denotes "B" suffix version available in the technology indicated.

FUNCTIONAL INDEX

DISPLAY DECODERS/DRIVERS, MEMORY/MICROPROCESSOR CONTROLLERS, AND VOLTAGE-CONTROLLED OSCILLATORS

OPEN-COLLECTOR DISPLAY DECODERS/DRIVERS

DESCRIPTION	OFF-STATE OUTPUT VOLTAGE	TYPE	TECHNOLOGY				VOLUME
			STD TTL	ALS	AS	L	LS	
BCD-To-Decimal	30 V	'45	●					
	60 V	'141	●					
	15 V	'145					●	
	7 V	'445					●	
BCD-To-Seven-Segment	30 V	'46	A			●		2
	15 V	'47	A			●	●	
	5.5 V	'48	●				●	
	5.5 V	'49	●				●	
	30 V	'246	●					
	15 V	'247	●				●	
	7 V	'347					●	
	7 V	'447	●				●	
	5.5 V	'248	●				●	
	5.5 V	'249	●				●	

OPEN COLLECTOR DISPLAY DECODERS/DRIVERS WITH COUNTERS/LATCH

DESCRIPTION	TYPE	TECHNOLOGY			VOLUME
		STD TTL	ALS	AS	
BCD Counter/4-Bit Latch/BCD-To-Decimal Decoder/Driver	'142	●			
BCD Counter/4-Bit Latch/BCD-To-Seven-Segment Decoder/Lad Driver	'143	●			2
BCD Counter/4-Bit Latch/BCD-To-Seven-Segment Decoder/Lamp Driver	'144	●			

VOLTAGE-CONTROLLED OSCILLATORS

DESCRIPTION					TYPE	TECHNOLOGY		VOLUME
No. VCO'S	COMP'L Z_{OUT}	ENABLE	RANGE INPUT	R_{ext}	f_{max} MHz		LS	S	
Single	Yes	Yes	Yes	No	20	'624	●		
Single	Yes	Yes	Yes	Yes	20	'628	●		
Dual	No	Yes	Yes	No	60	'124		●	2
Dual	Yes	Yes	No	No	20	'626	●		
Dual	No	No	No	No	20	'627	●		
Dual	No	Yes	Yes	No	20	'629	●		

MEMORY/MICROPROCESSOR CONTROLLERS

DESCRIPTION			TYPE	TECHNOLOGY				VOLUME
				ALS	AS	LS	S	
System Controllers For 8080A			'428				●	
System Controller, Universal			'482			●		
System Controllers, Universal			'890	▲				4
(or For '888, '889)			'891	▲				
Memory Refresh Controllers	Transparent, Burst Modes	4K, 16K	'600		A			
		64K	'601		A			
	Cycle Steal, Burst Modes	4K, 16K	'602		A			
		64K	'603		A			
Memory Cycle Controller			'608			●		2
Memory Mappers		3-State	'612			●		
		OC	'613			●		
Memory Mappers With Output Latches		3-State	'610			●		
		OC	'611			●		
Multi-Mode Latches (8080A Applications)			'412				●	

CLOCK GENERATOR CIRCUITS

DESCRIPTION	TYPE	TECHNOLOGY					VOLUME
		STD TTL	ALS	AS	LS	S	
Quadruple Complementary-Output Logic Elements	'265	●					
Dual Pulse Synchronizers/Drivers	'120	●					
Crystal-Controlled Oscillators	'320				●		2
	'321				●		
Digital Phase-Lock Loop	'297				●		
Programmable Frequency Dividers/Digital Timers	'292				●		
	'294				●		
Triple 4-Input AND/NAND Drivers	'800		▲				3
Triple 4-Input OR/NOR Drivers	'802		▲				
Dual VCO	'124					●	2

GENERAL INFORMATION 1

RESULTANT DISPLAYS USING '46A, '47A, '48, '49, 'L46, 'L47, 'LS47, 'LS48, 'LS49, 'LS347

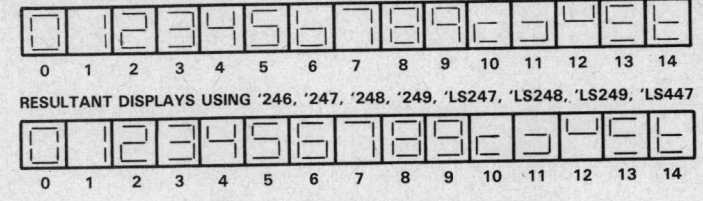

RESULTANT DISPLAYS USING '246, '247, '248, '249, 'LS247, 'LS248, 'LS249, 'LS447

RESULTANT DISPLAYS USING '143, '144

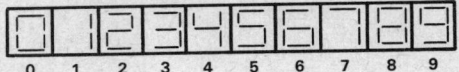

● Denotes available technology.
▲ Denotes planned new products.
A Denotes "A" suffix version available in the technology indicated.

FUNCTIONAL INDEX

ARITHMETIC CIRCUITS, ERROR DETECTION CIRCUITS, AND PROCESSOR ELEMENTS

4-BIT COMPARATORS

DESCRIPTION			OUTPUT ENABLE	TYPE	TECHNOLOGY						VOLUME
P=Q	P>Q	P<Q			STD TTL	ALS	AS	L	S	LS	
Yes	Yes	No	2-State	Yes	'85	●			●	●	●	2

8-BIT COMPARATORS

INPUTS	DESCRIPTION				OUTPUT ENABLE	TYPE	TECHNOLOGY			VOLUME
	P=Q	P>Q	P<Q	OUTPUT			ALS	AS	LS	
20-kΩ Pull-Up	Yes	No	No	OC	Yes	'518	●			3
	No	Yes	No	2-State	Yes	'520	●			
	No	No	Yes	OC	Yes	'522	●			
	Yes	No	Yes	2-State	No	'682			●	2
	Yes	No	Yes	OC	No	'683			●	
Standard	Yes	No	No	OC	Yes	'519	●			3
	No	Yes	No	2-State	Yes	'521	●			
	Yes	No	Yes	2-State	No	'684			●	
	Yes	No	Yes	OC	No	'685			●	2
	Yes	No	Yes	2-State	Yes	'686			●	
	Yes	No	Yes	OC	Yes	'687			●	
	No	Yes	No	2-State	Yes	'688	●		●	3
										2
	No	Yes	No	OC	Yes	'689	●		●	3
										2
Latched P	No	No	Yes	2-State	Yes	'885		●		
Latched P and Q	Yes	No	Yes	Latched	Yes	'866		●		3

ADDRESS COMPARATORS

DESCRIPTION	OUTPUT ENABLE	LATCHED OUTPUT	TYPE	TECHNOLOGY		VOLUME
				ALS	AS	
16-Bit to 4-Bit	Yes		'677	●		3
		Yes	'678	▲		
12-Bit to 4-Bit	Yes		'679	●		
		Yes	'680	▲		

PARITY GENERATORS/CHECKERS, ERROR DETECTION AND CORRECTION CIRCUITS

DESCRIPTION	NO. OF BITS	TYPE	TECHNOLOGY				VOLUME
			STD TTL	ALS	AS	LS	S	
Odd/Even Parity Generators/Checkers	8	'180	●				2
	9	'280			●	●	3
	9	'286		▲			3
Parallel Error Detection/Correction Circuits	3-State 8	'636				●	2
	OC 8	'637				●	
	3-State 16	'630				●	
	OC 16	'631				●	
	3-State 32	'632	●				3
	OC 32	'633	▲				
	3-State 32	'634	▲				
	OC 32	'635	▲				

FUSE-PROGRAMMABLE COMPARATORS

DESCRIPTION	TYPE	TECHNOLOGY				VOLUME
		STD TTL	ALS	AS	LS	S	
16-Bit Identity Comparator	'526		▲			3
12-Bit Identity Comparator	'528		▲			
8-Bit Identity Comparator and 4-Bit Comparator	'527		▲			

PARALLEL BINARY ADDERS

DESCRIPTION	TYPE	TECHNOLOGY						VOLUME
		STD TTL	ALS	AS	H	LS	S	
1-Bit Gated	'80	●						
2-Bit	'82	●						
4-Bit	'83	A			A			2
	'283	●				●	●	
Dual 1-Bit Carry-Save	'183	●				●	●	

ACCUMULATORS, ARITHMETIC LOGIC UNITS, LOOK-AHEAD CARRY GENERATORS

DESCRIPTION	TYPE	TECHNOLOGY					VOLUME
		STD TTL	ALS	AS	LS	S	
4-Bit Parallel Binary Accumulators	'281			●			2
	'681		●				
4-Bit Arithmetic Logic Units/ Function Generators	'181	●	A				3
	'381			A			2
	'881		A				3
4-Bit Arithmetic Logic Unit with Ripple Carry	'382	●			●		2
Look-Ahead Carry Generators	16-Bit '182	●	▲				2
	'282		▲				3
	32-Bit '882		▲				3
Quad Serial Adder/Subtractor	'385			●			2
4-Bit Slice Elements	'481			●	●		
8-Bit Slice Elements	'888		▲				4
	'889		▲				

MULTIPLIERS

DESCRIPTION	TYPE	TECHNOLOGY				VOLUME
		STD TTL	ALS	AS	LS	S	
2-Bit-by-4-Bit Parallel Binary Multipliers	'261			●			
4-Bit-by-4-Bit Parallel Binary Multipliers	'274					●	
	'284	●					
	'285	●					2
25-MHz 6-Bit Binary Rate Multipliers	'97	●					
25-MHz Decade Rate Multipliers	'167	●					
8-Bit × 1-Bit 2's Complement Multipliers	'384			●			
16-Bit Parallel Multiplier	'1616		▲				4

OTHER ARITHMETIC OPERATORS

DESCRIPTION	TYPE	TECHNOLOGY						VOLUME
		STD TTL	ALS	AS	H	L	LS	S	
Quad 2-Input Exclusive-OR Gates with Totem-Pole Outputs	'86	●			●	A	●		2
	'386		▲				A		3
Quad 2-Input Exclusive-OR Gates with Open-Collector Outputs	'136	●					●		2
Quad 2-Input Exclusive-NOR Gates	'266						●		2
Quad Exclusive OR/NOR Gates	'135							●	
4-Bit True/Complement Element	'87			●					

BIPOLAR BIT-SLICE PROCESSOR ELEMENTS

DESCRIPTION	CASCADABLE TO N-BITS	TYPE	TECHNOLOGY			VOLUME
			ALS	AS	LS	
4-Bit-Slice	Yes	'481		●	●	
8-Bit-Slice	Yes	'888	▲			4
	Yes	'889	▲			

● Denotes available technology.
▲ Denotes planned new products.
A Denotes "A" suffix version available in the technology indicated.

TEXAS INSTRUMENTS
POST OFFICE BOX 225012 • DALLAS, TEXAS 75265

FUNCTIONAL INDEX

MEMORIES

USER-PROGRAMMABLE READ-ONLY MEMORIES (PROM's)
STANDARD PROM's

DESCRIPTION	TYPE	ORGANIZATION	TYPE OUTPUT	VOLUME
16K-Bit Arrays	▲ TBP28S165	2048W × 8B	3-State	
	TBP28S166	2048W × 8B	3-State	
8K-Bit Arrays	TBP24S81	2048W × 4B	3-State	
	TBP24SA81	2048W × 4B	OC	
	TBP28S86A	1024W × 8B	3-State	
	TBP28SA86A	1024W × 8B	OC	
	▲ TBP28S85A	1024W × 8B	3-State	
4K-Bit Arrays	TBP28S42	512W × 8B	3-State	4
	TBP28SA42	512W × 8B	OC	
	▲ TBP28S45	512W × 8B	3-State	
	TBP28S46	512W × 8B	3-State	
	TBP28SA46	512W × 8B	OC	
	TBP24S41	1024W × 4B	3-State	
	TBP24SA41	1024W × 4B	OC	
1K-Bit Arrays	TBP24S10	256W × 4B	3-State	
	TBP24SA10	256W × 4B	OC	
256-Bit Arrays	TBP18S030	32W × 8B	3-State	
	TBP18SA030	32W × 8B	OC	

LOW-POWER PROM's

DESCRIPTION	TYPE	ORGANIZATION	TYPE OUTPUT	VOLUME
16K-Bit Arrays	▲ TBP28L165	2048W × 8B	3-State	
	▲ TBP28L166	2048W × 8B	3-State	
8K-Bit Arrays	▲ TBP28L85A	1024W × 8B	3-State	
	TBP28L86A	1024W × 8B	3-State	
4K-Bit Arrays	TBP28L42	512W × 8B	3-State	4
	▲ TBP28L45	512W × 8B	3-State	
	TBP28L46	512W × 8B	3-State	
2K-Bit Arrays	TBP28L22	256W × 8B	3-State	
	TBP28LA22	256W × 8B	OC	

READ-ONLY MEMORIES (ROM's)

DESCRIPTION	ORGANIZATION	TYPE OF OUTPUT	TYPE	STD TTL	ALS	AS	S	VOLUME
1024-Bit Arrays	256 × 4	OC	'187	●				4
256-Bit Arrays	32 × 8	OC	'88	A				

RANDOM-ACCESS READ-WRITE MEMORIES (RAM's)

DESCRIPTION	ORGANIZATION	TYPE OF OUTPUT	TYPE	STD TTL	ALS	AS	LS	S	VOLUME
256-Bit Arrays	256 × 1	3-State	'201					●	
		OC	'301					●	
64-Bit Arrays	16 × 4	OC	'89	●					4
		3-State	'189			A		B	
		3-State	'219			A			
		OC	'289			A		B	
		OC	'319			A			
16-Bit Multiple-Port Register File	8 × 2	3-State	'172	●					2
16-Bit Register File	4 × 4	OC	'170				●		
		3-State	'670				●		
Dual 64-Bit Register Files	16 × 4	3-State	'870				●		3
			'871				●		

FIRST-IN FIRST-OUT MEMORIES (FIFO'S)

DESCRIPTION	TYPE OF OUTPUT	TYPE	ALS	AS	LS	LS	VOLUME
Asynchronous 16 × 5	3-State	'225			●		
Asynchronous 16 × 4	3-State	'222			●		4
	3-State	'224			●		
	OC	'227			●		
	OC	'228			●		

● Denotes available technology.
▲ Denotes planned new products.
A Denotes ''A'' suffix version available in the technology indicated.

GENERAL INFORMATION

TEXAS INSTRUMENTS
POST OFFICE BOX 225012 ● DALLAS, TEXAS 75265

The TTL Data Book
Volume 3

General Information — 1

ALS and AS Circuits — 2

Mechanical Data — 3

2
ALS AND AS CIRCUITS

TYPES SN54ALS00A, SN54AS00, SN74ALS00A, SN74AS00
QUADRUPLE 2-INPUT POSITIVE-NAND GATES

D2661, APRIL 1982 – REVISED DECEMBER 1983

- Package Options Include Both Plastic and Ceramic Chip Carriers in Addition to Plastic and Ceramic DIPs
- Dependable Texas Instruments Quality and Reliability

description

These devices contain four independent 2-input NAND gates. They perform the Boolean functions $Y = \overline{A \cdot B}$ or $Y = \overline{A} + \overline{B}$ in positive logic.

The SN54ALS00A and SN54AS00 are characterized for operation over the full military temperature range of $-55\,°C$ to $125\,°C$. The SN74ALS00A and SN74AS00 are characterized for operation from $0\,°C$ to $70\,°C$.

SN54ALS00A, SN54AS00 . . . J PACKAGE
SN74ALS00A, SN74AS00 . . . N PACKAGE
(TOP VIEW)

```
1A [ 1   14 ] VCC
1B [ 2   13 ] 4B
1Y [ 3   12 ] 4A
2A [ 4   11 ] 4Y
2B [ 5   10 ] 3B
2Y [ 6    9 ] 3A
GND[ 7    8 ] 3Y
```

FUNCTION TABLE (each gate)

INPUTS		OUTPUT
A	B	Y
H	H	L
L	X	H
X	L	H

SN54ALS00A, SN54AS00 . . . FH PACKAGE
SN74ALS00A, SN74AS00 . . . FN PACKAGE
(TOP VIEW)

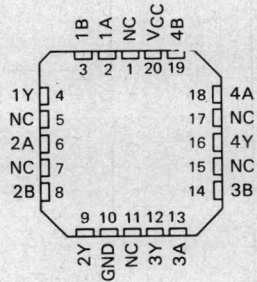

NC – No internal connection

logic symbol

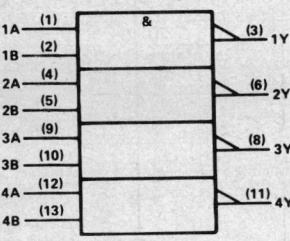

Pin numbers shown are for J and N packages.

TYPES SN54ALS00A, SN74ALS00A
QUADRUPLE 2-INPUT POSITIVE-NAND GATES

absolute maximum ratings over operating free-air temperature range (unless otherwise noted)

Supply voltage, V_{CC} .... 7 V
Input voltage .... 7 V
Operating free-air temperature range: SN54ALS00A .... $-55\,°C$ to $125\,°C$
                                     SN74ALS00A .... $0\,°C$ to $70\,°C$
Storage temperature range .... $-65\,°C$ to $150\,°C$

recommended operating conditions

		SN54ALS00A			SN74ALS00A			UNIT
		MIN	NOM	MAX	MIN	NOM	MAX	
V_{CC}	Supply voltage	4.5	5	5.5	4.5	5	5.5	V
V_{IH}	High-level input voltage	2			2			V
V_{IL}	Low-level input voltage			0.8			0.8	V
I_{OH}	High-level output current			-0.4			-0.4	mA
I_{OL}	Low-level output current			4			8	mA
T_A	Operating free-air temperature	-55		125	0		70	°C

electrical characteristics over recommended operating free-air temperature range (unless otherwise noted)

PARAMETER	TEST CONDITIONS		SN54ALS00A			SN74ALS00A			UNIT
			MIN	TYP†	MAX	MIN	TYP†	MAX	
V_{IK}	$V_{CC} = 4.5$ V,	$I_I = -18$ mA			-1.5			-1.5	V
V_{OH}	$V_{CC} = 4.5$ V to 5.5 V,	$I_{OH} = -0.4$ mA	$V_{CC}-2$			$V_{CC}-2$			V
V_{OL}	$V_{CC} = 4.5$ V,	$I_{OL} = 4$ mA		0.25	0.4		0.25	0.4	V
	$V_{CC} = 4.5$ V,	$I_{OL} = 8$ mA					0.35	0.5	
I_I	$V_{CC} = 5.5$ V,	$V_I = 7$ V			0.1			0.1	mA
I_{IH}	$V_{CC} = 5.5$ V,	$V_I = 2.7$ V			20			20	µA
I_{IL}	$V_{CC} = 5.5$ V,	$V_I = 0.4$ V			-0.1			-0.1	mA
I_O‡	$V_{CC} = 5.5$ V,	$V_O = 2.25$ V	-15		-70	-15		-70	mA
I_{CCH}	$V_{CC} = 5.5$ V,	$V_I = 0$ V		0.5	0.85		0.5	0.85	mA
I_{CCL}	$V_{CC} = 5.5$ V,	$V_I = 4.5$ V		1.5	3		1.5	3	mA

†All typical values are at $V_{CC} = 5$ V, $T_A = 25\,°C$.
‡The output conditions have been chosen to produce a current that closely approximates one half of the true short-circuit output current, I_{OS}.

switching characteristics (see Note 1)

PARAMETER	FROM (INPUT)	TO (OUTPUT)	$V_{CC} = 4.5$ V to 5.5 V, $C_L = 50$ pF, $R_L = 500\,\Omega$, $T_A = $ MIN to MAX				UNIT
			SN54ALS00A		SN74ALS00A		
			MIN	MAX	MIN	MAX	
t_{PLH}	A or B	Y	3	14	3	11	ns
t_{PHL}	A or B	Y	2	10	2	8	ns

NOTE 1: For load circuit and voltage waveforms, see page 1-12.

TEXAS INSTRUMENTS
POST OFFICE BOX 225012 • DALLAS, TEXAS 75265

TYPES SN54AS00, SN74AS00
QUADRUPLE 2-INPUT POSITIVE-NAND GATES

absolute maximum ratings over operating free-air temperature range (unless otherwise noted)

Supply voltage, V_{CC} .... 7 V
Input voltage .... 7 V
Operating free-air temperature range: SN54AS00 .... −55 °C to 125 °C
                                      SN74AS00 .... 0 °C to 70 °C
Storage temperature range .... −65 °C to 150 °C

recommended operating conditions

		SN54AS00			SN74AS00			UNIT
		MIN	NOM	MAX	MIN	NOM	MAX	
V_{CC}	Supply voltage	4.5	5	5.5	4.5	5	5.5	V
V_{IH}	High-level input voltage	2			2			V
V_{IL}	Low-level input voltage			0.8			0.8	V
I_{OH}	High-level output current			−2			−2	mA
I_{OL}	Low-level output current			20			20	mA
T_A	Operating free-air temperature	−55		125	0		70	°C

electrical characteristics over recommended operating free-air temperature range (unless otherwise noted)

PARAMETER	TEST CONDITIONS		SN54AS00			SN74AS00			UNIT
			MIN	TYP†	MAX	MIN	TYP†	MAX	
V_{IK}	$V_{CC} = 4.5$ V,	$I_I = -18$ mA			−1.2			−1.2	V
V_{OH}	$V_{CC} = 4.5$ V to 5.5 V,	$I_{OH} = -2$ mA	$V_{CC}-2$			$V_{CC}-2$			V
V_{OL}	$V_{CC} = 4.5$ V,	$I_{OL} = 20$ mA		0.35	0.5		0.35	0.5	V
I_I	$V_{CC} = 5.5$ V,	$V_I = 7$ V			0.1			0.1	mA
I_{IH}	$V_{CC} = 5.5$ V,	$V_I = 2.7$ V			20			20	µA
I_{IL}	$V_{CC} = 5.5$ V,	$V_I = 0.4$ V			−0.5			−0.5	mA
I_O‡	$V_{CC} = 5.5$ V,	$V_O = 2.25$ V	−30		−112	−30		−112	mA
I_{CCH}	$V_{CC} = 5.5$ V,	$V_I = 0$ V		2	3.2		2	3.2	mA
I_{CCL}	$V_{CC} = 5.5$ V,	$V_I = 4.5$ V		10.8	17.4		10.8	17.4	mA

†All typical values are at $V_{CC} = 5$ V, $T_A = 25$ °C.
‡The output conditions have been chosen to produce a current that closely approximates one half of the true short-circuit output current, I_{OS}.

switching characteristics (see Note 1)

PARAMETER	FROM (INPUT)	TO (OUTPUT)	$V_{CC} = 4.5$ V to 5.5 V, $C_L = 50$ pF, $R_L = 500$ Ω, T_A = MIN to MAX				UNIT
			SN54AS00		SN74AS00		
			MIN	MAX	MIN	MAX	
t_{PLH}	A or B	Y	1	5	1	4.5	ns
t_{PHL}	A or B	Y	1	5	1	4	ns

NOTE 1: For load circuit and voltage waveforms, see page 1-12.

ALS AND AS CIRCUITS

TEXAS INSTRUMENTS
POST OFFICE BOX 225012 • DALLAS, TEXAS 75265

2
ALS AND AS CIRCUITS

TYPES SN54ALS01, SN74ALS01
QUADRUPLE 2-INPUT POSITIVE-NAND GATES WITH OPEN-COLLECTOR OUTPUTS

D2661, APRIL 1982—REVISED DECEMBER 1983

- Package Options Include Both Plastic and Ceramic Chip Carriers in Addition to Plastic and Ceramic DIPs
- Dependable Texas Instruments Quality and Reliability

description

These devices contain four independent 2-input NAND gates. They perform the Boolean functions $Y = \overline{A \cdot B}$ or $Y = \overline{A} + \overline{B}$ in positive logic. The open-collector outputs require pull-up resistors to perform correctly. They may be connected to other open-collector outputs to implement active-low wired-OR or active-high wired-AND functions. Open-collector devices are often used to generate higher V_{OH} levels.

The SN54ALS01 is characterized for operation over the full military temperature range of $-55\,°C$ to $125\,°C$. The SN74ALS01 is characterized for operation from $0\,°C$ to $70\,°C$.

SN54ALS01 . . . J PACKAGE
SN74ALS01 . . . N PACKAGE
(TOP VIEW)

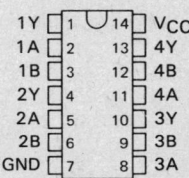

SN54ALS01 . . . FH PACKAGE
SN74ALS01 . . . FN PACKAGE
(TOP VIEW)

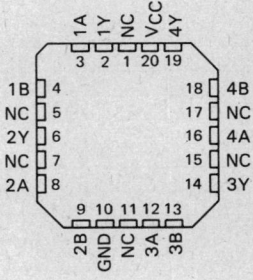

NC—No internal connection

FUNCTION TABLE (each gate)

INPUTS		OUTPUT
A	B	Y
H	H	L
L	X	H
X	L	H

logic symbol

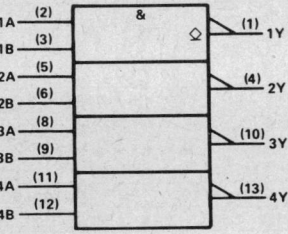

Pin numbers shown are for J and N packages.

ALS AND AS CIRCUITS

Copyright © 1982 by Texas Instruments Incorporated

TEXAS INSTRUMENTS
POST OFFICE BOX 225012 • DALLAS, TEXAS 75265

TYPES SN54ALS01, SN74ALS01
QUADRUPLE 2-INPUT POSITIVE-NAND GATES
WITH OPEN-COLLECTOR OUTPUTS

absolute maximum ratings over operating free-air temperature range (unless otherwise noted)

Supply voltage, V_{CC} .... 7 V
Input voltage .... 7 V
Off-state output voltage .... 7 V
Operating free-air temperature range: SN54ALS01 .... −55 °C to 125 °C
                                       SN74ALS01 .... 0 °C to 70 °C
Storage temperature range .... −65 °C to 150 °C

recommended operating conditions

		SN54ALS01			SN74ALS01			UNIT
		MIN	NOM	MAX	MIN	NOM	MAX	
V_{CC}	Supply voltage	4.5	5	5.5	4.5	5	5.5	V
V_{IH}	High-level input voltage	2			2			V
V_{IL}	Low-level input voltage			0.8			0.8	V
V_{OH}	High-level output voltage			5.5			5.5	V
I_{OL}	Low-level output current			4			8	mA
T_A	Operating free-air temperature	−55		125	0		70	°C

electrical characteristics over recommended operating free-air temperature range (unless otherwise noted)

PARAMETER	TEST CONDITIONS		SN54ALS01			SN74ALS01			UNIT
			MIN	TYP†	MAX	MIN	TYP†	MAX	
V_{IK}	V_{CC} = 4.5 V,	I_I = −18 mA			−1.5			−1.5	V
I_{OH}	V_{CC} = 4.5 V,	V_{OH} = 5.5 V			0.1			0.1	mA
V_{OL}	V_{CC} = 4.5 V,	I_{OL} = 4 mA		0.25	0.4		0.25	0.4	V
	V_{CC} = 4.5 V,	I_{OL} = 8 mA					0.35	0.5	
I_I	V_{CC} = 5.5 V,	V_I = 7 V			0.1			0.1	mA
I_{IH}	V_{CC} = 5.5 V,	V_I = 2.7 V			20			20	µA
I_{IL}	V_{CC} = 5.5 V,	V_I = 0.4 V			−0.1			−0.1	mA
I_{CCH}	V_{CC} = 5.5 V,	V_I = 0 V		0.43	0.85		0.43	0.85	mA
I_{CCL}	V_{CC} = 5.5 V,	V_I = 4.5 V		1.62	3		1.62	3	mA

† All typical values are at V_{CC} = 5 V, T_A = 25 °C

switching characteristics (see Note 1)

PARAMETER	FROM (INPUT)	TO (OUTPUT)	V_{CC} = 4.5 V to 5.5 V, C_L = 50 pF, R_L = 2 kΩ, T_A = MIN to MAX				UNIT
			SN54ALS01		SN74ALS01		
			MIN	MAX	MIN	MAX	
t_{PLH}	A or B	Y	23	59	23	54	ns
t_{PHL}	A or B	Y	8	29	8	28	ns

NOTE 1: For load circuit and voltage waveforms, see page 1-12.

TEXAS INSTRUMENTS
POST OFFICE BOX 225012 • DALLAS, TEXAS 75265

TYPES SN54ALS02, SN54AS02, SN74ALS02, SN74AS02
QUADRUPLE 2-INPUT POSITIVE-NOR GATES

D2661, APRIL 1982—REVISED DECEMBER 1983

- Package Options Include Both Plastic and Ceramic Chip Carriers in Addition to Plastic and Ceramic DIPs
- Dependable Texas Instruments Quality and Reliability

description

These devices contain four independent 2-input NOR gates. They perform the Boolean functions $Y = \overline{A+B}$ or $Y = \overline{A} \cdot \overline{B}$ in positive logic.

The SN54ALS02 and SN54AS02 are characterized for operation over the full military temperature range of $-55\,°C$ to $125\,°C$. The SN74ALS02 and SN74AS02 are characterized for operation from $0\,°C$ to $70\,°C$.

SN54ALS02, SN54AS02 . . . J PACKAGE
SN74ALS02, SN74AS02 . . . N PACKAGE
(TOP VIEW)

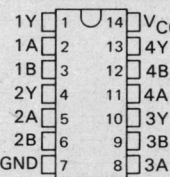

FUNCTION TABLE (each gate)

INPUTS		OUTPUT
A	B	Y
H	X	L
X	H	L
L	L	H

logic symbol

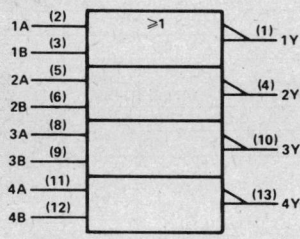

Pin numbers shown are for J and N packages.

SN54ALS02, SN54AS02 . . . FH PACKAGE
SN74ALS02, SN74AS02 . . . FN PACKAGE
(TOP VIEW)

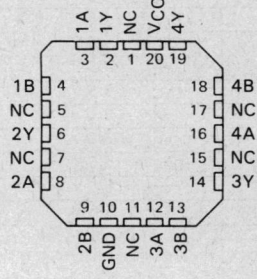

NC—No internal connection

Copyright © 1982 by Texas Instruments Incorporated

TEXAS INSTRUMENTS
POST OFFICE BOX 225012 • DALLAS, TEXAS 75265

TYPES SN54ALS02, SN74ALS02
QUADRUPLE 2-INPUT POSITIVE-NOR GATES

absolute maximum ratings over operating free-air temperature range (unless otherwise noted)

Supply voltage, V_{CC} .... 7 V
Input voltage .... 7 V
Operating free-air temperature range: SN54ALS02 .... −55°C to 125°C
                                       SN74ALS02 .... 0°C to 70°C
Storage temperature range .... −65°C to 150°C

recommended operating conditions

		SN54ALS02			SN74ALS02			UNIT
		MIN	NOM	MAX	MIN	NOM	MAX	
V_{CC}	Supply voltage	4.5	5	5.5	4.5	5	5.5	V
V_{IH}	High-level input voltage	2			2			V
V_{IL}	Low-level input voltage			0.8			0.8	V
I_{OH}	High-level output current			−0.4			−0.4	mA
I_{OL}	Low-level output current			4			8	mA
T_A	Operating free-air temperature	−55		125	0		70	°C

electrical characteristics over recommended operating free-air temperature range (unless otherwise noted)

PARAMETER	TEST CONDITIONS		SN54ALS02			SN74ALS02			UNIT
			MIN	TYP†	MAX	MIN	TYP†	MAX	
V_{IK}	V_{CC} = 4.5 V,	I_I = −18 mA			−1.5			−1.5	V
V_{OH}	V_{CC} = 4.5 V to 5.5 V,	I_{OH} = −0.4 mA	V_{CC}−2			V_{CC}−2			V
V_{OL}	V_{CC} = 4.5 V,	I_{OL} = 4 mA		0.25	0.4		0.25	0.4	V
	V_{CC} = 4.5 V,	I_{OL} = 8 mA					0.35	0.5	
I_I	V_{CC} = 5.5 V,	V_I = 7 V			0.1			0.1	mA
I_{IH}	V_{CC} = 5.5 V,	V_I = 2.7 V			20			20	µA
I_{IL}	V_{CC} = 5.5 V,	V_I = 0.4 V			−0.1			−0.1	mA
I_O‡	V_{CC} = 5.5 V,	V_O = 2.25 V	−30		−112	−30		−112	mA
I_{CCH}	V_{CC} = 5.5 V,	V_I = 0 V		0.86	2.2		0.86	2.2	mA
I_{CCL}	V_{CC} = 5.5 V,	V_I = 4.5 V		2.16	4		2.16	4	mA

†All typical values are at V_{CC} = 5 V, T_A = 25°C.
‡The output conditions have been chosen to produce a current that closely approximates one half of the true short-circuit output current, I_{OS}.

switching characteristics (see Note 1)

PARAMETER	FROM (INPUT)	TO (OUTPUT)	V_{CC} = 4.5 V to 5.5 V, C_L = 50 pF, R_L = 500 Ω, T_A = MIN to MAX				UNIT
			SN54ALS02		SN74ALS02		
			MIN	MAX	MIN	MAX	
t_{PLH}	A or B	Y	3	14	3	12	ns
t_{PHL}	A or B	Y	3	11	3	10	ns

NOTE 1: For load circuit and voltage waveforms, see page 1-12.

TEXAS INSTRUMENTS
POST OFFICE BOX 225012 • DALLAS, TEXAS 75265

TYPES SN54AS02, SN74AS02
QUADRUPLE 2-INPUT POSITIVE-NOR GATES

absolute maximum ratings over operating free-air temperature range (unless otherwise noted)

Supply voltage, V_{CC} .... 7 V
Input voltage .... 7 V
Operating free-air temperature range: SN54AS02 .... −55°C to 125°C
                                                       SN74AS02 .... 0°C to 70°C
Storage temperature range .... −65°C to 150°C

recommended operating conditions

		SN54AS02			SN74AS02			UNIT
		MIN	NOM	MAX	MIN	NOM	MAX	
V_{CC}	Supply voltage	4.5	5	5.5	4.5	5	5.5	V
V_{IH}	High-level input voltage	2			2			V
V_{IL}	Low-level input voltage			0.8			0.8	V
I_{OH}	High-level output current			−2			−2	mA
I_{OL}	Low-level output current			20			20	mA
T_A	Operating free-air temperature	−55		125	0		70	°C

electrical characteristics over recommended operating free-air temperature range (unless otherwise noted)

PARAMETER	TEST CONDITIONS		SN54AS02			SN74AS02			UNIT
			MIN	TYP†	MAX	MIN	TYP†	MAX	
V_{IK}	V_{CC} = 4.5 V,	I_I = −18 mA			−1.2			−1.2	V
V_{OH}	V_{CC} = 4.5 V to 5.5 V,	I_{OH} = −2 mA	V_{CC}−2			V_{CC}−2			V
V_{OL}	V_{CC} = 4.5 V,	I_{OL} = 20 mA		0.35	0.5		0.35	0.5	V
I_I	V_{CC} = 5.5 V,	V_I = 7 V			0.1			0.1	mA
I_{IH}	V_{CC} = 5.5 V,	V_I = 2.7 V			20			20	µA
I_{IL}	V_{CC} = 5.5 V,	V_I = 0.4 V			−0.5			−0.5	mA
I_O‡	V_{CC} = 5.5 V,	V_O = 2.25 V	−30		−112	−30		−112	mA
I_{CCH}	V_{CC} = 5.5 V,	V_I = 0 V		3.7	5.9		3.7	5.9	mA
I_{CCL}	V_{CC} = 5.5 V,	V_I = 4.5 V		12.5	20.1		12.5	20.1	mA

†All typical values are at V_{CC} = 5 V, T_A = 25°C.
‡The output conditions have been chosen to produce a current that closely approximates one half of the true short-circuit output current, I_{OS}.

switching characteristics (see Note 1)

PARAMETER	FROM (INPUT)	TO (OUTPUT)	V_{CC} = 4.5 V to 5.5 V, C_L = 50 pF, R_L = 500 Ω, T_A = MIN to MAX				UNIT
			SN54AS02		SN74AS02		
			MIN	MAX	MIN	MAX	
t_{PLH}	A or B	Y	1	5	1	4.5	ns
t_{PHL}	A or B	Y	1	5	1	4.5	ns

NOTE 1: For load circuit and voltage waveforms, see page 1-12.

Texas Instruments
POST OFFICE BOX 225012 • DALLAS, TEXAS 75265

2
ALS AND AS CIRCUITS

TYPES SN54ALS03A, SN74ALS03A
QUADRUPLE 2-INPUT POSITIVE-NAND GATES WITH OPEN-COLLECTOR OUTPUTS

D2661, APRIL 1982—REVISED DECEMBER 1983

- **Package Options Include Both Plastic and Ceramic Chip Carriers in Addition to Plastic and Ceramic DIPs**
- **Dependable Texas Instruments Quality and Reliability**

description

These devices contain four independent 2-input NAND gates. They perform the Boolean functions $Y = \overline{A \cdot B}$ or $Y = \overline{A} + \overline{B}$ in positive logic. The open-collector outputs require pull-up resistors to perform correctly. They may be connected to other open-collector outputs to implement active-low wired-OR or active-high wired-AND functions. Open-collector devices are often used to generate higher V_{OH} levels.

The SN54ALS03A is characterized for operation over the full military temperature range of −55°C to 125°C. The SN74ALS03A is characterized for operation from 0°C to 70°C.

SN54ALS03A . . . J PACKAGE
SN74ALS03A . . . N PACKAGE
(TOP VIEW)

1A	1		14	V_{CC}
1B	2		13	4B
1Y	3		12	4A
2A	4		11	4Y
2B	5		10	3B
2Y	6		9	3A
GND	7		8	3Y

SN54ALS03A . . . FH PACKAGE
SN74ALS03A . . . FN PACKAGE
(TOP VIEW)

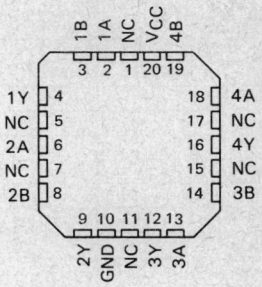

NC — No internal connection

FUNCTION TABLE (each gate)

INPUTS		OUTPUT
A	B	Y
H	H	L
L	X	H
X	L	H

logic symbol

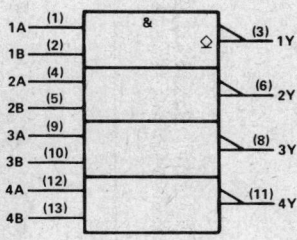

Pin numbers shown are for J and N packages.

ALS AND AS CIRCUITS

Copyright © 1982 by Texas Instruments Incorporated

TEXAS INSTRUMENTS
POST OFFICE BOX 225012 • DALLAS, TEXAS 75265

TYPES SN54ALS03A, SN74ALS03A
QUADRUPLE 2-INPUT POSITIVE-NAND GATES
WITH OPEN-COLLECTOR OUTPUTS

absolute maximum ratings over operating free-air temperature range (unless otherwise noted)

Supply voltage, V_{CC} .... 7 V
Input voltage .... 7 V
Off-state output voltage .... 7 V
Operating free-air temperature range: SN54ALS03A .... −55 °C to 125 °C
                                SN74ALS03A .... 0 °C to 70 °C
Storage temperature range .... −65 °C to 150 °C

recommended operating conditions

		SN54ALS03A			SN74ALS03A			UNIT
		MIN	NOM	MAX	MIN	NOM	MAX	
V_{CC}	Supply voltage	4.5	5	5.5	4.5	5	5.5	V
V_{IH}	High-level input voltage	2			2			V
V_{IL}	Low-level input voltage			0.8			0.8	V
V_{OH}	High-level output voltage			5.5			5.5	V
I_{OL}	Low-level output current			4			8	mA
T_A	Operating free-air temperature	−55		125	0		70	°C

electrical characteristics over recommended operating free-air temperature range (unless otherwise noted)

PARAMETER	TEST CONDITIONS		SN54ALS03A			SN74ALS03A			UNIT
			MIN	TYP†	MAX	MIN	TYP†	MAX	
V_{IK}	$V_{CC} = 4.5$ V,	$I_I = -18$ mA			−1.5			−1.5	V
I_{OH}	$V_{CC} = 4.5$ V,	$V_{OH} = 5.5$ V			0.1			0.1	mA
V_{OL}	$V_{CC} = 4.5$ V,	$I_{OL} = 4$ mA		0.25	0.4		0.25	0.4	V
	$V_{CC} = 4.5$ V,	$I_{OL} = 8$ mA					0.35	0.5	
I_I	$V_{CC} = 5.5$ V,	$V_I = 7$ V			0.1			0.1	mA
I_{IH}	$V_{CC} = 5.5$ V,	$V_I = 2.7$ V			20			20	µA
I_{IL}	$V_{CC} = 5.5$ V,	$V_I = 0.4$ V			−0.1			−0.1	mA
I_{CCH}	$V_{CC} = 5.5$ V,	$V_I = 0$ V		0.43	0.85		0.43	0.85	mA
I_{CCL}	$V_{CC} = 5.5$ V,	$V_I = 4.5$ V		1.62	3		1.62	3	mA

†All typical values are at $V_{CC} = 5$ V, $T_A = 25$ °C.

switching characteristics (see Note 1)

PARAMETER	FROM (INPUT)	TO (OUTPUT)	$V_{CC} = 4.5$ V to 5.5 V, $C_L = 50$ pF, $R_L = 2$ kΩ, $T_A = $ MIN to MAX				UNIT
			SN54ALS03A		SN74ALS03A		
			MIN	MAX	MIN	MAX	
t_{PLH}	A or B	Y	23	59	23	54	ns
t_{PHL}	A or B	Y	5	26	5	22	ns

NOTE 1: For load circuit and voltage waveforms, see page 1-12.

TEXAS INSTRUMENTS
POST OFFICE BOX 225012 • DALLAS, TEXAS 75265

TYPES SN54ALS04A, SN54AS04, SN74ALS04A, SN74AS04
HEX INVERTERS

D2661, APRIL 1982—REVISED DECEMBER 1983

- Package Options Include Both Plastic and Ceramic Chip Carriers in Addition to Plastic and Ceramic DIPs
- Dependable Texas Instruments Quality and Reliability

description

These devices contain six independent inverters. They perform the Boolean function $Y = \overline{A}$.

The SN54ALS04A and SN54AS04 are characterized for operation over the full military temperature range of $-55\,°C$ to $125\,°C$. The SN74ALS04A and SN74AS04 are characterized for operation from $0\,°C$ to $70\,°C$.

SN54ALS04A, SN54AS04 . . . J PACKAGE
SN74ALS04A, SN74AS04 . . . N PACKAGE
(TOP VIEW)

```
1A  [1    14] VCC
1Y  [2    13] 6A
2A  [3    12] 6Y
2Y  [4    11] 5A
3A  [5    10] 5Y
3Y  [6     9] 4A
GND [7     8] 4Y
```

FUNCTION TABLE
(each inverter)

INPUT A	OUTPUT Y
H	L
L	H

SN54ALS04A, SN54AS04 . . . FH PACKAGE
SN74ALS04A, SN74AS04 . . . FN PACKAGE
(TOP VIEW)

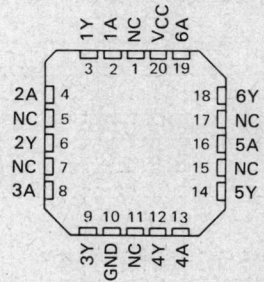

logic symbol

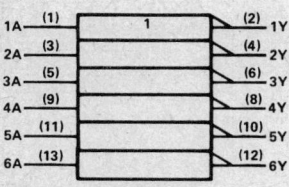

Pin numbers shown are for J and N packages.

NC—No internal connection

Copyright © 1982 by Texas Instruments Incorporated

2-15

TYPES SN54ALS04A, SN74ALS04A
HEX INVERTERS

absolute maximum ratings over operating free-air temperature range (unless otherwise noted)

Supply voltage, V_{CC} .... 7 V
Input voltage .... 7 V
Operating free-air temperature range: SN54ALS04A .... −55 °C to 125 °C
                                      SN74ALS04A .... 0 °C to 70 °C
Storage temperature range .... −65 °C to 150 °C

recommended operating conditions

		SN54ALS04A			SN74ALS04A			UNIT
		MIN	NOM	MAX	MIN	NOM	MAX	
V_{CC}	Supply voltage	4.5	5	5.5	4.5	5	5.5	V
V_{IH}	High-level input voltage	2			2			V
V_{IL}	Low-level input voltage			0.8			0.8	V
I_{OH}	High-level output current			−0.4			−0.4	mA
I_{OL}	Low-level output current			4			8	mA
T_A	Operating free-air temperature	−55		125	0		70	°C

electrical characteristics over recommended operating free-air temperature range (unless otherwise noted)

PARAMETER	TEST CONDITIONS		SN54ALS04A			SN74ALS04A			UNIT
			MIN	TYP†	MAX	MIN	TYP†	MAX	
V_{IK}	V_{CC} = 4.5 V,	I_I = −18 mA			−1.5			−1.5	V
V_{OH}	V_{CC} = 4.5 V to 5.5 V,	I_{OH} = −0.4 mA	V_{CC}−2			V_{CC}−2			V
V_{OL}	V_{CC} = 4.5 V,	I_{OL} = 4 mA		0.25	0.4		0.25	0.4	V
	V_{CC} = 4.5 V,	I_{OL} = 8 mA					0.35	0.5	
I_I	V_{CC} = 5.5 V,	V_I = 7 V			0.1			0.1	mA
I_{IH}	V_{CC} = 5.5 V,	V_I = 2.7 V			20			20	µA
I_{IL}	V_{CC} = 5.5 V,	V_I = 0.4 V			−0.1			−0.1	mA
I_O‡	V_{CC} = 5.5 V,	V_O = 2.25 V	−15		−70	−15		−70	mA
I_{CCH}	V_{CC} = 5.5 V,	V_I = 0 V		0.65	1.1		0.65	1.1	mA
I_{CCL}	V_{CC} = 5.5 V,	V_I = 4.5 V		2.9	4.2		2.9	4.2	mA

†All typical values are at V_{CC} = 5 V, T_A = 25°C.
‡The output conditions have been chosen to produce a current that closely approximates one half of the true short-circuit output current, I_{OS}.

switching characteristics (see Note 1)

PARAMETER	FROM (INPUT)	TO (OUTPUT)	V_{CC} = 4.5 V to 5.5 V, C_L = 50 pF, R_L = 500 Ω, T_A = MIN to MAX				UNIT
			SN54ALS04A		SN74ALS04A		
			MIN	MAX	MIN	MAX	
t_{PLH}	A	Y	3	14	3	11	ns
t_{PHL}	A	Y	2	12	2	8	ns

NOTE 1: For load circuit and voltage waveforms, see page 1-12.

TEXAS INSTRUMENTS
POST OFFICE BOX 225012 • DALLAS, TEXAS 75265

TYPES SN54AS04, SN74AS04
HEX INVERTERS

absolute maximum ratings over operating free-air temperature range (unless otherwise noted)

Supply voltage, V_{CC} .... 7 V
Input voltage .... 7 V
Operating free-air temperature range: SN54AS04 .... −55 °C to 125 °C
                                       SN74AS04 .... 0 °C to 70 °C
Storage temperature range .... −65 °C to 150 °C

recommended operating conditions

		SN54AS04			SN74AS04			UNIT
		MIN	NOM	MAX	MIN	NOM	MAX	
V_{CC}	Supply voltage	4.5	5	5.5	4.5	5	5.5	V
V_{IH}	High-level input voltage	2			2			V
V_{IL}	Low-level input voltage			0.8			0.8	V
I_{OH}	High-level output current			−2			−2	mA
I_{OL}	Low-level output current			20			20	mA
T_A	Operating free-air temperature	−55		125	0		70	°C

electrical characteristics over recommended operating free-air temperature range (unless otherwise noted)

PARAMETER	TEST CONDITIONS		SN54AS04			SN74AS04			UNIT
			MIN	TYP†	MAX	MIN	TYP†	MAX	
V_{IK}	V_{CC} = 4.5 V,	I_I = −18 mA			−1.2			−1.2	V
V_{OH}	V_{CC} = 4.5 V to 5.5 V,	I_{OH} = −2 mA	V_{CC}−2			V_{CC}−2			V
V_{OL}	V_{CC} = 4.5 V,	I_{OL} = 20 mA		0.35	0.5		0.35	0.5	V
I_I	V_{CC} = 5.5 V,	V_I = 7 V			0.1			0.1	mA
I_{IH}	V_{CC} = 5.5 V,	V_I = 2.7 V			20			20	µA
I_{IL}	V_{CC} = 5.5 V,	V_I = 0.4 V			−0.5			−0.5	mA
I_O‡	V_{CC} = 5.5 V,	V_O = 2.25 V	−30		−112	−30		−112	mA
I_{CCH}	V_{CC} = 5.5 V,	V_I = 0 V		3	4.8		3	4.8	mA
I_{CCL}	V_{CC} = 5.5 V,	V_I = 4.5 V		14	26.3		14	26.3	mA

†All typical values are at V_{CC} = 5 V, T_A = 25 °C.
‡The output conditions have been chosen to produce a current that closely approximates one half of the true short-circuit output current, I_{OS}.

switching characteristics (see Note 1)

PARAMETER	FROM (INPUT)	TO (OUTPUT)	V_{CC} = 4.5 V to 5.5 V, C_L = 50 pF, R_L = 500 Ω, T_A = MIN to MAX				UNIT
			SN54AS04		SN74AS04		
			MIN	MAX	MIN	MAX	
t_{PLH}	A	Y	1	6	1	5	ns
t_{PHL}	A	Y	1	4.5	1	4	ns

NOTE 1: For load circuit and voltage waveforms, see page 1-12.

TYPES SN54ALS05A, SN74ALS05A
HEX INVERTERS WITH OPEN-COLLECTOR OUTPUTS

D2661, APRIL 1982—REVISED DECEMBER 1983

- Package Options Include Both Plastic and Ceramic Chip Carriers in Addition to Plastic and Ceramic DIPs
- Dependable Texas Instruments Quality and Reliability

description

These devices contain six independent inverters. They perform the Boolean function $Y = \overline{A}$. The open-collector outputs require pull-up resistors to perform correctly. They may be connected to other open-collector outputs to implement active-low wired-OR or active-high wired-AND functions. Open-collector devices are often used to generate higher V_{OH} levels.

The SN54ALS05A is characterized for operation over the full military temperature range of $-55\,°C$ to $125\,°C$. The SN74ALS05A is characterized for operation from $0\,°C$ to $70\,°C$.

SN54ALS05A . . . J PACKAGE
SN74ALS05A . . . N PACKAGE
(TOP VIEW)

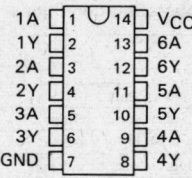

SN54ALS05A . . . FH PACKAGE
SN74ALS05A . . . FN PACKAGE
(TOP VIEW)

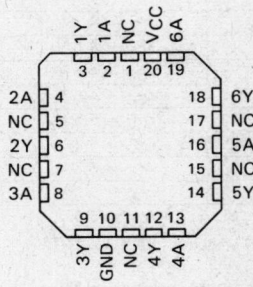

NC — No internal connection

FUNCTION TABLE (each inverter)

INPUT A	OUTPUT Y
H	L
L	H

logic symbol

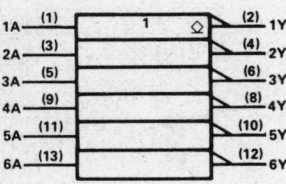

Pin numbers shown are for J and N packages.

Copyright © 1982 by Texas Instruments Incorporated

TEXAS INSTRUMENTS
POST OFFICE BOX 225012 • DALLAS, TEXAS 75265

TYPES SN54ALS05A, SN74ALS05A
HEX INVERTERS WITH OPEN-COLLECTOR OUTPUTS

absolute maximum ratings over operating free-air temperature range (unless otherwise noted)

Supply voltage, V_{CC} .... 7 V
Input voltage .... 7 V
Off-state output voltage .... 7 V
Operating free-air temperature range: SN54ALS05A .... −55 °C to 125 °C
                                      SN74ALS05A .... 0 °C to 70 °C
Storage temperature range .... −65 °C to 150 °C

recommended operating conditions

		SN54ALS05A			SN74ALS05A			UNIT
		MIN	NOM	MAX	MIN	NOM	MAX	
V_{CC}	Supply voltage	4.5	5	5.5	4.5	5	5.5	V
V_{IH}	High-level input voltage	2			2			V
V_{IL}	Low-level input voltage			0.8			0.8	V
V_{OH}	High-level output voltage			5.5			5.5	V
I_{OL}	Low-level output current			4			8	mA
T_A	Operating free-air temperature	−55		125	0		70	°C

electrical characteristics over recommended operating free-air temperature range (unless otherwise noted)

PARAMETER	TEST CONDITIONS		SN54ALS05A			SN74ALS05A			UNIT
			MIN	TYP†	MAX	MIN	TYP†	MAX	
V_{IK}	$V_{CC} = 4.5$ V,	$I_I = -18$ mA			−1.5			−1.5	V
I_{OH}	$V_{CC} = 4.5$ V,	$V_{OH} = 5.5$ V			0.1			0.1	mA
V_{OL}	$V_{CC} = 4.5$ V,	$I_{OL} = 4$ mA		0.25	0.4		0.25	0.4	V
	$V_{CC} = 4.5$ V,	$I_{OL} = 8$ mA					0.35	0.5	
I_I	$V_{CC} = 5.5$ V,	$V_I = 7$ V			0.1			0.1	mA
I_{IH}	$V_{CC} = 5.5$ V,	$V_I = 2.7$ V			20			20	µA
I_{IL}	$V_{CC} = 5.5$ V,	$V_I = 0.4$ V			−0.1			−0.1	mA
I_{CCH}	$V_{CC} = 5.5$ V,	$V_I = 0$ V		0.65	1.1		0.65	1.1	mA
I_{CCL}	$V_{CC} = 5.5$ V,	$V_I = 4.5$ V		2.9	4.2		2.9	4.2	mA

†All typical values are at $V_{CC} = 5$ V, $T_A = 25$ °C.

switching characteristics (see Note 1)

PARAMETER	FROM (INPUT)	TO (OUTPUT)	$V_{CC} = 4.5$ V to 5.5 V, $C_L = 50$ pF, $R_L = 2$ kΩ, T_A = MIN to MAX				UNIT
			SN54ALS05A		SN74ALS05A		
			MIN	MAX	MIN	MAX	
t_{PLH}	A or B	Y	23	59	23	54	ns
t_{PHL}	A or B	Y	4	19	4	14	ns

NOTE 1: For load circuit and voltage waveforms, see page 1-12.

TEXAS INSTRUMENTS
POST OFFICE BOX 225012 • DALLAS, TEXAS 75265

TYPES SN54ALS08, SN54AS08, SN74ALS08, SN74AS08
QUADRUPLE 2-INPUT POSITIVE-AND GATES

D2661, APRIL 1982—REVISED DECEMBER 1983

- Package Options Include Both Plastic and Ceramic Chip Carriers in Addition to Plastic and Ceramic DIPs
- Dependable Texas Instruments Quality and Reliability

description

These devices contain four independent 2-input AND gates. They perform the Boolean functions $Y = A \cdot B$ or $Y = \overline{\overline{A} + \overline{B}}$ in positive logic.

The SN54ALS08 and SN54AS08 are characterized for operation over the full military temperature range of $-55\,°C$ to $125\,°C$. The SN74ALS08 and SN74AS08 are characterized for operation from $0\,°C$ to $70\,°C$.

SN54ALS08, SN54AS08 . . . J PACKAGE
SN74ALS08, SN74AS08 . . . N PACKAGE
(TOP VIEW)

```
       ___
1A  [ 1   14 ] VCC
1B  [ 2   13 ] 4B
1Y  [ 3   12 ] 4A
2A  [ 4   11 ] 4Y
2B  [ 5   10 ] 3B
2Y  [ 6    9 ] 3A
GND [ 7    8 ] 3Y
```

FUNCTION TABLE
(each gate)

INPUTS		OUTPUT
A	B	Y
H	H	H
L	X	L
X	L	L

SN54ALS08, SN54AS08 . . . FH PACKAGE
SN74ALS08, SN74AS08 . . . FN PACKAGE
(TOP VIEW)

```
        1B 1A NC VCC 4B
         3  2  1 20 19
1Y  [ 4              18 ] 4A
NC  [ 5              17 ] NC
2A  [ 6              16 ] 4Y
NC  [ 7              15 ] NC
2B  [ 8              14 ] 3B
         9 10 11 12 13
        2Y GND NC 3Y 3A
```

NC—No internal connection

logic symbol

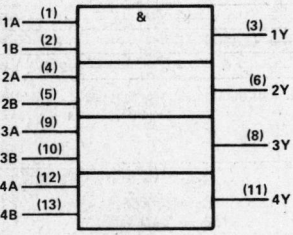

Pin numbers shown are for J and N packages.

ALS AND AS CIRCUITS

Copyright © 1982 by Texas Instruments Incorporated

TEXAS INSTRUMENTS
POST OFFICE BOX 225012 • DALLAS, TEXAS 75265

TYPES SN54ALS08, SN74ALS08
QUADRUPLE 2-INPUT POSITIVE-AND GATES

absolute maximum ratings over operating free-air temperature range (unless otherwise noted)

Supply voltage, V_{CC} .... 7 V
Input voltage .... 7 V
Operating free-air temperature range: SN54ALS08 .... $-55\,°C$ to $125\,°C$
SN74ALS08 .... $0\,°C$ to $70\,°C$
Storage temperature range .... $-65\,°C$ to $150\,°C$

recommended operating conditions

		SN54ALS08			SN74ALS08			UNIT
		MIN	NOM	MAX	MIN	NOM	MAX	
V_{CC}	Supply voltage	4.5	5	5.5	4.5	5	5.5	V
V_{IH}	High-level input voltage	2			2			V
V_{IL}	Low-level input voltage			0.8			0.8	V
I_{OH}	High-level output current			-0.4			-0.4	mA
I_{OL}	Low-level output current			4			8	mA
T_A	Operating free-air temperature	-55		125	0		70	$°C$

electrical characteristics over recommended operating free-air temperature range (unless otherwise noted)

PARAMETER	TEST CONDITIONS		SN54ALS08			SN74ALS08			UNIT
			MIN	TYP†	MAX	MIN	TYP†	MAX	
V_{IK}	$V_{CC} = 4.5$ V,	$I_I = -18$ mA			-1.5			-1.5	V
V_{OH}	$V_{CC} = 4.5$ V to 5.5 V,	$I_{OH} = -0.4$ mA	$V_{CC}-2$						V
V_{OL}	$V_{CC} = 4.5$ V,	$I_{OL} = 4$ mA		0.25	0.4		0.25	0.4	V
	$V_{CC} = 4.5$ V,	$I_{OL} = 8$ mA					0.35	0.5	
I_I	$V_{CC} = 5.5$ V,	$V_I = 7$ V			0.1			0.1	mA
I_{IH}	$V_{CC} = 5.5$ V,	$V_I = 2.7$ V			20			20	µA
I_{IL}	$V_{CC} = 5.5$ V,	$V_I = 0.4$ V			-0.1			-0.1	mA
I_O‡	$V_{CC} = 5.5$ V,	$V_O = 2.25$ V	-30		-112	-30		-112	mA
I_{CCH}	$V_{CC} = 5.5$ V,	$V_I = 4.5$ V		1.3	2.4		1.3	2.4	mA
I_{CCL}	$V_{CC} = 5.5$ V,	$V_I = 0$ V		2.2	4		2.2	4	mA

†All typical values are at $V_{CC} = 5$ V, $T_A = 25\,°C$.
‡The output conditions have been chosen to produce a current that closely approximates one half of the true short-circuit output current, I_{OS}.

switching characteristics (see Note 1)

PARAMETER	FROM (INPUT)	TO (OUTPUT)	$V_{CC} = 4.5$ V to 5.5 V, $C_L = 50$ pF, $R_L = 500\,\Omega$, T_A = MIN to MAX				UNIT
			SN54ALS08		SN74ALS08		
			MIN	MAX	MIN	MAX	
t_{PLH}	A or B	Y	4	16	4	14	ns
t_{PHL}	A or B	Y	3	12	3	10	ns

NOTE 1: For load circuit and voltage waveforms, see page 1-12.

TEXAS INSTRUMENTS
POST OFFICE BOX 225012 • DALLAS, TEXAS 75265

TYPES SN54AS08, SN74AS08
QUADRUPLE 2-INPUT POSITIVE-AND GATES

absolute maximum ratings over operating free-air temperature range (unless otherwise noted)

Supply voltage, V_{CC} .... 7 V
Input voltage .... 7 V
Operating free-air temperature range: SN54AS08 .... −55 °C to 125 °C
                                       SN74AS08 .... 0 °C to 70 °C
Storage temperature range .... −65 °C to 150 °C

recommended operating conditions

		SN54AS08			SN74AS08			UNIT
		MIN	NOM	MAX	MIN	NOM	MAX	
V_{CC}	Supply voltage	4.5	5	5.5	4.5	5	5.5	V
V_{IH}	High-level input voltage	2			2			V
V_{IL}	Low-level input voltage			0.8			0.8	V
I_{OH}	High-level output current			−2			−2	mA
I_{OL}	Low-level output current			20			20	mA
T_A	Operating free-air temperature	−55		125	0		70	°C

electrical characteristics over recommended operating free-air temperature range (unless otherwise noted)

PARAMETER	TEST CONDITIONS		SN54AS08			SN74AS08			UNIT
			MIN	TYP†	MAX	MIN	TYP†	MAX	
V_{IK}	V_{CC} = 4.5 V,	I_I = −18 mA			−1.2			−1.2	V
V_{OH}	V_{CC} = 4.5 V to 5.5 V,	I_{OH} = −2 mA	V_{CC}−2			V_{CC}−2			V
V_{OL}	V_{CC} = 4.5 V,	I_{OL} = 20 mA		0.35	0.5		0.35	0.5	V
I_I	V_{CC} = 5.5 V,	V_I = 7 V			0.1			0.1	mA
I_{IH}	V_{CC} = 5.5 V,	V_I = 2.7 V			20			20	µA
I_{IL}	V_{CC} = 5.5 V,	V_I = 0.4 V			−0.5			−0.5	mA
I_O‡	V_{CC} = 5.5 V,	V_O = 2.25 V	−30		−112	−30		−112	mA
I_{CCH}	V_{CC} = 5.5 V,	V_I = 4.5 V		5.8	9.3		5.8	9.3	mA
I_{CCL}	V_{CC} = 5.5 V,	V_I = 0 V		14.9	24		14.9	24	mA

†All typical values are at V_{CC} = 5 V, T_A = 25 °C.
‡The output conditions have been chosen to produce a current that closely approximates one half of the true short-circuit output current, I_{OS}.

switching characteristics (see Note 1)

PARAMETER	FROM (INPUT)	TO (OUTPUT)	V_{CC} = 4.5 V to 5.5 V, C_L = 50 pF, R_L = 500 Ω, T_A = MIN to MAX				UNIT
			SN54AS08		SN74AS08		
			MIN	MAX	MIN	MAX	
t_{PLH}	A or B	Y	1	6.5	1	5.5	ns
t_{PHL}	A or B	Y	1	6.5	1	5.5	ns

NOTE 1: For load circuit and voltage waveforms, see page 1-12.

ALS AND AS CIRCUITS

TEXAS INSTRUMENTS
POST OFFICE BOX 225012 • DALLAS, TEXAS 75265

2
ALS AND AS CIRCUITS

TYPES SN54ALS09, SN74ALS09
QUADRUPLE 2-INPUT POSITIVE-AND GATES WITH OPEN-COLLECTOR OUTPUTS

D2661, APRIL 1982—REVISED DECEMBER 1983

- Package Options Include Both Plastic and Ceramic Chip Carriers in Addition to Plastic and Ceramic DIPs
- Dependable Texas Instruments Quality and Reliability

description

These devices contain four independent 2-input AND gates. They perform the Boolean functions $Y = A \cdot B$ or $Y = \overline{A} + \overline{B}$ in positive logic. The open-collector outputs require pull-up resistors to perform correctly. They may be connected to other open-collector outputs to implement active-low wired-OR or active-high wired-AND functions. Open-collector devices are often used to generate higher V_{OH} levels.

The SN54ALS09 is characterized for operation over the full military temperature range of −55°C to 125°C. The SN74ALS09 is characterized for operation from 0°C to 70°C.

SN54ALS09 . . . J PACKAGE
SN74ALS09 . . . N PACKAGE
(TOP VIEW)

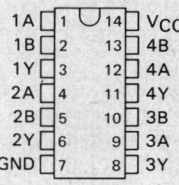

FUNCTION TABLE (each gate)

INPUTS		OUTPUT
A	B	Y
H	H	H
L	X	L
X	L	L

SN54ALS09 . . . FH PACKAGE
SN74ALS09 . . . FN PACKAGE
(TOP VIEW)

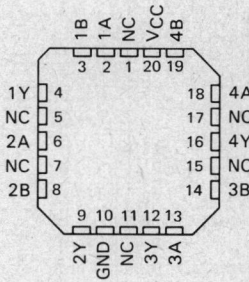

NC — No internal connection

logic symbol

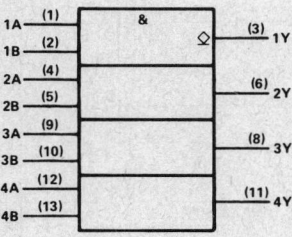

Pin numbers shown are for J and N packages.

ALS AND AS CIRCUITS

Copyright © 1982 by Texas Instruments Incorporated

TEXAS INSTRUMENTS
POST OFFICE BOX 225012 • DALLAS, TEXAS 75265

2-25

TYPES SN54ALS09, SN74ALS09
QUADRUPLE 2-INPUT POSITIVE-AND GATES
WITH OPEN-COLLECTOR OUTPUTS

absolute maximum ratings over operating free-air temperature range (unless otherwise noted)

Supply voltage, V_{CC} .... 7 V
Input voltage .... 7 V
Off-state output voltage .... 7 V
Operating free-air temperature range: SN54ALS09 .... −55 °C to 125 °C
                                       SN74ALS09 .... 0 °C to 70 °C
Storage temperature range .... −65 °C to 150 °C

recommended operating conditions

		SN54ALS09			SN74ALS09			UNIT
		MIN	NOM	MAX	MIN	NOM	MAX	
V_{CC}	Supply voltage	4.5	5	5.5	4.5	5	5.5	V
V_{IH}	High-level input voltage	2			2			V
V_{IL}	Low-level input voltage			0.8			0.8	V
V_{OH}	High-level output voltage			5.5			5.5	V
I_{OL}	Low-level output current			4			8	mA
T_A	Operating free-air temperature	−55		125	0		70	°C

electrical characteristics over recommended operating free-air temperature range (unless otherwise noted)

PARAMETER	TEST CONDITIONS		SN54ALS09			SN74ALS09			UNIT
			MIN	TYP†	MAX	MIN	TYP†	MAX	
V_{IK}	V_{CC} = 4.5 V,	I_I = −18 mA			−1.5			−1.5	V
I_{OH}	V_{CC} = 4.5 V,	V_{OH} = 5.5 V			0.1			0.1	mA
V_{OL}	V_{CC} = 4.5 V,	I_{OL} = 4 mA		0.25	0.4		0.25	0.4	V
	V_{CC} = 4.5 V,	I_{OL} = 8 mA					0.35	0.5	
I_I	V_{CC} = 5.5 V,	V_I = 7 V			0.1			0.1	mA
I_{IH}	V_{CC} = 5.5 V,	V_I = 2.7 V			20			20	µA
I_{IL}	V_{CC} = 5.5 V,	V_I = 0.4 V			−0.1			−0.1	mA
I_{CCH}	V_{CC} = 5.5 V,	V_I = 4.5 V	1.35		2.4	1.35		2.4	mA
I_{CCL}	V_{CC} = 5.5 V,	V_I = 0 V	2.2		4	2.2		4	mA

† All typical values are at V_{CC} = 5 V, T_A = 25 °C

switching characteristics (see Note 1)

PARAMETER	FROM (INPUT)	TO (OUTPUT)	V_{CC} = 4.5 V to 5.5 V, C_L = 50 pF, R_L = 2 kΩ, T_A = MIN to MAX				UNIT
			SN54ALS09		SN74ALS09		
			MIN	MAX	MIN	MAX	
t_{PLH}	A or B	Y	23	59	23	54	ns
t_{PHL}	A or B	Y	5	17	5	15	ns

NOTE 1: For load circuit and voltage waveforms, see page 1-12.

Texas Instruments
POST OFFICE BOX 225012 • DALLAS, TEXAS 75265

TYPES SN54ALS10, SN54AS10, SN74ALS10, SN74AS10
TRIPLE 3-INPUT POSITIVE-NAND GATES

D2661, APRIL 1982—REVISED DECEMBER 1983

- Package Options Include Both Plastic and Ceramic Chip Carriers in Addition to Plastic and Ceramic DIPs
- Dependable Texas Instruments Quality and Reliability

description

These devices contain three independent 3-input NAND gates. They perform the Boolean functions $Y = \overline{A \cdot B \cdot C}$ or $Y = \overline{A} + \overline{B} + \overline{C}$ in positive logic.

The SN54ALS10 and SN54AS10 are characterized for operation over the full military temperature range of $-55\,°C$ to $125\,°C$. The SN74ALS10 and SN74AS10 are characterized for operation from $0\,°C$ to $70\,°C$.

FUNCTION TABLE (each gate)

INPUTS			OUTPUT
A	B	C	Y
H	H	H	L
L	X	X	H
X	L	X	H
X	X	L	H

logic symbol

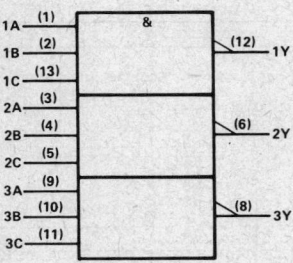

Pin numbers shown are for J and N packages.

SN54ALS10, SN54AS10 . . . J PACKAGE
SN74ALS10, SN74AS10 . . . N PACKAGE
(TOP VIEW)

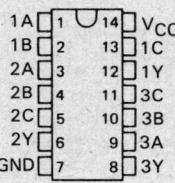

SN54ALS10, SN54AS10 . . . FH PACKAGE
SN74ALS10, SN74AS10 . . . FN PACKAGE
(TOP VIEW)

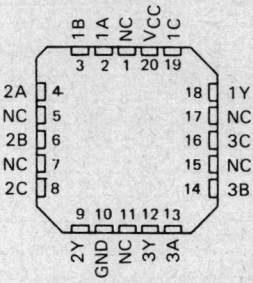

NC—No internal connection

ALS AND AS CIRCUITS

Copyright © 1982 by Texas Instruments Incorporated

TEXAS INSTRUMENTS
POST OFFICE BOX 225012 • DALLAS, TEXAS 75265

TYPES SN54ALS10, SN74ALS10
TRIPLE 3-INPUT POSITIVE-NAND GATES

absolute maximum ratings over operating free-air temperature range (unless otherwise noted)

Supply voltage, V_{CC} .... 7 V
Input voltage .... 7 V
Operating free-air temperature range: SN54ALS10 .... −55°C to 125°C
SN74ALS10 .... 0°C to 70°C
Storage temperature range .... −65°C to 150°C

recommended operating conditions

		SN54ALS10			SN74ALS10			UNIT
		MIN	NOM	MAX	MIN	NOM	MAX	
V_{CC}	Supply voltage	4.5	5	5.5	4.5	5	5.5	V
V_{IH}	High-level input voltage	2			2			V
V_{IL}	Low-level input voltage			0.8			0.8	V
I_{OH}	High-level output current			−0.4			−0.4	mA
I_{OL}	Low-level output current			4			8	mA
T_A	Operating free-air temperature	−55		125	0		70	°C

electrical characteristics over recommended operating free-air temperature range (unless otherwise noted)

PARAMETER	TEST CONDITIONS		SN54ALS10			SN74ALS10			UNIT
			MIN	TYP†	MAX	MIN	TYP†	MAX	
V_{IK}	V_{CC} = 4.5 V,	I_I = −18 mA			−1.5			−1.5	V
V_{OH}	V_{CC} = 4.5 V to 5.5 V,	I_{OH} = −0.4 mA	V_{CC}−2			V_{CC}−2			V
V_{OL}	V_{CC} = 4.5 V,	I_{OL} = 4 mA		0.25	0.4		0.25	0.4	V
	V_{CC} = 4.5 V,	I_{OL} = 8 mA					0.35	0.5	
I_I	V_{CC} = 5.5 V,	V_I = 7 V			0.1			0.1	mA
I_{IH}	V_{CC} = 5.5 V,	V_I = 2.7 V			20			20	µA
I_{IL}	V_{CC} = 5.5 V,	V_I = 0.4 V			−0.1			−0.1	mA
I_O‡	V_{CC} = 5.5 V,	V_O = 2.25 V	−30		−112	−30		−112	mA
I_{CCH}	V_{CC} = 5.5 V,	V_I = 0 V		0.32	0.6		0.32	0.6	mA
I_{CCL}	V_{CC} = 5.5 V,	V_I = 4.5 V		1.2	2.2		1.2	2.2	mA

† All typical values are at V_{CC} = 5 V, T_A = 25°C.
‡ The output conditions have been chosen to produce a current that closely approximates one half of the true short-circuit output current, I_{OS}.

switching characteristics (see Note 1)

PARAMETER	FROM (INPUT)	TO (OUTPUT)	V_{CC} = 4.5 V to 5.5 V, C_L = 50 pF, R_L = 500 Ω, T_A = MIN to MAX				UNIT
			SN54ALS10		SN74ALS10		
			MIN	MAX	MIN	MAX	
t_{PLH}	Any	Y	3	14	3	11	ns
t_{PHL}	Any	Y	4	21	4	18	ns

NOTE 1: For load circuit and voltage waveforms, see page 1-12.

TEXAS INSTRUMENTS
POST OFFICE BOX 225012 • DALLAS, TEXAS 75265

TYPES SN54AS10, SN74AS10
TRIPLE 3-INPUT POSITIVE-NAND GATES

absolute maximum ratings over operating free-air temperature range (unless otherwise noted)

Supply voltage, V_{CC} .... 7 V
Input voltage .... 7 V
Operating free-air temperature range: SN54AS10 .... −55 °C to 125 °C
　　　　　　　　　　　　　　　　　　 SN74AS10 .... 0 °C to 70 °C
Storage temperature range .... −65 °C to 150 °C

recommended operating conditions

		SN54AS10			SN74AS10			UNIT
		MIN	NOM	MAX	MIN	NOM	MAX	
V_{CC}	Supply voltage	4.5	5	5.5	4.5	5	5.5	V
V_{IH}	High-level input voltage	2			2			V
V_{IL}	Low-level input voltage			0.8			0.8	V
I_{OH}	High-level output current			−2			−2	mA
I_{OL}	Low-level output current			20			20	mA
T_A	Operating free-air temperature	−55		125	0		70	°C

electrical characteristics over recommended operating free-air temperature range (unless otherwise noted)

PARAMETER	TEST CONDITIONS		SN54AS10			SN74AS10			UNIT
			MIN	TYP†	MAX	MIN	TYP†	MAX	
V_{IK}	V_{CC} = 4.5 V,	I_I = −18 mA			−1.2			−1.2	V
V_{OH}	V_{CC} = 4.5 V to 5.5 V,	I_{OH} = −2 mA	$V_{CC}-2$			$V_{CC}-2$			V
V_{OL}	V_{CC} = 4.5 V,	I_{OL} = 20 mA		0.35	0.5		0.35	0.5	V
I_I	V_{CC} = 5.5 V,	V_I = 7 V			0.1			0.1	mA
I_{IH}	V_{CC} = 5.5 V,	V_I = 2.7 V			20			20	µA
I_{IL}	V_{CC} = 5.5 V,	V_I = 0.4 V			−0.5			−0.5	mA
I_O‡	V_{CC} = 5.5 V,	V_O = 2.25 V	−30		−112	−30		−112	mA
I_{CCH}	V_{CC} = 5.5 V,	V_I = 0 V		1.5	2.4		1.5	2.4	mA
I_{CCL}	V_{CC} = 5.5 V,	V_I = 4.5 V		8.1	13		8.1	13	mA

†All typical values are at V_{CC} = 5 V, T_A = 25 °C.
‡The output conditions have been chosen to produce a current that closely approximates one half of the true short-circuit output current, I_{OS}.

switching characteristics (see Note 1)

PARAMETER	FROM (INPUT)	TO (OUTPUT)	V_{CC} = 4.5 V to 5.5 V, C_L = 50 pF, R_L = 500 Ω, T_A = MIN to MAX				UNIT
			SN54AS10		SN74AS10		
			MIN	MAX	MIN	MAX	
t_{PLH}	Any	Y	1	5	1	4.5	ns
t_{PHL}	Any	Y	1	5	1	4.5	ns

NOTE 1: For load circuit and voltage waveforms, see page 1-12.

ALS AND AS CIRCUITS

TEXAS INSTRUMENTS
POST OFFICE BOX 225012 • DALLAS, TEXAS 75265

2 ALS AND AS CIRCUITS

TYPES SN54ALS11, SN54AS11, SN74ALS11, SN74AS11
TRIPLE 3-INPUT POSITIVE-AND GATES

D2661, APRIL 1982—REVISED DECEMBER 1983

- Package Options Include Both Plastic and Ceramic Chip Carriers in Addition to Plastic and Ceramic DIPs
- Dependable Texas Instruments Quality and Reliability

description

These devices contain three independent 3-input AND gates. They perform the Boolean functions $Y = A \cdot B \cdot C$ or $Y = \overline{\overline{A} + \overline{B} + \overline{C}}$ in positive logic.

The SN54ALS11 and SN54AS11 are characterized for operation over the full military temperature range of $-55°C$ to $125°C$. The SN74ALS11 and SN74AS11 are characterized for operation from $0°C$ to $70°C$.

SN54ALS11, SN54AS11 . . . J PACKAGE
SN74ALS11, SN74AS11 . . . N PACKAGE
(TOP VIEW)

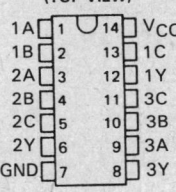

FUNCTION TABLE (each gate)

INPUTS			OUTPUT
A	B	C	Y
H	H	H	H
L	X	X	L
X	L	X	L
X	X	L	L

SN54ALS11, SN54AS11 . . . FH PACKAGE
SN74ALS11, SN74AS11 . . . FN PACKAGE
(TOP VIEW)

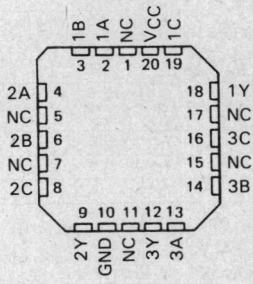

NC—No internal connection

logic symbol

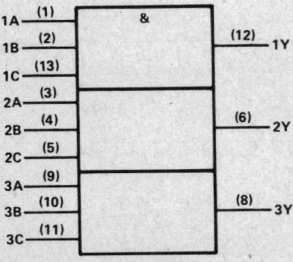

Pin numbers shown are for J and N packages.

ALS AND AS CIRCUITS

Copyright © 1982 by Texas Instruments Incorporated

TEXAS INSTRUMENTS
POST OFFICE BOX 225012 • DALLAS, TEXAS 75265

2-31

TYPES SN54ALS11, SN74ALS11
TRIPLE 3-INPUT POSITIVE-AND GATES

absolute maximum ratings over operating free-air temperature range (unless otherwise noted)

Supply voltage, V_{CC} .... 7 V
Input voltage .... 7 V
Operating free-air temperature range: SN54ALS11 .... $-55\,°C$ to $125\,°C$
              SN74ALS11 .... $0\,°C$ to $70\,°C$
Storage temperature range .... $-65\,°C$ to $150\,°C$

recommended operating conditions

		SN54ALS11			SN74ALS11			UNIT
		MIN	NOM	MAX	MIN	NOM	MAX	
V_{CC}	Supply voltage	4.5	5	5.5	4.5	5	5.5	V
V_{IH}	High-level input voltage	2			2			V
V_{IL}	Low-level input voltage			0.8			0.8	V
I_{OH}	High-level output current			-0.4			-0.4	mA
I_{OL}	Low-level output current			4			8	mA
T_A	Operating free-air temperature	-55		125	0		70	°C

electrical characteristics over recommended operating free-air temperature range (unless otherwise noted)

PARAMETER	TEST CONDITIONS		SN54ALS11			SN74ALS11			UNIT
			MIN	TYP†	MAX	MIN	TYP†	MAX	
V_{IK}	$V_{CC} = 4.5$ V,	$I_I = -18$ mA			-1.5			-1.5	V
V_{OH}	$V_{CC} = 4.5$ V to 5.5 V,	$I_{OH} = -0.4$ mA	$V_{CC}-2$			$V_{CC}-2$			V
V_{OL}	$V_{CC} = 4.5$ V,	$I_{OL} = 4$ mA		0.25	0.4		0.25	0.4	V
	$V_{CC} = 4.5$ V,	$I_{OL} = 8$ mA					0.35	0.5	
I_I	$V_{CC} = 5.5$ V,	$V_I = 7$ V			0.1			0.1	mA
I_{IH}	$V_{CC} = 5.5$ V,	$V_I = 2.7$ V			20			20	µA
I_{IL}	$V_{CC} = 5.5$ V,	$V_I = 0.4$ V			-0.1			-0.1	mA
I_O‡	$V_{CC} = 5.5$ V,	$V_O = 2.25$ V	-30		-112	-30		-112	mA
I_{CCH}	$V_{CC} = 5.5$ V,	$V_I = 4.5$ V		1	1.8		1	1.8	mA
I_{CCL}	$V_{CC} = 5.5$ V,	$V_I = 0$ V		1.6	3		1.6	3	mA

†All typical values are at $V_{CC} = 5$ V, $T_A = 25\,°C$.
‡The output conditions have been chosen to produce a current that closely approximates one half of the true short-circuit output current, I_{OS}.

switching characteristics (see Note 1)

PARAMETER	FROM (INPUT)	TO (OUTPUT)	$V_{CC} = 4.5$ V to 5.5 V, $C_L = 50$ pF, $R_L = 500\,\Omega$, $T_A =$ MIN to MAX				UNIT
			SN54ALS11		SN74ALS11		
			MIN	MAX	MIN	MAX	
t_{PLH}	Any	Y	5	23	5	20	ns
t_{PHL}	Any	Y	3	12	3	10	ns

NOTE 1: For load circuit and voltage waveforms, see page 1-12.

TEXAS INSTRUMENTS
POST OFFICE BOX 225012 • DALLAS, TEXAS 75265

TYPES SN54AS11, SN74AS11
TRIPLE 3-INPUT POSITIVE-AND GATES

absolute maximum ratings over operating free-air temperature range (unless otherwise noted)

Supply voltage, V_{CC} .... 7 V
Input voltage .... 7 V
Operating free-air temperature range: SN54AS11 .... −55 °C to 125 °C
                                       SN74AS11 .... 0 °C to 70 °C
Storage temperature range .... −65 °C to 150 °C

recommended operating conditions

		SN54AS11			SN74AS11			UNIT
		MIN	NOM	MAX	MIN	NOM	MAX	
V_{CC}	Supply voltage	4.5	5	5.5	4.5	5	5.5	V
V_{IH}	High-level input voltage	2			2			V
V_{IL}	Low-level input voltage			0.8			0.8	V
I_{OH}	High-level output current			−2			−2	mA
I_{OL}	Low-level output current			20			20	mA
T_A	Operating free-air temperature	−55		125	0		70	°C

electrical characteristics over recommended operating free-air temperature range (unless otherwise noted)

PARAMETER	TEST CONDITIONS		SN54AS11			SN74AS11			UNIT
			MIN	TYP†	MAX	MIN	TYP†	MAX	
V_{IK}	$V_{CC} = 4.5$ V,	$I_I = -18$ mA			−1.2			−1.2	V
V_{OH}	$V_{CC} = 4.5$ V to 5.5 V,	$I_{OH} = -2$ mA	$V_{CC}-2$			$V_{CC}-2$			V
V_{OL}	$V_{CC} = 4.5$ V,	$I_{OL} = 20$ mA		0.35	0.5		0.35	0.5	V
I_I	$V_{CC} = 5.5$ V,	$V_I = 7$ V			0.1			0.1	mA
I_{IH}	$V_{CC} = 5.5$ V,	$V_I = 2.7$ V			20			20	μA
I_{IL}	$V_{CC} = 5.5$ V,	$V_I = 0.4$ V			−0.5			−0.5	mA
I_O‡	$V_{CC} = 5.5$ V,	$V_O = 2.25$ V	−30		−112	−30		−112	mA
I_{CCH}	$V_{CC} = 5.5$ V,	$V_I = 4.5$ V		4.3	7		4.3	7	mA
I_{CCL}	$V_{CC} = 5.5$ V,	$V_I = 0$ V		11.2	18		11.2	18	mA

†All typical values are at $V_{CC} = 5$ V, $T_A = 25$ °C.
‡The output conditions have been chosen to produce a current that closely approximates one half of the true short-circuit output current, I_{OS}.

switching characteristics (see Note 1)

PARAMETER	FROM (INPUT)	TO (OUTPUT)	$V_{CC} = 4.5$ V to 5.5 V, $C_L = 50$ pF, $R_L = 500$ Ω, $T_A =$ MIN to MAX				UNIT
			SN54AS11		SN74AS11		
			MIN	MAX	MIN	MAX	
t_{PLH}	Any	Y	1	6.5	1	6	ns
t_{PHL}	Any	Y	1	6.5	1	5.5	ns

NOTE 1: For load circuit and voltage waveforms, see page 1-12.

TEXAS INSTRUMENTS
POST OFFICE BOX 225012 • DALLAS, TEXAS 75265

2 ALS AND AS CIRCUITS

TYPES SN54ALS12, SN74ALS12
TRIPLE 3-INPUT POSITIVE-NAND GATES WITH OPEN-COLLECTOR OUTPUTS

D2661, APRIL 1982—REVISED DECEMBER 1983

- Package Options Include Both Plastic and Ceramic Chip Carriers in Addition to Plastic and Ceramic DIPs
- Dependable Texas Instruments Quality and Reliability

description

These devices contain three independent 3-input NAND gates with open-collector outputs. These gates perform the Boolean functions $Y = \overline{A \cdot B \cdot C}$ or $Y = \overline{A} + \overline{B} + \overline{C}$ in positive logic. The open-collector outputs require pull-up resistors to perform correctly. They may be connected to other open-collector outputs to implement active-low wired-OR or active-high wired-AND functions. Open-collector devices are often used to generate higher V_{OH} levels.

The SN54ALS12 is characterized for operation over the full military temperature range of −55°C to 125°C. The SN74ALS12 is characterized for operation from 0°C to 70°C.

SN54ALS12 . . . J PACKAGE
SN74ALS12 . . . N PACKAGE
(TOP VIEW)

```
1A  [1  U 14]  Vcc
1B  [2    13]  1C
2A  [3    12]  1Y
2B  [4    11]  3C
2C  [5    10]  3B
2Y  [6     9]  3A
GND [7     8]  3Y
```

SN54ALS12 . . . FH PACKAGE
SN74ALS12 . . . FN PACKAGE
(TOP VIEW)

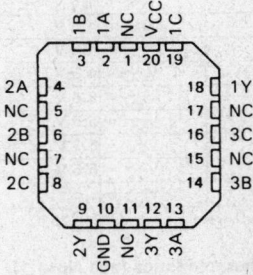

NC — No internal connection

FUNCTION TABLE (each gate)

INPUTS			OUTPUT
A	B	C	Y
H	H	H	L
L	X	X	H
X	L	X	H
X	X	L	H

logic symbol

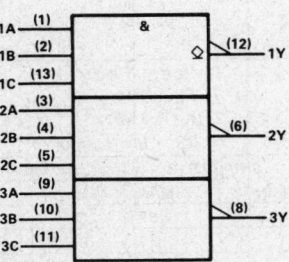

Pin numbers shown are for J and N packages.

ALS AND AS CIRCUITS

Copyright © 1982 by Texas Instruments Incorporated

2-35

TYPES SN54ALS12, SN74ALS12
TRIPLE 3-INPUT POSITIVE-NAND GATES
WITH OPEN-COLLECTOR OUTPUTS

absolute maximum ratings over operating free-air temperature range (unless otherwise noted)

Supply voltage, V_{CC} .... 7 V
Input voltage .... 7 V
Off-state output voltage .... 7 V
Operating free-air temperature range: SN54ALS12 .... −55 °C to 125 °C
             SN74ALS12 .... 0 °C to 70 °C
Storage temperature range .... −65 °C to 150 °C

recommended operating conditions

		SN54ALS12			SN74ALS12			UNIT
		MIN	NOM	MAX	MIN	NOM	MAX	
V_{CC}	Supply voltage	4.5	5	5.5	4.5	5	5.5	V
V_{IH}	High-level input voltage	2			2			V
V_{IL}	Low-level input voltage			0.8			0.8	V
V_{OH}	High-level output voltage			5.5			5.5	V
I_{OL}	Low-level output current			4			8	mA
T_A	Operating free-air temperature	−55		125	0		70	°C

electrical characteristics over recommended operating free-air temperature range (unless otherwise noted)

PARAMETER	TEST CONDITIONS		SN54ALS12			SN74ALS12			UNIT
			MIN	TYP†	MAX	MIN	TYP†	MAX	
V_{IK}	V_{CC} = 4.5 V,	I_I = −18 mA			−1.5			−1.5	V
I_{OH}	V_{CC} = 4.5 V,	V_{OH} = 5.5 V			0.1			0.1	mA
V_{OL}	V_{CC} = 4.5 V,	I_{OL} = 4 mA		0.25	0.4		0.25	0.4	V
	V_{CC} = 4.5 V,	I_{OL} = 8 mA					0.35	0.5	
I_I	V_{CC} = 5.5 V,	V_I = 7 V			0.1			0.1	mA
I_{IH}	V_{CC} = 5.5 V,	V_I = 2.7 V			20			20	µA
I_{IL}	V_{CC} = 5.5 V,	V_I = 0.4 V			−0.1			−0.1	mA
I_{CCH}	V_{CC} = 5.5 V,	V_I = 0 V		0.32	0.6		0.32	0.6	mA
I_{CCL}	V_{CC} = 5.5 V,	V_I = 4.5 V		1.2	2.2		1.2	2.2	mA

†All typical values are at V_{CC} = 5 V, T_A = 25 °C

switching characteristics (see Note 1)

PARAMETER	FROM (INPUT)	TO (OUTPUT)	V_{CC} = 4.5 V to 5.5 V, C_L = 50 pF, R_L = 2 kΩ, T_A = MIN to MAX				UNIT
			SN54ALS12		SN74ALS12		
			MIN	MAX	MIN	MAX	
t_{PLH}	Any	Y	23	59	23	54	ns
t_{PHL}	Any	Y	9	37	9	30	ns

NOTE 1: For load circuit and voltage waveforms, see page 1-12.

TYPES SN54ALS15, SN74ALS15
TRIPLE 3-INPUT POSITIVE-AND GATES
WITH OPEN-COLLECTOR OUTPUTS

D2661, APRIL 1982—REVISED DECEMBER 1983

- Package Options Include Both Plastic and Ceramic Chip Carriers in Addition to Plastic and Ceramic DIPs
- Dependable Texas Instruments Quality and Reliability

description

These devices contain three independent 3-input AND gates with open-collector outputs. These gates perform the Boolean functions $Y = A \cdot B \cdot C$ or $Y = \overline{\overline{A} + \overline{B} + \overline{C}}$ in positive logic. The open-collector outputs require pull-up resistors to perform correctly. They may be connected to other open-collector outputs to implement active-low wired-OR or active-high wired-AND functions. Open-collector devices are often used to generate higher V_{OH} levels.

The SN54ALS15 is characterized for operation over the full military temperature range of $-55\,°C$ to $125\,°C$. The SN74ALS15 is characterized for operation from $0\,°C$ to $70\,°C$.

SN54ALS15 . . . J PACKAGE
SN74ALS15 . . . N PACKAGE
(TOP VIEW)

```
1A [ 1  U 14 ] VCC
1B [ 2    13 ] 1C
2A [ 3    12 ] 1Y
2B [ 4    11 ] 3C
2C [ 5    10 ] 3B
2Y [ 6     9 ] 3A
GND[ 7     8 ] 3Y
```

SN54ALS15 . . . FH PACKAGE
SN74ALS15 . . . FN PACKAGE
(TOP VIEW)

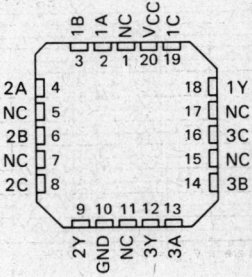

NC — No internal connection

FUNCTION TABLE (each gate)

INPUTS			OUTPUT
A	B	C	Y
H	H	H	H
L	X	X	L
X	L	X	L
X	X	L	L

logic symbol

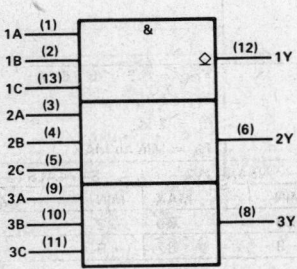

Pin numbers shown are for J and N packages.

ALS AND AS CIRCUITS

Copyright © 1982 by Texas Instruments Incorporated

TEXAS INSTRUMENTS
POST OFFICE BOX 225012 • DALLAS, TEXAS 75265

2-37

TYPES SN54ALS15, SN74ALS15
TRIPLE 3-INPUT POSITIVE-AND GATES WITH OPEN-COLLECTOR OUTPUTS

absolute maximum ratings over operating free-air temperature range (unless otherwise noted)

Supply voltage, V_{CC} .... 7 V
Input voltage .... 7 V
Off-state output voltage .... 7 V
Operating free-air temperature range: SN54ALS15 .... −55°C to 125°C
                                       SN74ALS15 .... 0°C to 70°C
Storage temperature range .... −65°C to 150°C

recommended operating conditions

		SN54ALS15			SN74ALS15			UNIT
		MIN	NOM	MAX	MIN	NOM	MAX	
V_{CC}	Supply voltage	4.5	5	5.5	4.5	5	5.5	V
V_{IH}	High-level input voltage	2			2			V
V_{IL}	Low-level input voltage			0.8			0.8	V
V_{OH}	High-level output voltage			5.5			5.5	V
I_{OL}	Low-level output current			4			8	mA
T_A	Operating free-air temperature	−55		125	0		70	°C

electrical characteristics over recommended operating free-air temperature range (unless otherwise noted)

PARAMETER	TEST CONDITIONS		SN54ALS15			SN74ALS15			UNIT
			MIN	TYP†	MAX	MIN	TYP†	MAX	
V_{IK}	V_{CC} = 4.5 V,	I_I = −18 mA			−1.5			−1.5	V
I_{OH}	V_{CC} = 4.5 V,	V_{OH} = 5.5 V			0.1			0.1	mA
V_{OL}	V_{CC} = 4.5 V,	I_{OL} = 4 mA		0.25	0.4		0.25	0.4	V
	V_{CC} = 4.5 V,	I_{OL} = 8 mA					0.35	0.5	
I_I	V_{CC} = 5.5 V,	V_I = 7 V			0.1			0.1	mA
I_{IH}	V_{CC} = 5.5 V,	V_I = 2.7 V			20			20	µA
I_{IL}	V_{CC} = 5.5 V,	V_I = 0.4 V			−0.1			−0.1	mA
I_{CCH}	V_{CC} = 5.5 V,	V_I = 4.5 V		1	1.8		1	1.8	mA
I_{CCL}	V_{CC} = 5.5 V,	V_I = 0 V		1.66	3		1.66	3	mA

†All typical values are at V_{CC} = 5 V, T_A = 25°C.

switching characteristics (see Note 1)

PARAMETER	FROM (INPUT)	TO (OUTPUT)	V_{CC} = 4.5 V to 5.5 V, C_L = 50 pF, R_L = 2 kΩ, T_A = MIN to MAX				UNIT
			SN54ALS15		SN74ALS15		
			MIN	MAX	MIN	MAX	
t_{PLH}	Any	Y	23	59	23	54	ns
t_{PHL}	Any	Y	6	14	6	13	ns

NOTE 1: For load circuit and voltage waveforms, see page 1-12.

TEXAS INSTRUMENTS
POST OFFICE BOX 225012 • DALLAS, TEXAS 75265

TYPES SN54ALS20A, SN54AS20, SN74ALS20A, SN74AS20
DUAL 4-INPUT POSITIVE-NAND GATES

D2661, APRIL 1982 – REVISED DECEMBER 1983

- Package Options Include Both Plastic and Ceramic Chip Carriers in Addition to Plastic and Ceramic DIPs
- Dependable Texas Instruments Quality and Reliability

description

These devices contain two independent 4-input NAND gates. They perform the Boolean functions $Y = \overline{A \cdot B \cdot C \cdot D}$ or $Y = \overline{A} + \overline{B} + \overline{C} + \overline{D}$ in positive logic.

The SN54ALS20A and SN54AS20 are characterized for operation over the full military temperature range of $-55\,°C$ to $125\,°C$. The SN74ALS20A and SN74AS20 are characterized for operation from $0\,°C$ to $70\,°C$.

SN54ALS20A, SN54AS20 . . . J PACKAGE
SN74ALS20A, SN74AS20 . . . N PACKAGE
(TOP VIEW)

```
1A  [ 1   14 ] VCC
1B  [ 2   13 ] 2D
NC  [ 3   12 ] 2C
1C  [ 4   11 ] NC
1D  [ 5   10 ] 2B
1Y  [ 6    9 ] 2A
GND [ 7    8 ] 2Y
```

FUNCTION TABLE (each gate)

INPUTS				OUTPUT
A	B	C	D	Y
H	H	H	H	L
L	X	X	X	H
X	L	X	X	H
X	X	L	X	H
X	X	X	L	H

SN54ALS20A, SN54AS20 . . . FH PACKAGE
SN74ALS20A, SN74AS20 . . . FN PACKAGE
(TOP VIEW)

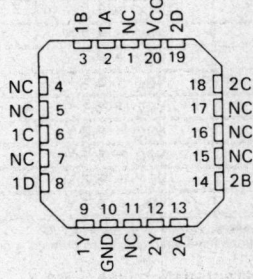

NC — No internal connection

logic symbol

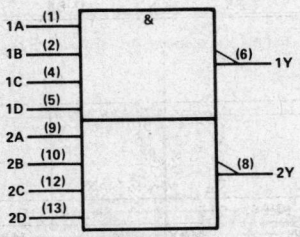

Pin numbers shown are for J and N packages.

ALS AND AS CIRCUITS

Copyright © 1982 by Texas Instruments Incorporated

TEXAS INSTRUMENTS
POST OFFICE BOX 225012 • DALLAS, TEXAS 75265

TYPES SN54ALS20A, SN74ALS20A
DUAL 4-INPUT POSITIVE-NAND GATES

absolute maximum ratings over operating free-air temperature range (unless otherwise noted)

Supply voltage, V_{CC} .... 7 V
Input voltage .... 7 V
Operating free-air temperature range: SN54ALS20A .... $-55\,°C$ to $125\,°C$
                                      SN74ALS20A .... $0\,°C$ to $70\,°C$
Storage temperature range .... $-65\,°C$ to $150\,°C$

recommended operating conditions

		SN54ALS20A			SN74ALS20A			UNIT
		MIN	NOM	MAX	MIN	NOM	MAX	
V_{CC}	Supply voltage	4.5	5	5.5	4.5	5	5.5	V
V_{IH}	High-level input voltage	2			2			V
V_{IL}	Low-level input voltage			0.8			0.8	V
I_{OH}	High-level output current			-0.4			-0.4	mA
I_{OL}	Low-level output current			4			8	mA
T_A	Operating free-air temperature	-55		125	0		70	°C

electrical characteristics over recommended operating free-air temperature range (unless otherwise noted)

PARAMETER	TEST CONDITIONS		SN54ALS20A			SN74ALS20A			UNIT
			MIN	TYP†	MAX	MIN	TYP†	MAX	
V_{IK}	$V_{CC} = 4.5$ V,	$I_I = -18$ mA			-1.5			-1.5	V
V_{OH}	$V_{CC} = 4.5$ V to 5.5 V,	$I_{OH} = -0.4$ mA	$V_{CC}-2$			$V_{CC}-2$			V
V_{OL}	$V_{CC} = 4.5$ V,	$I_{OL} = 4$ mA		0.25	0.4		0.25	0.4	V
	$V_{CC} = 4.5$ V,	$I_{OL} = 8$ mA					0.35	0.5	
I_I	$V_{CC} = 5.5$ V,	$V_I = 7$ V			0.1			0.1	mA
I_{IH}	$V_{CC} = 5.5$ V,	$V_I = 2.7$ V			20			20	µA
I_{IL}	$V_{CC} = 5.5$ V,	$V_I = 0.4$ V			-0.1			-0.1	mA
I_O‡	$V_{CC} = 5.5$ V,	$V_O = 2.25$ V	-15		-70	-15		-70	mA
I_{CCH}	$V_{CC} = 5.5$ V,	$V_I = 0$ V		0.22	0.4		0.22	0.4	mA
I_{CCL}	$V_{CC} = 5.5$ V,	$V_I = 4.5$ V		0.81	1.5		0.81	1.5	mA

† All typical values are at $V_{CC} = 5$ V, $T_A = 25\,°C$.
‡ The output conditions have been chosen to produce a current that closely approximates one half of the true short-circuit output current, I_{OS}.

switching characteristics (see Note 1)

PARAMETER	FROM (INPUT)	TO (OUTPUT)	$V_{CC} = 4.5$ V to 5.5 V, $C_L = 50$ pF, $R_L = 500\,\Omega$, $T_A = $ MIN to MAX				UNIT
			SN54ALS20A		SN74ALS20A		
			MIN	MAX	MIN	MAX	
t_{PLH}	Any	Y	3	13	3	11	ns
t_{PHL}	Any	Y	3	12	3	10	ns

NOTE 1: For load circuit and voltage waveforms, see page 1-12.

TYPES SN54AS20, SN74AS20
DUAL 4-INPUT POSITIVE-NAND GATES

absolute maximum ratings over operating free-air temperature range (unless otherwise noted)

Supply voltage, V_{CC} .... 7 V
Input voltage .... 7 V
Operating free-air temperature range: SN54AS20 .... $-55\,°C$ to $125\,°C$
                                             SN74AS20 .... $0\,°C$ to $70\,°C$
Storage temperature range .... $-65\,°C$ to $150\,°C$

recommended operating conditions

		SN54AS20			SN74AS20			UNIT
		MIN	NOM	MAX	MIN	NOM	MAX	
V_{CC}	Supply voltage	4.5	5	5.5	4.5	5	5.5	V
V_{IH}	High-level input voltage	2			2			V
V_{IL}	Low-level input voltage			0.8			0.8	V
I_{OH}	High-level output current			-2			-2	mA
I_{OL}	Low-level output current			20			20	mA
T_A	Operating free-air temperature	-55		125	0		70	°C

electrical characteristics over recommended operating free-air temperature range (unless otherwise noted)

PARAMETER	TEST CONDITIONS		SN54AS20			SN74AS20			UNIT
			MIN	TYP†	MAX	MIN	TYP†	MAX	
V_{IK}	$V_{CC} = 4.5$ V,	$I_I = -18$ mA			-1.2			-1.2	V
V_{OH}	$V_{CC} = 4.5$ V to 5.5 V,	$I_{OH} = -2$ mA	$V_{CC}-2$			$V_{CC}-2$			V
V_{OL}	$V_{CC} = 4.5$ V,	$I_{OL} = 20$ mA		0.35	0.5		0.35	0.5	V
I_I	$V_{CC} = 5.5$ V,	$V_I = 7$ V			0.1			0.1	mA
I_{IH}	$V_{CC} = 5.5$ V,	$V_I = 2.7$ V			20			20	µA
I_{IL}	$V_{CC} = 5.5$ V,	$V_I = 0.4$ V			-0.5			-0.5	mA
I_O‡	$V_{CC} = 5.5$ V,	$V_O = 2.25$ V	-30		-112	-30		-112	mA
I_{CCH}	$V_{CC} = 5.5$ V,	$V_I = 0$ V		1	1.6		1	1.6	mA
I_{CCL}	$V_{CC} = 5.5$ V,	$V_I = 4.5$ V		5.4	8.7		5.4	8.7	mA

†All typical values are at $V_{CC} = 5$ V, $T_A = 25\,°C$.
‡The output conditions have been chosen to produce a current that closely approximates one half of the true short-circuit output current, I_{OS}.

switching characteristics (see Note 1)

PARAMETER	FROM (INPUT)	TO (OUTPUT)	$V_{CC} = 4.5$ V to 5.5 V, $C_L = 50$ pF, $R_L = 500$ Ω, T_A = MIN to MAX				UNIT
			SN54AS20		SN74AS20		
			MIN	MAX	MIN	MAX	
t_{PLH}	Any	Y	1	5.5	1	5	ns
t_{PHL}	Any	Y	1	5	1	4.5	ns

NOTE 1: For load circuit and voltage waveforms, see page 1-12.

TEXAS INSTRUMENTS
POST OFFICE BOX 225012 • DALLAS, TEXAS 75265

ALS AND AS CIRCUITS

2

ALS AND AS CIRCUITS

TYPES SN54ALS21, SN54AS21, SN74ALS21, SN74AS21
DUAL 4-INPUT POSITIVE-AND GATES

D2661, APRIL 1982—REVISED DECEMBER 1983

- Package Options Include Both Plastic and Ceramic Chip Carriers in Addition to Plastic and Ceramic DIPs
- Dependable Texas Instruments Quality and Reliability

description

These devices contain two independent 4-input AND gates. They perform the Boolean functions $Y = A \cdot B \cdot C \cdot D$ or $Y = \overline{\overline{A} + \overline{B} + \overline{C} + \overline{D}}$ in positive logic.

The SN54ALS21 and SN54AS21 are characterized for operation over the full military temperature range of $-55\,°C$ to $125\,°C$. The SN74ALS21 and SN74AS21 are characterized for operation from $0\,°C$ to $70\,°C$.

SN54ALS21, SN54AS21 . . . J PACKAGE
SN74ALS21, SN74AS21 . . . N PACKAGE
(TOP VIEW)

```
1A  [1  U 14]  VCC
1B  [2    13]  2D
NC  [3    12]  2C
1C  [4    11]  NC
1D  [5    10]  2B
1Y  [6     9]  2A
GND [7     8]  2Y
```

FUNCTION TABLE (each gate)

INPUTS				OUTPUT
A	B	C	D	Y
H	H	H	H	H
L	X	X	X	L
X	L	X	X	L
X	X	L	X	L
X	X	X	L	L

SN54ALS21, SN54AS21 . . . FH PACKAGE
SN74ALS21, SN74AS21 . . . FN PACKAGE
(TOP VIEW)

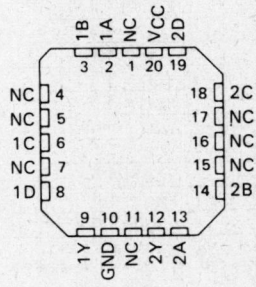

NC—No internal connection

logic symbol

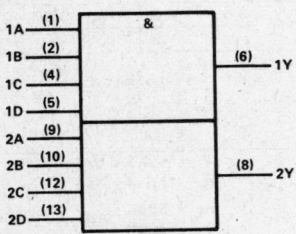

Pin numbers shown are for J and N packages.

ALS AND AS CIRCUITS

Copyright © 1982 by Texas Instruments Incorporated

TEXAS INSTRUMENTS
POST OFFICE BOX 225012 • DALLAS, TEXAS 75265

TYPES SN54ALS21, SN74ALS21
DUAL 4-INPUT POSITIVE-AND GATES

absolute maximum ratings over operating free-air temperature range (unless otherwise noted)

Supply voltage, V_{CC} .... 7 V
Input voltage .... 7 V
Operating free-air temperature range: SN54ALS21 .... −55°C to 125°C
                                     SN74ALS21 .... 0°C to 70°C
Storage temperature range .... −65°C to 150°C

recommended operating conditions

		SN54ALS21			SN74ALS21			UNIT
		MIN	NOM	MAX	MIN	NOM	MAX	
V_{CC}	Supply voltage	4.5	5	5.5	4.5	5	5.5	V
V_{IH}	High-level input voltage	2			2			V
V_{IL}	Low-level input voltage			0.8			0.8	V
I_{OH}	High-level output current			−0.4			−0.4	mA
I_{OL}	Low-level output current			4			8	mA
T_A	Operating free-air temperature	−55		125	0		70	°C

electrical characteristics over recommended operating free-air temperature range (unless otherwise noted)

PARAMETER	TEST CONDITIONS		SN54ALS21			SN74ALS21			UNIT
			MIN	TYP†	MAX	MIN	TYP†	MAX	
V_{IK}	V_{CC} = 4.5 V,	I_I = −18 mA			−1.5			−1.5	V
V_{OH}	V_{CC} = 4.5 V to 5.5 V,	I_{OH} = −0.4 mA	V_{CC}−2			V_{CC}−2			V
V_{OL}	V_{CC} = 4.5 V,	I_{OL} = 4 mA		0.25	0.4		0.25	0.4	V
	V_{CC} = 4.5 V,	I_{OL} = 8 mA					0.35	0.5	
I_I	V_{CC} = 5.5 V,	V_I = 7 V			0.1			0.1	mA
I_{IH}	V_{CC} = 5.5 V,	V_I = 2.7 V			20			20	µA
I_{IL}	V_{CC} = 5.5 V,	V_I = 0.4 V			−0.1			−0.1	mA
I_O‡	V_{CC} = 5.5 V,	V_O = 2.25 V	−30		−112	−30		−112	mA
I_{CCH}	V_{CC} = 5.5 V,	V_I = 4.5 V		0.67	1.2		0.67	1.2	mA
I_{CCL}	V_{CC} = 5.5 V,	V_I = 0 V		1.1	2		1.1	2	mA

† All typical values are at V_{CC} = 5 V, T_A = 25°C.
‡ The output conditions have been chosen to produce a current that closely approximates one half of the true short-circuit output current, I_{OS}.

switching characteristics (see Note 1)

PARAMETER	FROM (INPUT)	TO (OUTPUT)	V_{CC} = 4.5 V to 5.5 V, C_L = 50 pF, R_L = 500 Ω, T_A = MIN to MAX				UNIT
			SN54ALS21		SN74ALS21		
			MIN	MAX	MIN	MAX	
t_{PLH}	Any	Y	6	30	6	26	ns
t_{PHL}	Any	Y	3	12	3	10	ns

NOTE 1: For load circuit and voltage waveforms, see page 1-12.

Texas Instruments
POST OFFICE BOX 225012 • DALLAS, TEXAS 75265

TYPES SN54AS21, SN74AS21
DUAL 4-INPUT POSITIVE-AND GATES

absolute maximum ratings over operating free-air temperature range (unless otherwise noted)

Supply voltage, V_{CC}	7 V
Input voltage	7 V
Operating free-air temperature range: SN54AS21	$-55\,°C$ to $125\,°C$
SN74AS21	$0\,°C$ to $70\,°C$
Storage temperature range	$-65\,°C$ to $150\,°C$

recommended operating conditions

		SN54AS21			SN74AS21			UNIT
		MIN	NOM	MAX	MIN	NOM	MAX	
V_{CC}	Supply voltage	4.5	5	5.5	4.5	5	5.5	V
V_{IH}	High-level input voltage	2			2			V
V_{IL}	Low-level input voltage			0.8			0.8	V
I_{OH}	High-level output current			-2			-2	mA
I_{OL}	Low-level output current			20			20	mA
T_A	Operating free-air temperature	-55		125	0		70	°C

electrical characteristics over recommended operating free-air temperature range (unless otherwise noted)

PARAMETER	TEST CONDITIONS		SN54AS21			SN74AS21			UNIT
			MIN	TYP†	MAX	MIN	TYP†	MAX	
V_{IK}	$V_{CC} = 4.5$ V,	$I_I = -18$ mA			-1.2			-1.2	V
V_{OH}	$V_{CC} = 4.5$ V to 5.5 V,	$I_{OH} = -2$ mA	$V_{CC}-2$			$V_{CC}-2$			V
V_{OL}	$V_{CC} = 4.5$ V,	$I_{OL} = 20$ mA		0.35	0.5		0.35	0.5	V
I_I	$V_{CC} = 5.5$ V,	$V_I = 7$ V			0.1			0.1	mA
I_{IH}	$V_{CC} = 5.5$ V,	$V_I = 2.7$ V			20			20	µA
I_{IL}	$V_{CC} = 5.5$ V,	$V_I = 0.4$ V			-0.5			-0.5	mA
I_O‡	$V_{CC} = 5.5$ V,	$V_O = 2.25$ V	-30		-112	-30		-112	mA
I_{CCH}	$V_{CC} = 5.5$ V,	$V_I = 4.5$ V		2.9	4.6		2.9	4.6	mA
I_{CCL}	$V_{CC} = 5.5$ V,	$V_I = 0$ V		7.4	12		7.4	12	mA

†All typical values are at $V_{CC} = 5$ V, $T_A = 25\,°C$.
‡The output conditions have been chosen to produce a current that closely approximates one half of the true short-circuit output current, I_{OS}.

switching characteristics (see Note 1)

PARAMETER	FROM (INPUT)	TO (OUTPUT)	$V_{CC} = 4.5$ V to 5.5 V, $C_L = 50$ pF, $R_L = 500\,\Omega$, T_A = MIN to MAX				UNIT
			SN54AS21		SN74AS21		
			MIN	MAX	MIN	MAX	
t_{PLH}	Any	Y	1	6.5	1	6	ns
t_{PHL}	Any	Y	1	6.5	1	6	ns

NOTE 1: For load circuit and voltage waveforms, see page 1-12.

2
ALS AND AS CIRCUITS

TYPES SN54ALS22A, SN74ALS22A
DUAL 4-INPUT POSITIVE-NAND GATES
WITH OPEN-COLLECTOR OUTPUTS

D2661, APRIL 1982—REVISED DECEMBER 1983

- Package Options Include Both Plastic and Ceramic Chip Carriers in Addition to Plastic and Ceramic DIPs
- Dependable Texas Instruments Quality and Reliability

description

These devices contain two independent 4-input NAND gates. These gates perform the Boolean functions $Y = \overline{A \cdot B \cdot C \cdot D}$ or $Y = \overline{A} + \overline{B} + \overline{C} + \overline{D}$ in positive logic. The open-collector outputs require pull-up resistors to perform correctly. They may be connected to other open-collector outputs to implement active-low wired-OR or active-high wired-AND functions. Open-collector devices are often used to generate higher V_{OH} levels.

The SN54ALS22A is characterized for operation over the full military temperature range of −55°C to 125°C. The SN74ALS22A is characterized for operation from 0°C to 70°C.

SN54ALS22A . . . J PACKAGE
SN74ALS22A . . . N PACKAGE
(TOP VIEW)

```
1A  [ 1    14 ]  VCC
1B  [ 2    13 ]  2D
NC  [ 3    12 ]  2C
1C  [ 4    11 ]  NC
1D  [ 5    10 ]  2B
1Y  [ 6     9 ]  2A
GND [ 7     8 ]  2Y
```

SN54ALS22A . . . FH PACKAGE
SN74ALS22A . . . FN PACKAGE
(TOP VIEW)

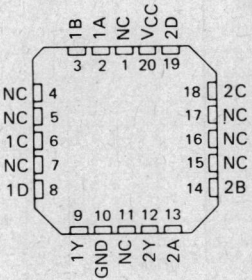

NC — No internal connection

FUNCTION TABLE (each gate)

INPUTS				OUTPUT
A	B	C	D	Y
H	H	H	H	L
L	X	X	X	H
X	L	X	X	H
X	X	L	X	H
X	X	X	L	H

logic symbol

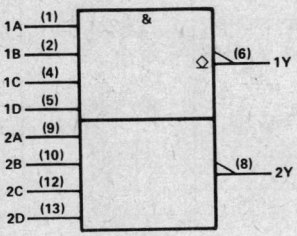

Pin numbers shown are for J and N packages.

ALS AND AS CIRCUITS

Copyright © 1982 by Texas Instruments Incorporated

2-47

TYPES SN54ALS22A, SN74ALS22A
DUAL 4-INPUT POSITIVE-NAND GATES
WITH OPEN-COLLECTOR OUTPUTS

absolute maximum ratings over operating free-air temperature range (unless otherwise noted)

Supply voltage, V_{CC} .... 7 V
Input voltage .... 7 V
Off-state output voltage .... 7 V
Operating free-air temperature range: SN54ALS22A .... −55 °C to 125 °C
                                           SN74ALS22A .... 0 °C to 70 °C
Storage temperature range .... −65 °C to 150 °C

recommended operating conditions

		SN54ALS22A			SN74ALS22A			UNIT
		MIN	NOM	MAX	MIN	NOM	MAX	
V_{CC}	Supply voltage	4.5	5	5.5	4.5	5	5.5	V
V_{IH}	High-level input voltage	2			2			V
V_{IL}	Low-level input voltage			0.8			0.8	V
V_{OH}	High-level output voltage			5.5			5.5	V
I_{OL}	Low-level output current			4			8	mA
T_A	Operating free-air temperature	−55		125	0		70	°C

electrical characteristics over recommended operating free-air temperature range (unless otherwise noted)

PARAMETER	TEST CONDITIONS		SN54ALS22A			SN74ALS22A			UNIT
			MIN	TYP†	MAX	MIN	TYP†	MAX	
V_{IK}	$V_{CC} = 4.5$ V,	$I_I = −18$ mA			−1.5			−1.5	V
I_{OH}	$V_{CC} = 4.5$ V,	$V_{OH} = 5.5$ V			0.1			0.1	mA
V_{OL}	$V_{CC} = 4.5$ V,	$I_{OL} = 4$ mA		0.25	0.4		0.25	0.4	V
	$V_{CC} = 4.5$ V,	$I_{OL} = 8$ mA					0.35	0.5	
I_I	$V_{CC} = 5.5$ V,	$V_I = 7$ V			0.1			0.1	mA
I_{IH}	$V_{CC} = 5.5$ V,	$V_I = 2.7$ V			20			20	µA
I_{IL}	$V_{CC} = 5.5$ V,	$V_I = 0.4$ V			−0.1			−0.1	mA
I_{CCH}	$V_{CC} = 5.5$ V,	$V_I = 0$ V		0.22	0.4		0.22	0.4	mA
I_{CCL}	$V_{CC} = 5.5$ V,	$V_I = 4.5$ V		0.8	1.5		0.8	1.5	mA

† All typical values are at $V_{CC} = 5$ V, $T_A = 25$ °C

switching characteristics (see Note 1)

PARAMETER	FROM (INPUT)	TO (OUTPUT)	$V_{CC} = 4.5$ V to 5.5 V, $C_L = 50$ pF, $R_L = 2$ kΩ, T_A = MIN to MAX				UNIT
			SN54ALS22A		SN74ALS22A		
			MIN	MAX	MIN	MAX	
t_{PLH}	Any	Y	23	59	23	54	ns
t_{PHL}	Any	Y	6	24	6	20	ns

NOTE 1: For load circuit and voltage waveforms, see page 1-12.

Texas Instruments
POST OFFICE BOX 225012 • DALLAS, TEXAS 75265

TYPES SN54ALS27, SN54AS27, SN74ALS27, SN74AS27
TRIPLE 3-INPUT POSITIVE-NOR GATES

D2661, APRIL 1982—REVISED DECEMBER 1983

- **Package Options Include Both Plastic and Ceramic Chip Carriers in Addition to Plastic and Ceramic DIPs**
- **Dependable Texas Instruments Quality and Reliability**

description

These devices contain three independent 3-input NOR gates. They perform the Boolean functions $Y = \overline{A + B + C}$ or $Y = \overline{A} \cdot \overline{B} \cdot \overline{C}$ in positive logic.

The SN54ALS27 and SN54AS27 are characterized for operation over the full military temperature range of $-55\,°C$ to $125\,°C$. The SN74ALS27 and SN74AS27 are characterized for operation from $0\,°C$ to $70\,°C$.

SN54ALS27, SN54AS27 . . . J PACKAGE
SN74ALS27, SN74AS27 . . . N PACKAGE
(TOP VIEW)

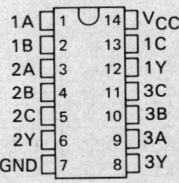

FUNCTION TABLE (each gate)

INPUTS			OUTPUT
A	B	C	Y
H	X	X	L
X	H	X	L
X	X	H	L
L	L	L	H

SN54ALS27, SN54AS27 . . . FH PACKAGE
SN74ALS27, SN74AS27 . . . FN PACKAGE
(TOP VIEW)

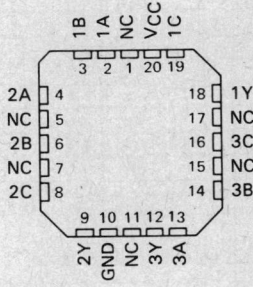

NC—No internal connection

logic symbol

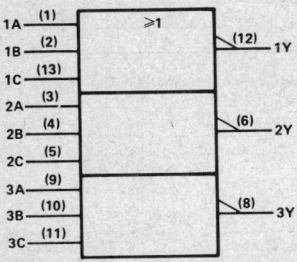

Pin numbers shown are for J and N packages.

Copyright © 1982 by Texas Instruments Incorporated

TEXAS INSTRUMENTS
POST OFFICE BOX 225012 • DALLAS, TEXAS 75265

2-49

TYPES SN54ALS27, SN74ALS27
TRIPLE 3-INPUT POSITIVE-NOR GATES

absolute maximum ratings over operating free-air temperature range (unless otherwise noted)

Supply voltage, V_{CC} .... 7 V
Input voltage .... 7 V
Operating free-air temperature range: SN54ALS27 .... −55 °C to 125 °C
SN74ALS27 .... 0 °C to 70 °C
Storage temperature range .... −65 °C to 150 °C

recommended operating conditions

		SN54ALS27			SN74ALS27			UNIT
		MIN	NOM	MAX	MIN	NOM	MAX	
V_{CC}	Supply voltage	4.5	5	5.5	4.5	5	5.5	V
V_{IH}	High-level input voltage	2			2			V
V_{IL}	Low-level input voltage			0.8			0.8	V
I_{OH}	High-level output current			−0.4			−0.4	mA
I_{OL}	Low-level output current			4			8	mA
T_A	Operating free-air temperature	−55		125	0		70	°C

electrical characteristics over recommended operating free-air temperature range (unless otherwise noted)

PARAMETER	TEST CONDITIONS		SN54ALS27			SN74ALS27			UNIT
			MIN	TYP†	MAX	MIN	TYP†	MAX	
V_{IK}	V_{CC} = 4.5 V,	I_I = −18 mA			−1.5			−1.5	V
V_{OH}	V_{CC} = 4.5 V to 5.5 V,	I_{OH} = −0.4 mA	V_{CC}−2			V_{CC}−2			V
V_{OL}	V_{CC} = 4.5 V,	I_{OL} = 4 mA		0.25	0.4		0.25	0.4	V
	V_{CC} = 4.5 V,	I_{OL} = 8 mA					0.35	0.5	
I_I	V_{CC} = 5.5 V,	V_I = 7 V			0.1			0.1	mA
I_{IH}	V_{CC} = 5.5 V,	V_I = 2.7 V			20			20	μA
I_{IL}	V_{CC} = 5.5 V,	V_I = 0.4 V			−0.1			−0.1	mA
I_O‡	V_{CC} = 5.5 V,	V_O = 2.25 V	−30		−112	−30		−112	mA
I_{CCH}	V_{CC} = 5.5 V,	V_I = 0 V		0.97	1.8		0.97	1.8	mA
I_{CCL}	V_{CC} = 5.5 V,	V_I = 4.5 V		2	4		2	4	mA

†All typical values are at V_{CC} = 5 V, T_A = 25°C.
‡The output conditions have been chosen to produce a current that closely approximates one half of the true short-circuit output current, I_{OS}.

switching characteristics (see Note 1)

PARAMETER	FROM (INPUT)	TO (OUTPUT)	V_{CC} = 4.5 V to 5.5 V, C_L = 50 pF, R_L = 500 Ω, T_A = MIN to MAX				UNIT
			SN54ALS27		SN74ALS27		
			MIN	MAX	MIN	MAX	
t_{PLH}	Any	Y	4	22	4	15	ns
t_{PHL}	Any	Y	3	10	3	9	ns

NOTE 1: For load circuit and voltage waveforms, see page 1-12.

TYPES SN54AS27, SN74AS27
TRIPLE 3-INPUT POSITIVE-NOR GATES

absolute maximum ratings over operating free-air temperature range (unless otherwise noted)

Supply voltage, V_{CC} .... 7 V
Input voltage .... 7 V
Operating free-air temperature range: SN54AS27 .... −55 °C to 125 °C
                                      SN74AS27 .... 0 °C to 70 °C
Storage temperature range .... −65 °C to 150 °C

recommended operating conditions

		SN54AS27			SN74AS27			UNIT
		MIN	NOM	MAX	MIN	NOM	MAX	
V_{CC}	Supply voltage	4.5	5	5.5	4.5	5	5.5	V
V_{IH}	High-level input voltage	2			2			V
V_{IL}	Low-level input voltage			0.8			0.8	V
I_{OH}	High-level output current			−2			−2	mA
I_{OL}	Low-level output current			20			20	mA
T_A	Operating free-air temperature	−55		125	0		70	°C

electrical characteristics over recommended operating free-air temperature range (unless otherwise noted)

PARAMETER	TEST CONDITIONS		SN54AS27			SN74AS27			UNIT
			MIN	TYP†	MAX	MIN	TYP†	MAX	
V_{IK}	V_{CC} = 4.5 V,	I_I = −18 mA			−1.2			−1.2	V
V_{OH}	V_{CC} = 4.5 V to 5.5 V,	I_{OH} = −2 mA	V_{CC}−2			V_{CC}−2			V
V_{OL}	V_{CC} = 4.5 V,	I_{OL} = 20 mA		0.35	0.5		0.35	0.5	V
I_I	V_{CC} = 5.5 V,	V_I = 7 V			0.1			0.1	mA
I_{IH}	V_{CC} = 5.5 V,	V_I = 2.7 V			20			20	µA
I_{IL}	V_{CC} = 5.5 V,	V_I = 0.4 V			−0.5			−0.5	mA
I_O‡	V_{CC} = 5.5 V,	V_O = 2.25 V	−30		−112	−30		−112	mA
I_{CCH}	V_{CC} = 5.5 V,	V_I = 0 V		4	6.4		4	6.4	mA
I_{CCL}	V_{CC} = 5.5 V,	V_I = 4.5 V		10.6	17.1		10.6	17.1	mA

† All typical values are at V_{CC} = 5 V, T_A = 25 °C.
‡ The output conditions have been chosen to produce a current that closely approximates one half of the true short-circuit output current, I_{OS}.

switching characteristics (see Note 1)

PARAMETER	FROM (INPUT)	TO (OUTPUT)	V_{CC} = 4.5 V to 5.5 V, C_L = 50 pF, R_L = 500 Ω, T_A = MIN to MAX				UNIT
			SN54AS27		SN74AS27		
			MIN	MAX	MIN	MAX	
t_{PLH}	Any	Y	1	6.5	1	5.5	ns
t_{PHL}	Any	Y	1	5	1	4.5	ns

NOTE 1: For load circuit and voltage waveforms, see page 1-12.

TEXAS
INSTRUMENTS
POST OFFICE BOX 225012 • DALLAS, TEXAS 75265

2
ALS AND AS CIRCUITS

TYPES SN54ALS28A, SN74ALS28A
QUADRUPLE 2-INPUT POSITIVE-NOR BUFFERS

D2661, APRIL 1982—REVISED DECEMBER 1983

- Package Options Include Both Plastic and Ceramic Chip Carriers in Addition to Plastic and Ceramic DIPs
- Dependable Texas Instruments Quality and Reliability

description

These devices contain four independent 2-input NOR buffer gates. They perform the Boolean functions $Y = \overline{A+B}$ or $Y = \overline{A} \cdot \overline{B}$ in positive logic.

The SN54ALS28A is characterized for operation over the full military temperature range of $-55\,°C$ to $125\,°C$. The SN74ALS28A is characterized for operation from $0\,°C$ to $70\,°C$.

SN54ALS28A . . . J PACKAGE
SN74ALS28A . . . N PACKAGE
(TOP VIEW)

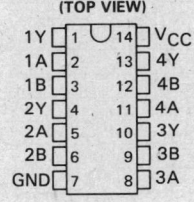

FUNCTION TABLE
(each gate)

INPUTS		OUTPUT
A	B	Y
H	X	L
X	H	L
L	L	H

SN54ALS28A . . . FH PACKAGE
SN74ALS28A . . . FN PACKAGE
(TOP VIEW)

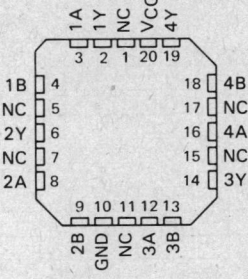

NC—No internal connection

logic symbol

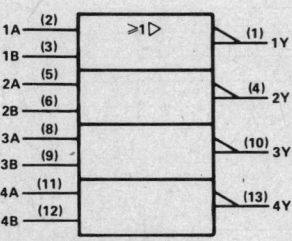

Pin numbers shown are for J and N packages.

Copyright © 1982 by Texas Instruments Incorporated

Texas Instruments
POST OFFICE BOX 225012 • DALLAS, TEXAS 75265

2-53

TYPES SN54ALS28A, SN74ALS28A
QUADRUPLE 2-INPUT POSITIVE-NOR BUFFERS

absolute maximum ratings over operating free-air temperature range (unless otherwise noted)

Supply voltage, V_{CC} .... 7 V
Input voltage .... 7 V
Operating free-air temperature range: SN54ALS28A .... −55 °C to 125 °C
              SN74ALS28A .... 0 °C to 70 °C
Storage temperature range .... −65 °C to 150 °C

recommended operating conditions

		SN54ALS28A			SN74ALS28A			UNIT
		MIN	NOM	MAX	MIN	NOM	MAX	
V_{CC}	Supply voltage	4.5	5	5.5	4.5	5	5.5	V
V_{IH}	High-level input voltage	2			2			V
V_{IL}	Low-level input voltage			0.8			0.8	V
I_{OH}	High-level output current			−1			−2.6	mA
I_{OL}	Low-level output current			12			24	mA
T_A	Operating free-air temperature	−55		125	0		70	°C

electrical characteristics over recommended operating free-air temperature range (unless otherwise noted)

PARAMETER	TEST CONDITIONS		SN54ALS28A			SN74ALS28A			UNIT
			MIN	TYP†	MAX	MIN	TYP†	MAX	
V_{IK}	V_{CC} = 4.5 V,	I_I = −18 mA			−1.5			−1.5	V
V_{OH}	V_{CC} = 4.5 V to 5.5 V,	I_{OH} = −0.4 mA	V_{CC}−2			V_{CC}−2			V
	V_{CC} = 4.5 V,	I_{OH} = −1 mA	2.4	3.3					
	V_{CC} = 4.5 V,	I_{OH} = −2.6 mA				2.4	3.2		
V_{OL}	V_{CC} = 4.5 V,	I_{OL} = 12 mA		0.25	0.4		0.25	0.4	V
	V_{CC} = 4.5 V,	I_{OL} = 24 mA					0.35	0.5	
I_I	V_{CC} = 5.5 V,	V_I = 7 V			0.1			0.1	mA
I_{IH}	V_{CC} = 5.5 V,	V_I = 2.7 V			20			20	μA
I_{IL}	V_{CC} = 5.5 V,	V_{IL} = 0.4 V			−0.1			−0.1	mA
I_O‡	V_{CC} = 5.5 V,	V_O = 2.25 V	−30		−112	−30		−112	mA
I_{CCH}	V_{CC} = 5.5 V,	V_I = 0 V		1.7	2.8		1.7	2.8	mA
I_{CCL}	V_{CC} = 5.5 V,	V_I = 4.5 V		5.6	9		5.6	9	mA

†All typical values are at V_{CC} = 5 V, T_A = 25 °C.
‡The output conditions have been chosen to produce a current that closely approximates one half of the true short-circuit output current, I_{OS}.

switching characteristics (see Note 1)

PARAMETER	FROM (INPUT)	TO (OUTPUT)	V_{CC} = 4.5 V to 5.5 V, C_L = 50 pF, R_L = 500 Ω, T_A = MIN to MAX				UNIT
			SN54ALS28A		SN74ALS28A		
			MIN	MAX	MIN	MAX	
t_{PLH}	A or B	Y	2	10	2	8	ns
t_{PHL}	A or B	Y	2	10	2	7	ns

NOTE 1: For load circuit and voltage waveforms, see page 1-12.

TYPES SN54ALS30, SN54AS30, SN74ALS30, SN74AS30
8-INPUT POSITIVE-NAND GATES

D2661, APRIL 1982—REVISED DECEMBER 1983

- Package Options Include Both Plastic and Ceramic Chip Carriers in Addition to Plastic and Ceramic DIPs
- Dependable Texas Instruments Quality and Reliability

description

These devices contain a single 8-input NAND gate and perform the following Boolean functions in positive logic:

$$Y = \overline{A \cdot B \cdot C \cdot D \cdot E \cdot F \cdot G \cdot H} \text{ OR}$$

$$Y = \overline{A} + \overline{B} + \overline{C} + \overline{D} + \overline{E} + \overline{F} + \overline{G} + \overline{H}$$

The SN54ALS30 and SN54AS30 are characterized for operation over the full military temperature range of −55°C to 125°C. The SN74ALS30 and SN74AS30 are characterized for operation from 0°C to 70°C.

SN54ALS30, SN54AS30 . . . J PACKAGE
SN74ALS30, SN74AS30 . . . N PACKAGE
(TOP VIEW)

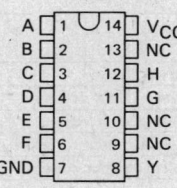

FUNCTION TABLE

INPUTS A THRU H	OUTPUT Y
All inputs H	L
One or more inputs L	H

SN54ALS30, SN54AS30 . . . FH PACKAGE
SN74ALS30, SN74AS30 . . . FN PACKAGE
(TOP VIEW)

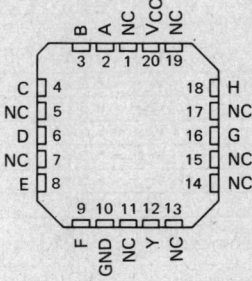

NC—No internal connection

logic symbol

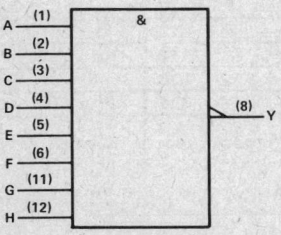

Pin numbers shown are for J and N packages.

Copyright © 1982 by Texas Instruments Incorporated

TEXAS INSTRUMENTS
POST OFFICE BOX 225012 • DALLAS, TEXAS 75265

2-55

TYPES SN54ALS30, SN74ALS30
8-INPUT POSITIVE-NAND GATES

absolute maximum ratings over operating free-air temperature range (unless otherwise noted)

Supply voltage, V_{CC} .... 7 V
Input voltage .... 7 V
Operating free-air temperature range: SN54ALS30 .... −55 °C to 125 °C
                                      SN74ALS30 .... 0 °C to 70 °C
Storage temperature range .... −65 °C to 150 °C

recommended operating conditions

		SN54ALS30			SN74ALS30			UNIT
		MIN	NOM	MAX	MIN	NOM	MAX	
V_{CC}	Supply voltage	4.5	5	5.5	4.5	5	5.5	V
V_{IH}	High-level input voltage	2			2			V
V_{IL}	Low-level input voltage			0.8			0.8	V
I_{OH}	High-level output current			−0.4			−0.4	mA
I_{OL}	Low-level output current			4			8	mA
T_A	Operating free-air temperature	−55		125	0		70	°C

electrical characteristics over recommended operating free-air temperature range (unless otherwise noted)

PARAMETER	TEST CONDITIONS		SN54ALS30			SN74ALS30			UNIT
			MIN	TYP†	MAX	MIN	TYP†	MAX	
V_{IK}	$V_{CC} = 4.5$ V,	$I_I = -18$ mA			−1.5			−1.5	V
V_{OH}	$V_{CC} = 4.5$ V to 5.5 V,	$I_{OH} = -0.4$ mA	$V_{CC}-2$			$V_{CC}-2$			V
V_{OL}	$V_{CC} = 4.5$ V,	$I_{OL} = 4$ mA		0.25	0.4		0.25	0.4	V
	$V_{CC} = 4.5$ V,	$I_{OL} = 8$ mA					0.35	0.5	
I_I	$V_{CC} = 5.5$ V,	$V_I = 7$ V			0.1			0.1	mA
I_{IH}	$V_{CC} = 5.5$ V,	$V_I = 2.7$ V			20			20	μA
I_{IL}	$V_{CC} = 5.5$ V,	$V_I = 0.4$ V			−0.1			−0.1	mA
I_O‡	$V_{CC} = 5.5$ V,	$V_O = 2.25$ V	−30		−112	−30		−112	mA
I_{CCH}	$V_{CC} = 5.5$ V,	$V_I = 0$ V		0.22	0.36		0.22	0.36	mA
I_{CCL}	$V_{CC} = 5.5$ V,	$V_I = 4.5$ V		0.54	0.9		0.54	0.9	mA

†All typical values are at $V_{CC} = 5$ V, $T_A = 25$ °C.
‡The output conditions have been chosen to produce a current that closely approximates one half of the true short-circuit output current, I_{OS}.

switching characteristics (see Note 1)

PARAMETER	FROM (INPUT)	TO (OUTPUT)	$V_{CC} = 4.5$ V to 5.5 V, $C_L = 50$ pF, $R_L = 500$ Ω, $T_A =$ MIN to MAX				UNIT
			SN54ALS30		SN74ALS30		
			MIN	MAX	MIN	MAX	
t_{PLH}	Any	Y	3	12	3	10	ns
t_{PHL}	Any	Y	5	22	5	20	ns

NOTE 1: For load circuit and voltage waveforms, see page 1-12.

Texas Instruments
POST OFFICE BOX 225012 • DALLAS, TEXAS 75265

TYPES SN54AS30, SN74AS30
8-INPUT POSITIVE-NAND GATES

absolute maximum ratings over operating free-air temperature range (unless otherwise noted)

Supply voltage, V_{CC} .... 7 V
Input voltage .... 7 V
Operating free-air temperature range: SN54AS30 .... −55 °C to 125 °C
　　　　　　　　　　　　　　　　　　　SN74AS30 .... 0 °C to 70 °C
Storage temperature range .... −65 °C to 150 °C

recommended operating conditions

		SN54AS30			SN74AS30			UNIT
		MIN	NOM	MAX	MIN	NOM	MAX	
V_{CC}	Supply voltage	4.5	5	5.5	4.5	5	5.5	V
V_{IH}	High-level input voltage	2			2			V
V_{IL}	Low-level input voltage			0.8			0.8	V
I_{OH}	High-level output current			−2			−2	mA
I_{OL}	Low-level output current			20			20	mA
T_A	Operating free-air temperature	−55		125	0		70	°C

electrical characteristics over recommended operating free-air temperature range (unless otherwise noted)

PARAMETER	TEST CONDITIONS		SN54AS30			SN74AS30			UNIT
			MIN	TYP†	MAX	MIN	TYP†	MAX	
V_{IK}	V_{CC} = 4.5 V,	I_I = −18 mA			−1.2			−1.2	V
V_{OH}	V_{CC} = 4.5 V to 5.5 V,	I_{OH} = −2 mA	V_{CC}−2			V_{CC}−2			V
V_{OL}	V_{CC} = 4.5 V,	I_{OL} = 20 mA		0.35	0.5		0.35	0.5	V
I_I	V_{CC} = 5.5 V,	V_I = 7 V			0.1			0.1	mA
I_{IH}	V_{CC} = 5.5 V,	V_I = 2.7 V			20			20	µA
I_{IL}	V_{CC} = 5.5 V,	V_I = 0.4 V			−0.5			−0.5	mA
I_O‡	V_{CC} = 5.5 V,	V_O = 2.25 V	−30		−112	−30		−112	mA
I_{CCH}	V_{CC} = 5.5 V,	V_I = 0 V		0.9	1.5		0.9	1.5	mA
I_{CCL}	V_{CC} = 5.5 V,	V_I = 4.5 V		3	4.9		3	4.9	mA

†All typical values are at V_{CC} = 5 V, T_A = 25 °C.
‡The output conditions have been chosen to produce a current that closely approximates one half of the true short-circuit output current, I_{OS}.

switching characteristics (see Note 1)

PARAMETER	FROM (INPUT)	TO (OUTPUT)	V_{CC} = 4.5 V to 5.5 V, C_L = 50 pF, R_L = 500 Ω, T_A = MIN to MAX				UNIT
			SN54AS30		SN74AS30		
			MIN	MAX	MIN	MAX	
t_{PLH}	Any	Y	1	5.5	1	5	ns
t_{PHL}	Any	Y	1	5	1	4.5	ns

NOTE 1: For load circuit and voltage waveforms, see page 1-12.

2

ALS AND AS CIRCUITS

TYPES SN54ALS32, SN54AS32, SN74ALS32, SN74AS32
QUADRUPLE 2-INPUT POSITIVE-OR GATES

D2661, APRIL 1982 — REVISED DECEMBER 1983

- Package Options Include Both Plastic and Ceramic Chip Carriers in Addition to Plastic and Ceramic DIPs
- Dependable Texas Instruments Quality and Reliability

description

These devices contain four independent 2-input OR gates. They perform the Boolean functions $Y = A + B$ or $Y = \overline{A} \cdot \overline{B}$ in positive logic.

The SN54ALS32 and SN54AS32 are characterized for operation over the full military temperature range of $-55\,°C$ to $125\,°C$. The SN74ALS32 and SN74AS32 are characterized for operation from $0\,°C$ to $70\,°C$.

FUNCTION TABLE
(each gate)

INPUTS		OUTPUT
A	B	Y
H	X	H
X	H	H
L	L	L

SN54ALS32, SN54AS32 . . . J PACKAGE
SN74ALS32, SN74AS32 . . . N PACKAGE
(TOP VIEW)

```
1A  [1   14] VCC
1B  [2   13] 4B
1Y  [3   12] 4A
2A  [4   11] 4Y
2B  [5   10] 3B
2Y  [6    9] 3A
GND [7    8] 3Y
```

SN54ALS32, SN54AS32 . . . FH PACKAGE
SN74ALS32, SN74AS32 . . . FN PACKAGE
(TOP VIEW)

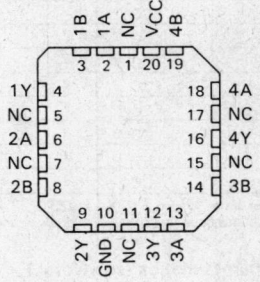

NC — No internal connection

logic symbol

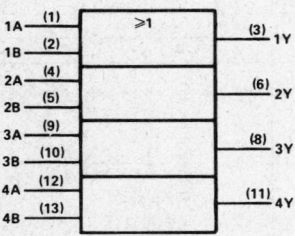

Pin numbers shown are for J and N packages.

Copyright © 1982 by Texas Instruments Incorporated

Texas Instruments
POST OFFICE BOX 225012 • DALLAS, TEXAS 75265

TYPES SN54ALS32, SN74ALS32
QUADRUPLE 2-INPUT POSITIVE-OR GATES

absolute maximum ratings over operating free-air temperature range (unless otherwise noted)

Supply voltage, V_{CC} .... 7 V
Input voltage .... 7 V
Operating free-air temperature range: SN54ALS32 .... −55 °C to 125 °C
                                      SN74ALS32 .... 0 °C to 70 °C
Storage temperature range .... −65 °C to 150 °C

recommended operating conditions

		SN54ALS32			SN74ALS32			UNIT
		MIN	NOM	MAX	MIN	NOM	MAX	
V_{CC}	Supply voltage	4.5	5	5.5	4.5	5	5.5	V
V_{IH}	High-level input voltage	2			2			V
V_{IL}	Low-level input voltage			0.8			0.8	V
I_{OH}	High-level output current			−0.4			−0.4	mA
I_{OL}	Low-level output current			4			8	mA
T_A	Operating free-air temperature	−55		125	0		70	°C

electrical characteristics over recommended operating free-air temperature range (unless otherwise noted)

PARAMETER	TEST CONDITIONS		SN54ALS32			SN74ALS32			UNIT
			MIN	TYP†	MAX	MIN	TYP†	MAX	
V_{IK}	V_{CC} = 4.5 V,	I_I = −18 mA			−1.5			−1.5	V
V_{OH}	V_{CC} = 4.5 V to 5.5 V,	I_{OH} = −0.4 mA	V_{CC}−2			V_{CC}−2			V
V_{OL}	V_{CC} = 4.5 V,	I_{OL} = 4 mA		0.25	0.4		0.25	0.4	V
	V_{CC} = 4.5 V,	I_{OL} = 8 mA					0.35	0.5	
I_I	V_{CC} = 5.5 V,	V_I = 7 V			0.1			0.1	mA
I_{IH}	V_{CC} = 5.5 V,	V_I = 2.7 V			20			20	µA
I_{IL}	V_{CC} = 5.5 V,	V_I = 0.4 V			−0.1			−0.1	mA
I_O‡	V_{CC} = 5.5 V,	V_O = 2.25 V	−30		−112	−30		−112	mA
I_{CCH}	V_{CC} = 5.5 V,	V_I = 4.5 V		1.9	4		1.9	4	mA
I_{CCL}	V_{CC} = 5.5 V,	V_I = 0 V		2.6	4.9		2.6	4.9	mA

†All typical values are at V_{CC} = 5 V, T_A = 25 °C.
‡The output conditions have been chosen to produce a current that closely approximates one half of the true short-circuit output current, I_{OS}.

switching characteristics (see Note 1)

PARAMETER	FROM (INPUT)	TO (OUTPUT)	V_{CC} = 4.5 V to 5.5 V, C_L = 50 pF, R_L = 500 Ω, T_A = MIN to MAX				UNIT
			SN54ALS32		SN74ALS32		
			MIN	MAX	MIN	MAX	
t_{PLH}	A or B	Y	3	16	3	14	ns
t_{PHL}	A or B	Y	3	13	3	12	ns

NOTE 1: For load circuit and voltage waveforms, see page 1-12.

TEXAS INSTRUMENTS
POST OFFICE BOX 225012 • DALLAS, TEXAS 75265

TYPES SN54AS32, SN74AS32
QUADRUPLE 2-INPUT POSITIVE-OR GATES

absolute maximum ratings over operating free-air temperature range (unless otherwise noted)

Supply voltage, V_{CC} .... 7 V
Input voltage .... 7 V
Operating free-air temperature range: SN54AS32 .... −55 °C to 125 °C
                                      SN74AS32 .... 0 °C to 70 °C
Storage temperature range .... −65 °C to 150 °C

recommended operating conditions

		SN54AS32			SN74AS32			UNIT
		MIN	NOM	MAX	MIN	NOM	MAX	
V_{CC}	Supply voltage	4.5	5	5.5	4.5	5	5.5	V
V_{IH}	High-level input voltage	2			2			V
V_{IL}	Low-level input voltage			0.8			0.8	V
I_{OH}	High-level output current			−2			−2	mA
I_{OL}	Low-level output current			20			20	mA
T_A	Operating free-air temperature	−55		125	0		70	°C

electrical characteristics over recommended operating free-air temperature range (unless otherwise noted)

PARAMETER	TEST CONDITIONS		SN54AS32			SN74AS32			UNIT
			MIN	TYP†	MAX	MIN	TYP†	MAX	
V_{IK}	V_{CC} = 4.5 V,	I_I = −18 mA			−1.2			−1.2	V
V_{OH}	V_{CC} = 4.5 V to 5.5 V,	I_{OH} = −2 mA	V_{CC}−2			V_{CC}−2			V
V_{OL}	V_{CC} = 4.5 V,	I_{OL} = 20 mA		0.35	0.5		0.35	0.5	V
I_I	V_{CC} = 5.5 V,	V_I = 7 V			0.1			0.1	mA
I_{IH}	V_{CC} = 5.5 V,	V_I = 2.7 V			20			20	µA
I_{IL}	V_{CC} = 5.5 V,	V_I = 0.4 V			−0.5			−0.5	mA
I_O‡	V_{CC} = 5.5 V,	V_O = 2.25 V	−30		−112	−30		−112	mA
I_{CCH}	V_{CC} = 5.5 V,	V_I = 4.5 V		7.3	12		7.3	12	mA
I_{CCL}	V_{CC} = 5.5 V,	V_I = 0 V		16.5	26.6		16.5	26.6	mA

†All typical values are at V_{CC} = 5 V, T_A = 25 °C.
‡The output conditions have been chosen to produce a current that closely approximates one half of the true short-circuit output current, I_{OS}.

switching characteristics (see Note 1)

PARAMETER	FROM (INPUT)	TO (OUTPUT)	V_{CC} = 4.5 V to 5.5 V, C_L = 50 pF, R_L = 500 Ω, T_A = MIN to MAX				UNIT
			SN54AS32		SN74AS32		
			MIN	MAX	MIN	MAX	
t_{PLH}	A or B	Y	1	7.5	1	5.8	ns
t_{PHL}	A or B	Y	1	6.5	1	5.8	ns

NOTE 1: For load circuit and voltage waveforms, see page 1-12.

2
ALS AND AS CIRCUITS

TYPES SN54ALS33A, SN74ALS33A
QUADRUPLE 2-INPUT POSITIVE-NOR BUFFERS WITH OPEN-COLLECTOR OUTPUTS

D2661, APRIL 1982—REVISED DECEMBER 1983

- Package Options Include Both Plastic and Ceramic Chip Carriers in Addition to Plastic and Ceramic DIPs
- Dependable Texas Instruments Quality and Reliability

description

These devices contain four independent 2-input NOR buffer gates with open-collector outputs. Open-collector outputs require resistive pull-up to perform logically but can deliver higher V_{OH} levels and are commonly used in wired-AND applications. These devices perform the Boolean functions $Y = \overline{A + B}$ or $Y = \overline{A} \cdot \overline{B}$ in positive logic.

The SN54ALS33A is characterized for operation over the full military temperature range of $-55\,°C$ to $125\,°C$. The SN74ALS33A is characterized for operation from $0\,°C$ to $70\,°C$.

SN54ALS33A . . . J PACKAGE
SN74ALS33A . . . N PACKAGE
(TOP VIEW)

```
1Y  [ 1    14 ] VCC
1A  [ 2    13 ] 4Y
1B  [ 3    12 ] 4B
2Y  [ 4    11 ] 4A
2A  [ 5    10 ] 3Y
2B  [ 6     9 ] 3B
GND [ 7     8 ] 3A
```

FUNCTION TABLE (each gate)

INPUTS		OUTPUT
A	B	Y
H	X	L
X	H	L
L	L	H

SN54ALS33A . . . FH PACKAGE
SN74ALS33A . . . FN PACKAGE
(TOP VIEW)

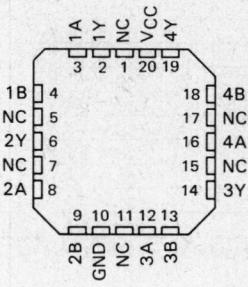

NC — No internal connection

logic symbol

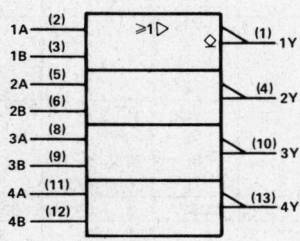

Pin numbers shown are for J and N packages.

TYPES SN54ALS33A, SN74ALS33A
QUADRUPLE 2-INPUT POSITIVE-NOR BUFFERS
WITH OPEN-COLLECTOR OUTPUTS

absolute maximum ratings over operating free-air temperature range (unless otherwise noted)

Supply voltage, V_{CC} .... 7 V
Input voltage .... 7 V
Off-state output voltage .... 7 V
Operating free-air temperature range: SN54ALS33A .... −55 °C to 125 °C
                                           SN74ALS33A .... 0 °C to 70 °C
Storage temperature range .... −65 °C to 150 °C

recommended operating conditions

		SN54ALS33A			SN74ALS33A			UNIT
		MIN	NOM	MAX	MIN	NOM	MAX	
V_{CC}	Supply voltage	4.5	5	5.5	4.5	5	5.5	V
V_{IH}	High-level input voltage	2			2			V
V_{IL}	Low-level input voltage			0.8			0.8	V
V_{OH}	High-level output voltage			5.5			5.5	V
I_{OL}	Low-level output current			12			24	mA
T_A	Operating free-air temperature	−55		125	0		70	°C

electrical characteristics over recommended operating free-air temperature range (unless otherwise noted)

PARAMETER	TEST CONDITIONS		SN54ALS33A			SN74ALS33A			UNIT
			MIN	TYP†	MAX	MIN	TYP†	MAX	
V_{IK}	V_{CC} = 4.5 V,	I_I = −18 mA			−1.5			−1.5	V
I_{OH}	V_{CC} = 4.5 V,	V_{OH} = 5.5 V			0.1			0.1	mA
V_{OL}	V_{CC} = 4.5 V,	I_{OL} = 12 mA		0.25	0.4		0.25	0.4	V
	V_{CC} = 4.5 V,	I_{OL} = 24 mA					0.35	0.5	
I_I	V_{CC} = 5.5 V,	V_I = 7 V			0.1			0.1	mA
I_{IH}	V_{CC} = 5.5 V,	V_I = 2.7 V			20			20	µA
I_{IL}	V_{CC} = 5.5 V,	V_I = 0.4 V			−0.1			−0.1	mA
I_{CCH}	V_{CC} = 5.5 V,	V_I = 0 V		1.7	2.8		1.7	2.8	mA
I_{CCL}	V_{CC} = 5.5 V,	V_I = 4.5 V		5.6	9		5.6	9	mA

† All typical values are at V_{CC} = 5 V, T_A = 25 °C

switching characteristics (see Note 1)

PARAMETER	FROM (INPUT)	TO (OUTPUT)	V_{CC} = 4.5 V to 5.5 V, C_L = 50 pF, R_L = 680 Ω, T_A = MIN to MAX				UNIT
			SN54ALS33A		SN74ALS33A		
			MIN	MAX	MIN	MAX	
t_{PLH}	A or B	Y	10	40	10	33	ns
t_{PHL}	A or B	Y	2	18	2	12	ns

NOTE 1: For load circuit and voltage waveforms, see page 1-12.

TEXAS INSTRUMENTS
POST OFFICE BOX 225012 • DALLAS, TEXAS 75265

TYPES SN54ALS34, SN54AS34, SN74ALS34, SN74AS34
HEX NONINVERTERS

D2261, DECEMBER 1983

- **Noninverters**
- **Package Options Include Both Plastic and Ceramic Chip Carriers in Addition to Plastic and Ceramic DIPs**
- **Dependable Texas Instruments Quality and Reliability**

SN54ALS34, SN54AS34 . . . J PACKAGE
SN74ALS34, SN74AS34 . . . N PACKAGE
(TOP VIEW)

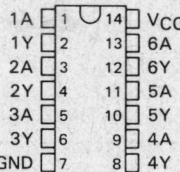

description

These devices contain six independent noninverters. They perform the Boolean functions Y = A.

The SN54ALS34 and SN54AS34 are characterized for operation over the full military temperature range of −55 °C to 125 °C. The SN74ALS34 and SN74AS34 are characterized for operation from 0 °C to 70 °C.

SN54ALS34, SN54AS34 . . . FH PACKAGE
SN74ALS34, SN74AS34 . . . FN PACKAGE
(TOP VIEW)

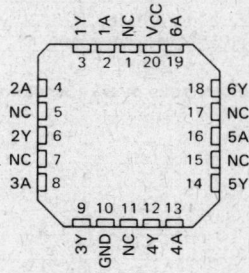

NC — No internal connection

FUNCTION TABLE (each buffer)

INPUT A	OUTPUT Y
H	H
L	L

logic symbol

```
1A ——(1)——| 1 |——(2)—— 1Y
2A ——(3)——|   |——(4)—— 2Y
3A ——(5)——|   |——(6)—— 3Y
4A ——(9)——|   |——(8)—— 4Y
5A ——(11)—|   |——(10)— 5Y
6A ——(13)—|   |——(12)— 6Y
```

Pin numbers shown are for J and N packages.

ALS AND AS CIRCUITS

Copyright © 1983 by Texas Instruments Incorporated

TYPES SN54ALS34, SN74ALS34
HEX NONINVERTERS

absolute maximum ratings over operating free-air temperature range (unless otherwise noted)

Supply voltage, V_{CC} .... 7 V
Input voltage .... 7 V
Operating free-air temperature range: SN54ALS34 .... −55 °C to 125 °C
                                      SN74ALS34 .... 0 °C to 70 °C
Storage temperature range .... −65 °C to 150 °C

recommended operating conditions

		SN54ALS34			SN74ALS34			UNIT
		MIN	NOM	MAX	MIN	NOM	MAX	
V_{CC}	Supply voltage	4.5	5	5.5	4.5	5	5.5	V
V_{IH}	High-level input voltage	2			2			V
V_{IL}	Low-level input voltage			0.8			0.8	V
I_{OH}	High-level output current			−0.4			−0.4	mA
I_{OL}	Low-level output current			4			8	mA
T_A	Operating free-air temperature	−55		125	0		70	°C

electrical characteristics over recommended operating free-air temperature range (unless otherwise noted)

PARAMETER	TEST CONDITIONS		SN54ALS34			SN74ALS34			UNIT
			MIN	TYP†	MAX	MIN	TYP†	MAX	
V_{IK}	$V_{CC} = 4.5$ V,	$I_I = -18$ mA			−1.5			−1.5	V
V_{OH}	$V_{CC} = 4.5$ V to 5.5 V,	$I_{OH} = -0.4$ mA	$V_{CC}-2$			$V_{CC}-2$			V
V_{OL}	$V_{CC} = 4.5$ V,	$I_{OL} = 4$ mA		0.25	0.4		0.25	0.4	V
	$V_{CC} = 4.5$ V,	$I_{OL} = 8$ mA					0.35	0.5	
I_I	$V_{CC} = 5.5$ V,	$V_I = 7$ V			0.1			0.1	mA
I_{IH}	$V_{CC} = 5.5$ V,	$V_I = 2.7$ V			20			20	µA
I_{IL}	$V_{CC} = 5.5$ V,	$V_I = 0.4$ V			−0.1			−0.1	mA
I_O‡	$V_{CC} = 5.5$ V,	$V_O = 2.25$ V	−30		−112	−30		−112	mA
I_{CCH}	$V_{CC} = 5.5$ V,	$V_I = 4.5$ V		1			1		mA
I_{CCL}	$V_{CC} = 5.5$ V,	$V_I = 0$ V		3.5			3.5		mA

†All typical values are at $V_{CC} = 5$ V, $T_A = 25$ °C.
‡The output conditions have been chosen to produce a current that closely approximates one half of the true short-circuit output current, I_{OS}.

switching characteristics (see Note 1)

PARAMETER	FROM (INPUT)	TO (OUTPUT)	$V_{CC} = 4.5$ V to 5.5 V, $C_L = 50$ pF, $R_L = 500$ Ω, T_A = MIN to MAX						UNIT
			SN54ALS34			SN74ALS34			
			MIN	TYP†	MAX	MIN	TYP†	MAX	
t_{PLH}	A	Y		8			8		ns
t_{PHL}				6			6		

†All typical values are at $V_{CC} = 5$ V, $T_A = 25$ °C.
NOTE 1: For load circuit and voltage waveforms, see page 1-12.

PRODUCT PREVIEW
This page contains information on a product under development. Texas Instruments reserves the right to change or discontinue this product without notice.

TEXAS INSTRUMENTS
POST OFFICE BOX 225012 • DALLAS, TEXAS 75265

TYPES SN54AS34, SN74AS34
HEX NONINVERTERS

absolute maximum ratings over operating free-air temperature range (unless otherwise noted)

Supply voltage, V_{CC} .... 7 V
Input voltage .... 7 V
Operating free-air temperature range: SN54AS34 .... −55 °C to 125 °C
                                        SN74AS34 .... 0 °C to 70 °C
Storage temperature range .... −65 °C to 150 °C

recommended operating conditions

		SN54AS34			SN74AS34			UNIT
		MIN	NOM	MAX	MIN	NOM	MAX	
V_{CC}	Supply voltage	4.5	5	5.5	4.5	5	5.5	V
V_{IH}	High-level input voltage	2			2			V
V_{IL}	Low-level input voltage			0.8			0.8	V
I_{OH}	High-level output current			−2			−2	mA
I_{OL}	Low-level output current			20			20	mA
T_A	Operating free-air temperature	−55		125	0		70	°C

electrical characteristics over recommended operating free-air temperature range (unless otherwise noted)

PARAMETER	TEST CONDITIONS		SN54AS34			SN74AS34			UNIT
			MIN	TYP†	MAX	MIN	TYP†	MAX	
V_{IK}	V_{CC} = 4.5 V,	I_I = −18 mA			−1.2			−1.2	V
V_{OH}	V_{CC} = 4.5 V to 5.5 V,	I_{OH} = −2 mA	V_{CC}−2			V_{CC}−2			V
V_{OL}	V_{CC} = 4.5 V,	I_{OL} = 20 mA		0.35	0.5		0.35	0.5	V
I_I	V_{CC} = 5.5 V,	V_I = 7 V			0.1			0.1	mA
I_{IH}	V_{CC} = 5.5 V,	V_I = 2.7 V			20			20	µA
I_{IL}	V_{CC} = 5.5 V,	V_I = 0.4 V			−0.1			−0.1	mA
I_O‡	V_{CC} = 5.5 V,	V_O = 2.25 V	−30		−112	−30		−112	mA
I_{CCH}	V_{CC} = 5.5 V,	V_I = 4.5 V		7.4	12		7.4	12	mA
I_{CCL}	V_{CC} = 5.5 V,	V_I = 0 V		21.3	34.6		21.3	34.6	mA

†All typical values are at V_{CC} = 5 V, T_A = 25 °C.
‡The output conditions have been chosen to produce a current that closely approximates one half of the true short-circuit output current, I_{OS}.

switching characteristics (see Note 1)

PARAMETER	FROM (INPUT)	TO (OUTPUT)	V_{CC} = 4.5 V to 5.5 V, C_L = 50 pF, R_L = 500 Ω, T_A = MIN to MAX				UNIT
			SN54AS34		SN74AS34		
			MIN	MAX	MIN	MAX	
t_{PLH}	A	Y	1	6.5	1	5.5	ns
t_{PHL}			1	7	1	6	

NOTE 1: For load circuit and voltage wavforms, see page 1-12.

ALS AND AS CIRCUITS

TYPES SN54ALS35, SN74ALS35
HEX NONINVERTERS WITH OPEN-COLLECTOR OUTPUTS

D2661, DECEMBER 1983

- **Noninverters with Open-Collector Outputs**
- **Package Options Include Both Plastic and Ceramic Chip Carriers in Addition to Plastic and Ceramic DIPs**
- **Dependable Texas Instruments Quality and Reliability**

description

These devices contain six independent noninverters. They perform the Boolean functions Y = A. The open-collector outputs require pull-up resistors to perform correctly. They may be connected to other open-collector outputs to implement active-low wired-OR or active-high wired-AND functions. Open-collector devices are often used to generate higher V_{OH} levels.

The SN54ALS35 is characterized for operation over the full military temperature range of −55°C to 125°C. The SN74ALS35 is characterized for operation from 0°C to 70°C.

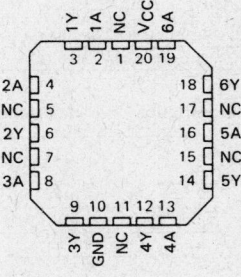

NC — No internal connection

FUNCTION TABLE (each buffer)

INPUT A	OUTPUT Y
H	H
L	L

logic symbol

```
1A ─(1)──┐ 1 ◇├──(2)── 1Y
2A ─(3)──┤        ├──(4)── 2Y
3A ─(5)──┤        ├──(6)── 3Y
4A ─(9)──┤        ├──(8)── 4Y
5A ─(11)─┤        ├──(10)─ 5Y
6A ─(13)─┤        ├──(12)─ 6Y
```

Pin numbers shown are for J and N packages.

PRODUCT PREVIEW

This document contains information on a product under development. Texas Instruments reserves the right to change or discontinue this product without notice.

Copyright © 1983 by Texas Instruments Incorporated

TEXAS INSTRUMENTS
POST OFFICE BOX 225012 • DALLAS, TEXAS 75265

TYPES SN54ALS35, SN74ALS35
HEX NONINVERTERS WITH OPEN-COLLECTOR OUTPUTS

absolute maximum ratings over operating free-air temperature range (unless otherwise noted)

Supply voltage, V_{CC} .... 7 V
Input voltage .... 7 V
Off-state output voltage .... 7 V
Operating free-air temperature range: SN54ALS35 .... −55 °C to 125 °C
                                  SN74ALS35 .... 0 °C to 70 °C
Storage temperature range .... −65 °C to 150 °C

recommended operating conditions

		SN54ALS35			SN74ALS35			UNIT
		MIN	NOM	MAX	MIN	NOM	MAX	
V_{CC}	Supply voltage	4.5	5	5.5	4.5	5	5.5	V
V_{IH}	High-level input voltage	2			2			V
V_{IL}	Low-level input voltage			0.8			0.8	V
V_{OH}	High-level output voltage			5.5			5.5	V
I_{OL}	Low-level output current			4			8	mA
T_A	Operating free-air temperature	−55		125	0		70	°C

electrical characteristics over recommended operating free-air temperature range (unless otherwise noted)

PARAMETER	TEST CONDITIONS		SN54ALS35			SN74ALS35			UNIT
			MIN	TYP†	MAX	MIN	TYP†	MAX	
V_{IK}	V_{CC} = 4.5 V,	I_I = −18 mA			−1.5			−1.5	V
I_{OH}	V_{CC} = 4.5 V,	V_{OH} = 5.5 V			0.1			0.1	mA
V_{OL}	V_{CC} = 4.5 V,	I_{OL} = 4 mA		0.25	0.4		0.25	0.4	V
	V_{CC} = 4.5 V,	I_{OL} = 8 mA					0.35	0.5	
I_I	V_{CC} = 5.5 V,	V_I = 7 V			0.1			0.1	mA
I_{IH}	V_{CC} = 5.5 V,	V_I = 2.7 V			20			20	µA
I_{IL}	V_{CC} = 5.5 V,	V_I = 0.4 V			−0.1			−0.1	mA
I_{CCH}	V_{CC} = 5.5 V,	V_I = 4.5 V		1			1		mA
I_{CCL}	V_{CC} = 5.5 V,	V_I = 0 V		3.5			3.5		mA

†All typical values are at V_{CC} = 5 V, T_A = 25 °C

switching characteristics (see Note 1)

PARAMETER	FROM (INPUT)	TO (OUTPUT)	V_{CC} = 4.5 V to 5.5 V, C_L = 50 pF, R_L = 680 Ω, T_A = MIN to MAX						UNIT
			SN54ALS35			SN74ALS35			
			MIN	TYP†	MAX	MIN	TYP†	MAX	
t_{PLH}	A	Y		25			25		ns
t_{PHL}	A	Y		8			8		ns

†All typical values are at V_{CC} = 5 V, T_A = 25 °C.
NOTE 1: For load circuit and voltage waveforms, see page 1-12.

TEXAS INSTRUMENTS
POST OFFICE BOX 225012 • DALLAS, TEXAS 75265

TYPES SN54ALS37A, SN74ALS37A
QUADRUPLE 2-INPUT POSITIVE-NAND BUFFERS

D2661, APRIL 1982—REVISED DECEMBER 1983

- Package Options Include Both Plastic and Ceramic Chip Carriers in Addition to Plastic and Ceramic DIPs
- Dependable Texas Instruments Quality and Reliability

description

These devices contain four independent 2-input NAND buffer gates. They perform the Boolean functions $Y = \overline{A \cdot B}$ or $Y = \overline{A} + \overline{B}$ in positive logic.

The SN54ALS37A is characterized for operation over the full military temperature range of $-55\,°C$ to $125\,°C$. The SN74ALS37A is characterized for operation from $0\,°C$ to $70\,°C$.

SN54ALS37A . . . J PACKAGE
SN74ALS37A . . . N PACKAGE
(TOP VIEW)

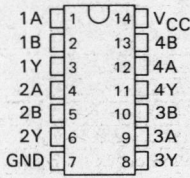

FUNCTION TABLE
(each gate)

INPUTS		OUTPUT
A	B	Y
H	H	L
L	X	H
X	L	H

SN54ALS37A . . . FH PACKAGE
SN74ALS37A . . . FN PACKAGE
(TOP VIEW)

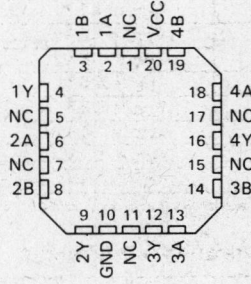

NC — No internal connection

logic symbol

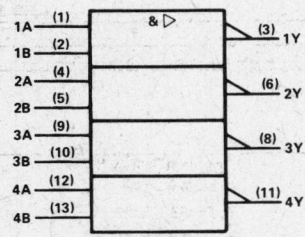

Pin numbers shown are for J and N packages.

Copyright © 1982 by Texas Instruments Incorporated

TEXAS INSTRUMENTS
POST OFFICE BOX 225012 • DALLAS, TEXAS 75265

ALS AND AS CIRCUITS

2-71

TYPES SN54ALS37A, SN74ALS37A
QUADRUPLE 2-INPUT POSITIVE-NAND BUFFERS

absolute maximum ratings over operating free-air temperature range (unless otherwise noted)

Supply voltage, V_{CC} .... 7 V
Input voltage .... 7 V
Operating free-air temperature range: SN54ALS37A .... −55 °C to 125 °C
                                       SN74ALS37A .... 0 °C to 70 °C
Storage temperature range .... −65 °C to 150 °C

recommended operating conditions

		SN54ALS37A			SN74ALS37A			UNIT
		MIN	NOM	MAX	MIN	NOM	MAX	
V_{CC}	Supply voltage	4.5	5	5.5	4.5	5	5.5	V
V_{IH}	High-level input voltage	2			2			V
V_{IL}	Low-level input voltage			0.8			0.8	V
I_{OH}	High-level output current			−1			−2.6	mA
I_{OL}	Low-level output current			12			24	mA
T_A	Operating free-air temperature	−55		125	0		70	°C

electrical characteristics over recommended operating free-air temperature range (unless otherwise noted)

PARAMETER	TEST CONDITIONS		SN54ALS37A			SN74ALS37A			UNIT
			MIN	TYP†	MAX	MIN	TYP†	MAX	
V_{IK}	V_{CC} = 4.5 V,	I_I = −18 mA			−1.5			−1.5	V
V_{OH}	V_{CC} = 4.5 V to 5.5 V,	I_{OH} = −0.4 mA	V_{CC}−2			V_{CC}−2			V
	V_{CC} = 4.5 V,	I_{OH} = −1 mA	2.4	3.3					
	V_{CC} = 4.5 V,	I_{OH} = −2.6 mA				2.4	3.2		
V_{OL}	V_{CC} = 4.5 V,	I_{OL} = 12 mA		0.25	0.4		0.25	0.4	V
	V_{CC} = 4.5 V,	I_{OL} = 24 mA					0.35	0.5	
I_I	V_{CC} = 5.5 V,	V_I = 7 V			0.1			0.1	mA
I_{IH}	V_{CC} = 5.5 V,	V_I = 2.7 V			20			20	µA
I_{IL}	V_{CC} = 5.5 V,	V_{IL} = 0.4 V			−0.1			−0.1	mA
I_O‡	V_{CC} = 5.5 V,	V_O = 2.25 V	−30		−112	−30		−112	mA
I_{CCH}	V_{CC} = 5.5 V,	V_I = 0 V		0.86	1.6		0.86	1.6	mA
I_{CCL}	V_{CC} = 5.5 V,	V_I = 4.5 V		4.8	7.8		4.8	7.8	mA

† All typical values are at V_{CC} = 5 V, T_A = 25 °C.
‡ The output conditions have been chosen to produce a current that closely approximates one half of the true short-circuit output current, I_{OS}.

switching characteristics (see Note 1)

PARAMETER	FROM (INPUT)	TO (OUTPUT)	V_{CC} = 4.5 V to 5.5 V, C_L = 50 pF, R_L = 500 Ω, T_A = MIN to MAX				UNIT
			SN54ALS37A		SN74ALS37A		
			MIN	MAX	MIN	MAX	
t_{PLH}	A or B	Y	2	10	2	8	ns
t_{PHL}	A or B	Y	2	10	2	7	ns

NOTE 1: For load circuit and voltage waveforms, see page 1-12.

TEXAS INSTRUMENTS
POST OFFICE BOX 225012 • DALLAS, TEXAS 75265

TYPES SN54ALS38A, SN74ALS38A
QUADRUPLE 2-INPUT POSITIVE-NAND BUFFERS WITH OPEN-COLLECTOR OUTPUTS

D2661, APRIL 1982—REVISED DECEMBER 1983

- Package Options Include Both Plastic and Ceramic Chip Carriers in Addition to Plastic and Ceramic DIPs
- Dependable Texas Instruments Quality and Reliability

description

These devices contain four independent 2-input NAND buffer gates with open-collector outputs. These NAND buffers perform the Boolean functions $Y = \overline{A \cdot B}$ or $Y = \overline{A} + \overline{B}$ in positive logic. The open-collector outputs require pull-up resistors to perform correctly. They may be connected to other open-collector outputs to implement active-low wired-OR or active-high wired-AND functions. Open-collector devices are often used to generate higher V_{OH} levels.

The SN54ALS38A is characterized for operation over the full military temperature range of $-55\,°C$ to $125\,°C$. The SN74ALS38A is characterized for operation from $0\,°C$ to $70\,°C$.

SN54ALS38A . . . J PACKAGE
SN74ALS38A . . . N PACKAGE
(TOP VIEW)

```
1A  [1   14] VCC
1B  [2   13] 4B
1Y  [3   12] 4A
2A  [4   11] 4Y
2B  [5   10] 3B
2Y  [6    9] 3A
GND [7    8] 3Y
```

SN54ALS38A . . . FH PACKAGE
SN74ALS38A . . . FN PACKAGE
(TOP VIEW)

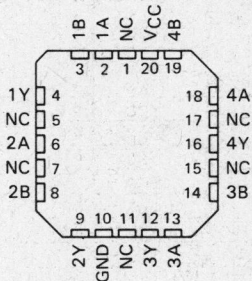

NC — No internal connection

FUNCTION TABLE (each gate)

INPUTS		OUTPUT
A	B	Y
H	H	L
L	X	H
X	L	H

logic symbol

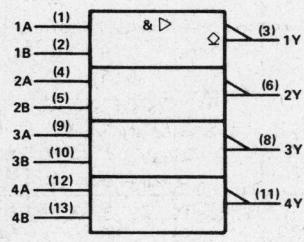

Pin numbers shown are for J and N packages.

ALS AND AS CIRCUITS

Copyright © 1982 by Texas Instruments Incorporated

TEXAS INSTRUMENTS
POST OFFICE BOX 225012 • DALLAS, TEXAS 75265

2-73

TYPES SN54ALS38A, SN74ALS38A
QUADRUPLE 2-INPUT POSITIVE-NAND BUFFERS WITH OPEN-COLLECTOR OUTPUTS

absolute maximum ratings over operating free-air temperature range (unless otherwise noted)

Supply voltage, V_{CC} .... 7 V
Input voltage .... 7 V
Off-state output voltage .... 7 V
Operating free-air temperature range: SN54ALS38A .... −55°C to 125°C
                                      SN74ALS38A .... 0°C to 70°C
Storage temperature range .... −65°C to 150°C

recommended operating conditions

		SN54ALS38A			SN74ALS38A			UNIT
		MIN	NOM	MAX	MIN	NOM	MAX	
V_{CC}	Supply voltage	4.5	5	5.5	4.5	5	5.5	V
V_{IH}	High-level input voltage	2			2			V
V_{IL}	Low-level input voltage			0.8			0.8	V
V_{OH}	High-level output voltage			5.5			5.5	V
I_{OL}	Low-level output current			12			24	mA
T_A	Operating free-air temperature	−55		125	0		70	°C

electrical characteristics over recommended operating free-air temperature range (unless otherwise noted)

PARAMETER	TEST CONDITIONS		SN54ALS38A			SN74ALS38A			UNIT
			MIN	TYP†	MAX	MIN	TYP†	MAX	
V_{IK}	V_{CC} = 4.5 V,	I_I = −18 mA			−1.5			−1.5	V
I_{OH}	V_{CC} = 4.5 V,	V_{OH} = 5.5 V			0.1			0.1	mA
V_{OL}	V_{CC} = 4.5 V,	I_{OL} = 12 mA		0.25	0.4		0.25	0.4	V
	V_{CC} = 4.5 V,	I_{OL} = 24 mA					0.35	0.5	
I_I	V_{CC} = 5.5 V,	V_I = 7 V			0.1			0.1	mA
I_{IH}	V_{CC} = 5.5 V,	V_I = 2.7 V			20			20	µA
I_{IL}	V_{CC} = 5.5 V,	V_I = 0.4 V			−0.1			−0.1	mA
I_{CCH}	V_{CC} = 5.5 V,	V_I = 0 V		0.86	1.6		0.86	1.6	mA
I_{CCL}	V_{CC} = 5.5 V,	V_I = 4.5 V		4.8	7.8		4.8	7.8	mA

† All typical values are at V_{CC} = 5 V, T_A = 25°C.

switching characteristics (see Note 1)

PARAMETER	FROM (INPUT)	TO (OUTPUT)	V_{CC} = 4.5 V to 5.5 V, C_L = 50 pF, R_L = 680 Ω, T_A = MIN to MAX				UNIT
			SN54ALS38A		SN74ALS38A		
			MIN	MAX	MIN	MAX	
t_{PLH}	A or B	Y	10	40	10	33	ns
t_{PHL}	A or B	Y	2	18	2	12	ns

NOTE 1: For load circuit and voltage waveforms, see page 1-12.

Texas Instruments
POST OFFICE BOX 225012 • DALLAS, TEXAS 75265

TYPES SN54ALS40A, SN74ALS40A
DUAL 4-INPUT POSITIVE-NAND BUFFERS

D2661, APRIL 1982—REVISED DECEMBER 1983

- Package Options Include Both Plastic and Ceramic Chip Carriers in Addition to Plastic and Ceramic DIPs
- Dependable Texas Instruments Quality and Reliability

description

These devices contain two independent 4-input NAND buffer gates. They perform the Boolean functions $Y = \overline{A \cdot B \cdot C \cdot D}$ or $Y = \overline{A} + \overline{B} + \overline{C} + \overline{D}$ in positive logic.

The SN54ALS40A is characterized for operation over the full military temperature range of $-55\,°C$ to $125\,°C$. The SN74ALS40A is characterized for operation from $0\,°C$ to $70\,°C$.

SN54ALS40A . . . J PACKAGE
SN74ALS40A . . . N PACKAGE
(TOP VIEW)

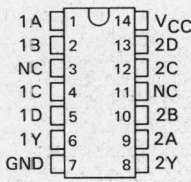

FUNCTION TABLE (each gate)

INPUTS				OUTPUT
A	B	C	D	Y
H	H	H	H	L
L	X	X	X	H
X	L	X	X	H
X	X	L	X	H
X	X	X	L	H

SN54ALS40A . . . FH PACKAGE
SN74ALS40A . . . FN PACKAGE
(TOP VIEW)

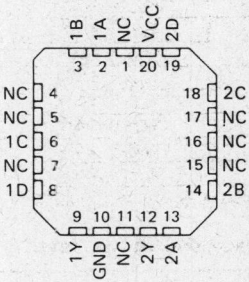

NC—No internal connection

logic symbol

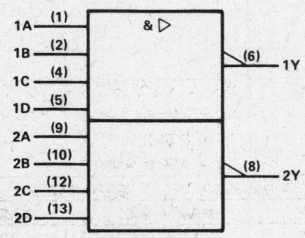

Pin numbers shown are for J and N packages.

ALS AND AS CIRCUITS

Copyright © 1982 by Texas Instruments Incorporated

TEXAS INSTRUMENTS
POST OFFICE BOX 225012 • DALLAS, TEXAS 75265

2-75

TYPES SN54ALS40A, SN74ALS40A
DUAL 4-INPUT POSITIVE-NAND BUFFERS

absolute maximum ratings over operating free-air temperature range (unless otherwise noted)

Supply voltage, V_{CC} .... 7 V
Input voltage .... 7 V
Operating free-air temperature range: SN54ALS40A .... −55 °C to 125 °C
                                       SN74ALS40A .... 0 °C to 70 °C
Storage temperature range .... −65 °C to 150 °C

recommended operating conditions

		SN54ALS40A			SN74ALS40A			UNIT
		MIN	NOM	MAX	MIN	NOM	MAX	
V_{CC}	Supply voltage	4.5	5	5.5	4.5	5	5.5	V
V_{IH}	High-level input voltage	2			2			V
V_{IL}	Low-level input voltage			0.8			0.8	V
I_{OH}	High-level output current			−1			−2.6	mA
I_{OL}	Low-level output current			12			24	mA
T_A	Operating free-air temperature	−55		125	0		70	°C

electrical characteristics over recommended operating free-air temperature range (unless otherwise noted)

PARAMETER	TEST CONDITIONS		SN54ALS40A			SN74ALS40A			UNIT
			MIN	TYP†	MAX	MIN	TYP†	MAX	
V_{IK}	V_{CC} = 4.5 V,	I_I = −18 mA			−1.5			−1.5	V
V_{OH}	V_{CC} = 4.5 V to 5.5 V,	I_{OH} = −0.4 mA	V_{CC}−2			V_{CC}−2			V
	V_{CC} = 4.5 V,	I_{OH} = −1 mA	2.4	3.3					
	V_{CC} = 4.5 V,	I_{OH} = −2.6 mA				2.4	3.2		
V_{OL}	V_{CC} = 4.5 V,	I_{OL} = 12 mA		0.25	0.4		0.25	0.4	V
	V_{CC} = 4.5 V,	I_{OL} = 24 mA					0.35	0.5	
I_I	V_{CC} = 5.5 V,	V_I = 7 V			0.1			0.1	mA
I_{IH}	V_{CC} = 5.5 V,	V_I = 2.7 V			20			20	µA
I_{IL}	V_{CC} = 5.5 V,	V_{IL} = 0.4 V			−0.1			−0.1	mA
I_O‡	V_{CC} = 5.5 V,	V_O = 2.25 V	−30		−112	−30		−112	mA
I_{CCH}	V_{CC} = 5.5 V,	V_I = 0 V		0.43	0.8		0.43	0.8	mA
I_{CCL}	V_{CC} = 5.5 V,	V_I = 4.5 V		2.4	3.9		2.4	3.9	mA

† All typical values are at V_{CC} = 5 V, T_A = 25 °C.
‡ The output conditions have been chosen to produce a current that closely approximates one half of the true short-circuit output current, I_{OS}.

switching characteristics (see Note 1)

PARAMETER	FROM (INPUT)	TO (OUTPUT)	V_{CC} = 4.5 V to 5.5 V, C_L = 50 pF, R_L = 500 Ω, T_A = MIN to MAX				UNIT
			SN54ALS40A		SN74ALS40A		
			MIN	MAX	MIN	MAX	
t_{PLH}	A or B	Y	2	10	2	8	ns
t_{PHL}	A or B	Y	2	10	2	7	ns

NOTE 1: For load circuit and voltage waveforms, see page 1-12.

TEXAS INSTRUMENTS
POST OFFICE BOX 225012 • DALLAS, TEXAS 75265

TYPES SN54ALS74, SN54AS74, SN74ALS74, SN74AS74
DUAL D-TYPE POSITIVE-EDGE-TRIGGERED FLIP-FLOPS WITH CLEAR AND PRESET

D2661, APRIL 1982—REVISED DECEMBER 1983

- Package Options Include Both Plastic and Ceramic Chip Carriers in Addition to Plastic and Ceramic DIPs
- Dependable Texas Instruments Quality and Reliability

TYPE	TYPICAL MAXIMUM CLOCK FREQUENCY (C_L = 50 pF)	TYPICAL POWER DISSIPATION PER FLIP-FLOP
'ALS74	50 MHz	6 mW
'AS74	134 MHz	26 mW

SN54ALS74, SN54AS74 . . . J PACKAGE
SN74ALS74, SN74AS74 . . . N PACKAGE
(TOP VIEW)

```
1CLR  [1   14] VCC
 1D   [2   13] 2CLR
1CLK  [3   12] 2D
1PRE  [4   11] 2CLK
 1Q   [5   10] 2PRE
 1Q̄   [6    9] 2Q
GND   [7    8] 2Q̄
```

description

These devices contain two independent D-type positive-edge-triggered flip-flops. A low level at the Preset or Clear inputs sets or resets the outputs regardless of the levels of the other inputs. When Preset and Clear are inactive (high), data at the D input meeting the setup time requirements are transferred to the outputs on the positive-going edge of the clock pulse. Clock triggering occurs at a voltage level and is not directly related to the rise time of the clock pulse. Following the hold time interval, data at the D input may be changed without affecting the levels at the outputs.

The SN54ALS74 and SN54AS74 are characterized for operation over the full military temperature range of −55 °C to 125 °C. The SN74ALS74 and SN74AS74 are characterized for operation from 0 °C to 70 °C.

SN54ALS74, SN54AS74 . . . FH PACKAGE
SN74ALS74, SN74AS74 . . . FN PACKAGE
(TOP VIEW)

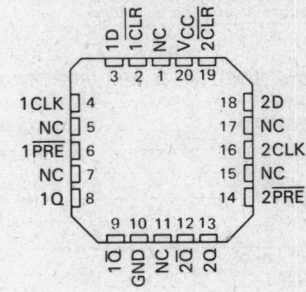

NC—No internal connection

FUNCTION TABLE

INPUTS				OUTPUTS	
PRESET	CLEAR	CLOCK	D	Q	Q̄
L	H	X	X	H	L
H	L	X	X	L	H
L	L	X	X	H*	H*
H	H	↑	H	H	L
H	H	↑	L	L	H
H	H	L	X	Q_0	$\bar{Q}_0$

*The output levels in this configuration are not guaranteed to meet the minimum levels for V_{OH} if the lows at Preset and Clear are near V_{IL} maximum. Furthermore, this configuration is nonstable; that is, it will not persist when either Preset or Clear returns to its inactive (high) level.

logic symbol

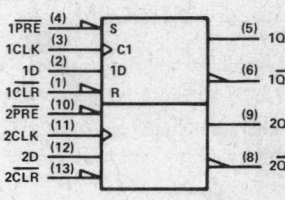

Pin numbers shown are for J and N packages.

absolute maximum ratings over operating free-air temperature range (unless otherwise noted)

Supply voltage, V_{CC} .... 7 V
Input voltage .... 7 V
Operating free-air temperature range: SN54ALS74, SN54AS74 .... −55 °C to 125 °C
SN74ALS74, SN74AS74 .... 0 °C to 70 °C
Storage temperature range .... −65 °C to 150 °C

Copyright © 1982 by Texas Instruments Incorporated

POST OFFICE BOX 225012 • DALLAS, TEXAS 75265

TYPES SN54ALS74, SN74ALS74
DUAL D-TYPE POSITIVE-EDGE-TRIGGERED FLIP-FLOPS WITH CLEAR AND PRESET

recommended operating conditions

			SN54ALS74			SN74ALS74			UNIT
			MIN	NOM	MAX	MIN	NOM	MAX	
V_{CC}	Supply voltage		4.5	5	5.5	4.5	5	5.5	V
V_{IH}	High-level input voltage		2			2			V
V_{IL}	Low-level input voltage				0.8			0.8	V
I_{OH}	High-level output current				−0.4			−0.4	mA
I_{OL}	Low-level output current				4			8	mA
f_{clock}	Clock frequency		0		30	0		34	MHz
t_w	Pulse duration	PRE or CLR low	15			15			ns
		CLK high	16.5			14.5			
		CLK low	16.5			14.5			
t_{su}	Setup time before CLK↑	Data	15			15			ns
		PRE or CLR inactive	10			10			
t_h	Hold time, data after CLK↑		0			0			ns
T_A	Operating free-air temperature		−55		125	0		70	°C

electrical characteristics over recommended operating free-air temperature range (unless otherwise noted)

PARAMETER		TEST CONDITIONS		SN54ALS74			SN74ALS74			UNIT
				MIN	TYP†	MAX	MIN	TYP†	MAX	
V_{IK}		V_{CC} = 4.5 V,	I_I = −18 mA			−1.5			−1.5	V
V_{OH}		V_{CC} = 4.5 V to 5.5 V,	I_{OH} = −0.4 mA	V_{CC}−2			V_{CC}−2			V
V_{OL}		V_{CC} = 4.5 V,	I_{OL} = 4 mA		0.25	0.4		0.25	0.4	V
		V_{CC} = 4.5 V,	I_{OL} = 8 mA					0.35	0.5	
I_I	CLK or D	V_{CC} = 5.5 V,	V_I = 7 V			0.1			0.1	mA
	PRE or CLR					0.2			0.2	
I_{IH}	CLK or D	V_{CC} = 5.5 V,	V_I = 2.7 V			20			20	µA
	PRE or CLR					40			40	
I_{IL}	CLK or D	V_{CC} = 5.5 V,	V_I = 0.4 V			−0.2			−0.2	mA
	PRE or CLR					−0.4			−0.4	
I_O‡		V_{CC} = 5.5 V,	V_O = 2.25 V	−10		−60	−10		−60	mA
I_{CC}		V_{CC} = 5.5 V,	See Note 1		2.4	4		2.4	4	mA

†All typical values are at V_{CC} = 5 V, T_A = 25 °C
‡The output conditions have been chosen to produce a current that closely approximates one half of the true short-circuit output current, I_{OS}.
NOTE 1: I_{CC} is measured with J, K, CLK, and PRE grounded, then with J, K, CLK, and CLR grounded.

switching characteristics (see Note 2)

PARAMETER	FROM (INPUT)	TO (OUTPUT)	V_{CC} = 4.5 V to 5.5 V, C_L = 50 pF, R_L = 500 Ω, T_A = MIN to MAX				UNIT
			SN54ALS74		SN74ALS74		
			MIN	MAX	MIN	MAX	
f_{max}			30		34		MHz
t_{PLH}	PRE or CLR	Q or $\bar{Q}$	3	15	3	13	ns
t_{PHL}			5	17	5	15	
t_{PLH}	CLK	Q or $\bar{Q}$	5	18	5	16	ns
t_{PHL}			7	20	7	18	

NOTE 2: For load circuit and voltage waveforms, see page 1-12.

TEXAS INSTRUMENTS
POST OFFICE BOX 225012 • DALLAS, TEXAS 75265

TYPES SN54AS74, SN74AS74
DUAL D-TYPE POSITIVE-EDGE-TRIGGERED FLIP-FLOPS WITH CLEAR AND PRESET

recommended operating conditions

			SN54AS74			SN74AS74			UNIT
			MIN	NOM	MAX	MIN	NOM	MAX	
V_{CC}	Supply voltage		4.5	5	5.5	4.5	5	5.5	V
V_{IH}	High-level input voltage		2			2			V
V_{IL}	Low-level input voltage				0.8			0.8	V
I_{OH}	High-level output current				−2			−2	mA
I_{OL}	Low-level output current				20			20	mA
f_{clock}	Clock frequency		0		90	0		105	MHz
t_w	Pulse duration	$\overline{PRE}$ or $\overline{CLR}$ low	4			4			ns
		CLK high	4			4			
		CLK low	5.5			5.5			
t_{su}	Setup time before CLK↑	Data	4.5			4.5			ns
		$\overline{PRE}$ or $\overline{CLR}$ inactive	2			2			
t_h	Hold time, data after CLK↑		0			0			ns
T_A	Operating free-air temperature		−55		125	0		70	°C

electrical characteristics over recommended operating free-air temperature range (unless otherwise noted)

PARAMETER	TEST CONDITIONS		SN54AS74			SN74AS74			UNIT
			MIN	TYP†	MAX	MIN	TYP†	MAX	
V_{IK}	V_{CC} = 4.5 V,	I_I = −18 mA			−1.2			−1.2	V
V_{OH}	V_{CC} = 4.5 V to 5.5 V,	I_{OH} = −2 mA	$V_{CC}-2$			$V_{CC}-2$			V
V_{OL}	V_{CC} = 4.5 V,	I_{OL} = 20 mA		0.25	0.5		0.25	0.5	V
I_I	V_{CC} = 5.5 V,	V_I = 7 V			0.1			0.1	mA
I_{IH}	V_{CC} = 5.5 V,	V_I = 2.7 V			20			20	μA
I_{IL} CLK or D	V_{CC} = 5.5 V,	V_I = 0.4 V			−0.5			−0.5	mA
$\overline{PRE}$ or $\overline{CLR}$					−1.5			−1.5	
I_O‡	V_{CC} = 5.5 V,	V_O = 2.25 V	−30		−112	−30		−112	mA
I_{CC}	V_{CC} = 5.5 V,	See Note 1		10.5	16		10.5	16	mA

†All typical values are at V_{CC} = 5 V, T_A = 25°C.
‡The output conditions have been chosen to produce a current that closely approximates one half of the true short-circuit output current, I_{OS}.
NOTE 1: I_{CC} is measured with D, CLK, and $\overline{PRE}$ grounded, then with D, CLK, and $\overline{CLR}$ grounded.

switching characteristics (see Note 2)

PARAMETER	FROM (INPUT)	TO (OUTPUT)	V_{CC} = 4.5 V to 5.5 V, C_L = 50 pF, R_L = 500 Ω, T_A = MIN to MAX				UNIT
			SN54AS74		SN74AS74		
			MIN	MAX	MIN	MAX	
f_{max}			90		105		MHz
t_{PLH}	$\overline{PRE}$ or $\overline{CLR}$	Q or $\overline{Q}$	3	8.5	3	7.5	ns
t_{PHL}			3.5	11.5	3.5	10.5	
t_{PLH}	CLK	Q or $\overline{Q}$	3.5	9	3.5	8	ns
t_{PHL}			4.5	10.5	4.5	9	

NOTE 2: For load circuit and voltage waveforms, see page 1-12.

2
ALS AND AS CIRCUITS

TYPES SN54ALS86, SN74ALS86
QUADRUPLE 2-INPUT EXCLUSIVE-OR GATES

D2661, APRIL 1982 – REVISED DECEMBER 1983

- Package Options Include Both Plastic and Ceramic Chip Carriers in Addition to Plastic and Ceramic DIPs
- Dependable Texas Instruments Quality and Reliability

SN54ALS86 . . . J PACKAGE
SN74ALS86 . . . N PACKAGE
(TOP VIEW)

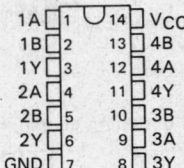

description

These devices contain four independent 2-input Exclusive-OR gates. They perform the Boolean functions $Y = A \oplus B = \overline{A}B + A\overline{B}$ in positive logic.

A common application is as a true/complement element. If one of the inputs is low, the other input will be reproduced in true form at the output. If one of the inputs is high, the signal on the other input will be reproduced inverted at the output.

The SN54ALS86 is characterized for operation over the full military temperature range of $-55\,°C$ to $125\,°C$. The SN74ALS86 is characterized for operation from $0\,°C$ to $70\,°C$.

SN54ALS86 . . . FH PACKAGE
SN74ALS86 . . . FN PACKAGE
(TOP VIEW)

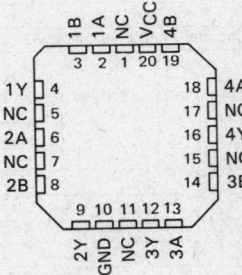

NC – No internal connection

logic symbol

FUNCTION TABLE
(each gate)

INPUTS		OUTPUT
A	B	Y
L	L	L
L	H	H
H	L	H
H	H	L

Pin numbers shown are for J and N packages.

exclusive-OR logic

An exclusive-OR gate has many applications, some of which can be represented better by alternative logic symbols.

EXCLUSIVE-OR

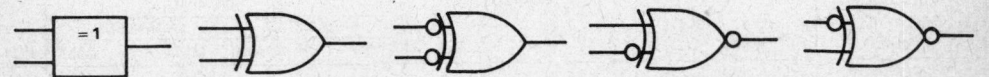

These are five equivalent Exclusive-OR symbols valid for an 'ALS86 gate in positive logic; negation may be shown at any two ports.

LOGIC IDENTITY ELEMENT	EVEN-PARITY	ODD-PARITY ELEMENT
	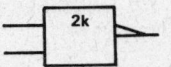	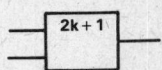
The output is active (low) if all inputs stand at the same logic level (i.e., A = B).	The output is active (low) if an even number of inputs (i.e., 0 or 2) are active.	The output is active (high) if an odd number of inputs (i.e., only 1 of the 2) are active.

PRODUCT PREVIEW

This document contains information on a product under development. Texas Instruments reserves the right to change or discontinue this product without notice.

Copyright © 1982 by Texas Instruments Incorporated

TEXAS INSTRUMENTS
POST OFFICE BOX 225012 • DALLAS, TEXAS 75265

TYPES SN54ALS86, SN74ALS86
QUADRUPLE 2-INPUT EXCLUSIVE-OR GATES

absolute maximum ratings over operating free-air temperature range (unless otherwise noted)

Supply voltage, V_{CC} .... 7 V
Input voltage .... 7 V
Operating free-air temperature range: SN54ALS86 .... −55 °C to 125 °C
                                       SN74ALS86 .... 0 °C to 70 °C
Storage temperature range .... −65 °C to 150 °C

recommended operating conditions

		SN54ALS86			SN74ALS86			UNIT
		MIN	NOM	MAX	MIN	NOM	MAX	
V_{CC}	Supply voltage	4.5	5	5.5	4.5	5	5.5	V
V_{IH}	High-level input voltage	2			2			V
V_{IL}	Low-level input voltage			0.8			0.8	V
I_{OH}	High-level output current			−0.4			−0.4	mA
I_{OL}	Low-level output current			4			8	mA
T_A	Operating free-air temperature	−55		125	0		70	°C

electrical characteristics over recommended operating free-air temperature range (unless otherwise noted)

PARAMETER	TEST CONDITIONS		SN54ALS86			SN74ALS86			UNIT
			MIN	TYP†	MAX	MIN	TYP†	MAX	
V_{IK}	V_{CC} = 4.5 V,	I_I = −18 mA			−1.5			−1.5	V
V_{OH}	V_{CC} = 4.5 V to 5.5 V,	I_{OH} = −0.4 mA	V_{CC}−2			V_{CC}−2			V
V_{OL}	V_{CC} = 4.5 V,	I_{OL} = 4 mA		0.25	0.4		0.25	0.4	V
	V_{CC} = 4.5 V,	I_{OL} = 8 mA					0.35	0.5	
I_I	V_{CC} = 5.5 V,	V_I = 7 V			0.1			0.1	mA
I_{IH}	V_{CC} = 5.5 V,	V_I = 2.7 V			20			20	μA
I_{IL}	V_{CC} = 5.5 V,	V_I = 0.4 V			0−0.1			−0.1	mA
I_O‡	V_{CC} = 5.5 V,	V_O = 2.25 V	−30		−112	−30		−112	mA
I_{CC}	V_{CC} = 5.5 V,	All inputs at 0 V		3			3		mA

† All typical values are at V_{CC} = 5 V, T_A = 25 °C.
‡ The output conditions have been chosen to produce a current that closely approximates one half of the true short-circuit output current, I_{OS}.

switching characteristics (see Note 1)

PARAMETER	FROM (INPUT)	TO (OUTPUT)	V_{CC} = 4.5 V to 5.5 V, C_L = 50 pF, R_L = 500 Ω, T_A = MIN to MAX						UNIT
			SN54ALS86			SN74ALS86			
			MIN	TYP†	MAX	MIN	TYP†	MAX	
t_{PLH}	A or B (other input low)	Y		7			7		ns
t_{PHL}				6			6		
t_{PLH}	A or B (other input high)	Y		8			8		ns
t_{PHL}				7			7		

† All typical values are at V_{CC} = 5 V, T_A = 25 °C.
NOTE 1: For load circuit and voltage waveforms, see page 1-12.

Additional information on these products can be obtained from the factory as it becomes available.

TEXAS INSTRUMENTS
POST OFFICE BOX 225012 • DALLAS, TEXAS 75265

TYPES SN54AS95, SN74AS95
4-BIT PARALLEL-ACCESS SHIFT REGISTER

D2661, DECEMBER 1983

- Serial-to-Parallel Conversions
- Parallel Synchronous Loading
- Right or Left Shifts
- Package Options Include Both Plastic and Ceramic Chip Carriers in Addition to Plastic and Ceramic DIPs
- Dependable Texas Instruments Quality and Reliability

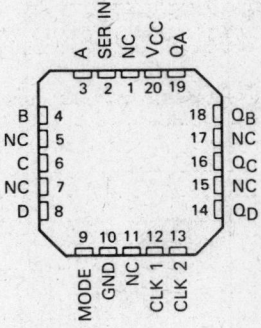

NC — No internal connection

description

These four-bit registers feature parallel and serial inputs, parallel outputs, mode control, and two clock inputs. The registers have three modes of operation:

Parallel (broadside) load
Shift right (the direction Q_A toward Q_D)
Shift left (the direction Q_D toward Q_A)

Parallel loading is accomplished by applying the four bits and taking the mode control input high. The data is loaded into the associated flip-flops and appears at the outputs after the high-to-low transition of the Clock-2 input. During loading, the entry of serial data is inhibited.

Shift right is accomplished on the high-to-low transition of Clock 1 when the mode control is low; shift left is accomplished on the high-to-low transition of Clock 2 when the mode control is high by connecting the output of each flip-flop to the parallel input of the previous flip-flop (Q_D to input C, etc.); and serial data is entered at input D. The clock input may be applied commonly to Clock 1 and Clock 2 if both modes can be clocked from the same source. Changes at the mode control input should normally be made while both clock inputs are low. However, conditions described in the last three lines of the function table will also ensure that the register contents are protected.

The SN54AS95 is characterized for operation over the full military temperature range of −55 °C to 125 °C. The SN74AS95 is characterized for operation from 0 °C to 70 °C.

PRODUCT PREVIEW
This document contains information on a product under development. Texas Instruments reserves the right to change or discontinue this product without notice.

Copyright © 1983 by Texas Instruments Incorporated

TYPES SN54AS95, SN74AS95
4-BIT PARALLEL-ACCESS SHIFT REGISTER

FUNCTION TABLE

INPUTS								OUTPUTS			
MODE CONTROL	CLOCKS		SERIAL	PARALLEL				Q_A	Q_B	Q_C	Q_D
	2 (L)	1 (R)		A	B	C	D				
H	H	X	X	X	X	X	X	Q_{A0}	Q_{B0}	Q_{C0}	Q_{D0}
H	↓	X	X	a	b	c	d	a	b	c	d
H	↓	X	X	Q_B†	Q_C†	Q_D†	d	Q_{Bn}	Q_{Cn}	Q_{Dn}	d
L	L	H	X	X	X	X	X	Q_{A0}	Q_{B0}	Q_{C0}	Q_{D0}
L	X	↓	H	X	X	X	X	H	Q_{An}	Q_{Bn}	Q_{Cn}
L	X	↓	L	X	X	X	X	L	Q_{An}	Q_{Bn}	Q_{Cn}
↑	L	L	X	X	X	X	X	Q_{A0}	Q_{B0}	Q_{C0}	Q_{D0}
↓	L	L	X	X	X	X	X	Q_{A0}	Q_{B0}	Q_{C0}	Q_{D0}
↓	L	H	X	X	X	X	X	Q_{A0}	Q_{B0}	Q_{C0}	Q_{D0}
↑	H	L	X	X	X	X	X	Q_{A0}	Q_{B0}	Q_{C0}	Q_{D0}
↑	H	H	X	X	X	X	X	Q_{A0}	Q_{B0}	Q_{C0}	Q_{D0}

†Shifting left requires external connection of Q_B to A, Q_C to B, and Q_D to C. Serial data is entered at input D.
H = high level (steady state), L = low level (steady state), X = irrelevant (any input, including transitions).
↓ = transition from high to low level, ↑ = transition from low to high level.
a, b, c, d = the level of steady-state input at inputs A, B, C, or D, respectively.
Q_{A0}, Q_{B0}, Q_{C0}, Q_{D0} = the level of Q_A, Q_B, Q_C, or Q_D, respectively, before the indicated steady-state input conditions were established.
Q_{An}, Q_{Bn}, Q_{Cn}, Q_{Dn} = the level of Q_A, Q_B, Q_C, or Q_D, respectively, before the most-recent ↓ transition of the clock.

logic symbol

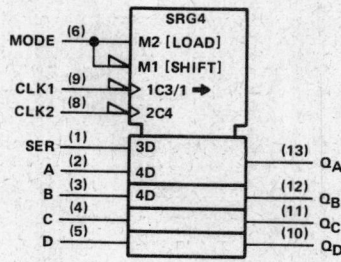

logic diagram (positive logic)

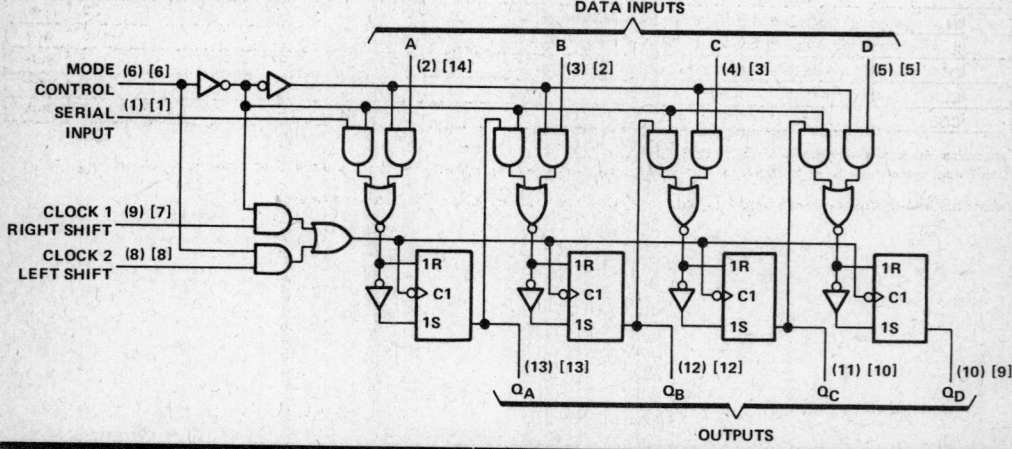

TYPES SN54AS95, SN74AS95
4-BIT PARALLEL-ACCESS SHIFT REGISTER

absolute maximum ratings over operating free-air temperature range (unless otherwise noted)

Supply voltage, V_{CC} .... 7 V
Input voltage .... 7 V
Operating free-air temperature range: SN54AS95 .... −55°C to 125°C
                                     SN74AS95 .... 0°C to 70°C
Storage temperature range .... −65°C to 150°C

recommended operating conditions

		SN54AS95			SN74AS95			UNIT
		MIN	NOM	MAX	MIN	NOM	MAX	
V_{CC}	Supply voltage	4.5	5	5.5	4.5	5	5.5	V
V_{IH}	High-level input voltage	2			2			V
V_{IL}	Low-level input voltage			0.8			0.8	V
I_{OH}	High-level output current			−2			−2	mA
I_{OL}	Low-level output current			20			20	mA
f_{clock}	Clock frequency	0			0			MHz
t_w	Pulse duration, CLK high or low							ns
t_{su}	Setup time, data before CLK ↓							ns
t_h	Hold time after CLK ↓ Data							ns
	Mode (see Figure 1)							
	Clock enable time CLK 1							ns
	(see Figure 1) CLK 2	0			0			
	Clock inhibit time CLK 1							ns
	(see Figure 1) CLK 2	0			0			
T_A	Operating free-air temperature	−55		125	0		70	°C

electrical characteristics over recommended operating free-air temperature range (unless otherwise noted)

PARAMETER	TEST CONDITIONS		SN54AS95			SN74AS95			UNIT
			MIN	TYP†	MAX	MIN	TYP†	MAX	
V_{IK}	$V_{CC} = 4.5$ V,	$I_I = −18$ mA			−1.2			−1.2	V
V_{OH}	$V_{CC} = 4.5$ V to 5.5 V,	$I_{OH} = −2$ mA	$V_{CC}-2$			$V_{CC}-2$			V
V_{OL}	$V_{CC} = 4.5$ V,	$I_{OL} = 20$ mA		0.35	0.5		0.35	0.5	V
I_I	$V_{CC} = 5.5$ V,	$V_I = 7$ V			0.1			0.1	mA
I_{IH}	$V_{CC} = 5.5$ V,	$V_I = 2.7$ V			20			20	μA
I_{IL}	$V_{CC} = 5.5$ V,	$V_{IL} = 0.4$ V			−0.5			−0.5	mA
I_O‡	$V_{CC} = 5.5$ V,	$V_O = 2.25$ V	−30		−112	−30		−112	mA
I_{CCH}	$V_{CC} = 5.5$ V								mA
I_{CCL}	$V_{CC} = 5.5$ V			26.1			26.1		mA

†All typical values are at $V_{CC} = 5$ V, $T_A = 25$°C.
‡The output conditions have been chosen to produce a current that closely approximates one half of the true short-circuit output current, I_{OS}.

Additional information on these products can be obtained from the factory as it becomes available.

TEXAS INSTRUMENTS
POST OFFICE BOX 225012 • DALLAS, TEXAS 75265

TYPES SN54AS95, SN74AS95
4-BIT PARALLEL-ACCESS SHIFT REGISTER

switching characteristics (see Note 1)

PARAMETER	FROM (INPUT)	TO (OUTPUT)	V_{CC} = 4.5 V to 5.5 V, C_L = 50 pF, R_L = 500 Ω, T_A = MIN to MAX						UNIT
			SN54AS95			SN74AS95			
			MIN	TYP[†]	MAX	MIN	TYP[†]	MAX	
f_{max}									MHz
t_{PLH}	CLK	Q		6.4			6.4		ns
t_{PHL}				7.1			7.1		

[†]All typical values are at V_{CC} = 5 V, T_A = 25°C.
NOTE 1: For load circuit and voltage waveforms, see page 1-12.

Additional information on these produces can be obtained from the factory as it becomes available.

PARAMETER MEASUREMENT INFORMATION

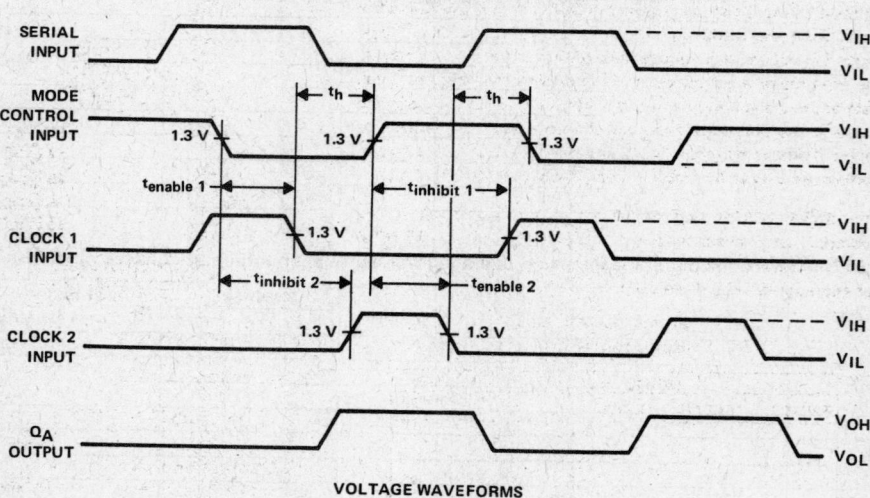

VOLTAGE WAVEFORMS

NOTES: A. Input A is at a low level.
       B. V_{IH} = 3.5 V, V_{IL} = 0.3 V.

FIGURE 1—CLOCK ENABLE, INHIBIT, AND HOLD TIMES

TYPES SN54ALS109, SN54AS109, SN74ALS109, SN74AS109
DUAL J-K̄ POSITIVE-EDGE-TRIGGERED FLIP-FLOPS WITH CLEAR AND PRESET

D2661, APRIL 1982 — REVISED DECEMBER 1983

- **Package Options Include Both Plastic and Ceramic Chip Carriers in Addition to Plastic and Ceramic DIPs**
- **Dependable Texas Instruments Quality and Reliability**

SN54ALS109, SN54AS109 . . . J PACKAGE
SN74ALS109, SN74AS109 . . . N PACKAGE
(TOP VIEW)

TYPE	TYPICAL MAXIMUM CLOCK FREQUENCY	TYPICAL POWER DISSIPATION PER FLIP-FLOP
'ALS109	50 MHz	6 mW
'AS109	129 MHz	29 mW

```
1CLR  [1   16] VCC
 1J   [2   15] 2CLR
 1K̄   [3   14] 2J
1CLK  [4   13] 2K̄
1PRE  [5   12] 2CLK
 1Q   [6   11] 2PRE
 1Q̄   [7   10] 2Q
 GND  [8    9] 2Q̄
```

description

These devices contain two independent J-K̄ positive-edge-triggered flip-flops. A low level at the Preset or Clear inputs sets or resets the outputs regardless of the levels of the other inputs. When Preset and Clear are inactive (high), data at the J and K̄ inputs meeting the setup time requirements are transferred to the outputs on the positive-going edge of the clock pulse. Clock triggering occurs at a voltage level and is not directly related to the rise time of the clock pulse. Following the hold time interval, data at the J and K̄ inputs may be changed without affecting the levels at the outputs. These versatile flip-flops can perform as toggle flip-flops by grounding K̄ and tying J high. They also can perform as D-type flip-flops if J and K̄ are tied together.

The SN54ALS109 and SN54AS109 are characterized for operation over the full military temperature range of −55 °C to 125 °C. The SN74ALS109 and SN74AS109 are characterized for operation from 0 °C to 70 °C.

SN54ALS109, SN54AS109 . . . FH PACKAGE
SN74ALS109, SN74AS109 . . . FN PACKAGE
(TOP VIEW)

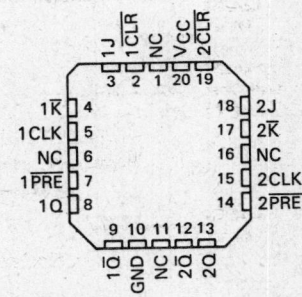

FUNCTION TABLE
(EACH FLIP-FLOP)

INPUTS					OUTPUTS	
PRESET	CLEAR	CLOCK	J	K̄	Q	Q̄
L	H	X	X	X	H	L
H	L	X	X	X	L	H
L	L	X	X	X	H*	H*
H	H	↑	L	L	L	H
H	H	↑	H	L	TOGGLE	
H	H	↑	L	H	Q_0	$\bar{Q}_0$
H	H	↑	H	H	H	L
H	H	L	X	X	Q_0	$\bar{Q}_0$

*The output levels in this configuration are not guaranteed to meet the minimum levels for V_{OH} if the lows at Preset and Clear are near V_{IL} maximum. Furthermore, this configuration is nonstable: that is, it will not persist when Preset or Clear; returns to their inactive (high) level.

logic symbol

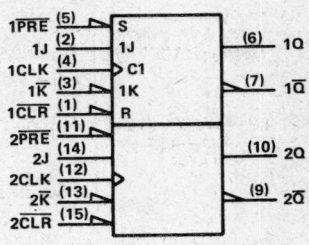

NC — No Internal connection

Pin numbers shown are for J and N packages.

absolute maximum ratings over operating free-air temperature range (unless otherwise noted)

Supply voltage, V_{CC} . . . . 7 V
Input voltage . . . . 7 V
Operating free-air temperature range: SN54ALS109, SN54AS109 . . . . −55 °C to 125 °C
SN74ALS109, SN74AS109 . . . . 0 °C to 70 °C
Storage temperature range . . . . −65 °C to 150 °C

Copyright © 1982 by Texas Instruments Incorporated

POST OFFICE BOX 225012 • DALLAS, TEXAS 75265

TYPES SN54ALS109, SN74ALS109
DUAL J-K POSITIVE-EDGE-TRIGGERED FLIP-FLOPS WITH CLEAR AND PRESET

recommended operating conditions

			SN54ALS109			SN74ALS109			UNIT
			MIN	NOM	MAX	MIN	NOM	MAX	
V_{CC}	Supply voltage		4.5	5	5.5	4.5	5	5.5	V
V_{IH}	High-level input voltage		2			2			V
V_{IL}	Low-level input voltage				0.8			0.8	V
I_{OH}	High-level output current				−0.4			−0.4	mA
I_{OL}	Low-level output current				4			8	mA
f_{clock}	Clock frequency		0		30	0		34	MHz
t_w	Pulse duration	$\overline{PRE}$ or $\overline{CLR}$ low	15			15			ns
		CLK high	16.5			14.5			
		CLK low	16.5			14.5			
t_{su}	Setup time before CLK↑	Data	15			15			ns
		$\overline{PRE}$ or $\overline{CLR}$ inactive	10			10			
t_h	Hold time, data after CLK↑		0			0			ns
T_A	Operating free-air temperature		−55		125	0		70	°C

electrical characteristics over recommended operating free-air temperature range (unless otherwise noted)

PARAMETER		TEST CONDITIONS		SN54ALS109			SN74ALS109			UNIT
				MIN	TYP†	MAX	MIN	TYP†	MAX	
V_{IK}		V_{CC} = 4.5 V,	I_I = −18 mA			−1.5			−1.5	V
V_{OH}		V_{CC} = 4.5 V to 5.5 V,	I_{OH} = −0.4 mA	V_{CC}−2			V_{CC}−2			V
V_{OL}		V_{CC} = 4.5 V,	I_{OL} = 4 mA		0.25	0.4		0.25	0.4	V
		V_{CC} = 4.5 V	I_{OL} = 8 mA					0.35	0.5	
I_I	CLK, J, or $\overline{K}$	V_{CC} = 5.5 V,	V_I = 7 V			0.1			0.1	mA
	$\overline{PRE}$ or $\overline{CLR}$					0.2			0.2	
I_{IH}	CLK, J, or $\overline{K}$	V_{CC} = 5.5 V,	V_I = 2.7 V			20			20	µA
	$\overline{PRE}$ or $\overline{CLR}$					40			40	
I_{IL}	CLK, J or $\overline{K}$	V_{CC} = 5.5 V,	V_I = 0.4 V			−0.2			−0.2	mA
	$\overline{PRE}$ or $\overline{CLR}$					−0.4			−0.4	
I_O‡		V_{CC} = 5.5 V,	V_O = 2.25 V	−10		−60	−10		−60	mA
I_{CC}		V_{CC} = 5.5 V,	See Note 1		2.4	4		2.4	4	mA

†All typical values are at V_{CC} = 5 V, T_A = 25°C.
‡The output conditions have been chosen to produce a current that closely approximates one half of the true short-circuit output current, I_{OS}.
NOTE 1: I_{CC} is measured with J, $\overline{K}$, CLK, and $\overline{PRE}$ grounded, then with K, $\overline{K}$, CLK, and $\overline{CLR}$ grounded.

switching characteristics (see Note 2)

PARAMETER	FROM (INPUT)	TO (OUTPUT)	V_{CC} = 4.5 V to 5.5 V, C_L = 50 pF, R_L = 500 Ω, T_A = MIN to MAX				UNIT
			SN54ALS109		SN74ALS109		
			MIN	MAX	MIN	MAX	
f_{max}			30		34		MHz
t_{PLH}	$\overline{PRE}$ or $\overline{CLR}$	Q or $\overline{Q}$	3	15	3	13	ns
t_{PHL}			5	17	5	15	
t_{PLH}	CLK	Q or $\overline{Q}$	5	18	5	16	ns
t_{PHL}			7	20	7	18	

NOTE 2: For load circuit and voltage waveforms, see page 1-12.

Texas Instruments
POST OFFICE BOX 225012 • DALLAS, TEXAS 75265

TYPES SN54AS109, SN74AS109
DUAL J-K POSITIVE-EDGE-TRIGGERED FLIP-FLOPS WITH CLEAR AND PRESET

recommended operating conditions

			SN54AS109			SN74AS109			UNIT
			MIN	NOM	MAX	MIN	NOM	MAX	
V_{CC}	Supply voltage		4.5	5	5.5	4.5	5	5.5	V
V_{IH}	High-level input voltage		2			2			V
V_{IL}	Low-level input voltage				0.8			0.8	V
I_{OH}	High-level output current				−2			−2	mA
I_{OL}	Low-level output current				20			20	mA
f_{clock}	Clock frequency		0		90	0		105	MHz
t_w	Pulse duration	$\overline{PRE}$ or $\overline{CLR}$ low	4			4			ns
		CLK high	4			4			
		CLK low	5.5			5.5			
t_{su}	Setup time before CLK↑	Data	5.5			5.5			ns
		$\overline{PRE}$ or $\overline{CLR}$ inactive	2			2			
t_h	Hold time, data after CLK↑		0			0			ns
T_A	Operating free-air temperature		−55		125	0		70	°C

electrical characteristics over recommended operating free-air temperature range (unless otherwise noted)

PARAMETER		TEST CONDITIONS		SN54AS109			SN74AS109			UNIT
				MIN	TYP†	MAX	MIN	TYP†	MAX	
V_{IK}		$V_{CC} = 4.5$ V,	$I_I = -18$ mA			−1.2			−1.2	V
V_{OH}		$V_{CC} = 4.5$ V to 5.5 V,	$I_{OH} = -2$ mA	$V_{CC}-2$			$V_{CC}-2$			V
V_{OL}		$V_{CC} = 4.5$ V,	$I_{OL} = 20$ mA		0.25	0.5		0.25	0.5	V
I_I		$V_{CC} = 5.5$ V,	$V_I = 7$ V			0.1			0.1	mA
I_{IH}		$V_{CC} = 5.5$ V,	$V_I = 2.7$ V			20			20	µA
I_{IL}	CLK or D	$V_{CC} = 5.5$ V,	$V_I = 0.4$ V			−0.5			−0.5	mA
	$\overline{PRE}$ or $\overline{CLR}$					−1.5			−1.5	
I_O‡		$V_{CC} = 5.5$ V,	$V_O = 2.25$ V	−30		−112	−30		−112	mA
I_{CC}		$V_{CC} = 5.5$ V,	See Note 1		11.5	17		11.5	17	mA

† All typical values are at $V_{CC} = 5$ V, $T_A = 25$°C.
‡ The output conditions have been chosen to produce a current that closely approximates one half of the true short-circuit output current, I_{OS}.
NOTE 1: I_{CC} is measured with D, CLK, and $\overline{PRE}$ grounded, then with D, CLK, and $\overline{CLR}$ grounded.

switching characteristics (see Note 2)

PARAMETER	FROM (INPUT)	TO (OUTPUT)	$V_{CC} = 4.5$ V to 5.5 V, $C_L = 50$ pF, $R_L = 500$ Ω, $T_A =$ MIN to MAX				UNIT
			SN54AS109		SN74AS109		
			MIN	MAX	MIN	MAX	
f_{max}			90		105		MHz
t_{PLH}	$\overline{PRE}$ or $\overline{CLR}$	Q or $\overline{Q}$	3	9	3	8	ns
t_{PHL}			3.5	11.5	3.5	10.5	
t_{PLH}	CLK	Q or $\overline{Q}$	3.5	10	3.5	9	ns
t_{PHL}			4.5	10.5	4.5	9	

NOTE 2: For load circuit and voltage waveforms, see page 1-12.

ALS AND AS CIRCUITS

TEXAS INSTRUMENTS
POST OFFICE BOX 225012 • DALLAS, TEXAS 75265

2
ALS AND AS CIRCUITS

TYPES SN54ALS112A, SN54AS112, SN74ALS112A, SN74AS112
DUAL J-K NEGATIVE-EDGE-TRIGGERED FLIP-FLOPS WITH CLEAR AND PRESET

D2661, APRIL 1982—REVISED DECEMBER 1983

- Fully Buffered to Offer Maximum Isolation from External Disturbance
- Package Options Include Both Plastic and Ceramic Carriers in Addition to Plastic and Ceramic DIPs.
- Dependable Texas Instruments Quality and Reliability

TYPE	TYPICAL MAXIMUM CLOCK FREQUENCY	TYPICAL POWER DISSIPATION PER FLIP-FLOP
'ALS112A	50 MHz	6 mW
'AS112	175 MHz	95 mW

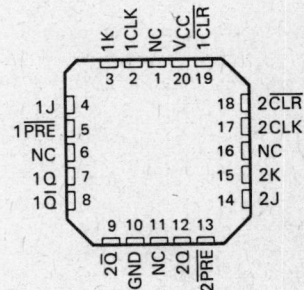

description

These devices contain two independent J-K negative-edge-triggered flip-flops. A low level at the Preset or Clear inputs sets or resets the outputs regardless of the levels of the other inputs. When Preset and Clear are inactive (high), data at the J and K inputs meeting the setup time requirements are transferred to the outputs on the negative-going edge of the clock pulse. Clock triggering occurs at a voltage level and is not directly related to the rise time of the clock pulse. Following the hold time interval, data at the J and K inputs may be changed without affecting the levels at the outputs. These versatile flip-flops can perform as toggle flip-flops by tying J and K high.

The SN54ALS112A and SN54AS112 are characterized for operation over the full military temperature range of $-55\,°C$ to $125\,°C$. The SN74ALS112A and SN74AS112 are characterized for operation from $0\,°C$ to $70\,°C$.

FUNCTION TABLE

INPUTS					OUTPUTS	
PRE	CLR	CLK	J	K	Q	Q̄
L	H	X	X	X	H	L
H	L	X	X	X	L	H
L	L	X	X	X	H*	H*
H	H	↓	L	L	Q₀	Q̄₀
H	H	↓	H	L	H	L
H	H	↓	L	H	L	H
H	H	↓	H	H	TOGGLE	
H	H	H	X	X	Q₀	Q̄₀

*The output levels in this configuration are not guaranteed to meet the minimum levels for V_{OH} if the lows at Preset and Clear are near V_{IL} maximum. Furthermore, this configuration is nonstable; that is, it will not persist when either Preset or Clear returns to its inactive (high) level.

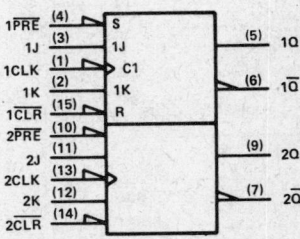

logic symbol

Pin numbers shown are for J and N packages.

absolute maximum ratings over operating free-air temperature range (unless otherwise noted)

Supply voltage, V_{CC} .... 7 V
Input voltage .... 7 V
Operating free-air temperature range: SN54ALS112A, SN54AS112 .... $-55\,°C$ to $125\,°C$
SN74ALS112A, SN74AS112 .... $0\,°C$ to $70\,°C$
Storage temperature range .... $-65\,°C$ to $150\,°C$

Copyright © 1982 by Texas Instruments Incorporated

Texas Instruments
POST OFFICE BOX 225012 • DALLAS, TEXAS 75265

TYPES SN54ALS112A, SN74ALS112A
DUAL J-K NEGATIVE-EDGE-TRIGGERED FLIP-FLOPS
WITH CLEAR AND PRESET

recommended operating conditions

			SN54ALS112A			SN74ALS112A			UNIT
			MIN	NOM	MAX	MIN	NOM	MAX	
V_{CC}	Supply voltage		4.5	5	5.5	4.5	5	5.5	V
V_{IH}	High-level input voltage		2			2			V
V_{IL}	Low-level input voltage				0.8			0.8	V
I_{OH}	High-level output current				−0.4			−0.4	mA
I_{OL}	Low-level output current				4			8	mA
f_{clock}	Clock frequency		0		25	0		30	MHz
t_w	Pulse duration	$\overline{PRE}$ or $\overline{CLR}$ low	15			10			ns
		CLK high	20			16.5			
		CLK low	20			16.5			
t_{su}	Setup time before CLK↓	Data	25			22			ns
		$\overline{PRE}$ or $\overline{CLR}$ inactive	22			20			
t_h	Hold time, data after CLK↓		0			0			ns
T_A	Operating free-air temperature		−55		125	0		70	°C

electrical characteristics over recommended operating free-air temperature range (unless otherwise noted)

PARAMETER	TEST CONDITIONS		SN54ALS112A			SN74ALS112A			UNIT
			MIN	TYP†	MAX	MIN	TYP†	MAX	
V_{IK}	V_{CC} = 4.5 V,	I_I = −18 mA			−1.5			−1.5	V
V_{OH}	V_{CC} = 4.5 V to 5.5 V,	I_{OH} = −0.4 mA	V_{CC}−2			V_{CC}−2			V
V_{OL}	V_{CC} = 4.5 V,	I_{OL} = 4 mA		0.25	0.4		0.25	0.4	V
	V_{CC} = 4.5 V	I_{OL} = 8 mA					0.35	0.5	
I_I	J, K, or CLK	V_{CC} = 5.5 V,	V_I = 7 V			0.1			0.1	mA
	$\overline{PRE}$ or $\overline{CLR}$					0.2			0.2	
I_{IH}	J, K, or CLK	V_{CC} = 5.5 V,	V_I = 2.7 V			20			20	µA
	$\overline{PRE}$ or $\overline{CLR}$					40			40	
I_{IL}	J, K, or CLK	V_{CC} = 5.5 V,	V_I = 0.4 V			−0.2			−0.2	mA
	$\overline{PRE}$ or $\overline{CLR}$					−0.4			−0.4	
I_O‡		V_{CC} = 5.5 V,	V_O = 2.25 V	−30		−112	−30		−112	mA
I_{CC}		V_{CC} = 5.5 V,	See Note 1		2.5	4.5		2.5	4.5	mA

†All typical values are at V_{CC} = 5 V, T_A = 25°C.
‡The output conditions have been chosen to produce a current that closely approximates one half of the true short-circuit output current, I_{OS}.
NOTE 1: I_{CC} is measured with J, K, CLK, and $\overline{PRE}$ grounded, then with J, K, CLK, and $\overline{CLR}$ grounded.

switching characteristics (see Note 2)

PARAMETER	FROM (INPUT)	TO (OUTPUT)	V_{CC} = 4.5 V to 5.5 V, C_L = 50 pF, R_L = 500 Ω, T_A = MIN to MAX				UNIT
			SN54ALS112A		SN74ALS112A		
			MIN	MAX	MIN	MAX	
f_{max}			25		30		MHz
t_{PLH}	$\overline{PRE}$ or $\overline{CLR}$	Q or $\overline{Q}$	3	20	3	15	ns
t_{PHL}			4	22	4	18	
t_{PLH}	CLK	Q or $\overline{Q}$	3	18	3	15	ns
t_{PHL}			5	23	5	19	

NOTE 2: For load circuit and voltage waveforms, see page 1-12.

TYPES SN54AS112, SN74AS112
DUAL J-K NEGATIVE-EDGE-TRIGGERED FLIP-FLOPS
WITH CLEAR AND PRESET

recommended operating conditions

			SN54AS112			SN74AS112			UNIT
			MIN	NOM	MAX	MIN	NOM	MAX	
V_{CC}	Supply voltage		4.5	5	5.5	4.5	5	5.5	V
V_{IH}	High-level input voltage		2			2			V
V_{IL}	Low-level input voltage				0.8			0.8	V
I_{OH}	High-level output current				−2			−2	mA
I_{OL}	Low-level output current				20			20	mA
f_{clock}	Clock frequency		0			0			MHz
t_w	Pulse duration	PRE or $\overline{CLR}$ low							ns
		CLK high							
		CLK low							
t_{su}	Setup time before CLK↓	Data							ns
		PRE or $\overline{CLR}$ inactive							
t_h	Hold time, data after CLK↓								ns
T_A	Operating free-air temperature		−55		125	0		70	°C

electrical characteristics over recommended operating free-air temperature range (unless otherwise noted)

PARAMETER		TEST CONDITIONS		SN54AS112			SN74AS112			UNIT
				MIN	TYP†	MAX	MIN	TYP†	MAX	
V_{IK}		V_{CC} = 4.5 V,	I_I = −18 mA			−1.2			−1.2	V
V_{OH}		V_{CC} = 4.5 V to 5.5 V,	I_{OH} = −2 mA	V_{CC}−2			V_{CC}−2			V
V_{OL}		V_{CC} = 4.5 V,	I_{OL} = 20 mA		0.35	0.5		0.35	0.5	V
I_I	J or K	V_{CC} = 5.5 V,	V_I = 7 V			0.1			0.1	mA
	PRE or $\overline{CLR}$					0.5			0.5	
	CLK					0.5			0.5	
I_{IH}	J or K	V_{CC} = 5.5 V,	V_I = 2.7 V			0.02			0.02	mA
	PRE or $\overline{CLR}$					0.1			0.1	
	CLK					0.1			0.1	
I_{IL}	J or K	V_{CC} = 5.5 V,	V_I = 0.4 V			−1			−1	mA
	PRE or $\overline{CLR}$					−5.5			−5.5	
	CLK					−5			−5	
I_O‡		V_{CC} = 5.5 V,	V_O = 2.25 V	−30		−112	−30		−112	mA
I_{CC}		V_{CC} = 5.5 V,	See Note 1		38			38		mA

NOTE 1: I_{CC} is measured with D, CLK, and PRE grounded, then with J, K, CLK, and PRE grounded, with K, K, CLK, and $\overline{CLR}$ grounded.

switching characteristics (see Note 2)

PARAMETER	FROM (INPUT)	TO (OUTPUT)	V_{CC} = 4.5 V to 5.5 V, C_L = 50 pF, R_L = 500 Ω, T_A = MIN to MAX						UNIT
			SN54AS112			SN74AS112			
			MIN	TYP†	MAX	MIN	TYP†	MAX	
f_{max}				175			175		MHz
t_{PLH}	PRE or $\overline{CLR}$	Q or $\overline{Q}$		3			3		ns
t_{PHL}				4			4		
t_{PLH}	CLK	Q or $\overline{Q}$		3			3		ns
t_{PHL}				4			4		

†All typical values are at V_{CC} = 5 V, T_A = 25°C.
‡The output conditions have been chosen to produce a current that closely approximates one half of the true short-circuit output current, I_{OS}.
NOTE 2: For load circuit and voltage waveforms, see page 1-12.

Additional information on these products can be obtained from the factory as it becomes available.

PRODUCT PREVIEW
This page contains information on a product under development. Texas Instruments reserves the right to change or discontinue this product without notice.

TEXAS INSTRUMENTS
POST OFFICE BOX 225012 • DALLAS, TEXAS 75265

ALS AND AS CIRCUITS

2
ALS AND AS CIRCUITS

TYPES SN54ALS113A, SN54AS113, SN74ALS113A, SN74AS113
DUAL J-K NEGATIVE-EDGE-TRIGGERED FLIP-FLOPS WITH PRESET

D2661, APRIL 1982 – REVISED DECEMBER 1983

- Fully Buffered to Offer Maximum Isolation from External Disturbance
- Package Options Include Both Plastic and Ceramic Carriers in Addition to Plastic and Ceramic DIPs.
- Dependable Texas Instruments Quality and Reliability

SN54ALS113A, SN54AS113 . . . J PACKAGE
SN74ALS113A, SN74AS113 . . . N PACKAGE
(TOP VIEW)

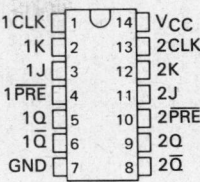

TYPE	TYPICAL MAXIMUM CLOCK FREQUENCY	TYPICAL POWER DISSIPATION PER FLIP-FLOP
'ALS113A	40 MHz (C_L = 15 pF)	6 mW
'AS113	175 MHz (C_L = 50 pF)	95 mW

description

These devices contain two independent J-K negative-edge-triggered flip-flops. A low level at the Preset input sets the outputs regardless of the levels of the other inputs. When Preset ($\overline{PRE}$) is inactive (high), data at the J and K inputs meeting the setup time requirements are transferred to the outputs on the negative-going edge of the clock pulse. Clock triggering occurs at a voltage level and is not directly related to the rise time of the clock pulse. Following the hold time interval, data at the J and K inputs may be changed without affecting the levels at the outputs. These versatile flip-flops can perform as toggle flip-flops by tying J and K high.

The SN54ALS113A and SN54AS113 are characterized for operation over the full military temperature range of $-55°C$ to $125°C$. The SN74ALS113A and SN74AS113 are characterized for operation from $0°C$ to $70°C$.

SN54ALS113A, SN54AS113 . . . FH PACKAGE
SN74ALS113A, SN74AS113 . . . FN PACKAGE
(TOP VIEW)

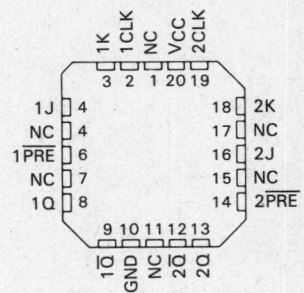

NC — No internal connection

FUNCTION TABLE

INPUTS				OUTPUTS	
$\overline{PRE}$	CLK	J	K	Q	$\overline{Q}$
L	X	X	X	H	L
H	↓	L	L	Q_0	$\overline{Q}_0$
H	↓	H	L	H	L
H	↓	L	H	L	H
H	↓	H	H	TOGGLE	
H	H	X	X	Q_0	$\overline{Q}_0$

logic symbol

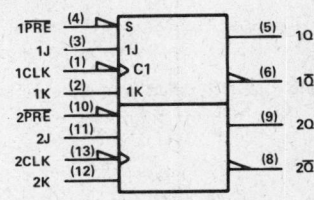

Pin numbers shown are for J and N packages.

absolute maximum ratings over operating free-air temperature range (unless otherwise noted)

Supply voltage, V_{CC} .... 7 V
Input voltage .... 7 V
Operating free-air temperature range: SN54ALS113A, SN54AS113 .... $-55°C$ to $125°C$
                                         SN74ALS113A, SN74AS113 .... $0°C$ to $70°C$
Storage temperature range .... $-65°C$ to $150°C$

Copyright © 1982 by Texas Instruments Incorporated

TEXAS INSTRUMENTS
POST OFFICE BOX 225012 • DALLAS, TEXAS 75265

TYPES SN54ALS113A, SN74ALS113A
DUAL J-K NEGATIVE-EDGE-TRIGGERED FLIP-FLOPS WITH PRESET

recommended operating conditions

			SN54ALS113A			SN74ALS113A			UNIT
			MIN	NOM	MAX	MIN	NOM	MAX	
V_{CC}	Supply voltage		4.5	5	5.5	4.5	5	5.5	V
V_{IH}	High-level input voltage		2			2			V
V_{IL}	Low-level input voltage				0.8			0.8	V
I_{OH}	High-level output current				−0.4			−0.4	mA
I_{OL}	Low-level output current				4			8	mA
f_{clock}	Clock frequency		0		25	0		30	MHz
t_w	Pulse duration	$\overline{PRE}$ low	15			10			ns
		CLK high	20			16.5			
		CLK low	20			16.5			
t_{su}	Setup time before CLK↓	Data	25			22			ns
		$\overline{PRE}$ inactive	22			20			
t_h	Hold time, data after CLK↓		0			0			ns
T_A	Operating free-air temperature		−55		125	0		70	°C

electrical characteristics over recommended operating free-air temperature range (unless otherwise noted)

PARAMETER		TEST CONDITIONS		SN54ALS113A			SN74ALS113A			UNIT
				MIN	TYP†	MAX	MIN	TYP†	MAX	
V_{IK}		V_{CC} = 4.5 V,	I_I = −18 mA			−1.5			−1.5	V
V_{OH}		V_{CC} = 4.5 V to 5.5 V,	I_{OH} = −0.4 mA	$V_{CC}-2$			$V_{CC}-2$			V
V_{OL}		V_{CC} = 4.5 V,	I_{OL} = 4 mA		0.25	0.4		0.25	0.4	V
		V_{CC} = 4.5 V,	I_{OL} = 8 mA					0.35	0.5	
I_I	J, K, or CLK	V_{CC} = 5.5 V,	V_I = 7 V			0.1			0.1	mA
	$\overline{PRE}$					0.2			0.2	
I_{IH}	J, K, or CLK	V_{CC} = 5.5 V,	V_I = 2.7 V			20			20	µA
	$\overline{PRE}$					40			40	
I_{IL}	J, K, or CLK	V_{CC} = 5.5 V,	V_I = 0.4 V			−0.2			−0.2	mA
	$\overline{PRE}$					−0.4			−0.4	
I_O‡		V_{CC} = 5.5 V,	V_O = 2.25 V	−30		−112	−30		−112	mA
I_{CC}		V_{CC} = 5.5 V,	See Note 1		2.5	4.5		2.5	4.5	mA

†All typical values are at V_{CC} = 5 V, T_A = 25°C.
‡The output conditions have been chosen to produce a current that closely approximates one half of the true short-circuit output current, I_{OS}.
NOTE 1: I_{CC} is measured with J, K, CLK, and $\overline{PRE}$ grounded, then with J, K, CLK, and $\overline{CLR}$ grounded.

switching characteristics (see Note 2)

PARAMETER	FROM (INPUT)	TO (OUTPUT)	V_{CC} = 4.5 V to 5.5 V, C_L = 50 pF, R_L = 500 Ω, T_A = MIN to MAX				UNIT
			SN54ALS113A		SN74ALS113A		
			MIN	MAX	MIN	MAX	
f_{max}			25		30		MHz
t_{PLH}	$\overline{PRE}$	Q or $\overline{Q}$	3	17	3	14	ns
t_{PHL}			4	20	4	16	
t_{PLH}	CLK	Q or $\overline{Q}$	3	18	3	15	ns
t_{PHL}			5	23	5	19	

NOTE 2: For load circuit and voltage waveforms, see page 1-12.

TEXAS INSTRUMENTS
POST OFFICE BOX 225012 • DALLAS, TEXAS 75265

TYPES SN54AS113, SN74AS113
DUAL J-K NEGATIVE-EDGE-TRIGGERED FLIP-FLOPS WITH PRESET

recommended operating conditions

		SN54AS113 MIN	NOM	MAX	SN74AS113 MIN	NOM	MAX	UNIT
V_{CC}	Supply voltage	4.5	5	5.5	4.5	5	5.5	V
V_{IH}	High-level input voltage	2			2			V
V_{IL}	Low-level input voltage			0.8			0.8	V
I_{OH}	High-level output current			−2			−2	mA
I_{OL}	Low-level output current			20			20	mA
f_{clock}	Clock frequency	0			0			MHz
t_w	Pulse duration — $\overline{PRE}$ low / CLK high / CLK low							ns
t_{su}	Setup time before CLK↓ — Data / $\overline{PRE}$ inactive							ns
t_h	Hold time, data after CLK↓							ns
T_A	Operating free-air temperature	−55		125	0		70	°C

electrical characteristics over recommended operating free-air temperature range (unless otherwise noted)

PARAMETER		TEST CONDITIONS		SN54AS113 MIN	TYP†	MAX	SN74AS113 MIN	TYP†	MAX	UNIT
V_{IK}		V_{CC} = 4.5 V,	I_I = −18 mA			−1.2			−1.2	V
V_{OH}		V_{CC} = 4.5 V to 5.5 V,	I_{OH} = −2 mA	V_{CC}−2			V_{CC}−2			V
V_{OL}		V_{CC} = 4.5 V,	I_{OL} = 20 mA		0.35	0.5		0.35	0.5	V
I_I	J or K	V_{CC} = 5.5 V,	V_I = 7 V			0.1			0.1	mA
	$\overline{PRE}$					0.5			0.5	
	CLK					0.5			0.5	
I_{IH}	J or K	V_{CC} = 5.5 V,	V_I = 2.7 V			0.02			0.02	mA
	$\overline{PRE}$					0.1			0.1	
	CLK					0.1			0.1	
I_{IL}	J or K	V_{CC} = 5.5 V,	V_I = 0.4 V			−1			−1	mA
	$\overline{PRE}$					−5.5			−5.5	
	CLR					−5			−5	
I_O‡		V_{CC} = 5.5 V,	V_O = 2.25 V	−30		−112	−30		−112	mA
I_{CC}		V_{CC} = 5.5 V,	See Note 1		38			38		mA

NOTE 1: I_{CC} is measured with D, CLK, and $\overline{PRE}$ grounded, then with D, CLK, and CLR grounded.

switching characteristics (see Note 2)

PARAMETER	FROM (INPUT)	TO (OUTPUT)	V_{CC} = 4.5 V to 5.5 V, C_L = 50 pF, R_L = 500 Ω, T_A = MIN to MAX						UNIT
			SN54AS113 MIN	TYP†	MAX	SN74AS113 MIN	TYP†	MAX	
f_{max}				175			175		MHz
t_{PLH}	$\overline{PRE}$	Q or $\overline{Q}$		3			3		ns
t_{PHL}				4			4		
t_{PLH}	CLK	Q or $\overline{Q}$		3			3		ns
t_{PHL}				4			4		

†All typical values are at V_{CC} = 5 V, T_A = 25°C.
‡The output conditions have been chosen to produce a current that closely approximates one half of the true short-circuit output current, I_{OS}.
NOTE 2: For load circuit and voltage waveforms, see page 1-12.
Additional information on these products can be obtained from the factory as it becomes available.

ALS AND AS CIRCUITS

PRODUCT PREVIEW

This page contains information on a product under development. Texas Instruments reserves the right to change or discontinue this product without notice.

Texas Instruments
POST OFFICE BOX 225012 • DALLAS, TEXAS 75265

2
ALS AND AS CIRCUITS

TYPES SN54ALS114A, SN54AS114, SN74ALS114A, SN74AS114
DUAL J-K NEGATIVE-EDGE-TRIGGERED FLIP-FLOPS
WITH PRESET, COMMON CLEAR, AND COMMON CLOCK

D2661, APRIL 1982—REVISED DECEMBER 1983

- Fully Buffered to Offer Maximum Isolation from External Disturbance
- Package Options Include Both Plastic and Ceramic Carriers in Addition to Plastic and Ceramic DIPs.
- Dependable Texas Instruments Quality and Reliability

SN54ALS114A, SN54AS114 . . . J PACKAGE
SN74ALS114A, SN74AS114 . . . N PACKAGE
(TOP VIEW)

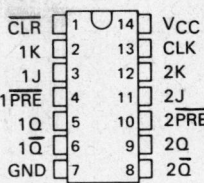

TYPE	TYPICAL MAXIMUM CLOCK FREQUENCY	TYPICAL POWER DISSIPATION PER FLIP-FLOP
'ALS114A	40 MHz (C_L = 15 pF)	6 mW
'AS114	175 MHz (C_L = 50 pF)	95 mW

description

These devices contain two independent J-K negative-edge-triggered flip-flops. A low level at the Preset or Clear inputs sets or resets the outputs regardless of the levels of the other inputs. When Preset and Clear are inactive (high), data at the J and K inputs meeting the setup time requirements are transferred to the outputs on the negative-going edge of the clock pulse. Clock triggering occurs at a voltage level and is not directly related to the rise time of the clock pulse. Following the hold time interval, data at the J and K inputs may be changed without affecting the levels at the outputs. These versatile flip-flops can perform as toggle flip-flops by tying J and K high.

The SN54ALS114A and SN54AS114 are characterized for operation over the full military temperature range of $-55\,°C$ to $125\,°C$. The SN74ALS114A and SN74AS114 are characterized for operation from $0\,°C$ to $70\,°C$.

SN54ALS114A, SN54AS114 . . . FH PACKAGE
SN74ALS114A, SN74AS114 . . . FN PACKAGE
(TOP VIEW)

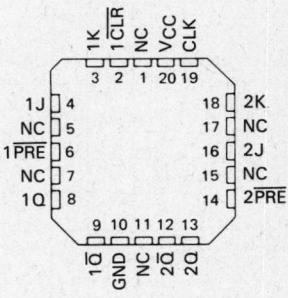

NC — No internal connection

FUNCTION TABLE

INPUTS					OUTPUTS	
PRE	CLR	CLK	J	K	Q	$\overline{Q}$
L	H	X	X	X	H	L
H	L	X	X	X	L	H
L	L	X	X	X	H*	H*
H	H	↓	L	L	Q_0	$\overline{Q}_0$
H	H	↓	H	L	H	L
H	H	↓	L	H	L	H
H	H	↓	H	H	TOGGLE	
H	H	X	X	X	Q_0	$\overline{Q}_0$

*The output levels in this configuration are not guaranteed to meet the minimum levels for V_{OH} if the lows at Preset and Clear are near V_{IL} maximum. Furthermore, this configuration is nonstable; that is, it will not persist when either Preset or Clear returns to its inactive (high) level.

logic symbol

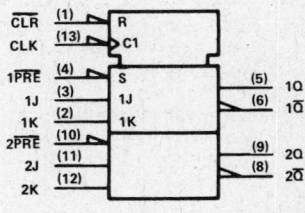

Pin numbers shown are for J and N packages.

absolute maximum ratings over operating free-air temperature range (unless otherwise noted)

Supply voltage, V_{CC} . . . . 7 V
Input voltage . . . . 7 V
Operating free-air temperature range: SN54ALS114A, SN54AS114 . . . . $-55\,°C$ to $125\,°C$
SN74ALS114A, SN74AS114 . . . . $0\,°C$ to $70\,°C$
Storage temperature range . . . . $-65\,°C$ to $150\,°C$

Copyright © 1982 by Texas Instruments Incorporated

TYPES SN54ALS114A, SN74ALS114A
DUAL J-K NEGATIVE-EDGE-TRIGGERED FLIP-FLOPS
WITH PRESET, COMMON CLEAR, AND COMMON CLOCK

recommended operating conditions

			SN54ALS114A			SN74ALS114A			UNIT
			MIN	NOM	MAX	MIN	NOM	MAX	
V_{CC}	Supply voltage		4.5	5	5.5	4.5	5	5.5	V
V_{IH}	High-level input voltage		2			2			V
V_{IL}	Low-level input voltage				0.8			0.8	V
I_{OH}	High-level output current				−0.4			−0.4	mA
I_{OL}	Low-level output current				4			8	mA
f_{clock}	Clock frequency		0		25	0		30	MHz
t_w	Pulse duration	$\overline{PRE}$ or $\overline{CLR}$ low	15			10			ns
		CLK high	20			16.5			
		CLK low	20			16.5			
t_{su}	Setup time before CLK↓	Data	25			22			ns
		$\overline{PRE}$ or $\overline{CLR}$ inactive	22			20			
t_h	Hold time, data after CLK↓		0			0			ns
T_A	Operating free-air temperature		−55		125	0		70	°C

electrical characteistics over recommended operating free-air temperature range (unless otherwise noted)

PARAMETER		TEST CONDITIONS		SN54ALS114A			SN74ALS114A			UNIT
				MIN	TYP†	MAX	MIN	TYP†	MAX	
V_{IK}		V_{CC} = 4.5 V,	I_I = −18 mA			−1.5			−1.5	V
V_{OH}		V_{CC} = 4.5 V to 5.5 V,	I_{OH} = −0.4 mA	V_{CC}−2			V_{CC}−2			V
V_{OL}		V_{CC} = 4.5 V,	I_{OL} = 4 mA		0.25	0.4				V
		V_{CC} = 4.5 V,	I_{OL} = 8 mA					0.35	0.5	
I_I	J, K, or CLK	V_{CC} = 5.5 V,	V_I = 7 V			0.1			0.1	mA
	$\overline{PRE}$ or $\overline{CLR}$					0.2			0.2	
I_{IH}	J, K, or CLK	V_{CC} = 5.5 V,	V_I = 2.7 V			20			20	µA
	$\overline{PRE}$ or $\overline{CLR}$					40			40	
I_{IL}	J, K, or CLK	V_{CC} = 5.5 V,	V_I = 0.4 V			−0.2			−0.2	mA
	$\overline{PRE}$ or $\overline{CLR}$					−0.4			−0.4	
I_O‡		V_{CC} = 5.5 V,	V_O = 2.25 V	−30		−112	−30		−112	mA
I_{CC}		V_{CC} = 5.5 V,	See Note 1		2.5	4.5		2.5	4.5	mA

†All typical values are at V_{CC} = 5 V, T_A = 25°C.
‡The output conditions have been chosen to produce a current that closely approximates one half of the true short-circuit output current, I_{OS}.
NOTE 1: I_{CC} is measured with J, K, CLK, and $\overline{PRE}$ grounded, then with J, K, CLK, and $\overline{CLR}$ grounded.

switching characteristics (see Note 2)

PARAMETER	FROM (INPUT)	TO (OUTPUT)	V_{CC} = 4.5 V to 5.5 V, C_L = 50 pF, R_L = 500 Ω, T_A = MIN to MAX				UNIT
			SN54ALS114A		SN74ALS114A		
			MIN	MAX	MIN	MAX	
f_{max}			25		30		MHz
t_{PLH}	$\overline{PRE}$ or $\overline{CLR}$	Q or $\overline{Q}$	3	20	3	15	ns
t_{PHL}			4	22	4	18	
t_{PLH}	CLK	Q or $\overline{Q}$	3	18	3	15	ns
t_{PHL}			5	23	5	19	

NOTE 2: For load circuit and voltage waveforms, see page 1-12.

TEXAS INSTRUMENTS
POST OFFICE BOX 225012 • DALLAS, TEXAS 75265

TYPES SN54AS114, SN74AS114
DUAL J-K NEGATIVE-EDGE-TRIGGERED FLIP-FLOPS
WITH PRESET, COMMON CLEAR, AND COMMON CLOCK

recommended operating conditions

			SN54AS114			SN74AS114			UNIT
			MIN	NOM	MAX	MIN	NOM	MAX	
V_{CC}	Supply voltage		4.5	5	5.5	4.5	5	5.5	V
V_{IH}	High-level input voltage		2			2			V
V_{IL}	Low-level input voltage				0.8			0.8	V
I_{OH}	High-level output current				−2			−2	mA
I_{OL}	Low-level output current				20			20	mA
f_{clock}	Clock frequency		0			0			MHz
t_w	Pulse duration	$\overline{PRE}$ or $\overline{CLR}$ low							ns
		CLK high							
		CLK low							
t_{su}	Setup time before CLK↓	Data							ns
		$\overline{PRE}$ or $\overline{CLR}$ inactive							
t_h	Hold time, data after CLK↓								ns
T_A	Operating free-air temperature		−55		125	0		70	°C

electrical characteristics over recommended operating free-air temperature range (unless otherwise noted)

PARAMETER		TEST CONDITIONS	SN54AS114			SN74AS114			UNIT
			MIN	TYP†	MAX	MIN	TYP†	MAX	
V_{IK}		V_{CC} = 4.5 V, I_I = −18 mA			−1.2			−1.2	V
V_{OH}		V_{CC} = 4.5 V to 5.5 V, I_{OH} = −2 mA	V_{CC}−2			V_{CC}−2			V
V_{OL}		V_{CC} = 4.5 V, I_{OL} = 20 mA		0.35	0.5		0.35	0.5	V
I_I	J or K	V_{CC} = 5.5 V, V_I = 7 V			0.1			0.1	mA
	$\overline{PRE}$				0.5			0.5	
	$\overline{CLR}$				1			1	
	CLK				1			1	
I_{IH}	J or K	V_{CC} = 5.5 V, V_I = 2.7 V			0.02			0.02	mA
	$\overline{PRE}$				0.1			0.1	
	$\overline{CLR}$				0.2			0.2	
	CLK				0.2			0.2	
I_{IL}	J or K	V_{CC} = 5.5 V, V_I = 0.4 V			−1			−1	mA
	$\overline{PRE}$				−5.5			−5.5	
	$\overline{CLR}$				−11.5			−11.5	
	CLK				−10.5			−10.5	
I_O‡		V_{CC} = 5.5 V, V_O = 2.25 V	−30		−112	−30		−112	mA
I_{CC}		V_{CC} = 5.5 V, See Note 1		38			38		mA

†All typical values are at V_{CC} = 5 V, T_A = 25°C.
‡The output conditions have been chosen to produce a current that closely approximates one half of the true short-circuit output current, I_{OS}.
NOTE 1: I_{CC} is measured with J, K, CLK, and $\overline{PRE}$ grounded, then with J, K, CLK, and $\overline{CLR}$ grounded.

Additional information on these products can be obtained from the factory as it becomes available.

PRODUCT PREVIEW

This page contains information on a product under development. Texas Instruments reserves the right to change or discontinue this product without notice.

TEXAS INSTRUMENTS
POST OFFICE BOX 225012 • DALLAS, TEXAS 75265

TYPES SN54AS114, SN74AS114
DUAL J-K NEGATIVE-EDGE-TRIGGERED FLIP FLOPS
WITH PRESET, COMMON CLEAR, AND COMMON CLOCK

switching characteristics (see Note 2)

PARAMETER	FROM (INPUT)	TO (OUTPUT)	V_{CC} = 4.5 V to 5.5 V, C_L = 50 pF, R_L = 500 Ω, T_A = MIN to MAX						UNIT
			SN54AS114			SN74AS114			
			MIN	TYP†	MAX	MIN	TYP†	MAX	
f_{max}				175			175		MHz
t_{PLH}	$\overline{PRE}$ or $\overline{CLR}$	Q or $\overline{Q}$		3			3		ns
t_{PHL}				4			4		
t_{PLH}	CLK	Q or $\overline{Q}$		3			3		ns
t_{PHL}				4			4		

†All typical values are at V_{CC} = 5 V, T_A = 25 °C.
NOTE 2: For load circuit and voltage waveforms, see page 1-12.

PRODUCT PREVIEW
This page contains information on a product under development. Texas Instruments reserves the right to change or discontinue this product without notice.

TYPES SN54ALS131, SN54AS131, SN74ALS131, SN74AS131
3-LINE TO 8-LINE DECODERS/DEMULTIPLEXERS WITH ADDRESS REGISTERS

D2661, APRIL 1982—REVISED DECEMBER 1983

- **Combines Decoder and 3-Bit Address Register**
- **Incorporates 2 Enable Inputs to Simplify Cascading**
- **Package Options Include Both Plastic and Ceramic Chip Carriers in Addition to Plastic and Ceramic DIPs**
- **Dependable Texas Instruments Quality and Reliability**

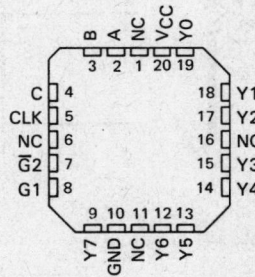

SN54ALS131, SN54AS131 . . . J PACKAGE
SN74ALS131, SN74AS131 . . . N PACKAGE
(TOP VIEW)

SN54ALS131, SN54AS131 . . . FH PACKAGE
SN74ALS131, SN74AS131 . . . FN PACKAGE
(TOP VIEW)

NC — No internal connection

description

The 'ALS131 and 'AS131 are three-line to eight-line decoders/demultiplexers with registers on the three address inputs. When the clock input (CLK) goes from low to high, the 'ALS131 and 'AS131 act as decoders/demultiplexers and the address present at the select inputs (A, B, and C) is stored in the registers. Further address changes are ignored until the next transition of CLK. The output enable controls, G1 and $\overline{G2}$, control the state of the outputs independently of the select or CLK inputs. All of the outputs are high unless G1 is high and $\overline{G2}$ is low. The 'ALS131 and 'AS131 are ideally suited for implementing glitch-free decoders in strobed (stored-address) applications in bus-oriented systems.

The SN54ALS131 and SN54AS131 are characterized for operation over the full military temperature range of −55°C to 125°C. The SN74ALS131 and SN74AS131 are characterized for operation from 0°C to 70°C.

logic symbols (alternatives)

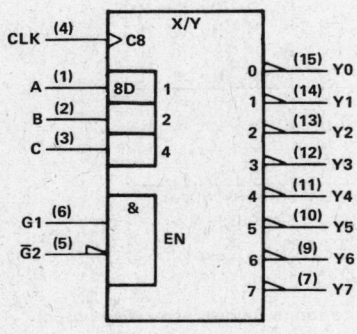

Pin numbers shown are for J and N packages.

Copyright © 1982 by Texas Instruments Incorporated

ALS AND AS CIRCUITS

TYPES SN54ALS131, SN54AS131, SN74ALS131, SN74AS131
3-LINE TO 8-LINE DECODERS/DEMULTIPLEXERS WITH ADDRESS REGISTERS

logic diagram (positive logic)

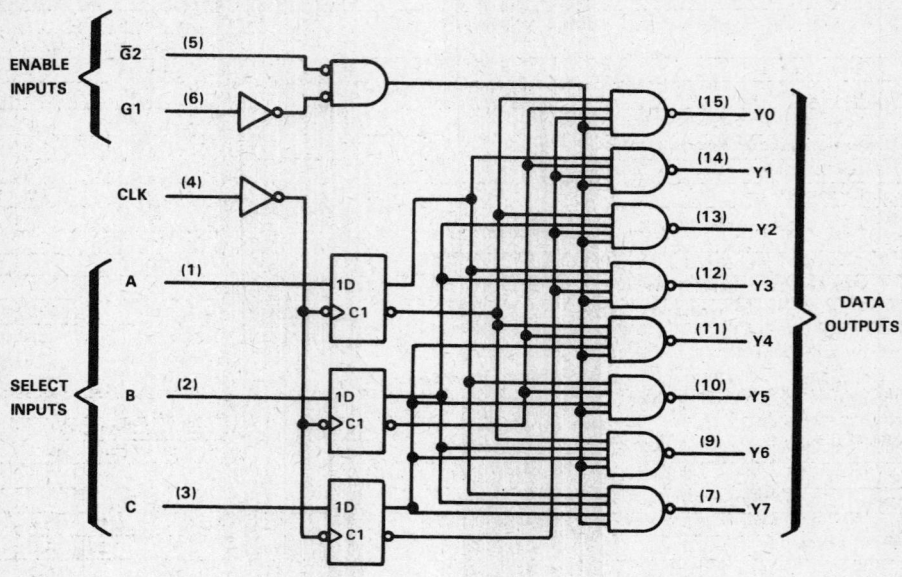

Pin numbers shown are for J and N packages.

FUNCTION TABLE

CLK	INPUTS			SELECT			OUTPUTS							
	ENABLE													
	G1	G2	C	B	A	Y0	Y1	Y2	Y3	Y4	Y5	Y6	Y7
X	X	H	X	X	X	H	H	H	H	H	H	H	H
X	L	X	X	X	X	H	H	H	H	H	H	H	H
↑	H	L	L	L	L	L	H	H	H	H	H	H	H
↑	H	L	L	L	H	H	L	H	H	H	H	H	H
↑	H	L	L	H	L	H	H	L	H	H	H	H	H
↑	H	L	L	H	H	H	H	H	L	H	H	H	H
↑	H	L	H	L	L	H	H	H	H	L	H	H	H
↑	H	L	H	L	H	H	H	H	H	H	L	H	H
↑	H	L	H	H	L	H	H	H	H	H	H	L	H
↑	H	L	H	H	H	H	H	H	H	H	H	H	L
L or H	H	L	X	X	X	OUTPUTS CORRESPONDING TO STORED ADDRESS, L; ALL OTHERS, H							

absolute maximum ratings over operating free-air temperature range (unless otherwise noted)

Supply voltage, V_CC .... 7 V
Input voltage .... 7 V
Operating free-air temperature range: SN54ALS131, SN54AS131 .... −55°C to 125°C
                                       SN74ALS131, SN74AS131 .... 0°C to 70°C
Storage temperature .... −65°C to 150°C

TYPES SN54ALS131, SN74ALS131
3-LINE TO 8-LINE DECODERS/DEMULTIPLEXERS WITH ADDRESS REGISTERS

recommended operating conditions

		SN54ALS131			SN74ALS131			UNIT
		MIN	NOM	MAX	MIN	NOM	MAX	
V_{CC}	Supply voltage	4.5	5	5.5	4.5	5	5.5	V
V_{IH}	High-level input voltage	2			2			V
V_{IL}	Low-level input voltage			0.8			0.8	V
I_{OH}	High-level output current			−0.4			−0.4	mA
I_{OL}	Low-level output current			4			8	mA
f_{clock}	Clock frequency	0		40	0		50	MHz
t_w	Pulse duration CLK high	12.5			10			ns
	Pulse duration CLK low	12.5			10			ns
t_{su}	Setup time at A, B, and C before CLK↑	15			10			ns
t_h	Hold time at A, B, and C after CLK↑	0			0			ns
T_A	Operating free-air temperature	−55		125	0		70	°C

electrical characteristics over recommended operating free-air temperature range (unless otherwise noted)

PARAMETER	TEST CONDITIONS		SN54ALS131			SN74ALS131			UNIT
			MIN	TYP†	MAX	MIN	TYP†	MAX	
V_{IK}	$V_{CC} = 4.5$ V,	$I_I = -18$ mA			−1.5			−1.5	V
V_{OH}	$V_{CC} = 4.5$ V to 5.5 V,	$I_{OH} = -0.4$ mA	$V_{CC}-2$			$V_{CC}-2$			V
V_{OL}	$V_{CC} = 4.5$ V,	$I_{OL} = 4$ mA		0.25	0.4		0.25	0.4	V
	$V_{CC} = 4.5$ V,	$I_{OL} = 8$ mA					0.35	0.5	
I_I	$V_{CC} = 5.5$ V,	$V_I = 7$ V			0.1			0.1	mA
I_{IH}	$V_{CC} = 5.5$ V,	$V_I = 2.7$ V			20			20	μA
I_{IL}	$V_{CC} = 5.5$ V,	$V_I = 0.4$ V			−0.1			−0.1	mA
I_O‡	$V_{CC} = 5.5$ V,	$V_O = 2.25$ V	−30		−112	−30		−112	mA
I_{CC}	$V_{CC} = 5.5$ V			5	11		5	11	mA

†All typical values are at $V_{CC} = 5$ V, $T_A = 25$°C.
‡The output conditions have been chosen to produce a current that closely approximates one half of the true short-circuit output current, I_{OS}.

switching characteristics (see Note 1)

PARAMETER	FROM (INPUT)	TO (OUTPUT)	$V_{CC} = 4.5$ V to 5.5 V, $C_L = 50$ pF, $R_L = 500$ Ω, $T_A =$ MIN to MAX				UNIT
			SN54ALS131		SN74ALS131		
			MIN	MAX	MIN	MAX	
f_{max}			40		50		MHz
t_{PLH}	CLK	Y	8	28	8	25	ns
t_{PHL}			7	24	7	20	
t_{PLH}	G1	Y	7	24	7	20	ns
t_{PHL}			6	20	6	17	
t_{PLH}	$\overline{G2}$	Y	5	18	5	15	ns
t_{PHL}			5	18	5	15	

NOTE 1: For load circuit and voltage waveforms, see page 1-12.

TEXAS INSTRUMENTS
POST OFFICE BOX 225012 • DALLAS, TEXAS 75265

TYPES SN54AS131, SN74AS131
3-LINE TO 8-LINE DECODERS/DEMULTIPLEXERS WITH ADDRESS REGISTERS

recommended operating conditions

		SN54AS131 MIN	NOM	MAX	SN74AS131 MIN	NOM	MAX	UNIT
V_{CC}	Supply voltage	4.5	5	5.5	4.5	5	5.5	V
V_{IH}	High-level input voltage	2			2			V
V_{IL}	Low-level input voltage			0.8			0.8	V
I_{OH}	High-level output current			-2			-2	mA
I_{OL}	Low-level output current			20			20	mA
f_{clock}	Clock frequency							MHz
t_w	Pulse duration CLK high							ns
	CLK low							
t_{su}	Setup time at A, B, and C before CLK↑							ns
t_h	Hold time at A, B, and C after CLK↑							ns
T_A	Operating free-air temperature	-55		125	0		70	°C

electrical characteristics over recommended operating free-air temperature range (unless otherwise noted)

PARAMETER	TEST CONDITIONS	SN54AS131 MIN	TYP†	MAX	SN74AS131 MIN	TYP†	MAX	UNIT
V_{IK}	V_{CC} = 4.5 V, I_I = -18 mA			-1.2			-1.2	V
V_{OH}	V_{CC} = 4.5 V to 5.5 V, I_{OH} = -2 mA	$V_{CC}-2$			$V_{CC}-2$			V
V_{OL}	V_{CC} = 4.5 V, I_{OL} = 20 mA		0.35	0.5		0.35	0.5	V
I_I	V_{CC} = 5.5 V, V_I = 7 V							mA
I_{IH}	V_{CC} = 5.5 V, V_I = 2.7 V							µA
I_{IL}	V_{CC} = 5.5 V, V_I = 0.4 V							mA
I_O‡	V_{CC} = 5.5 V, V_O = 2.25 V	-30		-112	-30		-112	mA
I_{CC}	V_{CC} = 5.5 V		16			16		mA

†All typical values are at V_{CC} = 5 V, T_A = 25°C.
‡The output conditions have been chosen to produce a current that closely approximates one half of the true short-circuit output current, I_{OS}.

switching characteristics (see Note 1)

PARAMETER	FROM (INPUT)	TO (OUTPUT)	V_{CC} = 4.5 V to 5.5 V, C_L = 50 pF, R_L = 500 Ω, T_A = MIN to MAX SN54AS131 MIN	TYP†	MAX	SN74AS131 MIN	TYP†	MAX	UNIT
f_{max}									MHz
t_{PLH}	CLK	Y		5.4			5.4		ns
t_{PHL}				5.3			5.3		
t_{PLH}	G1	Y		6.2			6.2		ns
t_{PHL}				5.6			5.6		
t_{PLH}	$\overline{G2}$	Y		5.4			5.4		ns
t_{PHL}				5.3			5.3		

†All typical values are at V_{CC} = 5 V, T_A = 25°C.
NOTE 1: For load circuit and voltage waveforms, see page 1-12.

PRODUCT PREVIEW
This page contains information on a product under development. Texas Instruments reserves the right to change or discontinue this product without notice.

TEXAS INSTRUMENTS
POST OFFICE BOX 225012 • DALLAS, TEXAS 75265

TYPES SN54ALS133, SN74ALS133
13-INPUT POSITIVE-NAND GATES

D2661, APRIL 1982—REVISION DECEMBER 1983

- Package Options Include Both Plastic and Ceramic Chip Carriers in Addition to Plastic and Ceramic DIPs
- Dependable Texas Instruments Quality and Reliability

description

These devices contain a single 13-input NAND gate. They perform the boolean functions in positive logic:

$$Y = \overline{A \cdot B \cdot C \cdot D \cdot E \cdot F \cdot G \cdot H \cdot I \cdot J \cdot K \cdot L \cdot M}$$

$$Y = \overline{A} + \overline{B} + \overline{C} + \overline{D} + \overline{E} + \overline{F} + \overline{G} + \overline{H} + \overline{I} + \overline{J} + \overline{K} + \overline{L} + \overline{M}$$

The SN54ALS133 is characterized for operation over the full military temperature range of −55°C to 125°C. The SN74ALS133 is characterized for operation from 0°C to 70°C.

FUNCTION TABLE

INPUTS A THRU M	OUTPUT Y
All inputs H	L
One or more inputs L	H

logic symbol

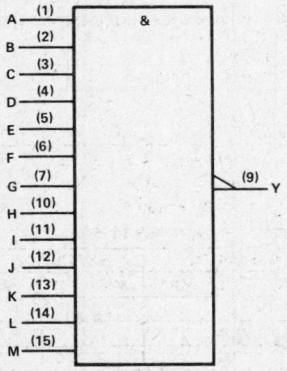

Pin numbers shown are for J and N packages.

SN54ALS133 . . . J PACKAGE
SN74ALS133 . . . N PACKAGE
(TOP VIEW)

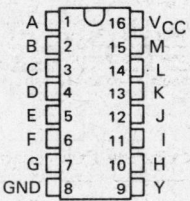

SN54ALS133 . . . FH PACKAGE
SN74ALS133 . . . FN PACKAGE
(TOP VIEW)

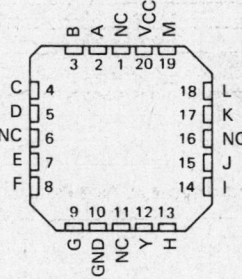

NC—No internal connection

ALS AND AS CIRCUITS

Copyright © 1982 by Texas Instruments Incorporated

TEXAS INSTRUMENTS
POST OFFICE BOX 225012 • DALLAS, TEXAS 75265

TYPES SN54ALS133, SN74ALS133
13-INPUT POSITIVE-NAND GATES

absolute maximum ratings over operating free-air temperature range (unless otherwise noted)

Supply voltage, V_{CC} .... 7 V
Input voltage .... 7 V
Operating free-air temperature range: SN54ALS133 .... $-55°C$ to $125°C$
                                      SN74ALS133 .... $0°C$ to $70°C$
Storage temperature range .... $-65°C$ to $150°C$

recommended operating conditions

		SN54ALS133			SN74ALS133			UNIT
		MIN	NOM	MAX	MIN	NOM	MAX	
V_{CC}	Supply voltage	4.5	5	5.5	4.5	5	5.5	V
V_{IH}	High-level input voltage	2			2			V
V_{IL}	Low-level input voltage			0.8			0.8	V
I_{OH}	High-level output current			-0.4			-0.4	mA
I_{OL}	Low-level output current			4			8	mA
T_A	Operating free-air temperature	-55		125	0		70	°C

electrical characteristics over recommended operating free-air temperature range (unless otherwise noted)

PARAMETER	TEST CONDITIONS		SN54ALS133			SN74ALS133			UNIT
			MIN	TYP†	MAX	MIN	TYP†	MAX	
V_{IK}	$V_{CC} = 4.5$ V,	$I_I = -18$ mA			-1.5			-1.5	V
V_{OH}	$V_{CC} = 4.5$ V to 5.5 V	$I_{OH} = -0.4$ mA	$V_{CC}-2$			$V_{CC}-2$			V
V_{OL}	$V_{CC} = 4.5$ V,	$I_{OL} = 4$ mA		0.25	0.4		0.25	0.4	V
	$V_{CC} = 4.5$ V	$I_{OL} = 8$ mA					0.35	0.5	
I_I	$V_{CC} = 5.5$ V,	$V_I = 7$ V			0.1			0.1	mA
I_{IH}	$V_{CC} = 5.5$ V,	$V_I = 2.7$ V			20			20	µA
I_{IL}	$V_{CC} = 5.5$ V,	$V_I = 0.4$ V			-0.1			-0.1	mA
I_O‡	$V_{CC} = 5.5$ V,	$V_O = 2.25$ V	-30		-112	-30		-112	mA
I_{CCH}	$V_{CC} = 5.5$ V,	$V_I = 0$ V		0.24	0.34		0.24	0.34	mA
I_{CCL}	$V_{CC} = 5.5$ V,	$V_I = 4.5$ V		0.56	0.8		0.56	0.8	mA

†All typical values aree at $V_{CC} = 5$ V, $T_A = 25°C$.
‡The output conditions have been chosen to produce a current that closely approximates one half of the true short-circuit output current, I_{OS}.

switching characteristics (see Note 1)

PARAMETER	FROM (INPUT)	TO (OUTPUT)	$V_{CC} = 4.5$ V to 5.5 V $C_L = 50$ pF, $R_L = 500$ Ω $T_A =$ MIN to MAX				UNIT
			SN54ALS133		SN74ALS133		
			MIN	MAX	MIN	MAX	
t_{PLH}	Any	Y	3	14	3	11	ns
t_{PHL}	Any	Y	5	28	5	25	ns

NOTE 2: For load circuit and voltage waveforms, see page 1-12.

TEXAS INSTRUMENTS
POST OFFICE BOX 225012 • DALLAS, TEXAS 75265

TYPES SN54ALS137, SN54AS137, SN74ALS137, SN74AS137
3-LINE TO 8-LINE DECODERS/DEMULTIPLEXERS WITH ADDRESS LATCHES

D2661, APRIL 1982—REVISED DECEMBER 1983

- **Combines Decoder and 3-Bit Address Latch**
- **Incorporates 2 Output Enables to Simplify Cascading**
- **Package Options Include Both Plastic and Ceramic Chip Carriers in Addition to Plastic and Ceramic DIPs**
- **Dependable Texas Instruments Quality and Reliability**

SN54ALS137, SN54AS137 . . . J PACKAGE
SN74ALS137, SN74AS137 . . . N PACKAGE
(TOP VIEW)

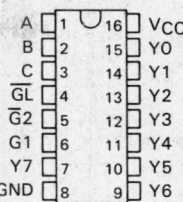

description

The 'ALS137 and 'AS137 are three-line to eight-line decoder/demultiplexer with latches on the three address inputs. When the latch-enable input ($\overline{GL}$) is low, the 'ALS137 and 'AS137 acts as a decoder/demultiplexer. When $\overline{GL}$ goes from low to high, the address present at the select inputs (A, B, and C) is stored in the latches. Further address changes are ignored as long as $\overline{GL}$ remains high. The output enable controls, G1 and $\overline{G2}$, control the outputs independently of the select or latch-enable inputs. All of the outputs are forced high if G1 is low or $\overline{G2}$ is high. The 'ALS137 and 'AS137 are ideally suited for implementing glitch-free decoders in strobed (stored-address) applications in bus-oriented systems.

The SN54ALS137 and SN54AS137 are characterized for operation over the full military temperature range of −55°C to 125°C. The SN74ALS137 and SN74AS137 are characterized for operation from 0°C to 70°C.

SN54ALS137, SN54AS137 . . . FH PACKAGE
SN74ALS137, SN74AS137 . . . FN PACKAGE
(TOP VIEW)

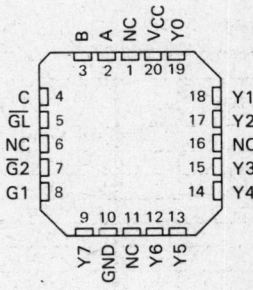

NC — No internal connection

logic symbols (alternatives)

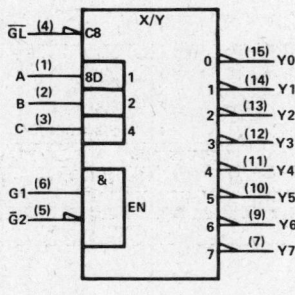
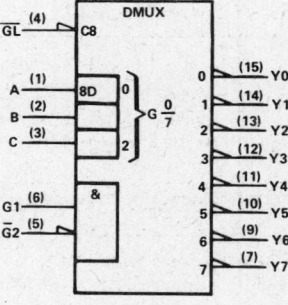

Pin numbers shown are for J and N packages.

Copyright © 1982 by Texas Instruments Incorporated

TEXAS INSTRUMENTS
POST OFFICE BOX 225012 • DALLAS, TEXAS 75265

2-109

TYPES SN54ALS137, SN54AS137, SN74ALS137, SN74AS137
3-LINE TO 8-LINE DECODERS/DEMULTIPLEXERS WITH ADDRESS LATCHES

logic diagram (positive logic)

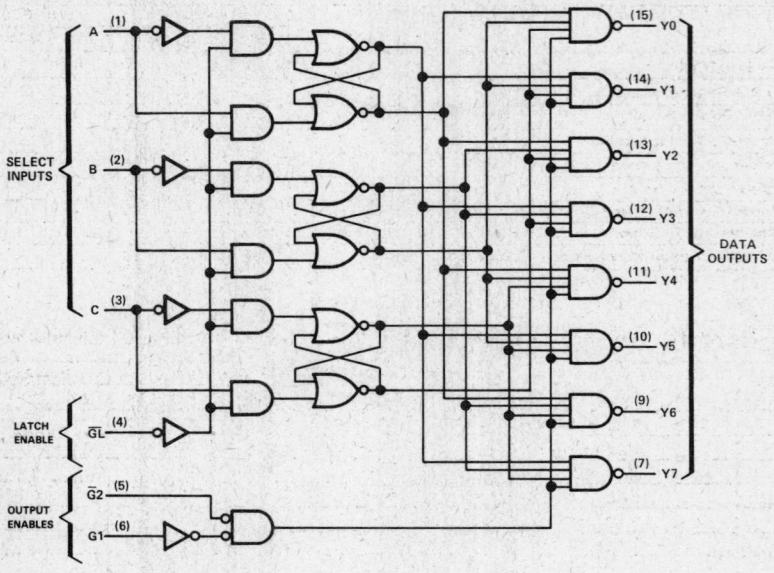

Pin numbers shown are for J and N packages.

FUNCTION TABLE

INPUTS					OUTPUTS							
ENABLE			SELECT									
$\overline{GL}$	$\overline{G1}$	G2	C	B	A	Y0	Y1	Y2	Y3	Y4	Y5	Y6	Y7
X	X	H	X	X	X	H	H	H	H	H	H	H	H
X	L	X	X	X	X	H	H	H	H	H	H	H	H
L	H	L	L	L	L	L	H	H	H	H	H	H	H
L	H	L	L	L	H	H	L	H	H	H	H	H	H
L	H	L	L	H	L	H	H	L	H	H	H	H	H
L	H	L	L	H	H	H	H	H	L	H	H	H	H
L	H	L	H	L	L	H	H	H	H	L	H	H	H
L	H	L	H	L	H	H	H	H	H	H	L	H	H
L	H	L	H	H	L	H	H	H	H	H	H	L	H
L	H	L	H	H	H	H	H	H	H	H	H	H	L
H	H	L	X	X	X	Output corresponding to stored address, L; all others, H						

absolute maximum ratings over operating free-air temperature range (unless otherwise noted)

Supply voltage, V_{CC} . . . . 7 V
Input voltage . . . . 7 V
Operating free-air temperature range: SN54ALS137, SN54AS137 . . . . −55°C to 125°C
                                   SN74ALS137, SN74AS137 . . . . 0°C to 70°C
Storage temperature . . . . −65°C to 150°C

TEXAS INSTRUMENTS
POST OFFICE BOX 225012 • DALLAS, TEXAS 75265

TYPES SN54ALS137, SN74ALS137
3-LINE TO 8-LINE DECODERS/DEMULTIPLEXERS WITH ADDRESS LATCHES

recommended operating conditions

		SN54ALS137			SN74ALS137			UNIT
		MIN	NOM	MAX	MIN	NOM	MAX	
V_{CC}	Supply voltage	4.5	5	5.5	4.5	5	5.5	V
V_{IH}	High-level input voltage	2			2			V
V_{IL}	Low-level input voltage			0.8			0.8	V
I_{OH}	High-level output current			−0.4			−0.4	mA
I_{OL}	Low-level output current			4			8	mA
t_w	Pulse duration, $\overline{GL}$ low	15			10			ns
t_{su}	Setup time at A, B, and C before $\overline{GL}\uparrow$	15			10			ns
t_h	Hold time at A, B, and C after $\overline{GL}\uparrow$	5			5			ns
T_A	Operating free-air temperature	−55		125	0		70	°C

electrical characteristics over recommended operating free-air temperature range (unless otherwise noted)

PARAMETER	TEST CONDITIONS		SN54ALS137			SN74ALS137			UNIT
			MIN	TYP[†]	MAX	MIN	TYP[†]	MAX	
V_{IK}	V_{CC} = 4.5 V,	I_I = −18 mA			−1.5			−1.5	V
V_{OH}	V_{CC} = 4.5 V to 5.5 V,	I_{OH} = −0.4 mA	$V_{CC}-2$			$V_{CC}-2$			V
V_{OL}	V_{CC} = 4.5 V,	I_{OL} = 4 mA		0.25	0.4		0.25	0.4	V
	V_{CC} = 4.5 V,	I_{OL} = 8 mA					0.35	0.5	
I_I	V_{CC} = 5.5 V,	V_I = 7 V			0.1			0.1	mA
I_{IH}	V_{CC} = 5.5 V,	V_I = 2.7 V			20			20	μA
I_{IL}	V_{CC} = 5.5 V,	V_I = 0.4 V			−0.1			−0.1	mA
I_O[‡]	V_{CC} = 5.5 V,	V_O = 2.25 V	−30		−112	−30		−112	mA
I_{CC}	V_{CC} = 5.5 V			5	11		5	11	mA

[†] All typical values are at V_{CC} = 5 V, T_A = 25°C.
[‡] The output conditions have been chosen to produce a current that closely approximates one half of the true short-circuit output current, I_{OS}.

switching characteristics (see Note 1)

PARAMETER	FROM (INPUT)	TO (OUTPUT)	V_{CC} = 4.5 V to 5.5 V, C_L = 50 pF, R_L = 500 Ω, T_A = MIN to MAX				UNIT
			SN54ALS137		SN74ALS137		
			MIN	MAX	MIN	MAX	
t_{PLH}	A, B, C	Y	5	25	5	20	ns
t_{PHL}			6	25	6	20	
t_{PLH}	$\overline{G2}$	Y	4	15	4	12	ns
t_{PHL}			5	18	5	15	
t_{PLH}	G1	Y	5	21	5	17	ns
t_{PHL}			5	19	5	15	
t_{PLH}	$\overline{GL}$	Y	7	27	7	22	ns
t_{PHL}			7	25	7	20	

NOTE 1: For load circuit and voltage waveforms, see page 1-12.

ALS AND AS CIRCUITS

TEXAS INSTRUMENTS
POST OFFICE BOX 225012 • DALLAS, TEXAS 75265

TYPES SN54AS137, SN74AS137
3-LINE TO 8-LINE DECODERS/MULTIPLEXERS WITH ADDRESS LATCHES

recommended operating conditions

		SN54AS137 MIN	SN54AS137 NOM	SN54AS137 MAX	SN74AS137 MIN	SN74AS137 NOM	SN74AS137 MAX	UNIT
V_{CC}	Supply voltage	4.5	5	5.5	4.5	5	5.5	V
V_{IH}	High-level input voltage	2			2			V
V_{IL}	Low-level input voltage			0.8			0.8	V
I_{OH}	High-level output current			−2			−2	mA
I_{OL}	Low-level output current			20			20	mA
t_w	Pulse duration, GL low							ns
t_{su}	Setup times at A, B, and C before $\overline{GL}\uparrow$							ns
t_h	Hold time at A, B, and C after $\overline{GL}\uparrow$							ns
T_A	Operating free-air temperature	−55		125	0		70	°C

electrical characteristics over recommended operating free-air temperature range (unless otherwise noted)

PARAMETER		TEST CONDITIONS		SN54AS137 MIN	SN54AS137 TYP†	SN54AS137 MAX	SN74AS137 MIN	SN74AS137 TYP†	SN74AS137 MAX	UNIT
V_{IK}		V_{CC} = 4.5 V,	I_I = −18 mA			−1.2			−1.2	V
V_{OH}		V_{CC} = 4.5 V to 5.5 V,	I_{OH} = −2 mA	V_{CC}−2			V_{CC}−2			V
V_{OL}		V_{CC} = 4.5 V,	I_{OL} = 20 mA		0.35	0.5		0.35	0.5	V
I_I	Enable	V_{CC} = 5.5 V,	V_I = 7 V							mA
	A, B, C									
I_{IH}	Enable	V_{CC} = 5.5 V,	V_I = 2.7 V							µA
	A, B, C									
I_{IL}	Enable	V_{CC} = 5.5 V,	V_I = 0.4 V			−0.05			−0.05	mA
	A, B, C					−0.05			−0.05	
I_O‡		V_{CC} = 5.5 V,	V_O = 2.25 V	−30		−112	−30		−112	mA
I_{CC}		V_{CC} = 5.5 V				16			16	mA

†All typical values are at V_{CC} = 5 V, T_A = 25°C.
‡The output conditions have been chosen to produce a current that closely approximates one half of the true short-circuit output current, I_{OS}.

switching characteristics (see Note 1)

PARAMETER	FROM (INPUT)	TO (OUTPUT)	V_{CC} = 4.5 V to 5.5 V, C_L = 50 pF, R_L = 500 Ω, T_A = MIN to MAX						UNIT
			SN54AS137 MIN	SN54AS137 TYP†	SN54AS137 MAX	SN74AS137 MIN	SN74AS137 TYP†	SN74AS137 MAX	
t_{PLH}	A, B, C	Y		6.6			6.6		ns
t_{PHL}				7.1			7.1		
t_{PLH}	$\overline{G2}$	Y		5.4			5.4		ns
t_{PHL}				5.3			5.3		
t_{PLH}	G1	Y		6.2			6.2		ns
t_{PHL}				5.6			5.6		
t_{PLH}	$\overline{GL}$	Y		5.4			5.4		ns
t_{PHL}				5.3			5.3		

†All typical values are at V_{CC} 5 V, T_A = 25°C.
NOTE 1: For load circuit and voltage waveforms, see page 1-12.

PRODUCT PREVIEW
This page contains information on a product under development. Texas Instruments reserves the right to change or discontinue this product without notice.

TEXAS INSTRUMENTS
POST OFFICE BOX 225012 • DALLAS, TEXAS 75265

ALS AND AS CIRCUITS

TYPES SN54ALS138, SN54AS138, SN74ALS138, SN74AS138
3-LINE TO 8-LINE DECODERS/DEMULTIPLEXERS

D2661, APRIL 1982—REVISED DECEMBER 1983

- Designed Specifically for High-Speed Memory Decoders and Data Transmission Systems
- Incorporates 3 Enable Inputs to Simplify Cascading and/or Data Reception
- Package Options Include Both Plastic and Ceramic Chip Carriers in Addition to Plastic and Ceramic DIPs
- Dependable Texas Instruments Quality and Reliability

SN54ALS138, SN54AS138 . . . J PACKAGE
SN74ALS138, SN74AS138 . . . N PACKAGE
(TOP VIEW)

```
     A  [ 1    16 ]  VCC
     B  [ 2    15 ]  Y0
     C  [ 3    14 ]  Y1
   G2A  [ 4    13 ]  Y2
   G2B  [ 5    12 ]  Y3
    G1  [ 6    11 ]  Y4
    Y7  [ 7    10 ]  Y5
   GND  [ 8     9 ]  Y6
```

description

The 'ALS138 and 'AS138 circuits are designed to be used in high-performance memory-decoding or data-routing applications requiring very short propagation delay times. In high-performance memory systems these decoders can be used to minimize the effects of system decoding. When employed with high-speed memories utilizing a fast enable circuit, the delay times of these decoders and the enable time of the memory are usually less than the typical access time of the memory. This means that the effective system delay introduced by the Schottky-clamped system decoder is negligible.

SN54ALS138, SN54AS138 . . . FH PACKAGE
SN74ALS138, SN74AS138 . . . FN PACKAGE
(TOP VIEW)

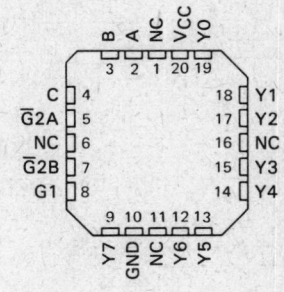

NC—No internal connection

The conditions at the binary select inputs and the three enable inputs select one of eight input lines. Two active-low and one active-high enable inputs reduce the need for external gates or inverters when expanding. A 24-line decoder can be implemented without external inverters and a 32-line decoder requires only one inverter. An enable input can be used as a data input for demultiplexing applications.

The SN54ALS138 and SN54AS138 are characterized for operation over the full military temperature range of $-55\,^{\circ}\text{C}$ to $125\,^{\circ}\text{C}$. The SN74ALS138 and SN74AS138 are characterized for operation from $0\,^{\circ}\text{C}$ to $70\,^{\circ}\text{C}$.

logic symbols (alternatives)

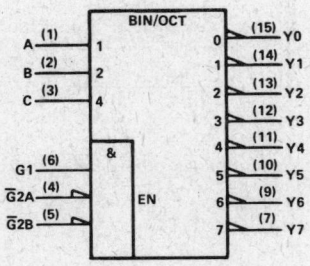
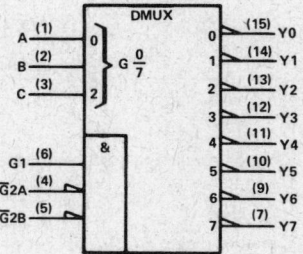

Pin numbers shown are for J and N packages.

Copyright © 1982 by Texas Instruments Incorporated

TYPES SN54ALS138, SN54AS138, SN74ALS138, SN74AS138
3-LINE TO 8-LINE DECODERS/DEMULTIPLEXERS

logic diagram (positive logic)

Pin numbers shown are for J and N packages.

FUNCTION TABLE

ENABLE INPUTS		SELECT INPUTS			OUTPUTS							
G1	$\overline{G2}$*	C	B	A	Y0	Y1	Y2	Y3	Y4	Y5	Y6	Y7
X	H	X	X	X	H	H	H	H	H	H	H	H
L	X	X	X	X	H	H	H	H	H	H	H	H
H	L	L	L	L	L	H	H	H	H	H	H	H
H	L	L	L	H	H	L	H	H	H	H	H	H
H	L	L	H	L	H	H	L	H	H	H	H	H
H	L	L	H	H	H	H	H	L	H	H	H	H
H	L	H	L	L	H	H	H	H	L	H	H	H
H	L	H	L	H	H	H	H	H	H	L	H	H
H	L	H	H	L	H	H	H	H	H	H	L	H
H	L	H	H	H	H	H	H	H	H	H	H	L

*$\overline{G2} = \overline{G2A} + \overline{G2B}$

absolute maximum ratings over operating free-air temperature range (unless otherwise noted)

Supply voltage, V_{CC} .... 7 V
Input voltage .... 7 V
Operating free-air temperature range: SN54ALS138, SN54AS138 .... −55°C to 125°C
                                             SN74ALS138, SN74AS138 .... 0°C to 70°C
Storage temperature range .... −65°C to 150°C

ALS AND AS CIRCUITS

TEXAS INSTRUMENTS
POST OFFICE BOX 225012 • DALLAS, TEXAS 75265

TYPES SN54ALS138, SN74ALS138
3-LINE TO 8-LINE DECODERS/DEMULTIPLEXERS

recommended operating conditions

		SN54ALS138			SN74ALS138			UNIT
		MIN	NOM	MAX	MIN	NOM	MAX	
V_{CC}	Supply voltage	4.5	5	5.5	4.5	5	5.5	V
V_{IH}	High-level input voltage	2			2			V
V_{IL}	Low-level input voltage			0.8			0.8	V
I_{OH}	High-level output current			−0.4			−0.4	mA
I_{OL}	Low-level output current			4			8	mA
T_A	Operating free-air temperature	−55		125	0		70	°C

electrical characteristics over recommended operating free-air temperature range (unless otherwise noted)

PARAMETER	TEST CONDITIONS		SN54ALS138			SN74ALS138			UNIT
			MIN	TYP†	MAX	MIN	TYP†	MAX	
V_{IK}	$V_{CC} = 4.5$ V,	$I_I = -18$ mA			−1.5			−1.5	V
V_{OH}	$V_{CC} = 4.5$ V to 5.5 V,	$I_{OH} = -0.4$ mA	$V_{CC}-2$			$V_{CC}-2$			V
V_{OL}	$V_{CC} = 4.5$ V,	$I_{OL} = 4$ mA		0.25	0.4		0.25	0.4	V
	$V_{CC} = 4.5$ V,	$I_{OL} = 8$ mA					0.35	0.5	
I_I	$V_{CC} = 5.5$ V,	$V_I = 7$ V			0.1			0.1	mA
I_{IH}	$V_{CC} = 5.5$ V,	$V_I = 2.7$ V			20			20	μA
I_{IL}	$V_{CC} = 5.5$ V,	$V_I = 0.4$ V			−0.1			−0.1	mA
I_O‡	$V_{CC} = 5.5$ V,	$V_O = 2.25$ V	−30		−112	−30		−112	mA
I_{CC}	$V_{CC} = 5.5$ V			5	10		5	10	mA

†All typical values are at $V_{CC} = 5$ V, $T_A = 25$°C.
‡The output conditions have been chosen to produce a current that closely approximates one half of the true short-circuit output current, I_{OS}.

switching characteristics (see Note 1)

PARAMETER	FROM (INPUT)	TO (OUTPUT)	$V_{CC} = 4.5$ V to 5.5 V, $C_L = 50$ pF, $R_L = 500$ Ω, T_A = MIN to MAX				UNIT
			SN54ALS138		SN74ALS138		
			MIN	MAX	MIN	MAX	
t_{PLH}	A, B, C	Any Y	6	27	6	22	ns
t_{PHL}			6	22	6	18	
t_{PLH}	Enable	Any Y	4	20	4	17	ns
t_{PHL}			5	20	5	17	

NOTE 1: For load circuit and voltage waveforms, see page 1-12.

TYPES SN54AS138, SN74AS138
3-LINE TO 8-LINE DECODERS/DEMULTIPLEXERS

recommended operating conditions

		SN54AS138			SN74AS138			UNIT
		MIN	NOM	MAX	MIN	NOM	MAX	
V_{CC}	Supply voltage	4.5	5	5.5	4.5	5	5.5	V
V_{IH}	High-level input voltage	2			2			V
V_{IL}	Low-level input voltage			0.8			0.8	V
I_{OH}	High-level output current			-2			-2	mA
I_{OL}	Low-level output current			20			20	mA
T_A	Operating free-air temperature	-55		125	0		70	°C

electrical characteristics over recommended operating free-air temperature range (unless otherwise noted)

PARAMETER	TEST CONDITIONS		SN54AS138			SN74AS138			UNIT
			MIN	TYP†	MAX	MIN	TYP†	MAX	
V_{IK}	$V_{CC} = 4.5$ V,	$I_I = -18$ mA							V
V_{OH}	$V_{CC} = 4.5$ V to 5.5 V,	$I_{OH} = -2$ mA	$V_{CC}-2$			$V_{CC}-2$			V
V_{OL}	$V_{CC} = 4.5$ V,	$I_{OL} = 20$ mA		0.35	0.5		0.35	0.5	V
I_I	$V_{CC} = 5.5$ V,	$V_I = 7$ V							mA
I_{IH}	$V_{CC} = 5.5$ V,	$V_I = 2.7$ V							µA
I_{IL}	$V_{CC} = 5.5$ V,	$V_I = 0.4$ V							mA
I_O‡	$V_{CC} = 5.5$ V,	$V_O = 2.25$ V	-30		-112	-30		-112	mA
I_{CC}	$V_{CC} = 5.5$ V			13			13		mA

†All typical values are at $V_{CC} = 5$ V, $T_A = 25$°C.
‡The output conditions have been chosen to produce a current that closely approximates one half of the true short-circuit output current, I_{OS}.

switching characteristics (see Note 1)

PARAMETER	FROM (INPUT)	TO (OUTPUT)	$V_{CC} = 4.5$ V to 5.5 V, $C_L = 50$ pF, $R_L = 500$ Ω, T_A = MIN to MAX						UNIT
			SN54AS138			SN74AS138			
			MIN	TYP†	MAX	MIN	TYP†	MAX	
t_{PLH}	A, B, C	Any Y		5.6			5.6		ns
t_{PHL}				6.1			6.1		
t_{PLH}	Enable	Any Y		5.8			5.8		ns
t_{PHL}				5.5			5.5		

†All typical values are at $V_{CC} = 5$ V, $T_A = 25$°C.
NOTE 1: For load circuit and voltage waveforms, see page 1-12.

PRODUCT PREVIEW
This page contains information on a product under development. Texas Instruments reserves the right to change or discontinue this product without notice.

TEXAS INSTRUMENTS
POST OFFICE BOX 225012 • DALLAS, TEXAS 75265

TYPES SN54ALS139, SN54AS139, SN74ALS139, SN74AS139
DUAL 2-LINE TO 4-LINE DECODERS/DEMULTIPLEXERS

D2661, APRIL 1982 – REVISED DECEMBER 1983

- Designed Specifically for High-Speed Memory Decoders and Data Transmission Systems
- Incorporates 2 Enable Inputs to Simplify Cascading and/or Data Reception
- Package Options Include Both Plastic and Ceramic Chip Carriers in Addition to Plastic and Ceramic DIPs
- Dependable Texas Instruments Quality and Reliability

SN54ALS139, SN54AS139 . . . J PACKAGE
SN74ALS139, SN74AS139 . . . N PACKAGE
(TOP VIEW)

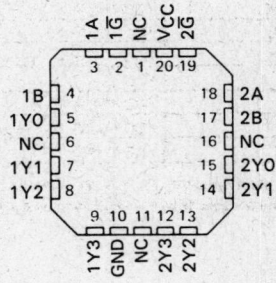

SN54ALS139, SN54AS139 . . . FH PACKAGE
SN74ALS139, SN74AS139 . . . FN PACKAGE
(TOP VIEW)

NC — No internal connection

description

The 'ALS139 and 'AS139 circuits are designed to be used in high-performance memory-decoding or data-routing applications requiring very short propagation delay times. In high-performance memory systems, these decoders can be used to minimize the effects of system decoding. When employed with high-speed memories utilizing a fast-enable circuit, the delay times of these decoders and the enable time of the memory are usually less than the typical access time of the memory. This means that the effective system delay introduced by the Schottky-clamped system decoder is negligible.

The 'ALS139 and 'AS139 are comprised of two individual two-line to four-line decoders in a single package. The active-low enable input can be used as a data line in demultiplexing applications. These decoders/demultiplexers feature fully buffered inputs, each of which represents only one normalized load to its driving circuit. All inputs are clamped with high-performance Schottky diodes to suppress line-ringing and simplify system design.

The SN54ALS139 and SN54AS139 are characterized for operation over the full military temperature range of −55°C to 125°C. The SN74ALS139 and SN74AS139 are characterized for operation from 0°C to 70°C.

logic symbols (alternatives)

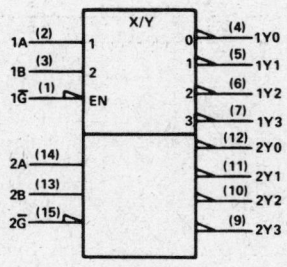

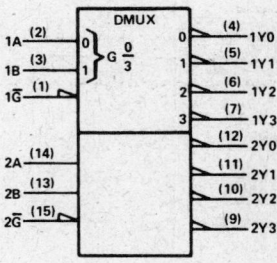

Pin numbers shown are for J and N packages.

PRODUCT PREVIEW

This document contains information on a product under development. Texas Instruments reserves the right to change or discontinue this product without notice.

Copyright © 1982 by Texas Instruments Incorporated

TEXAS INSTRUMENTS
POST OFFICE BOX 225012 • DALLAS, TEXAS 75265

TYPES SN54ALS139, SN54AS139, SN74ALS139, SN74AS139
DUAL 2-LINE TO 4-LINE DECODERS/DEMULTIPLEXERS

FUNCTION TABLE

INPUTS			OUTPUTS			
ENABLE	SELECT					
$\overline{G}$	B	A	Y0	Y1	Y2	Y3
H	X	X	H	H	H	H
L	L	L	L	H	H	H
L	L	H	H	L	H	H
L	H	L	H	H	L	H
L	H	H	H	H	H	L

functional block diagram (positive logic)

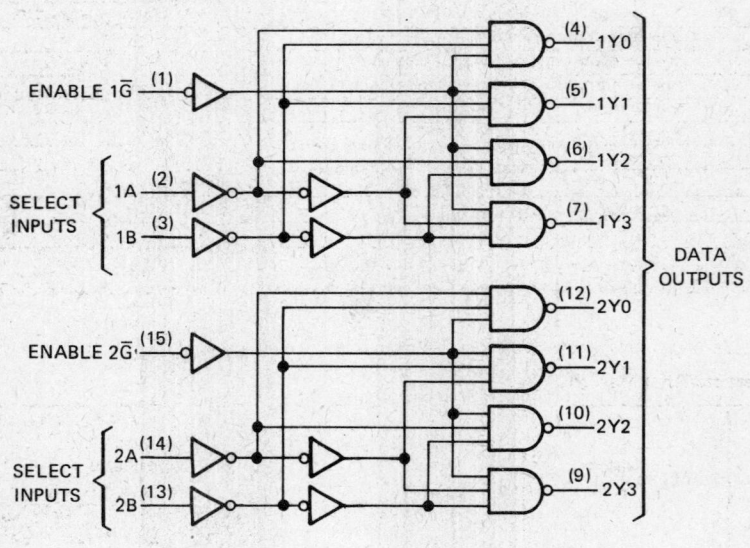

Pin numbers shown are for J and N packages.

absolute maximum ratings over operating free-air temperature range (unless otherwise noted)

Supply voltage, V_{CC} .... 7 V
Input voltage .... 7 V
Operating free-air temperature range: SN54ALS139, SN54AS139 .... −55°C to 125°C
　　　　　　　　　　　　　　　　　　　 SN74ALS139, SN74AS139 .... 0°C to 70°C
Storage temperature range .... −65°C to 150°C

TEXAS INSTRUMENTS
POST OFFICE BOX 225012 • DALLAS, TEXAS 75265

ALS AND AS CIRCUITS

TYPES SN54ALS139, SN74ALS139
DUAL 2-LINE TO 4-LINE DECODERS/DEMULTIPLEXERS

recommended operating conditions

		SN54ALS139			SN74ALS139			UNIT
		MIN	NOM	MAX	MIN	NOM	MAX	
V_{CC}	Supply voltage	4.5	5	5.5	4.5	5	5.5	V
V_{IH}	High-level input voltage	2			2			V
V_{IL}	Low-level input voltage			0.8			0.8	V
I_{OH}	High-level output current			−0.4			−0.4	mA
I_{OL}	Low-level output current			4			8	mA
T_A	Operating free-air temperature	−55		125	0		70	°C

electrical characteristics over recommended operating free-air temperature range (unless otherwise noted)

PARAMETER	TEST CONDITIONS		SN54ALS139			SN74ALS139			UNIT
			MIN	TYP[†]	MAX	MIN	TYP[†]	MAX	
V_{IK}	$V_{CC} = 4.5$ V,	$I_I = -18$ mA			−1.5			−1.5	V
V_{OH}	$V_{CC} = 4.5$ V to 5.5 V,	$I_{OH} = -0.4$ mA	$V_{CC}-2$			$V_{CC}-2$			V
V_{OL}	$V_{CC} = 4.5$ V,	$I_{OL} = 4$ mA		0.25	0.4		0.25	0.4	V
	$V_{CC} = 4.5$ V,	$I_{OL} = 8$ mA					0.35	0.5	
I_I	$V_{CC} = 5.5$ V,	$V_I = 7$ V			0.1			0.1	mA
I_{IH}	$V_{CC} = 5.5$ V,	$V_I = 2.7$ V			20			20	μA
I_{IL}	$V_{CC} = 5.5$ V,	$V_I = 0.4$ V			−0.1			−0.1	mA
I_O[‡]	$V_{CC} = 5.5$ V,	$V_O = 2.25$ V	−30		−112	−30		−112	mA
I_{CC}	$V_{CC} = 5.5$ V			4.5			4.5		mA

[†]All typical values are at $V_{CC} = 5$ V, $T_A = 25$ °C.
[‡]The output conditions have been chosen to produce a current that closely approximates one half of the true short-circuit output current, I_{OS}.

switching characteristics (see Note 1)

PARAMETER	FROM (INPUT)	TO (OUTPUT)	$V_{CC} = 4.5$ V to 5.5 V, $C_L = 50$ pF, $R_L = 500$ Ω, T_A = MIN to MAX						UNIT
			SN54ALS139			SN74ALS139			
			MIN	TYP[†]	MAX	MIN	TYP[†]	MAX	
t_{PLH}	A or B	Y		10			10		ns
t_{PHL}				10			10		
t_{PLH}	$\overline{G}$	Y		8			8		ns
t_{PHL}				8			8		

[†]All typical values are at $V_{CC} = 5$ V, $T_A = 25$ °C.
NOTE 1: For load circuit and voltage waveforms, see page 1-12.

Additional information on these products can be obtained from the factory as it becomes available.

ALS AND AS CIRCUITS

TEXAS INSTRUMENTS
POST OFFICE BOX 225012 • DALLAS, TEXAS 75265

TYPES SN54AS139, SN74AS139
DUAL 2-LINE TO 4-LINE DECODERS/DEMULTIPLEXERS

recommended operating conditions

		SN54AS139			SN74AS139			UNIT
		MIN	NOM	MAX	MIN	NOM	MAX	
V_{CC}	Supply voltage	4.5	5	5.5	4.5	5	5.5	V
V_{IH}	High-level input voltage	2			2			V
V_{IL}	Low-level input voltage			0.8			0.8	V
I_{OH}	High-level output current			−2			−2	mA
I_{OL}	Low-level output current			20			20	mA
T_A	Operating free-air temperature	−55		125	0		70	°C

electrical characteristics over recommended operating free-air temperature range (unless otherwise noted)

PARAMETER	TEST CONDITIONS	SN54AS139			SN74AS139			UNIT
		MIN	TYP†	MAX	MIN	TYP†	MAX	
V_{IK}	V_{CC} = 4.5 V, I_I = −18 mA			−1.2			−1.2	V
V_{OH}	V_{CC} = 4.5 V to 5.5 V, I_{OH} = −2 mA	V_{CC}−2			V_{CC}−2			V
V_{OL}	V_{CC} = 4.5 V, I_{OL} = 20 mA		0.35	0.5		0.35	0.5	V
I_I	V_{CC} = 5.5 V, V_I = 7 V			0.1			0.1	mA
I_{IH}	V_{CC} = 5.5 V, V_I = 2.7 V			20			20	μA
I_{IL}	V_{CC} = 5.5 V, V_I = 0.4 V			−0.5			−0.5	mA
I_O‡	V_{CC} = 5.5 V, V_O = 2.25 V	−30		−112	−30		−112	mA
I_{CC}	V_{CC} = 5.5 V		13			13		mA

†All typical values are at V_{CC} = 5 V, T_A = 25°C.
‡The output conditions have been chosen to produce a current that closely approximates one half of the true short-circuit output current, I_{OS}.

switching characteristics (see Note 1)

PARAMETER	FROM (INPUT)	TO (OUTPUT)	V_{CC} = 4.5 V to 5.5 V, C_L = 50 pF, R_L = 500 Ω, T_A = MIN to MAX						UNIT
			SN54AS139			SN74AS139			
			MIN	TYP†	MAX	MIN	TYP†	MAX	
t_{PLH}	A or B	Y		5.5			5.5		ns
t_{PHL}				6			6		
t_{PLH}	$\overline{G}$	Y		5.5			5.5		ns
t_{PHL}				5			5		

†All typical values are at V_{CC} = 5 V, T_A = 25°C.
NOTE 1: For load circuit and voltage waveforms, see page 1-12.

Additional information on these products can be obtained from the factory as it becomes available.

TYPES SN54ALS151, SN54AS151, SN74ALS151, SN74AS151
1 OF 8 DATA SELECTORS/MULTIPLEXERS

D2661, APRIL 1982—REVISED DECEMBER 1983

- **8-Line to 1-Line Multiplexers Can Perform As:**
  - Boolean Function Generators
  - Parallel-to-Serial Converters
  - Data Source Selectors
- **Input Clamping Diodes Simplify System Design**
- **Fully Compatible With Most TTL Circuits**
- **Package Options Include Both Plastic and Ceramic Chip Carriers in Addition to Plastic and Ceramic DIPs**
- **Dependable Texas Instruments Quality and Reliability**

SN54ALS151, SN54AS151 . . . J PACKAGE
SN74ALS151, SN74AS151 . . . N PACKAGE
(TOP VIEW)

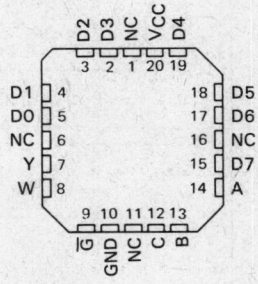

SN54ALS151, SN54AS151 . . . FH PACKAGE
SN74ALS151, SN74AS151 . . . FN PACKAGE
(TOP VIEW)

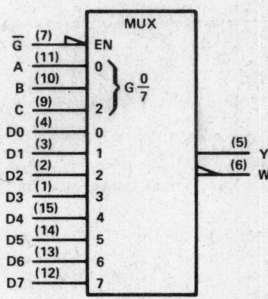

NC — No internal connection

description

These monolithic data selectors/multiplexers provide full binary decoding to select one of eight data sources. The strobe input ($\overline{G}$) must be at a low logic level to enable the inputs. A high level at the strobe terminal forces the W output high and the Y output low.

The SN54ALS151 and SN54AS151 are characterized for operation over the full military temperature range of $-55\,°C$ to $125\,°C$. The SN74ALS151 and SN74AS151 are characterized for operation from $0\,°C$ to $70\,°C$.

FUNCTION TABLE

INPUTS			OUTPUTS	
SELECT			STROBE	
C	B	A	$\overline{G}$	Y	W
X	X	X	H	L	H
L	L	L	L	D0	$\overline{D0}$
L	L	H	L	D1	$\overline{D1}$
L	H	L	L	D2	$\overline{D2}$
L	H	H	L	D3	$\overline{D3}$
H	L	L	L	D4	$\overline{D4}$
H	L	H	L	D5	$\overline{D5}$
H	H	L	L	D6	$\overline{D6}$
H	H	H	L	D7	$\overline{D7}$

H = high level, L = low level, X = irrelevant
D0, D1 . . . D7 = the level of the D respective input

logic symbol

```
             MUX
  G  (7)  ─►EN
  A  (11) ─┐ 0
  B  (10) ─┤G 
  C  (9)  ─┘ 7
  D0 (4)  ─ 0
  D1 (3)  ─ 1        (5)
  D2 (2)  ─ 2        ────► Y
  D3 (1)  ─ 3        (6)
  D4 (15) ─ 4        ────► W
  D5 (14) ─ 5
  D6 (13) ─ 6
  D7 (12) ─ 7
```

Pin numbers shown are for J and N packages.

Copyright © 1982 by Texas Instruments Incorporated

TEXAS INSTRUMENTS
POST OFFICE BOX 225012 • DALLAS, TEXAS 75265

TYPES SN54ALS151, SN54AS151, SN74ALS151, SN74AS151
1 OF 8 DATA SELECTORS/MULTIPLEXERS

logic diagram (positive logic)

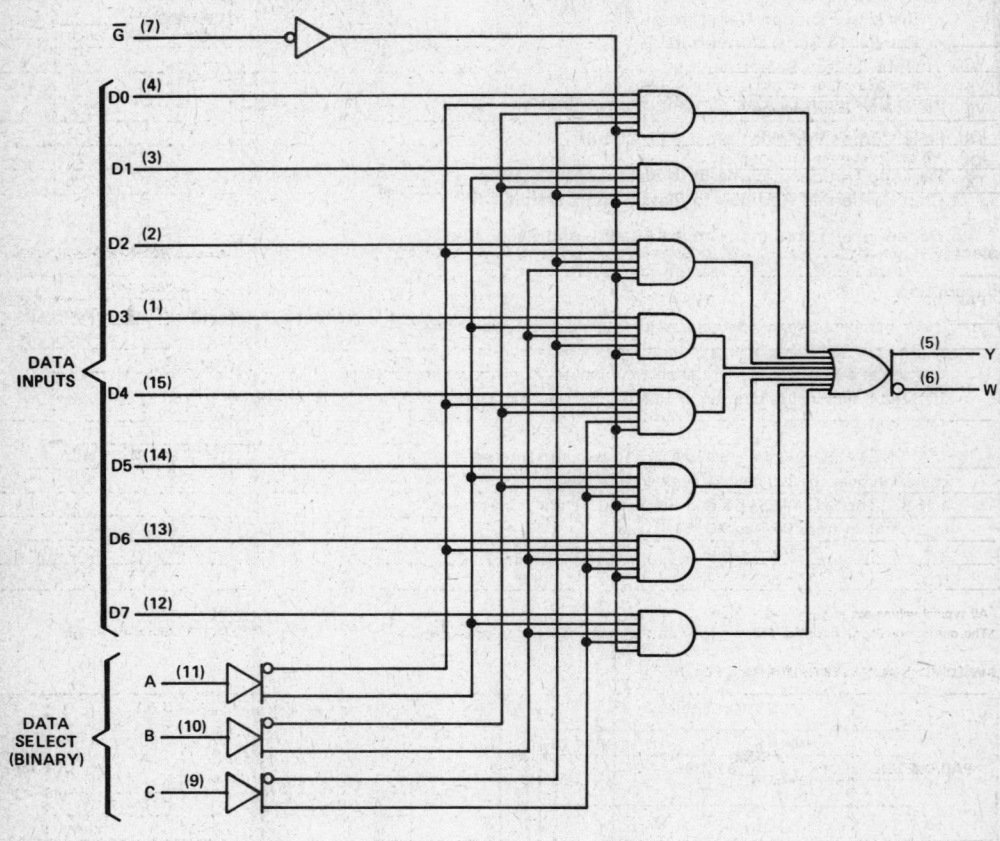

Pin numbers shown are for J and N packages.

absolute maximum ratings over operating free-air temperature range (unless otherwise noted)

Supply voltage, V_{CC} .... 7 V
Input voltage .... 7 V
Operating free-air temperature range: SN54ALS151, SN54AS151 .... $-55\,°C$ to $125\,°C$
  SN74ALS151, SN74AS151 .... $0\,°C$ to $70\,°C$
Storage temperature range .... $-65\,°C$ to $150\,°C$

Texas Instruments
POST OFFICE BOX 225012 • DALLAS, TEXAS 75265

TYPES SN54ALS151, SN74ALS151
1 OF 8 DATA SELECTORS/MULTIPLEXERS

recommended operating conditions

		SN54ALS151			SN74ALS151			UNIT
		MIN	NOM	MAX	MIN	NOM	MAX	
V_{CC}	Supply voltage	4.5	5	5.5	4.5	5	5.5	V
V_{IH}	High-level input voltage	2			2			V
V_{IL}	Low-level input voltage			0.8			0.8	V
I_{OH}	High-level output current			−1			−2.6	mA
I_{OL}	Low-level output current			12			24	mA
T_A	Operating free-air temperature	−55		125	0		70	°C

electrical characteristics over recommended operating free-air temperature range (unless otherwise noted)

PARAMETER	TEST CONDITIONS		SN54ALS151			SN74ALS151			UNIT
			MIN	TYP†	MAX	MIN	TYP†	MAX	
V_{IK}	V_{CC} = 4.5 V,	I_I = −18 mA			−1.5			−1.5	V
V_{OH}	V_{CC} = 4.5 V to 5.5 V,	I_{OH} = −0.4 mA	V_{CC}−2			V_{CC}−2			V
	V_{CC} = 4.5 V,	I_{OH} = −1 mA	2.4	3.3					
	V_{CC} = 4.5 V,	I_{OH} = −2.6 mA				2.4	3.2		
V_{OL}	V_{CC} = 4.5 V,	I_{OL} = 12 mA		0.25	0.4		0.25	0.4	V
	V_{CC} = 4.5 V	I_{OL} = 24 mA					0.35	0.5	
I_I	V_{CC} = 5.5 V,	V_I = 7 V			0.1			0.1	mA
I_{IH}	V_{CC} = 5.5 V,	V_I = 2.7 V			20			20	µA
I_{IL}	V_{CC} = 5.5 V,	V_I = 0.4 V			−0.1			−0.1	mA
I_O‡	V_{CC} = 5.5 V,	V_O = 2.25 V	−30		−112	−30		−112	mA
I_{CC}	V_{CC} = 5.5 V,	Inputs at 4.5 V		7.5	12		7.5	12	mA

†All typical values are at V_{CC} = 5 V, T_A = 25°C.
‡The output conditions have been chosen to produce a current that closely approximates one half of the true short-circuit output current, I_{OS}.

switching characteristics (see Note 1)

PARAMETER	FROM (INPUT)	TO (OUTPUT)	V_{CC} = 4.5 V to 5.5 V, C_L = 50 pF, R_L = 500 Ω, T_A = MIN to MAX				UNIT
			SN54ALS151		SN74ALS151		
			MIN	MAX	MIN	MAX	
t_{PLH}	A, B, or C	Y	4	21	4	18	ns
t_{PHL}			8	28	8	24	
t_{PLH}	A, B, or C	W	7	28	7	24	ns
t_{PHL}			7	26	7	23	
t_{PLH}	Any D	Y	3	12	3	10	ns
t_{PHL}			5	18	5	15	
t_{PLH}	Any D	W	3	18	3	15	ns
t_{PHL}			4	18	4	15	
t_{PLH}	$\overline{G}$	Y	4	21	4	18	ns
t_{PHL}			4	23	4	19	
t_{PLH}	$\overline{G}$	W	5	23	5	19	ns
t_{PHL}			5	26	5	23	

NOTE 1: For load circuit and voltage waveforms, see page 1-12.

TEXAS INSTRUMENTS
POST OFFICE BOX 225012 • DALLAS, TEXAS 75265

TYPES SN54AS151, SN74AS151
1 OF 8 DATA SELECTORS/MULTIPLEXERS

recommended operating conditions

		SN54AS151 MIN	NOM	MAX	SN74AS151 MIN	NOM	MAX	UNIT
V_{CC}	Supply voltage	4.5	5	5.5	4.5	5	5.5	V
V_{IH}	High-level input voltage	2			2			V
V_{IL}	Low-level input voltage			0.8			0.8	V
I_{OH}	High-level output current			−12			−15	mA
I_{OL}	Low-level output current			32			−48	mA
T_A	Operating free-air temperature	−55		125	0		70	°C

electrical characteristics over recommended operating free-air temperature range (unless otherwise noted)

PARAMETER		TEST CONDITIONS		SN54AS151 MIN	TYP†	MAX	SN74AS151 MIN	TYP†	MAX	UNIT
V_{IK}		V_{CC} = 4.5 V,	I_I = −18 mA			−1.2			−1.2	V
V_{OH}		V_{CC} = 4.5 V to 5.5 V,	I_{OH} = −2 mA	V_{CC}−2			V_{CC}−2			V
		V_{CC} = 4.5 V,	I_{OH} = −12 mA	2.4	3.2					
		V_{CC} = 4.5 V,	I_{OH} = −15 mA				2.4	3.3		
V_{OL}		V_{CC} = 4.5 V,	I_{OL} = 32 mA		0.25	0.5				V
		V_{CC} = 4.5 V	I_{OL} = 48 mA					0.35	0.5	
I_I	A, B, or C	V_{CC} = 5.5 V,	V_I = 7 V			0.2			0.2	mA
	All others					0.1			0.1	
I_{IH}	A, B, or C	V_{CC} = 5.5 V,	V_I = 2.7 V			40			40	μA
	All others					20			20	
I_{IL}	A, B, or C	V_{CC} = 5.5 V,	V_I = 0.4 V			−0.6			−0.6	mA
	All others					−0.3			−0.3	
I_O‡		V_{CC} = 5.5 V,	V_O = 2.25 V	−30		−112	−30		−112	mA
I_{CC}		V_{CC} = 5.5 V,			26			26		mA

†All typical values are at V_{CC} = 5 V, T_A = 25°C.
‡The output conditions have been chosen to produce a current that closely approximates one half of the true short-circuit output current, I_{OS}.

switching characteristics (see Note 1)

V_{CC} = 4.5 V to 5.5 V, C_L = 50 pF, R_L = 500 Ω, T_A = MIN to MAX

PARAMETER	FROM (INPUT)	TO (OUTPUT)	SN54AS151 MIN	TYP†	MAX	SN74AS151 MIN	TYP†	MAX	UNIT
t_{PLH}	A, B, or C	Y		5			5		ns
t_{PHL}				5			5		
t_{PLH}	A, B, or C	W		4.5			4.5		ns
t_{PHL}				4.5			4.5		
t_{PLH}	Any D	Y		3			3		ns
t_{PHL}				4			4		
t_{PLH}	Any D	W		3			3		ns
t_{PHL}				2.5			2.5		
t_{PLH}	$\overline{G}$	Y		5			5		ns
t_{PHL}				5			5		
t_{PLH}	$\overline{G}$	W		4.5			4.5		ns
t_{PHL}				4.5			4.5		

†All typical values are at V_{CC} = 5 V, T_A = 25°C.
NOTE 1: For load circuit and voltage waveforms, see page 1-12.

PRODUCT PREVIEW
This page contains information on a product under development. Texas Instruments reserves the right to change or discontinue this product without notice.

TEXAS INSTRUMENTS
POST OFFICE BOX 225012 • DALLAS, TEXAS 75265

TYPES SN54ALS153, SN54AS153, SN74ALS153, SN74AS153
DUAL 1 OF 4 DATA SELECTORS/MULTIPLEXERS

D2661, APRIL 1982—REVISED DECEMBER 1983

- Permits Multiplexing from N Lines to 1 Line
- Performs Parallel-to-Serial Conversion
- Strobe (Enable) Line Provided for Cascading (N lines to n lines)
- Fully Compatible with Most TTL Circuits
- 'ALS253 and 'AS253 Are 3-State Versions of These Parts
- Package Options Include Both Plastic and Ceramic Chip Carriers in Addition to Plastic and Ceramic DIPs
- Dependable Texas Instruments Quality and Reliability

SN54ALS153, SN54AS153 . . . J PACKAGE
SN74ALS153, SN74AS153 . . . N PACKAGE
(TOP VIEW)

```
1G   [ 1    16 ] VCC
B    [ 2    15 ] 2G
1C3  [ 3    14 ] A
1C2  [ 4    13 ] 2C3
1C1  [ 5    12 ] 2C2
1C0  [ 6    11 ] 2C1
1Y   [ 7    10 ] 2C0
GND  [ 8     9 ] 2Y
```

description

Each of these data selectors/multiplexers contains inverters and drivers to supply full binary decoding data selection to the AND-OR gates. Separate strobe inputs (G) are provided for each of the two four-line sections.

The SN54ALS153 and SN54AS153 are characterized for operation over the full military temperature range of −55°C to 125°C. The SN74ALS153 and SN74AS153 are characterized for operation from 0°C to 70°C.

SN54ALS153, SN54AS153 . . . FH PACKAGE
SN74ALS153, SN74AS153 . . . FN PACKAGE
(TOP VIEW)

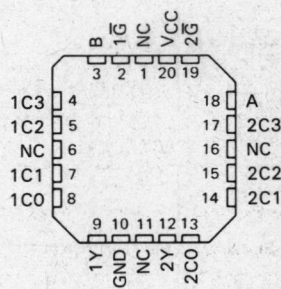

NC — No internal connection

logic symbol

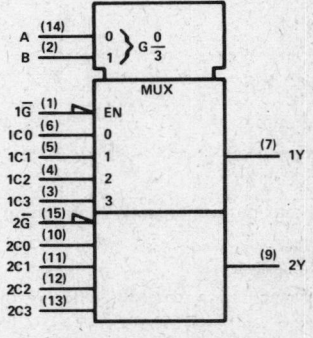

Pin numbers shown are for J and N packages.

FUNCTION TABLE

SELECT INPUTS		DATA INPUTS				STROBE	OUTPUT
B	A	C0	C1	C2	C3	G	Y
X	X	X	X	X	X	H	L
L	L	L	X	X	X	L	L
L	L	H	X	X	X	L	H
L	H	X	L	X	X	L	L
L	H	X	H	X	X	L	H
H	L	X	X	L	X	L	L
H	L	X	X	H	X	L	H
H	H	X	X	X	L	L	L
H	H	X	X	X	H	L	H

Select inputs A and B are common to both sections.

Copyright © 1982 by Texas Instruments Incorporated

TYPES SN54ALS153, SN54AS153, SN74ALS153, SN74AS153
DUAL 1 OF 4 DATA SELECTORS/MULTIPLEXERS

logic diagram (positive logic)

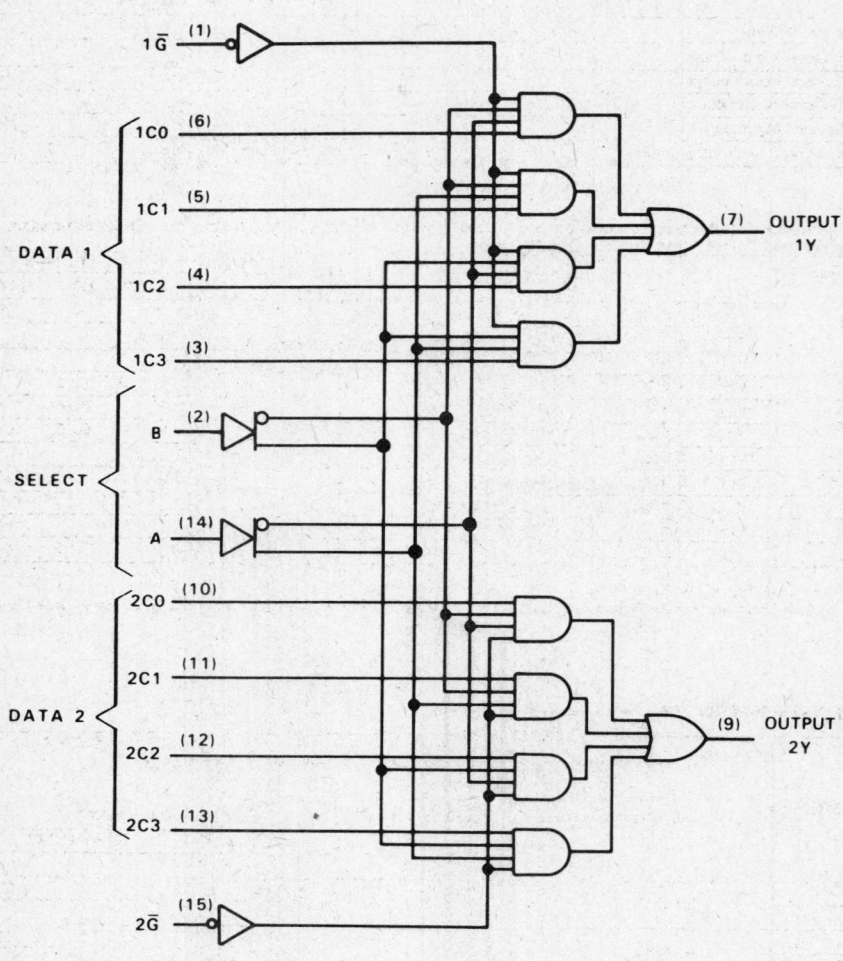

Pin numbers shown are for J and N packages.

absolute maximum ratings over operating free-air temperature range (unless otherwise noted)

Supply voltage, V_{CC} .... 7 V
Input voltage .... 7 V
Operating free-air temperature range: SN54ALS153, SN54AS153 .... −55 °C to 125 °C
                                       SN74ALS153, SN74AS153 .... 0 °C to 70 °C
Storage temperature range .... −65 °C to 150 °C

TEXAS INSTRUMENTS
POST OFFICE BOX 225012 • DALLAS, TEXAS 75265

TYPES SN54ALS153, SN74ALS153
DUAL 1 OF 4 DATA SELECTORS/MULTIPLEXERS

recommended operating conditions

		SN54ALS153			SN74ALS153			UNIT
		MIN	NOM	MAX	MIN	NOM	MAX	
V_{CC}	Supply voltage	4.5	5	5.5	4.5	5	5.5	V
V_{IH}	High-level input voltage	2			2			V
V_{IL}	Low-level input voltage			0.8			0.8	V
I_{OH}	High-level output current			−1			−2.6	mA
I_{OL}	Low-level output current			12			24	mA
T_A	Operating free-air temperature	−55		125	0		70	°C

electrical characteristics over recommended operating free-air temperature range (unless otherwise noted)

PARAMETER	TEST CONDITIONS		SN54ALS153			SN74ALS153			UNIT
			MIN	TYP†	MAX	MIN	TYP†	MAX	
V_{IK}	V_{CC} = 4.5 V,	I_I = −18 mA			−1.5			−1.5	V
V_{OH}	V_{CC} = 4.5 V to 5.5 V,	I_{OH} = −0.4 mA	V_{CC}−2			V_{CC}−2			V
	V_{CC} = 4.5 V,	I_{OH} = −1 mA	2.4	3.3					
	V_{CC} = 4.5 V,	I_{OH} = −2.6 mA				2.4	3.2		
V_{OL}	V_{CC} = 4.5 V,	I_{OL} = 12 mA		0.25	0.4		0.25	0.4	V
	V_{CC} = 4.5 V	I_{OL} = 24 mA					0.35	0.5	
I_I	V_{CC} = 5.5 V,	V_I = 7 V			0.1			0.1	mA
I_{IH}	V_{CC} = 5.5 V,	V_I = 2.7 V			20			20	µA
I_{IL}	V_{CC} = 5.5 V,	V_I = 0.4 V			−0.1			−0.1	mA
I_O‡	V_{CC} = 5.5 V,	V_O = 2.25 V	−30		−112	−30		−112	mA
I_{CC}	V_{CC} = 5.5 V,	All inputs at 4.5 V		7.5	14		7.5	14	mA

†All typical values are at V_{CC} = 5 V, T_A = 25°C.
‡The output conditions have been chosen to produce a current that closely approximates one half of the true short-circuit output current, I_{OS}.

switching characteristics (see Note 1)

PARAMETER	FROM (INPUT)	TO (OUTPUT)	V_{CC} = 4.5 V to 5.5 V, C_L = 50 pF, R_L = 500 Ω, T_A = MIN to MAX				UNIT
			SN54ALS153		SN74ALS153		
			MIN	MAX	MIN	MAX	
t_{PLH}	A or B	Y	5	25	5	21	ns
t_{PHL}			5	25	5	21	
t_{PLH}	Data (Any C)	Y	3	12	3	10	ns
t_{PHL}			4	18	4	15	
t_{PLH}	G	Y	5	22	5	18	ns
t_{PHL}			5	22	5	18	

NOTE 1: For load circuit and voltage waveforms, see page 1-12.

TYPES SN54AS153, SN74AS153
DUAL 1 OF 4 DATA SELECTORS/MULTIPLEXERS

recommended operating conditions

		SN54AS153			SN74AS153			UNIT
		MIN	NOM	MAX	MIN	NOM	MAX	
V_{CC}	Supply voltage	4.5	5	5.5	4.5	5	5.5	V
V_{IH}	High-level input voltage	2			2			V
V_{IL}	Low-level input voltage			0.8			0.8	V
I_{OH}	High-level output current			−12			−15	mA
I_{OL}	Low-level output current			32			48	mA
T_A	Operating free-air temperature	−55		125	0		70	°C

electrical characteristics over recommended operating free-air temperature range (unless otherwise noted)

PARAMETER		TEST CONDITIONS		SN54AS153			SN74AS153			UNIT
				MIN	TYP[†]	MAX	MIN	TYP[†]	MAX	
V_{IK}		V_{CC} = 4.5 V,	I_I = −18 mA			−1.2			−1.2	V
V_{OH}		V_{CC} = 4.5 V to 5.5 V,	I_{OH} = −2 mA	V_{CC}−2			V_{CC}−2			
		V_{CC} = 4.5 V,	I_{OH} = −12 mA	2.4	3.2					V
		V_{CC} = 4.5 V,	I_{OH} = −15 mA				2.4	3.3		
V_{OL}		V_{CC} = 4.5 V,	I_{OL} = 32 mA		0.25	0.5				V
		V_{CC} = 4.5 V	I_{OL} = 48 mA					0.35	0.5	
I_I	A, B	V_{CC} = 5.5 V,	V_I = 7 V			0.2			0.2	mA
	All others					0.1			0.1	
I_{IH}	A, B	V_{CC} = 5.5 V,	V_I = 2.7 V			40			40	μA
	All others					20			20	
I_{IL}	A, B	V_{CC} = 5.5 V,	V_I = 0.4 V			−1			−1	mA
	All others					−0.5			−0.5	
I_O[‡]		V_{CC} = 5.5 V,	V_O = 2.25 V	−30		−112	−30		−112	mA
I_{CC}		V_{CC} = 5.5 V,	Outputs high		16	26		16	26	mA
			Outputs low		21	33		21	33	

[†] All typical values are at V_{CC} = 5 V, T_A = 25°C.
[‡] The output conditions have been chosen to produce a current that closely approximates one half of the true short-circuit output current, I_{OS}.

switching characteristics (see Note 1)

PARAMETER	FROM (INPUT)	TO (OUTPUT)	V_{CC} = 4.5 V to 5.5 V, C_L = 50 pF, R_L = 500 Ω, T_A = MIN to MAX				UNIT
			SN54AS153		SN74AS153		
			MIN	MAX	MIN	MAX	
t_{PLH}	A or B	Y	3	14	3	12.5	ns
t_{PHL}			3	12.5	3	11	
t_{PLH}	Data (Any C)	Y	2	8	2	7	ns
t_{PHL}			2	8.5	2	8	
t_{PLH}	$\overline{G}$	Y	3	13	3	11.5	ns
t_{PHL}			2	10	2	9	

NOTE 1: For load circuit and voltage waveforms, see page 1-12.

Texas Instruments
POST OFFICE BOX 225012 • DALLAS, TEXAS 75265

TYPES SN54ALS157, SN54ALS158, SN54AS157, SN54AS158, SN74ALS157, SN74ALS158, SN74AS157, SN74AS158
QUADRUPLE 1 OF 2 DATA SELECTORS/MULTIPLEXERS

D2661, APRIL 1982—REVISED DECEMBER 1983

- Buffered Inputs and Outputs
- Package Options Include Both Plastic and Ceramic Chip Carriers in Addition to Plastic and Ceramic DIPs
- Dependable Texas Instruments Quality and Reliability

description

These monolithic data selectors/multiplexers contain inverters and drivers to supply full data selection to the four output gates. A separate strobe input ($\overline{G}$) is provided. A 4-bit word is selected from one of two sources and is routed to the four outputs. The 'ALS157 and 'AS157 present true data whereas the 'ALS158 and 'AS158 present inverted data to minimize propagation delay time.

The SN54' family is characterized for operation over the full military temperature range $-55\,°C$ to $125\,°C$. The SN74' family is characterized for operation from $0\,°C$ to $70\,°C$.

SN54ALS', SN54AS' . . . J PACKAGE
SN74ALS', SN74AS' . . . N PACKAGE
(TOP VIEW)

```
 $\overline{A/B}$ [ 1   16 ] V_CC
 1A    [ 2   15 ] $\overline{G}$
 1B    [ 3   14 ] 4A
 1Y    [ 4   13 ] 4B
 2A    [ 5   12 ] 4Y
 2B    [ 6   11 ] 3A
 2Y    [ 7   10 ] 3B
 GND   [ 8    9 ] 3Y
```

SN54ALS', SN54AS' . . . FH PACKAGE
SN74ALS', SN74AS' . . . FN PACKAGE
(TOP VIEW)

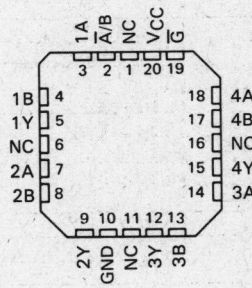

NC — No internal connection

logic symbols

'ALS157, 'AS157

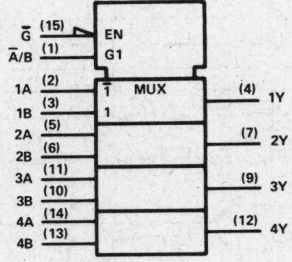

'ALS158, 'AS158

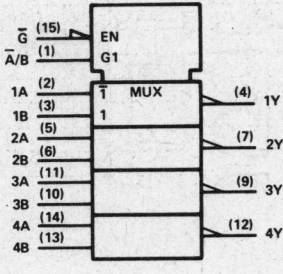

Pin numbers shown are for J and N packages.

FUNCTION TABLE

INPUTS				OUTPUT Y	
STROBE	SELECT	DATA		'ALS157	'ALS158
$\overline{G}$	$\overline{A/B}$	A	B	'AS157	'AS158
H	X	X	X	L	H
L	L	L	X	L	H
L	L	H	X	H	L
L	H	X	L	L	H
L	H	X	H	H	L

Copyright © 1982 by Texas Instruments Incorporated

TYPES SN54ALS157, SN54ALS158, SN74ALS157, SN74ALS158
QUADRUPLE 1 OF 2 DATA SELECTORS/MULTIPLEXERS

logic diagrams (positive logic)

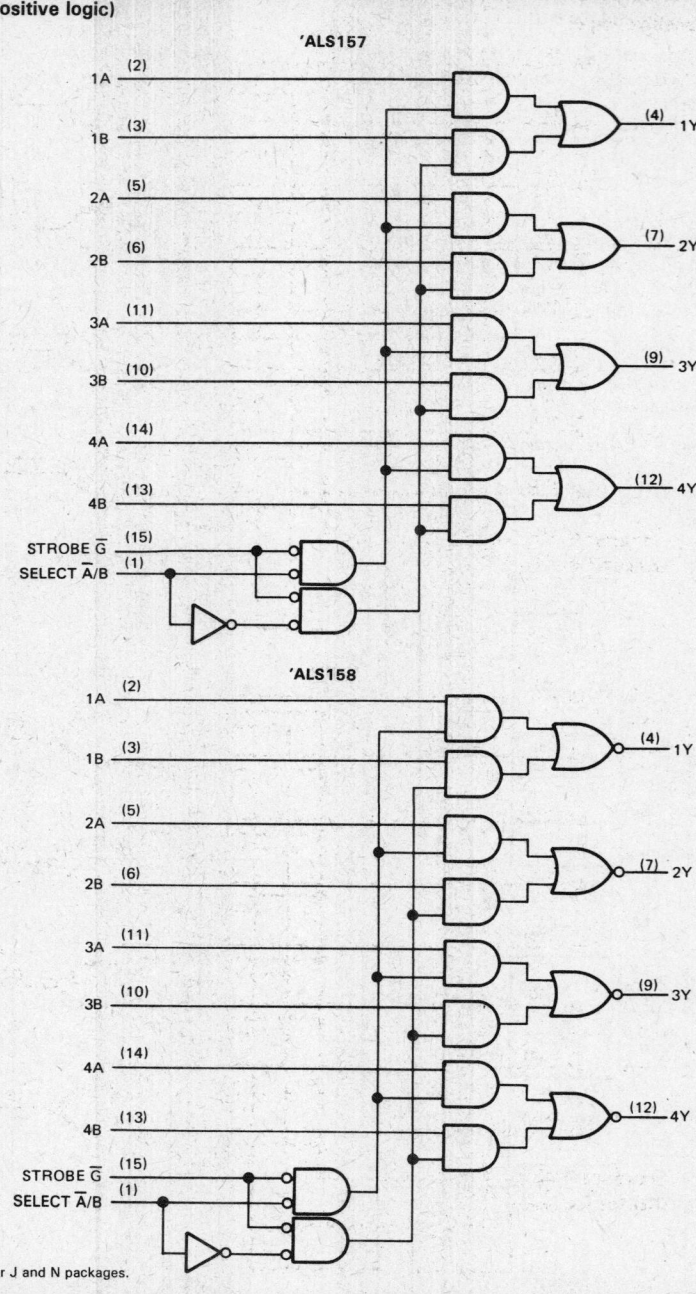

Pin numbers shown are for J and N packages.

TYPES SN54AS157, SN54AS158, SN74AS157, SN74AS158
QUADRUPLE 1 OF 2 DATA SELECTORS/MULTIPLEXERS

logic diagrams (positive logic)

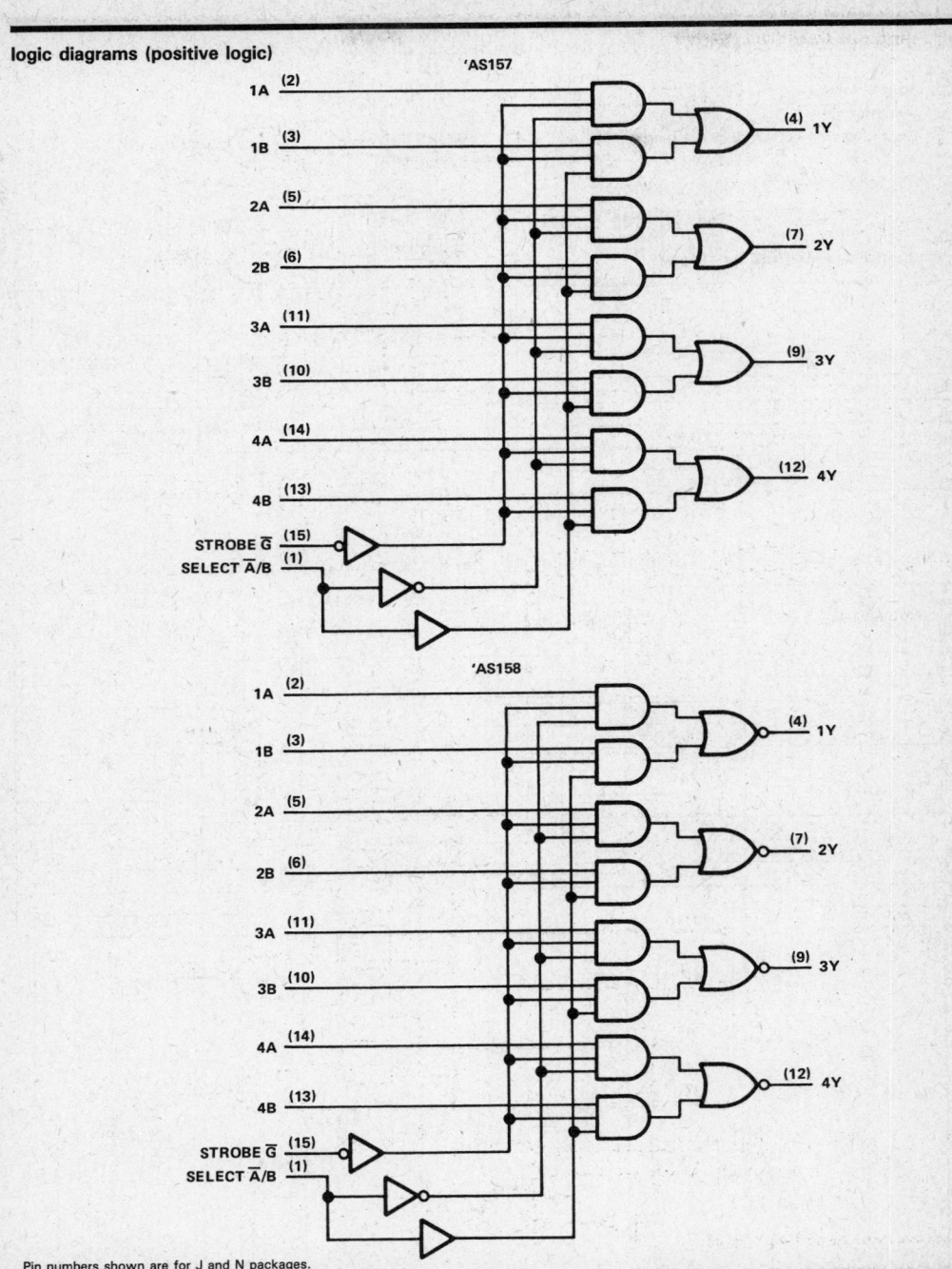

Pin numbers shown are for J and N packages.

TYPES SN54ALS157, SN54ALS158, SN74ALS157, SN74ALS158
QUADRUPLE 2-LINE TO 1-LINE DATA SELECTORS/MULTIPLEXERS

absolute maximum ratings over operating free-air temperature range (unless otherwise noted)

Supply voltage, V_{CC} .... 7 V
Input voltage .... 7 V
Operating free-air temperature range: SN54ALS157, SN54ALS158 .... −55 °C to 125 °C
                                        SN74ALS157, SN74ALS158 .... 0 °C to 70 °C
Storage temperature range .... −65 °C to 150 °C

recommended operating conditions

		SN54ALS157 SN54ALS158			SN74ALS157 SN74ALS158			UNIT
		MIN	NOM	MAX	MIN	NOM	MAX	
V_{CC}	Supply voltage	4.5	5	5.5	4.5	5	5.5	V
V_{IH}	High-level input voltage	2			2			V
V_{IL}	Low-level input voltage			0.8			0.8	V
I_{OH}	High-level output current			−1			−2.6	mA
I_{OL}	Low-level output current			12			24	mA
T_A	Operating free-air temperature	−55		125	0		70	°C

electrical characteristics over recommended operating free-air temperature range (unless otherwise noted)

PARAMETER		TEST CONDITIONS		SN54ALS157 SN54ALS158			SN74ALS157 SN74ALS158			UNIT
				MIN	TYP†	MAX	MIN	TYP†	MAX	
V_{IK}		V_{CC} = 4.5 V,	I_I = −18 mA			−1.5			−1.5	V
V_{OH}		V_{CC} = 4.5 V to 5.5 V,	I_{OH} = −0.4 mA	V_{CC}−2			V_{CC}−2			V
		V_{CC} = 4.5 V,	I_{OH} = −1 mA	2.4	3.3					
		V_{CC} = 4.5 V,	I_{OH} = −2.6 mA				2.4	3.2		
V_{OL}		V_{CC} = 4.5 V,	I_{OL} = 12 mA		0.25	0.4				V
		V_{CC} = 4.5 V	I_{OL} = 24 mA					0.35	0.5	
I_I		V_{CC} = 5.5 V,	V_I = 7 V			0.1			0.1	mA
I_{IH}		V_{CC} = 5.5 V,	V_I = 2.7 V			20			20	µA
I_{IL}		V_{CC} = 5.5 V,	V_I = 0.4 V			−0.1			−0.1	mA
I_O‡		V_{CC} = 5.5 V,	V_O = 2.25 V	−30		−112	−30		−112	mA
I_{CC}	'ALS157	V_{CC} = 5.5 V,			7.8			7.8		mA
	'ALS158				2.3			2.3		

†All typical values are at V_{CC} = 5 V, T_A = 25°C.
‡The output conditions have been chosen to produce a current that closely approximates one half of the true short-circuit current, I_{OS}.

ADVANCE INFORMATION
This page contains information on a new product.
Specifications are subject to change without notice.

ALS AND AS CIRCUITS

TEXAS INSTRUMENTS
POST OFFICE BOX 225012 • DALLAS, TEXAS 75265

TYPES SN54ALS157, SN54ALS158, SN74ALS157, SN74ALS158
QUADRUPLE 1 OF 2 DATA SELECTORS/MULTIPLEXERS

'ALS157 switching characteristics (see Note 1)

PARAMETER	FROM (INPUT)	TO (OUTPUT)	V_{CC} = 4.5 V to 5.5 V, C_L = 50 pF, R_L = 500 Ω, T_A = MIN to MAX						UNIT
			SN54ALS157			SN74ALS157			
			MIN	TYP†	MAX	MIN	TYP†	MAX	
t_{PLH}	A or B	Y		3.5			3.5		ns
t_{PHL}				5			5		
t_{PLH}	$\overline{A}/B$	Y		6			6		ns
t_{PHL}				6.5			6.5		
t_{PLH}	$\overline{G}$	Y		6			6		ns
t_{PHL}				6.5			6.5		

'ALS158 switching characteristics (see Note 1)

PARAMETER	FROM (INPUT)	TO (OUTPUT)	V_{CC} = 4.5 V to 5.5 V, C_L = 50 pF, R_L = 500 Ω, T_A = MIN to MAX						UNIT
			SN54ALS158			SN74ALS158			
			MIN	TYP†	MAX	MIN	TYP†	MAX	
t_{PLH}	A or B	Y		3.5			3.5		ns
t_{PHL}				5			5		
t_{PLH}	$\overline{A}/B$	Y		6			6		ns
t_{PHL}				6.5			6.5		
t_{PLH}	$\overline{G}$	Y		6			6		ns
t_{PHL}				6.5			6.5		

†All typical values are at V_{CC} = 5 V, T_A = 25°C.
NOTE 1: For load circuit and voltage waveforms, see page 1-12.

ADVANCE INFORMATION
This page contains information on a new product. Specifications are subject to change without notice.

TEXAS INSTRUMENTS
POST OFFICE BOX 225012 • DALLAS, TEXAS 75265

TYPES SN54AS157, SN54AS158, SN74AS157, SN74AS158
QUADRUPLE 1 OF 2 DATA SELECTORS/MULTIPLEXERS

absolute maximum ratings over operating free-air temperature range (unless otherwise noted)

Supply voltage, V_{CC} .... 7 V
Input voltage .... 7 V
Operating free-air temperature range: SN54AS157, SN54AS158 .... −55°C to 125°C
                                      SN74AS157, SN74AS158 .... 0°C to 70°C
Storage temperature range .... −65°C to 150°C

recommended operating conditions

		SN54AS157 SN54AS158			SN74AS157 SN74AS158			UNIT
		MIN	NOM	MAX	MIN	NOM	MAX	
V_{CC}	Supply voltage	4.5	5	5.5	4.5	5	5.5	V
V_{IH}	High-level input voltage	2			2			V
V_{IL}	Low-level input voltage			0.8			0.8	V
I_{OH}	High-level output current			−2			−2	mA
I_{OL}	Low-level output current			20			20	mA
T_A	Operating free-air temperature	−55		125	0		70	°C

electrical characteristics over recommended operating free-air temperature range (unless otherwise noted)

PARAMETER		TEST CONDITIONS		SN54AS157 SN54AS158			SN74AS157 SN74AS158			UNIT
				MIN	TYP†	MAX	MIN	TYP†	MAX	
V_{IK}		V_{CC} = 4.5 V,	I_I = −18 mA			−1.2			−1.2	V
V_{OH}		V_{CC} = 4.5 V to 5.5 V,	I_{OH} = −2 mA	$V_{CC}-2$			$V_{CC}-2$			V
V_{OL}		V_{CC} = 4.5 V,	I_{OL} = 20 mA		0.35	0.5		0.35	0.5	V
I_I	$\overline{A/B}$	V_{CC} = 5.5 V,	V_I = 7 V			0.2			0.2	mA
	A, B, or $\overline{G}$					0.1			0.1	
I_{IH}	$\overline{A/B}$	V_{CC} = 5.5 V,	V_I = 2.7 V			40			40	μA
	A, B, or $\overline{G}$					20			20	
I_{IL}	$\overline{A/B}$	V_{CC} = 5.5 V,	V_I = 0.4 V			−1			−1	mA
	A, B or $\overline{G}$					−0.5			−0.5	
I_O ‡		V_{CC} = 5.5 V,	V_O = 2.25 V	−30		−112	−30		−112	mA
I_{CC}	'AS157	V_{CC} = 5.5 V,			17.5	28		17.5	28	mA
	'AS158				15.6	22.5		15.6	22.5	

†All typical values are at V_{CC} = 5 V, T_A = 25°C.
‡The output conditions have been chosen to produce a current that closely approximates one half of the true short-circuit current, I_{OS}.

ALS AND AS CIRCUITS

TEXAS INSTRUMENTS
POST OFFICE BOX 225012 • DALLAS, TEXAS 75265

TYPES SN54AS157, SN54AS158, SN74AS157, SN74AS158
QUADRUPLE 1 OF 2 DATA SELECTORS/MULTIPLEXERS

'AS157 switching characteristics (see Note 1)

PARAMETER	FROM (INPUT)	TO (OUTPUT)	V_{CC} = 4.5 V to 5.5 V, C_L = 50 pF, R_L = 500 Ω, T_A = MIN to MAX				UNIT
			SN54AS157		SN74AS157		
			MIN	MAX	MIN	MAX	
t_{PLH}	A or B	Y	1	7.5	1	6	ns
t_{PHL}			1	6.5	1	5.5	
t_{PLH}	$\overline{A}/B$,	Y	2	12	2	11	ns
t_{PHL}			2	12	2	10	
t_{PLH}	$\overline{G}$	Y	2	12.5	2	10.5	ns
t_{PHL}			2	8.5	2	7.5	

'AS158 switching characteristics (see Note 1)

PARAMETER	FROM (INPUT)	TO (OUTPUT)	V_{CC} = 4.5 V to 5.5 V, C_L = 50 pF, R_L = 500 Ω, T_A = MIN to MAX				UNIT
			SN54AS158		SN74AS158		
			MIN	MAX	MIN	MAX	
t_{PLH}	A or B	Y	1	6	1	5	ns
t_{PHL}			1	5.5	1	4.5	
t_{PLH}	$\overline{A}/B$	Y	2	11	2	9.5	ns
t_{PHL}			2	11.5	2	10.5	
t_{PLH}	$\overline{G}$	Y	2	8	2	6.5	ns
t_{PHL}			2	11.5	2	10	

NOTE 1: For load circuit and voltage waveforms, see page 1-12.

2 ALS AND AS CIRCUITS

TYPES SN54ALS160A THRU SN54ALS163A, SN54AS160 THRU SN54AS163
SN74ALS160A THRU SN74ALS163A, SN74AS160 THRU SN74AS163
SYNCHRONOUS 4-BIT DECADE AND BINARY COUNTERS

D2661, APRIL 1982—REVISED DECEMBER 1983

- Internal Look-Ahead for Fast Counting
- Carry Output for n-Bit Cascading
- Synchronous Counting
- Synchronously Programmable
- Package Options Include Both Plastic and Ceramic Chip Carriers in Addition to Plastic and Ceramic DIPs
- Dependable Texas Instruments Quality and Reliability

SN54ALS', SN54AS' . . . J PACKAGE
SN74ALS', SN74AS' . . . N PACKAGE
(TOP VIEW)

```
CLR  [ 1   16 ] VCC
CLK  [ 2   15 ] RCO
  A  [ 3   14 ] QA
  B  [ 4   13 ] QB
  C  [ 5   12 ] QC
  D  [ 6   11 ] QD
ENP  [ 7   10 ] ENT
GND  [ 8    9 ] LOAD
```

SN54ALS', SN54AS . . . FH PACKAGE
SN74ALS', SN74AS' . . . FN PACKAGE
(TOP VIEW)

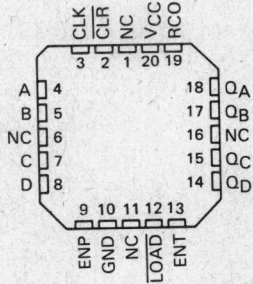

NC—No internal connection

description

These synchronous, presettable counters feature an internal carry look-ahead for application in high-speed counting designs. The 'ALS160A, 'ALS162A, 'AS160, and 'AS162 are decade counters, and the 'ALS161A, 'ALS163A, 'AS161, and 'AS163 are 4-bit binary counters. Synchronous operation is provided by having all flip-flops clocked simultaneously so that the outputs change coincident with each other when so instructed by the count-enable inputs and internal gating. This mode of operation eliminates the output counting spikes that are normally associated with asynchronous (ripple clock) counters. A buffered clock input triggers the four flip-flops on the rising (positive-going) edge of the clock input waveform.

These counters are fully programmable; that is, the outputs may be preset to either level. As presetting is synchronous, setting up a low level at the load input disables the counter and causes the outputs to agree with the setup data after the next clock pulse regardless of the levels of the enable inputs.

The clear function for the 'ALS160A, 'ALS161A, 'AS160, and 'AS161 is asynchronous and a low level at the clear input sets all four of the flip-flop outputs low regardless of the levels of the clock, load, or enable inputs.

The clear function for the 'ALS162A, 'ALS163A, 'AS162, and 'AS163 is synchronous and a low level at the clear input sets all four of the flip-flop outputs low after the next clock pulse, regardless of the levels of the enable inputs. This synchronous clear allows the count length to be modified easily as decoding the maximum count desired can be accomplished with one external NAND gate. The gate output is connected to the clear input to synchronously clear the counter to 0000 (LLLL).

The carry look-ahead circuitry provides for cascading counters for n-bit synchronous applications without additional gating. Instrumental in accomplishing this function are two count-enable inputs and a ripple carry output. Both count-enable inputs (ENP and ENT) must be high to count, and ENT is fed forward to enable the ripple carry output. The ripple carry output (RCO) thus enabled will produce a high-level pulse while the count is maximum (9 or 15 with Q_A high). This high-level overflow ripple carry pulse can be used to enable successive cascaded stages. Transitions at the ENP or ENT are allowed regardless of the level of the clock input.

These counters feature a fully independent clock circuit. Changes at control inputs (ENP, ENT, or $\overline{LOAD}$) that will modify the operating mode have no effect on the contents of the counter until clocking occurs. The function of the counter (whether enabled, disabled, loading, or counting) will be dictated solely by the conditions meeting the stable setup and hold times.

The SN54ALS160A through SN54ALS163A and SN54AS160 through SN54AS163 are characterized for operation over the full military temperature range of −55°C to 125°C. The SN74ALS160A through SN74ALS163A and SN74AS160 through SN74AS163 are characterized for operation from 0°C to 70°C.

Copyright © 1982 by Texas Instruments Incorporated

TEXAS INSTRUMENTS
POST OFFICE BOX 225012 • DALLAS, TEXAS 75265

TYPES SN54ALS160A, SN54ALS162A, SN54AS160, SN54AS162 SN74ALS160A, SN74ALS162A, SN74AS160, SN74AS162 SYNCHRONOUS 4-BIT DECADE COUNTERS

logic symbols

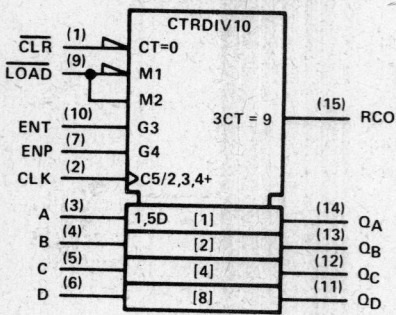

'ALS160A AND 'AS160 DECADE COUNTERS WITH DIRECT CLEAR

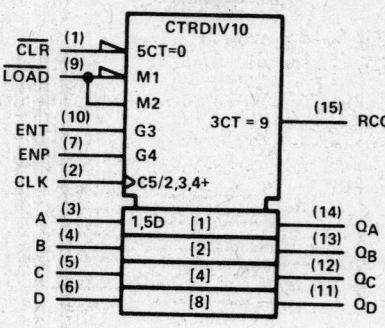

'ALS162A AND 'AS162 DECADE COUNTERS WITH SYNCHRONOUS CLEAR

'ALS160A and 'AS160 logic diagram (positive logic)

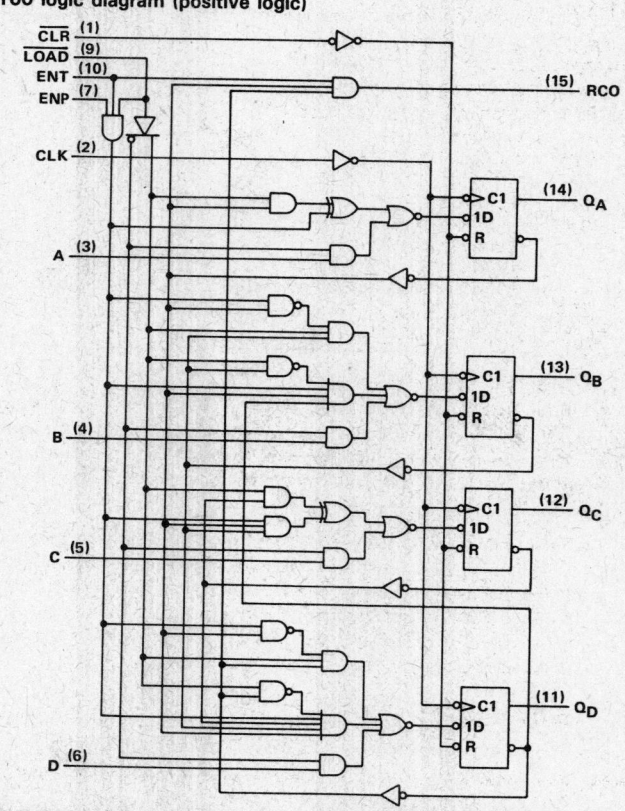

'ALS162A and 'AS162 decade counters are similar; however the clear is synchronous as shown for the 'ALS163A and 'AS163 binary counters.

Pin numbers shown are for J and N packages.

TYPES SN54ALS161A, SN54ALS163A, SN54AS161, SN54AS163 SN74ALS161A, SN74ALS163A, SN74AS161, SN74AS163
SYNCHRONOUS 4-BIT DECADE COUNTERS

logic symbols

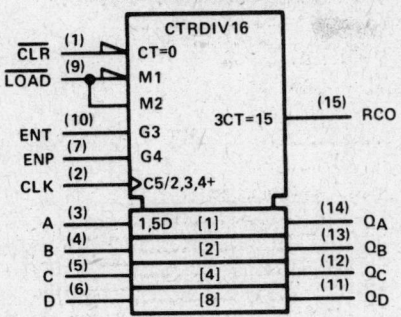

'ALS161A AND 'AS161 BINARY COUNTERS WITH DIRECT CLEAR

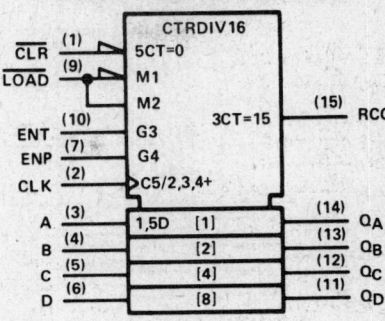

'ALS163A AND 'AS163 BINARY COUNTERS WITH SYNCHRONOUS CLEAR

'ALS163A and 'AS163 logic diagram (positive logic)

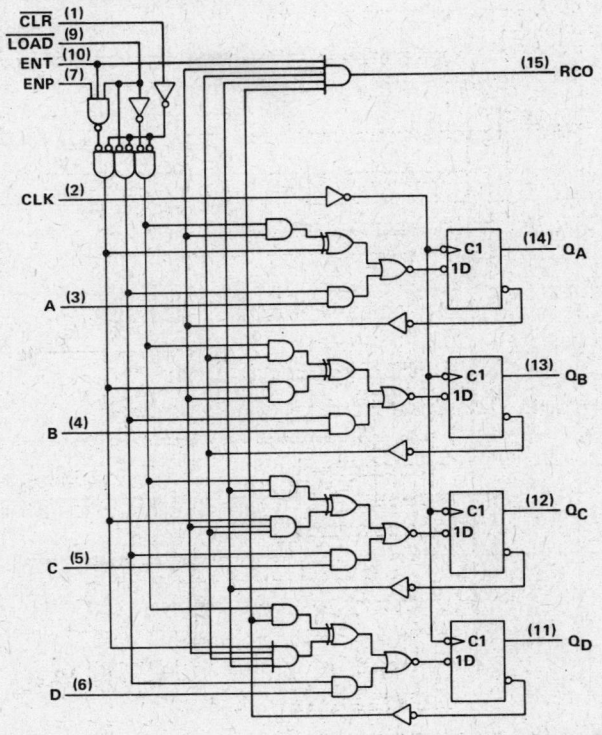

'ALS161A and 'AS161 synchronous binary counters are similar; however the clear is asynchronous as shown for the 'ALS160A and 'AS160 decade counters.

Pin numbers shown are for J and N packages.

TEXAS INSTRUMENTS
POST OFFICE BOX 225012 • DALLAS, TEXAS 75265

TYPES SN54ALS160A, SN54ALS162A, SN54AS160, SN54AS162, SN74ALS160A, SN74ALS162A, SN74AS160, SN74AS162 SYNCHRONOUS 4-BIT BINARY COUNTERS

typical clear, preset, count, and inhibit sequences

'ALS160A, 'AS160, 'ALS162A, 'AS162

Illustrated below is the following sequence:
1. Clear outputs to zero ('ALS160A and 'AS160 are asynchronous; 'ALS162A and 'AS162 are synchronous)
2. Preset to BCD seven
3. Count to eight, nine, zero, one, two, and three
4. Inhibit

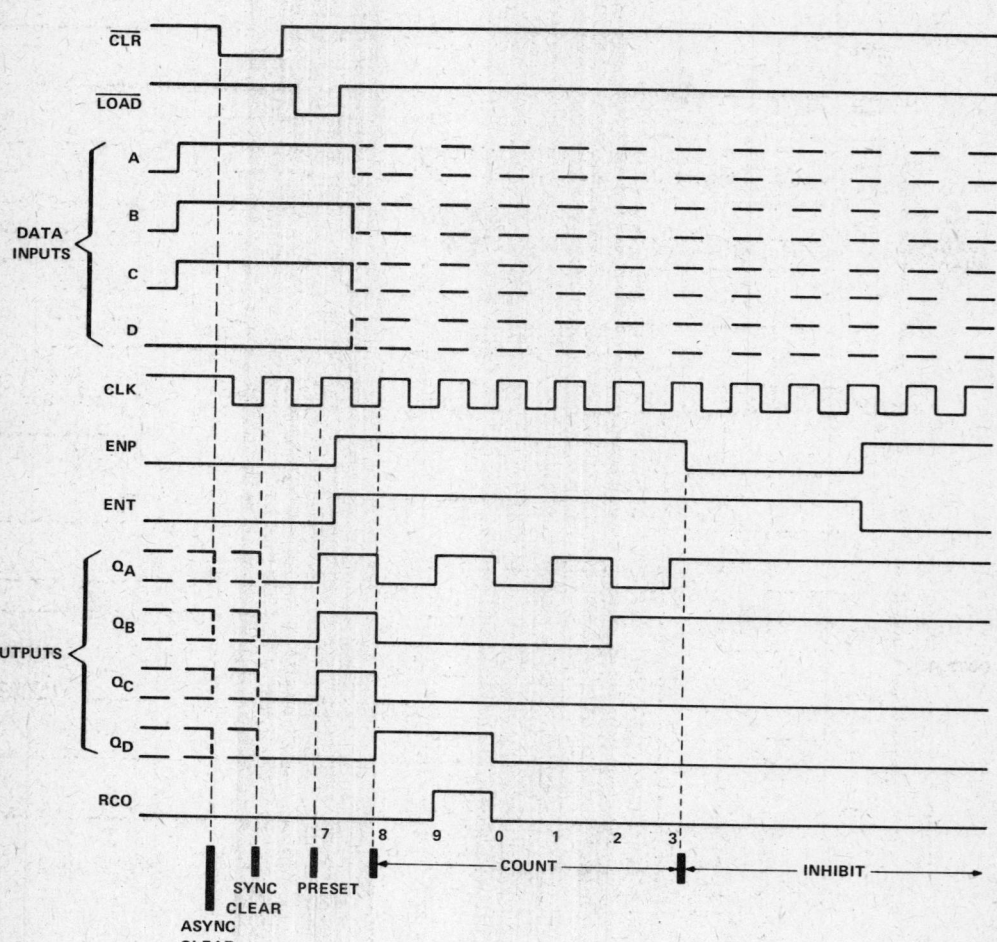

TYPES SN54ALS161A, SN54ALS163A, SN54AS161, SN54AS163 SN74ALS161A, SN74ALS163A, SN74AS161, SN74AS163 SYNCHRONOUS 4-BIT DECADE COUNTERS

typical clear, preset, count, and inhibit sequences

'ALS161A, 'AS161, 'ALS163A, 'AS163

Illustrated below is the following sequence:
1. Clear outputs to zero ('ALS161A and 'AS161 are asynchronous; 'ALS163A and 'AS163 are synchronous)
2. Preset to binary twelve
3. Count to thirteen, fourteen, fifteen, zero, one, and two
4. Inhibit

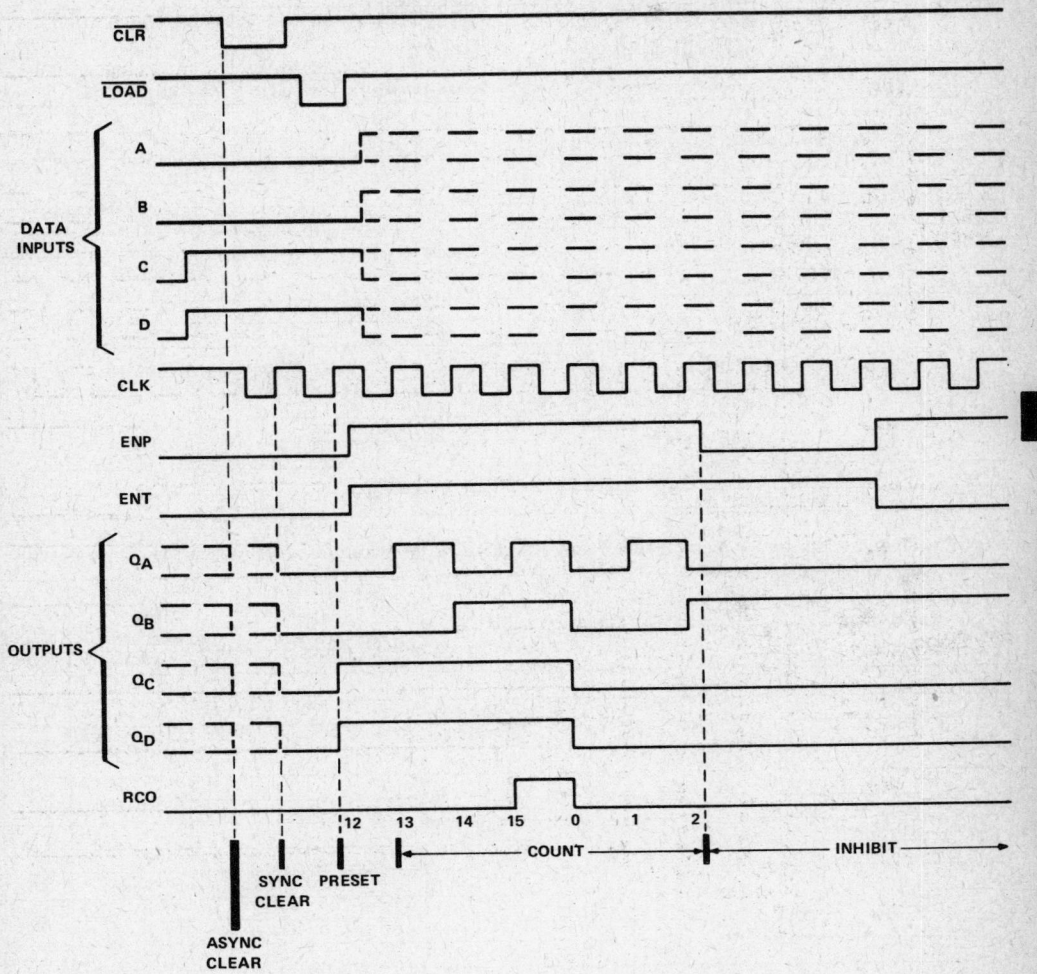

TYPES SN54ALS160A THRU SN54ALS163A
SN74ALS160A THRU SN74ALS163A
SYNCHRONOUS 4-BIT DECADE AND BINARY COUNTERS

absolute maximum ratings over operating free-air temperature range (unless otherwise noted)

Supply voltage, V_{CC} .... 7 V
Input voltage .... 7 V
Operating free-air temperature range: SN54ALS160A thru SN54ALS163A .... −55°C to 125°C
                                      SN74ALS160A thru SN74ALS163A .... 0°C to 70°C
Storage temperature range .... −65°C to 150°C

recommended operating conditions

			SN54ALS160A THRU SN54ALS163A			SN74ALS160A THRU SN74ALS163A			UNIT
			MIN	NOM	MAX	MIN	NOM	MAX	
V_{CC}	Supply voltage		4.5	5	5.5	4.5	5	5.5	V
V_{IH}	High-level input voltage		2			2			V
V_{IL}	Low-level input voltage				0.8			0.8	V
I_{OH}	High-level output current				−0.4			−0.4	mA
I_{OL}	Low-level output current				4			8	mA
f_{clock}	Clock frequency		0		25	0		30	MHz
t_w	Pulse duration	CLK high or low	20			16.5			ns
		'ALS160A, 'ALS161A, $\overline{CLR}$ low	20			15			
t_{su}	Setup time before CLK↑	A, B, C, D	20			15			ns
		$\overline{LOAD}$	20			15			
		ENP, ENT 'ALS160A, 'ALS161A	25			20			
		ENP, ENT 'ALS162A, 'ALS163A	30			25			
		'ALS160A, 'ALS161A $\overline{CLR}$ inactive	10			10			
		'ALS162A, 'ALS163A $\overline{CLR}$ low	20			15			
		'ALS162A, 'ALS163A $\overline{CLR}$ high (inactive)	10			10			
t_h	Hold time, all synchronous inputs after CLK↑		0			0			ns
T_A	Operating free-air temperature		−55		125	0		70	°C

electrical characteristics over recommended operating free-air temperature range (unless otherwise noted)

PARAMETER		TEST CONDITIONS		SN54ALS160A THRU SN54ALS163A			SN74ALS160A THRU SN74ALS163A			UNIT
				MIN	TYP†	MAX	MIN	TYP†	MAX	
V_{IK}		$V_{CC} = 4.5$ V,	$I_I = -18$ mA			−1.5			−1.5	V
V_{OH}		$V_{CC} = 4.5$ V to 5.5 V,	$I_{OH} = -0.4$ mA	$V_{CC} - 2$			$V_{CC} - 2$			V
V_{OL}		$V_{CC} = 4.5$ V,	$I_{OL} = 4$ mA		0.25	0.4		0.25	0.4	V
		$V_{CC} = 4.5$ V,	$I_{OL} = 8$ mA					0.35	0.5	
I_I	LOAD, CLK or ENT	$V_{CC} = 5.5$ V,	$V_I = 7$ V			0.2			0.2	mA
	All other					0.1			0.1	
I_{IH}	LOAD, CLK or ENT	$V_{CC} = 5.5$ V,	$V_I = 2.7$ V			40			40	µA
	All other					20			20	
I_{IL}		$V_{CC} = 5.5$ V,	$V_I = 0.4$ V			−0.2			−0.2	mA
I_O‡	RCO	$V_{CC} = 5.5$ V,	$V_O = 2.25$ V	−15		−70	−15		−70	mA
	Q			−30		−112	−30		−112	
I_{CC}		$V_{CC} = 5.5$ V			12	21		12	21	mA

†All typical values are at $V_{CC} = 5$ V, $T_A = 25$°C.
‡The output conditions have been chosen to produce a current that closely approximates one half of the true short-circuit output current, I_{OS}.

TYPES SN54ALS160A THRU SN54ALS163A, SN74ALS160A THRU SN74ALS163A SYNCHRONOUS 4-BIT DECADE AND BINARY COUNTERS

'ALS160A, 'ALS161A switching characteristics (see Note 1)

PARAMETER	FROM (INPUT)	TO (OUTPUT)	V_{CC} = 4.5 V to 5.5 V, C_L = 50 pF, R_L = 500 Ω, T_A = MIN to MAX				UNIT
			SN54ALS160A SN54ALS161A		SN74ALS160A SN74ALS161A		
			MIN	MAX	MIN	MAX	
f_{max}			25		30		MHz
t_{PLH}	CLK	RCO	8	30	8	26	ns
t_{PHL}			7	25	7	23	
t_{PLH}	CLK	Any Q	4	18	4	15	ns
t_{PHL}			6	20	6	17	
t_{PLH}	ENT	RCO	3	16	3	13	ns
t_{PHL}			3	16	3	13	
t_{PHL}	$\overline{CLR}$	Any Q	8	27	8	24	ns
t_{PHL}	$\overline{CLR}$	RCO	11	31	11	28	ns

'ALS162A, 'ALS163A switching characteristics (see Note 1)

PARAMETER	FROM (INPUT)	TO (OUTPUT)	V_{CC} = 4.5 V to 5.5 V, C_L = 50 pF, R_L = 500 Ω, T_A = MIN to MAX				UNIT
			SN54ALS162A SN54ALS163A		SN74ALS162A SN74ALS163A		
			MIN	MAX	MIN	MAX	
f_{max}			25		30		MHz
t_{PLH}	CLK	RCO	8	30	8	26	ns
t_{PHL}			7	25	7	23	
t_{PLH}	CLK	Any Q	4	18	4	15	ns
t_{PHL}			6	20	6	17	
t_{PLH}	ENT	RCO	3	20	3	17	ns
t_{PHL}			3	16	3	13	

NOTE 1: For load circuit and voltage waveforms, see page 1-12.

TYPES SN54AS160 THRU SN54AS163, SN74AS160 THRU SN74AS163 SYNCHRONOUS 4-BIT DECADE AND BINARY COUNTERS

absolute maximum ratings over operating free-air temperature range (unless otherwise noted)

Supply voltage, V_{CC} .... 7 V
Input voltage .... 7 V
Operating free-air temperature range: SN54AS160 thru SN54AS163 .... −55°C to 125°C
                                      SN74AS160 thru SN74AS163 .... 0°C to 70°C
Storage temperature range .... −65°C to 150°C

recommended operating conditions

				SN54AS160 THRU SN54AS163			SN74AS160 THRU SN74AS163			UNIT
				MIN	NOM	MAX	MIN	NOM	MAX	
V_{CC}	Supply voltage			4.5	5	5.5	4.5	5	5.5	V
V_{IH}	High-level input voltage			2			2			V
V_{IL}	Low-level input voltage					0.8			0.8	V
I_{OH}	High-level output current					−2			−2	mA
I_{OL}	Low-level output current					20			20	mA
f_{clock}	Clock frequency									MHz
t_w	Pulse duration	CLK high or low								ns
		'AS160, 'AS161 CLR low								
t_{su}	Setup time before CLK↑	A, B, C, D								ns
		LOAD								
		ENP, ENT								
		'AS160, 'AS161 CLR inactive								
		'AS162, 'AS163	CLR low							
			CLR high (inactive)							
t_h	Hold time, all synchronous inputs after CLK↑									ns
T_A	Operating free-air temperature			−55		125	0		70	°C

electrical characteristics over recommended operating free-air temperature range (unless otherwise noted)

PARAMETER		TEST CONDITIONS		SN54AS160 THRU SN54AS163			SN74AS160 THRU SN74AS163			UNIT
				MIN	TYP†	MAX	MIN	TYP†	MAX	
V_{IK}		V_{CC} = 4.5 V,	I_I = −18 mA			−1.2			−1.2	V
V_{OH}		V_{CC} = 4.5 V to 5.5 V,	I_{OH} = −2 mA	V_{CC}−2			V_{CC}−2			V
V_{OL}		V_{CC} = 4.5 V,	I_{OL} = 20 mA		0.25	0.5		0.25	0.5	V
I_I		V_{CC} = 5.5 V,	V_I = 7 V			0.1			0.1	mA
I_{IH}		V_{CC} = 5.5 V,	V_I = 2.7 V			20			20	μA
I_{IL}	LOAD, ENT	V_{CC} = 5.5 V,	V_I = 0.4 V			−1			−1	mA
	All other					−0.5			−0.5	
I_O‡		V_{CC} = 5.5 V,	V_O = 2.25 V	−30		−112	−30		−112	mA
I_{CC}		V_{CC} = 5.5 V				40			40	mA

†All typical values are at V_{CC} = 5 V, T_A = 25°C.
‡The output conditions have been chosen to produce a current that closely approximates one half of the true short-circuit output current, I_{OS}.

Additional information on these products can be obtained from the factory as it becomes available.

PRODUCT PREVIEW

This page contains information on a product under development. Texas Instruments reserves the right to change or discontinue this product without notice.

TEXAS INSTRUMENTS
POST OFFICE BOX 225012 • DALLAS, TEXAS 75265

TYPES SN54AS160 THRU SN54AS163, SN74AS160 THRU SN74AS163
SYNCHRONOUS 4-BIT DECADE AND BINARY COUNTERS

'AS160, 'AS161 switching characteristics (see Note 1)

PARAMETER	FROM (INPUT)	TO (OUTPUT)	V_{CC} = 4.5 V to 5.5 V, C_L = 50 pF, R_L = 500 Ω, T_A = MIN to MAX						UNIT
			SN54AS160 SN54AS161			SN74AS160 SN74AS161			
			MIN	TYP†	MAX	MIN	TYP†	MAX	
f_{max}									MHz
t_{PLH}	CLK	RCO		7			7		ns
t_{PHL}		RCO (with $\overline{LOAD}$ high)		6			6		
t_{PHL}		RCO (with $\overline{LOAD}$ low)		10			10		
t_{PLH}	CLK	Any Q		5			5		ns
t_{PHL}				6			6		
t_{PLH}	ENT	RCO		3			3		ns
t_{PHL}				4			4		
t_{PHL}	$\overline{CLR}$	Any Q		7			7		ns

'AS162, 'AS163 switching characteristics (see Note 1)

PARAMETER	FROM (INPUT)	TO (OUTPUT)	V_{CC} = 4.5 V to 5.5 V, C_L = 50 pF, R_L = 500 Ω, T_A = MIN to MAX						UNIT
			SN54AS162 SN54AS163			SN74AS162 SN74AS163			
			MIN	TYP†	MAX	MIN	TYP†	MAX	
f_{max}									MHz
t_{PLH}	CLK	RCO		7			7		ns
t_{PHL}		RCO (with $\overline{LOAD}$ high)		6			6		
t_{PHL}		RCO (with $\overline{LOAD}$ low)		10			10		
t_{PLH}	CLK	Any Q		5			5		ns
t_{PHL}				6			6		
t_{PLH}	ENT	RCO		3			3		ns
t_{PHL}				4			4		
t_{PHL}	$\overline{CLR}$	Any Q		7			7		ns

†All typical values are at V_{CC} = 5 V, T_A = 25°C.
NOTE 1: For load circuit and voltage waveforms, see page 1-12.

Additional information on these products can be obtained from the factory as it becomes available.

PRODUCT PREVIEW
This page contains information on a product under development. Texas Instruments reserves the right to change or discontinue this product without notice.

TEXAS INSTRUMENTS
POST OFFICE BOX 225012 • DALLAS, TEXAS 75265

TYPES SN54ALS160A THRU SN54ALS163A, SN54AS160 THRU SN54AS163
SN74ALS160A THRU SN74ALS163A, SN74AS160 THRU SN74AS163
SYNCHRONOUS 4-BIT DECADE AND BINARY COUNTERS

TYPICAL APPLICATION DATA

N-BIT SYNCHRONOUS COUNTERS

This application demonstrates how the look-ahead carry circuit can be used to implement a high-speed n-bit counter. The 'ALS160A, 'AS160, 'ALS162A, and 'AS162 will count in BCD and the 'ALS161A, 'AS161, 'ALS163A and 'AS163 will count in binary. Virtually any count mode (modulo-N, N_1-to-N_2, N_1-to-maximum) can be used with this fast look-ahead circuit.

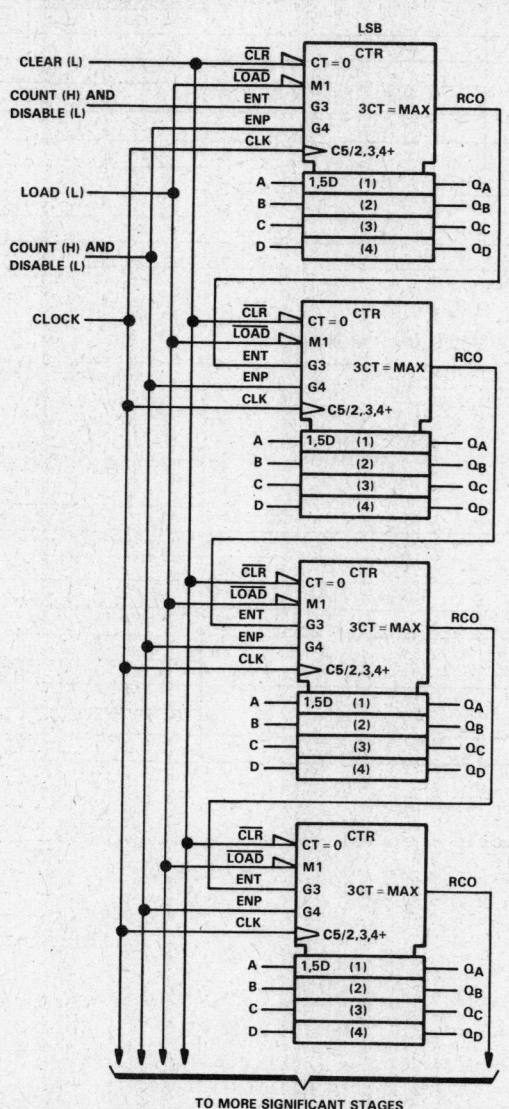

TYPES SN54ALS164, SN74ALS164
8-BIT PARALLEL-OUT SERIAL SHIFT REGISTERS

D2661, APRIL 1982 – REVISED DECEMBER 1983

- AND-Gated (Enable/Disable) Serial Inputs
- Fully Buffered Clock and Serial Inputs
- Direct Clear
- Package Options Include Both Plastic and Ceramic Chip Carriers in Addition to Plastic and Ceramic DIPs
- Dependable Texas Instruments Quality and Reliability

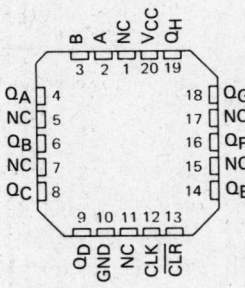

description

These 8-bit shift registers feature AND-gated serial inputs and an asynchronous clear. The gated serial inputs (A and B) permit complete control over incoming data as a low at either input inhibits entry of the new data and resets the first flip-flop to the low level at the next clock pulse. A high-level input enables the other input, which will then determine the state of the first flip-flop. Data at the serial inputs may be changed while the clock is high or low, provided the minimum setup time requirements are met. Clocking occurs on the low-to-high-level transition of the clock input. All inputs are diode-clamped to minimize transmission-line effects.

The SN54ALS164 is characterized for operation over the full military temperature range of −55°C to 125°C. The SN74ALS164 is characterized for operation from 0°C to 70°C.

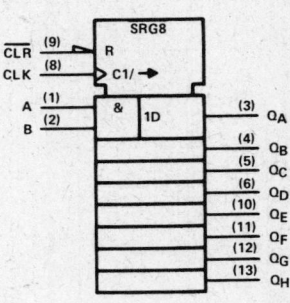

NC – No Internal connection

logic symbol

FUNCTION TABLE						
INPUTS				OUTPUTS		
CLEAR	CLOCK	A	B	Q_A	Q_B	... Q_H
L	X	X	X	L	L	L
H	L	X	X	Q_{A0}	Q_{B0}	Q_{H0}
H	↑	H	H	H	Q_{An}	Q_{Gn}
H	↑	L	X	L	Q_{An}	Q_{Gn}
H	↑	X	L	L	Q_{An}	Q_{Gn}

H = high level (steady state), L = low level (steady state)
X = irrelevant (any input, including transitions)
↑ = transition from low to high level.
Q_{A0}, Q_{B0}, Q_{H0} = the level of Q_A, Q_B, or Q_H, respectively, before the indicated steady-state input conditions were established.
Q_{An}, Q_{Gn} = the level of Q_A or Q_G before the most-recent ↑ transition of the clock; indicates a one-bit shift.

Pin numbers shown are for J and N packages.

PRODUCT PREVIEW

This document contains information on a product under development. Texas Instruments reserves the right to change or discontinue this product without notice.

Copyright © 1982 by Texas Instruments Incorporated

TEXAS INSTRUMENTS
POST OFFICE BOX 225012 • DALLAS, TEXAS 75265

ALS AND AS CIRCUITS

2-147

TYPES SN54ALS164, SN74ALS164
8-BIT-PARALLEL-OUT SERIAL SHIFT REGISTERS

logic diagram (positive logic)

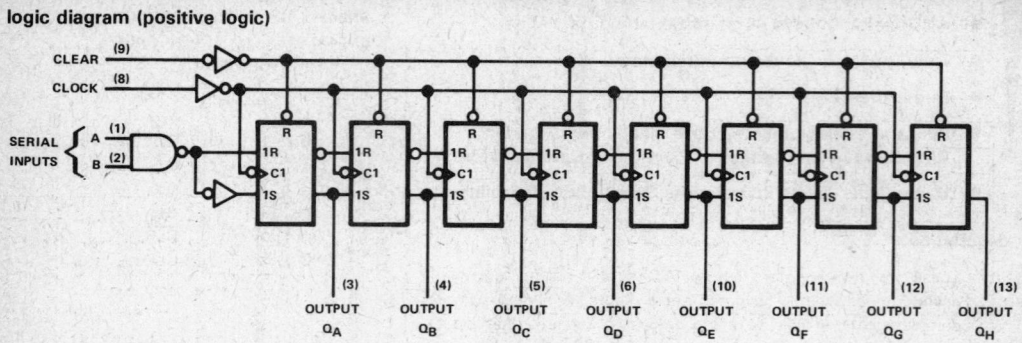

Pin numbers shown are for J and N packages.

typical clear, shift, and clear sequences

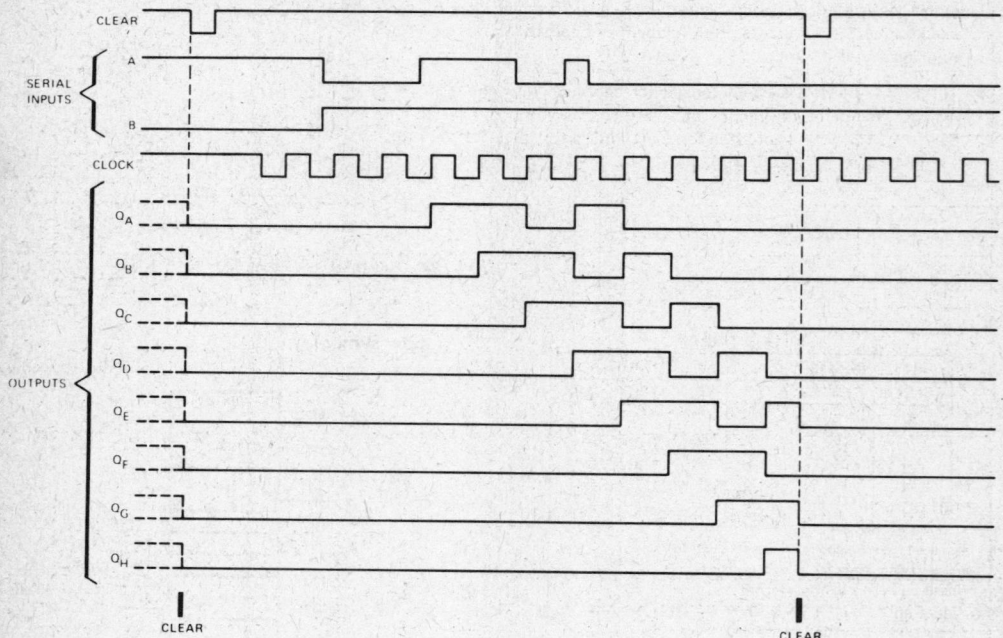

absolute maximum ratings over operating free-air temperature range (unless otherwise noted)

Supply voltage, V_{CC} .... 7 V
Input voltage .... 7 V
Operating free-air temperature range: SN54ALS164 .... −55 °C to 125 °C
                                       SN74ALS164 .... 0 °C to 70 °C
Storage temperature range .... −65 °C to 150 °C

TYPES SN54ALS164, SN74ALS164
8-BIT-PARALLEL-OUT SERIAL SHIFT REGISTERS

recommended operating conditions

		SN54ALS164			SN74ALS164			UNIT
		MIN	NOM	MAX	MIN	NOM	MAX	
V_{CC}	Supply voltage	4.5	5	5.5	4.5	5	5.5	V
V_{IH}	High-level input voltage	2			2			V
V_{IL}	Low-level input voltage			0.8			0.8	V
I_{OH}	High-level output current			−0.4			−0.4	mA
I_{OL}	Low-level output current			4			8	mA
f_{clock}	Clock frequency							MHz
t_w	Pulse duration CLR low / CLK high / CLK low							ns
t_{su}	Setup time before CLK↑ SH/LD / Data / CLR inactive							ns
t_h	Hold time, data after CLK↑	0			0			ns
T_A	Operating free-air temperature	−55		125	0		70	°C

electrical characteristics over recommended operating free-air temperature range (unless otherwise noted)

PARAMETER	TEST CONDITIONS		SN54ALS164			SN74ALS164			UNIT
			MIN	TYP†	MAX	MIN	TYP†	MAX	
V_{IK}	$V_{CC} = 4.5$ V,	$I_I = -18$ mA			−1.5			−1.5	V
V_{OH}	$V_{CC} = 4.5$ V to 5.5 V,	$I_{OH} = -0.4$ mA	$V_{CC}-2$			$V_{CC}-2$			V
V_{OL}	$V_{CC} = 4.5$ V,	$I_{OL} = 4$ mA		0.25	0.4		0.25	0.4	V
	$V_{CC} = 4.5$ V,	$I_{OL} = 8$ mA					0.35	0.5	
I_I	$V_{CC} = 5.5$ V,	$V_I = 7$ V			0.1			0.1	mA
I_{IH}	$V_{CC} = 5.5$ V,	$V_I = 2.7$ V			20			20	μA
I_{IL}	$V_{CC} = 5.5$ V,	$V_I = 0.4$ V			−0.1			−0.1	mA
I_O‡	$V_{CC} = 5.5$ V,	$V_O = 2.25$ V	−30		−112	−30		−112	mA
I_{CC}	$V_{CC} = 5.5$ V	See Note 1		10			10		mA

†All typical values are at $V_{CC} = 5$ V, $T_A = 25$°C.
‡The output conditions have been chosen to produce a current that closely approximates one half of the true short-circuit output current, I_{OS}.
NOTE 1: With 4.5 Volts applied to the serial input and all other inputs except the clock grounded, I_{CC} is measured after a clock transition from 0 to 4.5 volts.

switching characteristics (see Note 2)

PARAMETER	FROM (INPUT)	TO (OUTPUT)	$V_{CC} = 4.5$ V to 5.5 V, $C_L = 50$ pF, $R_L = 500$ Ω, T_A = MIN to MAX						UNIT
			SN54ALS164			SN74ALS164			
			MIN	TYP†	MAX	MIN	TYP†	MAX	
f_{max}				60			60		MHz
t_{PHL}	CLR	Any Q		12			12		ns
t_{PLH}	CLK	Any Q		10			10		ns
t_{PHL}				11			11		

†All typical values are at $V_{CC} = 5$ V, $T_A = 25$°C.
NOTE 2: For load circuit and voltage waveforms, see page 1-12.

Additional information on these products can be obtained from the factory as it becomes available.

ALS AND AS CIRCUITS

TEXAS INSTRUMENTS
POST OFFICE BOX 225012 • DALLAS, TEXAS 75265

2
ALS AND AS CIRCUITS

TYPES SN54ALS165, SN74ALS165
PARALLEL-LOAD 8-BIT SHIFT REGISTERS

D2661, JUNE 1982

- Complementary Outputs
- Direct Overriding Load (Data) Inputs
- Gated Clock Inputs
- Parallel-to-Serial Data Conversion
- Package Options Include Both Plastic and Ceramic Chip Carriers in Addition to Plastic and Ceramic DIPs
- Dependable Texas Instruments Quality and Reliability

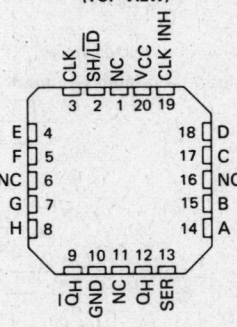

description

The 'ALS165 is an 8-bit serial shift register that, when clocked, shifts the data toward serial output $\overline{Q}_H$. Parallel-in access to each stage is provided by eight individual direct data inputs that are enabled by a low level at the SH/LD input. The 'ALS165 also features a clock inhibit function and a complemented serial output $\overline{Q}_H$.

Clocking is accomplished by a low-to-high transition of the CLK input while SH/$\overline{LD}$ is held high and CLK INH is held low. The functions of the CLK and CLK INH (clock inhibit) inputs are interchangeable. Since a low CLK input and a low-to-high transition of CLK INH will also accomplish clocking, CLK INH should be changed to the high level only while the CLK input is high. Parallel loading is inhibited when SH/$\overline{LD}$ is held high. The parallel inputs to the register are enabled while SH/$\overline{LD}$ is low independently of the levels of CLK, CLK INH, or SER inputs.

The SN54ALS165 is characterized for operation over the full military temperature range of −55°C to 125°C. The SN74ALS165 is characterized for operation from 0°C to 70°C.

FUNCTION TABLE

INPUTS			FUNCTION
SH/$\overline{LD}$	CLK	CLK INH	
L	X	X	PARALLEL LOAD
H	H	X	NO CHANGE
H	X	H	NO CHANGE
H	L	↑	SHIFT
H	↑	L	SHIFT

SHIFT — content of each internal register shifts toward serial output Q_H. Data at serial input is shifted into first register.

logic symbol

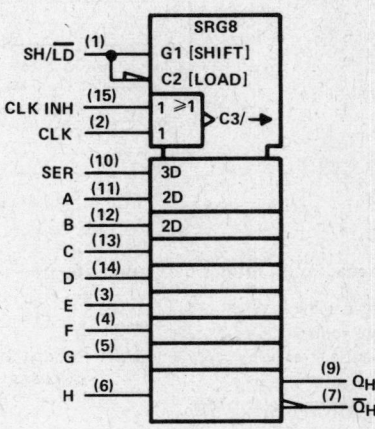

Pin numbers shown are for J and N packages.

PRODUCT PREVIEW

This document contains information on a product under development. Texas Instruments reserves the right to change or discontinue this product without notice.

Copyright © 1982 by Texas Instruments Incorporated

TEXAS INSTRUMENTS
POST OFFICE BOX 225012 • DALLAS, TEXAS 75265

TYPES SN54ALS165, SN74ALS165
PARALLEL-LOAD 8-BIT SHIFT REGISTERS

logic diagram (positive logic)

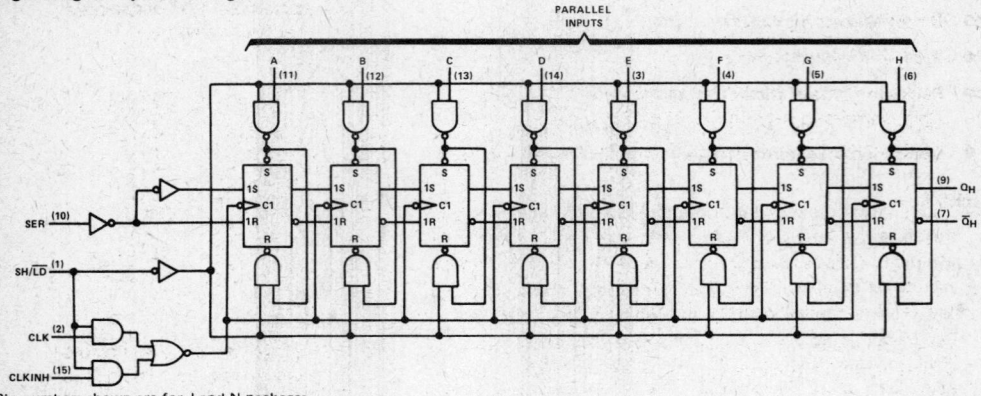

Pin numbers shown are for J and N packages

typical shift, load, and inhibit sequences

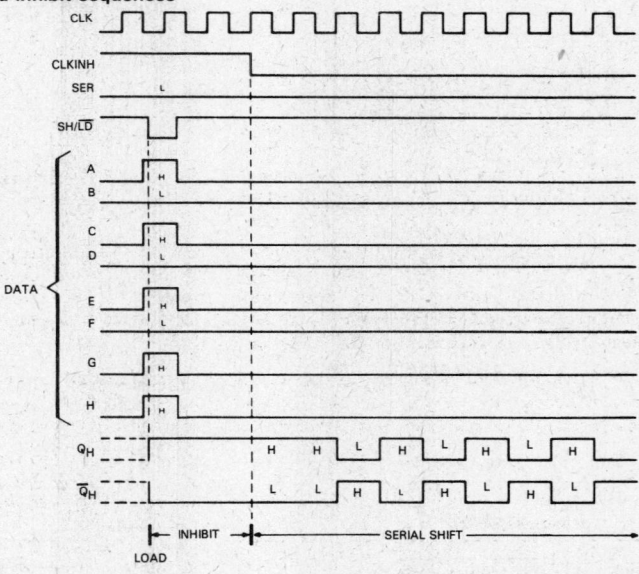

absolute maximum ratings over operating free-air temperature range (unless otherwise noted)

Supply voltage, V_{CC} .... 7 V
Input voltage .... 7 V
Operating free-air temperature range: SN54ALS165 .... −55°C to 125°C
                                          SN74ALS165 .... 0°C to 70°C
Storage temperature range .... −65°C to 150°C

TYPES SN54ALS166, SN74ALS166
PARALLEL-LOAD 8-BIT SHIFT REGISTERS

D2661, APRIL 1982—DECEMBER 1983

- Synchronous Load
- Direct Overriding Clear
- Parallel to Serial Conversion
- Package Options Include Both Plastic and Ceramic Chip Carriers in Addition to Plastic And Ceramic DIPs
- Dependable Texas Instruments Quality and Reliability

description

The 'ALS166 8-bit shift register is compatible with most other TTL logic families. All inputs are buffered to lower the drive requirements. Input clamping diodes minimize switching transients and simplify system design.

These parallel-in or serial-in, serial-out registers have a complexity of 77 equivalent gates on a monolithic chip. They feature gated clock inputs and an overriding clear input. The parallel-in or serial-in modes are established by the shift/load input. When high, this input enables the serial data input and couples the eight flip-flops for serial shifting with each clock pulse. When low, the parallel (broadside) data inputs are enabled and synchronous loading occurs on the next clock pulse. During parallel loading, serial data flow is inhibited. Clocking is accomplished on the low-to-high-level edge of the clock pulse through a two-input positive NOR gate permitting one input to be used as a clock-enable or clock-inhibit function. Holding either of the clock inputs high inhibits clocking; holding either low enables the other clock input. This, of course, allows the system clock to be free-running and the register can be stopped on command with the clock input. The clock-inhibit input should be changed to the high level only when the clock input is high. A buffered, direct clear input overrides all other inputs, including the clock, and sets all flip-flops to zero.

The SN54ALS166 is characterized for operation over the full military temperature range of −55°C to 125°C. The SN74ALS166 is characterized for operation from 0°C to 70°C.

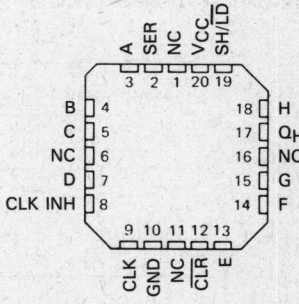

SN54ALS166 . . . J PACKAGE
SN74ALS166 . . . N PACKAGE
(TOP VIEW)

SN54ALS166 . . . FH PACKAGE
SN74ALS166 . . . FN PACKAGE
(TOP VIEW)

NC—No internal connection

logic symbol

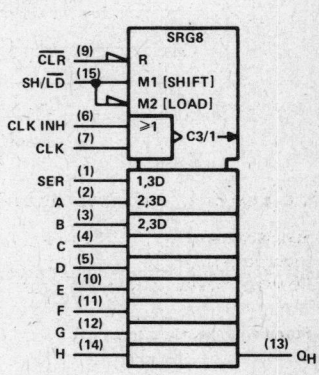

Pin numbers shown are for J and N packages.

FUNCTION TABLE

	INPUTS					INTERNAL OUTPUTS		OUTPUT
CLEAR	SHIFT/ LOAD	CLOCK INHIBIT	CLOCK	SERIAL	PARALLEL A...H	Q_A	Q_B	Q_H
L	X	X	X	X	X	L	L	L
H	X	L	L	X	X	Q_{A0}	Q_{B0}	Q_{H0}
H	L	L	↑	X	a...h	a	b	h
H	H	L	↑	H	X	H	Q_{An}	Q_{Gn}
H	H	L	↑	L	X	L	Q_{An}	Q_{Gn}
H	X	H	↑	X	X	Q_{A0}	Q_{B0}	Q_{H0}

PRODUCT PREVIEW

This document contains information on a product under development. Texas Instruments reserves the right to change or discontinue this product without notice.

Copyright © 1982 by Texas Instruments Incorporated

TYPES SN54ALS166, SN74ALS166
PARALLEL-LOAD 8-BIT SHIFT REGISTERS

logic diagram (positive logic)

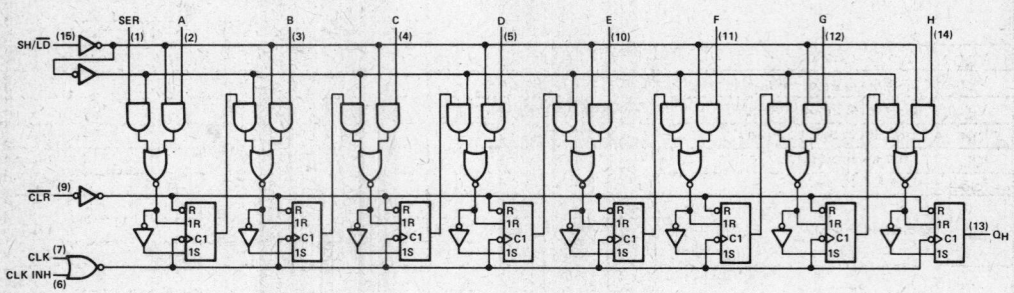

Pin numbers shown are for J and N packages.

typical clear, shift, load, inhibit, and shift sequences

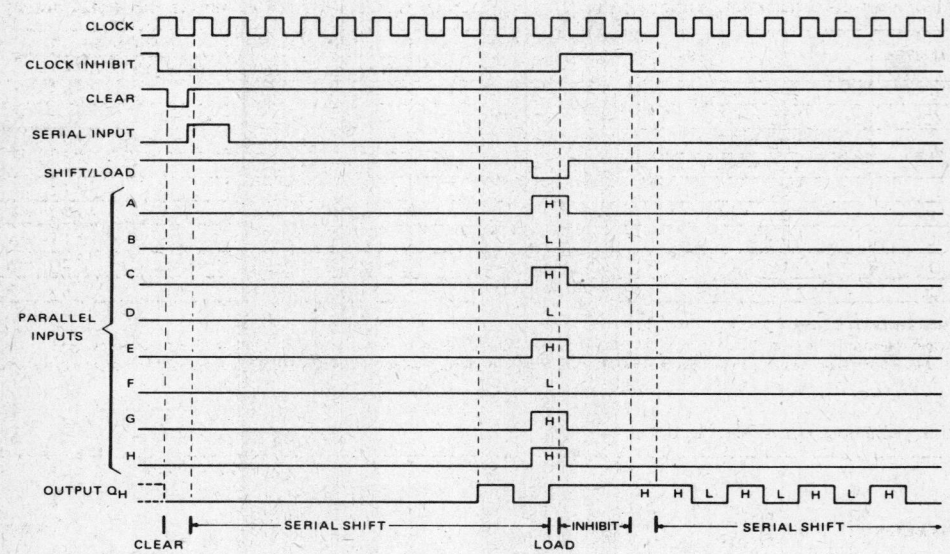

absolute maximum ratings over operating free-air temperature range (unless otherwise noted)

Supply voltage, V_{CC} .... 7 V
Input voltage .... 7 V
Operating free-air temperature range: SN54ALS166 .... −55 °C to 125 °C
                                        SN74ALS166 .... 0 °C to 70 °C
Storage temperature range .... −65 °C to 150 °C

TEXAS
INSTRUMENTS
POST OFFICE BOX 225012 • DALLAS, TEXAS 75265

TYPES SN54ALS166, SN74ALS166
PARALLEL-LOAD 8-BIT SHIFT REGISTERS

recommended operating conditions

			SN54ALS166			SN74ALS166			UNIT
			MIN	NOM	MAX	MIN	NOM	MAX	
V_{CC}	Supply voltage		4.5	5	5.5	4.5	5	5.5	V
V_{IH}	High-level input voltage		2			2			V
V_{IL}	Low-level input voltage				0.8			0.8	V
I_{OH}	High-level output current				−0.4			−0.4	mA
I_{OL}	Low-level output current				4			8	mA
f_{clock}	Clock frequency								MHz
t_w	Pulse duration	$\overline{CLR}$ low							ns
		CLK high							
		CLK low							
t_{su}	Setup time before CLK↑	SH/$\overline{LD}$							ns
		Data							
		$\overline{CLR}$ inactive							
t_h	Hold time, data after CLK↑								ns
T_A	Operating free-air temperature		−55		125	0		70	°C

electrical characteristics over recommended operating free-air temperature range (unless otherwise noted)

PARAMETER	TEST CONDITIONS		SN54ALS166			SN74ALS166			UNIT
			MIN	TYP†	MAX	MIN	TYP†	MAX	
V_{IK}	V_{CC} = 4.5 V,	I_I = −18 mA			−1.5			−1.5	V
V_{OH}	V_{CC} = 4.5 V to 5.5 V,	I_{OH} = −0.4 mA	$V_{CC}-2$			$V_{CC}-2$			V
V_{OL}	V_{CC} = 4.5 V,	I_{OL} = 4 mA		0.25	0.4		0.25	0.4	V
	V_{CC} = 4.5 V,	I_{OL} = 8 mA					0.35	0.5	
I_I	V_{CC} = 5.5 V,	V_I = 7 V			0.1			0.1	mA
I_{IH}	V_{CC} = 5.5 V,	V_I = 2.7 V			20			20	μA
I_{IL}	V_{CC} = 5.5 V,	V_I = 0.4 V			−0.1			−0.1	mA
I_O‡	V_{CC} = 5.5 V,	V_O = 2.25 V	−30		−112	−30		−112	mA
I_{CC}	V_{CC} = 5.5 V	See Note 1		16			16		mA

†All typical values are at V_{CC} = 5 V, T_A = 25°C.
‡The output conditions have been chosen to produce a current that closely approximates one half of the true short-circuit output current, I_{OS}.
NOTE 1: With 4.5 Volts applied to the serial input and all other inputs except the clock grounded, I_{CC} is measured after a clock transition from 0 to 4.5 volts.

switching characteristics (see Note 1)

PARAMETER	FROM (INPUT)	TO (OUTPUT)	V_{CC} = 4.5 V to 5.5 V, C_L = 50 pF, R1 = 500 Ω, T_A = MIN to MAX						UNIT
			SN54ALS166			SN74ALS166			
			MIN	TYP†	MAX	MIN	TYP†	MAX	
f_{max}				60			60		MHz
t_{PHL}	$\overline{CLR}$	Q_H		10			10		ns
t_{PLH}	CLK	Q_H		12			12		ns
t_{PHL}				13			13		

†All typical values are at V_{CC} = 5 V, T_A = 25°C.
NOTE 2: For load circuit and voltage waveforms, see page 1-12.
Additional information on these products can be obtained from the factory as it becomes available.

ALS AND AS CIRCUITS

POST OFFICE BOX 225012 • DALLAS, TEXAS 75265

2

ALS AND AS CIRCUITS

TYPES SN54ALS168A, SN54ALS169A, SN54AS168, SN54AS169 SN74ALS168A, SN74ALS169A, SN74AS168, AS74AS169 SYNCHRONOUS 4-BIT UP/DOWN DECADE AND BINARY COUNTERS

D2661, DECEMBER 1982—REVISED DECEMBER 1983

- Fully Synchronous Operation for Counting and Programming
- Internal Look-Ahead for Fast Counting
- Carry Output for n-Bit Cascading
- Fully Independent Clock Circuit
- Package Options Include Both Plastic and Ceramic Chip Carriers in Addition to Plastic and Ceramic DIPs
- Dependable Texas Instruments Quality and Reliability

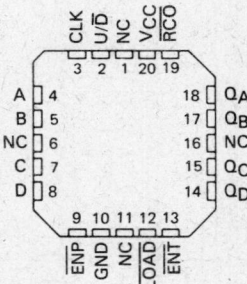

NC — no internal connection.

description

These synchronous presettable counters feature an internal carry look-ahead for cascading in high-speed counting applications. The 'ALS168A and 'AS168 are decade counters and the 'ALS169A and 'AS169 are 4-bit binary counters. Synchronous operation is provided by having all flip-flops clocked simultaneously so that the outputs change coincident with each other when so instructed by the count-enable inputs and internal gating. This mode of operation helps eliminate the output counting spikes that are normally associated with asynchronous (ripple clock) counters. A buffered clock input triggers the four flip-flops on the rising (positive-going) edge of the clock waveform.

These counters are fully programmable; that is, the outputs may each be preset to either level. The load input circuitry allows loading with the carry-enable output of cascaded counters. As loading is synchronous, setting up a low level at the load input disables the counter and causes the outputs to agree with the data inputs after the next clock pulse.

The carry look-ahead circuitry provides for cascading counters for n-bit synchronous application without additional gating. Instrumental in accomplishing this function are two count-enable inputs and a carry output. Both count enable inputs ($\overline{ENP}$ and $\overline{ENT}$) must be low to count. The direction of the count is determined by the level of the U/$\overline{D}$ input. When U/$\overline{D}$ is high, the counter counts up; when low, it counts down. Input $\overline{ENT}$ is fed forward to enable the carry output. The ripple carry output ($\overline{RCO}$) thus enabled will produce a low-level pulse while the count is zero (all inputs low) counting down or maximum (9 or 15) counting up. This low-level overflow carry pulse can be used to enable successive cascaded stages. Transitions at $\overline{ENP}$ or $\overline{ENT}$ are allowed regardless of the level of the clock input. All inputs are diode-clamped to minimize transmission-line effects, thereby simplifying system design.

These counters feature a fully independent clock circuit. Changes at control inputs ($\overline{ENP}$, $\overline{ENT}$, $\overline{LOAD}$, U/$\overline{D}$) that will modify the operating mode have no effect on the contents of the counter until clocking occurs. The function of the counter (whether enabled, disabled, loading, or counting) will be dictated solely by the conditions meeting the stable setup and hold times.

The SN54ALS168A, SN54AS168, SN54ALS169A, and SN54AS169 are characterized for operation over the full military temperature range of −55°C to 125°C. The SN74ALS168A, SN74AS168, SN74ALS169A, and SN74AS169 are characterized for operation from 0°C to 70°C.

Copyright © 1982 by Texas Instruments Incorporated

TYPES SN54ALS168A, SN54AS168, SN74ALS168A, SN74AS168
SYNCHRONOUS 4-BIT UP/DOWN DECADE COUNTERS

'ALS168A, 'AS168 logic diagram (positive logic)   'ALS168A, 'AS168 logic symbol

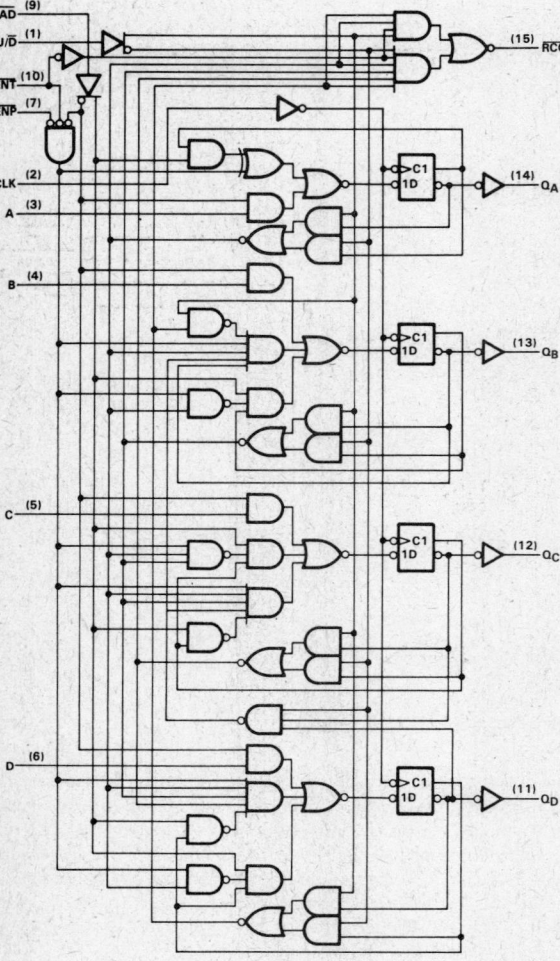
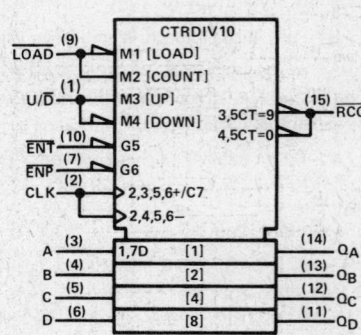

Pin numbers shown are for J and N packages.

2-158

ALS AND AS CIRCUITS

TEXAS INSTRUMENTS
POST OFFICE BOX 225012 • DALLAS, TEXAS 75265

TYPES SN54ALS169A, SN54AS169, SN74ALS169A, SN74AS169
SYNCHRONOUS 4-BIT UP/DOWN BINARY COUNTERS

'ALS169A, AS169 logic diagram (positive logic)

'ALS169A, AS169 logic symbol

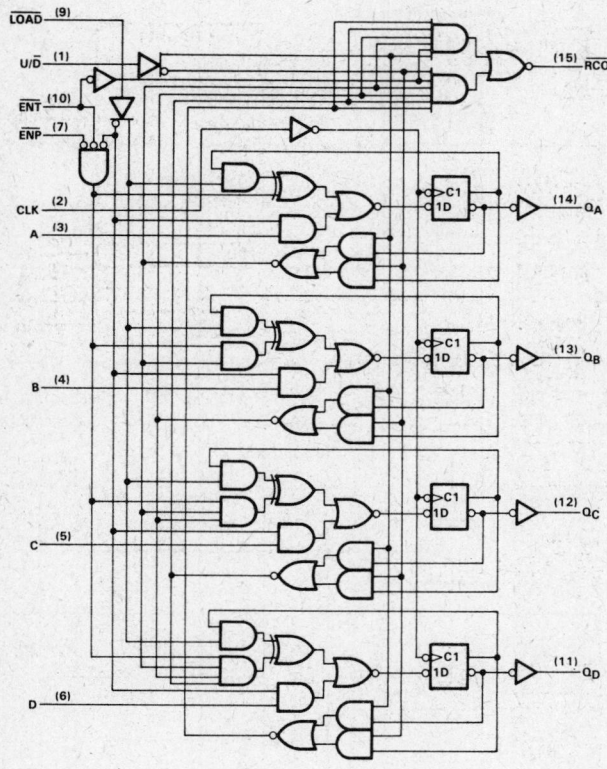

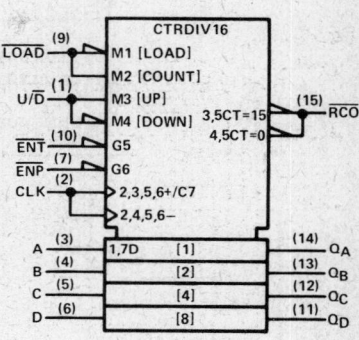

Pin numbers shown are for J and N packages.

TEXAS INSTRUMENTS
POST OFFICE BOX 225012 • DALLAS, TEXAS 75265

TYPES SN54ALS168A, SN54AS168, SN74ALS168A, SN74AS168
SYNCHRONOUS 4-BIT UP/DOWN DECADE COUNTERS

'ALS168A, 'AS168 typical load, count, and inhibit sequences

Illustrated below is the following sequence:

1. Load (preset) to BCD seven
2. Count up to eight, nine (maximum), zero, one, and two
3. Inhibit
4. Count down to one, zero (minimum), nine, eight, and seven

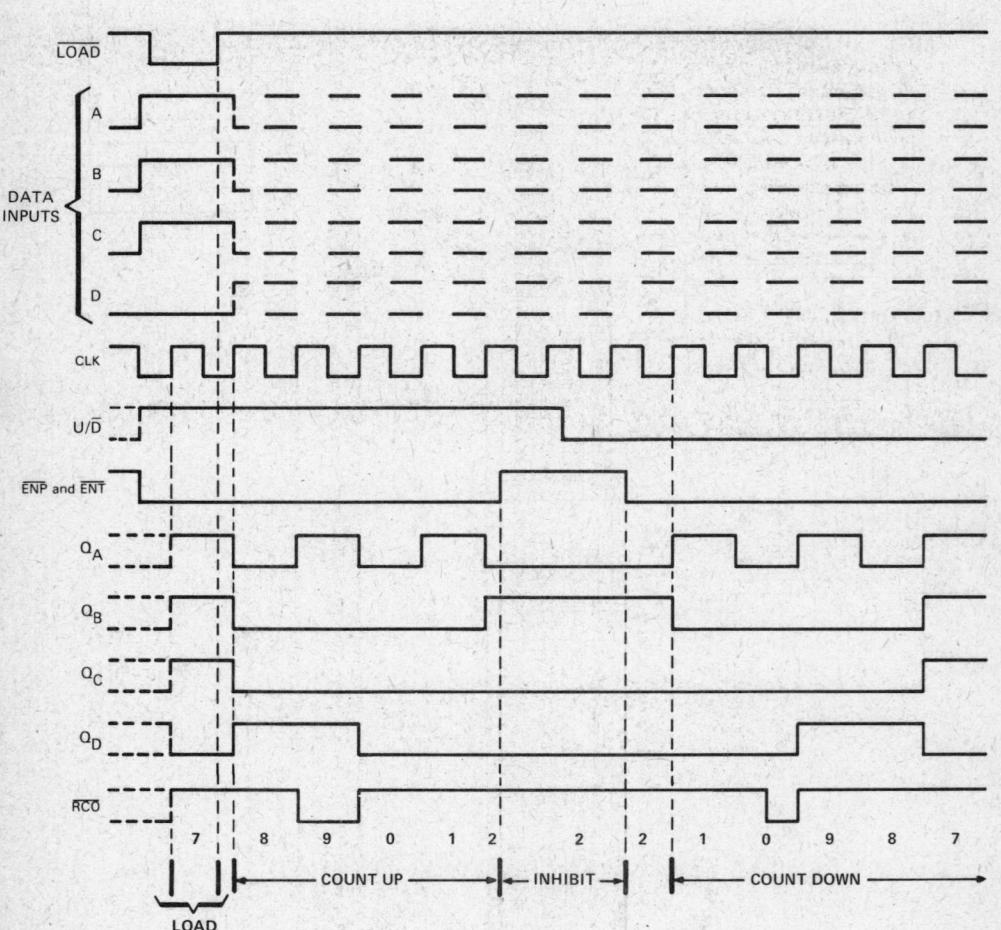

TYPES SN54ALS169A, SN54AS169, SN74ALS169A, SN74AS169
SYNCHRONOUS 4-BIT UP/DOWN BINARY COUNTERS

'ALS169A, 'AS169 typical load, count, and inhibit sequences

Illustrated below is the following sequence:

1. Load (preset) to binary thirteen
2. Count up to fourteen, fifteen (maximum), zero, one, and two
3. Inhibit
4. Count down to one, zero (minimum), fifteen, fourteen, and thirteen

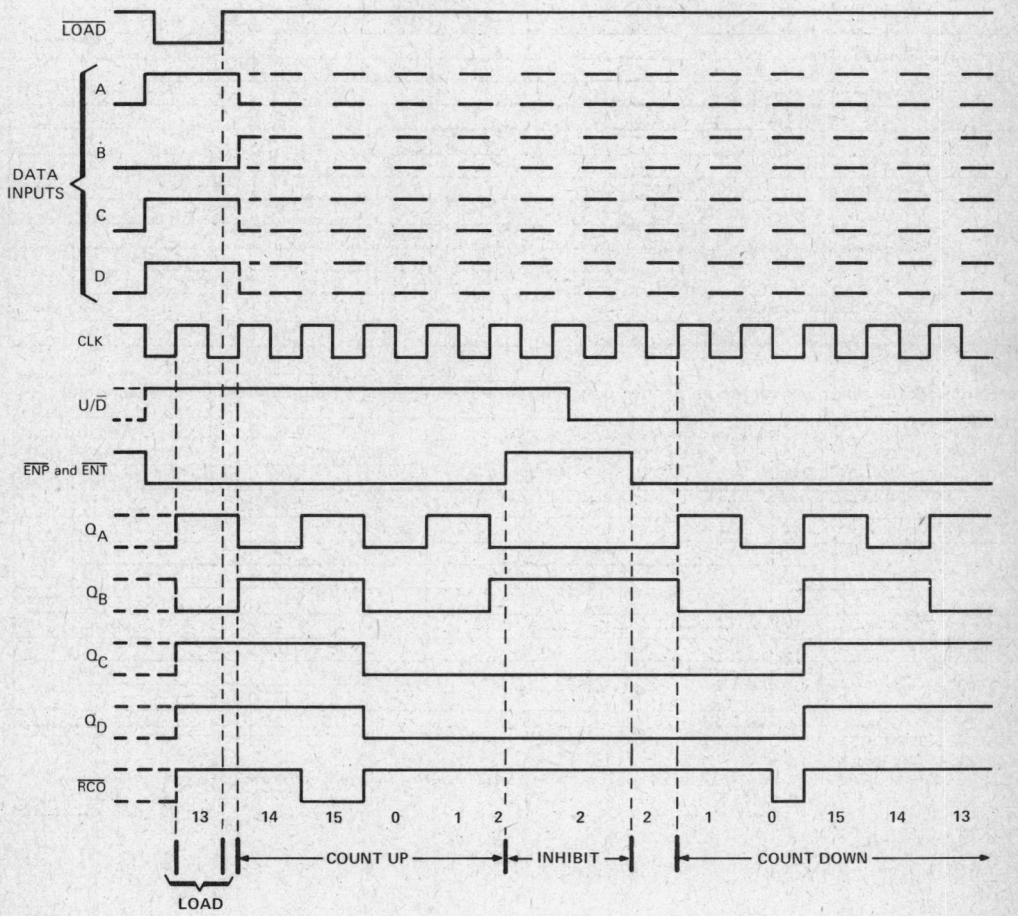

TEXAS INSTRUMENTS
POST OFFICE BOX 225012 • DALLAS, TEXAS 75265

TYPES SN54ALS168A, SN54ALS169A, SN74ALS168A, SN74ALS169A
SYNCHRONOUS 4-BIT UP/DOWN DECADE AND BINARY COUNTERS

absolute maximum ratings over operating free-air temperature range (unless otherwise noted)

Supply voltage, V_{CC} .... 7 V
Input voltage .... 7 V
Operating free-air temperature range: SN54ALS168A, SN54ALS169A .... −55°C to 125°C
                                       SN74ALS168A, SN74ALS169A .... 0°C to 70°C
Storage temperature range .... −65°C to 150°C

recommended operating conditions

		SN54ALS168A SN54ALS169A			SN74ALS168A SN74ALS169A			UNIT
		MIN	NOM	MAX	MIN	NOM	MAX	
V_{CC}	Supply voltage	4.5	5	5.5	4.5	5	5.5	V
V_{IH}	High-level input voltage	2			2			V
V_{IL}	Low-level input voltage			0.8			0.8	V
I_{OH}	High-level output current			−0.4			−0.4	mA
I_{OL}	Low-level output current			4			8	mA
f_{clock}	Clock frequency	0		25	0		30	MHz
t_w	Pulse duration CLK high or low	20			16.5			ns
t_{su}	Setup time before CLK↑ A, B, C, or D	20			15			ns
	ENP or ENT	25			20			
	LOAD	20			15			
	U/$\overline{D}$	20			15			
t_h	Hold time, data after CLK↑	0			0			ns
T_A	Operating free-air temperature	−55		125	0		70	°C

electrical characteristics over recommended operating free-air temperature range (unless otherwise noted)

PARAMETER		TEST CONDITIONS		SN54ALS168A SN54ALS169A			SN74ALS168A SN74ALS169A			UNIT
				MIN	TYP†	MAX	MIN	TYP†	MAX	
V_{IK}		V_{CC} = 4.5 V,	I_I = −18 mA			−1.5			−1.5	V
V_{OH}		V_{CC} = 4.5 V to 5.5 V,	I_{OH} = −0.4 mA	V_{CC}−2			V_{CC}−2			V
V_{OL}		V_{CC} = 4.5 V,	I_{OL} = 4 mA		0.25	0.4		0.25	0.4	V
		V_{CC} = 4.5 V,	I_{OL} = 8 mA					0.35	0.5	
I_I		V_{CC} = 5.5 V,	V_I = 7 V			0.1			0.1	mA
I_{IH}		V_{CC} = 5.5 V,	V_I = 2.7 V			20			20	μA
I_{IL}		V_{CC} = 5.5 V,	V_I = 0.4 V			−0.2			−0.2	mA
I_O‡	RCO	V_{CC} = 5.5 V,	V_O = 2.25 V	−15		−70	−15		−70	mA
	Q			−30		−112	−30		−112	
I_{CC}		V_{CC} = 5.5 V			15	25		15	25	mA

†All typical values are at V_{CC} = 5 V, T_A = 25°C.
‡The output conditions have been chosen to produce a current that closely approximates one half of the true short-circuit output current, I_{OS}.

ALS AND AS CIRCUITS

Texas Instruments
POST OFFICE BOX 225012 • DALLAS, TEXAS 75265

TYPES SN54ALS168A, SN54ALS169A, SN74ALS168A, SN74ALS169A
SYNCHRONOUS 4-BIT UP/DOWN DECADE AND BINARY COUNTERS

'ALS168A, 'ALS169A switching characteristics (see Note 1)

PARAMETER	FROM (INPUT)	TO (OUTPUT)	V_{CC} = 4.5 V to 5.5 V, C_L = 50 pF, R_L = 500 Ω, T_A = MIN to MAX				UNIT
			SN54ALS168A SN54ALS169A		SN74ALS168A SN74ALS169A		
			MIN	MAX	MIN	MAX	
f_{max}			25		30		MHz
t_{PLH}	CLK	$\overline{RCO}$	10	32	10	28	ns
t_{PHL}			6	22	6	18	
t_{PLH}	CLK	Any Q	5	19	5	16	ns
t_{PHL}			5	20	5	16	
t_{PLH}	ENT	$\overline{RCO}$	5	19	5	16	ns
t_{PHL}			3	16	3	13	
t_{PLH}	U/$\overline{D}$	$\overline{RCO}$	5	25	5	23	ns
t_{PHL}			5	23	5	19	

NOTE 1: For load circuit and voltage waveforms, see page 1-12.

ALS AND AS CIRCUITS

TEXAS INSTRUMENTS
POST OFFICE BOX 225012 • DALLAS, TEXAS 75265

TYPES SN54AS168, SN54AS169, SN74AS168, SN74AS169
SYNCHRONOUS 4-BIT UP/DOWN DECADE AND BINARY COUNTERS

absolute maximum ratings over operating free-air temperature range (unless otherwise noted)

Supply voltage, V_{CC} .... 7 V
Input voltage .... 7 V
Operating free-air temperature range: SN54AS168, SN54AS169 .... −55°C to 125°C
$\qquad\qquad\qquad\qquad\qquad\qquad\qquad$ SN74AS168, SN74AS169 .... 0°C to 70°C
Storage temperature range .... −65°C to 150°C

recommended operating conditions

			SN54AS168 SN54AS169			SN74AS168 SN74AS169			UNIT
			MIN	NOM	MAX	MIN	NOM	MAX	
V_{CC}	Supply voltage		4.5	5	5.5	4.5	5	5.5	V
V_{IH}	High-level input voltage		2			2			V
V_{IL}	Low-level input voltage				0.8			0.8	V
I_{OH}	High-level output current				−2			−2	mA
I_{OL}	Low-level output current				20			20	mA
f_{clock}	Clock frequency								MHz
t_w	Pulse duration	CLK high or low							ns
t_{su}	Setup time before CLK↑	A, B, C, or D							ns
		ENP or $\overline{ENT}$							
		$\overline{LOAD}$							
		$U/\overline{D}$							
t_h	Hold time, data after CLK↑								ns
T_A	Operating free-air temperature		−55		125	0		70	°C

electrical characteristics over recommended operating free-air temperature range (unless otherwise noted)

PARAMETER		TEST CONDITIONS		SN54AS168 SN54AS169			SN74AS168 SN74AS169			UNIT
				MIN	TYP[†]	MAX	MIN	TYP[†]	MAX	
V_{IK}		V_{CC} = 4.5 V,	I_I = −18 mA			−1.2			−1.2	V
V_{OH}		V_{CC} = 4.5 V to 5.5 V,	I_{OH} = −2 mA	V_{CC} − 2			V_{CC} − 2			V
V_{OL}		V_{CC} = 4.5 V,	I_{OL} = 20 mA		0.25	0.5		0.25	0.5	V
I_I		V_{CC} = 5.5 V,	V_I = 7 V			0.1			0.1	mA
I_{IH}		V_{CC} = 5.5 V,	V_I = 2.7 V			20			20	µA
I_{IL}	Load, $\overline{ENT}$, $U/\overline{D}$	V_{CC} = 5.5 V,	V_I = 0.4 V			−1			−1	mA
	All others					−0.5			−0.5	
I_O[‡]		V_{CC} = 5.5 V,	V_O = 2.25 V	−30		−112	−30		−112	mA
I_{CC}		V_{CC} = 5.5 V			46			46		mA

[†] All typical values are at V_{CC} = 5 V, T_A = 25°C.
[‡] The output conditions have been chosen to produce a current that closely approximates one half of the true short-circuit output current, I_{OS}.

Additional information on these products can be obtained from the factory as it becomes available.

PRODUCT PREVIEW

This page contains information on a product under development. Texas Instruments reserves the right to change or discontinue this product without notice.

Texas Instruments
POST OFFICE BOX 225012 • DALLAS, TEXAS 75265

TYPES SN54AS168, SN54AS169, SN74AS168, SN74AS169
SYNCHRONOUS 4-BIT UP/DOWN DECADE AND BINARY COUNTERS

'AS168, 'AS169 switching characteristics (see Note 1)

PARAMETER	FROM (INPUT)	TO (OUTPUT)	V_{CC} = 4.5 V to 5.5 V, C_L = 50 pF, R_L = 500 Ω, T_A = MIN to MAX						UNIT
			SN54AS168 SN54AS169			SN74AS168 SN74AS169			
			MIN	TYP†	MAX	MIN	TYP†	MAX	
f_{max}									MHz
t_{PLH}	CLK	$\overline{RCO}$ (with $\overline{LOAD}$ high)		9.5			9.5		ns
t_{PLH}	CLK	$\overline{RCO}$ (with $\overline{LOAD}$ low)		5.5			5.5		
t_{PHL}		$\overline{RCO}$		7.5			7.5		
t_{PLH}	CLK	Any Q		5			5		ns
t_{PHL}				6			6		
t_{PLH}	$\overline{ENT}$	$\overline{RCO}$		3			3		ns
t_{PHL}				4			4		
t_{PLH}	U/$\overline{D}$	$\overline{RCO}$		7.5			7.5		ns
t_{PHL}				10.5			10.5		

†All typical values are at V_{CC} = 5 V, T_A = 25°C.
NOTE 1: For load circuit and voltage waveforms, see page 1-12.

Additional information on these products can be obtained from the factory as it becomes available.

PRODUCT PREVIEW

This page contains information on a product under development. Texas Instruments reserves the right to change or discontinue this product without notice.

TEXAS INSTRUMENTS
POST OFFICE BOX 225012 • DALLAS, TEXAS 75265

2 ALS AND AS CIRCUITS

TYPES SN54ALS174, SN54ALS175, SN54AS174, SN54AS175 SN74ALS174, SN74ALS175, SN74AS174, SN74AS175
HEX/QUADRUPLE D-TYPE FLIP-FLOPS WITH CLEAR

D2661, APRIL 1982—REVISION DECEMBER 1983

- 'ALS174 and 'AS174 Contain Six Flip-Flops with Single-Rail Outputs
- 'ALS175 and 'AS175 Contain Four Flip-Flops with Double-Rail Outputs
- Buffered Clock and Direct Clear Inputs
- Applications Include:
  - Buffer/Storage Registers
  - Shift Registers
  - Pattern Generators
- Fully Buffered Outputs for Maximum Isolation from External Disturbance ('AS only)
- Package Options Include Both Plastic and Ceramic Chip Carriers in Addition to Plastic and Ceramic DIPs
- Dependable Texas Instruments Quality and Reliability

SN54ALS174, SN54AS174 . . . J PACKAGE
SN74ALS174, SN74AS174 . . . N PACKAGE
(TOP VIEW)

```
 CLR [ 1   16 ] VCC
  1Q [ 2   15 ] 6Q
  1D [ 3   14 ] 6D
  2D [ 4   13 ] 5D
  2Q [ 5   12 ] 5Q
  3D [ 6   11 ] 4D
  3Q [ 7   10 ] 4Q
 GND [ 8    9 ] CLK
```

SN54ALS174, SN54AS174 . . . FH PACKAGE
SN74ALS174, SN74AS174 . . . FN PACKAGE
(TOP VIEW)

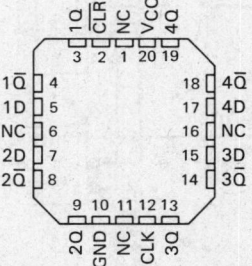

description

These monolithic, positive-edge-triggered flip-flops utilize TTL circuitry to implement D-type flip-flop logic. All have a direct clear input and the 'ALS175 and 'AS175 feature complementary outputs from each flip-flop.

Information at the D inputs meeting the setup time requirements is transferred to the outputs on the positive-going edge of the clock pulse. Clock triggering occurs at a particular voltage level and is not directly related to the transition time of the positive-going pulse. When the clock input is at either the high or low level, the D input signal has no effect at the output.

These circuits are fully compatible for use with most TTL circuits.

The SN54ALS174, SN54ALS175, SN54AS174, and SN54AS175 are characterized for operation over the full military temperature range of $-55\,°C$ to $125\,°C$. The SN74ALS174, SN74ALS175, SN74AS174, and SN74AS175 are characterized for operation from $0\,°C$ to $70\,°C$.

SN54ALS175, SN54AS175 . . . J PACKAGE
SN74ALS175, SN74AS175 . . . N PACKAGE
(TOP VIEW)

```
 CLR [ 1   16 ] VCC
  1Q [ 2   15 ] 4Q
  1Q̄ [ 3   14 ] 4Q̄
  1D [ 4   13 ] 4D
  2D [ 5   12 ] 3D
  2Q̄ [ 6   11 ] 3Q̄
  2Q [ 7   10 ] 3Q
 GND [ 8    9 ] CLK
```

SN54ALS175, SN54AS175 . . . FH PACKAGE
SN74ALS175, SN74AS175 . . . FN PACKAGE
(TOP VIEW)

FUNCTION TABLE
(EACH FLIP-FLOP)

INPUTS			OUTPUTS	
CLR	CLK	D	Q	Q̄†
L	X	X	L	H
H	↑	H	H	L
H	↑	L	L	H
H	L	X	Q₀	Q̄₀

† 'ALS175 and 'AS175 only

NC — No internal connection.

Copyright © 1983 by Texas Instruments Incorporated

TEXAS INSTRUMENTS
POST OFFICE BOX 225012 • DALLAS, TEXAS 75265

TYPES SN54ALS174, SN54ALS175, SN54AS174, SN54AS175, SN74ALS174, SN74ALS175, SN74AS174, SN74AS175
HEX/QUADRUPLE D-TYPE FLIP-FLOPS WITH CLEAR

logic symbols

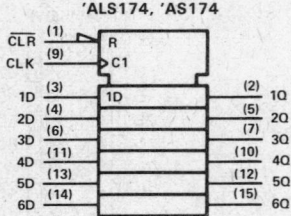
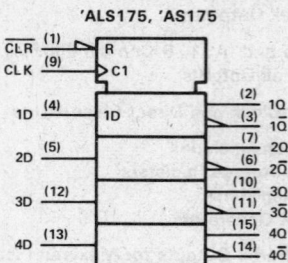

logic diagrams (positive logic)

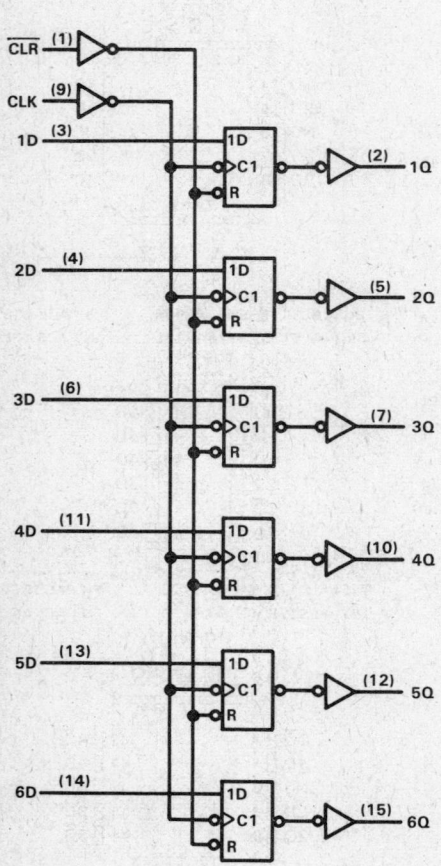
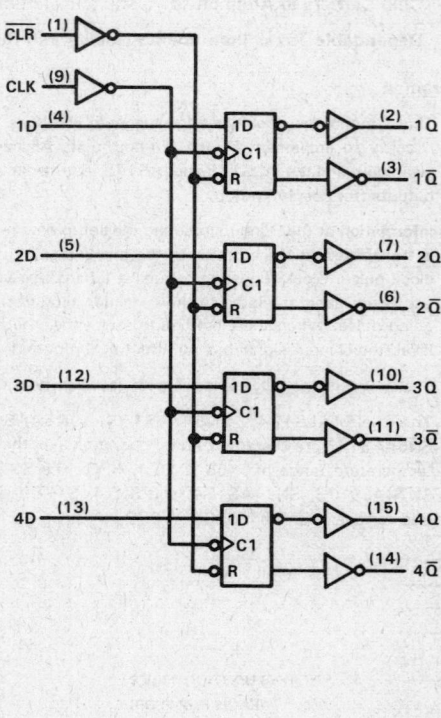

Pin numbers shown are for J and N packages.

Texas Instruments
POST OFFICE BOX 225012 • DALLAS, TEXAS 75265

TYPES SN54ALS174, SN54ALS175, SN74ALS174, SN74ALS175
HEX/QUADRUPLE D-TYPE FLIP-FLOPS WITH CLEAR

absolute maximum ratings over operating free-air temperature range (unless otherwise noted)

Supply voltage, V_{CC} .... 7 V
Input voltage .... 7 V
Operating free-air temperature range: SN54ALS174, SN54ALS175 .... −55°C to 125°C
 SN74ALS174, SN74ALS175 .... 0°C to 70°C
Storage temperature range .... −65°C to 150°C

recommended operating conditions

			SN54ALS174 SN54ALS175			SN74ALS174 SN74ALS175			UNIT
			MIN	NOM	MAX	MIN	NOM	MAX	
V_{CC}	Supply voltage		4.5	5	5.5	4.5	5	5.5	V
V_{IH}	High-level input voltage		2			2			V
V_{IL}	Low-level input voltage				0.8			0.8	V
I_{OH}	High-level output current				−0.4			−0.4	mA
I_{OL}	Low-level output current				4			8	mA
f_{clock}	Clock frequency		0		40	0		50	MHz
t_w	Pulse duration	$\overline{CLR}$ low	15			10			ns
		CLK high	12.5			10			
		CLK low	12.5			10			
t_{su}	Setup time before CLK↑	Data	15			10			ns
		$\overline{CLR}$ inactive	8			6			
t_h	Hold time, data after CLK↑		0			0			ns
T_A	Operating free-air temperature		−55		125	0		70	°C

electrical characteristics over recommended operating free-air temperature range (unless otherwise noted)

PARAMETER		TEST CONDITIONS		SN54ALS174 SN54ALS175			SN74ALS174 SN74ALS175			UNIT
				MIN	TYP[†]	MAX	MIN	TYP[†]	MAX	
V_{IK}		V_{CC} = 4.5 V,	I_I = −18 mA			−1.5			−1.5	V
V_{OH}		V_{CC} = 4.5 V to 5.5 V	I_{OH} = −0.4 mA	V_{CC}−2			V_{CC}−2			V
V_{OL}		V_{CC} = 4.5 V,	I_{OL} = 4 mA		0.25	0.4		0.25	0.4	V
		V_{CC} = 4.5 V	I_{OL} = 8 mA					0.35	0.5	
I_I		V_{CC} = 5.5 V,	V_I = 7 V			0.1			0.1	mA
I_{IH}		V_{CC} = 5.5 V,	V_I = 2.7 V			20			20	µA
I_{IL}		V_{CC} = 5.5 V,	V_I = 0.4 V			−0.1			−0.1	mA
I_O[‡]		V_{CC} = 5.5 V,	V_O = 2.25 V	−30		−112	−30		−112	mA
I_{CC}	'ALS174	V_{CC} = 5.5 V,	See Note 1		11	19		11	19	mA
	'ALS175				8	14		9	14	

[†] All typical values are at V_{CC} = 5 V, T_A = 25°C.
[‡] The output conditions have been chosen to produce a current that closely approximates one half of the true short-circuit output current, I_{OS}.
NOTE 1: I_{CC} is measured with D inputs and $\overline{CLR}$ grounded, and CLK at 4.5 V.

TYPES SN54ALS174, SN54ALS175, SN74ALS174, SN74ALS175
HEX/QUADRUPLE D-TYPE FLIP-FLOPS WITH CLEAR

switching characteristics (see Note 2)

PARAMETER	FROM (INPUT)	TO (OUTPUT)	V_{CC} = 4.5 V to 5.5 V, C_L = 50 pF, R_L = 500 Ω, T_A = MIN to MAX				UNIT
			SN54ALS174 SN54ALS175		SN74ALS174 SN74ALS175		
			MIN	MAX	MIN	MAX	
f_{max}			40		50		MHz
t_{PLH}	$\overline{CLR}$	Any $\overline{Q}$ ('ALS175)	5	20	5	18	ns
t_{PHL}		Any Q	8	26	8	23	
t_{PLH}	CLK	Any Q	3	17	3	15	ns
t_{PHL}		(or $\overline{Q}$, 'ALS175)	5	20	5	17	

NOTE 2: For load circuit and voltage waveforms, see page 1-12.

D flip-flop signal conventions

It is TI practice to name the outputs and other inputs of a D-type flip-flop and to draw its logic symbol based on the assumption of true data (D) inputs. Then outputs that produce data in phase with the data inputs are called Q and those producing complementary data are called $\overline{Q}$. An input that causes a Q output to go high or a $\overline{Q}$ output to go low is called Preset; an input that causes a $\overline{Q}$ output to go high or a Q output to go low is called Clear. Bars are used over these pin names ($\overline{PRE}$ and $\overline{CLR}$) if they are active low.

In some applications it may be advantageous to redesignate the data input $\overline{D}$. In that case all the other inputs and outputs should be renamed as shown below. Also shown are corresponding changes in the graphical symbol. Arbitrary pin numbers are shown in parentheses.

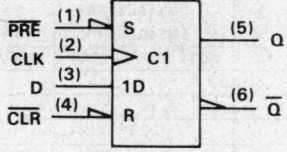 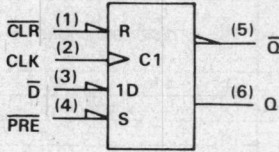

Notice that Q and $\overline{Q}$ exchange names, which causes Preset and Clear to do likewise. Also notice that the polarity indicators (◁) on $\overline{PRE}$ and $\overline{CLR}$ remain since these inputs are still active-low, but that the presence or absence of the polarity indicator changes at $\overline{D}$, Q, and $\overline{Q}$. Of course pin 5 ($\overline{Q}$) is still in phase with the data input $\overline{D}$, but now both are considered active-low.

TYPES SN54AS174, SN54AS175, SN74AS174, SN74AS175
HEX/QUADRUPLE D-TYPE FLIP-FLOPS WITH CLEAR

absolute maximum ratings over operating free-air temperature range (unless otherwise noted)

Supply voltage, V_{CC} .... 7 V
Input voltage .... 7 V
Operating free-air temperature range: SN54AS174, SN54AS175 .... $-55\,°C$ to $125\,°C$
                                      SN74AS174, SN74AS175 .... $0\,°C$ to $70\,°C$
Storage temperature range .... $-65\,°C$ to $150\,°C$

recommended operating conditions

			SN54AS174 SN54AS175			SN74AS174 SN74AS175			UNIT
			MIN	NOM	MAX	MIN	NOM	MAX	
V_{CC}	Supply voltage		4.5	5	5.5	4.5	5	5.5	V
V_{IH}	High-level input voltage		2			2			V
V_{IL}	Low-level input voltage				0.8			0.8	V
I_{OH}	High-level output current				-2			-2	mA
I_{OL}	Low-level output current				20			20	mA
f_{clock}	Clock frequency		0		100	0		100	MHz
t_w	Pulse duration	$\overline{CLR}$ low	5.5			5			ns
		CLK high	4			4			
		CLK low	6			6			
t_{su}	Setup time before CLK↑	Data	4			4			ns
		$\overline{CLR}$ inactive	6			6			
t_h	Hold time, data after CLK↑		1			1			ns
T_A	Operating free-air temperature		-55		125	0		70	°C

electrical characteristics over recommended operating free-air temperature range (unless otherwise noted)

PARAMETER		TEST CONDITIONS		SN54AS174 SN54AS175			SN74AS174 SN74AS175			UNIT
				MIN	TYP†	MAX	MIN	TYP†	MAX	
V_{IK}		$V_{CC} = 4.5$ V,	$I_I = -18$ mA			-1.2			-1.2	V
V_{OH}		$V_{CC} = 4.5$ V to 5.5 V	$I_{OH} = -2$ mA	$V_{CC}-2$			$V_{CC}-2$			V
V_{OL}		$V_{CC} = 4.5$ V,	$I_{OL} = 20$ mA		0.25	0.5		0.25	0.5	V
I_I		$V_{CC} = 5.5$ V,	$V_I = 7$ V			0.1			0.1	mA
I_{IH}		$V_{CC} = 5.5$ V,	$V_I = 2.7$ V			20			20	μA
I_{IL}		$V_{CC} = 5.5$ V,	$V_I = 0.4$ V			-0.5			-0.5	mA
I_O‡		$V_{CC} = 5.5$ V,	$V_O = 2.25$ V	-30		-112	-30		-112	mA
I_{CC}	'AS174	$V_{CC} = 5.5$ V,	See Note 1		30	45		30	45	mA
	'AS175				33			33		

†All typical values are at $V_{CC} = 5$ V, $T_A = 25\,°C$.
‡The output conditions have been chosen to produce a current that closely approximates one half of the true short-circuit output current, I_{OS}.
NOTE 1: I_{CC} is measured with D, CLK, and PRE grounded, then with D, CLK, and $\overline{CLR}$ grounded.

ADVANCE INFORMATION
This page contains information on a new product.
Specifications are subject to change without notice.

TEXAS INSTRUMENTS
POST OFFICE BOX 225012 • DALLAS, TEXAS 75265

TYPES SN54AS174, SN54AS175, SN74AS174, SN74AS175
HEX/QUADRUPLE D-TYPE FLIP-FLOPS WITH CLEAR

'AS174 switching characteristics (see Note 2)

PARAMETER	FROM (INPUT)	TO (OUTPUT)	V_{CC} = 4.5 V to 5.5 V, C_L = 50 pF, R_L = 500 Ω, T_A = MIN to MAX				UNIT
			SN54AS174		SN74AS174		
			MIN	MAX	MIN	MAX	
f_{max}			100		100		MHz
t_{PHL}	$\overline{CLR}$	Any Q	5	15	5	14	ns
t_{PLH}	CLK	Any Q	3.5	9.5	3.5	8	ns
t_{PHL}			4.5	11.5	4.5	10	

'AS175 switching characteristics (see Note 2)

PARAMETER	FROM (INPUT)	TO (OUTPUT)	V_{CC} = 4.5 V to 5.5 V, C_L = 50 pF, R_L = 500 Ω, T_A = MIN to MAX						UNIT
			SN54AS175			SN74AS175			
			MIN	TYP†	MAX	MIN	TYP†	MAX	
f_{max}				160			160		MHz
t_{PLH}	$\overline{CLR}$	Any Q or $\overline{Q}$		5			5		ns
t_{PHL}				5.5			5.5		
t_{PLH}	CLK	Any Q or $\overline{Q}$		4			4		ns
t_{PHL}				4			4		

†All typical values are at V_{CC} = 5 V, T_A = 25°C.
NOTE 2: For load circuit and voltage waveforms, see page 1-12.

ADVANCE INFORMATION
This page contains information on a new product.
Specifications are subject to change without notice.

Texas Instruments
POST OFFICE BOX 225012 • DALLAS, TEXAS 75265

TYPES SN54AS181A, SN54AS881A, SN74AS181A, SN74AS881A
ARITHMETIC LOGIC UNITS/FUNCTION GENERATORS

D2661, DECEMBER 1982 – REVISED DECEMBER 1983

- Package Options Include the 'AS181A in Compact 300-mil or Standard 600-mil DIPs. The 'AS881A Is Offered in 300-mil DIPs. Both Devices Are Available in Both Plastic and Ceramic Chip Carriers
- Full Look-Ahead for High-Speed Operations on Long Words
- Arithmetic Operating Modes:
    Addition
    Subtraction
    Shift Operand A One Position
    Magnitude Comparison
    Plus Twelve Other Arithmetic Operations
- Logic Function Modes
    Exclusive-OR
    Comparator
    AND, NAND, OR, NOR
    'AS881A Provides Status
    Register Checks
    Plus Ten Other Logic Operations
- Dependable Texas Instruments Quality and Reliability

SN54AS181A . . . . J OR JT PACKAGE
SN54AS881A . . . . JT PACKAGE
SN74AS181A . . . . N OR NT PACKAGE
SN74AS881A . . . . NT PACKAGE
(TOP VIEW)

```
     ___
     B0 [ 1    24 ] VCC
     ___           ___
     A0 [ 2    23 ] A1
     S3 [ 3    22 ] B1
                    ___
     S2 [ 4    21 ] A2
                    ___
     S1 [ 5    20 ] B2
                    ___
     S0 [ 6    19 ] A3
                    ___
     Cn [ 7    18 ] B3
                    _
     M  [ 8    17 ] G
     __
     F0 [ 9    16 ] Cn+4
     __
     F1 [10    15 ] P
     __
     F2 [11    14 ] A=B
                    __
     GND[12    13 ] F3
```

SN54AS181A, SN54AS881A . . . . FH PACKAGE
SN74AS181A, SN74AS881A . . . . FN PACKAGE

'AS181A, 'AS881A
(TOP VIEW)

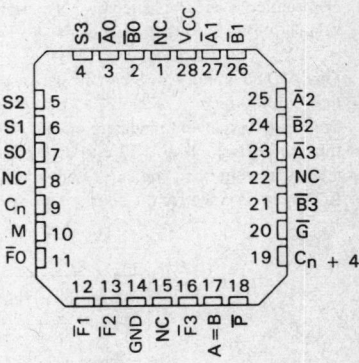

NC – no internal connection

logic symbol

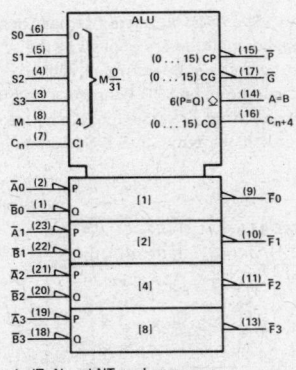

Pin numbers shown are J, JT, N and NT packages.

TYPICAL ADDITION TIMES (C_L = 15 pF, R_L = 280 Ω, T_A = 25°C)

NUMBER OF BITS	ADDITION TIMES			PACKAGE COUNT		CARRY METHOD BETWEEN ALU's
	USING 'AS881A AND 'AS882	USING 'AS181A AND 'AS882	USING 'S181 AND 'S182	ARITHMETIC LOGIC UNITS	LOOK-AHEAD CARRY GENERATORS	
1 to 4	5 ns	5 ns	11 ns	1		NONE
5 to 8	10 ns	10 ns	18 ns	2		RIPPLE
9 to 16	14 ns	14 ns	19 ns	3 or 4	1	FULL LOOK-AHEAD
17 to 64	19 ns	19 ns	28 ns	5 to 16	2 to 5	FULL LOOK-AHEAD

ADVANCE INFORMATION
This document contains information on a new product. Specifications are subject to change without notice.

Copyright © 1982 by Texas Instruments Incorporated

POST OFFICE BOX 225012 • DALLAS, TEXAS 75265

TYPES SN54AS181A, SN54AS881A, SN74AS181A, SN74AS881A
ARITHMETIC LOGIC UNITS/FUNCTION GENERATORS

description

The 'AS181A and 'AS881A are arithmetic logic units (ALU)/function generators that have a complexity of 75 and 77 equivalent gates respectively, on a monolithic chip. These circuits perform 16 binary arithmetic operations on two 4-bit words as shown in Tables 1 and 2. These operations are selected by the four function-select lines (S0, S1, S2, S3) and include addition, subtraction, decrement, and straight transfer. When performing arithmetic manipulations, the internal carries must be enabled by applying a low-level voltage to the mode control input (M). A full carry look-ahead scheme is made available in these devices for fast, simultaneous carry generation by means of two cascade-outputs (pins 15 and 17) for the four bits in the package. When used in conjunction with the SN54AS882 or SN74AS882 full carry look-ahead circuits, high-speed arithmetic operations can be performed. The typical addition times shown previously illustrate the little additional time required for addition of longer words when full carry look-ahead is employed. The method of cascading 'AS882 circuits with these ALU's to provide multi-level full carry look-ahead is illustrated under signal designations.

If high speed is not of importance, a ripple-carry input (C_n) and a ripple-carry output (C_{n+4}) are available. However, the ripple-carry delay has also been minimized so that arithmetic manipulations for small word lengths can be performed without external circuitry.

The 'AS181A and 'AS881A will accommodate active-high or active-low data if the pin designations are interpreted as follows:

PIN NUMBER	2	1	23	22	21	20	19	18	9	10	11	13	7	16	15	17
Active-low data (Table 1)	$\bar{A}0$	$\bar{B}0$	$\bar{A}1$	$\bar{B}1$	$\bar{A}2$	$\bar{B}2$	$\bar{A}3$	$\bar{B}3$	$\bar{F}0$	$\bar{F}1$	$\bar{F}2$	$\bar{F}3$	C_n	C_{n+4}	$\bar{P}$	$\bar{G}$
Active-high data (Table 2)	A0	B0	A1	B1	A2	B2	A3	B3	F0	F1	F2	F3	$\bar{C}_n$	$\bar{C}_{n+4}$	X	Y

Subtraction is accomplished by 1's complement addition where the 1's complement of the subtrahend is generated internally. The resultant output is $A - B - 1$, which requires an end-around or forced carry to provide $A - B$.

The 'AS181A and 'AS881A can also be utilized as a comparator. The A = B output is internally decoded from the function outputs (F0, F1, F2, F3) so that when two words of equal magnitude are applied at the A and B inputs, it will assume a high level to indicate equality (A = B). The ALU must be in the subtract mode with $C_n = H$ when performing this comparison. The A = B output is open-collector so that it can be wire-AND connected to give a comparison for more than four bits. The carry output (C_{n+4}) can also be used to supply relative magnitude information. Again, the ALU must be placed in the subtract mode by placing the function select inputs S3, S2, S1, S0 at L, H, H, L, respectively.

INPUT C_n	OUTPUT C_{n+4}	ACTIVE-LOW DATA (FIGURE 1)	ACTIVE-HIGH DATA (FIGURE 2)
H	H	A ⩾ B	A ⩽ B
H	L	A < B	A > B
L	H	A > B	A < B
L	L	A ⩽ B	A ⩾ B

These circuits have been designed to not only incorporate all of the designer's requirements for arithmetic operations, but also to provide 16 possible functions of two Boolean variables without the use of external circuitry. These logic functions are selected by use of the four function-select inputs (S0, S1, S2, S3) with the mode-control input (M) at a high level to disable the internal carry. The 16 logic functions are detailed in Tables 1 and 2 and include exclusive-OR, NAND, AND, NOR, and OR functions.

TYPES SN54AS181A, SN54AS881A, SN74AS181A, SN74AS881A
ARITHMETIC LOGIC UNITS/FUNCTION GENERATORS

description (continued)

The 'AS881A has the same pinout and same functionality as the 'AS181A except for the $\overline{P}$, $\overline{G}$, and C_{n+4} outputs when the device is in the logic mode (M = H).

In the logic mode the 'AS881 provides the user with a status check on the input words, A and B, and the output word F. While in the logic mode the $\overline{P}$, $\overline{G}$ and C_{n+4} outputs supply status information based upon the following logical combinations:

$\overline{P}$ = F0 + F1 + F2 + F3
$\overline{G}$ = H
C_{n+4} = PC_n

FUNCTION TABLE FOR INPUT BITS EQUAL/NOT EQUAL

S0 = S3 = H, S1 = S2 = L, and M = H

C_n	DATA INPUTS				OUTPUTS $\overline{G}$	$\overline{P}$	C_{n+4}
H	A0 = B0	A1 = B1	A2 = B2	A3 = B3	H	L	H
L	A0 = B0	A1 = B1	A2 = B2	A3 = B3	H	L	L
X	A0≠B0	X	X	X	H	H	L
X	X	A1≠B1	X	X	H	H	L
X	X	X	A2≠B2	X	H	H	L
X	X	X	X	A3≠B3	H	H	L

FUNCTION TABLE FOR INPUT PAIRS HIGH/NOT HIGH

S0 = S1 = S3 = L, S2 = H, and M = H

C_n	DATA INPUTS				OUTPUTS $\overline{G}$	$\overline{P}$	C_{n+4}
H	$\overline{A0}$ or $\overline{B0}$ = L	$\overline{A1}$ or $\overline{B1}$ = L	$\overline{A2}$ or $\overline{B2}$ = L	$\overline{A3}$ or $\overline{B3}$ = L	H	L	H
L	$\overline{A0}$ or $\overline{B0}$ = L	$\overline{A1}$ or $\overline{B1}$ = L	$\overline{A2}$ or $\overline{B2}$ = L	$\overline{A3}$ or $\overline{B3}$ = L	H	L	L
X	$\overline{A0}$ = $\overline{B0}$ = H	X	X	X	H	H	L
X	X	$\overline{A1}$ = $\overline{B1}$ = H	X	X	H	H	L
X	X	X	$\overline{A2}$ = $\overline{B2}$ = H	X	H	H	L
X	X	X	X	$\overline{A3}$ = $\overline{B3}$ = H	H	H	L

The combination of signals on the S3 through S0 control lines determine the operation performed on the data words to generate the output bits Fi. By monitoring the $\overline{P}$ and C_{n+4} outputs, the user can determine if all pairs of input bits are equal (see table above) or if any pair of inputs are both high (see table above). The 'AS881A has the unique feature of providing an A = B status while the exclusive-OR ($\oplus$) function is being utilized. When the control inputs (S3, S2, S1, S0) equal H, L, L, H; a status check is generated to determine whether all pairs (Ai, Bi) are equal in the following manner: $\overline{P}$ = (A0 $\oplus$ B0) + (A1 $\oplus$ B1) + (A2 $\oplus$ B2) + (A3 $\oplus$ B3). This unique bit-by-bit comparison of the data words which is available on the totem pole $\overline{P}$ output is particularly useful when cascading 'AS881's. As the A = B condition is sensed in the first stage the signal is propagated through the same ports used for carry generation in the arithmetic mode ($\overline{P}$ and $\overline{G}$). Thus the A = B status is transmitted to the second stage more quickly without the need for external multiplexing logic. The A = B open-collector output allows the user to check the validity of the bit-by-bit result by comparing the two signals for parity.

If the user wishes to check for any pair of data inputs ($\overline{Ai}$, $\overline{Bi}$) being high, it is necessary to set the control lines (S3,S2,S1,S0) to L, H, L, L. The data pairs will then be ANDed together and the results ORed in the following manner: $\overline{P}$ = $\overline{A0}\overline{B0}$ + $\overline{A1}\overline{B1}$ + $\overline{A2}\overline{B2}$ + $\overline{A3}\overline{B3}$.

S3	S2	S1	S0	M	
L	L	L	L	H	$\overline{P}$ = F0 + F1 + F2 + F3
L	H	L	L	H	$\overline{A0}\overline{B0}$ + $\overline{A1}\overline{B1}$ + $\overline{A2}\overline{B2}$ + $\overline{A3}\overline{B3}$
H	L	L	H	H	(A0 $\oplus$ B0) + (A1 $\oplus$ B1) + (A2 $\oplus$ B2) + (A3 $\oplus$ B3)

signal designations

In both Figures 1 and 2, the polarity indicators (▷) indicate that the associated input or output is active-low with respect to the function shown inside the symbol and the symbols are the same in both figures. The signal designations in Figure 1 agree with the indicated internal functions based on active-low data, and are for use with the logic functions and arithmetic operations shown in Table 1. The signal designations have been changed in Figure 2 to accommodate the logic functions and arithmetic operations for the active-high data given in Table 2. The 'AS181 and 'AS881 together with the 'AS882 and 'S182 can be used with the signal designation of either Figure 1 or Figure 2.

TYPES SN54AS181A, SN54AS881A, SN74AS181A, SN74AS881A
ARITHMETIC LOGIC UNITS/FUNCTION GENERATORS

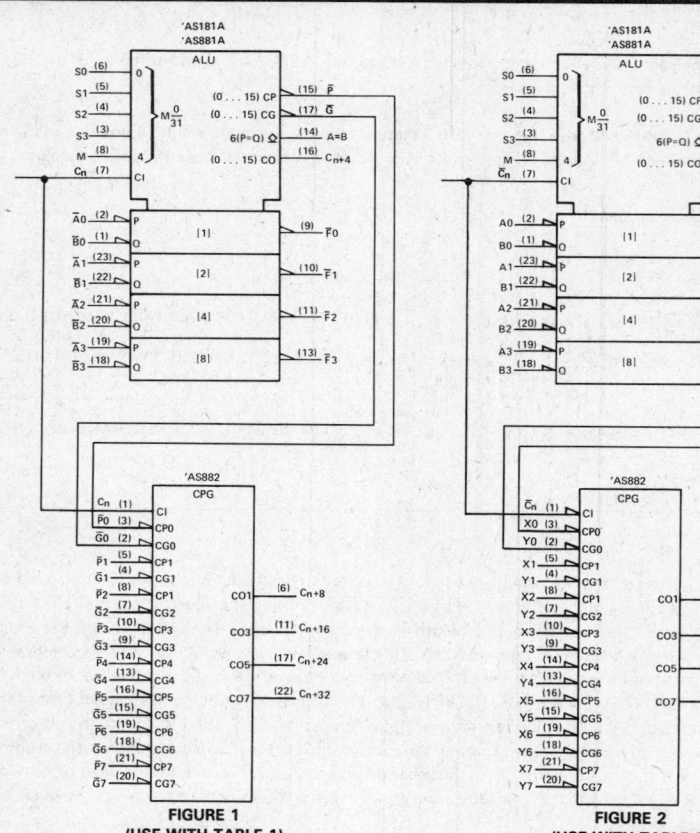

FIGURE 1
(USE WITH TABLE 1)

FIGURE 2
(USE WITH TABLE 2)

TABLE 1

SELECTION				ACTIVE-LOW DATA		
				M = H	M = L; ARITHMETIC OPERATIONS	
S3	S2	S1	S0	LOGIC FUNCTIONS	C_n = L (no carry)	C_n = H (with carry)
L	L	L	L	F = $\overline{A}$	F = A MINUS 1	F = A
L	L	L	H	F = $\overline{AB}$	F = AB MINUS 1	F = AB
L	L	H	L	F = $\overline{A}$ + B	F = $A\overline{B}$ MINUS 1	F = $A\overline{B}$
L	L	H	H	F = 1	F = MINUS 1 (2's COMP)	F = ZERO
L	H	L	L	F = $\overline{A + B}$	F = A PLUS (A + $\overline{B}$)	F = A PLUS (A + $\overline{B}$) PLUS 1
L	H	L	H	F = $\overline{B}$	F = AB PLUS (A + $\overline{B}$)	F = AB PLUS (A + $\overline{B}$) PLUS 1
L	H	H	L	F = A $\oplus$ $\overline{B}$	F = A MINUS B MINUS 1	F = A MINUS B
L	H	H	H	F = A + $\overline{B}$	F = A + $\overline{B}$	F = (A + $\overline{B}$) PLUS 1
H	L	L	L	F = $\overline{A}B$	F = A PLUS (A + B)	F = A PLUS (A + B) PLUS 1
H	L	L	H	F = A $\oplus$ B	F = A PLUS B	F = A PLUS B PLUS 1
H	L	H	L	F = B	F = $A\overline{B}$ PLUS (A + B)	F = $A\overline{B}$ PLUS (A + B) PLUS 1
H	L	H	H	F = A + B	F = (A + B)	F = (A + B) PLUS 1
H	H	L	L	F = 0	F = A PLUS A*	F = A PLUS A PLUS 1
H	H	L	H	F = $A\overline{B}$	F = AB PLUS A	F = AB PLUS A PLUS 1
H	H	H	L	F = AB	F = $A\overline{B}$ PLUS A	F = $A\overline{B}$ PLUS A PLUS 1
H	H	H	H	F = A	F = A	F = A PLUS 1

TABLE 2

SELECTION				ACTIVE-HIGH DATA		
				M = H	M = L; ARITHMETIC OPERATIONS	
S3	S2	S1	S0	LOGIC FUNCTIONS	$\overline{C_n}$ = H (no carry)	$\overline{C_n}$ = L (with carry)
L	L	L	L	F = $\overline{A}$	F = A	F = A PLUS 1
L	L	L	H	F = $\overline{A + B}$	F = A + B	F = (A + B) PLUS 1
L	L	H	L	F = $\overline{A}B$	F = A + $\overline{B}$	F = (A + $\overline{B}$) PLUS 1
L	L	H	H	F = 0	F = MINUS 1 (2's COMPL)	F = ZERO
L	H	L	L	F = $\overline{AB}$	F = A PLUS $A\overline{B}$	F = A PLUS $A\overline{B}$ PLUS 1
L	H	L	H	F = $\overline{B}$	F = (A + B) PLUS $A\overline{B}$	F = (A + B) PLUS $A\overline{B}$ PLUS 1
L	H	H	L	F = A $\oplus$ B	F = A MINUS B MINUS 1	F = A MINUS B
L	H	H	H	F = $A\overline{B}$	F = $A\overline{B}$ MINUS 1	F = $A\overline{B}$
H	L	L	L	F = $\overline{A}$ + B	F = A PLUS AB	F = A PLUS AB PLUS 1
H	L	L	H	F = A $\oplus$ $\overline{B}$	F = A PLUS B	F = A PLUS B PLUS 1
H	L	H	L	F = B	F = (A + $\overline{B}$) PLUS AB	F = (A + $\overline{B}$) PLUS AB PLUS 1
H	L	H	H	F = AB	F = AB MINUS 1	F = AB
H	H	L	L	F = 1	F = A PLUS A*	F = A PLUS A PLUS 1
H	H	L	H	F = A + $\overline{B}$	F = (A + B) PLUS A	F = (A + B) PLUS A PLUS 1
H	H	H	L	F = A + B	F = (A + $\overline{B}$) PLUS A	F = (A + $\overline{B}$) PLUS A PLUS 1
H	H	H	H	F = A	F = A MINUS 1	F = A

*Each bit is shifted to the next more significant position.

TYPES SN54AS181A, SN74AS181A
ARITHMETIC LOGIC UNITS/FUNCTION GENERATORS

logic diagram (positive logic)

'AS181A

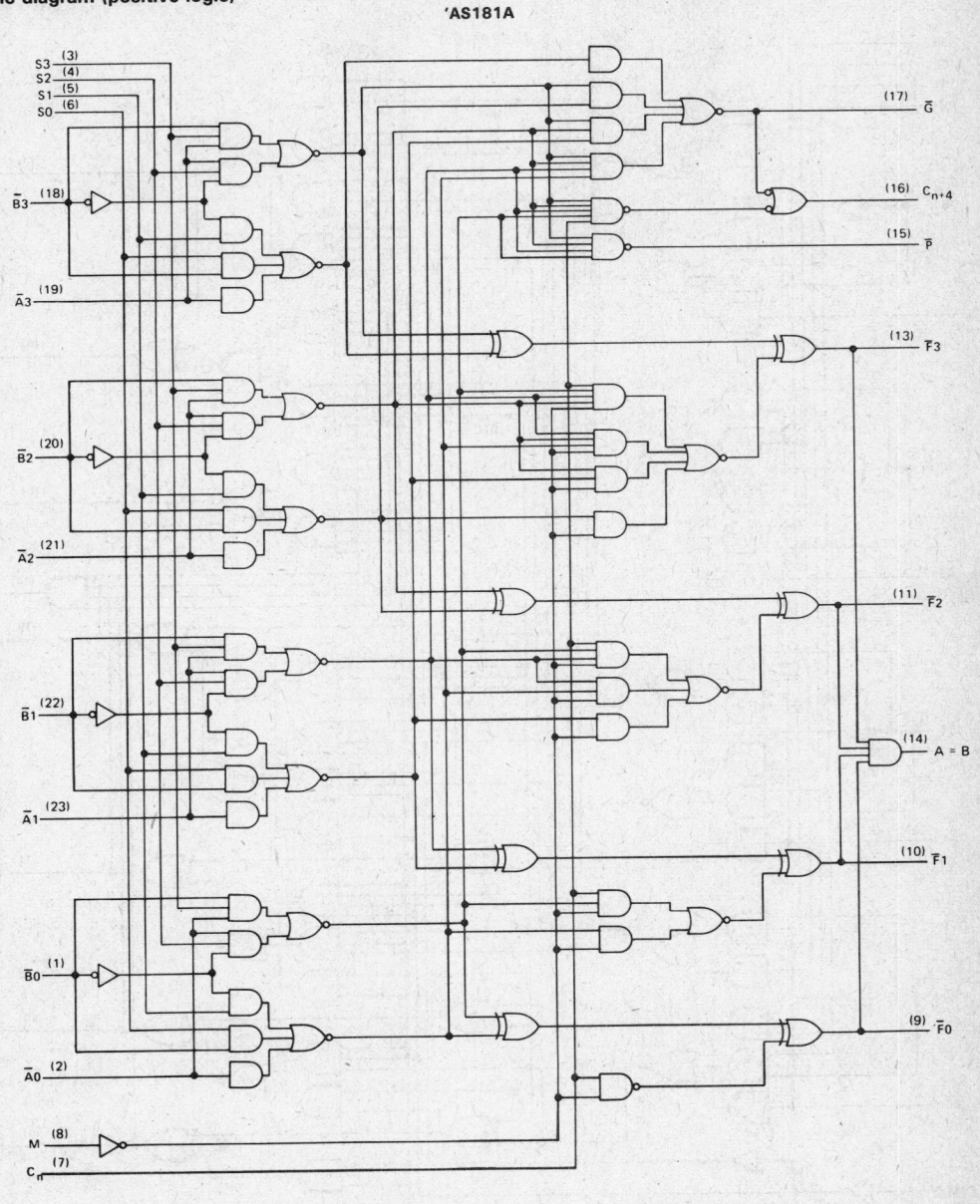

TYPES SN54AS881A, SN74AS881A
ARITHMETIC LOGIC UNITS/FUNCTION GENERATORS

logic diagram (positive logic)

TYPES SN54AS181A, SN54AS881A, SN74AS181A, SN74AS881A
ARITHMETIC LOGIC UNITS/FUNCTION GENERATORS

absolute maximum ratings over operating free-air temperature range (unless otherwise noted)

Supply voltage, V_{CC} .... 7 V
Input voltage .... 7 V
Off-state output voltage (A = B output only) .... 7 V
Operating free-air temperature range: SN54AS181A, SN54AS881A .... −55 °C to 125 °C
 SN74AS181A, SN74AS881A .... 0 °C to 70 °C
Storage temperature range .... −65 °C to 150 °C

recommended operating conditions

			SN54AS'			SN74AS'			UNIT
			MIN	NOM	MAX	MIN	NOM	MAX	
V_{CC}	Supply voltage		4.5	5	5.5	4.5	5	5.5	V
V_{IH}	High-level input voltage		2			2			V
V_{IL}	Low-level input voltage				0.8			0.8	V
V_{OH}	High-level output voltage	A = B output only			5.5			5.5	V
I_{OH}	High-level output current	All outputs except A = B and $\overline{G}$			−2			−2	mA
		$\overline{G}$			−3			−3	mA
I_{OL}	Low-level output current	All outputs except $\overline{G}$			20			20	mA
		$\overline{G}$			48			48	mA
T_A	Operating free-air temperature		−55		125	0		70	°C

TEXAS
INSTRUMENTS
POST OFFICE BOX 225012 • DALLAS, TEXAS 75265

TYPES SN54AS181A, SN54AS881A, SN74AS181A, SN74AS881A
ARITHMETIC LOGIC UNITS/FUNCTION GENERATORS

electrical characteristics over recommended operating free-air temperature range (unless otherwise noted)

PARAMETER		TEST CONDITIONS		SN54AS' MIN	TYP[†]	MAX	SN74AS' MIN	TYP[†]	MAX	UNIT
V_{IK}		V_{CC} = 4.5 V,	I_I = −18 mA			−1.2			−1.2	V
V_{OH}	Any output except A = B	V_{CC} = 4.5 V to 5.5 V,	I_{OH} = −2 mA	V_{CC}−2			V_{CC}−2			V
	$\overline{G}$	V_{CC} = 4.5 V,	I_{OH} = −3 mA	2.4	3.4		2.4	3.4		V
I_{OH}	A = B	V_{CC} = 4.5 V,	V_{OH} = 5.5 V			0.1			0.1	mA
V_{OL}	Any output except $\overline{G}$	V_{CC} = 4.5 V,	I_{OL} = 20 mA		0.3	0.5		0.3	0.5	V
	$\overline{G}$	V_{CC} = 4.5 V,	I_{OL} = 48 mA		0.4	0.5		0.4	0.5	V
I_I	M input	V_{CC} = 5.5 V,	V_I = 7 V			0.1			0.1	mA
	Any A or B input					0.3			0.3	
	Any S input					0.4			0.4	
	Carry input					0.6			0.6	
I_{IH}	M input	V_{CC} = 5.5 V,	V_I = 2.7 V			20			20	µA
	Any A or B input					60			60	
	Any S input					80			80	
	Carry input					120			120	
I_{IL}	M input	V_{CC} = 5.5 V,	V_I = 0.4 V			−2			−2	mA
	Any A or B input					−6			−6	
	Any S input					−8			−8	
	Carry input					−12			−12	
I_O[‡]	All outputs except A = B and $\overline{G}$	V_{CC} = 5.5 V,	V_O = 2.25 V	−30	−45	−112	−30	−45	−112	mA
	$\overline{G}$				−165			−165		
I_{CC}		V_{CC} = 5.5 V	'AS181A		135	200		135	200	mA
			'AS881A		135	210		135	210	

[†] All typical values are at V_{CC} = 5 V, T_A = 25°C.
[‡] The output conditions have been chosen to produce a current that closely approximates one-half of the true short-circuit current, I_{OS}.

TYPES SN54AS181A, SN54AS881A, SN74AS181A, SN74AS881A
ARITHMETIC LOGIC UNITS/FUNCTION GENERATORS

switching characteristics (see Note 1)

PARAMETER	FROM (INPUT)	TO (OUTPUT)	TEST CONDITIONS	$V_{CC} = 5$ V, $C_L = 15$ pF, $R_L = 500\ \Omega$ (280 Ω for A = B), $T_A = 25°C$ 'AS181A 'AS881A			$V_{CC} = 4.5$ V to 5.5 V, $C_L = 50$ pF (15 pF for A = B), $R_L = 500\ \Omega$ (280 Ω for A = B), $T_A = $ MIN to MAX SN54AS181A SN54AS881A						UNIT
				MIN	TYP†	MAX	MIN	TYP†	MAX	MIN	TYP†	MAX	
t_{pd}	C_n	C_{n+4}	M = 0 V, S1 = S2 = 0 V, S0 = S3 = 4.5 V		5			2	7	11	2	7	9	ns
t_{pd}	Any $\bar{A}$ or $\bar{B}$	C_{n+4}	M = 0 V, S1 = S2 = 0 V, S0 = S3 = 4.5 V ($\overline{SUM}$ mode)		6			2	8	14	2	8	12	ns
t_{pd}	Any $\bar{A}$ or $\bar{B}$	C_{n+4}	M = 0 V, S0 = S3 = 0 V, S1 = S2 = 4.5 V ($\overline{DIFF}$ mode)		7			2	8	20	2	8	16	ns
t_{pd}	C_n	Any $\bar{F}$	M = 0 V ($\overline{SUM}$ or $\overline{DIFF}$ mode)		5			3	6	11	3	6	9	ns
t_{pd}	Any $\bar{A}$ or $\bar{B}$	$\bar{G}$	M = 0 V, S1 = S2 = 0 V, S0 = S3 = 4.5 V ($\overline{SUM}$ mode)		4			2	5	9	2	5	7	ns
t_{pd}	Any $\bar{A}$ or $\bar{B}$	$\bar{G}$	M = 0 V, S0 = S3 = 0 V, S1 = S2 = 4.5 V ($\overline{DIFF}$ mode)		5			2	6	12	2	6	9	ns
t_{pd}	Any $\bar{A}$ or $\bar{B}$	$\bar{P}$	M = 0 V, S1 = S2 = 0 V, S0 = S3 = 4.5 V ($\overline{SUM}$ mode)		5			2	6	11	2	6	8	ns
t_{pd}	Any $\bar{A}$ or $\bar{B}$	$\bar{P}$	M = 0 V, S0 = S3 = 0 V, S1 = S2 = 4.5 V ($\overline{DIFF}$ mode)		5			2	6	13	2	6	10	ns
t_{pd}	$\bar{A}i$ or $\bar{B}i$	$\bar{F}i$	M = 0 V, S1 = S2 = 0 V, S0 = S3 = 4.5 V ($\overline{SUM}$ mode)		5			2	5	11	2	5	8	ns
t_{pd}	$\bar{A}i$ or $\bar{B}i$	$\bar{F}i$	M = 0 V, S0 = S1 = 0 V, S1 = S2 = 4.5 V ($\overline{DIFF}$ mode)		5			2	6	12	2	6	10	ns
t_{pd}	$\bar{A}i$ or $\bar{B}i$	$\bar{F}i$	M = 4.5 V (LOGIC mode)		6			2	6	16	2	6	11	ns
t_{pd}	Any A or B	A = B	M = 0 V, S0 = S3 = 0 V, S1 = S2 = 4.5 V ($\overline{DIFF}$ mode)		12			4	14	26	4	14	21	ns

addditional 'AS881A switching characteristics involving status checks (see Note 1)

PARAMETER	FROM (INPUT)	TO (OUTPUT)	TEST CONDITIONS	$V_{CC} = 5$ V, $C_L = 15$ pF, $R_L = 500\ \Omega$, $T_A = 25°C$ 'AS881A			$V_{CC} = 4.5$ V to 5.5 V, $C_L = 50$ pF, $R_L = 500\ \Omega$, $T_A = $ MIN to MAX SN54AS881A			SN74AS881A			UNIT
				MIN	TYP†	MAX	MIN	TYP†	MAX	MIN	TYP†	MAX	
t_{pd}	Any $\bar{A}$ or $\bar{B}$	$\bar{P}$	$C_n = 4.5$ V, M = 4.5 V, S0 = S3 = 4.5 V, S1 = S2 = 0 V, Equality ($\bar{A}i = \bar{B}i$ or $\bar{A}i \neq \bar{B}i$)		8			2	10	19	2	10	15	ns
t_{pd}	Any $\bar{A}$ or $\bar{B}$	C_{n+4}	$C_n = 4.5$ V, M = 4.5 V, S0 = S3 = 4.5 V, S1 = S2 = 0 V, Equality ($\bar{A}i = \bar{B}i$ or $\bar{A}i \neq \bar{B}i$)		10			2	12	24	2	12	18	ns
t_{pd}	Any $\bar{A}$ or $\bar{B}$	$\bar{P}$	$C_n = 4.5$ V, M = 4.5 V, S2 = 4.5 V, S0 = S1 = S3 = 0 V, ($\bar{A}i = \bar{B}i =$ H or $\bar{A}i$ or $\bar{B}i =$ L)		8			2	10	19	2	10	15	ns
t_{pd}	Any $\bar{A}$ or $\bar{B}$	C_{n+4}	$C_n = 4.5$ V, M = 4.5 V, S2 = 4.5 V, S0 = S1 = S3 = 0 V, ($\bar{A}i = \bar{B}i =$ H or $\bar{A}i$ or $\bar{B}i =$ L)		11			2	13	25	2	13	19	ns

$t_{pd} = t_{PHL}$ or t_{PLH}
†All typical values are at $V_{CC} = 5$ V, $T_A = 25°C$.
NOTE 1: For load circuit and voltage waveforms, see page 1-12.

ALS AND AS CIRCUITS

TEXAS INSTRUMENTS
POST OFFICE BOX 225012 • DALLAS, TEXAS 75265

TYPES SN54AS181A, SN54AS881A, SN74AS181A, SN74AS881A
ARITHMETIC LOGIC UNITS/FUNCTION GENERATORS

PARAMETER MEASUREMENT INFORMATION

$\overline{\text{SUM}}$ MODE TEST TABLE
FUNCTION INPUTS: $S0 = S3 = 4.5\text{ V}, S1 = S2 = M = 0\text{ V}$

PARAMETER	INPUT UNDER TEST	OTHER INPUT SAME BIT		OTHER DATA INPUTS		OUTPUT UNDER TEST	OUTPUT WAVEFORM (SEE NOTE A)
		APPLY 4.5 V	APPLY GND	APPLY 4.5 V	APPLY GND		
t_{PLH} / t_{PHL}	$\overline{A}_i$	$\overline{B}_i$	None	Remaining $\overline{A}$ and $\overline{B}$	C_n	$\overline{F}_i$	In-Phase
t_{PLH} / t_{PHL}	$\overline{B}_i$	$\overline{A}_i$	None	Remaining $\overline{A}$ and $\overline{B}$	C_n	$\overline{F}_i$	In-Phase
t_{PLH} / t_{PHL}	$\overline{A}_i$	$\overline{B}_i$	None	None	Remaining $\overline{A}$ and $\overline{B}$, C_n	$\overline{P}$	In-Phase
t_{PLH} / t_{PHL}	$\overline{B}_i$	$\overline{A}_i$	None	None	Remaining $\overline{A}$ and $\overline{B}$, C_n	$\overline{P}$	In-Phase
t_{PLH} / t_{PHL}	$\overline{A}_i$	None	$\overline{B}_i$	Remaining $\overline{B}$	Remaining $\overline{A}$, C_n	$\overline{G}$	In-Phase
t_{PLH} / t_{PHL}	$\overline{B}_i$	None	$\overline{A}_i$	Remaining $\overline{B}$	Remaining $\overline{A}$, C_n	$\overline{G}$	In-Phase
t_{PLH} / t_{PHL}	C_n	None	None	All $\overline{A}$	All $\overline{B}$	Any $\overline{F}$ or C_{n+4}	In-Phase
t_{PLH} / t_{PHL}	$\overline{A}_i$	None	$\overline{B}_i$	Remaining $\overline{B}$	Remaining $\overline{A}$, C_n	C_{n+4}	Out-of-Phase
t_{PLH} / t_{PHL}	$\overline{B}_i$	None	$\overline{A}_i$	Remaining $\overline{B}$	Remaining $\overline{A}$, C_n	C_{n+4}	Out-of-Phase

$\overline{\text{DIFF}}$ MODE TEST TABLE
FUNCTION INPUTS: $S1 = S2 = 4.5\text{ V}, S0 = S3 = M = 0\text{ V}$

PARAMETER	INPUT UNDER TEST	OTHER INPUT SAME BIT		OTHER DATA INPUTS		OUTPUT UNDER TEST	OUTPUT WAVEFORM (SEE NOTE A)
		APPLY 4.5 V	APPLY GND	APPLY 4.5 V	APPLY GND		
t_{PLH} / t_{PHL}	$\overline{A}_i$	None	$\overline{B}_i$	Remaining $\overline{A}$	Remaining $\overline{B}$, C_n	$\overline{F}_i$	In-Phase
t_{PLH} / t_{PHL}	$\overline{B}_i$	$\overline{A}_i$	None	Remaining $\overline{A}$	Remaining $\overline{B}$, C_n	$\overline{F}_i$	Out-of-Phase
t_{PLH} / t_{PHL}	$\overline{A}_i$	None	$\overline{B}_i$	None	Remaining $\overline{A}$ and $\overline{B}$, C_n	$\overline{P}$	In-Phase
t_{PLH} / t_{PHL}	$\overline{B}_i$	$\overline{A}_i$	None	None	Remaining $\overline{A}$ and $\overline{B}$, C_n	$\overline{P}$	Out-of-Phase
t_{PLH} / t_{PHL}	$\overline{A}_i$	$\overline{B}_i$	None	None	Remaining $\overline{A}$ and $\overline{B}$, C_n	$\overline{G}$	In-Phase
t_{PLH} / t_{PHL}	$\overline{B}_i$	None	$\overline{A}_i$	None	Remaining $\overline{A}$ and $\overline{B}$, C_n	$\overline{G}$	Out-of-Phase
t_{PLH} / t_{PHL}	$\overline{A}_i$	None	$\overline{B}_i$	Remaining $\overline{A}$	Remaining $\overline{B}$, C_n	$A = B$	In-Phase
t_{PLH} / t_{PHL}	$\overline{B}_i$	$\overline{A}_i$	None	Remaining $\overline{A}$	Remaining $\overline{B}$, C_n	$A = B$	Out-of-Phase
t_{PLH} / t_{PHL}	C_n	None	None	All $\overline{A}$ and $\overline{B}$	None	C_{n+4} or any $\overline{F}$	In-Phase
t_{PLH} / t_{PHL}	$\overline{A}_i$	$\overline{B}_i$	None	None	Remaining $\overline{A}$, $\overline{B}$, C_n	C_{n+4}	Out-of-Phase
t_{PLH} / t_{PHL}	$\overline{B}_i$	None	$\overline{A}_i$	None	Remaining $\overline{A}$, $\overline{B}$, C_n	C_{n+4}	In-Phase

NOTE A: For load circuit and voltage waveforms, see page 1-12.

TYPES SN54AS181A, SN54AS881A, SN74AS181A, SN74AS881A
ARITHMETIC LOGIC UNITS/FUNCTION GENERATORS

PARAMETER MEASUREMENT INFORMATION

LOGIC MODE TEST TABLE
FUNCTION INPUTS: S1 = S2 = M = 4.5 V, S0 = S3 = 0 V

PARAMETER	INPUT UNDER TEST	OTHER INPUT SAME BIT		OTHER DATA INPUTS		OUTPUT UNDER TEST	OUTPUT WAVEFORM (SEE NOTE A)
		APPLY 4.5 V	APPLY GND	APPLY 4.5 V	APPLY GND		
t_{PLH} t_{PHL}	$\overline{A}_i$	$\overline{B}$	None	None	Remaining $\overline{A}$ and $\overline{B}$, C_n	$\overline{F}_i$	Out-of-Phase
t_{PLH} t_{PHL}	$\overline{B}_i$	$\overline{A}_i$	None	None	Remaining $\overline{A}$ and $\overline{B}$, C_n	$\overline{F}_i$	Out-of-Phase

INPUT BITS EQUAL/NOT EQUAL TEST TABLE
FUNCTION INPUTS: S0 = S3 = M = 4.5 V, S1 = S2 = 0 V

PARAMETER	INPUT UNDER TEST	OTHER INPUT SAME BIT		OTHER DATA INPUTS		OUTPUT UNDER TEST	OUTPUT WAVEFORM (SEE NOTE A)
		APPLY 4.5 V	APPLY GND	APPLY 4.5 V	APPLY GND		
t_{PLH} t_{PHL}	$\overline{A}_i$	$\overline{B}_i$	None	Remaining $\overline{A}$ and $\overline{B}$, C_n	None	$\overline{P}$	Out-of-Phase
t_{PLH} t_{PHL}	$\overline{B}_i$	$\overline{A}_i$	None	Remaining $\overline{A}$ and $\overline{B}$, C_n	None	$\overline{P}$	Out-of-Phase
t_{PLH} t_{PHL}	$\overline{A}_i$	None	$\overline{B}_i$	Remaining $\overline{A}$ and $\overline{B}$, C_n	None	$\overline{P}$	In-Phase
t_{PLH} t_{PHL}	$\overline{B}_i$	None	$\overline{A}_i$	Remaining $\overline{A}$ and $\overline{B}$, C_n	None	$\overline{P}$	In-Phase
t_{PLH} t_{PHL}	$\overline{A}_i$	$\overline{B}_i$	None	Remaining $\overline{A}$ and $\overline{B}$, C_n	None	C_{n+4}	In-Phase
t_{PLH} t_{PHL}	$\overline{B}_i$	$\overline{A}_i$	None	Remaining $\overline{A}$ and $\overline{B}$, C_n	None	C_{n+4}	In-Phase
t_{PLH} t_{PHL}	$\overline{A}_i$	None	$\overline{B}_i$	Remaining $\overline{A}$ and $\overline{B}$, C_n	None	C_{n+4}	Out-of-Phase
t_{PLH} t_{PHL}	$\overline{B}_i$	None	$\overline{A}_i$	Remaining $\overline{A}$ and $\overline{B}$, C_n	None	C_{n+4}	Out-of-Phase

INPUT PAIRS HIGH/NOT HIGH TEST TABLE
FUNCTION INPUTS: S2 = M = 4.5 V, S0 = S1 = S3 = 0V

PARAMETER	INPUT UNDER TEST	OTHER INPUT SAME BIT		OTHER DATA INPUTS		OUTPUT UNDER TEST	OUTPUT WAVEFORM (SEE NOTE A)
		APPLY 4.5 V	APPLY GND	APPLY 4.5V	APPLY GND		
t_{PLH} t_{PHL}	$\overline{A}_i$	$\overline{B}_i$	None	Remaining $\overline{A}$, C_n	Remaining $\overline{B}$	$\overline{P}$	In-Phase
t_{PLH} t_{PHL}	$\overline{B}_i$	$\overline{A}_i$	None	Remaining $\overline{B}$, C_n	Remaining $\overline{A}$	$\overline{P}$	In-Phase
t_{PLH} t_{PHL}	$\overline{A}_i$	$\overline{B}_i$	None	Remaining $\overline{A}$, C_n	Remaining $\overline{B}$	C_{n+4}	Out-of-Phase
t_{PLH} t_{PHL}	$\overline{B}_i$	$\overline{A}_i$	None	Remaining $\overline{B}$, C_n	Remaining $\overline{A}$	C_{n+4}	Out-of-Phase

NOTE A: For load circuit and voltage waveforms, see page 1-12.

Texas Instruments
POST OFFICE BOX 225012 • DALLAS, TEXAS 75265

ALS AND AS CIRCUITS

2
ALS AND AS CIRCUITS

TYPES SN54AS182, SN74AS182
LOOK-AHEAD CARRY GENERATOR

D2661, DECEMBER 1983

- High-Speed Replacement for the 'S182
- Offers Carry Functions in a Compatible Form for Direct Connections to the ALU
- Cascadable to Perform Look-Ahead Across n-Bit Adders
- Dependable Texas Instruments Quality and Reliability

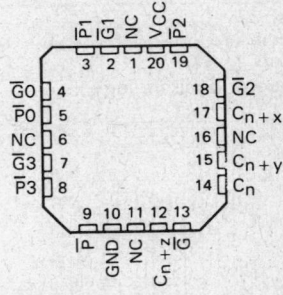

SN54AS182 ... J PACKAGE
SN74AS182 ... N PACKAGE
(TOP VIEW)

SN54AS182 ... FH PACKAGE
SN74AS182 ... FN PACKAGE
(TOP VIEW)

NC — No internal connection

PIN DESIGNATIONS

ALTERNATIVE	DESIGNATIONS†	FUNCTION
$\overline{G0}, \overline{G1}, \overline{G2}, \overline{G3}$	G0, G1, G2, G3	Carry Generate Inputs
$\overline{P0}, \overline{P1}, \overline{P2}, \overline{P3}$	P0, P1, P2, P3	Carry Propagate Inputs
C_n	$\overline{C_n}$	Carry Input
$C_{n+x}, C_{n+y}, C_{n+z}$	$\overline{C_{n+x}}, \overline{C_{n+y}}, \overline{C_{n+z}}$	Carry Outputs
$\overline{G}$	Y	Carry Generate Output
$\overline{P}$	X	Carry Propagate Output
V_{CC}		Supply Voltage
GND		Ground

† Interpretations are illustrated in connection with the Function Tables for the 'AS181A and 'AS881A.

description

The 'AS182 look-ahead carry generators are capable of anticipating a carry across four binary adders or group of adders. They are cascadable to perform full look-ahead across n-bit adders.

This generator, when used in conjunction with the 'AS181 or 'AS881 Arithmetic Logic Unit ALU, provides high-speed carry look-ahead capability for any word length. The 'AS182 generates the look-ahead (anticipated carry) across a group of four ALUs. In addition, other carry look-ahead circuits may be employed to anticipate carry-across sections of four look-ahead packages up to n-bits. The method of cascading 'AS182 circuits to perform multilevel look-ahead is illustrated under the typical application data.

The carry functions (inputs, outputs, generate, and propagate) of the look-ahead generators are implemented in the compatible forms for direct connections to the ALU. Reinterpretations of carry functions as explained on the 'AS181A and 'AS881A data sheet are also applicable to and compatible with the look-ahead generator. Logic equations for the 'AS182 are:

$C_{n+x} = G0 + P0\, C_n$
$C_{n+y} = G1 + P1\, G0 + P1\, P0\, C_n$
$C_{n+z} = G2 + P2\, G1 + P2\, P1\, G0 + P2\, P1\, P0\, C_n$
$\overline{G} = \overline{G3 + P3\, G2 + P3\, P2\, G1 + P3\, P2\, P1\, G0}$
$\overline{P} = \overline{P3\, P2\, P1\, P0}$

or

$\overline{C_{n+x}} = \overline{Y0\, (X0 + \overline{C_n})}$
$\overline{C_{n+y}} = \overline{Y1\, [X1 + Y0\, (X0 + \overline{C_n})]}$
$\overline{C_{n+z}} = \overline{Y2\, \{X2 + Y1\, [X1 + Y0\, (X0 + \overline{C_n})]\}}$
$Y = Y3\, (X3 + Y2)\, (X3 + X2 + Y1)\, (X3 + X2 + X1 + Y0)$
$X = X3 + X2 + X1 + X0$

PRODUCT PREVIEW
This page contains information on a product under development. Texas Instruments reserves the right to change or discontinue this product without notice.

Copyright © 1983 by Texas Instruments Incorporated

ALS AND AS CIRCUITS

TYPES SN54AS182, SN74AS182
LOOK-AHEAD CARRY GENERATOR

FUNCTION TABLE FOR $\overline{G}$ OUTPUT

INPUTS							OUTPUT
$\overline{G}3$	$\overline{G}2$	$\overline{G}1$	$\overline{G}0$	$\overline{P}3$	$\overline{P}2$	$\overline{P}1$	$\overline{G}$
L	X	X	X	X	X	X	L
X	L	X	X	L	X	X	L
X	X	L	X	L	L	X	L
X	X	X	L	L	L	L	L
All other combinations							H

FUNCTION TABLE FOR $\overline{P}$ OUTPUT

INPUTS				OUTPUT
$\overline{P}3$	$\overline{P}2$	$\overline{P}1$	$\overline{P}0$	$\overline{P}$
L	L	L	L	L
All other combinations				H

FUNCTION TABLE FOR C_{n+x} OUTPUT

INPUTS			OUTPUT
$\overline{G}0$	$\overline{P}0$	C_n	C_{n+x}
L	X	X	H
X	L	H	H
All other combinations			L

FUNCTION TABLE C_{n+y} OUTPUT

INPUTS					OUTPUT
$\overline{G}1$	$\overline{G}0$	$\overline{P}1$	$\overline{P}0$	C_n	C_{n+y}
L	X	X	X	X	H
X	L	L	X	X	H
X	X	L	L	H	H
All other combinations					L

FUNCTION TABLE FOR C_{n+z} OUTPUT

INPUTS							OUTPUT
$\overline{G}2$	$\overline{G}1$	$\overline{G}0$	$\overline{P}2$	$\overline{P}1$	$\overline{P}0$	C_n	C_{n+z}
L	X	X	X	X	X	X	H
X	L	X	L	X	X	X	H
X	X	L	L	L	X	X	H
X	X	X	L	L	L	H	H
All other combinations							L

H = High-level, L = Low-level, = Irrelevant
Any inputs not shown in a given table are irrelevant with respect to that output.

logic diagram (positive logic)

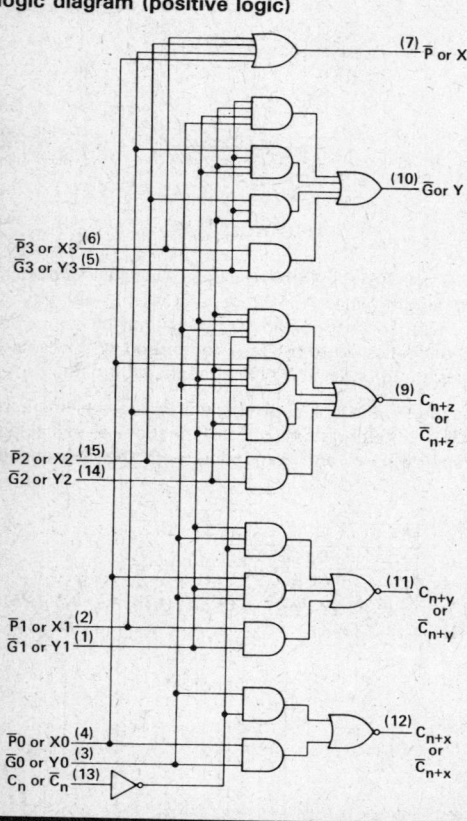

logic symbols

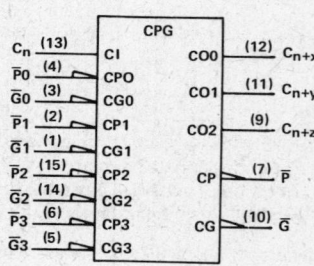

OR

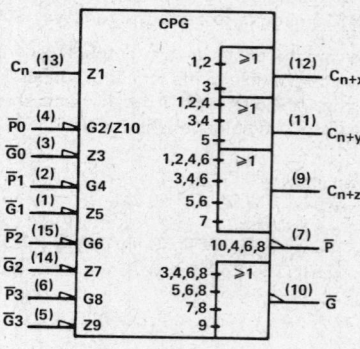

Pin numbers shown are for J and N packages only.

Texas Instruments
POST OFFICE BOX 225012 • DALLAS, TEXAS 75265

TYPES SN54AS182, SN74AS182
LOOK-AHEAD CARRY GENERATOR

absolute maximum ratings over operating free-air temperature range (unless otherwise noted)

Supply voltage, V_{CC} .... 7 V
Input voltage .... 7 V
Operating free-air temperature range: SN54AS182 .... $-55°C$ to $125°C$
                                      SN74AS182 .... $0°C$ to $70°C$
Storage temperature range .... $-65°C$ to $150°C$

recommended operating conditions

		SN54AS182			SN74AS182			UNIT
		MIN	NOM	MAX	MIN	NOM	MAX	
V_{CC}	Supply voltage	4.5	5	5.5	4.5	5	5.5	V
V_{IH}	High-level input voltage	2			2			V
V_{IL}	Low-level input voltage			0.8			0.8	V
I_{OH}	High-level output current			-2			-2	mA
I_{OL}	Low-level output current			20			20	mA
T_A	Operating free-air temperature	-55		125	0		70	°C

electrical characteristics over recommended operating free-air temperature range (unless otherwise noted)

PARAMETER		TEST CONDITIONS		SN54AS182			SN74AS182			UNIT
				MIN	TYP[†]	MAX	MIN	TYP[†]	MAX	
V_{IK}		$V_{CC} = 4.5$ V,	$I_I = -18$ mA			-1.2			-1.2	V
V_{OH}		$V_{CC} = 4.5$ V to 5.5 V,	$I_{OH} = -2$ mA	$V_{CC}-2$			$V_{CC}-2$			V
V_{OL}		$V_{CC} = 4.5$ V,	$I_{OL} = 20$ mA		0.3	0.5		0.3	0.5	V
I_I	C_n	$V_{CC} = 5.5$ V,	$V_I = 7.0$ V			0.1			0.1	mA
	$\overline{P3}$					0.2			0.2	
	$\overline{P2}$					0.3			0.3	
	$\overline{P0}, \overline{P1}, \overline{G3}$					0.4			0.4	
	$\overline{G0}, \overline{G2}$					0.7			0.7	
	$\overline{G1}$					0.8			0.8	
I_{IH}	C_n	$V_{CC} = 5.5$ V,	$V_I = 2.7$ V			0.02			0.02	mA
	$\overline{P3}$					0.04			0.04	
	$\overline{P2}$					0.06			0.06	
	$\overline{P0}, \overline{P1}, \overline{G3}$					0.08			0.08	
	$\overline{G1}, \overline{G2}$					0.14			0.14	
	$\overline{G1}$					0.16			0.16	
I_{IL}	C_n	$V_{CC} = 5.5$ V,	$V_I = 0.4$ V			-0.5			-0.5	mA
	$\overline{P3}$					-1			-1	
	$\overline{P2}$					-1.5			-1.5	
	$\overline{P2}, \overline{P1}, \overline{G3}$					-2			-2	
	$\overline{G0}, \overline{G2}$					-3.5			-3.5	
	$\overline{G1}$					-4			-4	
I_O[‡]		$V_{CC} = 5.5$ V,	$V_O = 2.25$ V	-30		-112	-30		-112	mA
I_{CCH}		$V_{CC} = 5.5$ V			17			17		mA
I_{CCL}					23			23		

[†] All typical values are at $V_{CC} = 5$ V, $T_A = 25°C$.
[‡] The output conditions have been chosen to produce a current that closely approximates one-half of the true short-circuit current, I_{OS}.

TEXAS INSTRUMENTS
POST OFFICE BOX 225012 • DALLAS, TEXAS 75265

TYPES SN54AS182, SN74AS182
LOOK-AHEAD CARRY GENERATOR

switching characteristics (see Note 1)

PARAMETER	FROM (INPUT)	TI (OUTPUT)	V_{CC} = 4.5 V to 5.5 V, C_L = 50 pF, R_L = 500 Ω, T_A = MIN to MAX						UNIT
			SN54AS182			SN74AS182			
			MIN	TYP‡	MAX	MIN	TYP‡	MAX	
t_{PLH}	C_n	C_{n+y}, C_{n+y} C_{n+z}		5			5		ns
t_{PHL}				5			5		
t_{PLH}	$\overline{P}$ or $\overline{G}$	C_{n+x}, C_{n+y} C_{n+z}		5			5		ns
t_{PHL}				5			5		
t_{PLH}	$\overline{P}$ or $\overline{G}$	$\overline{G}$		6			6		ns
t_{PHL}				5			5		
t_{PLH}	$\overline{P}$	$\overline{P}$		5			5		ns
t_{PHL}				5			5		

‡All typical values are at V_{CC} = 5 V, T_A = 25°C.
NOTE 1: For load circuit and voltage waveforms, see page 1-12.

TYPICAL APPLICATION DATA

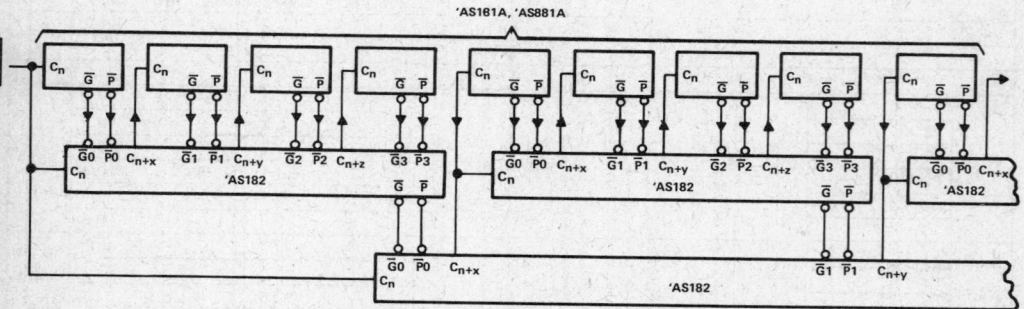

FIGURE 1—THE 'AS182 IN A 64-BIT LOOK-AHEAD CARRY CIRCUIT

TYPES SN54ALS190, SN54ALS191, SN74ALS190, SN74ALS191
SYNCHRONOUS 4-BIT UP/DOWN DECADE AND BINARY COUNTERS

D2661, DECEMBER 1982—REVISED DECEMBER 1983

- Single Down/Up Count Control Line
- Look-Ahead Circuitry Enhances Speed of Cascaded Counters
- Fully Synchronous in Count Modes
- Asynchronously Presettable with Load Control
- Package Options Include Both Plastic and Ceramic Chip Carriers in Addition to Plastic and Ceramic DIPS
- Dependable Texas Instruments Quality and Reliability

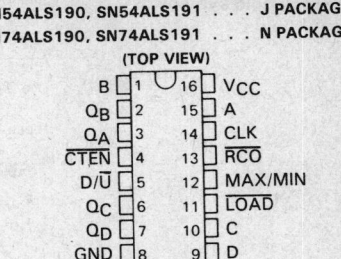

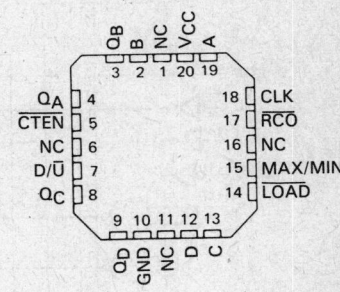

NC — no internal connection.

description

The 'ALS190 and 'ALS191 are synchronous, reversible up/down counters. The 'ALS190 is a 4-bit decade counter and the 'ALS191 is a 4-bit binary counter. Synchronous counting operation is provided by having all flip-flops clocked simultaneously so that the outputs change coincident with each other when so instructed by the steering logic. This mode of operation eliminates the output counting spikes normally associated with asynchronous (ripple clock) counters.

The outputs of the four flip-flops are triggered on a low-to-high-level transition of the clock input if the enable input ($\overline{\text{CTEN}}$) is low. A high at $\overline{\text{CTEN}}$ inhibits counting. The direction of the count is determined by the level of the down/up (D/$\overline{\text{U}}$) input. When D/$\overline{\text{U}}$ is low, the counter counts up and when D/$\overline{\text{U}}$ is high, it counts down.

These counters feature a fully independent clock circuit. Changes at the control inputs ($\overline{\text{CTEN}}$ and D/$\overline{\text{U}}$) that will modify the operating mode have no effect on the contents of the counter until clocking occurs. The function of the counter will be dictated solely by the condition meeting the stable setup and hold times.

These counters are fully programmable; that is, the outputs may each be preset to either level by placing a low on the load input and entering the desired data at the data inputs. The output will change to agree with the data inputs independently of the level of the clock input. This feature allows the counters to be used as modulo-N dividers by simply modifying the count length with the preset inputs.

The CLK, D/$\overline{\text{U}}$, and LOAD inputs are buffered to lower the drive requirement, which significantly reduces the loading on, or current required by, clock drivers, etc., for long parallel words.

Two outputs have been made available to perform the cascading function: ripple clock and maximum/minimum count. The latter output produces a high-level output pulse with a duration approximately equal to one complete cycle of the clock while the count is zero (all outputs low) counting down or maximum (9 or 15) counting up. The ripple clock output produces a low-level output pulse under those same conditions but only while the clock input is low. The counters can be easily cascaded by feeding the ripple clock output to the enable input of the succeeding counter if parallel clocking is used, or to the clock input if parallel enabling is used. The maximum/minimum count output can be used to accomplish look-ahead for high-speed operation.

The SN54ALS190 and SN54ALS191 are characterized for operation over the full military temperature range of −55 °C to 125 °C. The SN74ALS190 and SN74ALS191 are characterized for operation from 0 °C to 70 °C.

Copyright © 1982 by Texas Instruments Incorporated

TYPES SN54ALS190, SN74ALS190
SYNCHRONOUS 4-BIT UP/DOWN DECADE COUNTERS

'ALS190 logic symbol

'ALS190 logic diagram (positive logic)

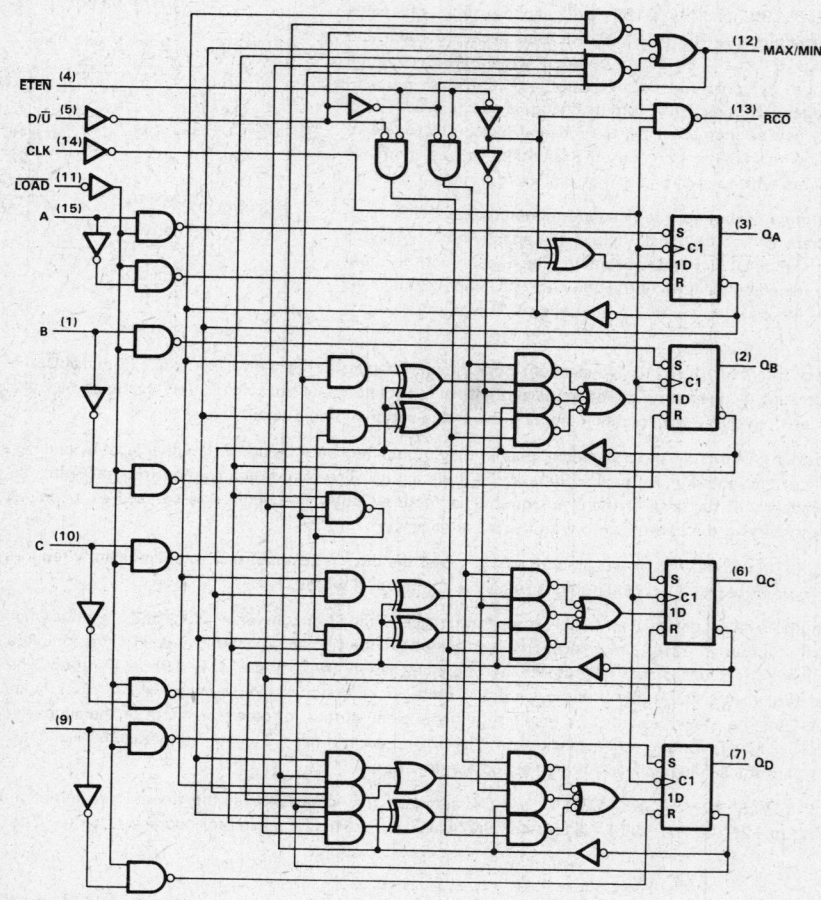

Pin numbers shown are for J and N packages.

2-190

Texas Instruments
POST OFFICE BOX 225012 • DALLAS, TEXAS 75265

TYPES SN54ALS191, SN74ALS191
SYNCHRONOUS 4-BIT UP/DOWN BINARY COUNTERS

'ALS191 logic symbol

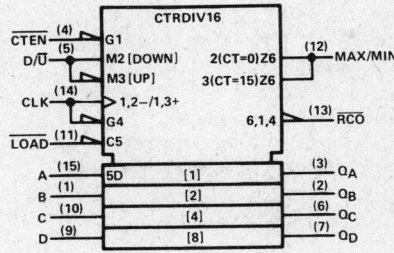

'ALS191 logic diagram (positive logic)

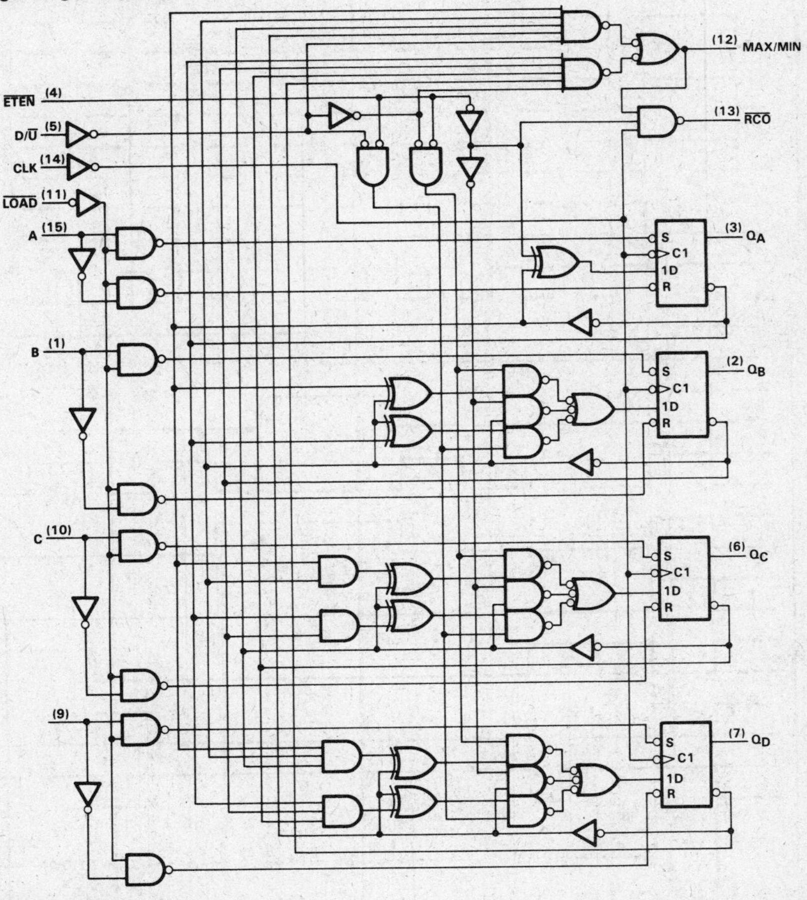

Pin numbers shown are for J and N packages.

TYPES SN54ALS190, SN74ALS190
SYNCHRONOUS 4-BIT UP/DOWN BINARY COUNTERS

typical load, count, and inhibit sequences

'ALS190

Illustrated below is the following sequence:

1. Load (preset) to BCD seven.
2. Count up to eight, nine (maximum), zero, one, and two.
3. Inhibit.
4. Count down to one, zero (minimum), nine, eight, and seven.

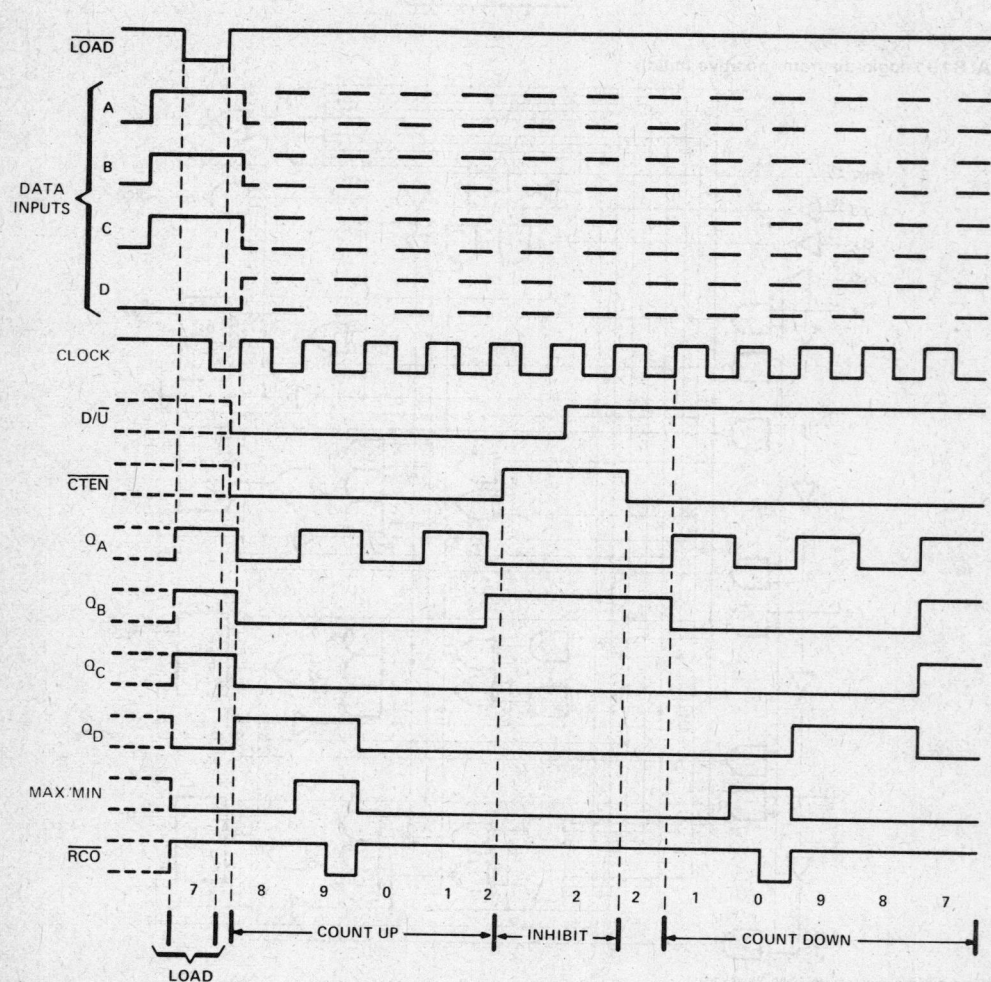

2-192

TEXAS INSTRUMENTS
POST OFFICE BOX 225012 • DALLAS, TEXAS 75265

TYPES SN54ALS191, SN74ALS191
SYNCHRONOUS 4-BIT UP/DOWN BINARY COUNTERS

typical load, count, and inhibit sequences

'ALS191

Illustrated below is the following sequence:

1. Load (preset) to binary thirteen.
2. Count up to fourteen, fifteen (maximum), zero, one, and two.
3. Inhibit.
4. Count down to one, zero (minimum), fifteen, fourteen, and thirteen.

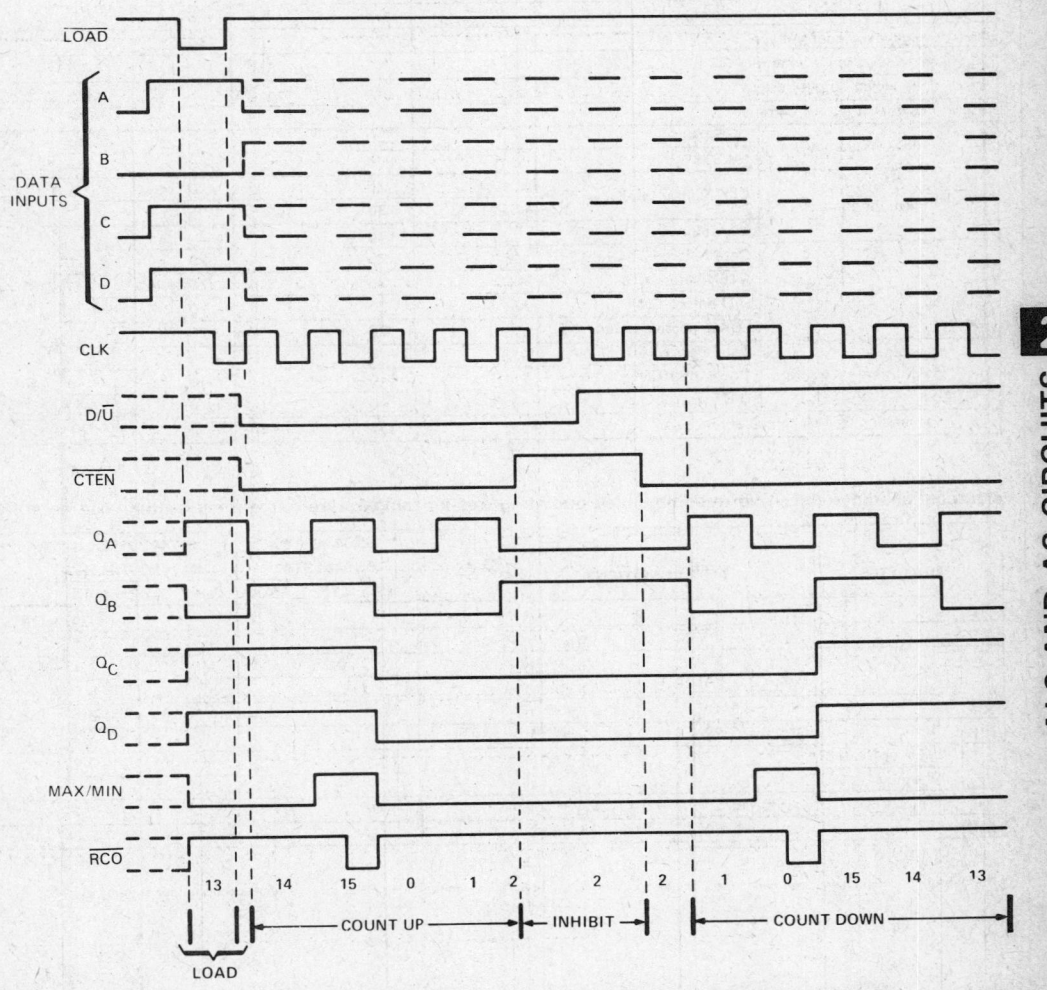

TYPES SN54ALS190, SN54ALS191, SN74ALS190, SN74ALS191
SYNCHRONOUS 4-BIT UP/DOWN DECADE AND BINARY COUNTERS

absolute maximum ratings over operating free-air temperature range (unless otherwise noted)

Supply voltage, V_{CC} .... 7 V
Input voltage .... 7 V
Operating free-air temperature range: SN54ALS190, SN54ALS191 .... −55°C to 125°C
                                      SN74ALS190, SN74ALS191 .... 0°C to 70°C
Storage temperature range .... −65°C to 150°C

recommended operating conditions

			SN54ALS190 SN54ALS191			SN74ALS190 SN74ALS191			UNIT
			MIN	NOM	MAX	MIN	NOM	MAX	
V_{CC}	Supply voltage		4.5	5	5.5	4.5	5	5.5	V
V_{IH}	High-level input voltage		2			2			V
V_{IL}	Low-level input voltage				0.8			0.8	V
I_{OH}	High-level output current				−0.4			−0.4	mA
I_{OL}	Low-level output current				4			8	mA
f_{clock}	Clock frequency	'ALS190	0		20	0		25	MHz
		'ALS191	0		25	0		30	
t_w	Pulse duration	CLK high or low 'ALS190	25			20			ns
		'ALS191	20			16.5			
		LOAD low	25			20			
t_{su}	Setup time	Data before LOAD↑	25			20			ns
		CTEN before CLK↑	25			20			
		D/Ū before CLK↑	20			20			
		LOAD inactive before CLK↑	20			20			
t_h	Hold time	Data after LOAD↑	5			5			ns
		CTEN after CLK↑	0			0			
		D/Ū after CLK↑	0			0			
T_A	Operating free-air temperature		−55		125	0		70	°C

electrical characteristics over recommended operating free-air temperature range (unless otherwise noted)

PARAMETER		TEST CONDITIONS		SN54ALS190 SN54ALS191			SN74ALS190 SN74ALS191			UNIT
				MIN	TYP[†]	MAX	MIN	TYP[†]	MAX	
V_{IK}		V_{CC} = 4.5 V,	I_I = −18 mA			−1.5			−1.5	V
V_{OH}		V_{CC} = 4.5 V to 5.5 V,	I_{OH} = −0.4 mA	V_{CC}−2			V_{CC}−2			V
V_{OL}		V_{CC} = 4.5 V,	I_{OL} = 4 mA		0.25	0.4		0.25	0.4	V
		V_{CC} = 4.5 V	I_{OL} = 8 mA					0.35	0.5	
I_I		V_{CC} = 5.5 V,	V_I = 7 V			0.1			0.1	mA
I_{IH}		V_{CC} = 5.5 V,	V_I = 2.7 V			20			20	µA
I_{IL}	CTEN or CLK	V_{CC} = 5.5 V,	V_I = 0.4 V			−0.2			−0.2	mA
	All others					−0.1			−0.1	
I_O[‡]		V_{CC} = 5.5 V,	V_O = 2.25 V	−30		−112	−30		−112	mA
I_{CC}		V_{CC} = 5.5 V,	All inputs at 0 V		12	22		12	22	mA

[†]All typical values are at V_{CC} = 5 V, T_A = 25°C.
[‡]The output conditions have been chosen to produce a current that closely approximates one half of the true short-circuit output current, I_{OS}.

TEXAS INSTRUMENTS
POST OFFICE BOX 225012 • DALLAS, TEXAS 75265

TYPES SN54ALS190, SN54ALS191, SN74ALS190, SN74ALS191
SYNCHRONOUS 4-BIT UP/DOWN DECADE AND BINARY COUNTERS

switching characteristics (see Note 1)

PARAMETER	FROM (INPUT)	TO (OUTPUT)	$V_{CC} = 4.5\ V$ to $5.5\ V$, $C_L = 50\ pF$, $R_L = 500\ \Omega$, $T_A = MIN$ to MAX				UNIT
			SN54ALS190 SN54ALS191		SN74ALS190 SN74ALS191		
			MIN	MAX	MIN	MAX	
f_{max}	'ALS190		20		25		MHz
	'ALS191		25		30		
t_{PLH}	$\overline{LOAD}$	Any Q	8	34	8	30	ns
t_{PHL}			8	34	8	30	
t_{PLH}	A, B, C, D	Any Q	4	25	4	21	ns
t_{PHL}			4	25	4	21	
t_{PLH}	CLK	$\overline{RCO}$	5	24	5	20	ns
t_{PHL}			5	24	5	20	
t_{PLH}	CLK	Any Q	3	22	3	18	ns
t_{PHL}			3	22	3	18	
t_{PLH}	CLK	MAX/MIN	8	34	8	31	ns
t_{PHL}			8	34	8	31	
t_{PLH}	$D/\overline{U}$	$\overline{RCO}$	15	42	15	37	ns
t_{PHL}			10	33	10	28	
t_{PLH}	$D/\overline{U}$	MAX/MIN	8	30	8	25	ns
t_{PHL}			8	30	8	25	
t_{PLH}	$\overline{CTEN}$	$\overline{RCO}$	4	21	4	18	ns
t_{PHL}			4	21	4	18	

NOTE 1: For load circuit and voltage waveforms, see page 1-12.

2 ALS AND AS CIRCUITS

TYPES SN54ALS192, SN54ALS193, SN74ALS192, SN74ALS193
SYNCHRONOUS 4-BIT UP/DOWN COUNTERS (DUAL CLOCK WITH CLEAR)

D2661, DECEMBER 1982 — REVISED DECEMBER 1983

- Look-Ahead Circuitry Enhances Cascaded Counters
- Fully Synchronous in Count Modes
- Parallel Asynchronous Load for Modulo-N Count Lengths
- Asynchronous Clear
- Package Options Include Both Plastic and Ceramic Chip Carriers in Addition to Plastic and Ceramic DIPs
- Dependable Texas Instruments Quality and Reliability

SN54ALS192, SN54ALS193 . . . J PACKAGE
SN74ALS192, SN74ALS193 . . . N PACKAGE
(TOP VIEW)

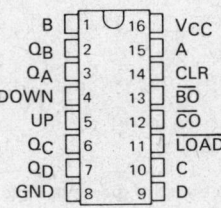

SN54ALS192, SN54ALS193 . . . FH PACKAGE
SN74ALS192, SN74ALS193 . . . FN PACKAGE
(TOP VIEW)

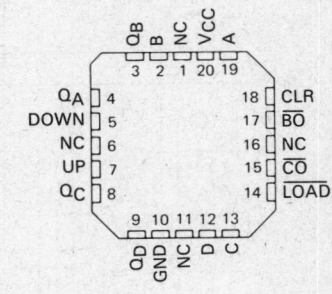

NC — no internal connection.

description

The 'ALS192 and 'ALS193 are synchronous, reversible up/down counters. The 'ALS192 is a 4-bit decade counter and the 'ALS193 is a 4-bit binary counter. Synchronous operation is provided by having all flip-flops clocked simultaneously so that the outputs change coincidently with each other when so instructed by the steering logic. This mode of operation eliminates the output counting spikes normally associated with asynchronous (ripple clock) counters.

The outputs of the four flip-flops are triggered by a low-to-high-level transition of either count (clock) input (Up or Down). The direction of counting is determined by which count input is pulsed while the other count input is high.

All four counters are fully programmable; that is, each output may be preset to either level by placing a low on the load input and entering the desired data at the data inputs. The output will change to agree with the data inputs independently of the count pulses. This feature allows the counters to be used as modulo-N dividers by simply modifying the count length with the preset inputs.

A clear input has been provided that forces all outputs to the low level when a high level is applied. The clear function is independent of the count and the load inputs. The clock, count, and load inputs are buffered to lower the drive requirements. This significantly reduces the loading on clock drivers, etc., for long parallel words.

These counters were designed to be cascaded without the need for external circuitry. The borrow output (BO) produces a low-level pulse while the count is zero (all outputs low) and the count-down input is low. Similarly, the carry output ($\overline{CO}$) produces a low-level pulse while the count is maximum (9 or 15) and the count-up input is low. The counters can then be easily cascaded by feeding the borrow and carry outputs to the count-down and count-up inputs, respectively, of the succeeding counter.

The SN54ALS192 and SN54ALS193 are characterized for operation over the full military temperature range of −55 °C to 125 °C. The SN74ALS192 and SN74ALS193 are characterized for operation from 0 °C to 70 °C.

Copyright © 1982 by Texas Instruments Incorporated

TYPES SN54ALS192, SN74ALS192
SYNCHRONOUS 4-BIT UP/DOWN DECADE COUNTERS (DUAL CLOCK WITH CLEAR)

'ALS192 logic symbol

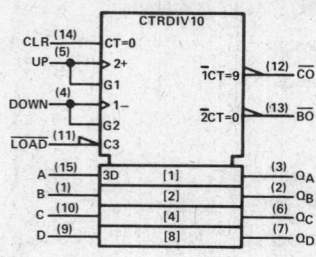

'ALS192 logic diagram (positive logic)

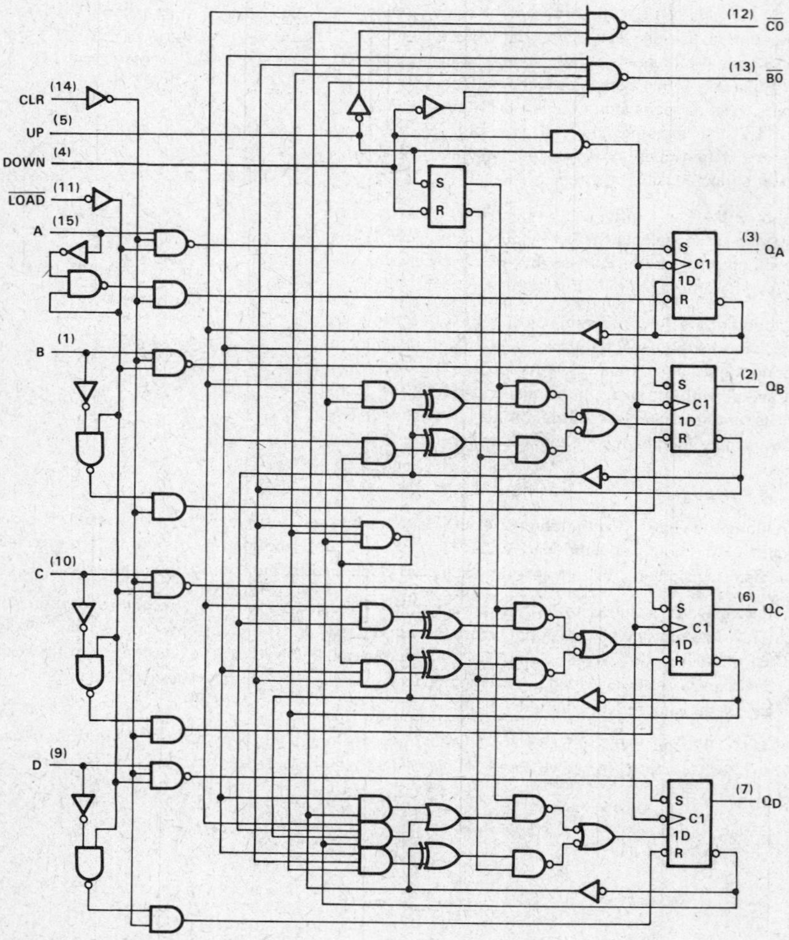

Pin numbers shown are for J and N packages.

TYPES SN54ALS193, SN74ALS193
SYNCHRONOUS 4-BIT UP/DOWN BINARY COUNTERS (DUAL CLOCK WITH CLEAR)

'ALS193 logic symbol

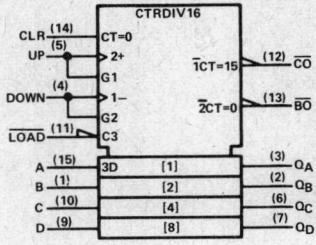

'ALS193 logic diagrams (positive logic)

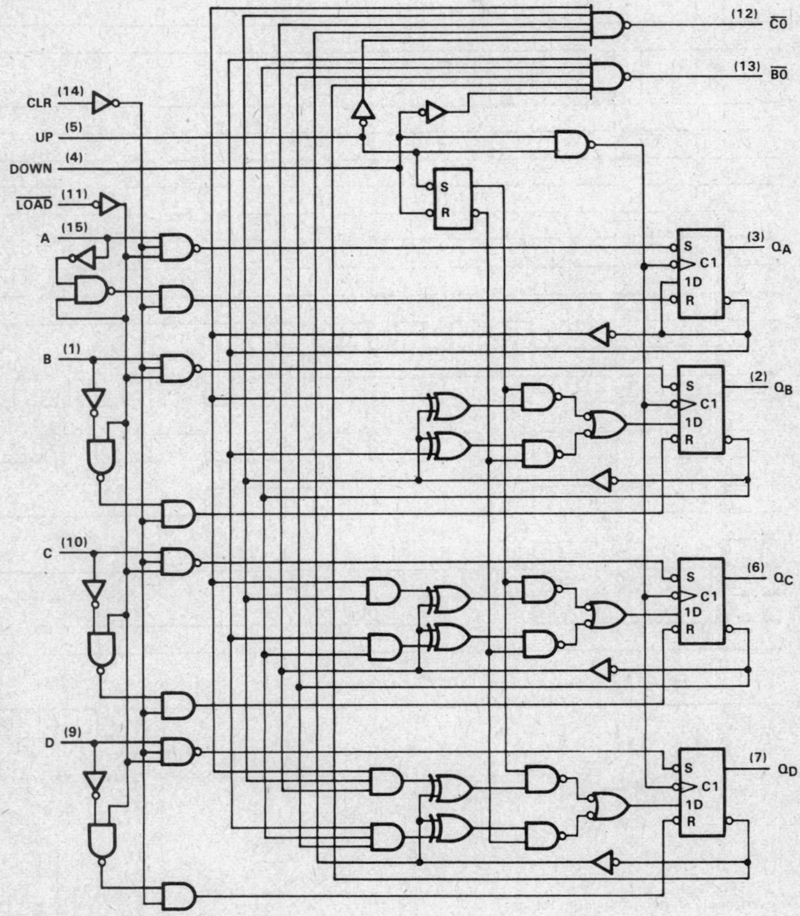

Pin numbers shown are for J and N packages.

TYPES SN54ALS192, SN74ALS192
SYNCHRONOUS 4-BIT UP/DOWN DECADE COUNTERS (DUAL CLOCK WITH CLEAR)

typical clear, load, and count sequence

'ALS192

Illustrated below is the following sequence:

1. Clear outputs to zero.
2. Load (preset) to BCD seven.
3. Count up to eight, nine, carry, zero, one, and two.
4. Count down to one, zero, borrow, nine, eight, and seven.

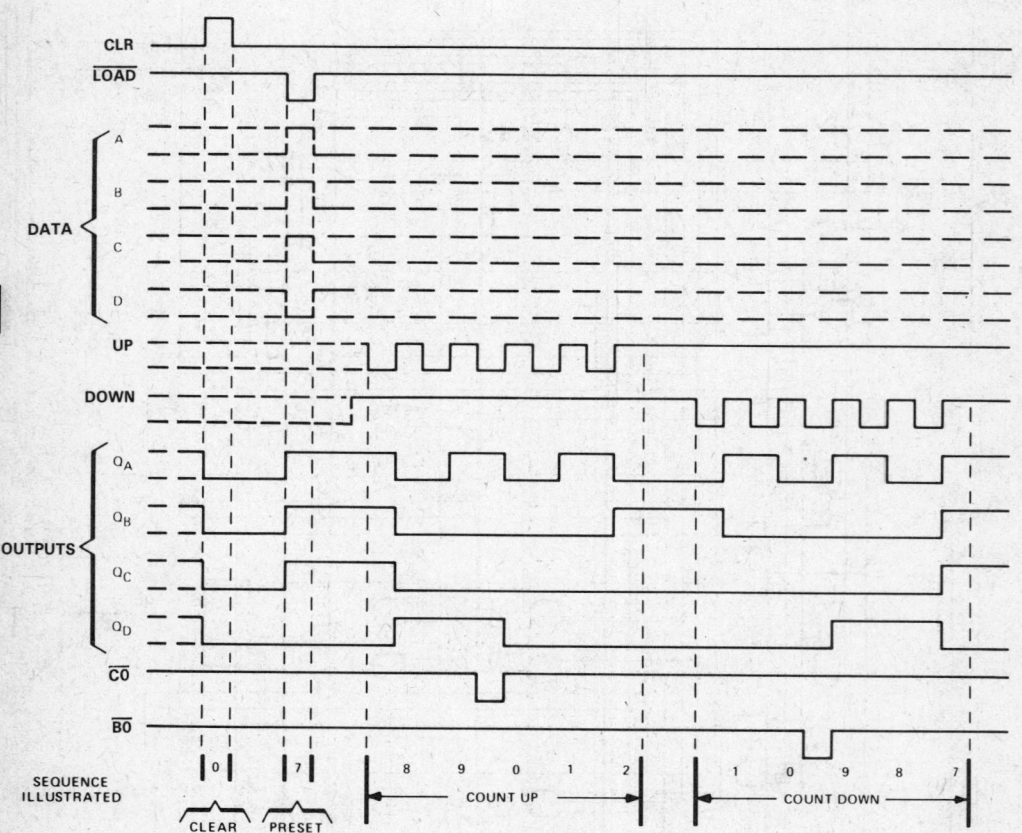

NOTES: A. Clear overrides load, data, and count inputs.
B. When counting up, count-down input must be high; when counting down, count-up input must be high.

TYPES SN54ALS193, SN74ALS193
SYNCHRONOUS 4-BIT UP/DOWN BINARY COUNTERS (DUAL CLOCK WITH CLEAR)

typical clear, load, and count sequences

'ALS193

Illustrated below is the following sequence:

1. Clear outputs to zero.
2. Load (preset) to binary thirteen.
3. Count up to fourteen, fifteen, carry, zero, one, and two.
4. Count down to one, zero, borrow, fifteen, fourteen, and thirteen.

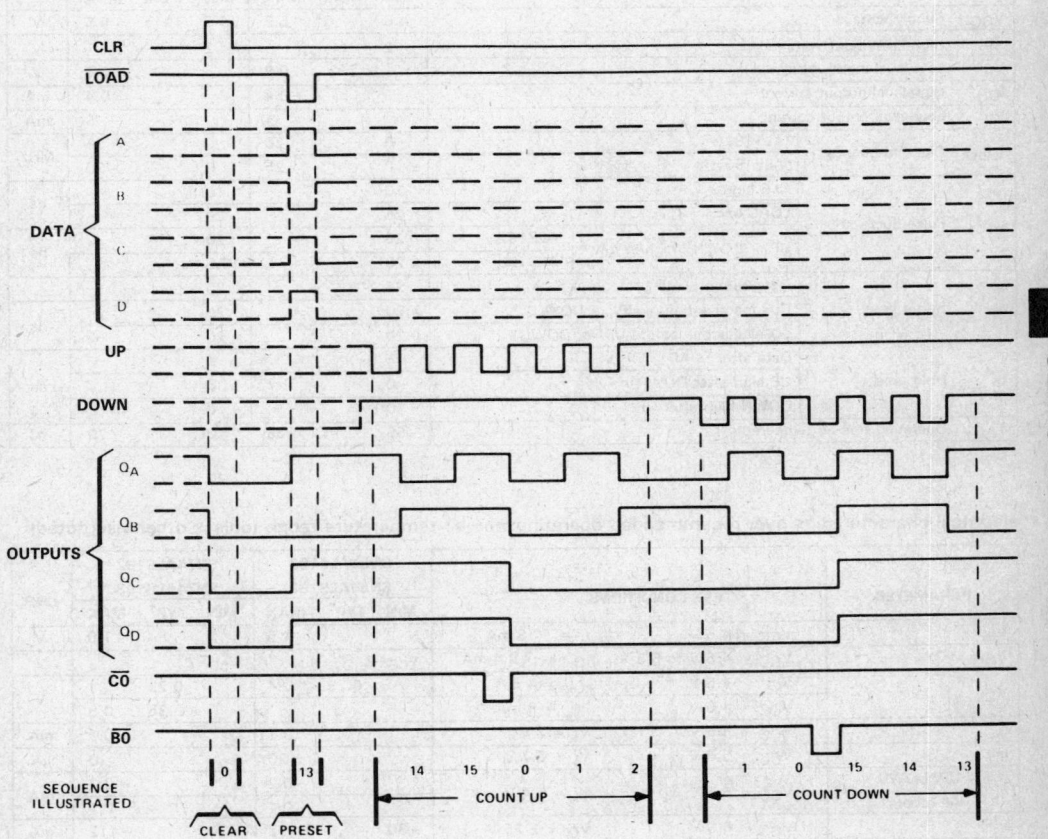

NOTES: A. Clear overrides load, data, and count inputs.
         B. When counting up, count-down input must be high; when counting down, count-up input must be high.

TYPES SN54ALS192, SN54ALS193, SN74ALS192, SN74ALS193
SYNCHRONOUS 4-BIT UP/DOWN COUNTERS (DUAL CLOCK WITH CLEAR)

absolute maximum ratings over operating free-air temperature range (unless otherwise noted)

Supply voltage, V_{CC} .... 7 V
Input voltage .... 7 V
Operating free-air temperature range: SN54ALS192, SN54ALS193 .... −55 °C to 125 °C
                                       SN74ALS192, SN74ALS193 .... 0 °C to 70 °C
Storage temperature range .... −65 °C to 150 °C

recommended operating conditions

			SN54ALS192 SN54ALS193			SN74ALS192 SN74ALS193			UNIT
			MIN	NOM	MAX	MIN	NOM	MAX	
V_{CC}	Supply voltage		4.5	5	5.5	4.5	5	5.5	V
V_{IH}	High-level input voltage		2			2			V
V_{IL}	Low-level input voltage				0.8			0.8	V
I_{OH}	High-level output current				−0.4			−0.4	mA
I_{OL}	Low-level output current				4			8	mA
f_{clock}	Clock frequency	'ALS192	0		20	0		25	MHz
		'ALS193	0		25	0		30	
t_w	Pulse duration	CLR high	10			10			ns
		$\overline{LOAD}$ low	25			20			
		UP or DOWN high or low 'ALS192	25			20			ns
		'ALS193	20			16.5			
t_{su}	Setup time	Data before $\overline{LOAD}\downarrow$	25			20			ns
		CLR inactive before UP↑ or DOWN↑	20			20			
		$\overline{LOAD}$ inactive before UP↑ or DOWN↑	20			20			
t_h	Hold time	Data after $\overline{LOAD}\uparrow$	5			5			ns
		UP high after DOWN↑	0			0			
		DOWN high after UP↑	0			0			
T_A	Operating free-air temperature		−55		125	0		70	°C

electrical characteristics over recommended operating free-air temperature range (unless otherwise noted)

PARAMETER		TEST CONDITIONS		SN54ALS192 SN54ALS193			SN74ALS192 SN74ALS193			UNIT
				MIN	TYP†	MAX	MIN	TYP†	MAX	
V_{IK}		V_{CC} = 4.5 V,	I_I = −18 mA			−1.5			−1.5	V
V_{OH}		V_{CC} = 4.5 V to 5.5 V,	I_{OH} = −0.4 mA	$V_{CC}-2$			$V_{CC}-2$			V
V_{OL}		V_{CC} = 4.5 V,	I_{OL} = 4 mA		0.25	0.4		0.25	0.4	V
		V_{CC} = 4.5 V	I_{OL} = 8 mA					0.35	0.5	
I_I		V_{CC} = 5.5 V,	V_I = 7 V			0.1			0.1	mA
I_{IH}		V_{CC} = 5.5 V,	V_I = 2.7 V			20			20	μA
I_{IL}	UP, DOWN	V_{CC} = 5.5 V,	V_I = 0.4 V			−0.2			−0.2	mA
	All others					−0.1			−0.1	
I_O‡		V_{CC} = 5.5 V,	V_O = 2.25 V	−30		−112	−30		−112	mA
I_{CC}		V_{CC} = 5.5 V,	See Note 1		12	22		12	22	mA

†All typical values are at V_{CC} = 5 V, T_A = 25 °C.
‡The output conditions have been chosen to produce a current that closely approximates one half of the true short-circuit output current, I_{OS}.
NOTE 1: I_{CC} is measured with the clear and load inputs grounded, and all other inputs at 4.5 V.

POST OFFICE BOX 225012 • DALLAS, TEXAS 75265

TYPES SN54ALS192, SN54ALS193, SN74ALS192, SN74ALS193
SYNCHRONOUS 4-BIT UP/DOWN COUNTERS (DUAL CLOCK WITH CLEAR)

switching characteristics (see Note 2)

PARAMETER	FROM (INPUT)	TO (OUTPUT)	V_{CC} = 4.5 V to 5.5 V, C_L = 50 pF, R_L = 500 Ω, T_A = MIN to MAX				UNIT
			SN54ALS192 SN54ALS193		SN74ALS192 SN74ALS193		
			MIN	MAX	MIN	MAX	
f_{max}		'ALS192	20		25		MHz
		'ALS193	25		30		
t_{PLH}	Up	CO	4	19	4	16	ns
t_{PHL}			5	21	5	18	
t_{PLH}	Down	BO	4	19	4	16	ns
t_{PHL}			5	21	5	18	
t_{PLH}	Up or Down	Any Q	4	23	4	19	ns
t_{PHL}			4	20	4	17	
t_{PLH}	$\overline{LOAD}$	Any Q	8	35	8	30	ns
t_{PHL}			8	31	8	28	
t_{PHL}	CLR	Any Q	5	20	5	17	ns

NOTE 2: For load circuit and voltage waveforms, see page 1-12.

2

ALS AND AS CIRCUITS

TYPES SN54AS194, SN74AS194
4-BIT BIDIRECTIONAL UNIVERSAL SHIFT REGISTERS

D2661, DECEMBER 1983

- Parallel-to-Serial, Serial-to-Parallel Conversions
- Left or Right Shifts
- Parallel Synchronous Loading
- Direct Overriding Clear
- Temporary Data Latching Capability
- Dependable Texas Instruments Quality and Reliability

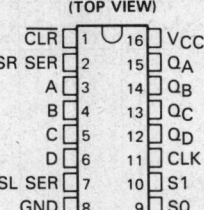

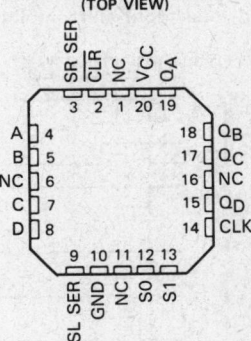

NC — No internal connection

description

These bidirectional shift registers feature parallel outputs, right-shift and left-shift serial inputs, operating-mode-control inputs, and a direct overriding clear line. The register has four distinct modes of operation, namely:

Inhibit clock (temporary data latch/do nothing
Shift-right (in the direction Q_A toward Q_D)
Shift-left (in the direction Q_D toward Q_A)
Parallel (broadside) load

Synchronous parallel loading is accomplished by applying the four bits of data and taking both mode control inputs, S0 and S1, high. The data are loaded into the associated flip-flops and appear at the outputs after the positive transition of the clock input. During loading, serial data flow is inhibited.

Shift-right is accomplished synchronously with the rising edge of the clock pulse when S0 is high and S1 is low. Serial data for this mode is entered at the shift-right data input. When S0 is low and S1 is high, data shifts left synchronously and new data is entered at the shift-left serial inputs. Clocking of the flip-flop is inhibited when both mode control inputs are low.

The SN54AS194 is characterized for operation over the full military temperature range of −55°C to 125°C. The SN74AS194 is characterized for operation from 0°C to 70°C.

PRODUCT PREVIEW

This page contains information on a product under development. Texas Instruments reserves the right to change or discontinue this product without notice.

Copyright © 1983 by Texas Instruments Incorporated

TEXAS INSTRUMENTS
POST OFFICE BOX 225012 • DALLAS, TEXAS 75265

TYPES SN54AS194, SN74AS194
4-BIT BIDIRECTIONAL UNIVERSAL SHIFT REGISTERS

FUNCTION TABLE

CLEAR	MODE		CLOCK	SERIAL		PARALLEL				OUTPUTS			
	S1	S0		LEFT	RIGHT	A	B	C	D	Q_A	Q_B	Q_C	Q_D
L	X	X	X	X	X	X	X	X	X	L	L	L	L
H	X	X	L	X	X	X	X	X	X	Q_{A0}	Q_{B0}	Q_{C0}	Q_{D0}
H	H	H	↑	X	X	a	b	c	d	a	b	c	d
H	L	H	↑	X	H	X	X	X	X	H	Q_{An}	Q_{Bn}	Q_{Cn}
H	L	H	↑	X	L	X	X	X	X	L	Q_{An}	Q_{Bn}	Q_{Cn}
H	H	L	↑	H	X	X	X	X	X	Q_{Bn}	Q_{Cn}	Q_{Dn}	H
H	H	L	↑	L	X	X	X	X	X	Q_{Bn}	Q_{Cn}	Q_{Dn}	L
H	L	L	X	X	X	X	X	X	X	Q_{A0}	Q_{B0}	Q_{C0}	Q_{D0}

H = high level (steady state)
L = low level (steady state)
X = irrelevant (any input, including transitions)
↑ = transition from low to high level
a, b, c, d = the level of steady-state input at inputs A, B, C, or D, respectively.
Q_{A0}, Q_{B0}, Q_{C0}, Q_{D0} = the level of Q_A, Q_B, Q_C, or Q_D, respectively, before the indicated steady-state input conditions were established.
Q_{An}, Q_{Bn}, Q_{Cn}, Q_{Dn} = the level of Q_A, Q_B, Q_C, respectively, before the most-recent ↑ transition of the clock.

typical clear, load, right-shift, inhibit, and clear sequences

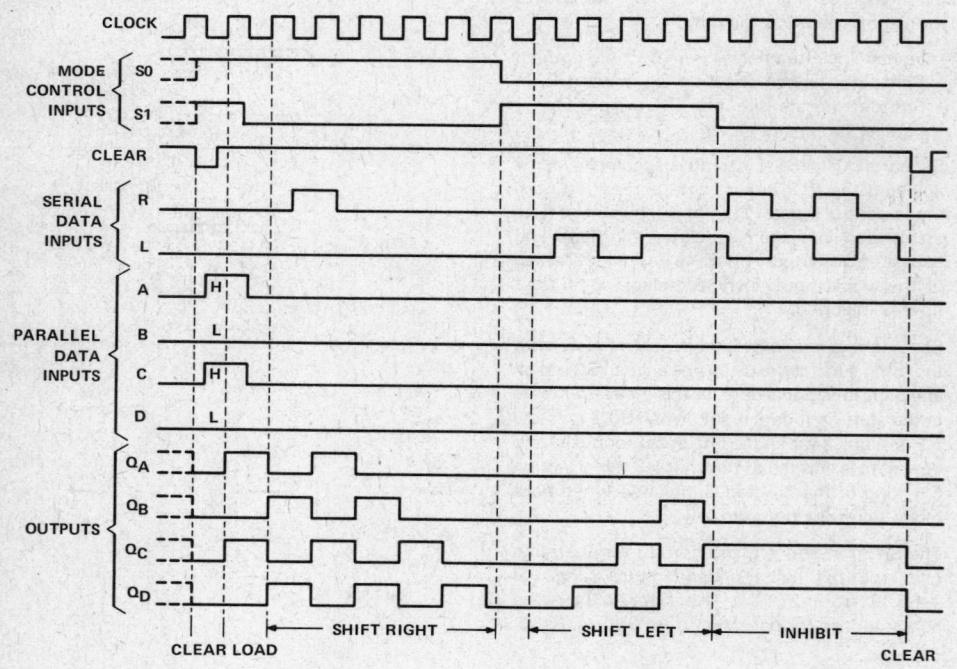

TYPES SN54AS194, SN74AS194
4-BIT BIDIRECTIONAL UNIVERSAL SHIFT REGISTERS

logic symbol

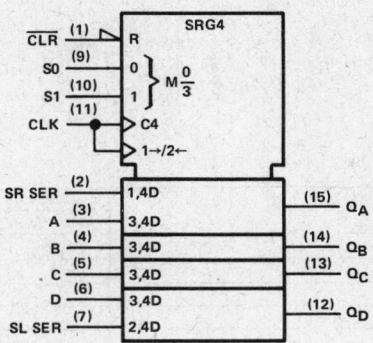

logic diagram (positive logic)

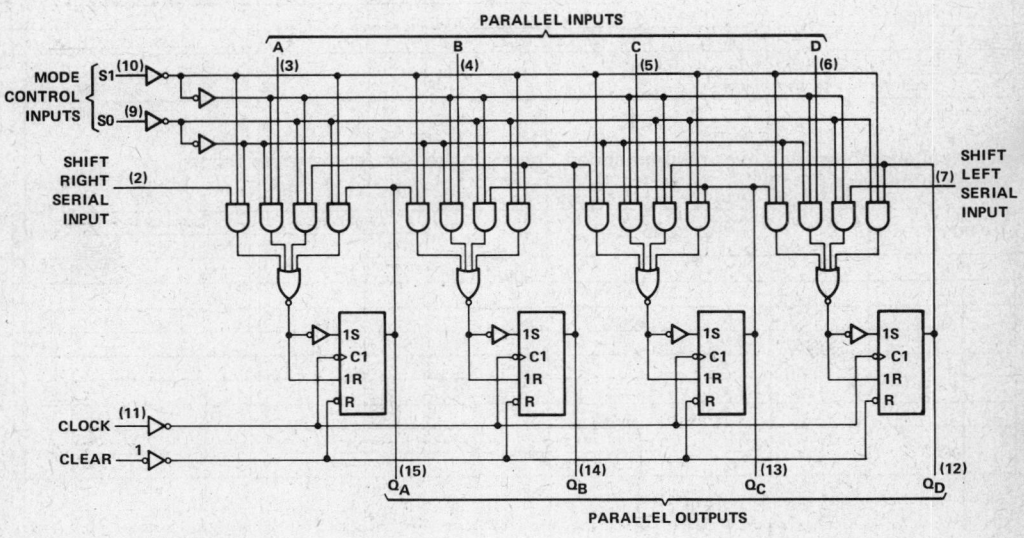

Pin numbers shown are for J and N package.

TYPES SN54AS194, SN74AS194
4-BIT BIDIRECTIONAL UNIVERSAL SHIFT REGISTERS

absolute maximum ratings over operating free-air temperature range (unless otherwise noted)

Supply voltage, V_{CC} .... 7 V
Input voltage .... 7 V
Operating free-air temperature range: SN54AS194 .... −55°C to 125°C
                                                                    SN74AS194 .... 0°C to 150°C
Storage temperature range .... −65°C to 150°C

recommended operating conditions

			SN54AS194			SN74AS194			UNIT
			MIN	NOM	MAX	MIN	NOM	MAX	
V_{CC}	Supply voltage		4.5	5	5.5	4.5	5	5.5	V
V_{IH}	High-level input voltage		2			2			V
V_{IL}	Low-level input voltage				0.8			0.8	V
I_{OH}	High-level output current				−2			−2	mA
I_{OL}	Low-level output current				20			20	mA
f_{clock}	Clock frequency								MHz
t_w	Pulse Duration	$\overline{CLR}$							
		CLK							
t_{su}	Set-up time before CLK↑	Select							
		Data							
t_{wr}		$\overline{CLR}$							
t_h	Hold time, data after CLK↑								ns
T_A	Operating free-air temperaure		−55		125	0		70	°C

electrical characteristics over recommended operating free-air temperature range (unless otherwise noted)

PARAMETER	TEST CONDITIONS		SN54AS194			SN74AS194			UNIT
			MIN	TYP†	MAX	MIN	TYP†	MAX	
V_{IK}	V_{CC} = 4.5 V,	I_I = −18 mA			−1.2			−1.2	V
V_{OH}	V_{CC} = 4.5 V to 5.5 V,	I_{OH} = −2 mA	V_{CC}−2			V_{CC}−2			V
V_{OL}	V_{CC} = 4.5 V,	I_{OL} = 20 mA		0.35	0.5		0.35	0.5	V
I_I	V_{CC} = 5.5 V,	V_I = 7 V							mA
I_{IH}	V_{CC} = 5.5 V,	V_I = 2.7 V							µA
I_{IL}	V_{CC} = 5.5 V,	V_I = 0.4 V							mA
I_O‡	V_{CC} = 5.5 V,	V_O = 2.25 V	−30		−112	−30		−112	mA
I_{CC}	V_{CC} = 5.5 V	Outputs high							mA
		Outputs low		27			27		

†All typical values are at V_{CC} = 5 V, T_A = 25°C.
‡The output conditions have been chosen to produce a current that closely approximates one half of the true short-circuit output current, I_{OS}.

ALS AND AS CIRCUITS

Additional information on these products can be obtained from the factory as it becomes available.

Texas Instruments
POST OFFICE BOX 225012 • DALLAS, TEXAS 75265

TYPES SN54AS194, SN74AS194
4-BIT BIDIRECTIONAL UNIVERSAL SHIFT REGISTERS

switching characteristics (see Note 1)

PARAMETER	FROM (INPUT)	TO L (OUTPUT)	V_{CC} = 4.5 V to 5.5 V, C_L = 50 pF, R1 = 500 Ω, R2 = 500 Ω, T_A = MIN to MAX						UNIT
			SN54AS194			SN74AS194			
			MIN	TYP†	MAX	MIN	TYP†	MAX	
f_{max}									MHz
t_{PLH}	CLK	Any Q		5			5		ns
t_{PHL}				5.5			5.5		
t_{PHL}	$\overline{CLR}$	Any Q		7.5			7.5		ns

†All typical values are at V_{CC} = 5 V, T_A = 25°C.
NOTE 1: For load circuit and voltage waveforms, see page 1-12.

2
ALS AND AS CIRCUITS

TYPES SN54AS195, SN74AS195
4-BIT BIDIRECTIONAL UNIVERSAL SHIFT REGISTERS

D2661, DECEMBER 1983

- Parallel-to-Serial, Serial-to-Parallel Conversions
- Parallel Synchronous Loading
- J and $\overline{K}$ Inputs to First Stage
- Right-Shift Only with Complementary Outputs on Last Stage
- Direct Overriding Clear
- Dependable Texas Instruments Quality and Reliability

SN54AS195 . . . J PACKAGE
SN74AS195 . . . N PACKAGE
(TOP VIEW)

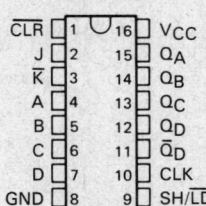

SN54AS195 . . . FH PACKAGE
SN74AS195 . . . FN PACKAGE
(TOP VIEW)

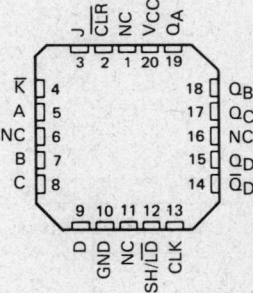

NC—No internal connection

description

These 4-bit registers feature parallel inputs, parallel outputs, J-$\overline{K}$ serial inputs, shift/load control input, and a direct overriding clear. The registers have two modes of operation:

Parallel (broadside) load
Shift (in the direction Q_A toward Q_D)

Parallel loading is accomplished by applying the four bits of data and taking the shift/load control input low. The data are loaded into the associated flip-flops and appear at the outputs after the positive transition of the clock input. During loading serial data flow is inhibited.

Shifting is accomplished synchronously when the shift/load control input is high. Serial data for this mode is entered at the J-$\overline{K}$ inputs. These inputs permit the first stage to perform as a J-$\overline{K}$, D-, or T-type flip-flop as shown in the function table.

The SN54AS195 is characterized for operation over the full military range of $-55\,°C$ to $125\,°C$. The SN74AS195 is characterized for operation from $0\,°C$ to $70\,°C$.

PRODUCT PREVIEW
This page contains information on a product under development. Texas Instruments reserves the right to change or discontinue this product without notice.

Copyright © 1983 by Texas Instruments Incorporated

TYPES SN54AS195, SN74AS195
4-BIT BIDIRECTIONAL UNIVERSAL SHIFT REGISTERS

logic symbol / logic diagram (positive logic)

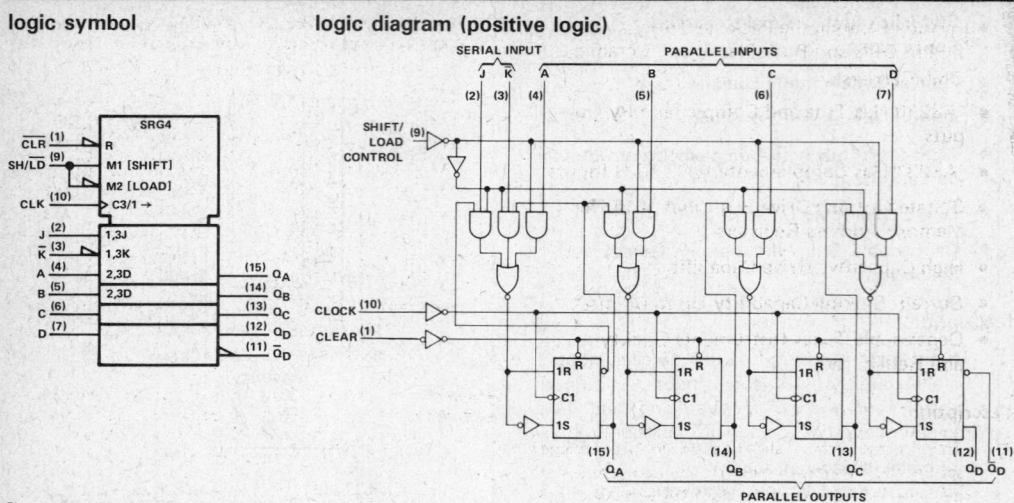

Pin numbers shown are for J and N packages.

FUNCTION TABLE

INPUTS									OUTPUTS				
CLEAR	SHIFT/LOAD	CLOCK	SERIAL		PARALLEL				Q_A	Q_B	Q_C	Q_D	$\overline{Q}_D$
			J	$\overline{K}$	A	B	C	D					
L	X	X	X	X	X	X	X	X	L	L	L	L	L
H	L	↑	X	X	a	b	c	d	a	b	c	d	$\overline{d}$
H	H	L	X	X	X	X	X	X	Q_{A0}	Q_{B0}	Q_{C0}	Q_{D0}	$\overline{Q}_{D0}$
H	H	↑	L	H	X	X	X	X	Q_{A0}	Q_{A0}	Q_{Bn}	Q_{Cn}	$\overline{Q}_{Cn}$
H	H	↑	L	L	X	X	X	X	L	Q_{An}	Q_{Bn}	Q_{Cn}	$\overline{Q}_{Cn}$
H	H	↑	H	H	X	X	X	X	H	Q_{An}	Q_{Bn}	Q_{Cn}	$\overline{Q}_{Cn}$
H	H	↑	H	L	X	X	X	X	$\overline{Q}_{An}$	Q_{An}	Q_{Bn}	Q_{Cn}	$\overline{Q}_{Cn}$

typical clear, shift, and load sequences

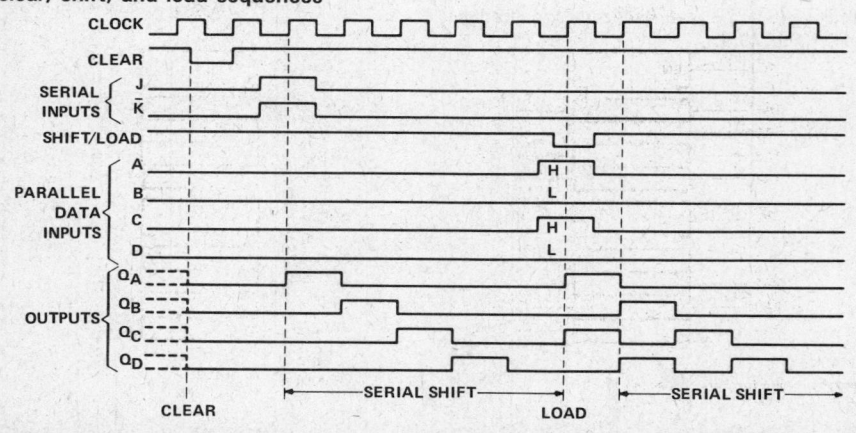

TEXAS INSTRUMENTS
POST OFFICE BOX 225012 • DALLAS, TEXAS 75265

TYPES SN54AS230, SN54AS231, SN74AS230, SN74AS231
OCTAL BUFFERS AND LINE DRIVERS WITH 3-STATE OUTPUTS

D2661, DECEMBER 1982—REVISED DECEMBER 1983

- Included among the Package Options Are 20-Pin DIPs and Both Plastic and Ceramic Chip Carriers
- 'AS230 Has True and Complementary Outputs
- 'AS231 Has Complementary G and $\overline{G}$ Inputs
- 3-State Outputs Drive Bus Lines or Buffer Memory Address Registers
- High Capacitive Drive Capability
- Current Sinking Capability Up to 64 mA
- Dependable Texas Instruments Quality and Reliability

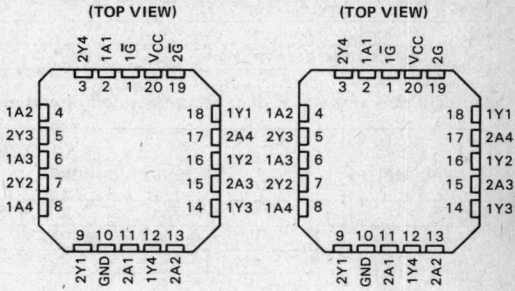

description

These octal buffers and line drivers are designed specifically to improve the performance of three-state memory address drivers, clock drivers, and bus-oriented receivers and transmitters. The designer has a choice of selected combinations of inverting and noninverting outputs, symmetrical $\overline{G}$ (active-low output control) inputs, and complementary G and $\overline{G}$ inputs.

The SN74AS230 and SN74AS231 can be used to drive terminated lines down to 133 ohms.

The SN54AS230 and SN54AS231 are characterized for operation over the full military temperature range of $-55°C$ to $125°C$. The SN74AS230 and SN74AS231 are characterized for operation from $0°C$ to $70°C$.

logic symbols

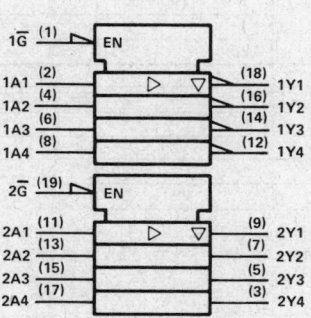

'AS230

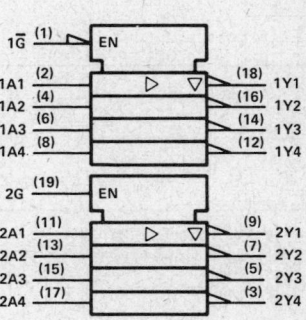

'AS231

Pin numbers shown are for J and N packages.

Copyright © 1982 by Texas Instruments Incorporated

TEXAS INSTRUMENTS
POST OFFICE BOX 225012 • DALLAS, TEXAS 75265

TYPES SN54AS230, SN54AS231, SN74AS230, SN74AS231
OCTAL BUFFERS AND LINE DRIVERS WITH 3-STATE OUTPUTS

absolute maximum ratings over operating free-air temperature range

Supply voltage, V_{CC} .... 7 V
Input voltage .... 7 V
Voltage applied to a disabled 3-state output .... 5.5 V
Operating free-air temperature range: SN54AS230, SN54AS231 .... $-55°C$ to $125°C$
                                 SN74AS230, SN74AS231 .... $0°C$ to $70°C$
Storage temperature range .... $-65°C$ to $150°C$

recommended operating conditions

		SN54AS230 SN54AS231			SN74AS230 SN74AS231			UNIT
		MIN	NOM	MAX	MIN	NOM	MAX	
V_{CC}	Supply voltage	4.5	5	5.5	4.5	5	5.5	V
V_{IH}	High-level input voltage	2			2			V
V_{IL}	Low-level input voltage			0.8			0.8	V
I_{OH}	High-level output current			-12			-15	mA
I_{OL}	Low-level output current			48			64	mA
T_A	Operating free-air temperature	-55		125	0		70	°C

electrical characteristics over recommended operating free-air temperature range (unless otherwise noted)

PARAMETER		TEST CONDITIONS		SN54AS230 SN54AS231			SN74AS230 SN74AS231			UNIT
				MIN	TYP†	MAX	MIN	TYP†	MAX	
V_{IK}		$V_{CC} = 4.5$ V,	$I_I = -18$ mA			-1.2			-1.2	V
V_{OH}		$V_{CC} = 4.5$ V to 5.5 V,	$I_{OH} = -2$ mA	$V_{CC}-2$			$V_{CC}-2$			V
		$V_{CC} = 4.5$ V,	$I_{OH} = -3$ mA	2.4	3.4		2.4	3.4		
		$V_{CC} = 4.5$ V,	$I_{OH} = -12$ mA	2.4						
		$V_{CC} = 4.5$ V,	$I_{OH} = -15$ mA				2.4			
V_{OL}		$V_{CC} = 4.5$ V,	$I_{OL} = 48$ mA		0.27	0.55				V
		$V_{CC} = 4.5$ V	$I_{OL} = 64$ mA					0.31	0.55	
I_{OZH}		$V_{CC} = 5.5$ V,	$V_O = 2.7$ V			50			50	µA
I_{OZL}		$V_{CC} = 5.5$ V,	$V_O = 0.4$ V			-50			-50	µA
I_I		$V_{CC} = 5.5$ V,	$V_I = 7$ V			0.1			0.1	mA
I_{IH}		$V_{CC} = 5.5$ V,	$V_I = 2.7$ V			20			20	µA
I_{IL}	'AS230 2A	$V_{CC} = 5.5$ V,	$V_I = 0.4$ V			-1			-1	mA
	All others					-0.5			-0.5	
I_O‡		$V_{CC} = 5.5$ V,	$V_O = 2.25$ V	-50		-150	-50		-150	mA
I_{CC}	'AS230	$V_{CC} = 5.5$ V	Outputs high		16	25		16	25	
			Outputs low		55	87		55	87	mA
			Outputs disabled		29	46		29	46	mA
	'AS231	$V_{CC} = 5.5$ V	Outputs high		12	18		12	18	
			Outputs low		52	82		52	82	mA
			Outputs disabled		25	39		25	39	

†All typical values are at $V_{CC} = 5$ V, $T_A = 25°C$.
‡The output conditions have been chosen to produce a current that closely approximates one half of the true short-circuit output current, I_{OS}.

TYPES SN54AS230, SN54AS231, SN74AS230, SN74AS231
OCTAL BUFFERS AND LINE DRIVERS WITH 3-STATE OUTPUTS

'AS230 switching characteristics (see Note 1)

PARAMETER	FROM (INPUT)	TO (OUTPUT)	V_{CC} = 4.5 V to 5.5 V, C_L = 50 pF, R1 = 500 Ω, R2 = 500 Ω, T_A = MIN to MAX				UNIT
			SN54AS230		SN74AS230		
			MIN	MAX	MIN	MAX	
t_{PLH}	1A	1Y	2.5	7	2.5	6.5	ns
t_{PHL}			2	6	2	5.7	
t_{PLH}	2A	2Y	2.5	9	2.5	6.2	ns
t_{PHL}			2	7	2	6.2	
t_{PZH}	$1\overline{G}$	1Y	2	7	2	6.4	ns
t_{PZL}			2	9	2	8.5	
t_{PHZ}			2	5.5	2	5	
t_{PLZ}			2	12.5	2	9.5	
t_{PZH}	$2\overline{G}$	2Y	2	10	2	9	ns
t_{PZL}			2	8	2	7.5	
t_{PHZ}			2	6.5	2	6	
t_{PLZ}			2	10.5	2	9	

'AS231 switching characteristics (see Note 1)

PARAMETER	FROM (INPUT)	TO (OUTPUT)	V_{CC} = 4.5 V to 5.5 V, C_L = 50 pF, R1 = 500 Ω, R2 = 500 Ω, T_A = MIN to MAX				UNIT
			SN54AS231		SN74AS231		
			MIN	MAX	MIN	MAX	
t_{PLH}	A	Y	2	7	2	6.5	ns
t_{PHL}			2	6	2	5.7	
t_{PZH}	$\overline{G}$	Y	2	7	2	6.4	ns
t_{PZL}			2	9	2	8.5	
t_{PHZ}			2	5.5	2	5	
t_{PLZ}			2	12.5	2	9.5	
t_{PZH}	G	Y	3	7	3	6	ns
t_{PZL}			3	10	3	9	
t_{PHZ}			3	6.5	3	6	
t_{PLZ}			3	13.5	3	7	

NOTE 1: For load circuit and voltage waveforms, see page 1-12.

2
ALS AND AS CIRCUITS

TYPES SN54ALS240A, SN54ALS241A, SN54AS240, SN54AS241
SN74ALS240A, SN74ALS241A, SN74AS240, SN74AS241
OCTAL BUFFERS AND LINE DRIVERS WITH 3-STATE OUTPUTS

D2661, DECEMBER 1982—REVISED DECEMBER 1983

- 3-State Outputs Drive Bus Lines or Buffer Memory Address Registers
- P-N-P Inputs Reduce DC Loading
- Dependable Texas Instruments Quality and Reliability

SN54ALS', SN54AS' . . . J PACKAGE
SN74ALS', SN74AS' . . . N PACKAGE
(TOP VIEW)

```
1G   [ 1    20 ] VCC
1A1  [ 2    19 ] 2G/2G*
2Y4  [ 3    18 ] 1Y1
1A2  [ 4    17 ] 2A4
2Y3  [ 5    16 ] 1Y2
1A3  [ 6    15 ] 2A3
2Y2  [ 7    14 ] 1Y3
1A4  [ 8    13 ] 2A2
2Y1  [ 9    12 ] 1Y4
GND  [10    11 ] 2A1
```

description

These octal buffers and line drivers are designed specifically to improve both the performance and density of three-state memory address drivers, clock drivers, and bus-oriented receivers and transmitters. The designer has a choice of selected combinations of inverting and noninverting outputs, symmetrical $\overline{G}$ (active-low output control) inputs, and complementary G and $\overline{G}$ inputs. These devices feature high fan-out and improved fan-in.

The -1 versions of the SN74ALS' parts are identical to their standard versions except that the recommended maximum I_{OL} is increased to 48 milliamperes. There are no -1 versions of the SN54ALS' parts.

The SN54' family is characterized for operation over the full military temperature range of −55°C to 125°C. The SN74' family is characterized for operation from 0°C to 70°C.

SN54ALS', SN54AS' . . . FH PACKAGE
SN74ALS', SN74AS' . . . FN PACKAGE
(TOP VIEW)

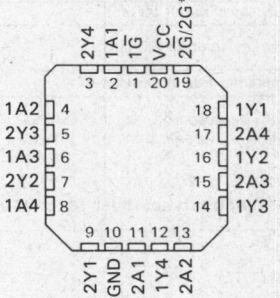

*2$\overline{G}$ for 'ALS240A, 'AS240 or 2G for 'ALS241A, 'AS241

logic symbols

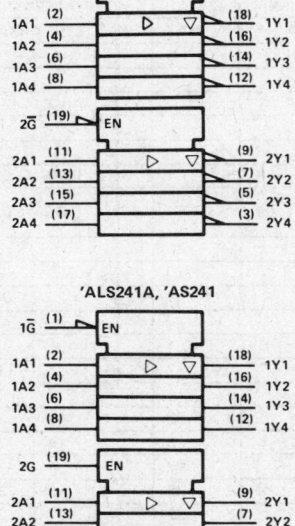

Pin numbers shown are for J and N packages.

logic diagrams (positive logic)

'ALS240A, 'AS240

```
1G   (1)  ─|>○─
1A1  (2)  ─|>○─  (18) 1Y1
1A2  (4)  ─|>○─  (16) 1Y2
1A3  (6)  ─|>○─  (14) 1Y3
1A4  (8)  ─|>○─  (12) 1Y4
2G   (19) ─|>○─
2A1  (11) ─|>○─  (9)  2Y1
2A2  (13) ─|>○─  (7)  2Y2
2A3  (15) ─|>○─  (5)  2Y3
2A4  (17) ─|>○─  (3)  2Y4
```

'ALS241A, 'AS241

```
1G   (1)  ─|>○─
1A1  (2)  ─|>─   (18) 1Y1
1A2  (4)  ─|>─   (16) 1Y2
1A3  (6)  ─|>─   (14) 1Y3
1A4  (8)  ─|>─   (12) 1Y4
2G   (19) ─|>─
2A1  (11) ─|>─   (9)  2Y1
2A2  (13) ─|>─   (7)  2Y2
2A3  (15) ─|>─   (5)  2Y3
2A4  (17) ─|>─   (3)  2Y4
```

Copyright © 1983 by Texas Instruments Incorporated

TYPES SN54ALS240A, SN54ALS241A, SN74ALS240A, SN74ALS241A
OCTAL BUFFERS AND LINE DRIVERS WITH 3-STATE OUTPUTS

absolute maximum ratings over operating free-air temperature range (unless otherwise noted)

Supply voltage, V_{CC} .... 7 V
Input voltage .... 7 V
Voltage applied to a disabled 3-state output .... 5.5 V
Operating free-air temperature range: SN54ALS240A, SN54ALS231A .... −55°C to 125°C
                                       SN74ALS240A, SN74ALS241A .... 0°C to 70°C
Storage temperature range .... −65°C to 150°C

recommended operating conditions

		SN54ALS240A SN54ALS241A			SN74ALS240A SN74ALS241A			UNIT
		MIN	NOM	MAX	MIN	NOM	MAX	
V_{CC}	Supply voltage	4.5	5	5.5	4.5	5	5.5	V
V_{IH}	High-level input voltage	2			2			V
V_{IL}	Low-level input voltage			0.8			0.8	V
I_{OH}	High-level output current			−12			−15	mA
I_{OL}	Low-level output current			12			24 48†	mA
T_A	Operating free-air temperature	−55		125	0		70	°C

†The extended limits apply only if V_{CC} is maintained between 4.75 V and 5.25 V.
The 48 mA limit applies for the SN74ALS240A-1 and SN74ALS241A-1 only.

electrical characteristics over recommended operating free-air temperature range (unless otherwise noted)

PARAMETER		TEST CONDITIONS		SN54ALS240A SN54ALS241A			SN74ALS240A SN74ALS241A			UNIT
				MIN	TYP‡	MAX	MIN	TYP‡	MAX	
V_{IK}		V_{CC} = 4.5 V,	I_I = −18 mA			−1.5			−1.5	V
V_{OH}		V_{CC} = 4.5 V to 5.5 V,	I_{OH} = −0.4 mA	V_{CC}−2			V_{CC}−2			V
		V_{CC} = 4.5 V,	I_{OH} = −3 mA	2.4	3.2		2.4	3.2		
		V_{CC} = 4.5 V,	I_{OH} = −12 mA	2						
		V_{CC} = 4.5 V,	I_{OH} = −15 mA				2			
V_{OL}		V_{CC} = 4.5 V,	I_{OL} = 12 mA		0.25	0.4		0.25	0.4	V
		V_{CC} = 4.5 V, (I_{OL} = 48 mA for −1 versions)	I_{OL} = 24 mA					0.35	0.5	
I_{OZH}		V_{CC} = 5.5 V,	V_O = 2.7 V			20			20	µA
I_{OZL}		V_{CC} = 5.5 V,	V_O = 0.4 V			−20			−20	µA
I_I		V_{CC} = 5.5 V,	V_I = 7 V			0.1			0.1	mA
I_{IH}		V_{CC} = 5.5 V,	V_I = 2.7 V			20			20	µA
I_{IL}		V_{CC} = 5.5 V,	V_I = 0.4 V			−0.1			−0.1	mA
I_O§		V_{CC} = 5.5 V,	V_O = 2.25 V	−30		−112	−30		−112	mA
I_{CC}	'ALS240A	V_{CC} = 5.5 V	Outputs high		4	10		4	10	mA
			Outputs low		13	23		13	23	
			Outputs disabled		14	25		14	25	
	'ALS241A		Outputs high		9	17		9	15	
			Outputs low		15	28		15	26	
			Outputs disabled		17	32		17	30	

†All typical values are at V_{CC} = 5 V, T_A = 25°C.
‡The output conditions have been chosen to produce a current that closely approximates one half of the true short-circuit output current, I_{OS}.

TEXAS INSTRUMENTS
POST OFFICE BOX 225012 • DALLAS, TEXAS 75265

TYPES SN54ALS240A, SN54ALS241A, SN74ALS240A, SN74ALS241A
OCTAL BUFFERS AND LINE DRIVERS WITH 3-STATE OUTPUTS

'ALS240A switching characteristics (see Note 1)

PARAMETER	FROM (INPUT)	TO (OUTPUT)	V_{CC} = 4.5 V to 5.5 V, C_L = 50 pF, $R1$ = 500 Ω, $R2$ = 500 Ω, T_A = MIN to MAX				UNIT
			SN54ALS240A		SN74ALS240A		
			MIN	MAX	MIN	MAX	
t_{PLH}	A	Y	2	12	2	9	ns
t_{PHL}			2	11	2	9	
t_{PZH}	$\overline{G}$	Y	5	15	5	13	ns
t_{PZL}			5	20	5	18	
t_{PHZ}	$\overline{G}$	Y	2	12	2	10	ns
t_{PLZ}			3	18	3	12	

'ALS241A switching characteristics (see Note 1)

PARAMETER	FROM (INPUT)	TO (OUTPUT)	V_{CC} = 4.5 V to 5.5 V, C_L = 50 pF, $R1$ = 500 Ω, $R2$ = 500 Ω, T_A = MIN to MAX				UNIT
			SN54ALS241A		SN74ALS241A		
			MIN	MAX	MIN	MAX	
t_{PLH}	A	Y	3	14	3	11	ns
t_{PHL}			3	13	3	10	
t_{PZH}	$1\overline{G}$	Y	7	25	7	21	ns
t_{PZL}			7	25	7	21	
t_{PHZ}	$1\overline{G}$	Y	2	12	2	10	ns
t_{PLZ}			3	20	3	15	
t_{PZH}	2G	Y	7	25	7	21	ns
t_{PZL}			7	25	7	21	
t_{PHZ}	2G	Y	2	12	2	10	ns
t_{PLZ}			3	20	3	15	

NOTE 1: For load circuit and voltage waveforms, see page 1-12.

TEXAS INSTRUMENTS
POST OFFICE BOX 225012 • DALLAS, TEXAS 75265

TYPES SN54AS240, SN54AS241, SN74AS240, SN74AS241
OCTAL BUFFERS AND LINE DRIVERS WITH 3-STATE OUTPUTS

absolute maximum ratings over operating free-air temperature range (unless otherwise noted)

Supply voltage, V_{CC} .... 7 V
Input voltage .... 7 V
Voltage applied to a disabled 3-state output .... 5.5 V
Operating free-air temperature range: SN54AS240, SN54AS241 .... −55°C to 125°C
                                      SN74AS240, SN74AS241 ....   0°C to 70°C
Storage temperature range .... −65°C to 150°C

recommended operating conditions

		SN54AS240 SN54AS241			SN74AS240 SN74AS241			UNIT
		MIN	NOM	MAX	MIN	NOM	MAX	
V_{CC}	Supply voltage	4.5	5	5.5	4.5	5	5.5	V
V_{IH}	High-level input voltage	2			2			V
V_{IL}	Low-level input voltage			0.8			0.8	V
I_{OH}	High-level output current			−12			−15	mA
I_{OL}	Low-level output current			48			64	mA
T_A	Operating free-air temperature	−55		125	0		70	°C

electrical characteristics over recommended operating free-air temperature range (unless otherwise noted)

PARAMETER		TEST CONDITIONS		SN54AS240 SN54AS241			SN74AS240 SN74AS241			UNIT
				MIN	TYP†	MAX	MIN	TYP†	MAX	
V_{IK}		V_{CC} = 4.5 V,	I_I = −18 mA			−1.2			−1.2	V
V_{OH}		V_{CC} = 4.5 V to 5.5 V,	I_{OH} = −2 mA	V_{CC}−2			V_{CC}−2			V
		V_{CC} = 4.5 V,	I_{OH} = −3 mA	2.4	3.4		2.4	3.4		
		V_{CC} = 4.5 V,	I_{OH} = −12 mA	2.4						
		V_{CC} = 4.5 V,	I_{OH} = −15 mA				2.4			
V_{OL}		V_{CC} = 4.5 V,	I_{OL} = 48 mA		0.27	0.55				V
		V_{CC} = 4.5 V,	I_{OL} = 64 mA					0.31	0.55	
I_{OZH}		V_{CC} = 5.5 V,	V_O = 2.7 V			50			50	μA
I_{OZL}		V_{CC} = 5.5 V,	V_O = 0.4 V			−50			−50	μA
I_I		V_{CC} = 5.5 V,	V_I = 7 V			0.1			0.1	mA
I_{IH}		V_{CC} = 5.5 V,	V_I = 2.7 V			20			20	μA
I_{IL}	'AS241 A inputs	V_{CC} = 5.5 V,	V_I = 0.4 V			−1			−1	mA
	All others					−0.5			−0.5	
I_O‡		V_{CC} = 5.5 V,	V_O = 2.25 V	−50		−150	−50		−150	mA
I_{CC}	'AS240	V_{CC} = 5.5 V	Outputs high		11	17		11	17	mA
			Outputs low		51	75		51	75	
			Outputs disabled		24	38		24	38	
	'AS241		Outputs high		22	35		22	35	
			Outputs low		61	90		61	90	
			Outputs disabled		35	56		35	56	

†All typical values are at V_{CC} = 5 V, T_A = 25°C.
‡The output conditions have been chosen to produce a current that closely approximates one half of the true short-circuit output current, I_{OS}.

TYPES SN54AS240, SN54AS241, SN74AS240, SN74AS241
OCTAL BUFFERS AND LINE DRIVERS WITH 3-STATE OUTPUTS

'AS240 switching characteristics (see Note 1)

PARAMETER	FROM (INPUT)	TO (OUTPUT)	V_{CC} = 4.5 V to 5.5 V, C_L = 50 pF, R1 = 500 Ω, R2 = 500 Ω, T_A = MIN to MAX				UNIT
			SN54AS240		SN74AS240		
			MIN	MAX	MIN	MAX	
t_{PLH}	A	Y	2	7	2	6.5	ns
t_{PHL}			2	6	2	5.7	
t_{PZH}	$\overline{G}$	Y	2	7	2	6.4	ns
t_{PZL}			2	9.5	2	9	
t_{PHZ}	$\overline{G}$	Y	2	5.5	2	5	ns
t_{PLZ}			2	12.5	2	9.5	

'AS241 switching characteristics (see Note 1)

PARAMETER	FROM (INPUT)	TO (OUTPUT)	V_{CC} = 4.5 V to 5.5 V, C_L = 50 pF, R1 = 500 Ω, R2 = 500 Ω, T_A = MIN to MAX				UNIT
			SN54AS241		SN74AS241		
			MIN	MAX	MIN	MAX	
t_{PLH}	A	Y	2	9	2	6.2	ns
t_{PHL}			2	7	2	6.2	
t_{PZH}	$1\overline{G}$	Y	2	10	2	9	ns
t_{PZL}			2	8	2	7.5	
t_{PHZ}	$1\overline{G}$	Y	2	6.5	2	6	ns
t_{PLZ}			2	10.5	2	9	
t_{PZH}	2G	Y	3	11	3	10.5	ns
t_{PZL}			3	9.5	3	8.5	
t_{PHZ}	2G	Y	3	7	3	7	ns
t_{PLZ}			3	12	3	12	

NOTE 1: For load circuit and voltage waveforms, see page 1-12.

2
ALS AND AS CIRCUITS

TYPES SN54ALS242A, SN54ALS243A, SN54AS242, SN54AS243 SN74ALS242A, SN74ALS243A, SN74AS242, SN74AS243 QUADRUPLE BUS TRANSCEIVERS WITH 3-STATE OUTPUTS

D2661, DECEMBER 1982—REVISED DECEMBER 1983

- 2-Way Asynchronous Communication Between Data Buses
- P-N-P Inputs Reduce Loading
- Dependable Texas Instruments Quality and Reliability

SN54′ . . . J PACKAGE
SN74′ . . . N PACKAGE
(TOP VIEW)

```
GAB  [ 1    14 ] VCC
NC   [ 2    13 ] GBA
A1   [ 3    12 ] NC
A2   [ 4    11 ] B1
A3   [ 5    10 ] B2
A4   [ 6     9 ] B3
GND  [ 7     8 ] B4
```

description

These four-data-line transceivers are designed for asynchronous two-way communications between data buses. The SN74ALS′ devices can be used to drive terminated lines down to 133 ohms.

The -1 versions of the SN74ALS′ parts are identical to the standard versions except that the recommended maximum I_{OL} is increased to 48 milliamperes. There are no -1 versions of the SN54ALS′ parts.

The SN54′ family is characterized for operation over the full military temperature range of −55°C to 125°C. The SN74′ family is characterized for operation from 0°C to 70°C.

SN54′ . . . FH PACKAGE
SN74′ . . . FN PACKAGE
(TOP VIEW)

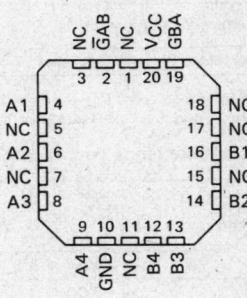

NC—No Internal connection

logic symbol

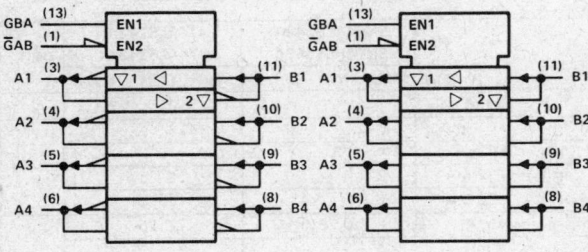

logic diagrams (positive logic)

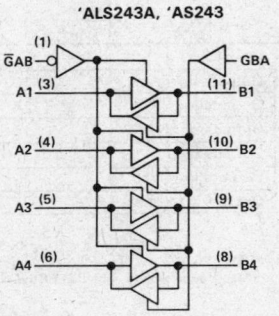

Pin numbers shown are for J and N packages.

FUNCTION TABLE

INPUTS		'ALS242A	'ALS243A
GAB	GBA	'AS242	'AS243
L	L	$\overline{A}$ to B	A to B
H	H	$\overline{B}$ to A	B to A
H	L	Isolation	Isolation
L	H	Latch A and B ($A = \overline{B}$)	Latch A and B ($A = B$)

Copyright © 1982 by Texas Instruments Incorporated

ALS AND AS CIRCUITS

TEXAS INSTRUMENTS
POST OFFICE BOX 225012 • DALLAS, TEXAS 75265

TYPES SN54ALS242A, SN54ALS243A, SN74ALS242A, SN74ALS243A
QUADRUPLE BUS TRANSCEIVERS WITH 3-STATE OUTPUTS

absolute maximum ratings over operating free-air temperature range (unless otherwise noted)

Supply voltage, V_{CC} .... 7 V
Input voltage: All inputs .... 7 V
                I/O ports .... 5.5 V
Operating free-air temperature range: SN54ALS242A, SN54ALS243A .... −55°C to 125°C
                             SN74ALS242A, SN74ALS243A .... 0°C to 70°C
Storage temperature range .... −65°C to 150°C

recommended operating conditions

		SN54ALS242A SN54ALS243A			SN74ALS242A SN74ALS243A			UNIT
		MIN	NOM	MAX	MIN	NOM	MAX	
V_{CC}	Supply voltage	4.5	5	5.5	4.5	5	5.5	V
V_{IH}	High-level input voltage	2			2			V
V_{IL}	Low-level input voltage			0.8			0.8	V
I_{OH}	High-level output current			−12			−15	mA
I_{OL}	Low-level output current			12			24 48†	mA
T_A	Operating free-air temperature	−55		125	0		70	°C

†The extended limits apply only if V_{CC} is maintained between 4.75 V and 5.25 V.
  The 48-mA limit applies for the SN74ALS242A-1 and SN74ALS243A-1 only.

electrical characteristics over recommended operating free-air temperature range (unless otherwise noted)

PARAMETER		TEST CONDITIONS		SN54ALS242A SN54ALS243A			SN74ALS242A SN74ALS243A			UNIT
				MIN	TYP‡	MAX	MIN	TYP‡	MAX	
V_{IK}		V_{CC} = 4.5 V,	I_I = −18 mA			−1.5			−1.5	V
V_{OH}		V_{CC} = 4.5 V to 5.5 V,	I_{OH} = −0.4 mA	V_{CC}−2			V_{CC}−2			V
		V_{CC} = 4.5 V	I_{OH} = −3 mA	2.4	3.2		2.4	3.2		
		V_{CC} = 4.5 V,	I_{OH} = −12 mA	2						
		V_{CC} = 4.5 V,	I_{OH} = −15 mA				2			
V_{OL}		V_{CC} = 4.5 V,	I_{OL} = 12 mA		0.25	0.4		0.25	0.4	V
		V_{CC} = 4.5 V,	I_{OL} = 24 mA					0.35	0.5	
		(I_{OL} = 48 mA for −1 versions)								
I_I	Control inputs	V_{CC} = 5.5 V,	V_I = 7 V			0.1			0.1	mA
	A or B ports	V_{CC} = 5.5 V,	V_I = 5.5 V			0.1			0.1	
I_{IH}	Control inputs	V_{CC} = 5.5 V,	V_I = 2.7 V			20			20	μA
	A or B ports§					20			20	
I_{IL}	Control inputs	V_{CC} = 5.5 V,	V_I = 0.4 V			−0.1			−0.1	mA
	A or B ports§					−0.1			−0.1	
I_O¶		V_{CC} = 5.5 V,	V_O = 2.25 V	−30		−112	−30		−112	mA
I_{CC}	'ALS242A	V_{CC} = 5.5 V	Outputs high		10	20		10	16	mA
			Outputs low		14	26		14	21	
			Outputs disabled		15	27		15	22	
	'ALS243A		Outputs high		15	30		15	25	
			Outputs low		20	35		20	30	
			Outputs disabled		21	37		21	32	

‡All typical values are at V_{CC} = 5 V, T_A = 25°C.
§For I/O ports, the parameters I_{IH} and I_{IL} include the off-state output current.
¶The output conditions have been chosen to produce a current that closely approximates one half of the true short-circuit output current, I_{OS}.

Texas Instruments
POST OFFICE BOX 225012 • DALLAS, TEXAS 75265

TYPES SN54ALS242A, SN54ALS243A, SN74ALS242A, SN74ALS243A
QUADRUPLE BUS TRANSCEIVERS WITH 3-STATE OUTPUTS

'ALS242A switching characteristics (see Note 1)

PARAMETER	FROM (INPUT)	TO (OUTPUT)	V_{CC} = 4.5 V to 5.5 V, C_L = 50 pF, R_1 = 500 Ω, R_2 = 500 Ω, T_A = MIN to MAX				UNIT
			SN54AS242A		SN74AS242A		
			MIN	MAX	MIN	MAX	
t_{PLH}	A or B	B or A	3	15	3	11	ns
t_{PHL}			2	14	2	10	
t_{PZH}	$\overline{G}AB$	B	4	22	4	18	ns
t_{PZL}			7	25	7	21	
t_{PHZ}	$\overline{G}AB$	B	2	16	2	14	ns
t_{PLZ}			4	28	4	22	
t_{PZH}	GBA	A	4	22	4	18	ns
t_{PZL}			7	25	7	21	
t_{PHZ}	GBA	A	2	16	2	14	ns
t_{PLZ}			4	28	4	22	

'ALS243 switching characteristics (see Note 1)

PARAMETER	FROM (INPUT)	TO (OUTPUT)	V_{CC} = 4.5 V to 5.5 V, C_L = 50 pF, R_1 = 500 Ω, R_2 = 500 Ω, T_A = MIN to MAX				UNIT
			SN54ALS243A		SN74ALS243A		
			MIN	MAX	MIN	MAX	
t_{PLH}	A or B	B or A	4	15	4	11	ns
t_{PHL}			4	15	4	11	
t_{PZH}	$\overline{G}AB$	B	7	25	7	20	ns
t_{PZL}			7	25	7	20	
t_{PHZ}	$\overline{G}AB$	B	2	16	2	14	ns
t_{PLZ}			3	27	3	22	
t_{PZH}	GBA	A	7	25	7	20	ns
t_{PZL}			7	25	7	20	
t_{PHZ}	GBA	A	2	16	2	14	ns
t_{PLZ}			3	27	3	22	

NOTE 1: For load circuit and voltage waveforms, see page 1-12.

TYPES SN54AS242, SN54AS243, SN74AS242, SN74AS243
QUADRUPLE BUS TRANSCEIVERS WITH 3-STATE OUTPUTS

absolute maximum ratings over operating free-air temperature range (unless otherwise noted)

Supply voltage, V_{CC} .... 7 V
Input voltage: All inputs .... 7 V
    I/O ports .... 5.5 V
Operating free-air temperature range: SN54AS242, SN54AS243 .... −55 °C to 125 °C
    SN74AS242, SN74AS243 .... 0 °C to 70 °C
Storage temperature range .... −65 °C to 150 °C

recommended operating conditions

		SN54AS242 SN54AS243			SN74AS242 SN74AS243			UNIT
		MIN	NOM	MAX	MIN	NOM	MAX	
V_{CC}	Supply voltage	4.5	5	5.5	4.5	5	5.5	V
V_{IH}	High-level input voltage	2			2			V
V_{IL}	Low-level input voltage			0.8			0.8	V
I_{OH}	High-level output current			−12			−15	mA
I_{OL}	Low-level output current			48			64	mA
T_A	Operating free-air temperature	−55		125	0		70	°C

electrical characteristics over recommended operating free-air temperature range (unless otherwise noted)

PARAMETER		TEST CONDITIONS		SN54AS242 SN54AS243			SN74AS242 SN74AS243			UNIT
				MIN	TYP[†]	MAX	MIN	TYP[†]	MAX	
V_{IK}		V_{CC} = 4.5 V,	I_I = −18 mA			−1.2			−1.2	V
V_{OH}		V_{CC} = 4.5 V to 5.5 V,	I_{OH} = −2 mA	V_{CC}−2			V_{CC}−2			V
		V_{CC} = 4.5 V	I_{OH} = −3 mA	2.4	3.4		2.4	3.4		
		V_{CC} = 4.5 V	I_{OH} = −12 mA	2.4						
		V_{CC} = 4.5 V,	I_{OH} = −15 mA				2.4			
V_{OL}		V_{CC} = 4.5 V,	I_{OL} = 48 mA			0.55				V
		V_{CC} = 4.5 V,	I_{OL} = 64 mA						0.55	
I_I	Control inputs	V_{CC} = 5.5 V,	V_I = 7 V			0.1			0.1	mA
	A or B ports	V_{CC} = 5.5 V,	V_I = 5.5 V			0.1			0.1	
I_{IH}	Control inputs	V_{CC} = 5.5 V,	V_I = 2.7 V			20			20	μA
	A or B ports[‡]					50			50	
I_{IL}	Control inputs	V_{CC} = 5.5 V,	V_I = 0.4 V			−0.5			−0.5	mA
	'AS242 A or B ports[‡]					−0.5			−0.5	
	'AS243 A or B ports[‡]					−1			−1	
I_O[§]		V_{CC} = 5.5 V,	V_O = 2.25 V	−50		−150	−50		−150	mA
I_{CC}	'AS242	V_{CC} = 5.5 V	Outputs high		18	28		18	28	mA
			Outputs low		38	60		38	60	
			Outputs disabled		25	39		25	39	
	'AS243		Outputs high		28	44		28	44	
			Outputs low		47	74		47	74	
			Outputs disabled		35	56		35	56	

[†] All typical values are at V_{CC} = 5 V, T_A = 25 °C.
[‡] For I/O ports, the parameters I_{IH} and I_{IL} include the off-state output current.
[§] The output conditions have been chosen to produce a current that closely approximates one half of the true short-circuit output current, I_{OS}.

ALS AND AS CIRCUITS

TEXAS INSTRUMENTS
POST OFFICE BOX 225012 • DALLAS, TEXAS 75265

TYPES SN54AS242, SN54AS243, SN74AS242, SN74AS243
QUADRUPLE BUS TRANSCEIVERS WITH 3-STATE OUTPUTS

'AS242 switching characteristics (see Note 1)

PARAMETER	FROM (INPUT)	TO (OUTPUT)	V_{CC} = 4.5 V to 5.5 V, C_L = 50 pF, R1 = 500 Ω, R2 = 500 Ω, T_A = MIN to MAX				UNIT
			SN54AS242		SN74AS242		
			MIN	MAX	MIN	MAX	
t_{PLH}	A or B	B or A	2	7	2	6.5	ns
t_{PHL}			2	6	2	5.7	
t_{PZH}	$\overline{GAB}$	B	2	9	2	5.5	ns
t_{PZL}			2	8.5	2	7.5	
t_{PHZ}	$\overline{GAB}$	B	2	7	2	6.5	ns
t_{PLZ}			2	12.5	2	9.5	
t_{PZH}	GAB	A	3	7	3	6	ns
t_{PZL}			3	9	3	8	
t_{PHZ}	GAB	A	3	8.5	3	6	ns
t_{PLZ}			3	13.5	3	10.5	

'AS243 switching characteristics (see Note 1)

PARAMETER	FROM (INPUT)	TO (OUTPUT)	V_{CC} = 4.5 V to 5.5 V, C_L = 50 pF, R1 = 500 Ω, R2 = 500 Ω, T_A = MIN to MAX				UNIT
			SN54AS243		SN74AS243		
			MIN	MAX	MIN	MAX	
t_{PLH}	A or B	B or A	3	9	3	7.5	ns
t_{PHL}			3	8	3	6.5	
t_{PZH}	$\overline{GAB}$	B	2	10	2	9	ns
t_{PZL}			2	9	2	7.5	
t_{PHZ}	$\overline{GAB}$	B	2	7	2	6.5	ns
t_{PLZ}			2	11	2	9	
t_{PZH}	GAB	A	3	11	3	10.5	ns
t_{PZL}			3	9.5	3	8.5	
t_{PHZ}	GAB	A	3	7.5	3	7	ns
t_{PLZ}			3	14	3	11	

NOTE 1: For load circuit and voltage waveforms, see page 1-12.

TEXAS INSTRUMENTS
POST OFFICE BOX 225012 • DALLAS, TEXAS 75265

2
ALS AND AS CIRCUITS

TYPES SN54ALS244A, SN54AS244, SN74ALS244A, SN74AS244
OCTAL BUFFERS AND LINE DRIVERS WITH 3-STATE OUTPUTS

D2661, DECEMBER 1982—REVISED DECEMBER 1983

- 3-State Outputs Drive Bus Lines or Buffer Memory Address Registers
- P-N-P Inputs Reduce DC Loading
- Package Options Include Both Plastic and Ceramic Chip Carriers in Addition to Plastic and Ceramic DIPs
- Dependable Texas Instruments Quality and Reliability

SN54ALS244A, SN54AS244 . . . J PACKAGE
SN74ALS244A, SN74AS244 . . . N PACKAGE
(TOP VIEW)

```
1G   [ 1    20 ] VCC
1A1  [ 2    19 ] 2G
2Y4  [ 3    18 ] 1Y1
1A2  [ 4    17 ] 2A4
2Y3  [ 5    16 ] 1Y2
1A3  [ 6    15 ] 2A3
2Y2  [ 7    14 ] 1Y3
1A4  [ 8    13 ] 2A2
2Y1  [ 9    12 ] 1Y4
GND  [ 10   11 ] 2A1
```

description

These octal buffers and line drivers are designed specifically to improve both the performance and density of three-state memory address drivers, clock drivers, and bus-oriented receivers and transmitters. Taken together with the 'ALS240A, 'ALS241A, 'AS240, and 'AS241, these devices provide the choice of selected combinations of inverting outputs, symmetrical G (active-low input control) inputs, and complementary G and G inputs.

The −1 version of the SN74ALS244A is identical to the standard version except that the recommended maximum I_{OL} is increased to 48 milliamperes. There is no −1 version of the SN54ALS244A.

The SN54ALS244A and SN54AS244 are characterized for operation over the full military temperature range of −55°C to 125°C. The SN74ALS244A and SN74AS244 are characterized for operation from 0°C to 70°C.

SN54ALS244A, SN54AS244 . . . FH PACKAGE
SN74ALS244A, SN74AS244 . . . FN PACKAGE
(TOP VIEW)

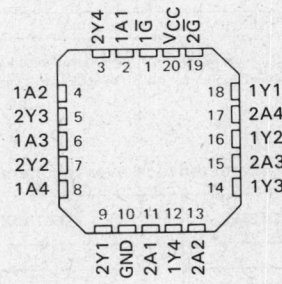

logic diagram (positive logic)

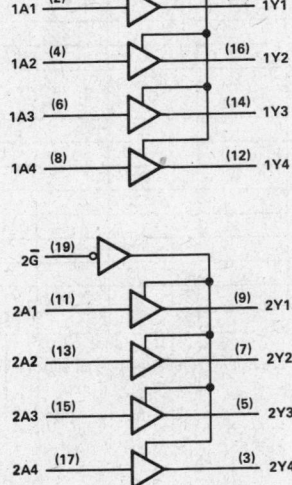

logic symbol

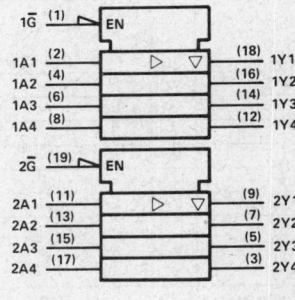

Pin numbers shown are for J and N packages.

Copyright © 1982 by Texas Instruments Incorporated

TEXAS INSTRUMENTS
POST OFFICE BOX 225012 • DALLAS, TEXAS 75265

TYPES SN54ALS244A, SN74ALS244A,
OCTAL BUFFERS AND LINE DRIVERS WITH 3-STATE OUTPUTS

absolute maximum ratings over operating free-air temperature range (unless otherwise noted)

Supply voltage, V_{CC} .... 7 V
Input voltage .... 7 V
Voltage applied to a disabled 3-state output .... 5.5 V
Operating free-air temperature range: SN54ALS244A .... $-55°C$ to $125°C$
                                      SN74ALS244A .... $0°C$ to $70°C$
Storage temperature range .... $-65°C$ to $150°C$

recommended operating conditions

		SN54ALS244A			SN74ALS244A			UNIT
		MIN	NOM	MAX	MIN	NOM	MAX	
V_{CC}	Supply voltage	4.5	5	5.5	4.5	5	5.5	V
V_{IH}	High-level input voltage	2			2			V
V_{IL}	Low-level input voltage			0.8			0.8	V
I_{OH}	High-level output current			-12			-15	mA
I_{OL}	Low-level output current			12			24	mA
							48†	
T_A	Operating free-air temperature	-55		125	0		70	°C

†The extended limits apply only if V_{CC} is maintained between 4.75 V and 5.25 V.
The 48-mA limit applies for the SN74ALS244A-1 only.

electrical characteristics over recommended operating free-air temperature range (unless otherwise noted)

PARAMETER	TEST CONDITIONS		SN54ALS244A			SN74ALS244A			UNIT
			MIN	TYP‡	MAX	MIN	TYP‡	MAX	
V_{IK}	$V_{CC} = 4.5$ V,	$I_I = -18$ mA			-1.5			-1.5	V
V_{OH}	$V_{CC} = 4.5$ V to 5.5 V,	$I_{OH} = -0.4$ mA	$V_{CC}-2$			$V_{CC}-2$			V
	$V_{CC} = 4.5$ V,	$I_{OH} = -3$ mA	2.4	3.2		2.4	3.2		
	$V_{CC} = 4.5$ V,	$I_{OH} = -12$ mA	2						
	$V_{CC} = 4.5$ V,	$I_{OH} = -15$ mA				2			
V_{OL}	$V_{CC} = 4.5$ V,	$I_{OL} = 12$ mA		0.25	0.4		0.25	0.4	V
	$V_{CC} = 4.5$ V	$I_{OL} = 24$ mA					0.35	0.5	
	($I_{OL} = 48$ mA for -1 version)								
I_{OZH}	$V_{CC} = 5.5$ V,	$V_O = 2.7$ V			20			20	µA
I_{OZL}	$V_{CC} = 5.5$ V,	$V_O = 0.4$ V			-20			-20	µA
I_I	$V_{CC} = 5.5$ V,	$V_I = 7$ V			0.1			0.1	mA
I_{IH}	$V_{CC} = 5.5$ V,	$V_I = 2.7$ V			20			20	µA
I_{IL}	$V_{CC} = 5.5$ V,	$V_I = 0.4$ V			-0.1			-0.1	mA
I_O §	$V_{CC} = 5.5$ V,	$V_O = 2.25$ V	-30		-112	-30		-112	mA
I_{CC}	$V_{CC} = 5.5$ V	Outputs high		9	15		9	15	mA
		Outputs low		15	24		15	24	
		Outputs disabled		17	27		17	27	

‡All typical values are at $V_{CC} = 5$ V, $T_A = 25°C$.
§The output conditions have been chosen to produce a current that closely approximates one half of the true short-circuit output current, I_{OS}.

TEXAS INSTRUMENTS
POST OFFICE BOX 225012 • DALLAS, TEXAS 75265

TYPES SN54ALS244A, SN74ALS244A
OCTAL BUFFERS AND LINE DRIVERS WITH 3-STATE OUTPUTS

switching characteristics (see Note 1)

PARAMETER	FROM (INPUT)	TO (OUTPUT)	V_{CC} = 4.5 V to 5.5 V, C_L = 50 pF, R1 = 500 Ω, R2 = 500 Ω, T_A = MIN to MAX				UNIT
			SN54ALS244A		SN74ALS244A		
			MIN	MAX	MIN	MAX	
t_{PLH}	A	Y	3	13	3	10	ns
t_{PHL}			3	13	3	10	
t_{PZH}	$\overline{G}$	Y	7	25	7	20	ns
t_{PZL}			7	25	7	20	
t_{PHZ}	$\overline{G}$	Y	2	12	2	10	ns
t_{PLZ}			3	18	3	13	

NOTE 1: For load circuit and voltage waveforms, see page 1-12.

TYPES SN54AS244, SN74AS244
OCTAL BUFFERS AND LINE DRIVERS WITH 3-STATE OUTPUTS

absolute maximum ratings over operating free-air temperature range (unless otherwise noted)

Supply voltage, V_{CC} .... 7 V
Input voltage .... 7 V
Voltage applied to a disabled 3-state output .... 5.5 V
Operating free-air temperature range: SN54AS244 .... −55 °C to 125 °C
                                         SN74AS244 .... 0 °C to 70 °C
Storage temperature range .... −65 °C to 150 °C

recommended operating conditions

		SN54AS244			SN74AS244			UNIT
		MIN	NOM	MAX	MIN	NOM	MAX	
V_{CC}	Supply voltage	4.5	5	5.5	4.5	5	5.5	V
V_{IH}	High-level input voltage	2			2			V
V_{IL}	Low-level input voltage			0.8			0.8	V
I_{OH}	High-level output current			−12			−15	mA
I_{OL}	Low-level output current			48			64	mA
T_A	Operating free-air temperature	−55		125	0		70	°C

electrical characteristics over recommended operating free-air temperature range (unless otherwise noted)

PARAMETER		TEST CONDITIONS		SN54AS244			SN74AS244			UNIT
				MIN	TYP[†]	MAX	MIN	TYP[†]	MAX	
V_{IK}		V_{CC} = 4.5 V,	I_I = −18 mA			−1.2			−1.2	V
V_{OH}		V_{CC} = 4.5 V to 5.5 V,	I_{OH} = −2 mA	V_{CC}−2			V_{CC}−2			V
		V_{CC} = 4.5 V,	I_{OH} = −3 mA	2.4	3.4		2.4	3.4		
		V_{CC} = 4.5 V,	I_{OH} = −12 mA	2.4						
		V_{CC} = 4.5 V,	I_{OH} = −15 mA				2.4			
V_{OL}		V_{CC} = 4.5 V,	I_{OL} = 48 mA			0.55				V
		V_{CC} = 4.5 V	I_{OL} = 64 mA						0.55	
I_{OZH}		V_{CC} = 5.5 V,	V_O = 2.7 V			50			50	µA
I_{OZL}		V_{CC} = 5.5 V,	V_O = 0.4 V			−50			−50	µA
I_I		V_{CC} = 5.5 V,	V_I = 7 V			0.1			0.1	mA
I_{IH}		V_{CC} = 5.5 V,	V_I = 2.7 V			20			20	µA
I_{IL}	$\overline{G}$	V_{CC} = 5.5 V,	V_I = 0.4 V			−0.5			−0.5	mA
	A					−1			−1	
I_O[‡]		V_{CC} = 5.5 V,	V_O = 2.25 V	−50		−150	−50		−150	mA
I_{CC}		V_{CC} = 5.5 V	Outputs high		22	34		22	34	mA
			Outputs low		60	90		60	90	
			Outputs disabled		34	54		34	54	

[†] All typical values are at V_{CC} = 5 V, T_A = 25 °C.
[‡] The output conditions have been chosen to produce a current that closely approximates one half of the true short-circuit output current, I_{OS}.

TEXAS INSTRUMENTS
POST OFFICE BOX 225012 • DALLAS, TEXAS 75265

TYPES SN54AS244, SN74AS244
OCTAL BUFFERS AND LINE DRIVERS WITH 3-STATE OUTPUTS

switching characteristics (see Note 1)

PARAMETER	FROM (INPUT)	TO (OUTPUT)	V_{CC} = 4.5 V to 5.5 V, C_L = 50 pF, R1 = 500 Ω, R2 = 500 Ω, T_A = MIN to MAX				UNIT
			SN54AS244		SN74AS244		
			MIN	MAX	MIN	MAX	
t_{PLH}	A	Y	2	9	2	6.2	ns
t_{PHL}			2	7	2	6.2	
t_{PZH}	$\overline{G}$	Y	2	10	2	9	ns
t_{PZL}			2	8	2	7.5	
t_{PHZ}	$\overline{G}$	Y	2	6.5	2	6	ns
t_{PLZ}			2	10.5	2	9	

NOTE 1: For load circuit and voltage waveforms, see page 1-12.

2

ALS AND AS CIRCUITS

TYPES SN54ALS245A, SN54AS245, SN74ALS245A, SN74AS245
OCTAL BUS TRANSCEIVERS WITH 3-STATE OUTPUTS

D2661, DECEMBER 1982 – REVISED DECEMBER 1983

- 3-State Outputs Drive Bus Lines Directly
- P-N-P Inputs Reduce Dc Loading
- 'AS Version in Development. Data Will Be Provided As It Becomes Available. Contact the Factory for Latest Information
- Package Options Include Both Plastic and Ceramic Chip Carriers in Addition to Plastic and Ceramic DIPs
- Dependable Texas Instruments Quality and Reliability

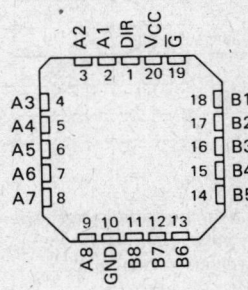

description

These octal bus transceivers are designed for synchronous two-way communication between data buses The control function implementation minimizes external timing requirements.

The devices allow data transmission from the A bus to the B bus or from the B bus to the A bus depending upon the logic level at the direction control (DIR) input. The enable input ($\overline{G}$) can be used to disable the device so that the buses are effectively isolated.

The –1 version of the SN74ALS245A is identical to the standard version except that the recommended maximum I_{OL} is increased to 48 milliamperes. There is no –1 version of the SN54ALS245A.

The SN54ALS245A and SN54AS245 are characterized for operation over the full military temperature range of –55°C to 125°C. The SN74ALS245A and SN74AS245 are characterized for operation from 0°C to 70°C.

FUNCTION TABLE

ENABLE $\overline{G}$	DIRECTION CONTROL DIR	OPERATION
L	L	B data to A bus
L	H	A data to B bus
H	X	Isolation

Copyright © 1982 by Texas Instruments Incorporated

TYPES SN54ALS245A, SN74ALS245A
OCTAL BUS TRANSCEIVERS WITH 3-STATE OUTPUTS

logic diagram (positive logic)

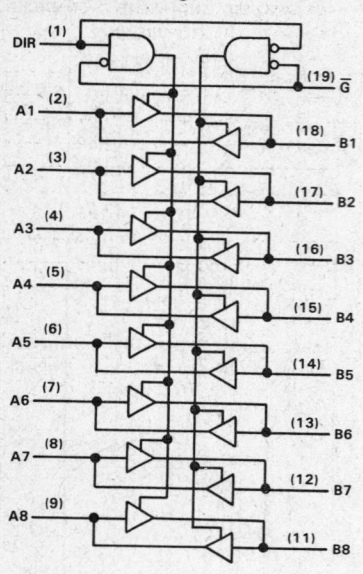

logic symbol

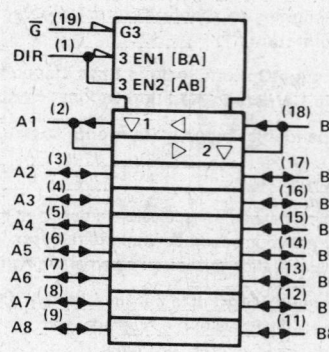

Pin numbers shown are for J and N packages.

absolute maximum ratings over operating free-air temperature range (unless otherwise noted)

Supply voltage, V_{CC} .... 7 V
Input voltage: All inputs .... 7 V
              I/O ports .... 5.5 V
Operating free-air temperature range: SN54ALS245A .... $-55\,°C$ to $125\,°C$
                                  SN74ALS245A .... $0\,°C$ to $70\,°C$
Storage temperature range .... $-65\,°C$ to $150\,°C$

recommended operating conditions

		SN54ALS245A			SN74ALS245A			UNIT
		MIN	NOM	MAX	MIN	NOM	MAX	
V_{CC}	Supply voltage	4.5	5	5.5	4.5	5	5.5	V
V_{IH}	High-level input voltage	2			2			V
V_{IL}	Low-level input voltage			0.8			0.8	V
I_{OH}	High-level output current			-12			-15	mA
I_{OL}	Low-level output current			12			24	mA
							48†	
T_A	Operating free-air temperature	-55		125	0		70	°C

†The extended limits apply only if V_{CC} is maintained between 4.75 V and 5.25 V.
The 48-mA limit applies for the SN74ALS245A-1 only.

TYPES SN54ALS245A, SN74ALS245A
OCTAL BUS TRANSCEIVERS WITH 3-STATE OUTPUTS

electrical characteristics over recommended operating free-air temperature range (unless otherwise noted)

PARAMETER		TEST CONDITIONS		SN54ALS245A MIN	SN54ALS245A TYP†	SN54ALS245A MAX	SN74ALS245A MIN	SN74ALS245A TYP†	SN74ALS245A MAX	UNIT
V_{IK}		V_{CC} = 4.5 V,	I_I = −18 mA			−1.5			−1.5	V
V_{OH}		V_{CC} = 4.5 V to 5.5 V,	I_{OH} = −0.4 mA	V_{CC}−2			V_{CC}−2			V
		V_{CC} = 4.5 V,	I_{OH} = −3 mA	2.4	3.2		2.4	3.2		
		V_{CC} = 4.5 V,	I_{OH} = −12 mA	2						
		V_{CC} = 4.5 V,	I_{OH} = −15 mA				2			
V_{OL}		V_{CC} = 4.5 V,	I_{OL} = 12 mA		0.25	0.4		0.25	0.4	V
		V_{CC} = 4.5 V,	I_{OL} = 24 mA					0.35	0.5	
		(I_{OL} = 48 mA for −1 versions)								
I_I	Control inputs	V_{CC} = 5.5 V,	V_I = 7 V			0.1			0.1	mA
	A or B ports	V_{CC} = 5.5 V,	V_I = 5.5 V			0.1			0.1	
I_{IH}	Control inputs	V_{CC} = 5.5 V,	V_I = 2.7 V			20			20	μA
	A or B ports‡					20			20	
I_{IL}	Control inputs	V_{CC} = 5.5 V,	V_I = 0.4 V			−0.1			−0.1	mA
	A or B ports‡					−0.1			−0.1	
I_O§		V_{CC} = 5.5 V,	V_O = 2.25 V	−30		−112	−30		−112	mA
I_{CC}		V_{CC} = 5.5 V	Outputs high		30	48		30	45	mA
			Outputs low		36	60		36	55	
			Outputs disabled		38	63		38	58	

†All typical values are at V_{CC} = 5 V, T_A = 25°C.
‡For I/O ports, the parameters I_{IH} and I_{IL} include the off-state output current.
§The output conditions have been chosen to produce a current that closely approximates one half of the true short-circuit output current, I_{OS}.

switching characteristics (see Note 1)

PARAMETER	FROM (INPUT)	TO (OUTPUT)	V_{CC} = 4.5 V to 5.5 V, C_L = 50 pF, R1 = 500 Ω, R2 = 500 Ω, T_A = MIN to MAX				UNIT
			SN54ALS245A MIN	SN54ALS245A MAX	SN74ALS245A MIN	SN74ALS245A MAX	
t_{PLH}	A or B	B or A	3	15	3	10	ns
t_{PHL}			3	13	3	10	
t_{PZH}	$\overline{G}$	A or B	5	25	5	20	ns
t_{PZL}			5	25	5	20	
t_{PHZ}	$\overline{G}$	A or B	2	12	2	10	ns
t_{PLZ}			4	18	4	15	

NOTE 1: For load circuit and voltage waveforms, see page 1-12.

TEXAS INSTRUMENTS
POST OFFICE BOX 225012 • DALLAS, TEXAS 75265

TYPES SN54AS245, SN74AS245
OCTAL BUS TRANSCEIVERS WITH 3-STATEE OUTPUTS

absolute maximum ratings over operating free-air temperature range (unless otherwise noted)

Supply voltage, V_{CC} .... 7 V
Input voltage: All inputs .... 7 V
              I/O ports .... 5.5 V
Operating free-air temperature range: SN54AS245 .... −55°C to 125°C
                                       SN74AS245 .... 0°C to 70°C
Storage temperature range .... −65°C to 150°C

recommended operating conditions

		SN54AS245			SN74AS245			UNIT
		MIN	NOM	MAX	MIN	NOM	MAX	
V_{CC}	Supply voltage	4.5	5	5.5	4.5	5	5.5	V
V_{IH}	High-level input voltage	2			2			V
V_{IL}	Low-level input voltage			0.8			0.8	V
I_{OH}	High-level output current			−12			−15	mA
I_{OL}	Low-level output current			32			48	mA
T_A	Operating free-air temperature	−55		125	0		70	°C

electrical characteristics over recommended operating free-air temperature range (unless otherwise noted)

PARAMETER		TEST CONDITIONS		SN54AS245			SN74AS245			UNIT
				MIN	TYP[†]	MAX	MIN	TYP[†]	MAX	
V_{IK}		V_{CC} = 4.5 V,	I_I = −18 mA			−1.2			−1.2	V
V_{OH}		V_{CC} = 4.5 V to 5.5 V,	I_{OH} = −2 mA	V_{CC}−2			V_{CC}−2			V
		V_{CC} = 4.5 V,	I_{OH} = −3 mA	2.4	3.2		2.4	3.2		
		V_{CC} = 4.5 V,	I_{OH} = −12 mA	2.4						
		V_{CC} = 4.5 V,	I_{OH} = −15 mA				2.4			
V_{OL}		V_{CC} = 4.5 V,	I_{OL} = 32 mA		0.25	0.5				V
		V_{CC} = 4.5 V,	I_{OL} = 48 mA					0.35	0.5	
I_I	Control inputs	V_{CC} = 5.5 V,	V_I = 7 V			0.1			0.1	mA
	A or B ports	V_{CC} = 5.5 V,	V_I = 5.5 V			0.1			0.1	
I_{IH}	Control inputs	V_{CC} = 5.5 V,	V_I = 2.7 V			20			20	µA
	A or B ports[‡]					50			50	
I_{IL}	Control inputs	V_{CC} = 5.5 V,	V_I = 0.4 V			−0.1			−0.1	mA
	A or B ports[‡]					−0.75			−0.75	
I_{OS}[§]		V_{CC} = 5.5 V,	V_O = 2.25 V	−30		−112	−30		−112	mA
I_{CC}		V_{CC} = 5.5 V	Outputs high		62			62		mA
			Outputs low		95			95		
			Outputs disabled		79			79		

[†] All typical values are at V_{CC} = 5 V, T_A = 25°C.
[‡] For I/O ports, the parameters I_{IH} and I_{IL} include the off-state output current.
[§] The output conditions have been chosen to produce a current that closely approximates one half of the true short-circuit output current, I_{OS}.

PRODUCT PREVIEW

This page contains information on a product under development. Texas Instruments reserves the right to change or discontinue this product without notice.

TEXAS INSTRUMENTS
POST OFFICE BOX 225012 • DALLAS, TEXAS 75265

TYPES SN54AS245, SN74AS245
OCTAL BUS TRANSCEIVERS WITH 3-STATE OUTPUTS

switching characteristics (see Note 1)

PARAMETER	FROM (INPUT)	TO (OUTPUT)	V_{CC} = 4.5 V to 5.5 V, C_L = 50 pF, $R1$ = 500 Ω, $R2$ = 500 Ω, T_A = MIN to MAX						UNIT
			SN54AS245			SN74AS245			
			MIN	TYP†	MAX	MIN	TYP†	MAX	
t_{PLH}	A or B	B or A		6			6		ns
t_{PHL}				5			5		
t_{PZH}	$\overline{G}$	A or B		8			8		ns
t_{PZL}				8			8		
t_{PHZ}	$\overline{G}$	A or B		4.5			4.5		ns
t_{PLZ}				5			5		

†All typical values are at V_{CC} = 5 V, T_A = 25°C.
NOTE 1: For load circuit and voltage waveforms, see page 1-12.

PRODUCT PREVIEW

This page contains information on a product under development. Texas Instruments reserves the right to change or discontinue this product without notice.

TEXAS INSTRUMENTS
POST OFFICE BOX 225012 • DALLAS, TEXAS 75265

ALS AND AS CIRCUITS

TYPES SN54AS250, SN74AS250
1-OF-16 DATA GENERATORS/MULTIPLEXERS
WITH 3-STATE OUTPUTS

DECEMBER 1983

- 4-Line to 1-Line Multiplexer that can Select 1 of 16 Data Inputs
- Applications:
  Boolean Function Generator
  Parallel-to-Serial Converter
  Data Source Selector
- Buffered 3-State Bus Driver Inputs Permit Multiplexing from N Lines to One Line
- Dependable Texas Instruments Quality and Reliability

SN54AS250 . . . J DUAL-IN-LINE PACKAGE
SN74AS250 . . . N DUAL-IN-LINE PACKAGE
(TOP VIEW)

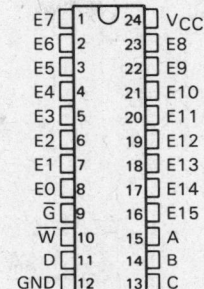

description

The 'AS250 provides full binary decoding to select one of sixteen data sources with an inverting $\overline{W}$ output. The selected sources are buffered with symmetrical propagation delay times. This reduces the possibility of transients occurring at the output.

A buffered enable output ($\overline{G}$) may be used for n-line-to-one-line cascading. Taking the $\overline{G}$ high will place the output in a high-impedance state. In the high-impedance state, the output neither loads nor drives the bus lines significantly.

The enable ($\overline{G}$) does not affect the internal operations of the data selector/multiplexer. New data can be set up while the outputs are disabled.

The SN54AS250 is characterized for operation over the full military temperature range of −55°C to 125°C. The SN74AS250 is characterized for operation from 0°C to 70°C.

SN54AS250 . . . FH CHIP CARRIER PACKAGE
SN74AS250 . . . FN CHIP CARRIER PACKAGE
(TOP VIEW)

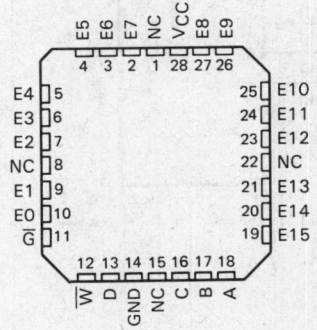

NC − No internal connection

PRODUCT PREVIEW

This document contains information on a product under development. Texas Instruments reserves the right to change or discontinue this product without notice.

Copyright © 1983 by Texas Instruments Incorporated

TEXAS INSTRUMENTS
POST OFFICE BOX 225012 • DALLAS, TEXAS 75265

TYPES SN54AS250, SN74AS250
1-OF-16 DATA GENERATORS/MULTIPLEXERS WITH 3-STATE OUTPUTS

logic symbol

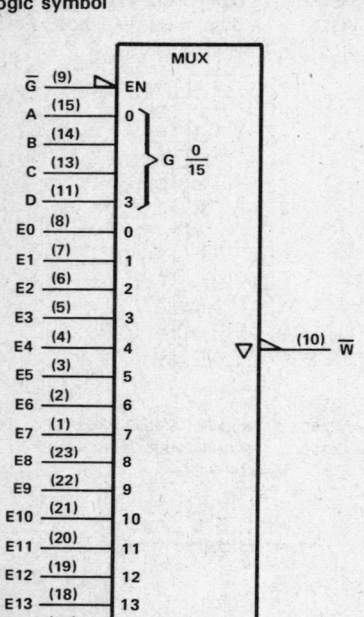

Pin numbers shown are for J or N packages.

logic diagram (positive logic)

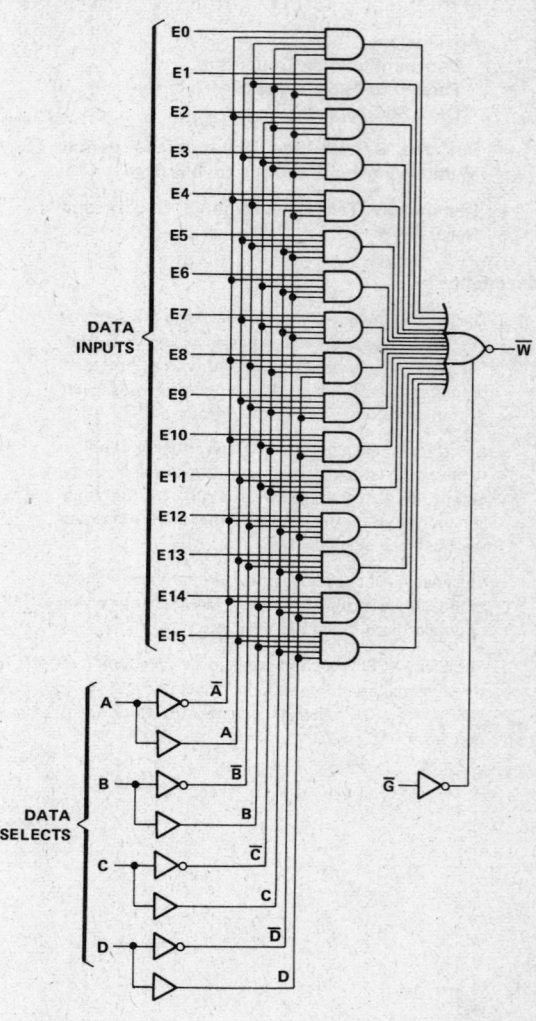

TYPES SN54AS250, SN74AS250
1-OF-16 DATA GENERATORS/MULTIPLEXERS
WITH 3-STATE OUTPUTS

FUNCTION TABLE

INPUT						OUTPUT
$\overline{G}$	A	B	C	D	Ei	$\overline{W}$
L	L	L	L	L	E0	E0
L	H	L	L	L	E1	E1
L	L	H	L	L	E2	E2
L	H	H	L	L	E3	E3
L	L	L	H	L	E4	E4
L	H	L	H	L	E5	E5
L	L	H	H	L	E6	E6
L	H	H	H	L	E7	E7
L	L	L	L	H	E8	E8
L	H	L	L	H	E9	E9
L	L	H	L	H	E10	E10
L	H	H	L	H	E11	E11
L	L	L	H	H	E12	E12
L	H	L	H	H	E13	E13
L	L	H	H	H	E14	E14
L	H	H	H	H	E15	E15
H	X	X	X	X	X	Z

absolute maximum ratings over operating free-air temperature range (unless otherwise noted)

Supply voltage, V_{CC} .... 7 V
Input voltage .... 7 V
Operating free-air temperature range: SN54AS250 .... −55°C to 125°C
                                     SN74AS250 .... 0°C to 70°C
Storage temperature range .... −65°C to 150°C

recommended operating conditions

		SN54AS250			SN74AS250			UNIT
		MIN	NOM	MAX	MIN	NOM	MAX	
V_{CC}	Supply voltage	4.5	5	5.5	4.5	5	5.5	V
V_{IH}	High-level input voltage	2			2			V
V_{IL}	Low-level input voltage			0.8			0.8	V
I_{OH}	High-level output current			−12			−15	mA
I_{OL}	Low-level output current			32			48	mA
T_A	Operating free-air temperature	−55		125	0		70	°C

TYPES SN54AS250, SN74AS250
1-OF-16 DATA GENERATORS/MULTIPLEXERS
WITH 3-STATE OUTPUTS

electrical characteristics over recommended operating free-air temperature range (unless otherwise noted)

PARAMETER	TEST CONDITIONS		SN54AS250			SN74AS250			UNIT
			MIN	TYP†	MAX	MIN	TYP†	MAX	
V_{IK}	V_{CC} = 4.5 V,	I_I = −18 mA			−1.2			−1.2	V
V_{OH}	V_{CC} = 4.5 V to 5.5 V,	I_{OH} = −2 mA	V_{CC}−2			V_{CC}−2			V
	V_{CC} = 4.5 V,	I_{OH} = −12 mA	2.4	3.2					
	V_{CC} = 4.5 V,	I_{OH} = −15 mA				2.4	3.3		
V_{OL}	V_{CC} = 4.5 V,	I_{OL} = 32 mA		0.25	0.5				V
	V_{CC} = 4.5 V,	I_{OL} = 48 mA					0.35	0.5	
I_{OZH}	V_{CC} = 5.5 V,	V_O = 2.7 V			50			50	µA
I_{OZL}	V_{CC} = 5.5 V,	V_O = 0.4 V			−50			−50	µA
I_I	V_{CC} = 5.5 V,	V_I = 7 V							mA
I_{IH}	V_{CC} = 5.5 V,	V_I = 2.7 V							µA
I_{IL}	V_{CC} = 5.5 V,	V_I = 0.4 V							mA
I_O ‡	V_{CC} = 5.5 V,	V_O = 2.25 V	−30		−112	−30		−112	mA
I_{CC}	V_{CC} = 5.5 V	Outputs high							mA
		Outputs low							
		Outputs disabled							

†All typical values are at V_{CC} = 5 V, T_A = 25°C.
‡The output conditions have been chosen to produce a current that closely approximates one half of the true short-circuit output current, I_{OS}.

switching characteristics (see Note 1)

PARAMETER	FROM (INPUT)	TO (OUTPUT)	V_{CC} = 4.5 V to 5.5 V, C_L = 50 pF, R1 = 500 Ω, R2 = 500 Ω, T_A = MIN to MAX						UNIT
			SN54AS250			SN74AS250			
			MIN	TYP†	MAX	MIN	TYP†	MAX	
t_{PLH}	DATA	$\overline{W}$		4.3			4.3		ns
t_{PHL}				4.7			4.7		
t_{PLH}	SELECT	$\overline{W}$		4.6			4.6		ns
t_{PHL}				7.7			7.7		
t_{PZH}	$\overline{G}$	$\overline{W}$		4			4		ns
t_{PZL}				4.9			4.9		
t_{PHZ}	$\overline{G}$	$\overline{W}$		3.1			3.1		ns
t_{PLZ}				3.9			3.9		

†All typical values are at V_{CC} = 5 V, T_A = 25°C.
NOTE 1: For load circuit and voltage waveforms, see page 1-12.

ALS AND AS CIRCUITS

TYPES SN54ALS251, SN54AS251, SN74ALS251, SN74AS251
1 OF 8 DATA SELECTORS/MULTIPLEXERS WITH 3-STATE OUTPUTS

D2661, APRIL 1982–REVISED DECEMBER 1983

- Three-State Versions of 'ALS151 and 'AS151
- Three-State Outputs Interface Directly with System Bus
- Performs Parallel-to-Serial Conversion
- Complementary Outputs Provide True and Inverted Data
- Fully Compatible with Most TTL Circuits
- Package Options Include Both Plastic and Ceramic Chip Carriers in Addition to Plastic and Ceramic DIPs
- Dependable Texas Instruments Quality and Reliability

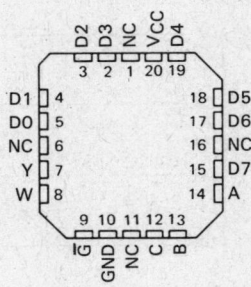

description

These data selectors/multiplexers contain full binary decoding to select one-of-eight data sources and feature strobe-controlled complementary three-state outputs.

The three-state outputs can interface with and drive data lines of bus-organized systems. With all but one of the common outputs disabled (at a high-impedance state), the low-impedance of the single enabled output will drive the bus line to a high or low logic level. Both outputs are controlled by the strobe ($\overline{G}$). The outputs are disabled when $\overline{G}$ is high.

The SN54ALS251 and SN54AS251 are characterized for operation over the full military temperature range of $-55\,°C$ to $125\,°C$. The SN74ALS251 and SN74AS251 are characterized for operation from $0\,°C$ to $70\,°C$.

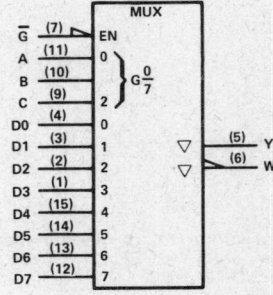

NC — No internal connection.

FUNCTION TABLE

INPUTS				OUTPUTS	
SELECT			STROBE	Y	W
C	B	A	$\overline{G}$		
X	X	X	H	Z	Z
L	L	L	L	D0	$\overline{D0}$
L	L	H	L	D1	$\overline{D1}$
L	H	L	L	D2	$\overline{D2}$
L	H	H	L	D3	$\overline{D3}$
H	L	L	L	D4	$\overline{D4}$
H	L	H	L	D5	$\overline{D5}$
H	H	L	L	D6	$\overline{D6}$
H	H	H	L	D7	$\overline{D7}$

D0, D1 . . . D7 = the level of the respective D input

logic symbol

```
        MUX
$\overline{G}$ (7) ──┤EN
A   (11) ──┐
B   (10) ──┤ G$^0_7$
C   (9)  ──┘
D0  (4)  ──┐ 0
D1  (3)  ──┤ 1      (5)  Y
D2  (2)  ──┤ 2      (6)  W
D3  (1)  ──┤ 3
D4  (15) ──┤ 4
D5  (14) ──┤ 5
D6  (13) ──┤ 6
D7  (12) ──┘ 7
```

Pin numbers shown are for J and N packages.

Copyright © 1982 by Texas Instruments Incorporated.

TYPES SN54ALS251, SN54AS251, SN74ALS251, SN74AS251
1 OF 8 DATA SELECTORS/MULTIPLEXERS WITH 3-STATE OUTPUTS

logic diagram (positive logic)

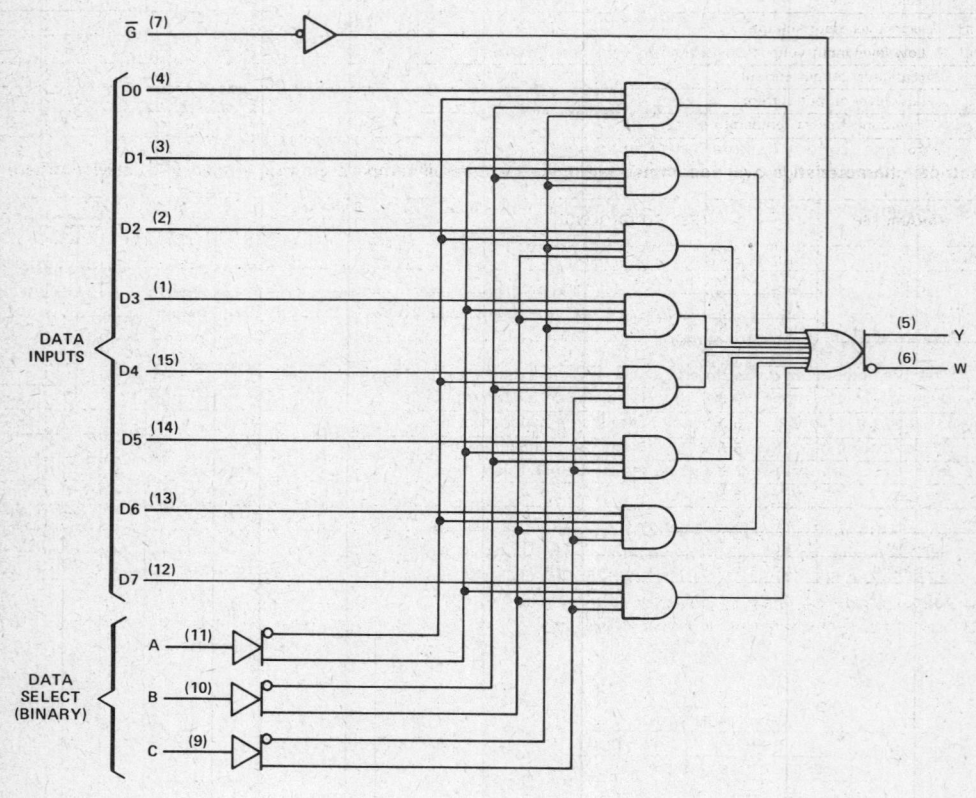

Pin numbers shown are for J and N packages.

absolute maximum ratings over operating free-air temperature range (unless otherwise noted)

Supply voltage, V_{CC} .... 7 V
Input voltage .... 7 V
Voltage applied to a disabled 3-state output .... 5.5 V
Operating free-air temperature range: SN54ALS251, SN54AS251 .... −55 °C to 125 °C
  SN74ALS251, SN74AS251 .... 0 °C to 70 °C
Storage temperature range .... −65 °C to 150 °C

TEXAS
INSTRUMENTS
POST OFFICE BOX 225012 • DALLAS, TEXAS 75265

TYPES SN54ALS251, SN74ALS251
1 OF 8 DATA SELECTORS/MULTIPLEXERS WITH 3-STATE OUTPUTS

recommended operating conditions

		SN54ALS251			SN74ALS251			UNIT
		MIN	NOM	MAX	MIN	NOM	MAX	
V_{CC}	Supply voltage	4.5	5	5.5	4.5	5	5.5	V
V_{IH}	High-level input voltage	2			2			V
V_{IL}	Low-level input voltage			0.8			0.8	V
I_{OH}	High-level output current			−1			−2.6	mA
I_{OL}	Low-level output current			12			24	mA
T_A	Operating free-air temperature	−55		125	0		70	°C

electrical characteristics over recommended operating free-air temperature range (unless otherwise noted)

PARAMETER		TEST CONDITIONS		SN54ALS251			SN74ALS251			UNIT
				MIN	TYP†	MAX	MIN	TYP†	MAX	
V_{IK}		$V_{CC} = 4.5$ V,	$I_I = -18$ mA			−1.5			−1.5	V
V_{OH}		$V_{CC} = 4.5$ V to 5.5 V,	$I_{OH} = -0.4$ mA	$V_{CC}-2$			$V_{CC}-2$			V
		$V_{CC} = 4.5$ V,	$I_{OH} = -1$ mA	2.4	3.3					
		$V_{CC} = 4.5$ V,	$I_{OH} = -2.6$ mA				2.4	3.2		
V_{OL}		$V_{CC} = 4.5$ V,	$I_{OL} = 12$ mA		0.25	0.4		0.25	0.4	V
		$V_{CC} = 4.5$ V,	$I_{OL} = 24$ mA					0.35	0.5	
I_{OZH}		$V_{CC} = 5.5$ V,	$V_O = 2.7$ V			20			20	µA
I_{OZL}		$V_{CC} = 5.5$ V,	$V_I = 0.4$ V			−20			−20	µA
I_I		$V_{CC} = 5.5$ V,	$V_I = 7$ V			0.1			0.1	mA
I_{IH}		$V_{CC} = 5.5$ V,	$V_I = 2.7$ V			20			20	µA
I_{IL}		$V_{CC} = 5.5$ V,	$V_I = 0.4$ V			−0.1			−0.1	mA
I_O‡		$V_{CC} = 5.5$ V,	$V_O = 2.25$ V	−30		−112	−30		−112	mA
I_{CC}	Enabled	$V_{CC} = 5.5$ V,	Inputs at Gnd.		7	10		7	10	mA
	Disabled	$V_{CC} = 5.5$ V,	Inputs at 4.5 V		9.4	14		9.4	14	

†All typical values are at $V_{CC} = 5$ V, $T_A = 25$ °C.
‡The output conditions have been chosen to produce a current that closely approximates one half of the true short-circuit output current, I_{OS}.

ALS AND AS CIRCUITS

TEXAS INSTRUMENTS
POST OFFICE BOX 225012 • DALLAS, TEXAS 75265

TYPES SN54ALS251, SN74ALS251
1 OF 8 DATA SELECTORS/MULTIPLEXERS WITH 3-STATE OUTPUTS

switching characteristics (see Note 1)

PARAMETER	FROM (INPUT)	TO (OUTPUT)	V_{CC} = 4.5 V to 5.5 V, C_L = 50 pF, R1 = 500 Ω, R2 = 500 Ω, T_A = MIN to MAX				UNIT
			SN54ALS251		SN74ALS251		
			MIN	MAX	MIN	MAX	
t_{PLH}	A, B or C	Y	5	21	5	18	ns
t_{PHL}			8	28	8	24	
t_{PLH}	A, B or C	W	8	28	8	24	ns
t_{PHL}			7	26	7	23	
t_{PLH}	Any D	Y	2	12	2	10	ns
t_{PHL}			3	18	3	15	
t_{PLH}	Any D	W	3	18	3	15	ns
t_{PHL}			3	18	3	15	
t_{PZH}	$\overline{G}$	Y	3	18	3	15	ns
t_{PZL}			3	18	3	15	
t_{PZH}	$\overline{G}$	W	3	18	3	15	ns
t_{PZL}			3	18	3	15	
t_{PHZ}	$\overline{G}$	Y	2	12	2	10	ns
t_{PLZ}			1	12	1	10	
t_{PHZ}	$\overline{G}$	W	2	12	2	10	ns
t_{PLZ}			1	12	1	10	

NOTE 1: For load circuit and voltage waveforms, see page 1-12.

TYPES SN54AS251, SN74AS251
1 OF 8 DATA SELECTORS/MULTIPLEXERS WITH 3-STATE OUTPUTS

recommended operating conditions

		SN54AS251			SN74AS251			UNIT
		MIN	NOM	MAX	MIN	NOM	MAX	
V_{CC}	Supply voltage	4.5	5	5.5	4.5	5	5.5	V
V_{IH}	High-level input voltage	2			2			V
V_{IL}	Low-level input voltage			0.8			0.8	V
I_{OH}	High-level output current			−12			−15	mA
I_{OL}	Low-level output current			32			48	mA
T_A	Operating free-air temperature	−55		125	0		70	°C

electrical characteristics over recommended operating free-air temperature range (unless otherwise noted)

PARAMETER		TEST CONDITIONS		SN54AS251			SN74AS251			UNIT
				MIN	TYP†	MAX	MIN	TYP†	MAX	
V_{IK}		V_{CC} = 4.5 V,	I_I = −18 mA			−1.2			−1.2	V
V_{OH}		V_{CC} = 4.5 V to 5.5 V,	I_{OH} = −2 mA	V_{CC}−2			V_{CC}−2			V
		V_{CC} = 4.5 V,	I_{OH} = −12 mA	2.4	3.2					
		V_{CC} = 4.5 V,	I_{OH} = −15 mA				2.4	3.3		
V_{OL}		V_{CC} = 4.5 V,	I_{OL} = 32 mA		0.25	0.5				V
		V_{CC} = 4.5 V,	I_{OL} = 48 mA					0.35	0.5	
I_{OZH}		V_{CC} = 5.5 V,	V_O = 2.7 V			50			50	µA
I_{OZL}		V_{CC} = 5.5 V,	V_I = 0.4 V			−50			−50	µA
I_I	A, B, C	V_{CC} = 5.5 V,	V_I = 7 V			0.2			0.2	mA
	All other					0.1			0.1	
I_{IH}	A, B, C	V_{CC} = 5.5 V,	V_I = 2.7 V			40			40	µA
	All other					20			20	
I_{IL}	A, B, C	V_{CC} = 5.5 V,	V_I = 0.4 V			−0.6			−0.6	mA
	All other					−0.3			−0.3	
I_O‡		V_{CC} = 5.5 V,	V_O = 2.25 V	−30		−112	−30		−112	mA
I_{CC}		V_{CC} = 5.5 V,			28			28		mA

†All typical values are at V_{CC} = 5 V, T_A = 25°C.
‡The output conditions have been chosen to produce a current that closely approximates one half of the true short-circuit output current, I_{OS}.

PRODUCT PREVIEW

This page contains information on a product under development. Texas Instruments reserves the right to change or discontinue this product without notice.

TEXAS INSTRUMENTS
POST OFFICE BOX 225012 • DALLAS, TEXAS 75265

TYPES SN54AS251, SN74AS251
1 OF 8 DATA SELECTORS/MULTIPLEXERS WITH 3-STATE OUTPUTS

switching characteristics (see Note 1)

PARAMETER	FROM (INPUT)	TO (OUTPUT)	V_{CC} = 4.5 V to 5.5 V, C_L = 50 pF, R1 = 500 Ω, R2 = 500 Ω, T_A = MIN to MAX						UNIT
			SN54AS251			SN74AS251			
			MIN	TYP†	MAX	MIN	TYP†	MAX	
t_{PLH}	A, B, or C	Y		5			5		ns
t_{PHL}				5			5		
t_{PLH}	A, B, or C	W		4.5			4.5		ns
t_{PHL}				4.5			4.5		
t_{PLH}	Any D	Y		3			3		ns
t_{PHL}				4			4		
t_{PLH}	Any D	W		3			3		ns
t_{PHL}				2.5			2.5		
t_{PZH}	$\overline{G}$	Y		5			5		ns
t_{PZL}				6			6		
t_{PZH}	$\overline{G}$	W		5			5		ns
t_{PZL}				6			6		
t_{PHZ}	$\overline{G}$	Y		3			3		ns
t_{PLZ}				4			4		
t_{PHZ}	$\overline{G}$	W		3			3		ns
t_{PLZ}				4			4		

†All typical values are at V_{CC} = 5 V, T_A = 25°C.
NOTE 1: For load circuit and voltage waveforms, see page 1-12.

Additional information on these products can be obtained from the factory as it becomes available.

PRODUCT PREVIEW

This page contains information on a product under development. Texas Instruments reserves the right to change or discontinue this product without notice.

TEXAS INSTRUMENTS
POST OFFICE BOX 225012 • DALLAS, TEXAS 75265

TYPES SN54ALS253, SN54AS253, SN74ALS253, SN74AS253
DUAL 1 OF 4 DATA SELECTORS/MULTIPLEXERS WITH 3-STATE OUTPUTS

D2661, APRIL 1982—REVISED DECEMBER 1983

- Three-State Versions of 'ALS153 and 'AS153
- Permits Multiplexing from N Lines to 1 Line
- Performs Parallel-to-Serial Conversion
- Fully Compatible with Most TTL Circuits
- Package Options Include Both Plastic and Ceramic Chip Carriers in Addition to Plastic and Ceramic DIPs
- Dependable Texas Instruments Quality and Reliability

description

Each of these data selectors/multiplexers contains inverters and drivers to supply full binary decoding data selection to the AND-OR gates. Separate output control inputs are provided for each of the two four-line sections.

The three-state outputs can interface with and drive data lines of bus-organized systems. With all but one of the common outputs disabled (at a high-impedance state) the low-impedance of the single enabled output will drive the bus line to a high or low logic level. Each output has its own strobe ($\overline{G}$). The output is disabled when its strobe is high.

The SN54ALS253 and SN54AS253 are characterized for operation over the full military temperature range of $-55\,°C$ to $125\,°C$. The SN74ALS253 and SN74AS253 are characterized for operation from $0\,°C$ to $70\,°C$.

SN54ALS253, SN54AS253 . . . J PACKAGE
SN74ALS253, SN74AS253 . . . N PACKAGE
(TOP VIEW)

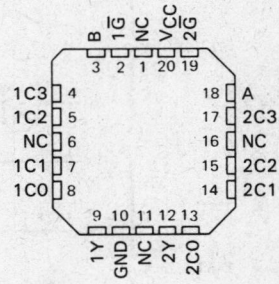

SN54ALS253, SN54AS253 . . . FH PACKAGE
SN74ALS253, SN74AS253 . . . FN PACKAGE
(TOP VIEW)

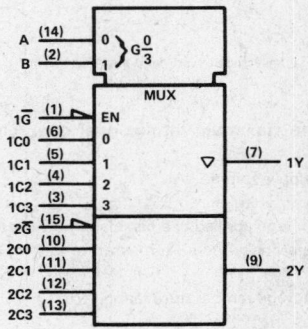

NC—No internal connection

FUNCTION TABLE

SELECT INPUTS		DATA INPUTS				OUTPUT CONTROL	OUTPUT
B	A	C0	C1	C2	C3	$\overline{G}$	Y
X	X	X	X	X	X	H	Z
L	L	L	X	X	X	L	L
L	L	H	X	X	X	L	H
L	H	X	L	X	X	L	L
L	H	X	H	X	X	L	H
H	L	X	X	L	X	L	L
H	L	X	X	H	X	L	H
H	H	X	X	X	L	L	L
H	H	X	X	X	H	L	H

Address inputs A and B are common to both sections.

logic symbol

Pin numbers shown are for J and N packages.

Copyright © 1982 by Texas Instruments Incorporated

Texas Instruments
POST OFFICE BOX 225012 • DALLAS, TEXAS 75265

TYPES SN54ALS253, SN54AS253, SN74ALS253, SN74AS253
DUAL 1 OF 4 DATA SELECTORS/MULTIPLEXERS
WITH 3-STATE OUTPUTS

logic diagram (positive logic)

Pin numbers shown are for J and N packages.

absolute maximum ratings over operating free-air temperature range (unless otherwise noted)

Supply voltage, V_{CC} .... 7 V
Input voltage .... 7 V
Voltage applied to a disabled 3-state output .... 5.5 V
Operating free-air temperature range: SN54ALS253, SN54AS253 .... −55 °C to 125 °C
                                      SN74ALS253, SN74AS253 .... 0 °C to 70 °C
Storage temperature range .... −65 °C to 150 °C

TEXAS
INSTRUMENTS
POST OFFICE BOX 225012 • DALLAS, TEXAS 75265

TYPES SN54ALS253, SN74ALS253
DUAL 1 OF 4 DATA SELECTORS/MULTIPLEXERS
WITH 3-STATE OUTPUTS

recommended operating conditions

		SN54ALS253			SN74ALS253			UNIT
		MIN	NOM	MAX	MIN	NOM	MAX	
V_{CC}	Supply voltage	4.5	5	5.5	4.5	5	5.5	V
V_{IH}	High-level input voltage	2			2			V
V_{IL}	Low-level input voltage			0.8			0.8	V
I_{OH}	High-level output current			−1			−2.6	mA
I_{OL}	Low-level output current			12			24	mA
T_A	Operating free-air temperature	−55		125	0		70	°C

electrical characteristics over recommended operating free-air temperature range (unless otherwise noted)

PARAMETER	TEST CONDITIONS		SN54ALS253			SN74ALS253			UNIT
			MIN	TYP†	MAX	MIN	TYP†	MAX	
V_{IK}	$V_{CC} = 4.5$ V,	$I_I = -18$ mA			−1.5			−1.5	V
V_{OH}	$V_{CC} = 4.5$ V to 5.5 V,	$I_{OH} = -0.4$ mA	$V_{CC}-2$			$V_{CC}-2$			V
	$V_{CC} = 4.5$ V,	$I_{OH} = -1$ mA	2.4	3.3					
	$V_{CC} = 4.5$ V,	$I_{OH} = -2.6$ mA				2.4	3.2		
V_{OL}	$V_{CC} = 4.5$ V,	$I_{OL} = 12$ mA		0.25	0.4		0.25	0.4	V
	$V_{CC} = 4.5$ V,	$I_{OL} = 24$ mA					0.35	0.5	
I_{OZH}	$V_{CC} = 5.5$ V,	$V_O = 2.7$ V			20			20	µA
I_{OZL}	$V_{CC} = 5.5$ V,	$V_O = 0.4$ V			−20			−20	µA
I_I	$V_{CC} = 5.5$ V,	$V_I = 7$ V			0.1			0.1	mA
I_{IH}	$V_{CC} = 5.5$ V,	$V_I = 2.7$ V			20			20	µA
I_{IL}	$V_{CC} = 5.5$ V,	$V_I = 0.4$ V			−0.1			−0.1	mA
I_O‡	$V_{CC} = 5.5$ V,	$V_O = 2.25$ V	−30		−112	−30		−112	mA
I_{CC}	$V_{CC} = 5.5$ V	Outputs enabled		6.5	12		6.5	12	mA
		Outputs disabled		7.5	14		7.5	14	

†All typical values are at $V_{CC} = 5$ V, $T_A = 25$ °C.
‡The output conditions have been chosen to produce a current that closely approximates one half of the true short-circuit output current, I_{OS}.

switching characteristics (see Note 1)

PARAMETER	FROM (INPUT)	TO (OUTPUT)	$V_{CC} = 4.5$ V to 5.5 V, $C_L = 50$ pF, R1 = 500 Ω, R2 = 500 Ω, T_A = MIN to MAX				UNIT
			SN54ALS253		SN74ALS253		
			MIN	MAX	MIN	MAX	
t_{PLH}	A or B	Any Y	5	25	5	21	ns
t_{PHL}			5	25	5	21	
t_{PLH}	Data (Any C)	Any Y	2	12	2	10	ns
t_{PHL}			3	17	3	14	
t_{PZH}	$\overline{G}$	Any Y	3	17	3	14	ns
t_{PZL}			4	19	4	16	
t_{PHZ}	$\overline{G}$	Any Y	2	12	2	10	ns
t_{PLZ}			2	16	2	14	

NOTE 1: For load circuit and voltage waveforms, see page 1-12.

ALS AND AS CIRCUITS

TEXAS INSTRUMENTS
POST OFFICE BOX 225012 • DALLAS, TEXAS 75265

TYPES SN54AS253, SN74AS253
DUAL 1 OF 4 DATA SELECTORS/MULTIPLEXERS
WITH 3-STATE OUTPUTS

recommended operating conditions

		SN54AS253 MIN	SN54AS253 NOM	SN54AS253 MAX	SN74AS253 MIN	SN74AS253 NOM	SN74AS253 MAX	UNIT
V_{CC}	Supply voltage	4.5	5	5.5	4.5	5	5.5	V
V_{IH}	High-level input voltage	2			2			V
V_{IL}	Low-level input voltage			0.8			0.8	V
I_{OH}	High-level output current			−12			−15	mA
I_{OL}	Low-level output current			32			48	mA
T_A	Operating free-air temperature	−55		125	0		70	°C

electrical characteristics over recommended operating free-air temperature range (unless otherwise noted)

PARAMETER		TEST CONDITIONS		SN54AS253 MIN	SN54AS253 TYP†	SN54AS253 MAX	SN74AS253 MIN	SN74AS253 TYP†	SN74AS253 MAX	UNIT
V_{IK}		V_{CC} = 4.5 V,	I_I = −18 mA			−1.2			−1.2	V
V_{OH}		V_{CC} = 4.5 V to 5.5 V,	I_{OH} = −2 mA	V_{CC}−2			V_{CC}−2			V
		V_{CC} = 4.5 V,	I_{OH} = −12 mA	2.4	3.2					
		V_{CC} = 4.5 V,	I_{OH} = −15 mA				2.4	3.2		
V_{OL}		V_{CC} = 4.5 V,	I_{OL} = 32 mA		0.25	0.5				V
		V_{CC} = 4.5 V,	I_{OL} = 48 mA					0.35	0.5	
I_{OZH}		V_{CC} = 5.5 V,	V_O = 2.7 V			50			50	µA
I_{OZL}		V_{CC} = 5.5 V,	V_O = 0.4 V			−50			−50	µA
I_I	A, B	V_{CC} = 5.5 V,	V_I = 7 V			0.2			0.2	mA
	All others					0.1			0.1	
I_{IH}	A, B	V_{CC} = 5.5 V,	V_I = 2.7 V			40			40	µA
	All others					20			20	
I_{IL}	A, B	V_{CC} = 5.5 V,	V_I = 0.4 V			−1			−1	mA
	All others					−0.5			−0.5	
I_O‡		V_{CC} = 5.5 V,	V_O = 2.25 V	−30		−112	−30		−112	mA
I_{CC}		V_{CC} = 5.5 V	Outputs high		18	29		18	29	mA
			Outputs low		20	32		20	32	
			Outputs disabled		21	33		21	33	

†All typical values are at V_{CC} = 5 V, T_A = 25°C.
‡The output conditions have been chosen to produce a current that closely approximates one half of the true short-circuit output current, I_{OS}.

switching characteristics (see Note 1)

PARAMETER	FROM (INPUT)	TO (OUTPUT)	V_{CC} = 4.5 V to 5.5 V, C_L = 50 pF, R1 = 500 Ω, R2 = 500 Ω, T_A = MIN to MAX				UNIT
			SN54AS253 MIN	SN54AS253 MAX	SN74AS253 MIN	SN74AS253 MAX	
t_{PLH}	A or B	Y	4	14.5	4	13.5	ns
t_{PHL}			4	12	4	11.5	
t_{PLH}	Data (Any C)	Y	3	8.5	3	7.5	ns
t_{PHL}			3	8.5	3	8	
t_{PZH}	G	Any Y	4	13	4	12.5	ns
t_{PZL}			4	12	4	11.5	
t_{PHZ}	G	Any Y	2	6.5	2	6	ns
t_{PLZ}			2	8	2	7	

NOTE 1: For load circuit and voltage waveforms, see page 1-12.

TYPES SN54ALS257, SN54ALS258, SN54AS257, SN54AS258, SN74ALS257, SN74ALS258, SN74AS257, SN74AS258
QUADRUPLE 1 OF 8 DATA SELECTORS/MULTIPLEXERS WITH 3-STATE OUTPUTS

D2661, APRIL 1982—REVISED DECEMBER 1983

- Three-State Outputs Interface Directly with System Bus
- Provides Bus Interface from Multiple Sources in High-Performance Systems
- Package Options Include Both Plastic and Ceramic Chip Carriers in Addition to Plastic and Ceramic DIPs
- Dependable Texas Instruments Quality and Reliability

description

These devices are designed to multiplex signals from four-bit data sources to four-output data lines in bus-organized systems. The 3-state outputs will not load the data lines when the output control pin ($\overline{G}$) is at a high-logic level.

The SN54' family is characterized for operation over the full military temperature range of −55°C to 125°C. The SN74' family is characterized for operation from 0°C to 70°C.

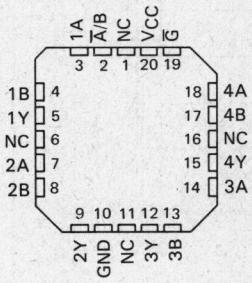

SN54ALS', SN54AS' . . . J PACKAGE
SN74ALS', SN74AS' . . . N PACKAGE
(TOP VIEW)

SN54ALS', SN54AS' . . . FH PACKAGE
SN74ALS', SN74AS' . . . FN PACKAGE
(TOP VIEW)

logic symbols

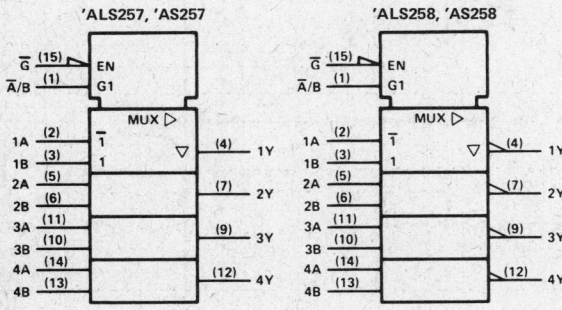

'ALS257, 'AS257   'ALS258, 'AS258

Pin numbers shown are for J and N packages.

FUNCTION TABLE

INPUTS			OUTPUT Y	
OUTPUT CONTROL $\overline{G}$	SELECT $\overline{A}/B$	DATA A B	'ALS257 'AS257	'ALS258 'AS258
H	X	X X	Z	Z
L	L	L X	L	H
L	L	H X	H	L
L	H	X L	L	H
L	H	X H	H	L

TYPES SN54ALS257, SN54ALS258, SN54AS257, SN54AS258, SN74ALS257, SN74ALS258, SN74AS257, SN74AS258
QUADRUPLE 1 OF 8 DATA SELECTORS/MULTIPLEXERS WITH 3-STATE OUTPUTS

logic diagrams (positive logic)

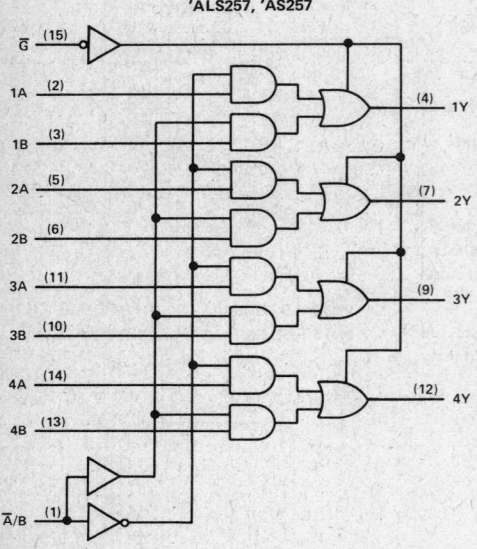

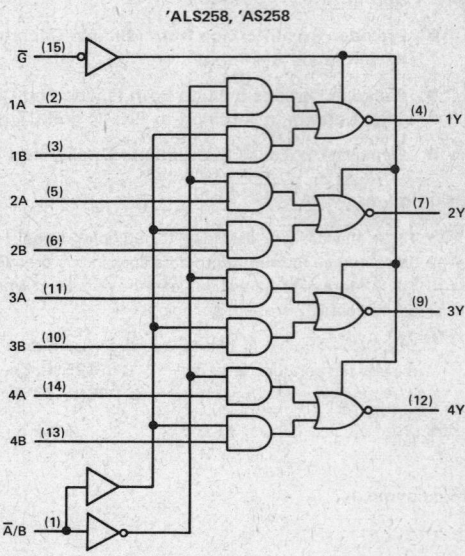

Pin numbers shown are for J and N packages.

absolute maximum ratings over operating free-air temperature range (unless otherwise noted)

Supply voltage, V_{CC} .... 7 V
Input voltage .... 7 V
Voltage applied to a disabled 3-state output .... 5.5 V
Operating free-air temperature range: SN54ALS′, SN54AS′ .... −55°C to 125°C
                                     SN74ALS′, SN74AS′ .... 0°C to 70°
Storage temperature range .... −65°C to 150°C

TYPES SN54ALS257, SN54ALS258, SN74ALS257, SN74ALS258
QUADRUPLE 1 OF 8 DATA SELECTORS/MULTIPLEXERS WITH 3-STATE OUTPUTS

recommended operating conditions

		SN54ALS257 SN54ALS258			SN74ALS257 SN74ALS258			UNIT
		MIN	NOM	MAX	MIN	NOM	MAX	
V_{CC}	Supply voltage	4.5	5	5.5	4.5	5	5.5	V
V_{IH}	High-level input voltage	2			2			V
V_{IL}	Low-level input voltage			0.8			0.8	V
I_{OH}	High-level output current			−1			−2.6	mA
I_{OL}	Low-level output current			12			24	mA
T_A	Operating free-air temperature	−55		125	0		70	°C

electrical characteristics over recommended operating free-air temperature range (unless otherwise noted)

PARAMETER		TEST CONDITIONS		SN54ALS257 SN54ALS258			SN74ALS257 SN74ALS258			UNIT
				MIN	TYP†	MAX	MIN	TYP†	MAX	
V_{IK}		V_{CC} = 4.5 V,	I_I = −18 mA			−1.5			−1.5	V
V_{OH}		V_{CC} = 4.5 V to 5.5 V,	I_{OH} = −0.4 mA	V_{CC}−2			V_{CC}−2			V
		V_{CC} = 4.5 V,	I_{OH} = −1 mA	2.4	3.3					
		V_{CC} = 4.5 V,	I_{OH} = −2.6 mA				2.4	3.2		
V_{OL}		V_{CC} = 4.5 V,	I_{OL} = 12 mA		0.25	0.4		0.25	0.4	V
		V_{CC} = 4.5 V	I_{OL} = 24 mA					0.35	0.5	
I_{OZH}		V_{CC} = 5.5 V,	V_O = 2.7 V			20			20	µA
I_{OZL}		V_{CC} = 5.5 V,	V_O = 0.4 V			−20			−20	µA
I_I		V_{CC} = 5.5 V,	V_I = 7 V			0.1			0.1	mA
I_{IH}		V_{CC} = 5.5 V,	V_I = 2.7 V			20			20	µA
I_{IL}		V_{CC} = 5.5 V,	V_I = 0.4 V			−0.1			−0.1	mA
I_O‡		V_{CC} = 5.5 V,	V_O = 2.25 V	−30		−112	−30		−112	mA
I_{CC}	'ALS257	V_{CC} = 5.5 V	Outputs high		3	6		3	6	mA
			Outputs low		8	12		8	12	
			Outputs disabled		9	14		9	14	
	'ALS258	V_{CC} = 5.5 V	Outputs high		2.5	4		2.5	4	
			Outputs low		7	11		7	11	
			Outputs disabled		8	13		8	13	

†All typical values are at V_{CC} = 5 V, T_A = 25°C.
‡The output conditions have been chosen to produce a current that closely approximates one half of the true short-circuit output current, I_{OS}.

TYPES SN54ALS257, SN54ALS258, SN74ALS257, SN74ALS258
QUADRUPLE 1 OF 8 DATA SELECTORS/MULTIPLEXERS WITH 3-STATE OUTPUTS

'ALS257 switching characteristics (see Note 1)

PARAMETER	FROM (INPUT)	TO (OUTPUT)	V_{CC} = 4.5 V to 5.5 V, C_L = 50 pF, R1 = 500 Ω, R2 = 500 Ω, T_A = MIN to MAX				UNIT
			SN54ALS257		SN74ALS257		
			MIN	MAX	MIN	MAX	
t_{PLH}	A or B	Any Y	2	12	2	10	ns
t_{PHL}			3	14	3	12	
t_{PLH}	$\overline{A}$/B	Any Y	7	21	7	18	ns
t_{PHL}			6	25	6	22	
t_{PZH}	$\overline{G}$	Any Y	4	20	4	16	ns
t_{PZL}			5	22	5	18	
t_{PHZ}	$\overline{G}$	Any Y	2	12	2	10	ns
t_{PLZ}			4	18	4	15	

'ALS258 switching characteristics (see Note 1)

PARAMETER	FROM (INPUT)	TO (OUTPUT)	V_{CC} = 4.5 V to 5.5 V, C_L = 50 pF, R1 = 500 Ω, R2 = 500 Ω, T_A = MIN to MAX				UNIT
			SN54ALS258		SN74ALS258		
			MIN	MAX	MIN	MAX	
t_{PLH}	A or B	Any Y	2	10	2	8	ns
t_{PHL}			2	9	2	7	
t_{PLH}	$\overline{A}$/B	Any Y	5	28	5	25	ns
t_{PHL}			8	23	8	20	
t_{PZH}	$\overline{G}$	Any Y	5	20	5	18	ns
t_{PZL}			5	20	5	18	
t_{PHZ}	$\overline{G}$	Any Y	2	12	2	10	ns
t_{PLZ}			5	20	5	18	

NOTE 1: For load circuit and voltage waveforms, see page 1-12.

TYPES SN54AS257, SN54AS258, SN74AS257, SN74AS258
QUADRUPLE 1 OF 8 DATA SELECTORS/MULTIPLEXERS WITH 3-STATE OUTPUTS

recommended operating conditions

		SN54AS257 SN54AS258			SN74AS257 SN74AS258			UNIT
		MIN	NOM	MAX	MIN	NOM	MAX	
V_{CC}	Supply voltage	4.5	5	5.5	4.5	5	5.5	V
V_{IH}	High-level input voltage	2			2			V
V_{IL}	Low-level input voltage			0.8			0.8	V
I_{OH}	High-level output current			−12			−15	mA
I_{OL}	Low-level output current			32			48	mA
T_A	Operating free-air temperature	−55		125	0		70	°C

electrical characteristics over recommended operating free-air temperature range (unless otherwise noted)

PARAMETER		TEST CONDITIONS		SN54AS257 SN54AS258			SN74AS257 SN74AS258			UNIT
				MIN	TYP†	MAX	MIN	TYP†	MAX	
V_{IK}		V_{CC} = 4.5 V,	I_I = −18 mA			−1.2			−1.2	V
V_{OH}		V_{CC} = 4.5 V to 5.5 V,	I_{OH} = −2 mA	V_{CC}−2			V_{CC}−2			V
		V_{CC} = 4.5 V,	I_{OH} = −12 mA	2.4	3.3					
		V_{CC} = 4.5 V,	I_{OH} = −15 mA				2.4	3.2		
V_{OL}		V_{CC} = 4.5 V,	I_{OL} = 32 mA		0.25	0.5				V
		V_{CC} = 4.5 V	I_{OL} = 48 mA					0.35	0.5	
I_{OZH}		V_{CC} = 5.5 V,	V_O = 2.7 V			50			50	µA
I_{OZL}		V_{CC} = 5.5 V,	V_O = 0.4 V			−50			−50	µA
I_I	A, B or $\overline{G}$	V_{CC} = 5.5 V,	V_I = 7 V			0.1			0.1	mA
	$\overline{A}/B$					0.2			0.2	
I_{IH}	A, B, or $\overline{G}$	V_{CC} = 5.5 V,	V_I = 2.7 V			20			20	µA
	$\overline{A}/B$					40			40	
I_{IL}	A, B, or $\overline{G}$	V_{CC} = 5.5 V,	V_I = 0.4 V			−0.5			−0.5	mA
	$\overline{A}/B$					−1			−1	
I_O‡		V_{CC} = 5.5 V,	V_O = 2.25 V	−30		−112	−30		−112	mA
I_{CC}	'AS257	V_{CC} = 5.5 V	Outputs high		12.1	19.7		12.1	19.7	mA
			Outputs low		19	30.6		19	30.6	
			Outputs disabled		19.7	31.9		19.7	31.9	
	'AS258	V_{CC} = 5.5 V	Outputs high		8.4	13.5		8.4	13.5	
			Outputs low		15.2	24.6		15.2	24.6	
			Outputs disabled		15.5	25.2		15.5	25.2	

†All typical values are at V_{CC} = 5 V, T_A = 25°C.
‡The output conditions have been chosen to produce a current that closely approximates one half of the true short-circuit output current, I_{OS}.

TEXAS INSTRUMENTS
POST OFFICE BOX 225012 • DALLAS, TEXAS 75265

TYPES SN54AS257, SN54AS258, SN74AS257, SN74AS258
QUADRUPLE 1 OF 8 DATA SELECTORS/MULTIPLEXERS WITH 3-STATE OUTPUTS

'AS257 switching characteristics (see Note 1)

PARAMETER	FROM (INPUT)	TO (OUTPUT)	V_{CC} = 4.5 V to 5.5 V, C_L = 50 pF, R1 = 500 Ω, R2 = 500 Ω, T_A = MIN to MAX				UNIT
			SN54AS257		SN74AS257		
			MIN	MAX	MIN	MAX	
t_{PLH}	A or B	Any Y	1	6.5	1	5.5	ns
t_{PHL}			1	7	1	6	
t_{PLH}	$\overline{A}/B$	Any Y	2	12	2	11	ns
t_{PHL}			2	10.5	2	10	
t_{PZH}	$\overline{G}$	Any Y	2	8.5	2	7.5	ns
t_{PZL}			2	10.5	2	9.5	
t_{PHZ}	$\overline{G}$	Any Y	1.5	8	1.5	6.5	ns
t_{PLZ}			2	8	2	7	

'AS258 switching characteristics (see Note 1)

PARAMETER	FROM (INPUT)	TO (OUTPUT)	V_{CC} = 4.5 V to 5.5 V, C_L = 50 pF, R1 = 500 Ω, R2 = 500 Ω, T_A = MIN to MAX				UNIT
			SN54AS258		SN74AS258		
			MIN	MAX	MIN	MAX	
t_{PLH}	A or B	Any Y	1	5.5	1	5	ns
t_{PHL}			1	5	1	4	
t_{PLH}	$\overline{A}/B$	Any Y	2	11	2	9.5	ns
t_{PHL}			2	11	2	10	
t_{PZH}	$\overline{G}$	Any Y	2	8.5	2	8	ns
t_{PZL}			2	11	2	10	
t_{PHZ}	$\overline{G}$	Any Y	1.5	7	1.5	6	ns
t_{PLZ}			2	8.5	2	6.5	

NOTE 1: For load circuit and voltage waveforms, see page 1-12.

TYPES SN54ALS259, SN74ALS259
8-BIT ADDRESSABLE LATCHES

D2661, DECEMBER 1982

- 8-Bit Parallel-Out Storage Register Performs Serial-to-Parallel Conversion with Storage
- Asynchronous Parallel Clear
- Active High Decoder
- Enable/Disable Input Simplifies Expansion
- Expandable for N-Bit Applications
- Four Distinct Functional Modes
- Package Options Include Both Plastic and Ceramic Chip Carriers in Addition to Plastic and Ceramic DIPS
- Dependable Texas Instruments Quality and Reliability

description

These 8-bit addressable latches are designed for general purpose storage applications in digital systems. Specific uses include working registers, serial-holding registers, and active-high decoders or demultiplexers. They are multifunctional devices capable of storing single-line data in eight addressable latches, and being a 1-of-8 decoder or demultiplexer with active-high outputs.

Four distinct modes of operation are selectable by controlling the clear ($\overline{CLR}$) and enable ($\overline{G}$) inputs as enumerated in the function table. In the addressable-latch mode, data at the data-in terminal is written into the addressed latch. The addressed latch will follow the data input with all unaddressed latches remaining in their previous states. In the memory mode, all latches remain in their previous states and are unaffected by the data or address inputs. To eliminate the possibility of entering erroneous data in the latches, enable $\overline{G}$ should be held high (inactive) while the address lines are changing. In the 1-of-8 decoding or demultiplexing mode, the addressed output will follow the level of the D input with all other outputs low. In the clear mode, all outputs are low and unaffected by the address and data inputs.

The SN54ALS259 will be characterized for operation over the full military temperature range of −55°C to 125°C. The SN74ALS259 will be characterized for operation from 0°C to 70°C.

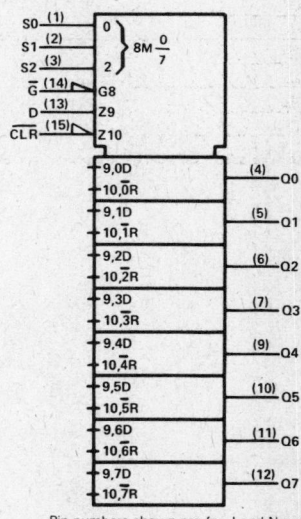

NC — No internal connection

FUNCTION TABLE

INPUTS		OUTPUT OF ADDRESSED LATCH	EACH OTHER OUTPUT	FUNCTION
$\overline{CLR}$	$\overline{G}$			
H	L	D	L	Addressable Latch
H	H	Q_{i0}	Q_{i0}	Memory
L	L	D	L	8-Line Demultiplexer
L	H	L	L	Clear

D = the level at the data input.
Q_{i0} = the level of Q_i (i = 0, 1, ...7, as appropriate) before the indicated steady-state input conditions were established.

LATCH SELECTION TABLE

SELECT INPUTS			LATCH ADDRESSED
S2	S1	S0	
L	L	L	0
L	L	H	1
L	H	L	2
L	H	H	3
H	L	L	4
H	L	H	5
H	H	L	6
H	H	H	7

Pin numbers shown are for J and N packages.

PRODUCT PREVIEW

This document contains information on a product under development. Texas Instruments reserves the right to change or discontinue this product without notice.

Copyright © 1982 by Texas Instruments Incorporated

TEXAS INSTRUMENTS
POST OFFICE BOX 225012 • DALLAS, TEXAS 75265

2
ALS AND AS CIRCUITS

TYPES SN54AS264, SN74AS264
LOOK-AHEAD CARRY GENERATORS FOR COUNTERS

D2824, DECEMBER 1983

- Performs Look-Ahead Carry Across n-Bit Counters
- Accommodates Active-High or Active-Low Carry
- Improves Cascaded Counters System Performance
- Dependable Texas Instruments Quality and Reliability

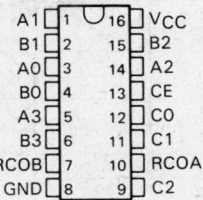

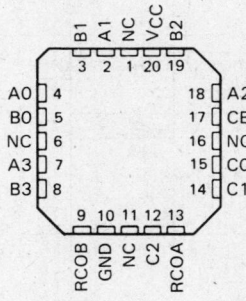

NC — No internal connection

description

This look-ahead generator was designed specifically to perform a carry-anticipate across any number of n-bit counters, thus increasing system clock frequency. A carry enable CE, and carry outputs RCOA and RCOB are provided for n-bit cascading.

The counter can be used with either active-high-carry or active-low-carry counters. For active-high-carry counters, CE is active high, the A set of inputs and output RCOA are used, and the B set of inputs are connected to a low logic level. For active-low-carry counters, CE is active low, the B set of inputs and output RCOB are used, and the A set of inputs are connected to a high logic level. See Figures 1 and 2 for typical applications.

The SN54AS264 is characterized for operation over the full military temperature range of $-55\,°C$ to $125\,°C$. The SN74AS264 is characterized for operation in the temperature range of $0\,°C$ to $70\,°C$.

positive logic equations

ACTIVE-HIGH-CARRY COUNTERS
(CE is high, all B inputs are low)
$C0 = A0$
$C1 = A0 \cdot A1$
$C2 = A0 \cdot A1 \cdot A2$
$RCOA = A0 \cdot A1 \cdot A2 \cdot A3$
RCOB is high

ACTIVE-LOW-CARRY COUNTERS
(CE is low, all A inputs are high)
$C0 = \overline{B0}$
$C1 = \overline{B0} \cdot \overline{B1}$
$C2 = \overline{B0} \cdot \overline{B1} \cdot \overline{B2}$
$RCOA = \overline{\overline{B1} \cdot \overline{B2} \cdot \overline{B3}}$
$RCOB = \overline{B0} \cdot \overline{B1} \cdot \overline{B2} \cdot \overline{B3}$

PRODUCT PREVIEW
This document contains information on a product under development. Texas Instruments reserves the right to change or discontinue this product without notice.

Copyright © 1983 by Texas Instruments Incorporated

TEXAS INSTRUMENTS
POST OFFICE BOX 225012 • DALLAS, TEXAS 75265

TYPES SN54AS264, SN74AS264
LOOK-AHEAD CARRY GENERATORS FOR COUNTERS

FUNCTION TABLE FOR C0 OUTPUT

INPUTS			OUTPUT
A0	B0	CE	C0
H	H	X	H
H	X	H	H
L	X	X	L
X	L	L	L

FUNCTION TABLE FOR C1 OUTPUT

INPUTS				OUTPUT
A1	A0	B1	B0	CE	C1
H	X	H	X	X	H
H	H	X	X	X	H
H	H	X	X	H	H
L	X	X	X	X	L
X	L	L	X	X	L
X	X	L	L	X	L

FUNCTION TABLE FOR C2 OUTPUT

INPUTS						OUTPUT
A2	A1	A0	B2	B1	B0	CE	C2
H	X	X	H	X	X	X	H
H	H	X	X	H	X	X	H
H	H	H	X	X	H	X	H
H	H	H	X	X	X	H	H
L	X	X	X	X	X	X	L
X	L	X	L	X	X	X	L
X	X	L	L	L	X	X	L
X	X	X	L	L	L	L	L

FUNCTION TABLE FOR RCOA OUTPUT

INPUTS						OUTPUT
A3	A2	A1	A0	B3	B2	B1	CE	RCOA
H	X	X	X	H	X	X	X	H
H	H	X	X	X	H	X	X	H
H	H	H	X	X	X	H	X	H
H	H	H	H	X	X	X	H	H
L	X	X	X	X	X	X	X	L
X	L	X	X	L	X	X	X	L
X	X	L	X	L	L	X	X	L
X	X	X	L	L	L	L	X	L
X	X	X	X	L	L	L	L	L

FUNCTION TABLE FOR RCOB OUTPUT

INPUTS				OUTPUT
B3	B2	B1	B0	CE	RCOB
H	X	X	X	X	H
X	H	X	X	X	H
X	X	H	X	X	H
X	X	X	H	X	H
X	X	X	X	H	H
L	L	L	L	L	L

logic symbols

ACTIVE-HIGH INPUTS

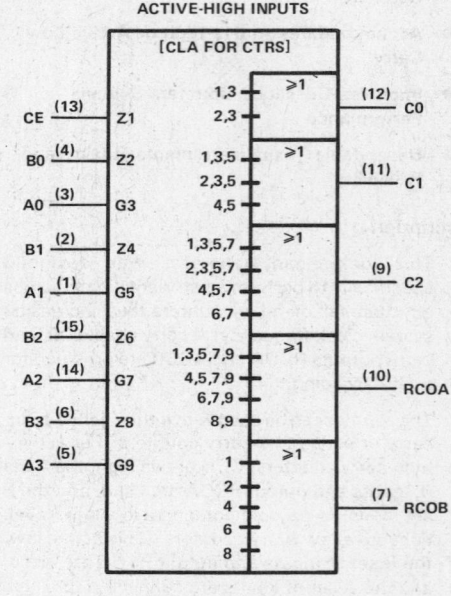

ACTIVE-LOW INPUTS

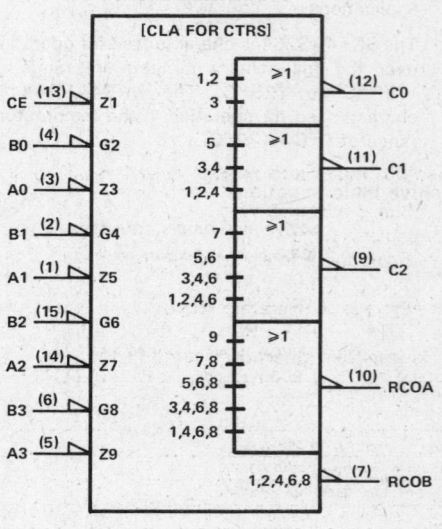

TYPES SN54AS264, SN74AS264
LOOK-AHEAD CARRY GENERATORS FOR COUNTERS

logic diagram (positive logic)

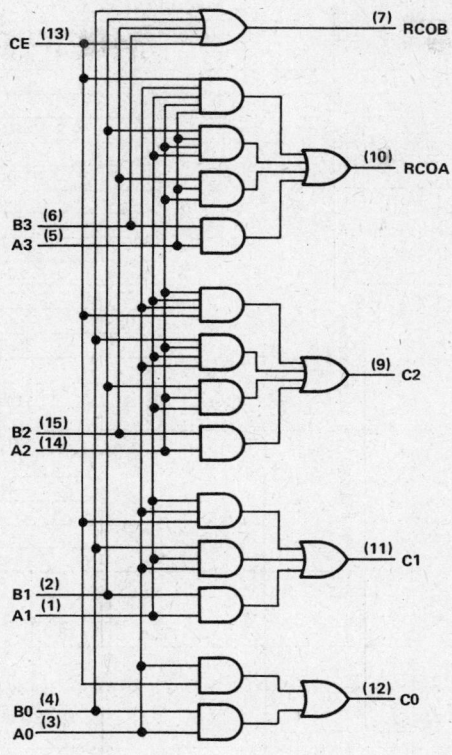

absolute maximum ratings over free-air temperature (unless otherwise noted)

Supply voltage, V_{CC} .... 7 V
Input voltage .... 7 V
Operating free-air temperature range: SN54AS264 .... −55°C to 125°C
                                      SN74AS264 .... 0°C to 70°C
Storage temperature range .... −65°C to 150°C

recommended operating conditions

		SN54AS264			SN74AS264			UNIT
		MIN	NOM	MAX	MIN	NOM	MAX	
V_{CC}	Supply voltage	4.5	5	5.5	4.5	5	5.5	V
V_{IH}	High-level input voltage	2			2			V
V_{IL}	Low-level input voltage			0.8			0.8	V
I_{OH}	High-level output current			−2			−2	mA
I_{OL}	Low-level output current			20			20	mA
T_A	Operating free-air temperature	−55		125	0		70	°C

TEXAS INSTRUMENTS
POST OFFICE BOX 225012 • DALLAS, TEXAS 75265

TYPES SN54AS264, SN74AS264
LOOK-AHEAD CARRY GENERATORS FOR COUNTERS

electrical characteristics over recommended operating free-air temperature range (unless otherwise noted)

PARAMETER		TEST CONDITIONS		SN54AS264 MIN	SN54AS264 TYP†	SN54AS264 MAX	SN74AS264 MIN	SN74AS264 TYP†	SN74AS264 MAX	UNIT
V_{IK}		$V_{CC} = 4.5$ V,	$I_I = -18$ mA			-1.2			-1.2	V
V_{OH}		$V_{CC} = 4.5$ V to 5.5 V,	$I_{OH} = -2$ mA	$V_{CC}-2$			$V_{CC}-2$			V
V_{OL}		$V_{CC} = 4.5$ V,	$I_{OL} = 20$ mA		0.3	0.5		0.3	0.5	V
I_I	CE	$V_{CC} = 5.5$ V,	$V_I = 7$ V			500			500	µA
	A0, A2					700			700	
	A1					800			800	
	A3, B0, B1					400			400	
	B2					300			300	
	B3					200			200	
I_{IH}	CE	$V_{CC} = 5.5$ V,	$V_I = 2.7$ V			100			100	µA
	A0, A2					140			140	
	A1					160			160	
	A3, B0, B1					80			80	
	B2					60			60	
	B3					40			40	
I_{IL}	CE	$V_{CC} = 5.5$ V,	$V_I = 0.4$ V			-2.5			-2.5	mA
	A0					-3.5			-3.5	
	A1, A2					-4			-4	
	A3, B0, B1					-2			-2	
	B2					-1			-1	
	B3					-1.5			-1.5	
I_O‡		$V_{CC} = 5.5$ V	$V_O = 2.25$ V	-30		-112	-30		-112	mA
I_{CCH}		$V_{CC} = 5.5$ V			26			26		mA
I_{CCL}					28			28		

†All typical values are at $V_{CC} = 5$ V, $T_A = 25°C$.
‡The output conditions have been chosen to produce a current that closely approximates one-half of the true short-circuit current, I_{OS}.

switching characteristics (see Note 1)

PARAMETER	FROM (INPUT)	TO (OUTPUT)	$V_{CC} = 4.5$ V to 5.5 V, $C_L = 50$ pF, $R_L = 50$ Ω, $T_A = $ MIN to MAX						UNIT
			SN54AS264 MIN	SN54AS264 TYP†	SN54AS264 MAX	SN74AS264 MIN	SN74AS264 TYP†	SN74AS264 MAX	
t_{PLH}	CE	C0, C1, C2		6			6		ns
t_{PHL}				5			5		
t_{PLH}	An or Bn	C0, C1, C2		5			5		ns
t_{PHL}				5			5		
t_{PLH}	An, Bn, or CE	RCOA		5			5		ns
t_{PHL}				5			5		
t_{PLH}	Bn or CE	RCOB		5			5		ns
t_{PHL}				5			5		

†All typical values are at $V_{CC} = 5$ V, $T_A = 25°C$.
NOTE 1: For load circuit and voltage waveforms, see page 1-12.

TYPES SN54AS264, SN74AS264
LOOK-AHEAD CARRY GENERATORS FOR COUNTERS

TYPICAL APPLICATION INFORMATION

The circuit shown in Figure 1 illustrates how the 'AS624 can implement look-ahead carry for the active-high-carry 'AS163, while Figure 2 shows the look-ahead carry for the active-low-carry 'AS169.

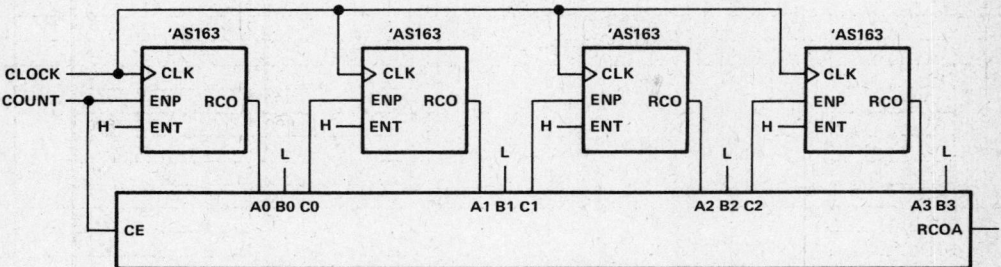

FIGURE 1—ACTIVE-HIGH-CARRY

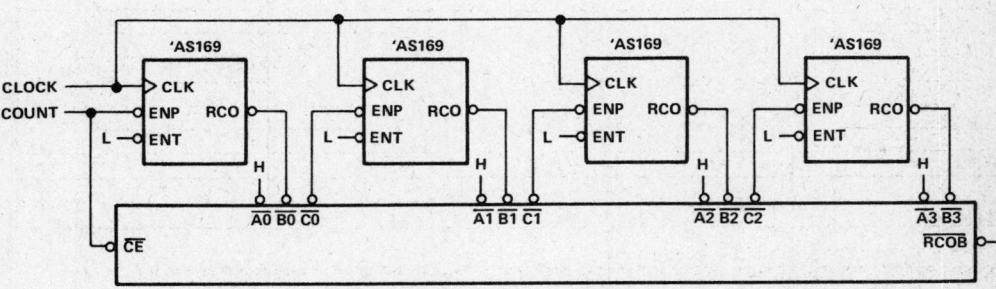

FIGURE 2—ACTIVE-LOW-CARRY

TYPES SN54ALS273, SN74ALS273
OCTAL D-TYPE FLIP-FLOPS WITH CLEAR

D2661, APRIL 1982—REVISED DECEMBER 1983

- Contains Eight Flip-Flops with Single-Rail Outputs
- Buffered Clock and Direct Clear Inputs
- Individual Data Input to Each Flip-Flop
- Applications Include:
    Buffer/Storage Registers
    Shift Registers
    Pattern Generators
- Package Options Include Both Plastic and Ceramic Chip Carriers in Addition to Plastic and Ceramic DIPs
- Dependable Texas Instruments Quality and Reliability

SN54ALS273 . . . J PACKAGE
SN74ALS273 . . . N PACKAGE
(TOP VIEW)

```
 CLR  [ 1   20 ] VCC
  1Q  [ 2   19 ] 8Q
  1D  [ 3   18 ] 8D
  2D  [ 4   17 ] 7D
  2Q  [ 5   16 ] 7Q
  3Q  [ 6   15 ] 6Q
  3D  [ 7   14 ] 6D
  4D  [ 8   13 ] 5D
  4Q  [ 9   12 ] 5Q
 GND  [ 10  11 ] CLK
```

description

These monolithic, positive-edge-triggered flip-flops utilize TTL circuitry to implement D-type flip-flop logic with a direct clear input.

Information at the D inputs meeting the setup time requirements is transferred to the Q outputs on the positive-going edge of the clock pulse. Clock triggering occurs at a particular voltage level and is not directly related to the transition time of the positive-going pulse. When the clock input is at either the high or low level, the D input signal has no effect at the output.

The SN54ALS273 is characterized for operation over the full military temperature range of $-55\,°C$ to $125\,°C$. The SN74ALS273 is characterized for operation from $0\,°C$ to $70\,°C$.

SN54ALS273 . . . FH PACKAGE
SN74ALS273 . . . FN PACKAGE
(TOP VIEW)

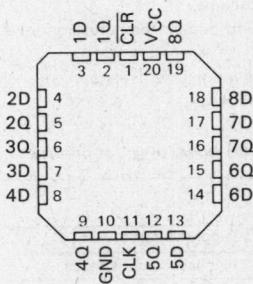

FUNCTION TABLE
(EACH FLIP-FLOP)

INPUTS			OUTPUT
CLEAR	CLOCK	D	Q
L	X	X	L
H	↑	H	H
H	↑	L	L
H	L	X	Q_0

logic symbol

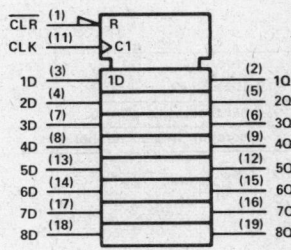

Pin numbers shown are for J and N packages.

Copyright © 1982 by Texas Instruments Incorporated

TEXAS INSTRUMENTS
POST OFFICE BOX 225012 • DALLAS, TEXAS 75265

TYPES SN54ALS273, SN74ALS273
OCTAL D-TYPE FLIP-FLOPS WITH CLEAR

logic diagram (positive logic)

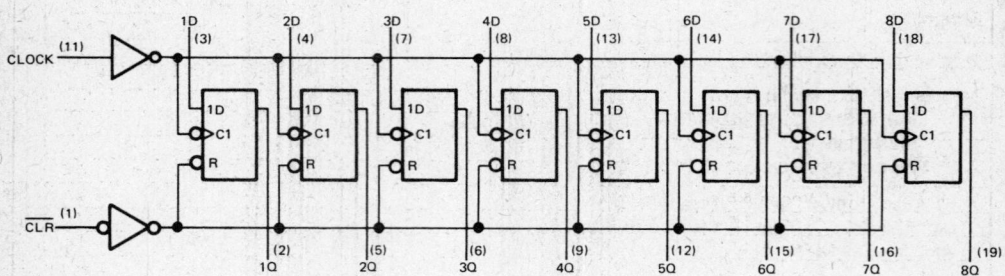

absolute maximum ratings over operating free-air temperature range (unless otherwise noted)

Supply voltage, V_{CC} .... 7 V
Input voltage .... 7 V
Operating free-air temperature range: SN54ALS273 .... −55°C to 125°C
                                       SN74ALS273 .... 0°C to 70°C
Storage temperature range .... −65°C to 150°C

recommended operating conditions

			SN54ALS273			SN74ALS273			UNIT
			MIN	NOM	MAX	MIN	NOM	MAX	
V_{CC}	Supply voltage		4.5	5	5.5	4.5	5	5.5	V
V_{IH}	High-level input voltage		2			2			V
V_{IL}	Low-level input voltage				0.8			0.8	V
I_{OH}	High-level output current				−1			−2.6	mA
I_{OL}	Low-level output current				12			24	mA
f_{clock}	Clock frequency		0		30	0		35	MHz
t_w	Pulse duration	CLR low	10			10			ns
		CLK high	16.5			14			
		CLK low	16.5			14			
t_{su}	Setup time before CLK↑	Data	10			10			ns
		Clear inactive state	15			15			
t_h	Hold time, data after CLK↑		0			0			ns
T_A	Operating free-air temperature		−55		125	0		70	°C

TEXAS INSTRUMENTS
POST OFFICE BOX 225012 • DALLAS, TEXAS 75265

TYPES SN54ALS273, SN74ALS273
OCTAL D-TYPE FLIP-FLOPS WITH CLEAR

electrical characteristics over recommended operating free-air temperature range (unless otherwise noted)

PARAMETER	TEST CONDITIONS		SN54ALS273			SN74ALS273			UNIT
			MIN	TYP†	MAX	MIN	TYP†	MAX	
V_{IK}	$V_{CC} = 4.5$ V,	$I_I = -18$ mA			-1.5			-1.5	V
V_{OH}	$V_{CC} = 4.5$ V to 5.5 V,	$I_{OH} = -0.4$ mA	$V_{CC}-2$			$V_{CC}-2$			V
	$V_{CC} = 4.5$ V,	$I_{OH} = -1$ mA	2.4	3.3					
	$V_{CC} = 4.5$ V,	$I_{OH} = -2.6$ mA				2.4	3.2		
V_{OL}	$V_{CC} = 4.5$ V,	$I_{OL} = 12$ mA		0.25	0.4		0.25	0.4	V
	$V_{CC} = 4.5$ V,	$I_{OL} = 24$ mA					0.35	0.5	
I_I	$V_{CC} = 5.5$ V,	$V_I = 7$ V			0.1			0.1	mA
I_{IH}	$V_{CC} = 5.5$ V,	$V_I = 2.7$ V			20			20	µA
I_{IL}	$V_{CC} = 5.5$ V,	$V_I = 0.4$ V			-0.2			-0.2	mA
I_O‡	$V_{CC} = 5.5$ V,	$V_O = 2.25$ V	-30		-112	-30		-112	mA
I_{CCH}	$V_{CC} = 5.5$ V			11	20		11	20	mA
I_{CCL}	$V_{CC} = 5.5$ V			19	29		19	29	

†All typical values are at $V_{CC} = 5$ V, $T_A = 25°C$.
‡The output conditions have been chosen to produce a current that closely approximates one half of the true short-circuit output current, I_{OS}.

switching characteristics (see Note 1)

PARAMETER	FROM (INPUT)	TO (OUTPUT)	$V_{CC} = 4.5$ V to 5.5 V, $C_L = 50$ pF, $R_L = 500$ Ω, T_A = MIN to MAX				UNIT
			SN54ALS273		SN74ALS273		
			MIN	MAX	MIN	MAX	
f_{max}			30		35		MHz
t_{PHL}	CLR	Any Q	4	21	4	18	ns
t_{PLH}	CLR	Any Q	2	16	2	12	ns
t_{PHL}			3	17	3	15	

NOTE 1: For load circuit and voltage waveforms, see page 1-12.

2
ALS AND AS CIRCUITS

TYPES SN54AS280, SN74AS280
9-BIT PARITY GENERATORS/CHECKERS

D2661, DECEMBER 1982 – REVISED DECEMBER 1983

- Generates Either Odd or Even Parity for Nine Data Lines
- Cascadable for n-Bits Parity
- Can Be Used to Upgrade Existing Systems using MSI Parity Circuits
- Package Options Include Both Plastic and Ceramic Chip Carriers in Addition to Plastic and Ceramic DIPs
- Dependable Texas Instruments Quality and Reliability

SN54AS280 . . . J PACKAGE
SN74AS280 . . . N PACKAGE
(TOP VIEW)

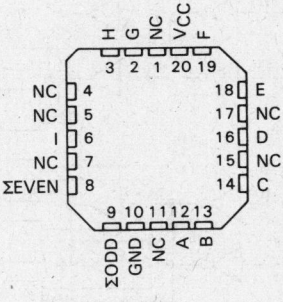

SN54AS280 . . . FH PACKAGE
SN74AS280 . . . FN PACKAGE
(TOP VIEW)

NC – No internal connection

FUNCTION TABLE

NUMBER OF INPUTS A THRU I THAT ARE HIGH	OUTPUTS Σ EVEN	Σ ODD
0,2,4,6,8	H	L
1,3,5,7,9	L	H

logic symbol

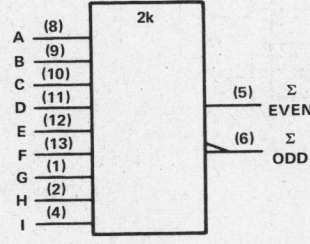

Pin numbers shown are for J and N packages.

description

These universal, monolithic, nine-bit parity generators/checkers utilize Advanced Schottky high-performance circuitry and feature odd and even outputs to facilitate operation of either odd or even parity application. The word-length capability is easily expanded by cascading.

These devices can be used to upgrade the performance of most systems utilizing the '180 parity generator/checker. Although the 'AS280 is implemented without expander inputs, the corresponding function is provided by the availability of an input at pin 4 and the absence of any internal connection at pin 3. This permits the 'AS280 to be substituted for the '180 in existing designs to produce an identical function even if 'AS280's are mixed with existing '180's.

All 'AS280 inputs are buffered to lower the drive requirements.

The SN54AS280 is characterized for operation over the full military temperature range of −55°C to 125°C. The SN74AS280 is characterized for operation from 0°C to 70°C.

PRODUCT PREVIEW

This document contains information on a product under development. Texas Instruments reserves the right to change or discontinue this product without notice.

Copyright © 1982 Texas Instruments Incorporated

TEXAS INSTRUMENTS
POST OFFICE BOX 225012 • DALLAS, TEXAS 75265

TYPES SN54AS280, SN74AS280
9-BIT PARITY GENERATORS/CHECKERS

logic diagram

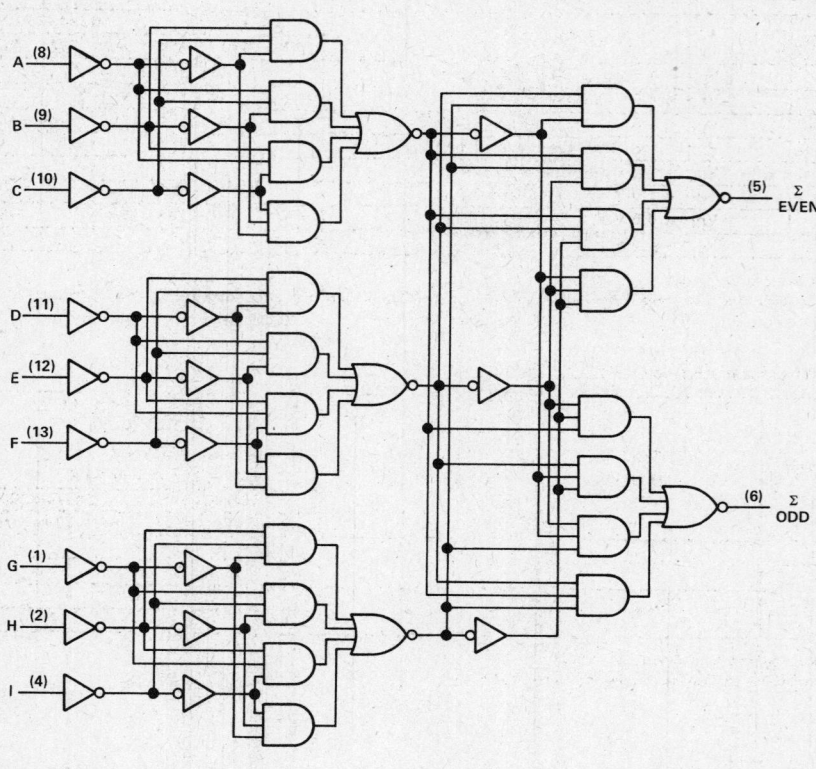

absolute maximum ratings over operating free-air temperature range (unless otherwise noted)

Supply voltage, V_{CC} .... 7 V
Input voltage .... 7 V
Operating free-air temperature range: SN54AS280 .... −55°C to 125°C
                                       SN74AS280 .... 0°C to 70°C
Storage temperature range .... −65°C to 150°C

recommended operating conditions

		SN54AS280			SN74AS280			UNIT
		MIN	NOM	MAX	MIN	NOM	MAX	
V_{CC}	Supply voltage	4.5	5	5.5	4.5	5	5.5	V
V_{IH}	High-level input voltage	2			2			V
V_{IL}	Low-level input voltage			0.8			0.8	V
I_{OH}	High-level output current			−2			−2	mA
I_{OL}	Low-level output current			20			20	mA
T_A	Operating free-air temperature	−55		125	0		70	°C

TEXAS INSTRUMENTS
POST OFFICE BOX 225012 • DALLAS, TEXAS 75265

TYPES SN54AS280, SN74AS280
9-BIT PARITY GENERATORS/CHECKERS

electrical characteristics over recommended operating free-air temperature range (unless otherwise noted)

PARAMETER	TEST CONDITIONS		SN54AS280 MIN	SN54AS280 TYP[†]	SN54AS280 MAX	SN74AS280 MIN	SN74AS280 TYP[†]	SN74AS280 MAX	UNIT
V_{IK}	$V_{CC} = 4.5$ V,	$I_I = -18$ mA			-1.2			-1.2	V
V_{OH}	$V_{CC} = 4.5$ V to 5.5 V,	$I_{OH} = -2$ mA	$V_{CC}-2$			$V_{CC}-2$			V
V_{OL}	$V_{CC} = 4.5$ V,	$I_{OL} = 20$ mA		0.35	0.5		0.35	0.5	V
I_I	$V_{CC} = 5.5$ V,	$V_I = 7$ V			0.1			0.1	mA
I_{IH}	$V_{CC} = 5.5$ V,	$V_I = 2.7$ V			20			20	μA
I_{IL}	$V_{CC} = 5.5$ V,	$V_I = 0.4$ V			-0.5			-0.5	mA
I_O[‡]	$V_{CC} = 5.5$ V,	$V_O = 2.25$ V	-30		-112	-30		-112	mA
I_{CC}	$V_{CC} = 5.5$ V			27			27		mA

[†] All typical values are at $V_{CC} = 5$ V, $T_A = 25$°C.
[‡] The output conditions have been chosen to produce a current that closely approximates one half of the true short-circuit output current, I_{OS}.

switching characteristics (see Note 1)

PARAMETER	FROM (INPUT)	TO (OUTPUT)	$V_{CC} = 4.5$ V to 5.5 V, $C_L = 50$ pF, $R_L = 500$ Ω, $T_A = $ MIN to MAX						UNIT
			SN54AS280 MIN	SN54AS280 TYP[‡]	SN54AS280 MAX	SN74AS280 MIN	SN74AS280 TYP[‡]	SN74AS280 MAX	
t_{PLH}	Any	Σ Even		7			7		ns
t_{PHL}				7.5			7.5		
t_{PLH}	Any	Σ Odd		7			7		ns
t_{PHL}				7.5			7.5		

[‡] All typical values are at $V_{CC} = 5$ V, $T_A = 25$°C.
NOTE 1: For load circuit and voltage waveforms, see page 1-12.

Additional information on these products can be obtained from the factory as it becomes available.

TEXAS INSTRUMENTS
POST OFFICE BOX 225012 • DALLAS, TEXAS 75265

TYPES SN54AS280, SN74AS280
9-BIT PARITY GENERATORS/CHECKERS

TYPICAL APPLICATION DATA

25-LINE PARITY/GENERATOR CHECKER

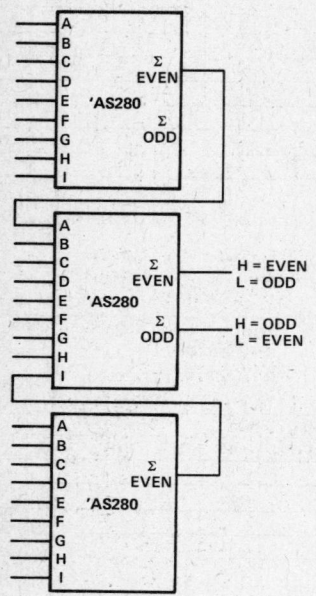

81-LINE PARITY/GENERATOR CHECKER

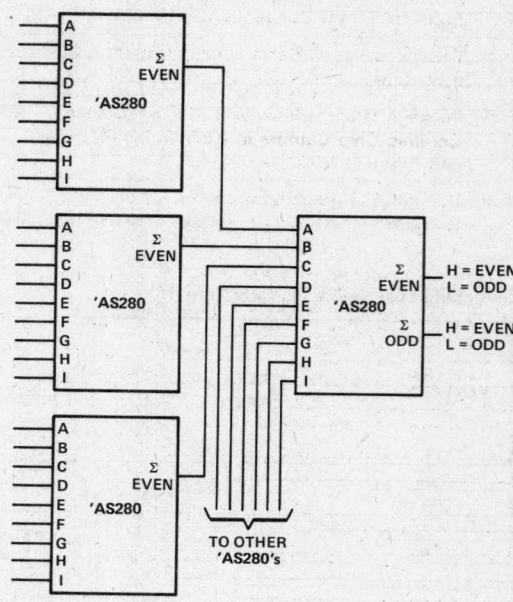

Three 'AS280 can be used to implement a 25-line parity generator/checker.

As an alternative, the outputs of two or three parity generators/checkers can be decoded with a 2-input ('S86 or 'LS86) or 3-input ('S135) exclusive-OR gate for 18- or 27-line parity applications.

Longer word lengths can be implemented by cascading 'AS280. As shown here, parity can be generated for word lengths up to 81 bits.

Texas Instruments
POST OFFICE BOX 225012 • DALLAS, TEXAS 75265

TYPES SN54AS282, SN74AS282
LOOK-AHEAD CARRY GENERATOR WITH SELECTABLE CARRY INPUTS

D2811, DECEMBER 1983

- Selectable Carry Inputs Version of the Popular 'S182 Allows Double Precision Carry
- Offers Carry Functions in a Compatible Form for Direct Connection to the ALU
- Cascadable to Perform Look-Ahead Across n-Bit Adders
- Package Options Include Both Plastic and Ceramic Chip Carriers in Addition to Plastic and Ceramic DIPs
- Dependable Texas Instruments Quality and Reliability

SN54AS282 . . . J PACKAGE
SN74AS282 . . . N PACKAGE
(TOP VIEW)

```
   ___
  G1  [ 1    20 ] VCC
  __          __
  P1  [ 2    19 ] P2
  __          __
  G0  [ 3    18 ] G2
  __
  P0  [ 4    17 ] CnA
  __
  G3  [ 5    16 ] CnB
  __
  P3  [ 6    15 ] Cn+x
  S0  [ 7    14 ] Cn+y
  S1  [ 8    13 ] Cn'
  __          _
  P   [ 9    12 ] G
  GND [ 10   11 ] Cn+z
```

SN54AS282 . . . FH PACKAGE
SN74AS282 . . . FN PACKAGE
(TOP VIEW)

PIN DESIGNATIONS

ALTERNATIVE DESIGNATIONS[†]		FUNCTION
$\overline{G0}, \overline{G1}, \overline{G2}, \overline{G3}$	G0, G1, G2, G3	Carry Generate Inputs
$\overline{P0}, \overline{P1}, \overline{P2}, \overline{P3}$	P0, P1, P2, P3	Carry Propagate Inputs
C_{nA}, C_{nB}	$\overline{C}_{nA}, \overline{C}_{nB}$	Carry Inputs
C_n'	$\overline{C}_n'$	Selected Carry
$C_{n+x}, C_{n+y},$ C_{n+z}	$\overline{C}_{n+x}, \overline{C}_{n+y},$ $\overline{C}_{n+z}$	Carry Outputs
$\overline{G}$	Y	Carry Generate Outputs
$\overline{P}$	X	Carry Propagate Outputs
S0, S1		Carry Select Inputs
VCC		Supply Voltage
GND		Ground

[†] Interpretations are illustrated in connection with the Function Tables for the 'AS181A and 'AS881A.

description

The 'AS282 look-ahead carry generator is capable of anticipating a carry across four binary adders or group of adders. They are cascadable to perform full look-ahead across n-bit adders. The 'AS282 is functionally the same as the SN54AS182/SN74AS182 except that the carry input (C_n) is selected from C_{nA}, C_{nB}, and their complements $\overline{C}_{nA}$ and $\overline{C}_{nB}$. The logic equations are written in terms of the selected carry C_n. This signal is also available as an output at C_n'.

When used in conjunction with the 'AS181A, 'AS881A, or 'AS888 arithmetic logic unit (ALU), this generator provides high-speed carry look-ahead capability for any word length. The 'AS282 generates the look-ahead (anticipated carry) across a group of four ALU's and, in addition, other carry across sections of four look-ahead circuits may be employed to anticipated carry across sections of four look-ahead packages up to n-bits. The method of cascading 'AS282 circuits to perform multi-level look-ahead is illustrated under typical application data.

logic equations

$C_{n+x} = G0 + P0\,C_n$
$C_{n+y} = G1 + P1\,G0 + P1\,P0\,C_n$
$C_{n+z} = G2 + P2\,G1 + P2\,P1\,G0 + P2\,P1\,P0\,C_n$
$\overline{G} = \overline{G3 + P3\,G2 + P3\,P2\,G1 + P3\,P2\,P1\,G0}$
$\overline{P} = \overline{P3\,P2\,P1\,P0}$

$\overline{C}_{n+x} = \overline{Y0\,(X0 + C_n)}$
$\overline{C}_{n+y} = \overline{Y1\,[X1 + Y0\,(X0 + C_n)]}$
$\overline{C}_{n+z} = \overline{Y2\,\{X2 + Y1\,[X1 + Y0\,(X0 + C_n)]\}}$
$Y = Y3\,(X3 + Y2)\,(X3 + X2 + Y1)\,(X3 + X2 + X1 + Y0)$
$X = X3 + X2 + X1 + X0$

Copyright © 1983 by Texas Instruments Incorporated

TEXAS INSTRUMENTS
POST OFFICE BOX 225012 • DALLAS, TEXAS 75265

TYPES SN54AS282, SN74AS282
LOOK-AHEAD CARRY GENERATOR WITH SELECTABLE CARRY INPUTS

FUNCTION TABLE FOR $\overline{G}$ OUTPUT

INPUTS							OUTPUT
$\overline{G}3$	$\overline{G}2$	$\overline{G}1$	$\overline{G}0$	$\overline{P}3$	$\overline{P}2$	$\overline{P}1$	$\overline{G}$
L	X	X	X	X	X	X	L
X	L	X	X	L	X	X	L
X	X	L	X	L	L	X	L
X	X	X	L	L	L	L	L
All other combinations							H

FUNCTION TABLE FOR $\overline{P}$ OUTPUT

INPUTS				OUTPUT
$\overline{P}3$	$\overline{P}2$	$\overline{P}1$	$\overline{P}0$	$\overline{P}$
L	L	L	L	L
All other combinations				H

FUNCTION TABLE FOR $C_n{'}$ OUTPUT

INPUTS		OUTPUT
S1	S0	$C_n{'}$
L	L	C_{nA}
L	H	$\overline{C}_{nA}$
H	L	C_{nB}
H	H	$\overline{C}_{nB}$

FUNCTION TABLE FOR C_{n+x} OUTPUT

INPUTS			OUTPUT
$\overline{G}0$	$\overline{P}0$	$C_n{'}$	C_{n+x}
L	X	X	H
X	L	H	H
All other combinations			L

FUNCTION TABLE C_{n+y} OUTPUT

INPUTS				OUTPUT
$\overline{G}1$	$\overline{G}0$	$\overline{P}1$	$\overline{P}0$	$C_n{'}$	C_{n+y}
L	X	X	X	X	H
X	L	L	X	X	H
X	X	L	L	H	H
All other combinations					L

FUNCTION TABLE FOR C_{n+z} OUTPUT

INPUTS						OUTPUT
$\overline{G}2$	$\overline{G}1$	$\overline{G}0$	$\overline{P}2$	$\overline{P}1$	$\overline{P}0$	$C_n{'}$	C_{n+z}
L	X	X	X	X	X	X	H
X	L	X	L	X	X	X	H
X	X	L	L	L	X	X	H
X	X	X	L	L	L	H	H
All other combinations							L

H = high-level, L = low level, X = irrelevant.
Any inputs not shown in a given table are irrelevant with respect to that output.

logic diagram (positive logic)

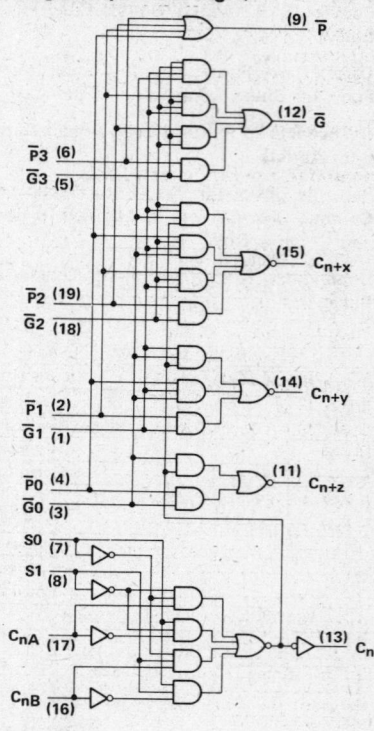

logic symbol

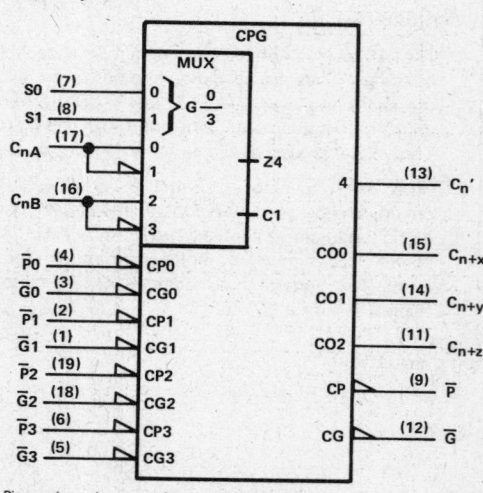

Pin numbers shown are for J and N packages.

Texas Instruments
POST OFFICE BOX 225012 • DALLAS, TEXAS 75265

TYPES SN54AS282, SN74AS282
LOOK-AHEAD CARRY GENERATOR WITH SELECTABLE CARRY INPUTS

absolute maximum ratings over operating free-air temperature range (unless otherwise noted)

Supply voltage, V_{CC} .... 7 V
Input voltage .... 7 V
Operating free-air temperature range: SN54AS282 .... −55°C to 125°C
                                          SN74AS282 .... 0°C to 70°C
Storage temperature range .... −65°C to 150°C

recommended operating conditions

		SN54AS282			SN74AS282			UNIT
		MIN	NOM	MAX	MIN	NOM	MAX	
V_{CC}	Supply voltage	4.5	5	5.5	4.5	5	5.5	V
V_{IH}	High-level input voltage	2			2			V
V_{IL}	Low-level input voltage			0.8			0.8	V
I_{OH}	High-level output current			−2			−2	mA
I_{OL}	Low-level output current			20			20	mA
T_A	Operating free-air temperature	−55		125	0		70	°C

electrical characteristics over recommended operating free-air temperature range (unless otherwise noted)

PARAMETER		TEST CONDITIONS		SN54AS282			SN74AS282			UNIT
				MIN	TYP†	MAX	MIN	TYP†	MAX	
V_{IK}		V_{CC} = 4.5 V,	I_I = −18 mA			−1.2			−1.2	V
V_{OH}		V_{CC} = 4.5 V to 5.5 V,	I_{OH} = −2 mA	V_{CC}−2			V_{CC}−2			V
V_{OL}		V_{CC} = 4.5 V,	I_{OL} = 20 mA		0.3	0.5		0.3	0.5	V
I_I	C_{nA}, C_{nB}	V_{CC} = 5.5 V,	V_I = 7 V			200			200	µA
	S0, S1, $\overline{P3}$					200			200	
	$\overline{P2}$					300			300	
	$\overline{P0}$, $\overline{P1}$, $\overline{G3}$					400			400	
	$\overline{G0}$, $\overline{G2}$					700			700	
	$\overline{G1}$					800			800	
I_{IH}	C_{nA}, C_{nB}	V_{CC} = 5.5 V,	V_I = 2.7 V			40			40	µA
	S0, S1, $\overline{P3}$					40			40	
	$\overline{P2}$					60			60	
	$\overline{P0}$, $\overline{P1}$, $\overline{G3}$					80			80	
	$\overline{G0}$, $\overline{G2}$					140			140	
	$\overline{G1}$					160			160	
I_{IL}	C_{nA}, C_{nB}	V_{CC} = 5.5 V,	V_I = 0.4 V			−1			−1	mA
	S0, S1, $\overline{P3}$					−1			−1	
	$\overline{P2}$					−1.5			−1.5	
	$\overline{P0}$, $\overline{P1}$, $\overline{G3}$					−2			−2	
	$\overline{G0}$, $\overline{G2}$					−3.5			−3.5	
	$\overline{G1}$					−4			−4	
I_O‡		V_{CC} = 5.5 V,	V_O = 2.25 V	−30		−112	−30		−112	mA
I_{CCH}		V_{CC} = 5.5 V			22			22		mA
I_{CCL}					26			26		

†All typical values are at V_{CC} = 5 V, T_A = 25°C.
‡The output conditions have been chosen to produce a current that closely approximates one-half of the true short-circuit current, I_{OS}.

Texas Instruments
POST OFFICE BOX 225012 • DALLAS, TEXAS 75265

TYPES SN54AS282, SN74AS282
LOOK-AHEAD CARRY GENERATOR WITH SELECTABLE CARRY INPUTS

switching characteristics (see Note 1)

PARAMETER	FROM (INPUT)	TO (OUTPUT)	V_{CC} = 4.5 V to 5.5 V, C_L = 50 pF, R_L = 500 Ω, T_A = MIN to MAX						UNIT
			SN54AS282			SN74AS282			
			MIN	TYP†	MAX	MIN	TYP†	MAX	
t_{PLH}	S0, S1, C_{nA}, or C_{nB}	C_n'		6			6		ns
t_{PHL}				6			6		
t_{PLH}	S0, S1, C_{nA}, or C_{nB}	C_{n+x}, C_{n+y}, C_{n+z}		6			6		ns
t_{PHL}				6			6		
t_{PLH}	$\overline{P}$ or $\overline{G}$	C_{n+x}, C_{n+y}, C_{n+z}		5			5		ns
t_{PHL}				5			5		
t_{PLH}	$\overline{P}$ or $\overline{G}$	$\overline{G}$		6			6		ns
t_{PHL}				5			5		
t_{PLH}	$\overline{P}$	$\overline{P}$		5			5		ns
t_{PHL}				5			5		

†All typical values are at V_{CC} = 5 V, T_A = 25°C.
NOTE 1: For load circuit and voltage waveforms, see page 1-12.

TYPICAL APPLICATION DATA

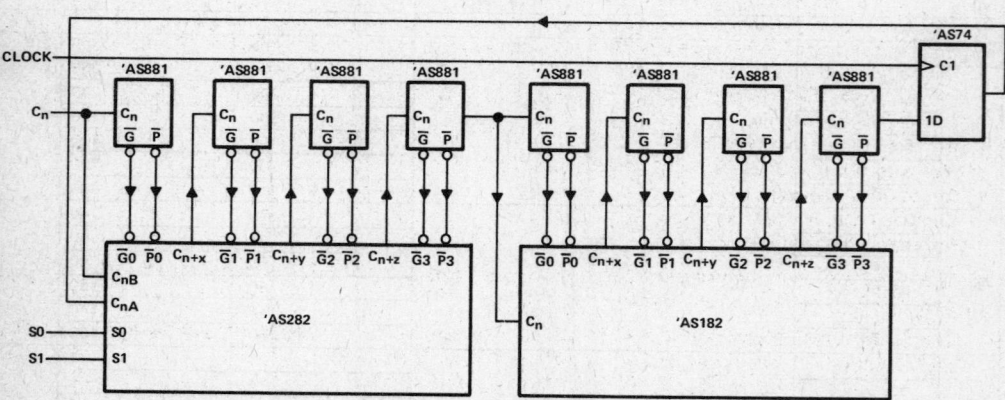

FIGURE 1–32-BIT LOOK-AHEAD CARRY WITH DOUBLE-PRECISION CARRY IN 'AS282 AND 'AS182

TYPES SN54AS286, SN74AS286
9-BIT PARITY GENERATORS/CHECKER WITH BUS DRIVER PARITY I/O PORT

D2809, DECEMBER 1983

- Generates Either Odd or Even Parity for Nine Data Lines
- Cascadable for n-Bits Parity
- Direct Bus Connection for Parity Generation or for Checking by Using the Parity I/O Port
- Glitch-Free Bus During Power Up/Down
- Package Options Include both Plastic and Ceramic Carriers in Addition to Plastic and Ceramic DIPs
- Dependable Texas Instruments Quality and Reliability

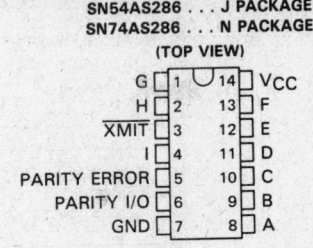

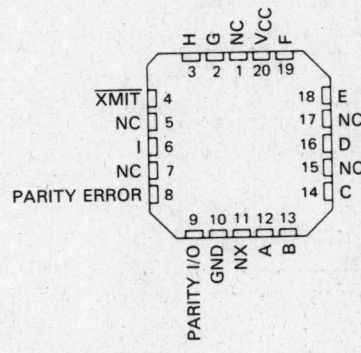

NC — No internal connection

description

The SN54AS286 and SN74AS286 universal nine-bit parity generators/checkers feature a local output for parity checking and a 48-milliampere bus-driving parity I/O port for parity generation/checking. The word-length capability is easily expanded by cascading.

The $\overline{XMIT}$ control input is implemented specifically to accommodate cascading. When $\overline{XMIT}$ is low the parity tree is disabled and PE will remain at a high logic level regardless of the input levels. When $\overline{XMIT}$ is high the parity tree is enabled. The Parity Error output will indicate a parity error when either an even number of inputs (A through I) are high and Parity I/O is forced to a low logic level, or when an odd number of inputs are high and Parity I/O is forced to a high logic level.

The I/O control circuitry was designed so that the I/O port will remain in the high-impedance state during power-up or power-down to prevent bus glitches.

The SN54AS286 is characterized for operation over the full military range of $-55\,°C$ to $125\,°C$. The SN74AS286 is characterized for operation from $0\,°C$ to $70\,°C$.

PRODUCT PREVIEW

This document contains information on a product under development. Texas Instruments reserves the right to change or discontinue this product without notice.

TYPES SN54AS286, SN74AS286
9-BIT PARITY GENERATORS/CHECKER
WITH BUS DRIVER PARITY I/O PORT

FUNCTION TABLE

NUMBER OF INPUTS (A THRU I) THAT ARE HIGH	XMIT	PARITY I/O	PARITY ERROR
0, 2, 4, 6, 8	l	H	H
1, 3, 5, 7, 9	l	L	H
0, 2, 4, 6, 8	h	h	H
	h	l	L
1, 3, 5, 7, 9	h	h	L
	h	l	H

h — high input level l — low input level
H — high output level L — low output level

logic symbol

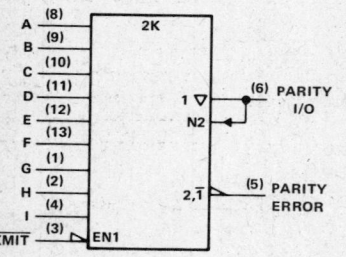

logic diagram (positive logic)

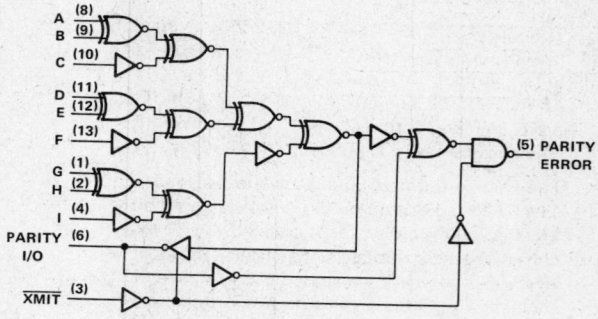

absolute maximum ratings over operating free-air temperature range

Supply voltage, V_{CC} .... 7 V
Input voltage .... 7 V
Voltage applied to a disabled 3-state output .... 5.5 V
Operating free-air temperature range: SN54AS286 .... −55°C to 125°C
                                      SN74AS286 .... 0°C to 70°C
Storage temperature .... −65°C to 140°C

recommended operating conditions

			SN54AS286			SN74AS286			UNIT
			MIN	NOM	MAX	MIN	NOM	MAX	
V_{CC}	Supply voltage		4.5	5	5.5	4.5	5	5.5	V
V_{IH}	High-level input voltage		2			2			V
V_{IL}	Low-level input voltage				0.8			0.8	V
I_{OH}	High-level output current	Parity error			−2			−2	mA
		Parity I/O			−12			−15	
I_{OL}	Low-level output current	Parity error			20			10	mA
		Parity I/O			32			48	
T_A	Operating free-air temperature		−55		125	0		70	°C

TYPES SN54AS286, SN74AS286
9-BIT PARITY GENERATORS/CHECKER
WITH BUS DRIVER PARITY I/O PORT

electrical characteristics over recommended free-air temperature range (unless otherwise noted)

PARAMETER		TEST CONDITIONS		SN54AS286 MIN	SN54AS286 TYP[†]	SN54AS286 MAX	SN74AS286 MIN	SN74AS286 TYP[†]	SN74AS286 MAX	UNIT
V_{IK}		$V_{CC} = 4.5$ V,	$I_I = -18$ mA			-1.2			-1.2	V
V_{OH}	All outputs	$V_{CC} = 4.5$ V to 5.5 V,	$I_{OH} = -2$ mA	$V_{CC}-2$			$V_{CC}-2$			V
		$V_{CC} = 4.5$ V,	$I_{OH} = -3$ mA	2.4	3.2		2.4	3.2		
	Parity I/O	$V_{CC} = 4.5$ V,	$I_{OH} = -12$ mA	2.4						
		$V_{CC} = 4.5$ V,	$I_{OH} = -15$ mA				2.4			
V_{OL}	Parity error	$V_{CC} = 4.5$ V,	$I_{OL} = 20$ mA		0.35	0.5		0.35	0.5	V
	Parity I/O	$V_{CC} = 4.5$ V,	$I_{OL} = 32$ mA			0.5				
		$V_{CC} = 4.5$ V,	$I_{OL} = 48$ mA						0.5	
I_I	Parity I/O	$V_{CC} = 5.5$ V,	$V_I = 5.5$ V			0.1			0.1	mA
	All other inputs	$V_{CC} = 5.5$ V,	$V_I = 7$ V			0.1			0.1	
I_{IH}	Parity I/O[‡]	$V_{CC} = 5.5$ V,	$V_I = 2.7$ V			50			50	μA
	All other inputs					20			20	
I_{IL}	Parity I/O[‡]	$V_{CC} = 5.5$ V,	$V_I = 0.4$ V			-0.5			-0.5	mA
	All other inputs					-0.5			-0.5	
I_O[§]		$V_{CC} = 5.5$ V,	$V_O = 2.25$ V	-30		-112	-30		-112	mA
I_{CC}	Transmit	$V_{CC} = 5.5$ V			29			29		mA
	Receive				34			34		

[†] All typical values are at $V_{CC} = 5$ V, $T_A = 25°C$.
[‡] For I/O ports, the parameters I_{IH} and I_{IL} include the off-state current.
[§] The output conditions have been chosen to produce a current that closely approximates one half of the true short-circuit output current, I_{OS}.

switching characteristics (see Note 1)

PARAMETER	FROM (INPUT)	TO (OUTPUT)	$V_{CC} = 4.5$ V to 5.5 V, $C_L = 50$ pF, R1 = 500 Ω, R2 = 500 Ω, T_A = MIN to MAX						UNIT
			SN54AS286 MIN	SN54AS286 TYP[†]	SN54AS286 MAX	SN74AS286 MIN	SN74AS286 TYP[†]	SN74AS286 MAX	
t_{PLH}	Any A thru I	Parity I/O		8			8		ns
t_{PHL}				8.5			8.5		
t_{PLH}	Any A thru I	Parity error		9			9		ns
t_{PHL}				9.5			9.5		
t_{PLH}	Parity I/O	Parity error		5			5		ns
t_{PHL}				6			6		
t_{PZH}	$\overline{XMIT}$	Parity I/O		5			5		ns
t_{PZL}				6			6		
t_{PHZ}				10			10		
t_{PLZ}				9.5			9.5		

[†] All typical values are at $V_{CC} = 5$ V, $T_A = 25°C$.
NOTE 1: For load circuit and voltage waveforms, see page 1-12.

ALS AND AS CIRCUITS

TEXAS INSTRUMENTS
POST OFFICE BOX 225012 • DALLAS, TEXAS 75265

TYPES SN54AS286, SN74AS286
9-BIT PARITY GENERATORS/CHECKER
WITH BUS DRIVER PARITY I/O PORT

TYPICAL APPLICATION DATA

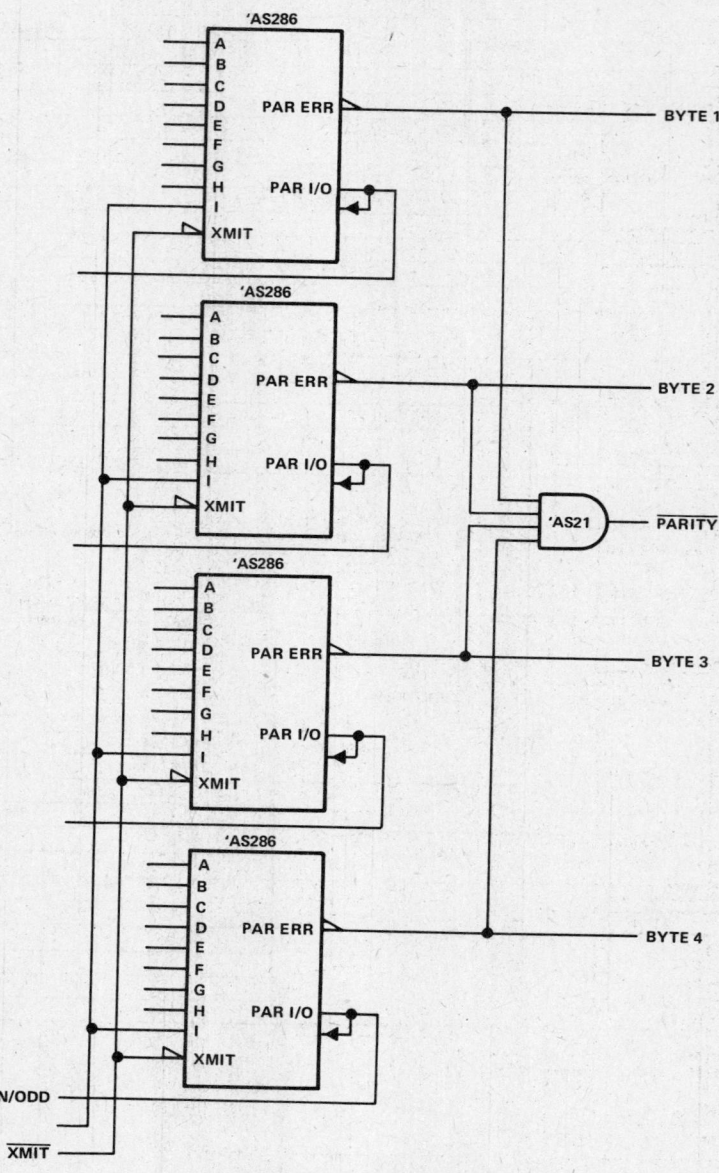

FIGURE 1—32-BIT PARITY GENERATOR/CHECKER

Figure 1 shows a 32-bit parity generator/checker with output polarity-switching, parity error detection, and parity on every byte.

TYPES SN54AS286, SN74AS286
9-BIT PARITY GENERATORS/CHECKER WITH BUS DRIVER PARITY I/O PORT

TYPICAL APPLICATION DATA

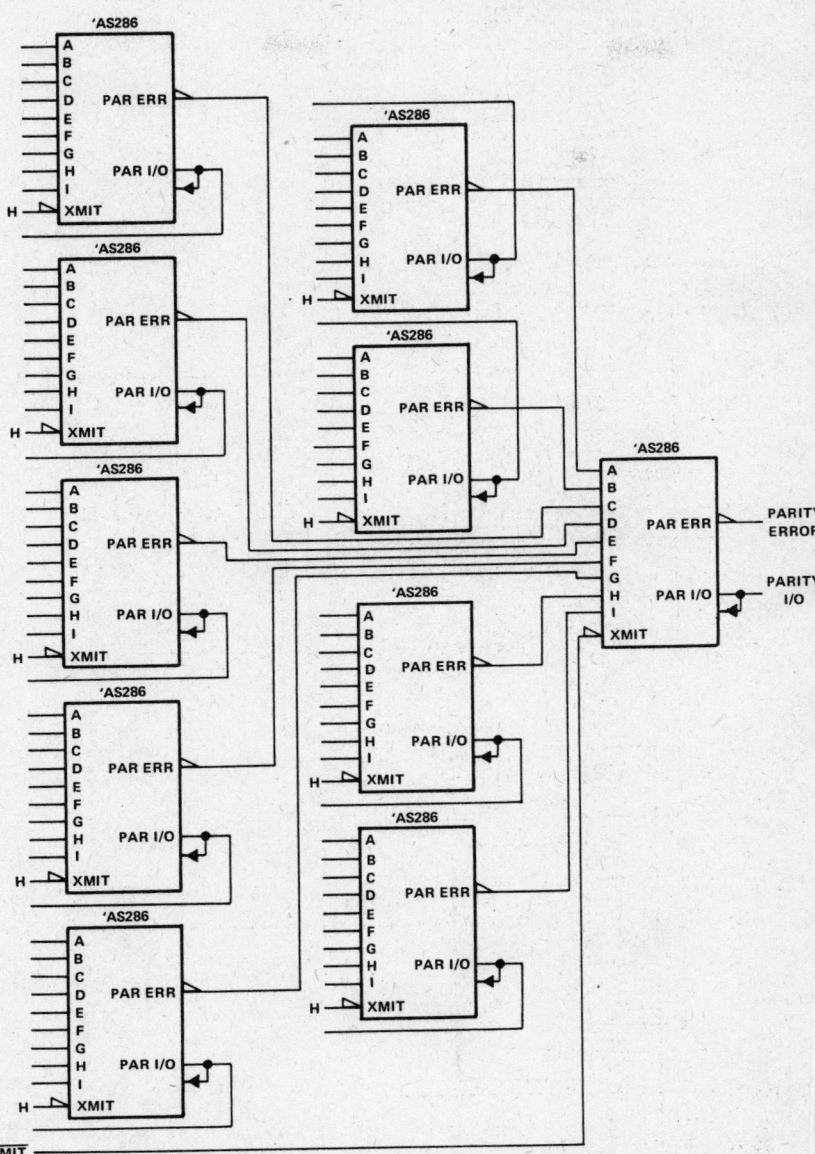

FIGURE 2 – 90-BIT PARITY GENERATOR/CHECKER WITH PARITY ERROR DETECTION

In Figure 2, a 90-bit parity generator/checker with the $\overline{\text{XMIT}}$ on the last stage is available for use with parity detection.

2 ALS AND AS CIRCUITS

TYPES SN54AS298, SN74AS298
QUADRUPLE 2-INPUT MULTIPLEXER WITH STORAGE

D2661, DECEMBER 1983

- Selects One of Two 4-Bit Data Sources and Stores Data Synchronously with System Clock
- Applications:
  Dual Source for Operands and Constants in Arithmetic Processor; Can Release Processor Register Files for Acquiring New Data

  Implements Separate Registers Capable of Parallel Exchange of Contents, yet Retains External Load Capability

  Has Universal-Type Register for Implementing Various Shift Patterns; even Has Compound Left-Right Capability
- Dependable Texas Instruments Quality and Reliability

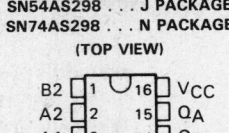

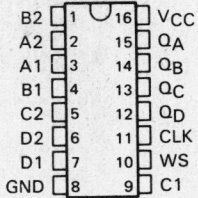

NC—No internal connection

description

This quadruple two-input multiplexer with storage provides essentially the equivalent functional capabilities of two separate MSI functions (SN54AS157/SN74AS157 and SN54AS175/SN74AS175) in a single 16-pin package.

When the word-select (WS) input is low, Word 1 (A1, B1, C1, D1 is applied to the flip-flops. A high input to the word-select (WS) will cause the selection of Word 2 (A2, B2, C2, D2). The selected word is clocked to the output terminals on the negative-going edge of the clock pulse.

The SN54AS298 is characterized for operation over the full military range of −55°C to 125°C. The SN74AS298 is characterized for operation from 0°C to 70°C.

FUNCTION TABLE

INPUTS		OUTPUTS			
WORD SELECT	CLOCK	Q_A	Q_B	Q_C	Q_D
L	↓	a1	b1	c1	d1
H	↓	a2	b2	c2	d2
X	H	Q_{A0}	Q_{B0}	Q_{C0}	Q_{D0}

H = high level (steady state)
L = low level (steady state)
X = irrelevant (any input, including transitions)
↓ = transition from high to low level
a1, a2, etc. = the level of steady-state input at A1, A2, etc.
Q_{A0}, Q_{B0}, etc. = the level of Q_A, Q_B, etc. entered on the most-recent ↓ transition of the clock input.

PRODUCT PREVIEW

This document contains information on a product under development. Texas Instruments reserves the right to change or discontinue this product without notice.

Copyright © 1983 by Texas Instruments Incorporated

TEXAS INSTRUMENTS
POST OFFICE BOX 225012 • DALLAS, TEXAS 75265

TYPES SN54AS298, SN74AS298
QUADRUPLE 2-INPUT MULTIPLEXER WITH STORAGE

logic symbol

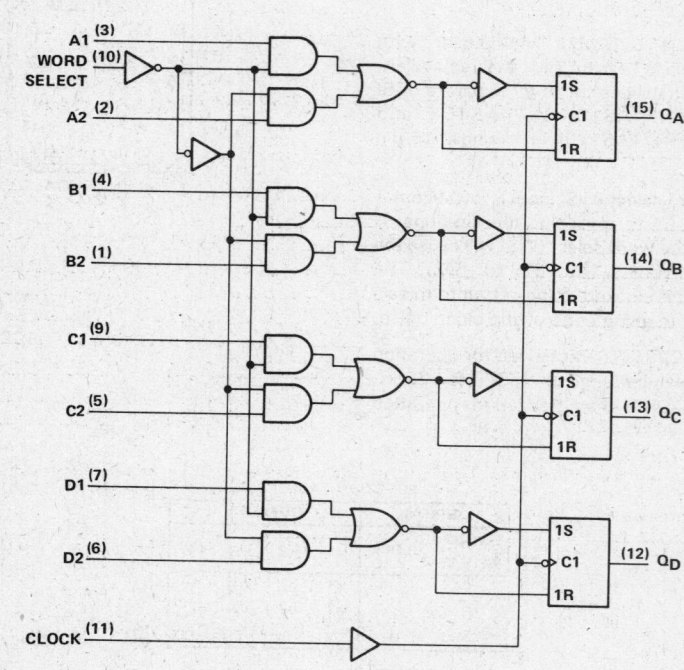

logic diagram (positive logic)

Pin numbers shown are for J and N packages.

TYPES SN54AS298, SN74AS298
QUADRUPLE 2-INPUT MULTIPLEXER WITH STORAGE

absolute maximum ratings over operating free-air temperature range (unless otherwise noted)

Supply voltage, V_{CC} .... 7 V
Input voltage .... 7 V
Operating free-air temperature range: SN54AS298 .... −55°C to 125°C
　　　　　　　　　　　　　　　　　　SN74AS298 .... 0°C to 70°C
Storage temperature range .... −65°C to 150°C

recommended operating conditions

		\multicolumn{3}{c}{SN54AS298}	\multicolumn{3}{c}{SN74AS298}	UNIT
		MIN	NOM	MAX	MIN	NOM	MAX	
V_{CC}	Supply voltage	4.5	5	5.5	4.5	5	5.5	V
V_{IH}	High-level input voltage	2			2			V
V_{IL}	Low-level input voltage			0.8			0.8	V
I_{OH}	High-level output current			−2			−2	mA
I_{OL}	Low-level output current			20			20	mA
f_{clock}	Clock frequency	0			0			MHz
t_w	Pulse duration, CLK high or low							ns
t_{su}	Setup time before CLK ↓							ns
t_h	Hold time after CLK ↓							ns
T_A	Operating free-air temperature	−55		125	0		70	°C

electrical characteristics over recommended operating free-air temperature range (unless otherwise noted)

PARAMETER	TEST CONDITIONS		SN54AS298			SN74AS298			UNIT
			MIN	TYP†	MAX	MIN	TYP†	MAX	
V_{IK}	$V_{CC} = 4.5$ V,	$I_I = -18$ mA			−1			−1	V
V_{OH}	$V_{CC} = 4.5$ V to 5.5 V,	$I_{OH} = -2$ mA	$V_{CC}-2$			$V_{CC}-2$			V
V_{OL}	$V_{CC} = 4.5$ V,	$I_{OL} = 20$ mA		0.35	0.5		0.35	0.5	V
I_I	$V_{CC} = 5.5$ V,	$V_I = 7$ V			0.1			0.1	mA
I_{IH}	$V_{CC} = 5.5$ V,	$V_I = 2.7$ V			20			20	μA
I_{IL}	$V_{CC} = 5.5$ V,	$V_I = 0.4$ V			−0.5			−0.5	mA
I_O‡	$V_{CC} = 5.5$ V,	$V_O = 2.25$ V	−30		−112	−30		−112	mA
I_{CCH}	$V_{CC} = 5.5$ V								mA
I_{CCL}	$V_{CC} = 5.5$ V			23			23		mA

†All typical values are at $V_{CC} = 5$ V, $T_A = 25$°C.
‡The output conditions have been chosen to produce a current that closely approximates one half of the true short-circuit output current, I_{OS}.

switching characteristics (see Note 1)

PARAMETER	FROM (INPUT)	TO (OUTPUT)	$V_{CC} = 4.5$ V to 5.5 V, $C_L = 50$ pF, $R_L = 500$ Ω, T_A = MIN to MAX						UNIT
			SN54AS298			SN74AS298			
			MIN	TYP†	MAX	MIN	TYP†	MAX	
f_{max}	CLK	Q							MHz
t_{PLH}				6.5			6.5		ns
t_{PHL}				7			7		

†All typical values are at $V_{CC} = 5$ V, $T_A = 25$°C.
NOTE 1: For load circuit and voltage waveforms, see page 1-12.

Additional information on these products can be obtained from the factory as it becomes available.

TEXAS INSTRUMENTS
POST OFFICE BOX 225012 • DALLAS, TEXAS 75265

TYPES SN54AS298, SN74AS298
QUADRUPLE 2-INPUT MULTIPLEXER WITH STORAGE

TYPICAL APPLICATION DATA

This versatile multiplexer/register can be connected to operate as a shift register that can shift N-places in a single clock pulse.

The following figure illustrates a BCD shift register that will shift an entire 4-bit BCD digit in one clock pulse.

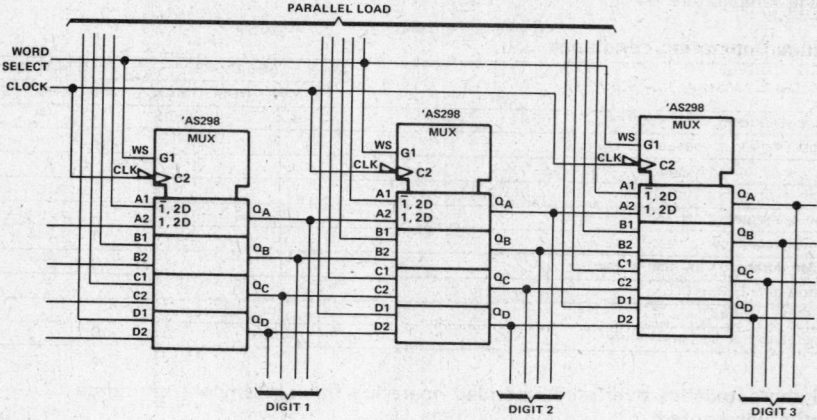

When the word-select input is high and the registers are clocked, the contents of register 1 is transferred (shifted) to register 2 and etc. In effect, the BCD digits are shifted one position. In addition, this application retains a parallel-load capability which means that new BCD data can be entered in the entire register with one clock pulse. This arrangement can be modified to perform the shifting of binary data for any number of bit locations.

Another function that can be implemented with the 'AS298 is a register that can be designed specifically for supporting multiplier or division operations. The example below is a one-place/two-place shift register.

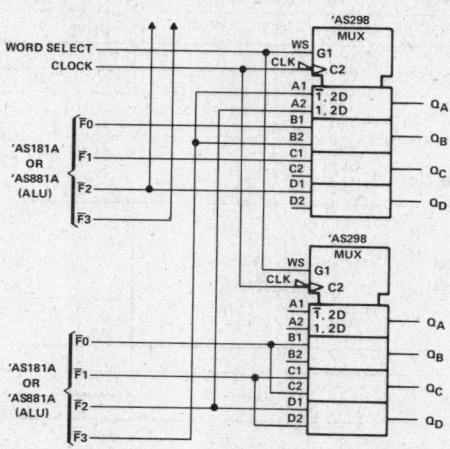

When word select is low and the register is clocked, the outputs of the arithmetic/logic units (ALU's) are shifted one place. When word select is high and the registers are clocked, the data is shifted two places.

2-290

Texas Instruments
POST OFFICE BOX 225012 • DALLAS, TEXAS 75265

TYPES SN54ALS299, SN54ALS323, SN54AS299, SN54AS323, SN74ALS299, SN74ALS323, SN74AS299, SN74AS323
8-BIT UNIVERSAL SHIFT/STORAGE REGISTERS WITH 3-STATE OUTPUTS

D2661, DECEMBER 1982—REVISED DECEMBER 1983

- Multiplexed I/O Ports Provides Improved Bit Density
- Four Modes of Operation: Hold (Store), Shift Right, Shift Left, and Load Data
- Operates with Outputs Enabled or at High Impedance
- 3-State Outputs Drive Bus Lines Directly
- Can Be Cascaded for N-Bit Word Lengths
- 'ALS299 and AS299 Have Direct Overriding Clear
- 'ALS323 and AS323 Have Synchronous Clear
- Application:
  Stacked or Push-Down Registers, Buffer Storage, and Accumulator Registers
- Package Options Include Both Plastic and Ceramic Chip Carriers in Addition to Plastic and Ceramic DIPs
- Dependable Texas Instruments Quality and Reliability

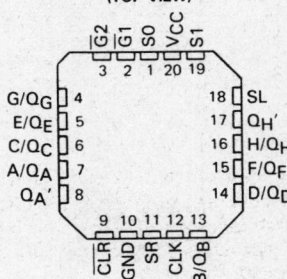

SN54ALS', SN54AS' . . . J PACKAGE
SN74ALS', SN74AS' . . . N PACKAGE
(TOP VIEW)

SN54ALS', SN54AS' . . . FH PACKAGE
SN74ALS', SN74AS' . . . FN PACKAGE
(TOP VIEW)

description

These eight-bit universal registers feature multiplexed I/O ports to achieve full eight-bit data handling in a single 20-pin package. Two function-select inputs and two output-control inputs can be used to choose the modes of operation listed in the function table.

Synchronous parallel loading is accomplished by taking both function-select lines S0 and S1 high. This places the three-state outputs in a high-impedance state and permits data that is applied on the I/O ports to be clocked into the register. Reading out of the register can be accomplished while the outputs are enabled in any mode. Clearing occurs asynchronously on 'ALS299, 'AS299 and synchronously on 'ALS323, 'AS323 when $\overline{CLR}$ is low. Taking either of the output controls, $\overline{G1}$ or $\overline{G2}$, high disables the outputs but this has no effect on clearing, shifting, or storage of data.

The SN54' family is characterized for operation over the full military range of −55°C to 125°C. The SN74' family is characterized for operation from 0°C to 70°C.

TYPES SN54ALS299, SN54ALS323, SN54AS299, SN54AS323, SN74ALS299, SN74ALS323, SN74AS299, SN74AS323
8-BIT UNIVERSAL SHIFT/STORAGE REGISTERS WITH 3-STATE OUTPUTS

FUNCTION TABLE

MODE	INPUTS							I/O PORTS								OUTPUTS	
	$\overline{CLR}$	S1	S0	OUTPUT CONTROL		CLK	SL	SR	A/Q_A	B/Q_B	C/Q_C	D/Q_D	E/Q_E	F/Q_F	G/Q_G	H/Q_H	Q_A'	Q_H'
				$\overline{G1}$	$\overline{G2}$													
Clear	L	X	L	L	L	X	X	X	L	L	L	L	L	L	L	L	L	L
('ALS299)	L	L	X	L	L	X	X	X	L	L	L	L	L	L	L	L	L	L
('AS299)	L	H	H	X	X	X	X	X	X	X	X	X	X	X	X	X	L	L
Clear	L	X	L	L	L	↑	X	X	L	L	L	L	L	L	L	L	L	L
('ALS323)	L	L	X	L	L	↑	X	X	L	L	L	L	L	L	L	L	L	L
('AS323)	L	H	H	X	X	↑	X	X	X	X	X	X	X	X	X	X	L	L
Hold	H	L	L	L	L	X	X	X	Q_{A0}	Q_{B0}	Q_{C0}	Q_{D0}	Q_{E0}	Q_{F0}	Q_{G0}	Q_{H0}	Q_{A0}	Q_{H0}
	H	X	X	L	L	L	X	X	Q_{A0}	Q_{B0}	Q_{C0}	Q_{D0}	Q_{E0}	Q_{F0}	Q_{G0}	Q_{H0}	Q_{A0}	Q_{H0}
Shift Right	H	L	H	L	L	↑	X	H	H	Q_{An}	Q_{Bn}	Q_{Cn}	Q_{Dn}	Q_{En}	Q_{Fn}	Q_{Gn}	H	Q_{Gn}
	H	L	H	L	L	↑	X	L	L	Q_{An}	Q_{Bn}	Q_{Cn}	Q_{Dn}	Q_{En}	Q_{Fn}	Q_{Gn}	L	Q_{Gn}
Shift Left	H	H	L	L	L	↑	H	X	Q_{Bn}	Q_{Cn}	Q_{Dn}	Q_{En}	Q_{Fn}	Q_{Gn}	Q_{Hn}	H	Q_{Bn}	H
	H	H	L	L	L	↑	L	X	Q_{Bn}	Q_{Cn}	Q_{Dn}	Q_{En}	Q_{Fn}	Q_{Gn}	Q_{Hn}	L	Q_{Bn}	L
Load	H	H	H	X	X	↑	X	X	a	b	c	d	e	f	g	h	a	h

When one or both output controls are high the eight input/output terminals are disabled to the high-impedance state; however, sequential operation or clearing of the register is not affected.

logic symbols

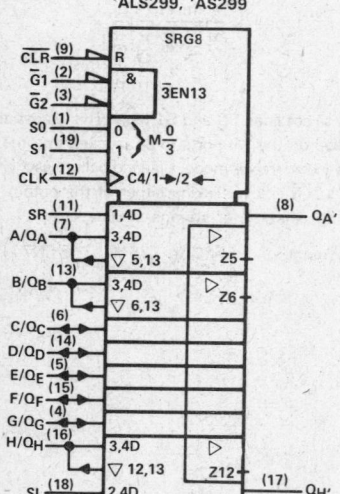

Pin numbers shown are for J and N packages.

TEXAS INSTRUMENTS
POST OFFICE BOX 225012 • DALLAS, TEXAS 75265

TYPES SN54ALS299, SN54ALS323, SN54AS299, SN54AS323, SN74ALS299, SN74ALS323, SN74AS299, SN74AS323
8-BIT UNIVERSAL SHIFT/STORAGE REGISTERS WITH 3-STATE OUTPUTS

logic diagrams (positive logic)

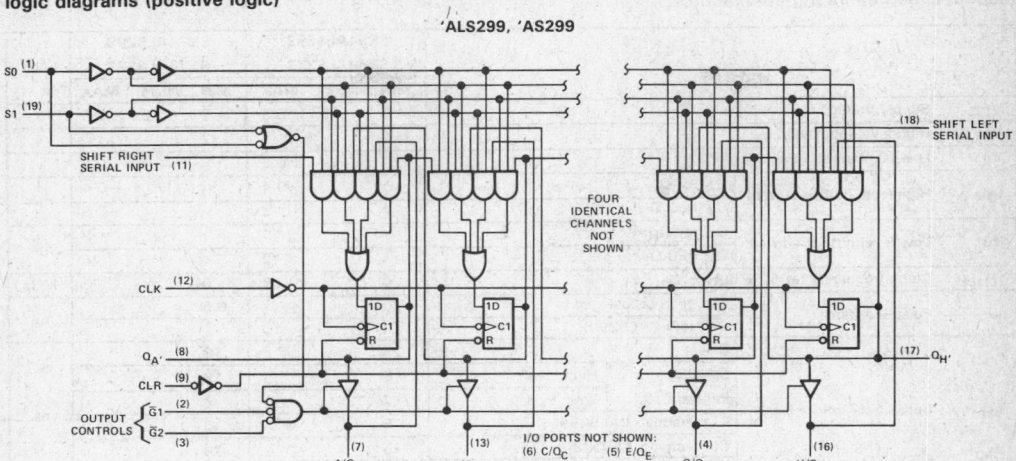

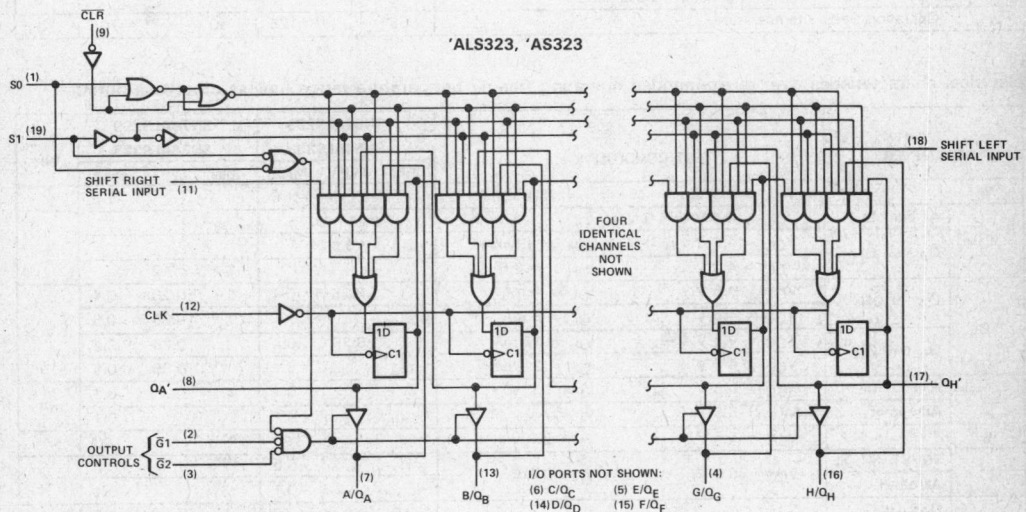

Pin numbers shown are for J and N packages.

absolute maximum ratings over operating free-air temperature range (unless otherwise noted)

Supply voltage, V_{CC} .... 7 V
Input voltage: All inputs .... 7 V
             I/O ports .... 5.5 V
Operating free-air temperature range: SN54ALS', SN54AS' .... −55°C to 125°C
                                 SN74ALS', SN74AS' .... 0°C to 70°C
Storage temperature range .... −65°C to 150°C

ALS AND AS CIRCUITS

Texas Instruments
POST OFFICE BOX 225012 • DALLAS, TEXAS 75265

TYPES SN54ALS299, SN54ALS323, SN74ALS299, SN74ALS323
8-BIT UNIVERSAL SHIFT/STORAGE REGISTERS WITH 3-STATE OUTPUTS

recommended operating conditions

				SN54ALS299 SN54ALS323			SN74ALS299 SN74ALS323			UNIT
				MIN	NOM	MAX	MIN	NOM	MAX	
V_{CC}	Supply voltage			4.5	5	5.5	4.5	5	5.5	V
V_{IH}	High-level input voltage			2			2			V
V_{IL}	Low-level input voltage					0.8			0.8	V
I_{OH}	High-level output current	$Q_{A'}$ or $Q_{H'}$				−0.4			−0.4	mA
		Q_A thru Q_H				−1			−2.6	
I_{OL}	Low-level output current	$Q_{A'}$ or $Q_{H'}$				4			8	mA
		Q_A thru Q_H				12			24	
f_{clock}	Clock frequency (at 50% duty cycle)			0		25	0		30	MHz
t_w	Pulse duration	CLK high or low		20			16.5			ns
		$\overline{CLR}$ low ('ALS299)		10			10			
t_{su}	Setup time before CLK↑	Select		25			20			ns
		Serial or Parallel data	High level	18			16			
			Low level	7			6			
		$\overline{CLR}$ inactive ('ALS299)		15			15			
		$\overline{CLR}$ active ('ALS323)		25			20			
		$\overline{CLR}$ inactive ('ALS323)		18			16			
t_h	Hold time after CLK↑	Select		0			0			ns
		Serial or parallel data		0			0			
T_A	Operating free-air temperature			−55		125	0		70	°C

electrical characteristics over recommended operating free-air temperature range (unless otherwise noted)

PARAMETER		TEST CONDITIONS		SN54ALS299 SN54ALS323			SN74ALS299 SN74ALS323			UNIT
				MIN	TYP†	MAX	MIN	TYP†	MAX	
V_{IK}		V_{CC} = 4.5 V,	I_I = −18 mA			−1.5			−1.5	V
V_{OH}	All outputs	V_{CC} = 4.5 V to 5.5 V,	I_{OH} = −0.4 mA	$V_{CC}-2$			$V_{CC}-2$			V
	Q_A thru Q_H	V_{CC} = 4.5 V,	I_{OH} = −1 mA	2.4	3.3					
		V_{CC} = 4.5 V,	I_{OH} = −2.6 mA				2.4	3.2		
V_{OL}	$Q_{A'}$ or $Q_{H'}$	V_{CC} = 4.5 V,	I_{OL} = 4 mA		0.25	0.4		0.25	0.4	V
		V_{CC} = 4.5 V,	I_{OL} = 8 mA					0.35	0.5	
	Q_A thru Q_H	V_{CC} = 4.5 V,	I_{OL} = 12 mA		0.25	0.4		0.25	0.4	
		V_{CC} = 4.5 V,	I_{OL} = 24 mA					0.35	0.5	
I_I	A thru H	V_{CC} = 5.5 V,	V_I = 5.5 V			0.1			0.1	mA
	Any other	V_{CC} = 5.5 V,	V_I = 7 V			0.1			0.1	
I_{IH}‡		V_{CC} = 5.5 V,	V_I = 2.7 V			20			20	µA
I_{IL}‡	S0, S1, SR, SL	V_{CC} = 5.5 V,	V_I = 0.4 V			−0.2			−0.2	mA
	All others					−0.1			−0.1	
I_O§	$Q_{A'}$, $Q_{H'}$	V_{CC} = 5.5 V,	V_O = 2.25 V	−15		−70	−15		−70	mA
	Q_A thru Q_H			−30		−112	−30		−112	
I_{CC}		V_{CC} = 5.5 V	Outputs high		15	28		15	28	mA
			Outputs low		22	38		22	38	
			Outputs disabled		23	40		23	40	

†All typical values are at V_{CC} = 5 V, T_A = 25°C.
‡For I/O ports (Q_A through Q_H), the parameters I_{IH} and I_{IL} include the off-state output current.
§The output conditions have been chosen to produce a current that closely approximates one half of the true short-circuit output current, I_{OS}.

Texas Instruments
POST OFFICE BOX 225012 • DALLAS, TEXAS 75265

TYPES SN54ALS299, SN54ALS323, SN74ALS299, SN74ALS323
8-BIT UNIVERSAL SHIFT/STORAGE REGISTERS WITH 3-STATE OUTPUTS

switching characteristics (see Note 1)

PARAMETER	FROM (INPUT)	TO (OUTPUT)	V_{CC} = 4.5 V to 5.5 V, C_L = 50 pF, $R1$ = 500 Ω, $R2$ = 500 Ω, T_A = MIN to MAX				UNIT
			SN54ALS299 SN54ALS323		SN74ALS299 SN74ALS323		
			MIN	MAX	MIN	MAX	
f_{max}			25		30		MHz
t_{PLH}	CLK	Q_A thru Q_H	4	15	4	13	ns
t_{PHL}			7	25	7	19	
t_{PLH}	CLK	Q_A' or Q_H'	5	20	5	15	ns
t_{PHL}			8	21	8	18	
t_{PHL}	$\overline{CLR}$ ('ALS299 only)	Q_A thru Q_H	6	29	6	22	ns
		Q_A' or Q_H'	6	29	6	22	
t_{PZH}	$\overline{G1}, \overline{G2}$	Q_A thru Q_H	6	21	6	16	ns
t_{PZL}			8	26	8	22	
t_{PZH}	S0, S1	Q_A thru Q_H	7	21	7	17	ns
t_{PZL}			8	26	8	22	
t_{PHZ}	$\overline{G1}, \overline{G2}$	Q_A thru Q_H	1	10	1	8	ns
t_{PLZ}			5	23	5	15	
t_{PHZ}	S0, S1	Q_A thru Q_H	1	16	1	12	ns
t_{PLZ}			8	30	8	25	

NOTE 1: For load circuit and voltage waveforms, see page 1-12.

TYPES SN54AS299, SN54AS323, SN74AS299, SN74AS323
8-BIT UNIVERSAL SHIFT/STORAGE REGISTERS WITH 3-STATE OUTPUTS

recommended operating conditions

			SN54AS299 SN54AS323			SN74AS299 SN74AS323			UNIT
			MIN	NOM	MAX	MIN	NOM	MAX	
V_{CC}	Supply voltage		4.5	5	5.5	4.5	5	5.5	V
V_{IH}	High-level input voltage		2			2			V
V_{IL}	Low-level input voltage				0.8			0.8	V
I_{OH}	High-level output current	Q_A' or Q_H'			-2			-2	mA
		Q_A thru Q_H			-12			-15	
I_{OL}	Low-level output current	Q_A' or Q_H'			20			20	mA
		Q_A thru Q_H			32			48	
f_{clock}	Clock frequency (at 50% duty cycle)								MHz
t_w	Pulse duration	CLK high or low							ns
		$\overline{CLR}$ low ('AS299)							
t_{su}	Setup time before CLK↑	Select							ns
		Serial or Parallel data — High level							
		Serial or Parallel data — Low level							
		$\overline{CLR}$ inactive ('AS299)							
		$\overline{CLR}$ active ('AS323)							
		$\overline{CLR}$ inactive ('AS323)							
t_h	Hold time after CLK↑	Select							ns
		Serial or parallel data							
T_A	Operating free-air temperature		-55		125	0		70	°C

electrical characteristics over recommended operating free-air temperature range (unless otherwise noted)

PARAMETER		TEST CONDITIONS		SN54AS299 SN54AS323			SN74AS299 SN74AS323			UNIT
				MIN	TYP†	MAX	MIN	TYP†	MAX	
V_{IK}		$V_{CC} = 4.5$ V,	$I_I = -18$ mA			-1.2			-1.2	V
V_{OH}	All outputs	$V_{CC} = 4.5$ V to 5.5 V,	$I_{OH} = -2$ mA	$V_{CC}-2$			$V_{CC}-2$			V
	Q_A thru Q_H	$V_{CC} = 4.5$ V,	$I_{OH} = -12$ mA	2.4	3.2					
		$V_{CC} = 4.5$ V,	$I_{OH} = -15$ mA				2.4	3.2		
V_{OL}	Q_A' or Q_H'	$V_{CC} = 4.5$ V,	$I_{OL} = 20$ mA		0.25	0.5		0.25	0.5	V
	Q_A thru Q_H	$V_{CC} = 4.5$ V,	$I_{OL} = 32$ mA		0.25	0.5				
		$V_{CC} = 4.5$ V,	$I_{OL} = 48$ mA					0.35	0.5	
I_I	A thru H	$V_{CC} = 5.5$ V,	$V_I = 5.5$ V							mA
	Any other	$V_{CC} = 5.5$ V,	$V_I = 7$ V							
I_{IH}‡		$V_{CC} = 5.5$ V,	$V_I = 2.7$ V							µA
I_{IL}‡		$V_{CC} = 5.5$ V,	$V_I = 0.4$ V							mA
I_O§		$V_{CC} = 5.5$ V,	$V_O = 2.25$ V	-30		-112	-30		-112	mA
I_{CC}		$V_{CC} = 5.5$ V	Outputs high							mA
			Outputs low							
			Outputs disabled		95			95		

†All typical values are at $V_{CC} = 5$ V, $T_A = 25$°C.
‡For I/O ports (Q_A through Q_H), the parameters I_{IH} and I_{IL} include the off-state output current.
§The output conditions have been chosen to produce a current that closely approximates one half of the true short-circuit output current, I_{OS}.

PRODUCT PREVIEW
This page contains information on a product under development. Texas Instruments reserves the right to change or discontinue this product without notice.

Texas Instruments
POST OFFICE BOX 225012 • DALLAS, TEXAS 75265

TYPES SN54AS299, SN54AS323, SN74AS299, SN74AS323
8-BIT UNIVERSAL SHIFT/STORAGE REGISTERS WITH 3-STATE OUTPUTS

switching characteristics (see Note 1)

PARAMETER	FROM (INPUT)	TO (OUTPUT)	V_{CC} = 4.5 V to 5.5 V, C_L = 50 pF, $R1$ = 500 Ω, $R2$ = 500 Ω, T_A = MIN to MAX						UNIT
			SN54AS299 SN54AS323			SN74AS299 SN74AS323			
			MIN	TYP†	MAX	MIN	TYP†	MAX	
f_{max}									MHz
t_{PLH}	CLK	Q_A thru Q_H		10			10		ns
t_{PHL}				10			10		
t_{PLH}	CLK	Q_A' or Q_H'		10			10		ns
t_{PHL}				10			10		
t_{PHL}	$\overline{CLR}$	Q_A thru Q_H		12			12		ns
		Q_A' or Q_H'		12			12		
t_{PZH}	$\overline{G1}, \overline{G2}$	Q_A thru Q_H		10			10		ns
t_{PZL}				10			10		
t_{PZH}	S0, S1	Q_A thru Q_H		10			10		ns
t_{PZL}				10			10		
t_{PHZ}	$\overline{G1}, \overline{G2}$	Q_A thru Q_H		7			7		ns
t_{PLZ}				7			7		
t_{PHZ}	S0, S1	Q_A thru Q_H		7			7		ns
t_{PLZ}				7			7		

†All typical values are at V_{CC} = 5 V, T_A = 25°C.
NOTE 1: For load circuit and voltage waveforms, see page 1-12.

PRODUCT PREVIEW
This page contains information on a product under development. Texas Instruments reserves the right to change or discontinue this product without notice.

TEXAS INSTRUMENTS
POST OFFICE BOX 225012 • DALLAS, TEXAS 75265

2
ALS AND AS CIRCUITS

TYPES SN54ALS323, SN54AS323, SN74ALS323, SN74AS323
8-BIT UNIVERSAL SHIFT/STORAGE REGISTERS WITH 3-STATE OUTPUTS

D2661, DECEMBER 1982—REVISED DECEMBER 1983

- Multiplexed I/O Ports Provides Improved Bit Density
- Four Modes of Operation: Hold (Store), Shift Right, Shift Left, and Load Data
- Operates with Outputs Enabled or at High Impedance
- 3-State Outputs Drive Bus Lines Directly
- Can Be Cascaded for N-Bit Word Lengths
- 'ALS323 and 'AS323 Have Synchronous Clear
- Application:
   Stacked or Push-Down Registers, Buffer Storage, and Accumulator Registers
- Package Options Include Both Plastic and Ceramic Chip Carriers in Addition to Plastic and Ceramic DIPs
- Dependable Texas Instruments Quality and Reliability

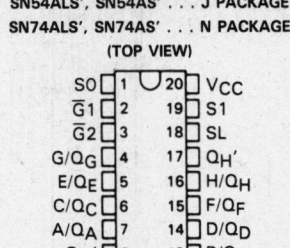

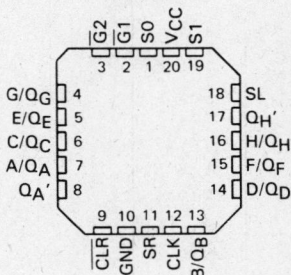

For complete information on the SN54ALS323, SN54AS323, SN74ALS323, SN74AS323, see page

ALS AND AS CIRCUITS

Copyright © 1982 by Texas Instruments Incorporated

2-299

2 ALS AND AS CIRCUITS

TYPES SN54ALS352, SN54AS352, SN74ALS352, SN74AS352
DUAL 4-LINE TO 1-LINE DATA SELECTORS/MULTIPLEXERS

D2661, APRIL 1982—REVISED DECEMBER 1983

- Inverting Versions of 'ALS153 and 'AS153
- Permits Multiplexing from N Lines to 1 Line
- Performs Parallel-to-Serial Conversion
- Strobe (Enable) Line Provided for Cascading (N Lines to n Lines)
- Typical 'ALS352 Power per Multiplexer . . . 16 mW
- Typical 'AS352 Average Propagation Delay Times
  Data Input to Output . . . 2.7 ns
  Strobe Input to Output . . . 4.5 ns
  Select Input to Output . . . 4.5 ns
- Fully Compatible with Most TTL Circuits
- Package Options Include Both Plastic and Ceramic Chip Carriers in Addition to Plastic and Ceramic DIPs
- Dependable Texas Instruments Quality and Reliability

SN54ALS352, SN54AS352 . . . J PACKAGE
SN74ALS352, SN74AS352 . . . N PACKAGE
(TOP VIEW)

SN54ALS352, SN54AS352 . . . FH PACKAGE
SN74ALS352, SN74AS352 . . . FN PACKAGE
(TOP VIEW)

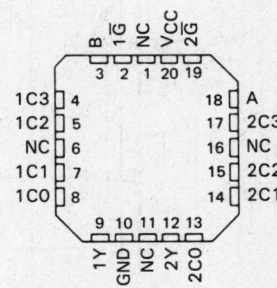

description

Each of these data selectors/multiplexers contains inverters and drivers to supply fully complementary binary decoding data selection to the AND-OR-invert gates. Separate strobe inputs ($\overline{G}$) are provided for each of the two four-line sections.

The SN54ALS352 and SN54AS352 are characterized for operation over the full military temperature range of $-55\,°C$ to $125\,°C$. The SN74ALS352 and SN74AS352 are characterized for operation from $0\,°C$ to $70\,°C$.

NC—No internal connection

logic symbol

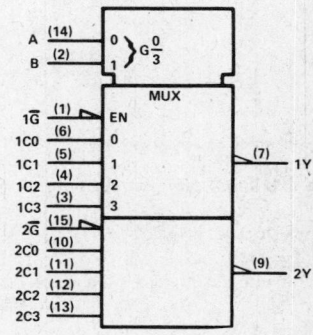

Pin numbers shown are for J and N packages.

FUNCTION TABLE

SELECT INPUTS		DATA INPUTS				STROBE	OUTPUT
B	A	C0	C1	C2	C3	$\overline{G}$	Y
X	X	X	X	X	X	H	H
L	L	L	X	X	X	L	H
L	L	H	X	X	X	L	L
L	H	X	L	X	X	L	H
L	H	X	H	X	X	L	L
H	L	X	X	L	X	L	H
H	L	X	X	H	X	L	L
H	H	X	X	X	L	L	H
H	H	X	X	X	H	L	L

Select inputs A and B are common to both sections.

Copyright © 1982 by Texas Instruments Incorporated

TEXAS INSTRUMENTS
POST OFFICE BOX 225012 • DALLAS, TEXAS 75265

TYPE SN54ALS352, SN54AS352, SN74ALS352, SN74AS352
DUAL 4-LINE TO 1-LINE DATA SELECTORS/MULTIPLEXERS

logic diagram (positive logic)

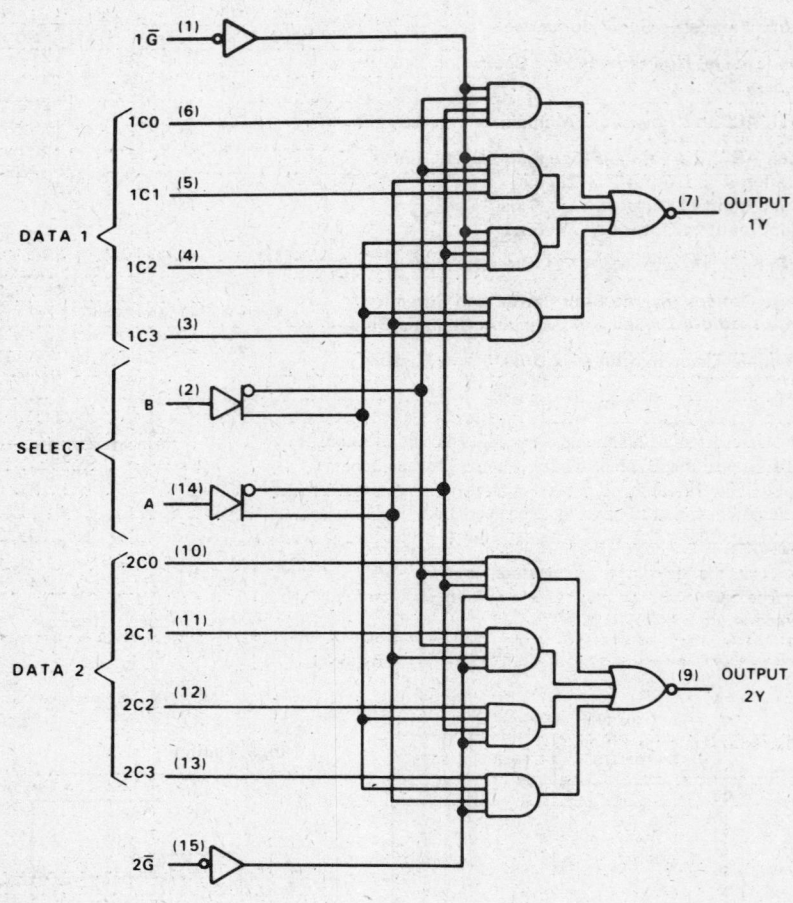

Pin numbers shown are for J and N packages.

absolute maximum ratings over operating free-air temperature range (unless otherwise noted)

Supply voltage, V_{CC} .... 7 V
Input voltage .... 7 V
Operating free-air temperature range: SN54ALS352, SN54AS352 .... −55 °C to 125 °C
                               SN74ALS352, SN74AS352 .... 0 °C to 70 °C
Storage temperature range .... −65 °C to 150 °C

TEXAS INSTRUMENTS
POST OFFICE BOX 225012 • DALLAS, TEXAS 75265

TYPE SN54ALS352, SN74ALS352
DUAL 4-LINE TO 1-LINE DATA SELECTORS/MULTIPLEXERS

recommended operating conditions

		SN54ALS352			SN74ALS352			UNIT
		MIN	NOM	MAX	MIN	NOM	MAX	
V_{CC}	Supply voltage	4.5	5	5.5	4.5	5	5.5	V
V_{IH}	High-level input voltage	2			2			V
V_{IL}	Low-level input voltage			0.8			0.8	V
I_{OH}	High-level output current			−1			−2.6	mA
I_{OL}	Low-level output current			12			24	mA
T_A	Operating free-air temperature	−55		125	0		70	°C

electrical characteristics over recommended operating free-air temperature range (unless otherwise noted)

PARAMETER	TEST CONDITIONS		SN54ALS352			SN74ALS352			UNIT
			MIN	TYP†	MAX	MIN	TYP†	MAX	
V_{IK}	V_{CC} = 4.5 V,	I_I = −18 mA			−1.5			−1.5	V
V_{OH}	V_{CC} = 4.5 V to 5.5 V,	I_{OH} = −0.4 mA	$V_{CC}-2$			$V_{CC}-2$			V
	V_{CC} = 4.5 V,	I_{OH} = −1 mA	2.4	3.3					
	V_{CC} = 4.5 V,	I_{OH} = −2.6 mA				2.4	3.2		
V_{OL}	V_{CC} = 4.5 V,	I_{OL} = 12 mA		0.25	0.4		0.25	0.4	V
	V_{CC} = 4.5 V,	I_{OL} = 24 mA					0.35	0.5	
I_I	V_{CC} = 5.5 V,	V_I = 7 V			0.1			0.1	mA
I_{IH}	V_{CC} = 5.5 V,	V_I = 2.7 V			20			20	µA
I_{IL}	V_{CC} = 5.5 V,	V_I = 0.4 V			−0.1			−0.1	mA
I_O‡	V_{CC} = 5.5 V,	V_O = 2.25 V	−30		−112	−30		−112	mA
I_{CC}	V_{CC} = 5.5 V,	See Note 1		6.5	10		6.5	10	mA

†All typical values are at V_{CC} = 5 V, T_A = 25°C.
‡The output conditions have been chosen to produce a current that closely approximates one half of the true short-circuit output current, I_{OS}.
NOTE 1: I_{CC} is measured with data and select inputs at 4.5 V, and G inputs grounded.

switching characteristics (see Note 2)

PARAMETER	FROM (INPUT)	TO (OUTPUT)	V_{CC} = 4.5 V to 5.5 V, C_L = 50 pF, R_L = 500 Ω, T_A = MIN to MAX				UNIT
			SN54ALS352		SN74ALS352		
			MIN	MAX	MIN	MAX	
t_{PLH}	A or B	Y	5	28	5	24	ns
t_{PHL}			5	24	5	21	
t_{PLH}	Data (Any C)	Y	3	21	3	18	ns
t_{PHL}			2	15	2	13	
t_{PLH}	$\overline{G}$	Y	4	22	4	18	ns
t_{PHL}			4	24	4	20	

NOTE 2: For load circuit and voltage waveforms, see page 1-12.

ALS AND AS CIRCUITS

TEXAS INSTRUMENTS
POST OFFICE BOX 225012 • DALLAS, TEXAS 75265

TYPE SN54AS352, SN74AS352
DUAL 4-LINE TO 1-LINE DATA SELECTORS/MULTIPLEXERS

recommended operating conditions

		SN54AS352 MIN	NOM	MAX	SN74AS352 MIN	NOM	MAX	UNIT
V_{CC}	Supply voltage	4.5	5	5.5	4.5	5	5.5	V
V_{IH}	High-level input voltage	2			2			V
V_{IL}	Low-level input voltage			0.8			0.8	V
I_{OH}	High-level output current			-12			-15	mA
I_{OL}	Low-level output current			32			48	mA
T_A	Operating free-air temperature	-55		125	0		70	°C

electrical characteristics over recommended operating free-air temperature range (unless otherwise noted)

PARAMETER		TEST CONDITIONS		SN54AS352 MIN	TYP†	MAX	SN74AS352 MIN	TYP†	MAX	UNIT
V_{IK}		$V_{CC} = 4.5$ V,	$I_I = -18$ mA			-1.2			-1.2	V
V_{OH}		$V_{CC} = 4.5$ V to 5.5 V,	$I_{OH} = -2$ mA	$V_{CC}-2$			$V_{CC}-2$			V
		$V_{CC} = 4.5$ V,	$I_{OH} = -12$ mA	2.4	3.2					
		$V_{CC} = 4.5$ V,	$I_{OH} = -15$ mA				2.4	3.3		
V_{OL}		$V_{CC} = 4.5$ V,	$I_{OL} = 32$ mA		0.25	0.5				V
		$V_{CC} = 4.5$ V,	$I_{OL} = 48$ mA					0.35	0.5	
I_I	A, B	$V_{CC} = 5.5$ V,	$V_I = 7$ V			0.2			0.2	mA
	All others					0.1			0.1	
I_{IH}	A, B	$V_{CC} = 5.5$ V,	$V_I = 2.7$ V			40			40	µA
	All others					20			20	
I_{IL}	A, B	$V_{CC} = 5.5$ V,	$V_I = 0.4$ V			-1			-1	mA
	All others					-0.5			-0.5	
I_O‡		$V_{CC} = 5.5$ V,	$V_O = 2.25$ V	-30		-112	-30		-112	mA
I_{CC}		$V_{CC} = 5.5$ V	Outputs high		15.5	25		15.5	25	mA
			Outputs low		17.5	28		17.5	28	

†All typical values are at $V_{CC} = 5$ V, $T_A = 25$°C.
‡The output conditions have been chosen to produce a current that closely approximates one half of the true short-circuit output current, I_{OS}.

switching characteristics (see Note 1)

PARAMETER	FROM (INPUT)	TO (OUTPUT)	$V_{CC} = 4.5$ V to 5.5 V, $C_L = 50$ pF, $R_L = 500$ Ω, $T_A =$ MIN to MAX				UNIT
			SN54AS352 MIN	MAX	SN74AS352 MIN	MAX	
t_{PLH}	A or B	Y	4	12.5	4	11	ns
t_{PHL}			4	14	4	13	
t_{PLH}	Data (Any C)	Y	2	7.5	2	6.5	ns
t_{PHL}			2	7	2	6	
t_{PLH}	$\overline{G}$	Y	3	8	3	7	ns
t_{PHL}			4	13.5	4	12	

NOTE 1: For load circuit and voltage waveforms, see page 1-12.

TYPES SN54ALS353, SN54AS353, SN74ALS353, SN74AS353
DUAL 1 OF 4 DATA SELECTORS/MULTIPLEXERS WITH 3-STATE OUTPUTS

D2661, APRIL 1982–REVISED DECEMBER 1983

- Inverting Versions of 'ALS253 and 'AS253
- Permits Multiplexing from N Lines to 1 Line
- Performs Parallel-to-Serial Conversion
- Typical 'ALS353 Power per Multiplexer . . . 20 mW
- Fully Compatible with Most TTL Circuits
- Package Options Include Both Plastic and Ceramic Chip Carriers in Addition to Plastic and Ceramic DIPs
- Dependable Texas Instruments Quality and Reliability

description

Each of these data selectors/multiplexers contains inverters and drivers to supply full binary decoding data selection to the AND-OR-invert gates. Separate strobe inputs ($\overline{G}$) are provided for each of the two four-line sections.

The three-state outputs can interface with and drive data lines of bus-organized systems. With all but one of the common outputs disabled (at a high-impedance state) the low-impedance of the single enabled output will drive the bus line to a high or low logic level. Each output has its own strobe ($\overline{G}$). The output is disabled when its strobe is high.

The SN54ALS353 and SN54AS353 are characterized for operation over the full military temperature range of $-55\,°C$ to $125\,°C$. The SN74ALS353 and SN74AS353 are characterized for operation from $0\,°C$ to $70\,°C$.

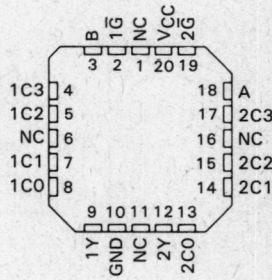

NC—No internal connection

FUNCTION TABLE

SELECT INPUTS		DATA INPUTS				OUTPUT CONTROL	OUTPUT
B	A	C0	C1	C2	C3	$\overline{G}$	Y
X	X	X	X	X	X	H	Z
L	L	L	X	X	X	L	H
L	L	H	X	X	X	L	L
L	H	X	L	X	X	L	H
L	H	X	H	X	X	L	L
H	L	X	X	L	X	L	H
H	L	X	X	H	X	L	L
H	H	X	X	X	L	L	H
H	H	X	X	X	H	L	L

Select inputs A and B are common to both sections.

logic symbol

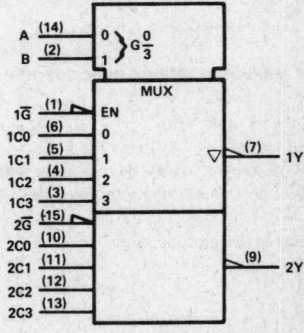

Pin numbers shown are for J and N packages.

Copyright © 1982 by Texas Instruments Incorporated

TYPES SN54ALS353, SN54AS353, SN74ALS353, SN74AS353
DUAL 1 OF 4 DATA SELECTORS/MULTIPLEXERS WITH 3-STATE OUTPUTS

logic diagram (positive logic)

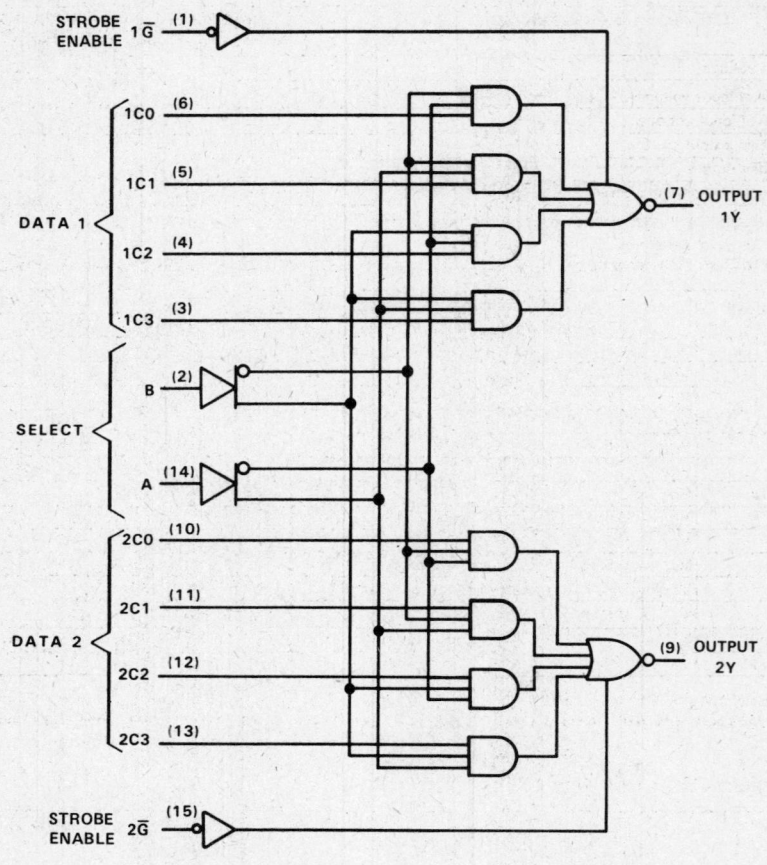

Pin numbers shown are for J and N packages.

absolute maximum ratings over operating free-air temperature range (unless otherwise noted)

Supply voltage, V_{CC} .... 7 V
Input voltage .... 7 V
Voltage applied to a disabled 3-state output .... 5.5 V
Operating free-air temperature range: SN54ALS353, SN54AS353 .... −55 °C to 125 °C
                                      SN74ALS353, SN74AS353 .... 0 °C to 70 °C
Storage temperature range .... −65 °C to 150 °C

Texas Instruments
POST OFFICE BOX 225012 • DALLAS, TEXAS 75265

TYPES SN54ALS353, SN74ALS353
DUAL 1 OF 4 DATA SELECTORS/MULTIPLEXERS
WITH 3-STATE OUTPUTS

recommended operating conditions

		SN54ALS353			SN74ALS353			UNIT
		MIN	NOM	MAX	MIN	NOM	MAX	
V_{CC}	Supply voltage	4.5	5	5.5	4.5	5	5.5	V
V_{IH}	High-level input voltage	2			2			V
V_{IL}	Low-level input voltage			0.8			0.8	V
I_{OH}	High-level output current			−1			−2.6	mA
I_{OL}	Low-level output current			12			24	mA
T_A	Operating free-air temperature	−55		125	0		70	°C

electrical characteristics over recommended operating free-air temperature range (unless otherwise noted)

PARAMETER		TEST CONDITIONS		SN54ALS353			SN74ALS353			UNIT
				MIN	TYP[†]	MAX	MIN	TYP[†]	MAX	
V_{IK}		V_{CC} = 4.5 V,	I_I = −18 mA			−1.5			−1.5	V
V_{OH}		V_{CC} = 4.5 V to 5.5 V,	I_{OH} = −0.4 mA	V_{CC}−2			V_{CC}−2			V
		V_{CC} = 4.5 V,	I_{OH} = −1 mA	2.4	3.3					
		V_{CC} = 4.5 V,	I_{OH} = −2.6 mA				2.4	3.2		
V_{OL}		V_{CC} = 4.5 V,	I_{OL} = 12 mA		0.25	0.4		0.25	0.4	V
		V_{CC} = 4.5 V,	I_{OL} = 24 mA					0.35	0.5	
I_{OZH}		V_{CC} = 5.5 V,	V_O = 2.7 V			20			20	µA
I_{OZL}		V_{CC} = 5.5 V,	V_O = 0.4 V,			−20			−20	µA
I_I		V_{CC} = 5.5 V,	V_I = 7 V			0.1			0.1	mA
I_{IH}		V_{CC} = 5.5 V,	V_I = 2.7 V			20			20	µA
I_{IL}		V_{CC} = 5.5 V,	V_I = 0.4 V			−0.1			−0.1	mA
I_O[‡]		V_{CC} = 5.5 V,	V_O = 2.25 V	−30		−112	−30		−112	mA
I_{CC}	disabled	V_{CC} = 5.5 V	All inputs, at 4.5 V		8	13		8	13	mA
	enabled		All inputs at Gnd		7	12		7	12	

[†] All typical values are at V_{CC} = 5 V, T_A = 25°C.
[‡] The output conditions have been chosen to produce a current that closely approximates one half of the true short-circuit output current, I_{OS}.

switching characteristics (see Note 1)

PARAMETER	FROM (INPUT)	TO (OUTPUT)	V_{CC} = 4.5 V to 5.5 V, C_L = 50 pF, R1 = 500 Ω, R2 = 500 Ω, T_A = MIN to MAX				UNIT
			SN54ALS353		SN74ALS353		
			MIN	MAX	MIN	MAX	
t_{PLH}	A or B	Y	5	28	5	24	ns
t_{PHL}			5	24	5	21	
t_{PLH}	Data (Any C)	Y	4	21	4	18	ns
t_{PHL}			3	15	3	13	
t_{PZH}	$\overline{G}$	Y	3	15	3	13	ns
t_{PZL}			3	19	2	16	
t_{PHZ}	$\overline{G}$	Y	2	12	2	10	ns
t_{PLZ}			2	16	2	14	

NOTE 1: For load circuit and voltage waveforms, see page 1-12.

TEXAS INSTRUMENTS
POST OFFICE BOX 225012 • DALLAS, TEXAS 75265

TYPES SN54AS353, SN74AS353
DUAL 1 OF 4 DATA SELECTORS/MULTIPLEXERS
WITH 3-STATE OUTPUTS

recommended operating conditions

		SN54AS353			SN74AS353			UNIT
		MIN	NOM	MAX	MIN	NOM	MAX	
V_{CC}	Supply voltage	4.5	5	5.5	4.5	5	5.5	V
V_{IH}	High-level input voltage	2			2			V
V_{IL}	Low-level input voltage			0.8			0.8	V
I_{OH}	High-level output current			−12			−15	mA
I_{OL}	Low-level output current			32			48	mA
T_A	Operating free-air temperature	−55		125	0		70	°C

electrical characteristics over recommended operating free-air temperature range (unless otherwise noted)

PARAMETER		TEST CONDITIONS		SN54AS353			SN74AS353			UNIT
				MIN	TYP†	MAX	MIN	TYP†	MAX	
V_{IK}		$V_{CC} = 4.5$ V,	$I_I = -18$ mA			−1.2			−1.2	V
V_{OH}		$V_{CC} = 4.5$ V to 5.5 V,	$I_{OH} = -2$ mA	$V_{CC}-2$			$V_{CC}-2$			V
		$V_{CC} = 4.5$ V,	$I_{OH} = -12$ mA	2.4	3.2					
		$V_{CC} = 4.5$ V,	$I_{OH} = -15$ mA				2.4	3.3		
V_{OL}		$V_{CC} = 4.5$ V,	$I_{OL} = 32$ mA		0.25	0.5				V
		$V_{CC} = 4.5$ V,	$I_{OL} = 48$ mA					0.35	0.5	
I_{OZH}		$V_{CC} = 5.5$ V,	$V_O = 2.7$ V			50			50	µA
I_{OZL}		$V_{CC} = 5.5$ V,	$V_O = 0.4$ V,			−50			−50	µA
I_I	A, B	$V_{CC} = 5.5$ V,	$V_I = 7$ V			0.2			0.2	mA
	All others					0.1			0.1	
I_{IH}	A, B	$V_{CC} = 5.5$ V,	$V_I = 2.7$ V			40			40	µA
	All others					20			20	
I_{IL}	A, B	$V_{CC} = 5.5$ V,	$V_I = 0.4$ V			−1			−1	mA
	All others					−0.5			−0.5	
I_O‡		$V_{CC} = 5.5$ V,	$V_O = 2.25$ V	−30		−112	−30		−112	mA
I_{CC}		$V_{CC} = 5.5$ V	Outputs high		15	24		15	24	mA
			Outputs low		19	31		19	31	
			Outputs disabled		18	30		18	30	

†All typical values are at $V_{CC} = 5$ V, $T_A = 25$°C.
‡The output conditions have been chosen to produce a current that closely approximates one half of the true short-circuit output current, I_{OS}.

switching characteristics (see Note 1)

PARAMETER	FROM (INPUT)	TO (OUTPUT)	$V_{CC} = 4.5$ V to 5.5 V, $C_L = 50$ pF, R1 = 500 Ω, R2 = 500 Ω, T_A = MIN to MAX				UNIT
			SN54AS353		SN74AS353		
			MIN	MAX	MIN	MAX	
t_{PLH}	A or B	Y	3	10	3	9	ns
t_{PHL}			4	14	4	12	
t_{PLH}	Data (Any C)	Y	3	8.5	3	7.5	ns
t_{PHL}			2	6.5	2	6	
t_{PZH}	Strobe	Y	3	8.5	3	7.5	ns
t_{PZL}			4	12	4	11	
t_{PHZ}	Strobe	Y	2	6.5	2	5.5	ns
t_{PLZ}			3	9	3	7.5	

NOTE 1: For load circuit and voltage waveforms, see page 1-12.

TYPES SN54ALS365 THRU SN54ALS368, SN74ALS365 THRU SN74ALS368
HEX BUS DRIVERS WITH 3-STATE OUTPUTS

D2661, DECEMBER 1982 – REVISED DECEMBER 1983

- 3-State Outputs Drive Bus Lines Or Buffer Memory Address Registers
- Choice of True or Inverting Outputs
- Package Options Include Both Plastic and Ceramic Chip Carriers in Addition to Plastic and Ceramic DIPs
- Dependable Texas Instruments Quality and Reliability

| 'ALS365, 'ALS367 | True Outputs |
| 'ALS366, 'ALS368 | Inverting Outputs |

description

These Hex buffers and line drivers are designed specifically to improve both the performance and density of three state memory address drivers, clock drivers, and bus oriented receivers and transmitters. The designer has a choice of selected combinations of inverting and noninverting outputs, symmetrical $\overline{G}$ (active-low control) inputs.

These devices feature high fan-out, and improved fan-in. The SN74ALS365 through SN74ALS368 can be used to drive terminated lines down to 133 ohms.

The -1 versions of the SN74ALS' parts are identical to the standard versions except that the recommended maximum I_{OL} is increased to 48 milliamperes. There are no -1 versions of the SN54ALS' parts.

The SN54' family is characterized for operation over the full military temperature range of −55 °C to 125 °C. The SN74' family is characterized for operation from 0 °C to 70 °C.

SN54ALS365, SN54ALS366 . . . J PACKAGE
SN74ALS365, SN74ALS366 . . . N PACKAGE
(TOP VIEW)

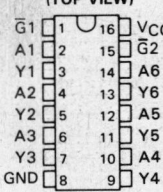

SN54ALS365, SN54ALS366 . . . FH PACKAGE
SN74ALS365, SN74ALS366 . . . FN PACKAGE
(TOP VIEW)

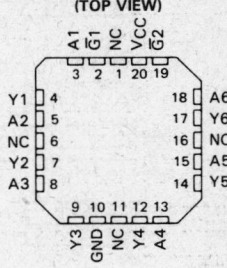

SN54ALS367, SN54ALS368 . . . J PACKAGE
SN74ALS367, SN74ALS368 . . . N PACKAGE
(TOP VIEW)

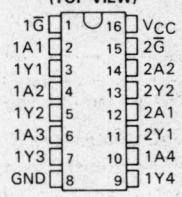

SN54ALS367, SN54ALS368 . . . FH PACKAGE
SN74ALS367, SN74ALS368 . . . FN PACKAGE
(TOP VIEW)

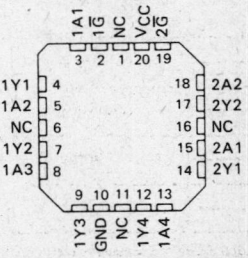

ALS AND AS CIRCUITS

ADVANCE INFORMATION
This document contains information on a new product. Specifications are subject to change without notice.

Copyright © 1982 by Texas Instruments Incorporated

TEXAS INSTRUMENTS
POST OFFICE BOX 225012 • DALLAS, TEXAS 75265

TYPES SN54ALS365 THRU SN54ALS368, SN74ALS365 THRU SN74ALS368
HEX BUS DRIVERS WITH 3-STATE OUTPUTS

logic symbols

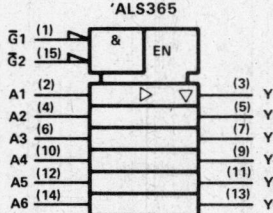
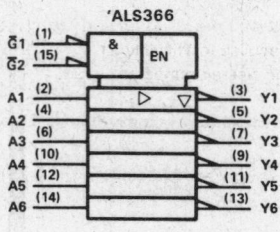
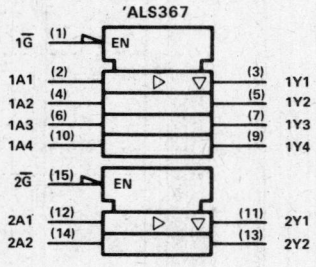
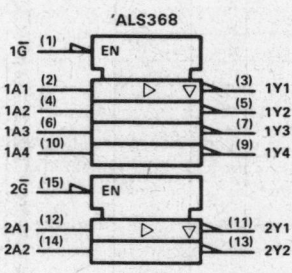

logic diagrams (positive logic)

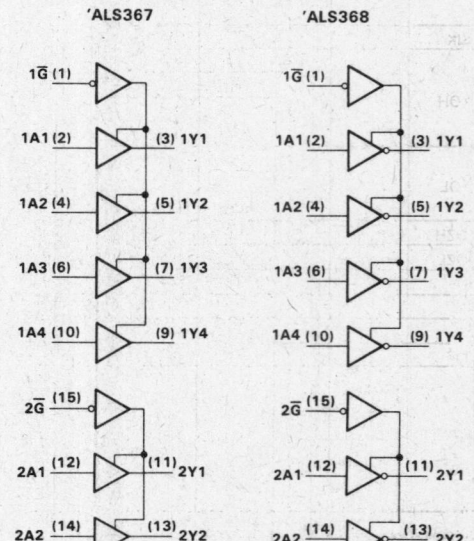

Pin numbers shown are for J and N packages.

TEXAS INSTRUMENTS
POST OFFICE BOX 225012 • DALLAS, TEXAS 75265

TYPES SN54ALS365 THRU SN54ALS368, SN74ALS365 THRU SN74ALS368
HEX BUS DRIVERS WITH 3-STATE OUTPUTS

absolute maximum ratings over operating free-air temperature range (unless otherwise noted)

Supply voltage, V_{CC} .... 7 V
Input voltage .... 7 V
Voltage applied to a disabled 3-state output .... 5.5 V
Operating free-air temperature range: SN54ALS365 thru SN54ALS368 .... −55 °C to 125 °C
                                    SN74ALS365 thru SN74ALS368 .... 0 °C to 70 °C
Storage temperature range .... −65 °C to 150 °C

recommended operating conditions

		SN54ALS365 THRU SN54ALS368			SN74ALS365 THRU SN74ALS368			UNIT
		MIN	NOM	MAX	MIN	NOM	MAX	
V_{CC}	Supply voltage	4.5	5	5.5	4.5	5	5.5	V
V_{IH}	High-level input voltage	2			2			V
V_{IL}	Low-level input voltage			0.8			0.8	V
I_{OH}	High-level output current			−12			−15	mA
I_{OL}	Low-level output current			12			24 48†	mA
T_A	Operating free-air temperature	−55		125	0		70	°C

†The extended limits apply only if V_{CC} is maintained between 4.75 V and 5.25 V.
  The 48-mA limit applies for the SN74ALS365-1 thru SN74ALS368-1 only.

electrical characteristics over recommended operating free-air temperature range (unless otherwise noted)

PARAMETER		TEST CONDITIONS		SN54ALS365 THRU SN54ALS368			SN74ALS365 THRU SN74ALS368			UNIT
				MIN	TYP‡	MAX	MIN	TYP‡	MAX	
V_{IK}		V_{CC} = 4.5 V,	I_I = −18 mA			−1.5			−1.5	V
V_{OH}		V_{CC} = 4.5 V to 5.5 V,	I_{OH} = −0.4 mA	V_{CC} − 2			V_{CC} − 2			V
		V_{CC} = 4.5 V,	I_{OH} = −3 mA	2.4	3.2		2.4	3.2		
		V_{CC} = 4.5 V,	I_{OH} = −12 mA	2						
		V_{CC} = 4.5 V,	I_{OH} = −15 mA				2			
V_{OL}		V_{CC} = 4.5 V,	I_{OL} = 12 mA		0.25	0.4		0.25	0.4	V
		V_{CC} = 4.5 V,	I_{OL} = 24 mA					0.35	0.5	
		(I_{OL} = 48 mA for −1 versions)								
I_{OZH}		V_{CC} = 5.5 V,	V_O = 2.7 V			20			20	µA
I_{OZL}		V_{CC} = 5.5 V,	V_O = 0.4 V			−20			−20	µA
I_I		V_{CC} = 5.5 V,	V_I = 7 V			0.1			0.1	mA
I_{IH}		V_{CC} = 5.5 V,	V_I = 2.7 V			20			20	µA
I_{IL}		V_{CC} = 5.5 V,	V_I = 0.4 V			−0.1			−0.1	mA
I_O §		V_{CC} = 5.5 V,	V_O = 2.25 V	−30		−112	−30		−112	mA
I_{CC}	'ALS365 'ALS367	V_{CC} = 5.5 V	Outputs high		7			7		mA
			Outputs low		12			12		
			Outputs disabled		13			13		
	'ALS366 'ALS368		Outputs high		3			3		
			Outputs low		10			10		
			Outputs disabled		11			11		

‡All typical values are at V_{CC} = 5 V, T_A = 25 °C.
§The output conditions have been chosen to produce a current that closely approximates one half of the true short-circuit output current, I_{OS}.

TYPES SN54ALS365 THRU SN54ALS368, SN74ALS365 THRU SN74ALS368
HEX BUS DRIVERS WITH 3-STATE OUTPUTS

'ALS365, 'ALS367 switching characteristics (see Note 1)

PARAMETER	FROM (INPUT)	TO (OUTPUT)	V_{CC} = 4.5 V to 5.5 V, C_L = 50 pF, R1 = 500 Ω, R2 = 500 Ω, T_A = MIN to MAX						UNIT
			SN54ALS365 SN54ALS367			SN74ALS365 SN74ALS367			
			MIN	TYP†	MAX	MIN	TYP†	MAX	
t_{PLH}	A	Y		7			7		ns
t_{PHL}				7			7		
t_{PZH}	$\overline{G}$	Y		14			14		ns
t_{PZL}				14			14		
t_{PHZ}	$\overline{G}$	Y		5			5		ns
t_{PLZ}				8			8		

'ALS366, 'ALS368 switching characteristics (see Note 1)

PARAMETER	FROM (INPUT)	TO (OUTPUT)	V_{CC} = 4.5 V to 5.5 V, C_L = 50 pF, R1 = 500 Ω, R2 = 500 Ω, T_A = MIN to MAX						UNIT
			SN54ALS366 SN54ALS368			SN74ALS366 SN74ALS368			
			MIN	TYP†	MAX	MIN	TYP†	MAX	
t_{PLH}	A	Y		6			6		ns
t_{PHL}				5			5		
t_{PZH}	$\overline{G}$	Y		10			10		ns
t_{PZL}				17			17		
t_{PHZ}	$\overline{G}$	Y		6			6		ns
t_{PLZ}				6			6		

†All typical values are at V_{CC} = 5 V, T_A = 25°C.
NOTE 1: For load circuit and voltage waveforms, see page 1-12.

TYPES SN54ALS373, SN54AS373, SN74ALS373, SN74AS373
OCTAL D-TYPE TRANSPARENT LATCHES WITH 3-STATE OUTPUTS

D2661, APRIL 1982—REVISED DECEMBER 1983

- 8 Latches in a Single Package
- 3-State Bus-Driving True Outputs
- Full Parallel Access for Loading
- Buffered Control Inputs
- P-N-P Inputs Reduce D-C Loading on Data Lines
- Package Options Include Both Plastic and Ceramic Chip Carriers in Addition to Plastic and Ceramic DIPs
- Dependable Texas Instruments Quality and Reliability

SN54ALS373, SN54AS373 . . . J PACKAGE
SN74ALS373, SN74AS373 . . . N PACKAGE
(TOP VIEW)

```
 OC  [ 1   U  20 ] VCC
 1Q  [ 2      19 ] 8Q
 1D  [ 3      18 ] 8D
 2D  [ 4      17 ] 7D
 2Q  [ 5      16 ] 7Q
 3Q  [ 6      15 ] 6Q
 3D  [ 7      14 ] 6D
 4D  [ 8      13 ] 5D
 4Q  [ 9      12 ] 5Q
 GND [10      11 ] C
```

SN54ALS373, SN54AS373 . . . FH PACKAGE
SN74ALS373, SN74AS373 . . . FN PACKAGE
(TOP VIEW)

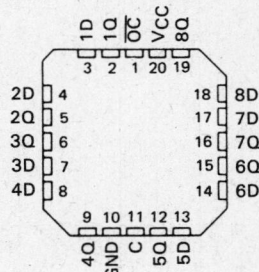

description

These 8-bit latches feature three-state outputs designed specifically for driving highly capacitive or relatively low-impedance loads. They are particularly suitable for implementing buffer registers, I/O ports, bidirectional bus drivers, and working registers.

The eight latches of the 'ALS373 and 'AS373 are transparent D-type latches. While the enable (C) is high the Q outputs will follow the data (D) inputs. When the enable is taken low, the Q outputs will be latched at the levels that were set up at the D inputs.

A buffered output-control input ($\overline{OC}$) can be used to place the eight outputs in either a normal logic state (high or low logic levels) or a high-impedance state. In the high-impedance state the outputs neither load nor drive the bus lines significantly. The high-impedance third state and increased drive provide the capability to drive the bus lines in a bus-organized system without need for interface or pull-up components.

The output control $\overline{OC}$ does not affect the internal operations of the latches. Old data can be retained or new data can be entered while the outputs are off.

The SN54ALS373 and SN54AS373 are characterized for operation over the full military temperature range of $-55\,°C$ to $125\,°C$. The SN74ALS373 and SN74AS373 are characterized for operation from $0\,°C$ to $70\,°C$.

FUNCTION TABLE (EACH LATCH)

INPUTS			OUTPUT
$\overline{OC}$	ENABLE C	D	Q
L	H	H	H
L	H	L	L
L	L	X	Q_0
H	X	X	Z

Copyright © 1982 by Texas Instruments Incorporated

TEXAS INSTRUMENTS
POST OFFICE BOX 225012 • DALLAS, TEXAS 75265

TYPES SN54ALS373, SN54AS373, SN74ALS373, SN74AS373
OCTAL D-TYPE TRANSPARENT LATCHES WITH 3-STATE OUTPUTS

logic symbol

logic diagram (positive logic)

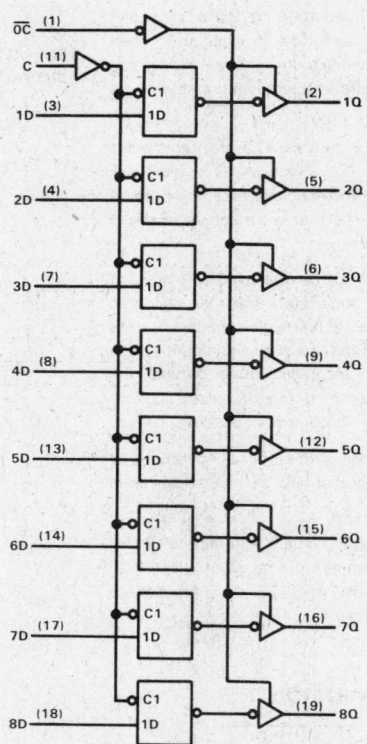

Pin numbers shown are for J and N packages.

absolute maximum ratings over operating free-air temperature range (unless otherwise noted)

Supply voltage, V_{CC} .... 7 V
Input voltage .... 7 V
Voltage applied to a disabled 3-state output .... 5.5 V
Operating free-air temperature range: SN54ALS373, SN54AS373 .... −55 °C to 125 °C
                              SN74ALS373, SN74AS373 .... 0 °C to 70 °C
Storage temperature range .... −65 °C to 150 °C

Texas Instruments
POST OFFICE BOX 225012 • DALLAS, TEXAS 75265

TYPES SN54ALS373, SN74ALS373
OCTAL D-TYPE TRANSPARENT LATCHES WITH 3-STATE OUTPUTS

recommended operating conditions

		SN54ALS373			SN74ALS373			UNIT
		MIN	NOM	MAX	MIN	NOM	MAX	
V_{CC}	Supply voltage	4.5	5	5.5	4.5	5	5.5	V
V_{IH}	High-level input voltage	2			2			V
V_{IL}	Low-level input voltage			0.8			0.8	V
I_{OH}	High-level output current			-1			-2.6	mA
I_{OL}	Low-level output current			12			24	mA
t_w	Pulse duration, enable C high	10			10			ns
t_{su}	Setup time, data before enable C↓	10			10			ns
t_h	Hold time, data after enable C↓	7			7			ns
T_A	Operating free-air temperature	-55		125	0		70	°C

electrical characteristics over recommended operating free-air temperature range (unless otherwise noted)

PARAMETER	TEST CONDITIONS		SN54ALS373			SN74ALS373			UNIT
			MIN	TYP†	MAX	MIN	TYP†	MAX	
V_{IK}	$V_{CC} = 4.5$ V,	$I_I = -18$ mA			-1.5			-1.5	V
V_{OH}	$V_{CC} = 4.5$ V to 5.5 V,	$I_{OH} = -0.4$ mA	$V_{CC}-2$			$V_{CC}-2$			V
	$V_{CC} = 4.5$ V,	$I_{OH} = -1$ mA	2.4	3.3					
	$V_{CC} = 4.5$ V,	$I_{OH} = -2.6$ mA				2.4	3.2		
V_{OL}	$V_{CC} = 4.5$ V,	$I_{OL} = 12$ mA		0.25	0.4		0.25	0.4	V
	$V_{CC} = 4.5$ V,	$I_{OL} = 24$ mA					0.35	0.5	
I_{OZH}	$V_{CC} = 5.5$ V,	$V_O = 2.7$ V			20			20	µA
I_{OZL}	$V_{CC} = 5.5$ V,	$V_O = 0.4$ V			-20			-20	µA
I_I	$V_{CC} = 5.5$ V,	$V_I = 7$ V			0.1			0.1	mA
I_{IH}	$V_{CC} = 5.5$ V,	$V_I = 2.7$ V			20			20	µA
I_{IL}	$V_{CC} = 5.5$ V,	$V_I = 0.4$ V			-0.1			-0.1	mA
I_O‡	$V_{CC} = 5.5$ V,	$V_O = 2.25$ V	-30		-112	-30		-112	mA
I_{CC}	$V_{CC} = 5.5$ V	Outputs high		9	16		9	16	mA
		Outputs low		16	25		16	25	
		Outputs disabled		17	27		17	27	

†All typical values are at $V_{CC} = 5$ V, $T_A = 25°C$.
‡The output conditions have been chosen to produce a current that closely approximates one half of the true short-circuit output current, I_{OS}.

TEXAS
INSTRUMENTS
POST OFFICE BOX 225012 • DALLAS, TEXAS 75265

TYPES SN54ALS373, SN74ALS373
OCTAL D-TYPE TRANSPARENT LATCHES WITH 3-STATE OUTPUTS

switching characteristics (see Note 1)

PARAMETER	FROM (INPUT)	TO (OUTPUT)	V_{CC} = 4.5 V to 5.5 V, C_L = 50 pF, R1 = 500 Ω, R2 = 500 Ω, T_A = MIN to MAX				UNIT
			SN54ALS373		SN74ALS373		
			MIN	MAX	MIN	MAX	
t_{PLH}	D	Q	2	14	2	12	ns
t_{PHL}			4	19	4	16	
t_{PLH}	C	Any Q	6	26	6	22	ns
t_{PHL}			7	27	7	23	
t_{PZH}	$\overline{OC}$	Any Q	5	24	5	20	ns
t_{PZL}			6	22	6	18	
t_{PHZ}	$\overline{OC}$	Any Q	2	16	2	12	ns
t_{PLZ}			2	12	2	10	

NOTE 1: For load circuit and voltage waveforms, see page 1-12.

D latch signal conventions

It is TI practice to name the outputs and other inputs of a D-type latch and to draw its logic symbol based on the assumption of true data (D) inputs. Then outputs that produce data in phase with the data inputs are called Q and those producing complementary data are called $\overline{Q}$. An input that causes a Q output to go high or a $\overline{Q}$ output to go low is called Preset; an input that causes a $\overline{Q}$ output to go high or a Q output to go low is called Clear. Bars are used over these pin names ($\overline{PRE}$ and $\overline{CLR}$) if they are active low.

In some applications it may be advantageous to redesignate the data input $\overline{D}$. In that case all the other inputs and outputs should be renamed as shown below. Also shown are corresponding changes in the graphical symbol. Arbitrary pin numbers are shown in parentheses.

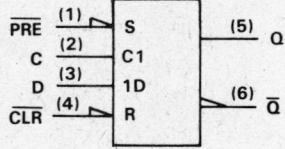 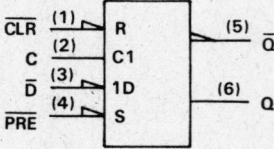

Notice that Q and $\overline{Q}$ exchange names, which causes Preset and Clear to do likewise. Also notice that the polarity indicators (▷) on $\overline{PRE}$ and $\overline{CLR}$ remain since these inputs are still active-low, but that the presence or absence of the polarity indicator changes at $\overline{D}$, Q, and $\overline{Q}$. Of course pin 5 ($\overline{Q}$) is still in phase with the data input $\overline{D}$, but now both are considered active-low.

TYPES SN54AS373, SN74AS373
OCTAL D-TYPE TRANSPARENT LATCHES WITH 3-STATE OUTPUTS

recommended operating conditions

		SN54AS373			SN74AS373			UNIT
		MIN	NOM	MAX	MIN	NOM	MAX	
V_{CC}	Supply voltage	4.5	5	5.5	4.5	5	5.5	V
V_{IH}	High-level input voltage	2			2			V
V_{IL}	Low-level input voltage			0.8			0.8	V
I_{OH}	High-level output current			−12			−15	mA
I_{OL}	Low-level output current			32			48	mA
t_w	Pulse duration, enable C high	5.5			4.5			ns
t_{su}	Setup time, data before enable C↓	2			2			ns
t_h	Hold time, data after enable C↓	3			3			ns
T_A	Operating free-air temperature	−55		125	0		70	°C

electrical characteristics over recommended operating free-air temperature range (unless otherwise noted)

PARAMETER	TEST CONDITIONS		SN54AS373			SN74AS373			UNIT
			MIN	TYP†	MAX	MIN	TYP†	MAX	
V_{IK}	$V_{CC} = 4.5$ V,	$I_I = -18$ mA			−1.2			−1.2	V
V_{OH}	$V_{CC} = 4.5$ V to 5.5 V,	$I_{OH} = -2$ mA	$V_{CC}-2$			$V_{CC}-2$			V
	$V_{CC} = 4.5$ V,	$I_{OH} = -12$ mA	2.4	3.2					
	$V_{CC} = 4.5$ V,	$I_{OH} = -15$ mA				2.4	3.3		
V_{OL}	$V_{CC} = 4.5$ V,	$I_{OL} = 32$ mA		0.27	0.5				V
	$V_{CC} = 4.5$ V,	$I_{OL} = 48$ mA					0.32	0.5	
I_{OZH}	$V_{CC} = 5.5$ V,	$V_O = 2.7$ V			50			50	µA
I_{OZL}	$V_{CC} = 5.5$ V,	$V_O = 0.4$ V			−50			−50	µA
I_I	$V_{CC} = 5.5$ V,	$V_I = 7$ V			0.1			0.1	mA
I_{IH}	$V_{CC} = 5.5$ V,	$V_I = 2.7$ V			20			20	µA
I_{IL}	$V_{CC} = 5.5$ V,	$V_I = 0.4$ V		−0.02	−0.5		−0.02	−0.5	mA
I_O‡	$V_{CC} = 5.5$ V,	$V_O = 2.25$ V	−30		−112	−30		−112	mA
I_{CC}	$V_{CC} = 5.5$ V	Outputs high		55	90		55	90	mA
		Outputs low		55	85		55	85	
		Outputs disabled		65	100		65	100	

†All typical values are at $V_{CC} = 5$ V, $T_A = 25$°C.
‡The output conditions have been chosen to produce a current that closely approximates one half of the true short-circuit output current, I_{OS}.

ALS AND AS CIRCUITS

TEXAS INSTRUMENTS
POST OFFICE BOX 225012 • DALLAS, TEXAS 75265

TYPES SN54AS373, SN74AS373
OCTAL D-TYPE TRANSPARENT LATCHES WITH 3-STATE OUTPUTS

switching characteristics (see Note 1)

PARAMETER	FROM (INPUT)	TO (OUTPUT)	V_{CC} = 4.5 V to 5.5 V, C_L = 50 pF, $R1$ = 500 Ω, $R2$ = 500 Ω, T_A = MIN to MAX				UNIT
			SN54AS373		SN74AS373		
			MIN	MAX	MIN	MAX	
t_{PLH}	D	Q	3.5	8	3.5	6	ns
t_{PHL}			3.5	7	3.5	6	
t_{PLH}	C	Any Q	6.5	14	6.5	11.5	ns
t_{PHL}			5	8	5	7.5	
t_{PZH}	$\overline{OC}$	Any Q	2	7.5	2	6.5	ns
t_{PZL}			4.5	10.5	4.5	9.5	
t_{PHZ}	$\overline{OC}$	Any Q	3	7.5	3	6.5	ns
t_{PLZ}			3	8	3	7	

NOTE 1: For load circuits and voltage waveforms, see page 1-12.

TEXAS INSTRUMENTS
POST OFFICE BOX 225012 • DALLAS, TEXAS 75265

TYPES SN54ALS374, SN54AS374, SN74ALS374, SN74AS374
OCTAL D-TYPE EDGE-TRIGGERED FLIP-FLOPS

D2661, APRIL 1982—REVISED DECEMBER 1983

- D-Type Flip-Flops In a Single Package
- 3-State Bus-Driving True Outputs
- Full Parallel Access for Loading
- Buffered Control Inputs
- Package Options Include Both Plastic and Ceramic Chip Carriers in Addition to Plastic and Ceramic DIPs
- Dependable Texas Instruments Quality and Reliability

SN54ALS374, SN54AS374 . . . J PACKAGE
SN74ALS374, SN74AS374 . . . N PACKAGE
(TOP VIEW)

```
OC  [ 1   20 ] VCC
1Q  [ 2   19 ] 8Q
1D  [ 3   18 ] 8D
2D  [ 4   17 ] 7D
2Q  [ 5   16 ] 7Q
3Q  [ 6   15 ] 6Q
3D  [ 7   14 ] 6D
4D  [ 8   13 ] 5D
4Q  [ 9   12 ] 5Q
GND [ 10  11 ] CLK
```

description

These 8-bit flip-flops feature three-state outputs designed specifically for driving highly capacitive or relatively low-impedance loads. They are particularly suitable for implementing buffer registers, I/O ports, bidirectional bus drivers, and working registers.

The eight flip-flops of the 'ALS374 and 'AS374 are edge-triggered D-type flip-flops. On the positive transition of the clock the Q outputs will be set to the logic levels that were set up at the D inputs.

A buffered output-control input can be used to place the eight outputs in either a normal logic state (high or low logic levels) or a high-impedance state. In the high-impedance state the outputs neither load nor drive the bus lines significantly. The high-impedance third state and increased drive provide the capability to drive the bus lines in a bus-organized system without need for interface or pull-up components.

The output control ($\overline{OC}$) does not affect the internal operation of the flip-flops. Old data can be retained or new data can be entered while the outputs are in the high-impedance state.

The SN54ALS374 and SN54AS374 are characterized for operation over the full military temperature range of -55°C to 125°C. The SN74ALS374 and SN74AS374 are characterized for operation from 0°C to 70°C.

SN54ALS374, SN54AS374 . . . FH PACKAGE
SN74ALS374, SN74AS374 . . . FN PACKAGE
(TOP VIEW)

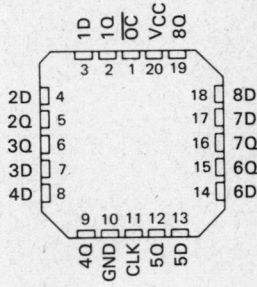

FUNCTION TABLE (EACH FLIP-FLOP)

INPUTS			OUTPUT
$\overline{OC}$	CLK	D	Q
L	↑	H	H
L	↑	L	L
L	L	X	Q_0
H	X	X	Z

ALS AND AS CIRCUITS

Copyright © 1983 by Texas Instruments Incorporated

TEXAS INSTRUMENTS
POST OFFICE BOX 225012 • DALLAS, TEXAS 75265

TYPES SN54ALS374, SN54AS374, SN74ALS374, SN74AS374
OCTAL D-TYPE EDGE-TRIGGERED FLIP-FLOPS

logic symbol

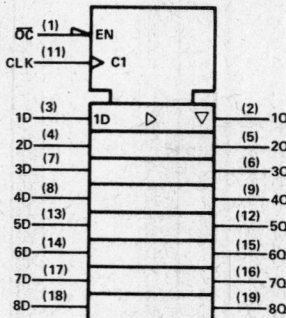

logic diagram (positive logic)

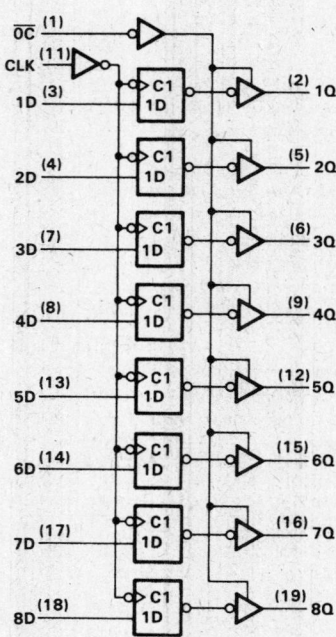

Pin numbers shown are for J and N packages.

absolute maximum ratings over operating free-air temperature range (unless otherwise noted)

Supply voltage, V_{CC}	7 V
Input voltage	7 V
Voltage applied to a disabled 3-state output	5.5 V
Operating free-air temperature range: SN54ALS374, SN54AS374	−55°C to 125°C
SN74ALS374, SN74AS374	0°C to 70°C
Storage temperature range	−65°C to 150°C

Texas Instruments
POST OFFICE BOX 225012 • DALLAS, TEXAS 75265

TYPES SN54ALS374, SN74ALS374
OCTAL D-TYPE EDGE-TRIGGERED FLIP-FLOPS

recommended operating conditions

		SN54ALS374			SN74ALS374			UNIT
		MIN	NOM	MAX	MIN	NOM	MAX	
V_{CC}	Supply voltage	4.5	5	5.5	4.5	5	5.5	V
V_{IH}	High-level input voltage	2			2			V
V_{IL}	Low-level input voltage			0.8			0.8	V
I_{OH}	high-level output current			−1			−2.6	mA
I_{OL}	Low-level output current			12			24	mA
f_{clock}	Clock frequency	0		30	0		35	MHz
t_w	Pulse duration CLK high	16.5			14			ns
	Pulse duration CLK low	16.5			14			
t_{su}	Setup time, data before CLK↑	10			10			ns
t_h	Hold time, data after CLK↑	4			0			ns
T_A	Operating free-air temperature	−55		125	0		70	°C

electrical characteristics over recommended operating free-air temperature range (unless otherwise noted)

PARAMETER	TEST CONDITIONS		SN54ALS374			SN74ALS374			UNIT
			MIN	TYP[†]	MAX	MIN	TYP[†]	MAX	
V_{IK}	$V_{CC} = 4.5$ V,	$I_I = -18$ mA			−1.5			−1.5	V
V_{OH}	$V_{CC} = 4.5$ V to 5.5 V,	$I_{OH} = -0.4$ mA	$V_{CC}-2$			$V_{CC}-2$			V
	$V_{CC} = 4.5$ V,	$I_{OH} = -1$ mA	2.4	3.3					
	$V_{CC} = 4.5$ V,	$I_{OH} = -2.6$ mA				2.4	3.2		
V_{OL}	$V_{CC} = 4.5$ V	$I_{OL} = 12$ mA		0.25	0.4		0.25	0.4	V
	$V_{CC} = 4.5$ V	$I_{OL} = 24$ mA					0.35	0.5	
I_{OZH}	$V_{CC} = 5.5$ V,	$V_O = 2.7$ V			20			20	µA
I_{OZL}	$V_{CC} = 5.5$ V,	$V_I = 0.4$ V			−20			−20	µA
I_I	$V_{CC} = 5.5$ V,	$V_I = 7$ V			0.1			0.1	mA
I_{IH}	$V_{CC} = 5.5$ V,	$V_I = 2.7$ V			20			20	µA
I_{IL}	$V_{CC} = 5.5$ V,	$V_I = 0.4$ V			−0.2			−0.2	mA
I_O[‡]	$V_{CC} = 5.5$ V,	$V_O = 2.25$ V	−30		−112	−30		−112	mA
I_{CC}	$V_{CC} = 5.5$ V	Outputs high		11	19		11	19	mA
		Outputs low		19	28		19	28	
		Outputs disabled		20	31		20	31	

[†] All typical values are at $V_{CC} = 5$ V, $T_A = 25$°C.
[‡] The output conditions have been chosen to produce a current that closely approximates one half of the true short-circuit output current, I_{OS}.

ALS AND AS CIRCUITS

TEXAS INSTRUMENTS
POST OFFICE BOX 225012 • DALLAS, TEXAS 75265

TYPES SN54ALS374, SN74ALS374
OCTAL D-TYPE EDGE-TRIGGERED FLIP-FLOPS

switching characteristics (see Note 1)

PARAMETER	FROM (INPUT)	TO (OUTPUT)	V_{CC} = 4.5 V to 5.5 V, C_L = 50 pF, R_1 = 500 Ω, R_2 = 500 Ω, T_A = MIN to MAX				UNIT
			SN54ALS374		SN74ALS374		
			MIN	MAX	MIN	MAX	
f_{max}			30		35		MHz
t_{PLH}	CLK	Q	3	15	3	12	ns
t_{PHL}			5	18	5	16	
t_{PZH}	$\overline{OC}$	Q	5	19	5	17	ns
t_{PZL}			7	20	7	18	
t_{PHZ}	$\overline{OC}$	Q	2	12	2	10	ns
t_{PLZ}			3	24	3	18	

NOTE 1: For load circuit and voltage waveforms, see page 1-12.

TYPICAL APPLICATION DATA

EXPANDABLE 4-WORD BY 8-BIT GENERAL REGISTER FILE

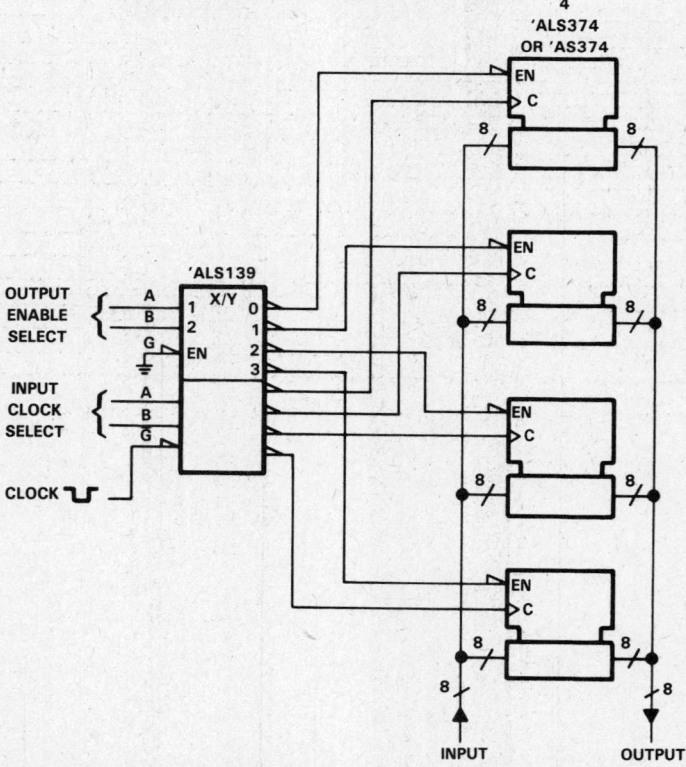

TEXAS INSTRUMENTS
POST OFFICE BOX 225012 • DALLAS, TEXAS 75265

TYPES SN54AS374, SN74AS374
OCTAL D-TYPE EDGE-TRIGGERED FLIP-FLOPS

recommended operating conditions

		SN54AS374			SN74AS374			UNIT
		MIN	NOM	MAX	MIN	NOM	MAX	
V_{CC}	Supply voltage	4.5	5	5.5	4.5	5	5.5	V
V_{IH}	High-level input voltage	2			2			V
V_{IL}	Low-level input voltage			0.8			0.8	V
I_{OH}	High-level output current			−12			−15	mA
I_{OL}	Low-level output current			32			48	mA
f_{clock}	Clock frequency	0		100	0		125	MHz
t_w	Pulse duration CLK high	5.5			4			ns
	Pulse duration CLK low	5			3			ns
t_{su}	Setup time data before CLK↑	3			2			ns
t_h	Hold time, data after CLK↑	3			2			ns
T_A	Operating free-air temperature	−55		125	0		70	°C

electrical characteristics over recommended operating free-air temperature range (unless otherwise noted)

PARAMETER		TEST CONDITIONS		SN54AS374			SN74AS374			UNIT
				MIN	TYP†	MAX	MIN	TYP†	MAX	
V_{IK}		V_{CC} = 4.5 V,	I_I = −18 mA			−1.2			−1.2	V
V_{OH}		V_{CC} = 4.5 V to 5.5 V,	I_{OH} = −2 mA	V_{CC}−2			V_{CC}−2			V
		V_{CC} = 4.5 V,	I_{OH} = −12 mA	2.4	3.2					
		V_{CC} = 4.5 V,	I_{OH} = −15 mA				2.4	3.3		
V_{OL}		V_{CC} = 4.5 V,	I_{OL} = 32 mA		0.29	0.5				V
		V_{CC} = 4.5 V	I_{OL} = 48 mA					0.34	0.5	
I_{OZH}		V_{CC} = 5.5 V,	V_O = 2.7 V			50			50	µA
I_{OZL}		V_{CC} = 5.5 V,	V_I = 0.4 V			−50			−50	µA
I_I		V_{CC} = 5.5 V,	V_I = 7 V			0.1			0.1	mA
I_{IH}		V_{CC} = 5.5 V,	V_I = 2.7 V			20			20	µA
I_{IL}	OC, CLK	V_{CC} = 5.5 V,	V_I = 0.4 V			−0.5			−0.5	mA
	Data					−3			−2	
I_O‡		V_{CC} = 5.5 V,	V_O = 2.25 V	−30		−112	−30		−112	mA
I_{CC}		V_{CC} = 5.5 V	Outputs high		77	120		77	120	mA
			Outputs low		84	128		84	128	
			Outputs disabled		84	128		84	128	

†All typical values are at V_{CC} = 5 V, T_A = 25°C.
‡The output conditions have been chosen to produce a current that closely approximates one half of the true short-circuit output current, I_{OS}.

ALS AND AS CIRCUITS

TEXAS
INSTRUMENTS
POST OFFICE BOX 225012 • DALLAS, TEXAS 75265

TYPES SN54AS374, SN74AS374
OCTAL D-TYPE EDGE-TRIGGERED FLIP-FLOPS

switching characteristics (see Note 1)

PARAMETER	FROM (INPUT)	TO (OUTPUT)	V_{CC} = 4.5 V to 5.5 V, C_L = 50 pF, R1 = 500 Ω, R2 = 500 Ω, T_A = MIN to MAX				UNIT
			SN54AS374		SN74AS374		
			MIN	MAX	MIN	MAX	
f_{max}			100		125		MHz
t_{PLH}	CLK	Q	3	11	3	8	ns
t_{PHL}			4	11.5	4	9	
t_{PZH}	$\overline{OC}$	Q	2	7	2	6	ns
t_{PZL}			3	11	3	10	
t_{PHZ}	$\overline{OC}$	Q	2	7	2	6	ns
t_{PLZ}			2	7	2	6	

NOTE 1: For load circuit and voltage waveforms, see page 1-12.

TYPICAL APPLICATION DATA

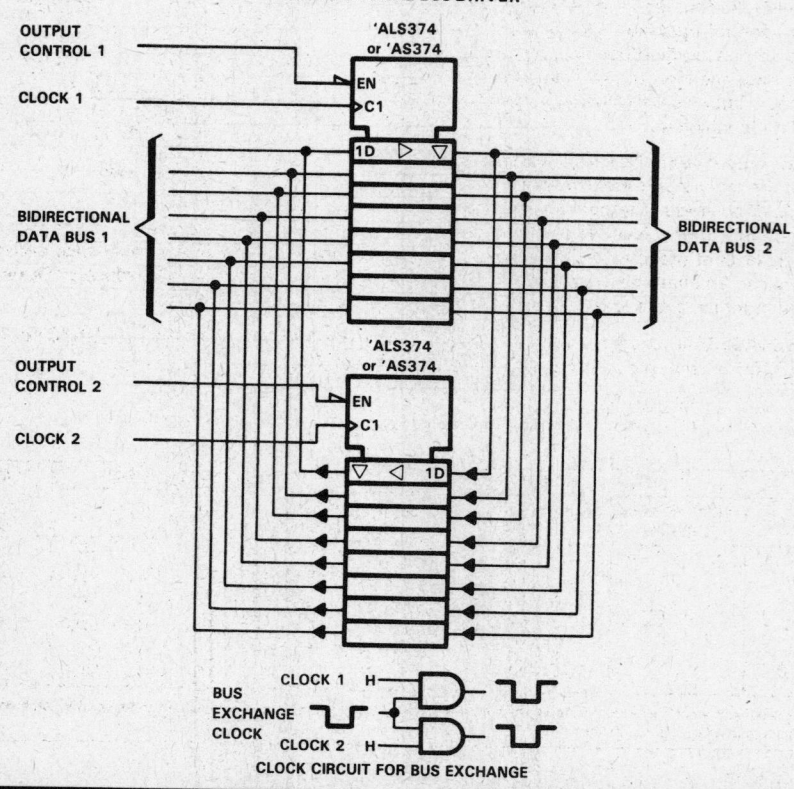

BIDIRECTIONAL BUS DRIVER

TYPES SN54AS395, SN74AS395
4-BIT CASCADABLE SHIFT REGISTERS WITH 3-STATE OUTPUTS

D2661, DECEMBER 1983

- Cascadable, 4-Bit, Three-State Parallel-to-Serial, Serial-to-Parallel Conversions
- Parallel Synchronous Loading
- Direct Overriding Clear
- Applications:
  N-Bit Serial-to-Parallel Converter
  N-Bit Parallel-to-Serial Converter
  N-Bit Storage Register
- Dependable Texas Instruments Quality and Reliability

SN54AS395 . . . J PACKAGE
SN74AS395 . . . N PACKAGE
(TOP VIEW)

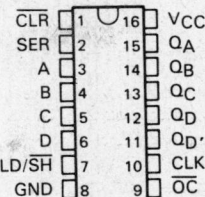

SN54AS395 . . . FH PACKAGE
SN74AS395 . . . FN PACKAGE
(TOP VIEW)

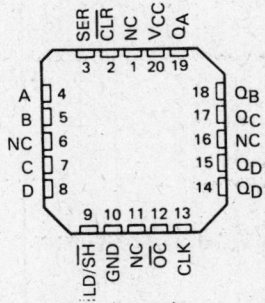

NC—No internal connection

description

These 4-bit registers feature parallel inputs, parallel outputs, cascadable output, and clock, serial load/shift, output control, and direct overriding clear inputs.

Shifting is accomplished when the load/shift control is low. Parallel loading is accomplished by applying the four bits of data and taking the load/shift control input high. The data are loaded into the associated flip-flops and appear at the outputs after the high-to-low transition of the clock input. During parallel loading, the entry of serial data is inhibited.

When the output control is low, the normal logic levels of the four outputs are available for driving the loads or bus lines. The outputs are disabled independently of the level of the clock by a high logic level at the output control input. The outputs then present a high impedance and neither load nor drive the bus line; however, sequential operation of the registers is not affected. During the high-impedance mode, the output at $Q_{D'}$ is still available for cascading.

The SN54AS395 is characterized for operation over the full military range of −55°C to 125°C. The SN74AS395 is characterized for operation from 0°C to 70°C.

FUNCTION TABLE

INPUTS								3-STATE OUTPUTS				CASCADE OUTPUT
CLEAR	LOAD/SHIFT CONTROL	CLOCK	SERIAL	PARALLEL				Q_A	Q_B	Q_C	Q_D	$Q_{D'}$
				A	B	B	B					
L	X	X	X	X	X	X	X	L	L	L	L	L
H	H	H	X	X	X	X	X	Q_{A0}	Q_{B0}	Q_{C0}	Q_{D0}	Q_{D0}
H	H	↓	X	a	b	c	d	a	b	c	d	d
H	L	H	X	X	X	X	X	Q_{A0}	Q_{B0}	Q_{C0}	Q_{D0}	Q_{D0}
H	L	↓	H	X	X	X	X	H	Q_{An}	Q_{Bn}	Q_{Cn}	Q_{Cn}
H	L	↓	L	X	X	X	X	L	Q_{An}	Q_{Bn}	Q_{Cn}	Q_{Cn}

When the output control is high, the 3-state outputs are disabled to the high-impedance state; however, sequential operation of the registers and the output at $Q_{D'}$ are not affected.

PRODUCT PREVIEW

This document contains information on a product under development. Texas Instruments reserves the right to change or discontinue this product without notice.

Copyright © 1983 by Texas Instruments Incorporated

TEXAS INSTRUMENTS
POST OFFICE BOX 225012 • DALLAS, TEXAS 75265

TYPES SN54AS395, SN74AS395
4-BIT CASCADABLE SHIFT REGISTERS WITH 3-STATES OUTPUTS

logic symbol

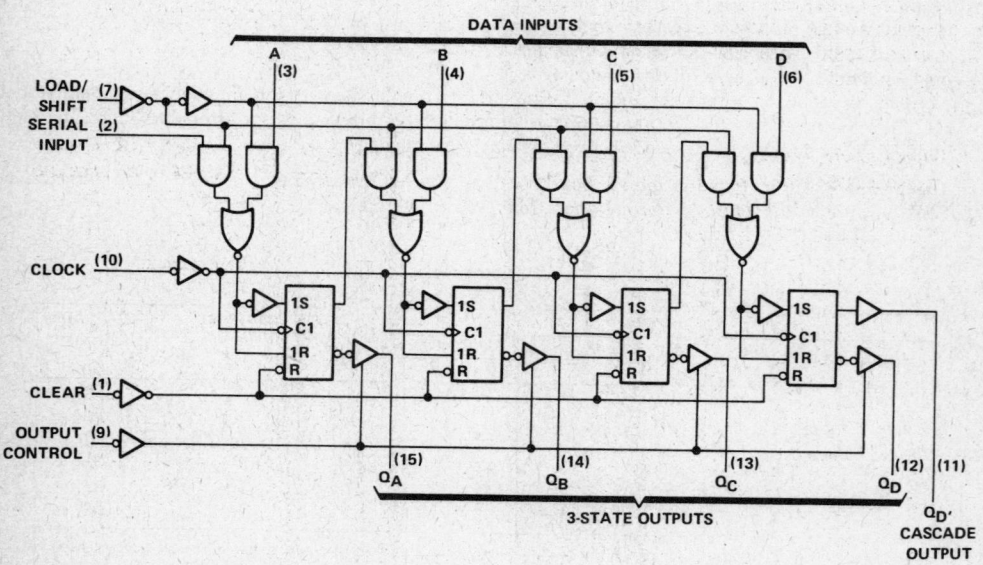

Pin numbers shown are for J and N packages.

logic diagram (positive logic)

Pin numbers shown are for J and N packages.

TYPES SN54ALS465A THRU SN54ALS468A, SN74ALS465A THRU SN74ALS468A
OCTAL BUFFERS WITH 3-STATE OUTPUTS

D2661, APRIL 1982—REVISED DECEMBER 1983

- Mechanically and Functionally Interchangeable with DM71/81LS97 and DM71/81LS98
- P-N-P Inputs Reduce Bus Loading
- 3-State Outputs Rated at I_{OL} of 12 mA and 24 mA for SN54ALS' and SN74ALS', Respectively
- Package Options Include Both Plastic and Ceramic Chip Carriers in Addition to Plastic and Ceramic DIPs
- Dependable Texas Instruments Quality and Reliability

DEVICE	DATA PATH
'ALS465A	True
'ALS466A	Inverting
'ALS467A	True
'ALS468A	Inverting

SN54ALS465A, SN54ALS466A . . . J PACKAGE
SN74ALS465A, SN74ALS466A . . . N PACKAGE
(TOP VIEW)

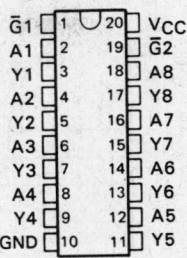

SN54ALS465A, SN54ALS466A . . . FH PACKAGE
SN74ALS465A, SN74ALS466A . . . FN PACKAGE
(TOP VIEW)

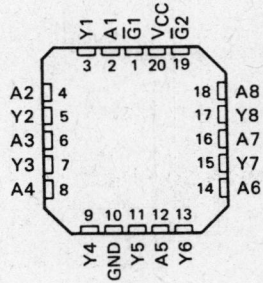

description

These octal buffers utilize the latest advanced low-power Schottky technology. The 'ALS465A and 'ALS466A have a two-input active-low AND enable gate controlling all eight data buffers. The 'ALS467A and 'ALS468A have two separate active-low enable inputs each controlling four data buffers. In each case, a high level on any $\overline{G}$ places the affected outputs at high impedance.

The SN54ALS465A, SN54ALS466A, SN54ALS467A, and SN54ALS468A are characterized for operation over the full military temperature range of −55°C to 125°C. The SN74ALS465A, SN74ALS466A, SN74ALS467A, and SN74ALS468A are characterized for operation from 0°C to 70°C.

SN54ALS467A, SN54ALS468A . . . J PACKAGE
SN74ALS467A, SN74ALS468A . . . N PACKAGE
(TOP VIEW)

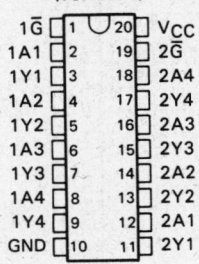

SN54ALS467A, SN54ALS468A . . . FH PACKAGE
SN74ALS467A, SN74ALS468A . . . FN PACKAGE
(TOP VIEW)

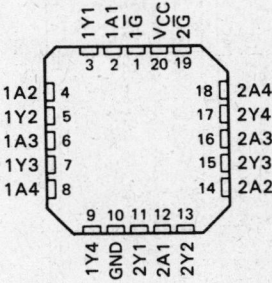

Copyright © 1982 by Texas Instruments Incorporated

Texas Instruments
POST OFFICE BOX 225012 • DALLAS, TEXAS 75265

ALS AND AS CIRCUITS

TYPES SN54ALS465A THRU SN54ALS468A, SN74ALS465A THRU SN74ALS468A
OCTAL BUFFERS WITH 3-STATE OUTPUTS

logic symbols

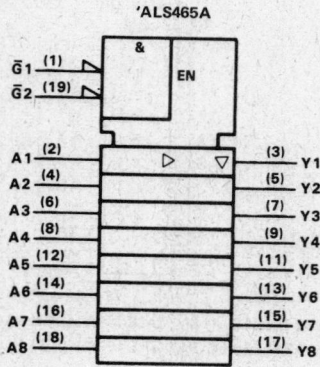
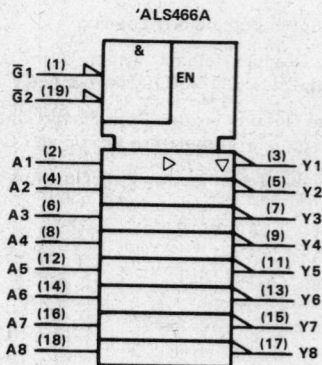

logic diagrams (positive logic)

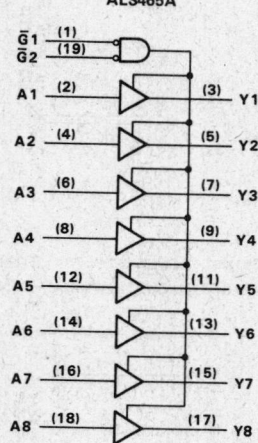
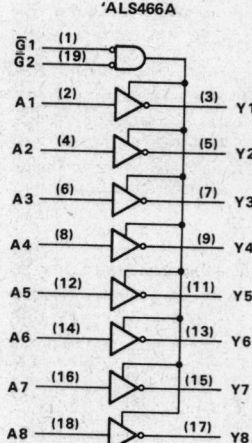

Pin numbers shown are for J and N packages.

TYPES SN54ALS465A THRU SN54ALS468A, SN74ALS465A THRU SN74ALS468A
OCTAL BUFFERS WITH 3-STATE OUTPUTS

logic symbols

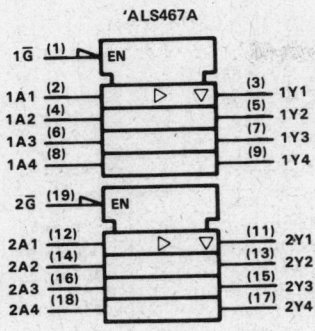
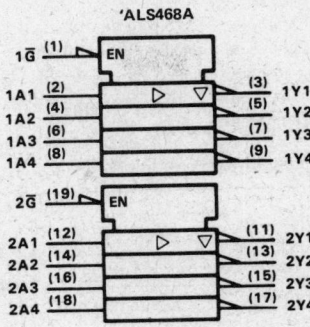

logic diagrams (positive logic)

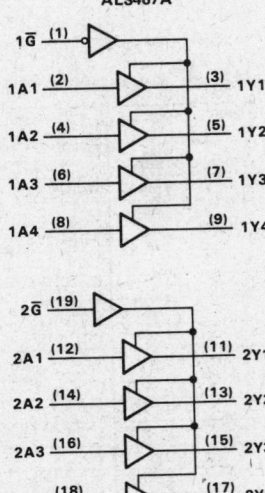
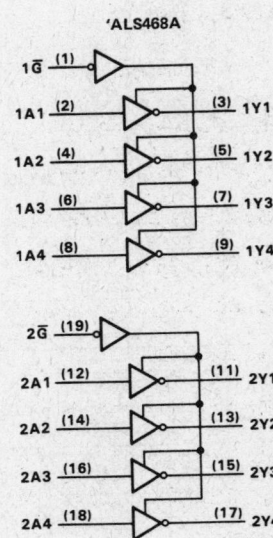

Pin numbers shown are for J and N packages.

TEXAS INSTRUMENTS
POST OFFICE BOX 225012 • DALLAS, TEXAS 75265

TYPES SN54ALS465A THRU SN54ALS468A, SN74ALS465A THRU SN74ALS468A OCTAL BUFFERS WITH 3-STATE OUTPUTS

absolute maximum ratings over operating free-air temperature range (unless otherwise noted)

Supply voltage, V_{CC} .... 7 V
Input voltage .... 7 V
Voltage applied to a disabled 3-state output .... 5.5 V
Operating free-air temperature range: SN54ALS465A THRU SN54ALS468A .... $-55\,°C$ to $125\,°C$
                                    SN74ALS465A THRU SN74ALS468A .... $0\,°C$ to $70\,°C$
Storage temperature range .... $-65\,°C$ to $150\,°C$

recommended operating conditions

		SN54ALS465A THRU SN54ALS468A			SN74ALS465A THRU SN74ALS468A			UNIT
		MIN	NOM	MAX	MIN	NOM	MAX	
V_{CC}	Supply voltage	4.5	5	5.5	4.5	5	5.5	V
V_{IH}	High-level input voltage	2			2			V
V_{IL}	Low-level input voltage			0.8			0.8	V
I_{OH}	High-level output current			-12			-15	mA
I_{OL}	Low-level output current			12			24	mA
							48†	
T_A	Operating free-air temperature	-55		125	0		70	°C

†The extended limit applies only if V_{CC} is maintained between 4.75 V and 5.25 V.
  The 48 mA limit applies for SN74ALS465A-1, SN74ALS466A-1, SN74ALS467A-1, and SN74ALS468A-1 only.

electrical characteristics over recommended operating free-air temperature range (unless otherwise noted)

PARAMETER		TEST CONDITIONS		SN54ALS465A THRU SN54ALS468A			SN74ALS465A THRU SN74ALS468A			UNIT
				MIN	TYP‡	MAX	MIN	TYP‡	MAX	
V_{IK}		$V_{CC} = 4.5$ V,	$I_I = -18$ mA			-1.5			-1.5	V
V_{OH}		$V_{CC} = 4.5$ V to 5.5 V,	$I_{OH} = -0.4$ mA	$V_{CC}-2$			$V_{CC}-2$			
		$V_{CC} = 4.5$ V,	$I_{OH} = -3$ mA	2.4	3.2		2.4	3.2		V
		$V_{CC} = 4.5$ V,	$I_{OH} = -12$ mA	2						
		$V_{CC} = 4.5$ V,	$I_{OH} = -15$ mA				2			
V_{OL}		$V_{CC} = 4.5$ V,	$I_{OL} = 12$ mA		0.25	0.4		0.25	0.4	V
		$V_{CC} = 4.5$ V,	$I_{OL} = 24$ mA							
		($I_{OL} = 48$ mA for -1 versions)						0.35	0.5	
I_{OZH}		$V_{CC} = 5.5$ V,	$V_O = 2.7$ V			20			20	µA
I_{OZL}		$V_{CC} = 5.5$ V,	$V_O = 0.4$ V			-20			-20	µA
I_I		$V_{CC} = 5.5$ V,	$V_I = 7$ V			0.1			0.1	mA
I_{IH}		$V_{CC} = 5.5$ V,	$V_I = 2.7$ V			20			20	µA
I_{IL}		$V_{CC} = 5.5$ V,	$V_I = 0.4$ V			-0.1			-0.1	mA
I_O‡		$V_{CC} = 5.5$ V,	$V_O = 2.25$ V	-30		-112	-30		-112	mA
I_{CC}	'ALS465A 'ALS467A	$V_{CC} = 5.5$ V	Outputs high		11	21		11	16	mA
			Outputs low		19	33		19	28	
			Outputs disabled		23	38		23	33	
	'ALS466A 'ALS468A	$V_{CC} = 5.5$ V	Outputs high		7	15		7	10	mA
			Outputs low		16	29		16	24	
			Outputs disabled		19	32		19	27	

‡All typical values are at $V_{CC} = 5$ V, $T_A = 25\,°C$.
§The output conditions have been chosen to produce a current that closely approximates one half of the true short-circuit output current, I_{OS}.

TYPES SN54ALS465A THRU SN54ALS468A, SN74ALS465A THRU SN74ALS468A
OCTAL BUFFERS WITH 3-STATE OUTPUTS

'ALS465A, 'ALS467A switching characteristics (see Note 1)

PARAMETER	FROM (INPUT)	TO (OUTPUT)	V_{CC} = 4.5 V to 5.5 V, C_L = 50 pF, R1 = 500 Ω, R2 = 500 Ω, T_A = MIN to MAX				UNIT
			SN54ALS465A SN54ALS467A		SN74ALS465A SN74ALS467A		
			MIN	MAX	MIN	MAX	
t_{PLH}	A	Y	2	16	2	13	ns
t_{PHL}			4	15	4	12	
t_{PZH}	$\overline{G}$	Any Y	4	27	4	23	ns
t_{PZL}			5	30	5	25	
t_{PHZ}	$\overline{G}$	Any Y	2	12	2	10	ns
t_{PLZ}			3	21	3	18	

'ALS466A, 'ALS468A switching characteristics (see Note 1)

PARAMETER	FROM (INPUT)	TO (OUTPUT)	V_{CC} = 4.5 V to 5.5 V, C_L = 50 pF, R1 = 500 Ω, R2 = 500 Ω, T_A = MIN to MAX				UNIT
			SN54ALS466A SN54ALS468A		SN74ALS466A SN74ALS468A		
			MIN	MAX	MIN	MAX	
t_{PLH}	A	Y	3	14	3	12	ns
t_{PHL}			2	11	2	9	
t_{PZH}	$\overline{G}$	Any Y	4	21	4	16	ns
t_{PZL}			7	25	7	23	
t_{PHZ}	$\overline{G}$	Any Y	2	12	2	10	ns
t_{PLZ}			2	20	2	17	

NOTE 1: For load circuit and voltage waveforms, see page 1-12.

ALS AND AS CIRCUITS

TEXAS INSTRUMENTS
POST OFFICE BOX 225012 • DALLAS, TEXAS 75265

2 ALS AND AS CIRCUITS

TYPES SN54ALS518 THRU SN54ALS522, SN74ALS518 THRU SN74ALS522
8-BIT IDENTITY COMPARATORS

D2661, JUNE 1982—REVISED DECEMBER 1983

- **Compares Two 8-Bit Words**
- **Choice of Totem-Pole or Open-Collector Outputs**
- **'ALS518, 'ALS520, and 'ALS522 Have 20-kΩ Pull-up Resistors on Q Inputs**
- **Package Options Include Both Plastic and Ceramic Chip Carriers in Addition to Plastic and Ceramic DIPs**
- **Dependable Texas Instruments Quality and Reliability**

TYPE	INPUT PULL-UP RESISTOR	OUTPUT FUNCTION AND CONFIGURATION
'ALS518	Yes	P=Q open-collector
'ALS519	No	P=Q open-collector
'ALS520	Yes	$\overline{P=Q}$ totem-pole
'ALS521†	No	$\overline{P=Q}$ totem-pole
'ALS522	Yes	$\overline{P=Q}$ open-collector

†'ALS521 is identical to 'ALS688

description

These identity comparators perfrom comparisons on two eight-bit binary or BCD words. The 'ALS518 and 'ALS519 provide P=Q outputs, while the 'ALS520, 'ALS521, and 'ALS522 provide $\overline{P=Q}$ outputs. The 'ALS518, 'ALS519, and 'ALS522 have open-collector outputs. The 'ALS518, 'ALS520, and 'ALS522 feature 20-kΩ pull-up termination resistors on the Q inputs for analog or switch data.

The SN54ALS518 through SN54ALS522 are characterized for operation over the full military temperature range of −55°C to 125°C. The SN74ALS518 through SN74ALS522 are characterized for operation from 0°C to 70°C.

SN54ALS' . . . J PACKAGE
SN74ALS' . . . N PACKAGE
(TOP VIEW)

```
 G  [ 1   20 ] VCC
P0  [ 2   19 ] P=Q/P=Q†
Q0  [ 3   18 ] Q7
P1  [ 4   17 ] P7
Q1  [ 5   16 ] Q6
P2  [ 6   15 ] P6
Q2  [ 7   14 ] Q5
P3  [ 8   13 ] P5
Q3  [ 9   12 ] Q4
GND [10   11 ] P4
```

SN54ALS' . . . FH PACKAGE
SN74ALS' . . . FN PACKAGE
(TOP VIEW)

†P=Q for 'ALS518 and 'ALS519, and $\overline{P=Q}$ for 'ALS520, 'ALS521, and 'ALS522.

FUNCTION TABLE

INPUTS		OUTPUTS	
DATA P, Q	ENABLE G	P=Q	$\overline{P=Q}$
P=Q	L	H	L
P>Q	L	L	H
P<Q	L	L	H
X	H	L	H

Copyright © 1982 by Texas Instruments Incorporated.

TEXAS INSTRUMENTS
POST OFFICE BOX 225012 • DALLAS, TEXAS 75265

ALS AND AS CIRCUITS

TYPES SN54ALS518 THRU SN54ALS522, SN74ALS518 THRU SN74ALS522
8-BIT IDENTITY COMPARATORS

logic symbols

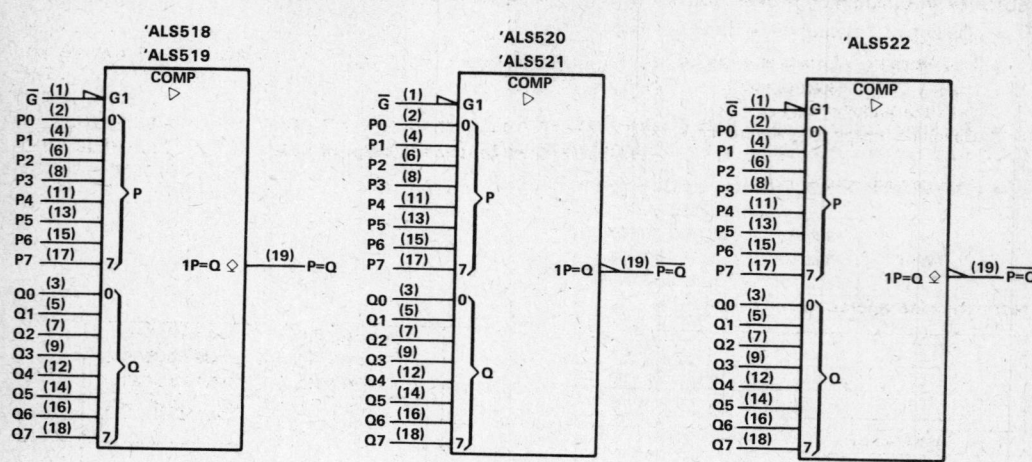

logic diagrams (positive logic)

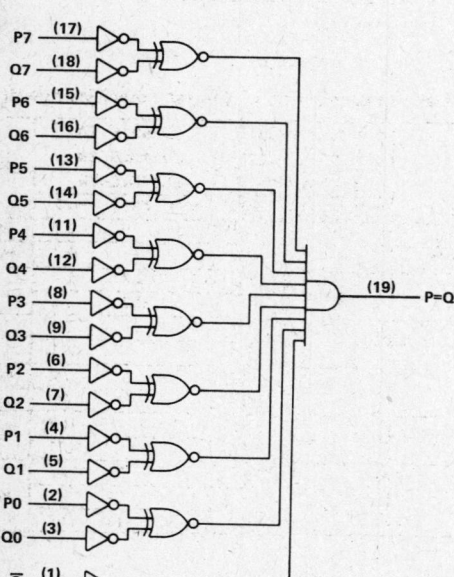

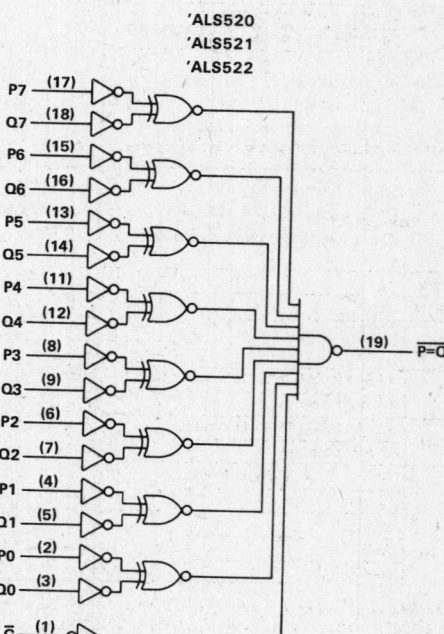

Pin numbers shown are for J and N packages.

TEXAS INSTRUMENTS
POST OFFICE BOX 225012 • DALLAS, TEXAS 75265

TYPES SN54ALS518 THRU SN54ALS522, SN74ALS518 THRU SN74ALS522
8-BIT IDENTITY COMPARATORS

absolute maximum ratings over operating free-air temperature range (unless otherwise noted)

Supply voltage, V_{CC} .... 7 V
Input voltage: Q inputs of 'ALS518, 'ALS522 .... V_{CC} + 0.5 V or 5.5 V, whichever is less
                 All other inputs .... 7 V
Off-state output voltage .... 7 V
Operating free-air temperature range: SN54ALS518, SN54ALS519, SN54ALS522 .... −55°C to 125°C
                                        SN74ALS518, SN74ALS519, SN74ALS522 .... 0°C to 70°C
Storage temperature range .... −65°C to 150°C

recommended operating conditions

		SN54ALS518 / SN54ALS519 / SN54ALS522			SN74ALS518 / SN74ALS519 / SN74ALS522			UNIT
		MIN	NOM	MAX	MIN	NOM	MAX	
V_{CC}	Supply voltage	4.5	5	5.5	4.5	5	5.5	V
V_{IH}	High-level input voltage	2			2			V
V_{IL}	Low-level input voltage			0.8			0.8	V
V_{OH}	High-level output voltage			5.5			5.5	V
I_{OL}	Low-level output current			12			24	mA
T_A	Operating free-air temperature	−55		125	0		70	°C

electrical characteristics over recommended operating free-air temperature range (unless otherwise noted)

PARAMETER		TEST CONDITIONS		SN54ALS518 / SN54ALS519 / SN54ALS522			SN74ALS518 / SN74ALS519 / SN74ALS522			UNIT
				MIN	TYP†	MAX	MIN	TYP†	MAX	
V_{IK}		V_{CC} = 4.5 V,	I_I = −18 mA			−1.5			−1.5	V
I_{OH}		V_{CC} = 4.5 V,	V_{OH} = 5.5 V			0.1			0.1	mA
V_{OL}		V_{CC} = 4.5 V,	I_{OL} = 12 mA		0.25	0.4		0.25	0.4	V
		V_{CC} = 4.5 V,	I_{OL} = 24 mA					0.35	0.5	
I_I	'ALS518, 'ALS522 Q inputs	V_{CC} = 5.5 V,	V_I = 5.5 V			0.1			0.1	mA
	All other inputs	V_{CC} = 5.5 V,	V_I = 7 V			0.1			0.1	
I_{IH}	'ALS518, 'ALS522 Q inputs	V_{CC} = 5.5 V,	V_I = 2.7 V			−0.2			−0.2	mA
	All other inputs					20			20	µA
I_{IL}	'ALS518, 'ALS522 Q inputs	V_{CC} = 5.5 V,	V_I = 0.4 V			−0.6			−0.6	mA
	All other inputs					−0.1			−0.1	
I_{CC}	'ALS518	V_{CC} = 5.5 V,	See Note 1		11	17		11	17	mA
	'ALS519				11	17		11	17	
	'ALS522				11	17		11	17	

†All typical values are at V_{CC} = 5 V, T_A = 25°C.
NOTE 1: I_{CC} is measured with $\overline{G}$ grounded, P and Q at 4.5 V.

Texas Instruments
POST OFFICE BOX 225012 • DALLAS, TEXAS 75265

TYPES SN54ALS518, THRU SN54ALS522, SN74ALS518 THRU SN74ALS522
8-BIT IDENTITY COMPARATORS

'ALS518, 'ALS519 switching characteristics (see Note 1)

PARAMETER	FROM (INPUT)	TO (OUTPUT)	V_{CC} = 4.5 V to 5.5 V, C_L = 50 pF, R_L = 680 Ω, T_A = MIN to MAX				UNIT
			SN54ALS518 SN54ALS519		SN74ALS518 SN74ALS519		
			MIN	MAX	MIN	MAX	
t_{PLH}	P or Q	P = Q	15	37	15	33	ns
t_{PHL}			3	18	3	15	
t_{PLH}	$\overline{G}$	P = Q	15	37	15	33	ns
t_{PHL}			3	18	3	15	

'ALS522 switching characteristics (see Note 1)

PARAMETER	FROM (INPUT)	TO (OUTPUT)	V_{CC} = 4.5 V to 5.5 V, C_L = 50 pF, R_L = 680 Ω, T_A = MIN to MAX				UNIT
			SN54ALS522		SN74ALS522		
			MIN	MAX	MIN	MAX	
t_{PLH}	P or Q	$\overline{P=Q}$	10	30	10	25	ns
t_{PHL}			5	25	5	23	
t_{PLH}	$\overline{G}$	$\overline{P=Q}$	8	30	8	25	ns
t_{PHL}			8	30	8	23	

NOTE 1: For load circuit and voltage waveforms, see page 1-12.

TYPES SN54ALS518, THRU SN54ALS522, SN74ALS518 THRU SN74ALS522
8-BIT IDENTITY COMPARATORS

absolute maximum ratings over operating free-air temperature range (unless otherwise noted)

Supply voltage, V_{CC} .... 7 V
Input voltage: Q inputs of 'ALS520 .... V_{CC} + 0.5 V or 5.5 V, whichever is less
               All other inputs .... 7 V
Operating free-air temperature range: SN54ALS520, SN54ALS521 .... −55 °C to 125 °C
                              SN74ALS520, SN74ALS521 .... 0 °C to 70 °C
Storage temperature range .... −65 °C to 150 °C

recommended operating conditions

		SN54ALS520 SN54ALS521			SN74ALS520 SN74ALS521			UNIT
		MIN	NOM	MAX	MIN	NOM	MAX	
V_{CC}	Supply voltage	4.5	5	5.5	4.5	5	5.5	V
V_{IH}	High-level input voltage	2			2			V
V_{IL}	Low-level input voltage			0.8			0.8	V
I_{OH}	High-level output current			−1			−2.6	mA
I_{OL}	Low-level output current			12			24	mA
T_A	Operating free-air temperature	−55		125	0		70	°C

electrical characteristics over recommended operating free-air temperature range (unless otherwise noted)

PARAMETER		TEST CONDITIONS		SN54ALS520 SN54ALS521			SN74ALS520 SN74ALS521			UNIT
				MIN	TYP†	MAX	MIN	TYP†	MAX	
V_{IK}		V_{CC} = 4.5 V,	I_I = −18 mA			−1.5			−1.5	V
V_{OH}		V_{CC} = 4.5 V to 5.5 V,	I_{OH} = −0.4 mA	$V_{CC}-2$			$V_{CC}-2$			V
		V_{CC} = 4.5 V,	I_{OH} = −1 mA	2.4	3.3					
		V_{CC} = 4.5 V,	I_{OH} = −2.6 mA				2.4	3.2		
V_{OL}		V_{CC} = 4.5 V,	I_{OL} = 12 mA		0.25	0.4		0.25	0.4	V
		V_{CC} = 4.5 V	I_{OL} = 24 mA					0.35	0.5	
I_I	'ALS520 Q inputs	V_{CC} = 5.5 V,	V_I = 5.5 V			0.1			0.1	mA
	All other inputs	V_{CC} = 5.5 V,	V_I = 7 V			0.1			0.1	
I_{IH}	'ALS520 Q inputs	V_{CC} = 5.5 V,	V_I = 2.7 V			−0.2			−0.2	mA
	All other inputs					20			20	μA
I_{IL}	'ALS520 Q inputs	V_{CC} = 5.5 V,	V_I = 0.4 V			−0.6			−0.6	mA
	All other inputs					−0.1			−0.1	
I_O‡		V_{CC} = 5.5 V,	V_O = 2.25 V	−30		−112	−30		−112	mA
I_{CC}	'ALS520	V_{CC} = 5.5 V,	See Note 1		12	19		12	19	mA
	'ALS521				12	19		12	19	

†All typical values are at V_{CC} = 5 V, T_A = 25°C.
‡The output conditions have been chosen to produce a current that closely approximates one half of the true short-circuit output current, I_{OS}.
NOTE 1: I_{CC} is measured with $\overline{G}$ grounded and P and Q inputs at 4.5 V.

Texas Instruments
POST OFFICE BOX 225012 • DALLAS, TEXAS 75265

TYPES SN54ALS518 THRU SN54ALS522, SN74ALS518 THRU SN74ALS522
8-BIT IDENTITY COMPARATORS

switching characteristics (see Note 1)

PARAMETER	FROM (INPUT)	TO (OUTPUT)	$V_{CC} = 4.5$ V to 5.5 V, $C_L = 50$ pF, $R_L = 500$ Ω, $T_A =$ MIN to MAX				UNIT
			SN54ALS520 SN54ALS521		SN74ALS520 SN74ALS521		
			MIN	MAX	MIN	MAX	
t_{PLH}	P or Q	$\overline{P=Q}$	3	16	3	12	ns
t_{PHL}			5	25	5	20	
t_{PLH}	$\overline{G}$	$\overline{P=Q}$	2	15	2	12	ns
t_{PHL}			5	23	5	22	

NOTE 1. For load circuit and voltage waveforms, see page 1-12.

ALS AND AS CIRCUITS

TEXAS INSTRUMENTS
POST OFFICE BOX 225012 • DALLAS, TEXAS 75265

TYPES SN54ALS526, SN54ALS527, SN54ALS528, SN74ALS526, SN74ALS527, SN74ALS528
FUSE-PROGRAMMABLE IDENTITY COMPARATORS

D2826, DECEMBER 1983

- Can Be Programmed and Verified on Most Incoming Test Equipment
- Reduces Board and Package Size for Similar Fixed Comparator Functions
- High-Speed Address Recognition
- Package Options Include Both Plastic and Ceramic Chip Carriers in Addition to Plastic and Ceramic DIPs
- Dependable Texas Instruments Quality and Reliability

Programming Capabilities
- 'ALS526 — Fuse Programmable 16-Bit Identity Comparator
- 'ALS527 — Fuse Programmable 8-Bit Identity Comparator and 4-Bit Comparator
- 'ALS528 — Fuse Programmable 12-Bit Identity Comparator

SN54ALS526 . . . J PACKAGE
SN74ALS526 . . . N PACKAGE
(TOP VIEW)

```
 G   [ 1   20] VCC
 P0  [ 2   19] P=Q
 P1  [ 3   18] P15
 P2  [ 4   17] P14
 P3  [ 5   16] P13
 P4  [ 6   15] P12
 P5  [ 7   14] P11
 P6  [ 8   13] P10
 P7  [ 9   12] P9
 GND [10   11] P8
```

SN54ALS527 . . . J PACKAGE
SN74ALS527 . . . N PACKAGE
(TOP VIEW)

```
 G   [ 1   20] VCC
 P0  [ 2   19] P=Q
 P1  [ 3   18] Q11
 P2  [ 4   17] P11
 P3  [ 5   16] Q10
 P4  [ 6   15] P10
 P5  [ 7   14] Q9
 P6  [ 8   13] P9
 P7  [ 9   12] Q8
 GND [10   11] P8
```

SN54ALS528 . . . J PACKAGE
SN74ALS528 . . . N PACKAGE
(TOP VIEW)

```
 G   [ 1   16] VCC
 P0  [ 2   15] P=Q
 P1  [ 3   14] P11
 P2  [ 4   13] P10
 P3  [ 5   12] P9
 P4  [ 6   11] P8
 P5  [ 7   10] P7
 GND [ 8    9] P6
```

SN54ALS526 . . . FH PACKAGE
SN54ALS526 . . . FN PACKAGE
(TOP VIEW)

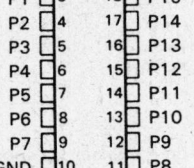

SN54ALS527 . . . FH PACKAGE
SN74ALS527 . . . FN PACKAGE
(TOP VIEW)

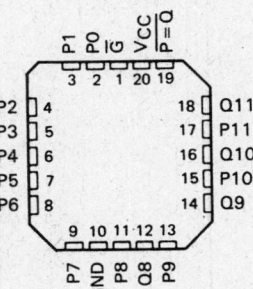

SN54ALS528 . . . FH PACKAGE
SN74ALS528 . . . FN PACKAGE
(TOP VIEW)

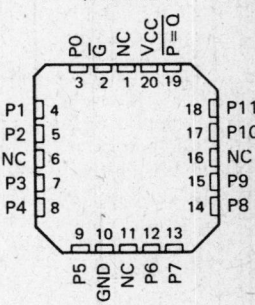

NC—No internal connection

PRODUCT PREVIEW
This document contains information on a product under development. Texas Instruments reserves the right to change or discontinue this product without notice.

Copyright © 1983 by Texas Instruments Incorporated

TEXAS INSTRUMENTS
POST OFFICE BOX 225012 • DALLAS, TEXAS 75265

TYPES SN54ALS526, SN54ALS527, SN54ALS528 SN74ALS526, SN74ALS527, SN54ALS528
FUSE-PROGRAMMABLE IDENTITY COMPARATORS

description

The 'ALS526 and 'ALS528 are fuse-programmable identity comparators designed for easy programming in fixed-comparator applications. The 'ALS526 compares a 16-bit data word against a preprogrammed 16-bit data word while the 'ALS528 compares a 12-bit data word against a preprogrammed 12-bit data word. The $\overline{P=Q}$ output will go low when the applied data word (P inputs) matches the preprogrammed data word (Q represents the preprogrammed data word). Programming is easily accomplished on the bench or with conventional automatic test equipment. Special equipment such as PROM-programmers are not required.

The 'ALS527 is a combination of an 8-bit fuse-programmable comparator and a conventional 4-bit comparator. For the $\overline{P=Q}$ output to go low, the applied data word P0 through P7 must match the preprogrammed data word Q0 through Q7, and the applied data word P8 through P11 must match the applied data word Q8 through Q11.

The SN54ALS526, SN54ALS527, and SN54ALS528 are characterized for operation over the full military temperature range of $-55\,°C$ to $125\,°C$. The SN74ALS526, SN74ALS527, and SN74ALS528 are characterized for operation from $0\,°C$ to $70\,°C$.

programming details

Before any fuses are blown, the inputs are programmed to recognize a low logic level. Therefore, only the bits that are to be programmed to recognize a high logic level require a fuse to be blown. A fuse is easily blown by applying 12 volts (V_{IHH}) to the desired P input pin and also to the $\overline{G}$ input. This permanently programs the pin to recognize a high. Only one input pin should be programmed at a time.

verification details

Before the device is programmed, all of the fuses are intact. In this condition, the $\overline{P=Q}$ output should go low only if lows are applied to all the P inputs. On the 'ALS527, the same is true for the P0 and P7 inputs, but in addition, the P8 through P11 inputs must match the Q8 through Q11 inputs.

It is possible to check the fuse circuitry before actually blowing it in the following manner. By placing a high (V_{IH}) at the desired P input pin while leaving a low on the $\overline{G}$ input, the $\overline{P=Q}$ output should be high. If the P input is then taken to V_{IHH}, the $\overline{P=Q}$ output should go low assuming all other P inputs are at a high level (V_{IH}).

In this condition, the fuse will not be blown as long as $\overline{G}$ is at V_{IL}. When $\overline{G}$ is taken to V_{IHH}, the fuse will be blown and the input will be permanently programmed to recognize a high logic level. The timing diagram in Figure 1 shows the recommended programming sequence. After all desired input pins have been programmed, it is easy to verify the device by applying the programmed data word and checking to be sure that the $\overline{P=Q}$ output is low.

TYPES SN54ALS526, SN54ALS527, SN54ALS528 SN74ALS526, SN74ALS527, SN74ALS528
FUSE-PROGRAMMABLE IDENTITY COMPARATORS

logic diagrams (positive logic)

'ALS526

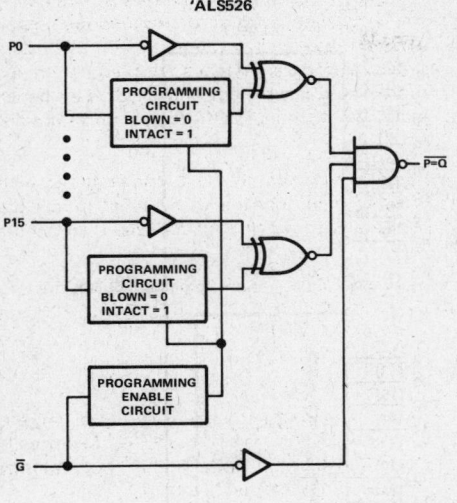

'ALS527

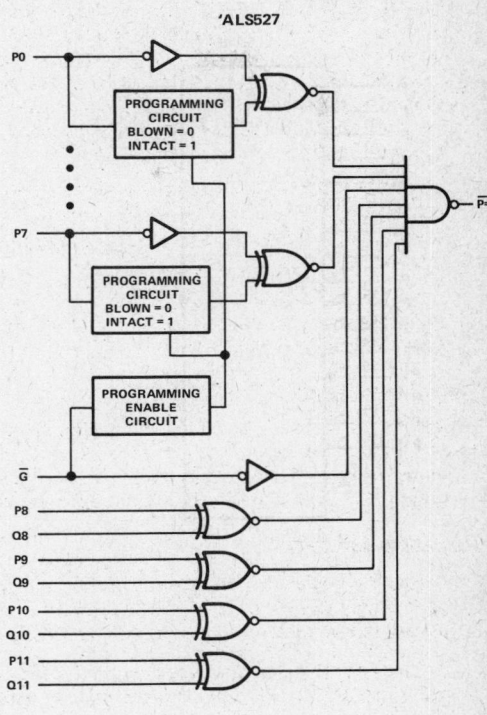

'ALS528

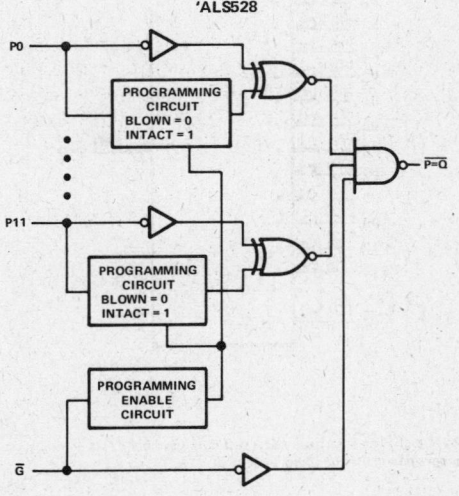

Pin numbers shown are for J and N packages.

TEXAS INSTRUMENTS
POST OFFICE BOX 225012 • DALLAS, TEXAS 75265

TYPES SN54ALS526, SN54ALS527, SN54ALS528
SN74ALS526, SN74ALS527, SN54ALS528
FUSE-PROGRAMMABLE IDENTITY COMPARATORS

logic symbols

'ALS526

Pin	(No.)
P0	(2)
P1	(3)
P2	(4)
P3	(5)
P4	(6)
P5	(7)
P6	(8)
P7	(9)
P8	(11)
P9	(12)
P10	(13)
P11	(14)
P12	(15)
P13	(16)
P14	(17)
P15	(18)
$\overline{G}$	(1)

Output: (19) $\overline{P=Q}$

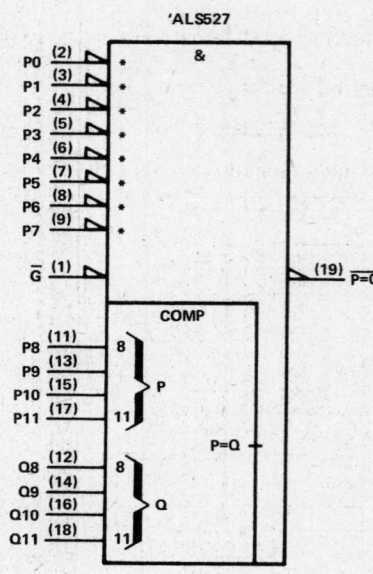

'ALS528

Pin	(No.)
P0	(2)
P1	(3)
P2	(4)
P3	(5)
P4	(6)
P5	(7)
P6	(9)
P7	(10)
P8	(11)
P9	(12)
P10	(13)
P11	(14)
$\overline{G}$	(1)

Output: (15) $\overline{P=Q}$

*These inputs can be programmed to be active high. The asterisk is not a part of the symbol. For a correct symbol for the programmed device, delete the polarity symbol (▷) at any input whose programming fuse has been blown.

Pin numbers shown are for J and N packages.

TEXAS INSTRUMENTS
POST OFFICE BOX 225012 • DALLAS, TEXAS 75265

TYPES SN54ALS526, SN54ALS527, SN54ALS528, SN74ALS526, SN74ALS527, SN74ALS528 FUSE-PROGRAMMABLE IDENTITY COMPARATORS

absolute maximum ratings over operating free-air temperature range (unless otherwise noted)

Supply voltage, V_{CC} (see Note 1) .... 7 V
Input voltage (see Note 1) .... 5.5 V
Operating free-air temperature range: SN54ALS' .... $-55\,°C$ to $125\,°C$
                                      SN74ALS' .... $0\,°CC$ to $70\,°C$
Storage temperature range .... $-65\,°C$ to $150\,°C$

NOTE 1: These ratings apply except for programming pins during a programming cycle.

recommended operating conditions

		SN54ALS'			SN74ALS'			UNIT
		MIN	NOM	MAX	MIN	NOM	MAX	
V_{CC}	Supply voltage	4.5	5	5.5	4.5	5	5.5	V
V_{IH}	High-level input voltage	2		5.5	2		5.5	V
V_{IL}	Low-level input voltage			0.8			0.8	V
I_{OH}	High-level output current			-1			-2.6	mA
I_{OL}	Low-level output current			12			24	mA
T_A	Operating free-air temperature	-55		125	0		70	°C

electrical characteristics over recommended operating free-air temperature range (unless otherwise noted)

PARAMETER		TEST CONDITIONS		SN54ALS'			SN74ALS'			UNIT
				MIN	TYP†	MAX	MIN	TYP†	MAX	
V_{IK}		$V_{CC} = 4.5$ V,	$I_I = -18$ mA			-1.5			-1.5	V
V_{OH}		$V_{CC} = 4.5$ V to 5.5 V,	$I_{OH} = -0.4$ mA	$V_{CC}-2$			$V_{CC}-2$			V
		$V_{CC} = 4.5$ V,	$I_{OH} = -1$ mA	2.4	3					
		$V_{CC} = 4.5$ V,	$I_{OH} = -2.6$ mA				2.4	2.9		
V_{OL}		$V_{CC} = 4.5$ V,	$I_{OL} = 12$ mA		0.25	0.4		0.25	0.4	V
		$V_{CC} = 4.5$ V,	$I_{OL} = 24$ mA					0.36	0.5	
I_I		$V_{CC} = 5.5$ V,	$V_I = 5.5$ V			0.1			0.1	mA
I_{IH}		$V_{CC} = 5.5$ V,	$V_I = 2.7$ V			20			20	μA
I_{IL}		$V_{CC} = 5.5$ V,	$V_{IL} = 0.4$ V			-0.2			-0.2	mA
I_O‡		$V_{CC} = 5.5$ V,	$V_O = 2.25$ V	-30		-130	-30		-130	mA
I_{CC}	'ALS526	$V_{CC} = 5.5$ V, All P inputs at 4.5 V, $\overline{G}$ input at GND			14			14		mA
	'ALS527				13			13		
	'ALS528				13			13		

†All typical values are at $V_{CC} = 5$ V, $T_A = 25\,°C$.
‡The output conditions have been chosen to produce a current that closely approximates one half of the tue short-circuit output current, I_{OS}.

switching characteristics (see Note 2)

PARAMETER	FROM (INPUT)	TO (OUTPUT)	$V_{CC} = 4.5$ V to 5.5 V, $C_L = 50$ pF, $R_L = 500\,\Omega$, $T_A =$ MIN to MAX						UNIT
			SN54ALS'			SN74ALS'			
			MIN	TYP†	MAX	MIN	TYP†	MAX	
t_{PLH}	P or Q	$\overline{P=Q}$		8			8		ns
t_{PHL}				9			9		
t_{PLH}	$\overline{G}$	$\overline{P=Q}$		6			6		ns
t_{PHL}				6			6		

†All typical values are at $V_{CC} = 5$ V, $T_A = 25\,°C$.
NOTE 2: For load circuit and voltage waveforms, see page 1-12.

TEXAS INSTRUMENTS
POST OFFICE BOX 225012 • DALLAS, TEXAS 75265

TYPES SN54ALS526, SN54ALS527, SN54ALS528
SN74ALS526, SN74ALS527, SN54ALS528
FUSE-PROGRAMMABLE IDENTITY COMPARATORS

programming parameters

	PARAMETER		MIN	MAX	UNIT
V_{IH}	High-level input voltage		2	5.5	V
V_{IL}	Low-level input voltage			0.8	V
V_{IHH}	Program-pulse input voltage		10.5	12	V
V_{CC}	Supply voltage		5	7	V
I_{IHH}	Program-pulse input current	Pn ($\overline{G}$ low)		2.08	mA
		$\overline{G}$		1.24	
I_{CCHH}	Supply current with V_{IHH} applied	'ALS526			mA
		'ALS527			
		'ALS528			
t_w	Pulse duration, program		10	50	µs
t_r	Rise time, program voltage			10	µs

programming waveforms

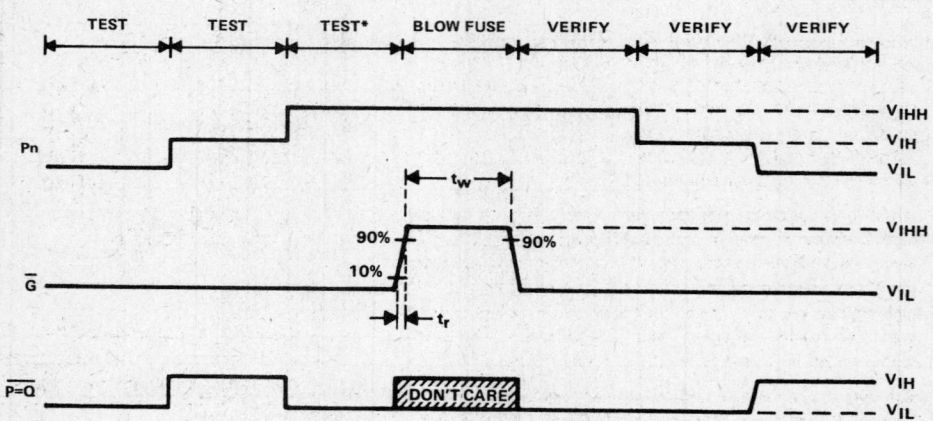

*This test is only true if all other P inputs are at V_{IH}.

ALS AND AS CIRCUITS

TYPES SN54ALS533, SN54AS533, SN74ALS533, SN74AS533
OCTAL D-TYPE TRANSPARENT LATCHES WITH 3-STATE OUTPUTS

D2661, APRIL 1982 — REVISED DECEMBER 1983

- 8-Latches In a Single Package
- 3-State Bus-Driving Inverting Outputs
- Full Parallel Access for Loading
- Buffered Control Inputs
- P-N-P Inputs Reduce D-C Loading on Data Lines
- Package Options Include Both Plastic and Ceramic Chip Carriers in Addition to Plastic and Ceramic DIPs
- Dependable Texas Instruments Quality and Reliability

SN54ALS533, SN54AS533 . . . J PACKAGE
SN74ALS533, SN74AS533 . . . N PACKAGE
(TOP VIEW)

```
       OC  [ 1   U  20 ] VCC
       1Q  [ 2      19 ] 8Q
       1D  [ 3      18 ] 8D
       2D  [ 4      17 ] 7D
       2Q  [ 5      16 ] 7Q
       3Q  [ 6      15 ] 6Q
       3D  [ 7      14 ] 6D
       4D  [ 8      13 ] 5D
       4Q  [ 9      12 ] 5Q
       GND [10      11 ] C
```

SN54ALS533, SN54AS533 . . . FH PACKAGE
SN74ALS533, SN74AS533 . . . FN PACKAGE
(TOP VIEW)

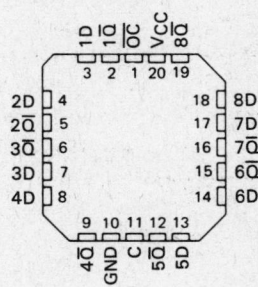

description

These 8-bit latches feature three-state outputs designed specifically for driving highly capacitive or relatively low-impedance loads. They are particularly suitable for implementing buffer registers, I/O ports, bidirectional bus drivers, and working registers.

The eight latches of the 'ALS533 and 'AS533 are transparent D-type latches. While the enable (C) is high, the $\overline{Q}$ outputs will follow the complements of the D inputs. When the enable is taken low, the $\overline{Q}$ outputs will be latched at the inverses of the levels that were set up at the D inputs. The 'ALS533 and 'AS533 are functionally equivalent to the 'ALS373 and 'AS373 except for having inverted outputs.

A buffered output-control ($\overline{OC}$) input can be used to place the eight outputs in either a normal logic state (high or low logic levels) or a high-impedance state. In the high-impedance state the outputs neither load nor drive the bus lines significantly. The high-impedance third state and increased drive provide the capability to drive the bus lines in a bus-organized system without need for interface or pull-up components.

The output control does not affect the internal operation of the latches. Old data can be retained or new data can be entered while the outputs are off.

The SN54ALS533 and SN54AS533 are characterized for operation over the full military temperature range of −55°C to 125°C. The SN74ALS533 and SN74AS533 are characterized for operation from 0°C to 70°C.

FUNCTION TABLE (EACH LATCH)

INPUTS			OUTPUT
$\overline{OC}$	ENABLE C	D	$\overline{Q}$
L	H	H	L
L	H	L	H
L	L	X	$\overline{Q}_0$
H	X	X	Z

ALS AND AS CIRCUITS

Copyright © 1982 by Texas Instruments Incorporated

TEXAS INSTRUMENTS
POST OFFICE BOX 225012 • DALLAS, TEXAS 75265

TYPES SN54ALS533, SN54AS533, SN74ALS533, SN74AS533
OCTAL D-TYPE TRANSPARENT LATCHES WITH 3-STATE OUTPUTS

logic symbol

logic diagram (positive logic)

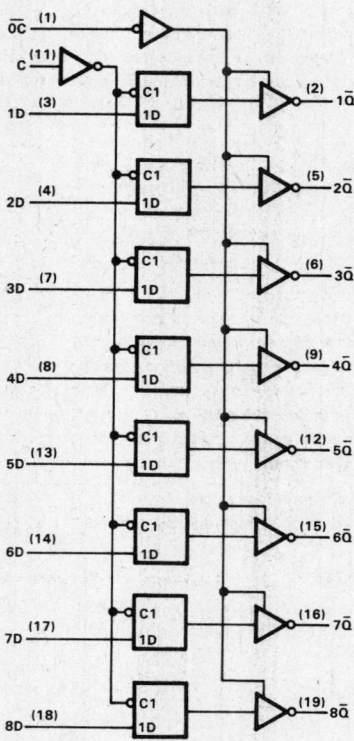

Pin numbers shown are for J and N packages.

absolute maximum ratings over operating free-air temperature range (unless otherwise noted)

Supply voltage, V_{CC} .... 7 V
Input voltage .... 7 V
Voltage applied to a disabled 3-state output .... 5.5 V
Operating free-air temperature range: SN54ALS533, SN54AS533 .... −55°C to 125°C
                                      SN74ALS533, SN74AS533 .... 0°C to 70°C
Storage temperature range .... −65°C to 150°C

TEXAS
INSTRUMENTS
POST OFFICE BOX 225012 • DALLAS, TEXAS 75265

TYPES SN54ALS533, SN74ALS533
OCTAL D-TYPE TRANSPARENT LATCHES WITH 3-STATE OUTPUTS

recommended operating conditions

		SN54ALS533			SN74ALS533			UNIT
		MIN	NOM	MAX	MIN	NOM	MAX	
V_{CC}	Supply voltage	4.5	5	5.5	4.5	5	5.5	V
V_{IH}	High-level input voltage	2			2			V
V_{IL}	Low-level input voltage			0.8			0.8	V
I_{OH}	High-level output current			−1			−2.6	mA
I_{OL}	Low-level output current			12			24	mA
t_w	Pulse duration, enable C high	15			15			ns
t_{su}	Setup time, data before enable C↓	15			15			ns
t_h	Hold time, data after enable C↓	7			7			ns
T_A	Operating free-air temperature	−55		125	0		70	°C

electrical characteristics over recommended operating free-air temperature range (unless otherwise noted)

PARAMETER	TEST CONDITIONS		SN54ALS533			SN74ALS533			UNIT
			MIN	TYP†	MAX	MIN	TYP†	MAX	
V_{IK}	V_{CC} = 4.5 V,	I_I = −18 mA			−1.5			−1.5	V
V_{OH}	V_{CC} = 4.5 V to 5.5 V,	I_{OH} = −0.4 mA	V_{CC}−2			V_{CC}−2			V
	V_{CC} = 4.5 V,	I_{OH} = −1 mA	2.4	3.3					
	V_{CC} = 4.5 V,	I_{OH} = −2.6 mA				2.4	3.2		
V_{OL}	V_{CC} = 4.5 V,	I_{OL} = 12 mA		0.25	0.4		0.25	0.4	V
	V_{CC} = 4.5 V,	I_{OL} = 24 mA					0.35	0.5	
I_{OZH}	V_{CC} = 5.5 V,	V_O = 2.7 V			20			20	μA
I_{OZL}	V_{CC} = 5.5 V,	V_I = 0.4 V			−20			−20	μA
I_I	V_{CC} = 5.5 V,	V_I = 7 V			0.1			0.1	mA
I_{IH}	V_{CC} = 5.5 V,	V_I = 2.7 V			20			20	μA
I_{IL}	V_{CC} = 5.5 V,	V_I = 0.4 V			−0.1			−0.1	mA
I_O‡	V_{CC} = 5.5 V,	V_O = 2.25 V	−30		−112	−30		−112	mA
I_{CC}	V_{CC} = 5.5 V	Outputs high		10	17		10	17	mA
		Outputs low		17	26		17	26	
		Outputs disabled		18.5	28		18.5	28	

†All typical values are at V_{CC} = 5 V, T_A = 25°C.
‡The output conditions have been chosen to produce a current that closely approximates one half of the true short-circuit output current, I_{OS}.

ALS AND AS CIRCUITS

TEXAS INSTRUMENTS
POST OFFICE BOX 225012 • DALLAS, TEXAS 75265

TYPES SN54ALS533, SN74ALS533
OCTAL D-TYPE TRANSPARENT LATCHES WITH 3-STATE OUTPUTS

switching characteristics (see Note 1)

PARAMETER	FROM (INPUT)	TO (OUTPUT)	V_{CC} = 4.5 V to 5.5 V C_L = 50 pF, R1 = 500 Ω, R2 = 500 Ω, T_A = MIN to MAX				UNIT
			SN54ALS533		SN74ALS533		
			MIN	MAX	MIN	MAX	
t_{PLH}	D	$\overline{Q}$	4	24	4	19	ns
t_{PHL}			4	14	4	13	
t_{PLH}	C	Any $\overline{Q}$	5	28	5	23	ns
t_{PHL}			4	21	4	18	
t_{PZH}	$\overline{OC}$	Any $\overline{Q}$	4	19	4	17	ns
t_{PZL}			4	20	4	18	
t_{PHZ}	$\overline{OC}$	Any $\overline{Q}$	2	12	2	10	ns
t_{PLZ}			3	22	3	16	

NOTE 1: For load circuit and voltage waveforms, see page 1-12.

D latch signal conventions

It is TI practice to name the outputs and other inputs of a D-type latch and to draw its logic symbol based on the assumption of true data (D) inputs. Then outputs that produce data in phase with the data inputs are called Q and those producing complementary data are called $\overline{Q}$. An input that causes a Q output to go high or a $\overline{Q}$ output to go low is called Preset; an input that causes a $\overline{Q}$ output to go high or a Q output to go low is called Clear. Bars are used over these pin names ($\overline{PRE}$ and $\overline{CLR}$) if they are active low.

In some applications it may be advantageous to redesignate the data input $\overline{D}$. In that case all the other inputs and outputs should be renamed as shown below. Also shown are corresponding changes in the graphical symbol. Arbitrary pin numbers are shown in parentheses.

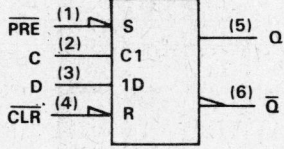 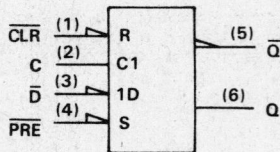

Notice that Q and $\overline{Q}$ exchange names, which causes Preset and Clear to do likewise. Also notice that the polarity indicators ($\triangleright$) on $\overline{PRE}$ and $\overline{CLR}$ remain since these inputs are still active-low, but that the presence or absence of the polarity indicator changes at $\overline{D}$, Q, and $\overline{Q}$. Of course pin 5 ($\overline{Q}$) is still in phase with the data input $\overline{D}$, but now both are considered active-low.

TYPES SN54AS533, SN74AS533
OCTAL D-TYPE TRANSPARENT LATCHES WITH 3-STATE OUTPUTS

recommended operating conditions

		SN54AS533			SN74AS533			UNIT
		MIN	NOM	MAX	MIN	NOM	MAX	
V_{CC}	Supply voltage	4.5	5	5.5	4.5	5	5.5	V
V_{IH}	High-level input voltage	2			2			V
V_{IL}	Low-level input voltage			0.8			0.8	V
I_{OH}	High-level output current			−12			−15	mA
I_{OL}	Low-level output current			32			48	mA
t_w	Pulse duration, enable C high	3			2			ns
t_{su}	Setup time, data before enable C↓	2			2			ns
t_h	Hold time, data after enable C↓	3			3			ns
T_A	Operating free-air temperature	−55		125	0		70	°C

electrical characteristics over recommended operating free-air temperature range (unless otherwise noted)

PARAMETER	TEST CONDITIONS		SN54AS533			SN74AS533			UNIT
			MIN	TYP†	MAX	MIN	TYP†	MAX	
V_{IK}	V_{CC} = 4.5 V,	I_I = −18 mA			−1.2			−1.2	V
V_{OH}	V_{CC} = 4.5 V to 5.5 V,	I_{OH} = −2 mA	$V_{CC}-2$			$V_{CC}-2$			V
	V_{CC} = 4.5 V,	I_{OH} = −12 mA	2.4	3.2					
	V_{CC} = 4.5 V,	I_{OH} = −15 mA				2.4	3.3		
V_{OL}	V_{CC} = 4.5 V,	I_{OL} = 32 mA		0.29	0.5				V
	V_{CC} = 4.5 V,	I_{OL} = 48 mA					0.34	0.5	
I_{OZH}	V_{CC} = 5.5 V,	V_O = 2.7 V			50			50	µA
I_{OZL}	V_{CC} = 5.5 V,	V_I = 0.4 V			−50			−50	µA
I_I	V_{CC} = 5.5 V,	V_I = 7 V			0.1			0.1	mA
I_{IH}	V_{CC} = 5.5 V,	V_I = 2.7 V			20			20	µA
I_{IL}	V_{CC} = 5.5 V,	V_I = 0.5 V		−0.02	−0.5		−0.02	−0.5	mA
I_O‡	V_{CC} = 5.5 V,	V_O = 2.25 V	−30		−112	−30		−112	mA
I_{CC}	V_{CC} = 5.5 V	Outputs high		62	100		62	100	mA
		Outputs low		64	100		64	100	
		Outputs disabled		71	110		71	110	

†All typical values are at V_{CC} = 5 V, T_A = 25°C.
‡The output conditions have been chosen to produce a current that closely approximates one half of the true short-circuit output current, I_{OS}.

TEXAS INSTRUMENTS
POST OFFICE BOX 225012 • DALLAS, TEXAS 75265

TYPES SN54AS533, SN74AS533
OCTAL D-TYPE TRANSPARENT LATCHES WITH 3-STATE OUTPUTS

switching characteristics (see Note 1)

PARAMETER	FROM (INPUT)	TO (OUTPUT)	V_{CC} = 4.5 V to 5.5 V, C_L = 50 pF, R1 = 500 Ω, R2 = 500 Ω, T_A = MIN to MAX				UNIT
			SN54AS533		SN74AS533		
			MIN	MAX	MIN	MAX	
t_{PLH}	D	$\overline{Q}$	4	10	4	7.5	ns
t_{PHL}			4	8	4	7	
t_{PLH}	C	Any $\overline{Q}$	5	11	5	9	ns
t_{PHL}			4.5	8.5	4.5	8	
t_{PZH}	$\overline{OC}$	Any $\overline{Q}$	2	7.5	2	6.5	ns
t_{PZL}			4.5	10.5	4.5	9.5	
t_{PHZ}	$\overline{OC}$	Any $\overline{Q}$	3	7.5	3	6.5	ns
t_{PLZ}			3	8	3	7	

NOTE 1: For load circuit and voltage waveforms, see page 1-12.

ALS AND AS CIRCUITS

TEXAS INSTRUMENTS
POST OFFICE BOX 225012 • DALLAS, TEXAS 75265

TYPES SN54ALS534, SN54AS534, SN74ALS534, SN74AS534
OCTAL D-TYPE EDGE-TRIGGERED FLIP-FLOPS WITH 3-STATE OUTPUTS

D2661, APRIL 1982—REVISED DECEMBER 1983

- 3-State Bus-Driving Inverting Outputs
- Buffered Control Inputs
- Package Options Include Both Plastic and Ceramic Chip Carriers in Addition to Plastic and Ceramic DIPs
- Dependable Texas Instruments Quality and Reliability

description

These 8-bit flip-flops feature three-state outputs designed specifically for driving highly capacitive or relatively low-impedance loads. They are particularly attractive for implementing buffer registers, I/O ports, bidirectional bus drivers, and working registers.

The eight flip-flops of the 'ALS534 and 'AS534 are edge-triggered D-type flip-flops. On the positive transition of the clock, the Q outputs will be set to the complement of the logic states that were set up at the D inputs. The 'ALS534 and 'AS534 are functionally equivalent to the 'ALS374 and 'AS374 except for having inverted outputs.

A buffered output-control input can be used to place the eight outputs in either a normal logic state (high or low logic levels) or a high-impedance state. In the high-impedance state the outputs neither load nor drive the bus lines significantly. The high-impedance third state and increased drive provide the capability to drive the bus lines in a bus-organized system without need for interface or pull-up components.

The output control does not affect the internal operation of the flip-flops. Old data can be retained or new data can be entered while the outputs are off.

The SN54ALS534 and SN54AS534 are characterized for operation over the full military temperature range of $-55\,°C$ to $125\,°C$. The SN74ALS534 and SN74AS534 are characterized for operation from $0\,°C$ to $70\,°C$.

SN54ALS534, SN54AS534 . . . J PACKAGE
SN74ALS534, SN74AS534 . . . N PACKAGE
(TOP VIEW)

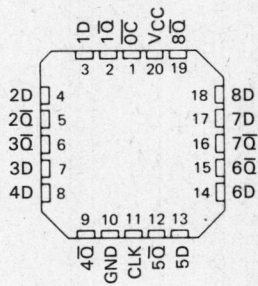

SN54ALS534, SN54AS534 . . . FH PACKAGE
SN74ALS534, SN74AS534 . . . FN PACKAGE
(TOP VIEW)

FUNCTION TABLE (EACH FLIP-FLOP)

INPUTS			OUTPUT
$\overline{OC}$	CLK	D	$\overline{Q}$
L	↑	H	L
L	↑	L	H
L	L	X	$\overline{Q}_0$
H	X	X	Z

Copyright © 1983 by Texas Instruments Incorporated

TYPES SN54ALS534, SN54AS534, SN74ALS534, SN74AS534
OCTAL D-TYPE EDGE-TRIGGERED FLIP-FLOPS WITH 3-STATE OUTPUTS

logic symbol

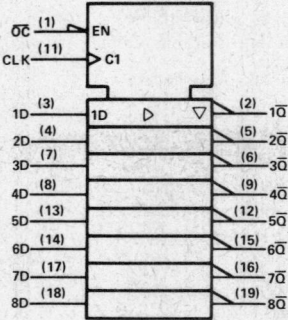

logic diagram (positive logic)

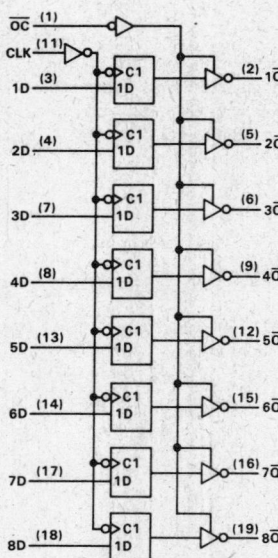

Pin numbers shown are for J and N packages.

absolute maximum ratings over operating free-air temperature range (unless otherwise noted)

Supply voltage, V_{CC} .... 7 V
Input voltage .... 7 V
Voltage applied to a disabled 3-state output .... 5.5 V
Operating free-air temperature range: SN54ALS534, SN54AS534 .... −55°C to 125°C
                                            SN74ALS534, SN74AS534 .... 0°C to 70°C
Storage temperature range .... −65°C to 150°C

ALS AND AS CIRCUITS

TEXAS INSTRUMENTS
POST OFFICE BOX 225012 • DALLAS, TEXAS 75265

TYPES SN54ALS534, SN74ALS534
OCTAL D-TYPE EDGE-TRIGGERED FLIP-FLOPS WITH 3-STATE OUTPUTS

recommended operating conditions

			SN54ALS534			SN74ALS534			UNIT
			MIN	NOM	MAX	MIN	NOM	MAX	
V_{CC}	Supply voltage		4.5	5	5.5	4.5	5	5.5	V
V_{IH}	High-level input voltage		2			2			V
V_{IL}	Low-level input voltage				0.8			0.8	V
I_{OH}	High-level output current				−1			−2.6	mA
I_{OL}	Low-level output current				12			24	mA
f_{clock}	Clock frequency		0		30	0		35	MHz
t_w	Pulse duration	CLK high	16.5			14			ns
		CLK low	16.5			14			
t_{su}	Setup time, data before CLK↑		10			10			ns
t_h	Hold time, data after CLK↑		0			0			ns
T_A	Operating free-air temperature		−55		125	0		70	°C

electrical characteristics over recommended operating free-air temperature range (unless otherwise noted)

PARAMETER		TEST CONDITIONS		SN54ALS534			SN74ALS534			UNIT
				MIN	TYP†	MAX	MIN	TYP†	MAX	
V_{IK}		$V_{CC} = 4.5$ V,	$I_I = -18$ mA			−1.5			−1.5	V
V_{OH}		$V_{CC} = 4.5$ V to 5.5 V,	$I_{OH} = -0.4$ mA	$V_{CC}-2$			$V_{CC}-2$			V
		$V_{CC} = 4.5$ V,	$I_{OH} = -1$ mA	2.4	3.3					
		$V_{CC} = 4.5$ V,	$I_{OH} = -2.6$ mA				2.4	3.2		
V_{OL}		$V_{CC} = 4.5$ V,	$I_{OL} = 12$ mA		0.25	0.4		0.25	0.4	V
		$V_{CC} = 4.5$ V,	$I_{OL} = 24$ mA					0.35	0.5	
I_{OZH}		$V_{CC} = 5.5$ V,	$V_O = 2.7$ V			20			20	µA
I_{OZL}		$V_{CC} = 5.5$ V,	$V_I = 0.4$ V			−20			−20	µA
I_I		$V_{CC} = 5.5$ V,	$V_I = 7$ V			0.1			0.1	mA
I_{IH}		$V_{CC} = 5.5$ V,	$V_I = 2.7$ V			20			20	µA
I_{IL}	CLK, $\overline{OC}$	$V_{CC} = 5.5$ V,	$V_I = 0.4$ V			−0.1			−0.1	mA
	D					−0.2			−0.2	
I_O‡		$V_{CC} = 5.5$ V,	$V_O = 2.25$ V	−30		−112	−30		−112	mA
I_{CC}		$V_{CC} = 5.5$ V	Outputs high		11	19		11	19	mA
			Outputs low		19	28		19	28	
			Outputs disabled		10	31		20	31	

†All typical values are at $V_{CC} = 5$ V, $T_A = 25$°C.
‡The output conditions have been chosen to produce a current that closely approximates one half of the true short-circuit output current, I_{OS}.

ALS AND AS CIRCUITS

Texas Instruments
POST OFFICE BOX 225012 • DALLAS, TEXAS 75265

TYPES SN54ALS534, SN74ALS534
OCTAL D-TYPE EDGE-TRIGGERED FLIP-FLOPS WITH 3-STATE OUTPUTS

switching characteristics (see Note 1)

PARAMETER	FROM (INPUT)	TO (OUTPUT)	V_{CC} = 4.5 V to 5.5 V, C_L = 50 pF, R1 = 500 Ω, R2 = 500 Ω, T_A = MIN to MAX				UNIT
			SN54ALS534		SN74ALS534		
			MIN	MAX	MIN	MAX	
f_{max}			30		35		MHz
t_{PLH}	CLK	Any $\overline{Q}$	3	15	3	12	ns
t_{PHL}			5	18	5	16	
t_{PZH}	$\overline{OC}$	Any $\overline{Q}$	5	19	5	17	ns
t_{PZL}			7	20	7	18	
t_{PHZ}	$\overline{OC}$	Any $\overline{Q}$	2	12	2	10	ns
t_{PLZ}			2	16	2	14	

NOTE 1: For load circuit and voltage waveforms, see page 1-12.

D flip-flop signal conventions

It is TI practice to name the outputs and other inputs of a D-type flip-flop and to draw its logic symbol based on the assumption of true data (D) inputs. Then outputs that produce data in phase with the data inputs are called Q and those producing complementary data are called $\overline{Q}$. An input that causes a Q output to go high or a $\overline{Q}$ output to go low is called Preset; an input that causes a $\overline{Q}$ output to go high or a Q output to go low is called Clear. Bars are used over these pin names ($\overline{PRE}$ and $\overline{CLR}$) if they are active low.

In some applications it may be advantageous to redesignate the data input $\overline{D}$. In that case all the other inputs and outputs should be renamed as shown below. Also shown are corresponding changes in the graphical symbol. Arbitrary pin numbers are shown in parentheses.

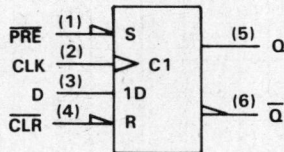
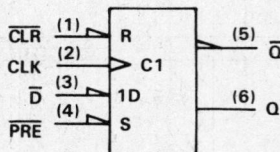

Notice that Q and $\overline{Q}$ exchange names, which causes Preset and Clear to do likewise. Also notice that the polarity indicators (▷) on $\overline{PRE}$ and $\overline{CLR}$ remain since these inputs are still active-low, but that the presence or absence of the polarity indicator changes at $\overline{D}$, Q, and $\overline{Q}$. Of course pin 5 ($\overline{Q}$) is still in phase with the data input $\overline{D}$, but now both are considered active-low.

TYPES SN54AS534, SN74AS534
OCTAL D-TYPE EDGE-TRIGGERED FLIP-FLOPS WITH 3-STATE OUTPUTS

recommended operating conditions

			SN54AS534			SN74AS534			UNIT
			MIN	NOM	MAX	MIN	NOM	MAX	
V_{CC}	Supply voltage		4.5	5	5.5	4.5	5	5.5	V
V_{IH}	High-level input voltage		2			2			V
V_{IL}	Low-level input voltage				0.8			0.8	V
I_{OH}	High-level output current				−12			−15	mA
I_{OL}	Low-level output current				32			48	mA
f_{clock}	Clock frequency		0		100	0		125	MHz
t_w	Pulse duration	CLK high	5.5			4			ns
		CLK low	5			3			
t_{su}	Setup time, data before CLK↑		3			2			ns
t_h	Hold time, data after CLK↑		3			2			ns
T_A	Operating free-air temperature		−55		125	0		70	°C

electrical characteristics over recommended operating free-air temperature range (unless otherwise noted)

PARAMETER		TEST CONDITIONS		SN54AS534			SN74AS534			UNIT
				MIN	TYP[†]	MAX	MIN	TYP[†]	MAX	
V_{IK}		V_{CC} = 4.5 V,	I_I = −18 mA			−1.2			−1.2	V
V_{OH}		V_{CC} = 4.5 V to 5.5 V,	I_{OH} = −2 mA	V_{CC}−2			V_{CC}−2			V
		V_{CC} = 4.5 V,	I_{OH} = −12 mA	2.4	3.2					
		V_{CC} = 4.5 V,	I_{OH} = −15 mA				2.4	3.3		
V_{OL}		V_{CC} = 4.5 V,	I_{OL} = 32 mA		0.29	0.5				V
		V_{CC} = 4.5 V,	I_{OL} = 48 mA					0.34	0.5	
I_{OZH}		V_{CC} = 5.5 V,	V_O = 2.7 V			50			50	µA
I_{OZL}		V_{CC} = 5.5 V,	V_I = 0.4 V			−50			−50	µA
I_I		V_{CC} = 5.5 V,	V_I = 7 V			0.1			0.1	mA
I_{IH}		V_{CC} = 5.5 V,	V_I = 2.7 V			20			20	µA
I_{IL}	OC, CLK	V_{CC} = 5.5 V,	V_I = 0.4 V			−0.5			−0.5	mA
	D					−3			−2	
I_O[‡]		V_{CC} = 5.5 V,	V_O = 2.25 V	−30		−112	−30		−112	mA
I_{CC} mA		V_{CC} = 5.5 V	Outputs high		77	120		77	120	mA
			Outputs low		84	128		84	128	
			Outputs disabled		84	128		84	128	

[†] All typical values are at V_{CC} = 5 V, T_A = 25°C.
[‡] The output conditions have been chosen to produce a current that closely approximates one half of the true short-circuit output current, I_{OS}.

ALS AND AS CIRCUITS

TEXAS INSTRUMENTS
POST OFFICE BOX 225012 • DALLAS, TEXAS 75265

TYPES SN54AS534, SN74AS534
OCTAL D-TYPE EDGE-TRIGGERED FLIP-FLOPS WITH 3-STATE OUTPUTS

switching characteristics (see Note 1)

PARAMETER	FROM (INPUT)	TO (OUTPUT)	V_{CC} = 4.5 V to 5.5 V, C_L = 50 pF, R1 = 500 Ω, R2 = 500 Ω, T_A = MIN to MAX				UNIT
			SN54AS534		SN74AS534		
			MIN	MAX	MIN	MAX	
f_{max}			100		125		MHz
t_{PLH}	CLK	Any $\overline{Q}$	3	11	3	8	ns
t_{PHL}			4	11.5	4	9	
t_{PZH}	$\overline{OC}$	Any $\overline{Q}$	2	7	2	6	ns
t_{PZL}			3	11	3	10	
t_{PHZ}	$\overline{OC}$	Any $\overline{Q}$	2	7	2	6	ns
t_{PLZ}			2	7	2	6	

NOTE 1: For load circuit and voltage waveforms, see page 1-12.

TYPES SN54ALS538, SN74ALS538
3-LINE TO 8-LINE DECODERS/DEMULTIPLEXERS WITH 3-STATE OUTPUTS

D2661, APRIL 1982 – REVISED DECEMBER 1983

- 3-State Outputs
- Output Polarity Control
- Data Multiplexing Capability
- Multiple Enables for Expansion
- Package Options Include Both Plastic and Ceramic Chip Carriers in Addition to Plastic and Ceramic DIPs
- Dependable Texas Instruments Quality and Reliability

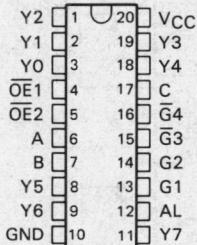

SN54ALS538 . . . J PACKAGE
SN74ALS538 . . . N PACKAGE
(TOP VIEW)

description

The 'ALS538 decoder/demultiplexer accepts three address input signals (A, B, C) and decodes them to select one-of-eight mutually exclusive outputs. If the polarity control input (AL) is high, the outputs are active-low; if AL is low, the outputs are active-high. Two active-high and two active-low input enables are available for easy expansion to 1-of-32 decoding with four packages, or for data demultiplexing to 1-of-8 or 1-of-16 destinations. A high signal on either of the output enables ($\overline{OE1}$ and $\overline{OE2}$) forces all outputs to the high-impedance state.

The SN54ALS538 is characterized for operation over the full military temperature range of $-55\,°C$ to $125\,°C$. The SN74ALS538 is characterized for operation from $0\,°C$ to $70\,°C$.

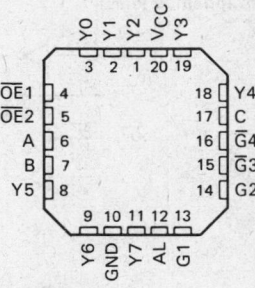

SN54ALS538 . . . FH PACKAGE
SN74ALS538 . . . FN PACKAGE
(TOP VIEW)

ALS AND AS CIRCUITS

FUNCTION TABLE (EACH DECODER/DEMULTIPLEXER)

FUNCTION	INPUTS									OUTPUTS							
	$\overline{OE1}$	$\overline{OE2}$	G1	G2	$\overline{G3}$	$\overline{G4}$	C	B	A	Y0	Y1	Y2	Y3	Y4	Y5	Y6	Y7
High impedance (AL = X)	H	X	X	X	X	X	X	X	X	Z	Z	Z	Z	Z	Z	Z	Z
	X	H	X	X	X	X	X	X	X	Z	Z	Z	Z	Z	Z	Z	Z
Disable (AL = X)	L	L	L	X	X	X	X	X	X	All outputs same level as AL							
	L	L	X	L	X	X	X	X	X								
	L	L	X	X	H	X	X	X	X								
	L	L	X	X	X	H	X	X	X								
Active-High Output (AL = L)	L	L	H	H	L	L	L	L	L	H	L	L	L	L	L	L	L
	L	L	H	H	L	L	L	L	H	L	H	L	L	L	L	L	L
	L	L	H	H	L	L	L	H	L	L	L	H	L	L	L	L	L
	L	L	H	H	L	L	L	H	H	L	L	L	H	L	L	L	L
	L	L	H	H	L	L	H	L	L	L	L	L	L	H	L	L	L
	L	L	H	H	L	L	H	L	H	L	L	L	L	L	H	L	L
	L	L	H	H	L	L	H	H	L	L	L	L	L	L	L	H	L
	L	L	H	H	L	L	H	H	H	L	L	L	L	L	L	L	H
Active-Low Output (AL = H)	L	L	H	H	L	L	L	L	L	L	H	H	H	H	H	H	H
	L	L	H	H	L	L	L	L	H	H	L	H	H	H	H	H	H
	L	L	H	H	L	L	L	H	L	H	H	L	H	H	H	H	H
	L	L	H	H	L	L	L	H	H	H	H	H	L	H	H	H	H
	L	L	H	H	L	L	H	L	L	H	H	H	H	L	H	H	H
	L	L	H	H	L	L	H	L	H	H	H	H	H	H	L	H	H
	L	L	H	H	L	L	H	H	L	H	H	H	H	H	H	L	H
	L	L	H	H	L	L	H	H	H	H	H	H	H	H	H	H	L

PRODUCT PREVIEW

This document contains information on a product under development. Texas Instruments reserves the right to change or discontinue this product without notice.

TEXAS INSTRUMENTS
POST OFFICE BOX 225012 • DALLAS, TEXAS 75265

TYPES SN54ALS538, SN74ALS538
3-LINE TO 8-LINE DECODERS/DEMULTIPLEXERS WITH 3-STATE OUTPUTS

logic symbols (alternatives)

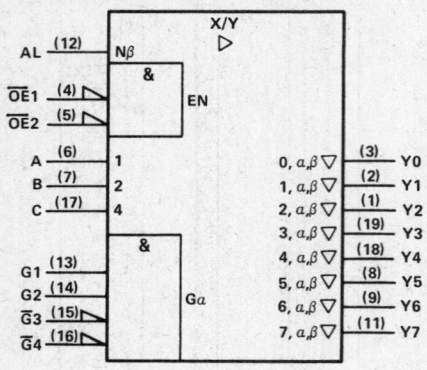
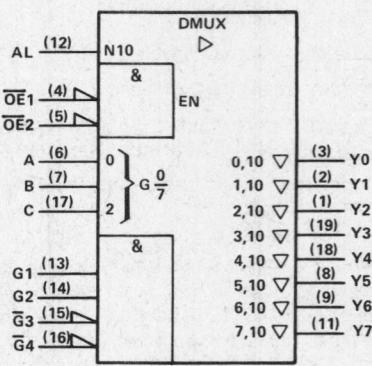

logic diagram (positive logic)

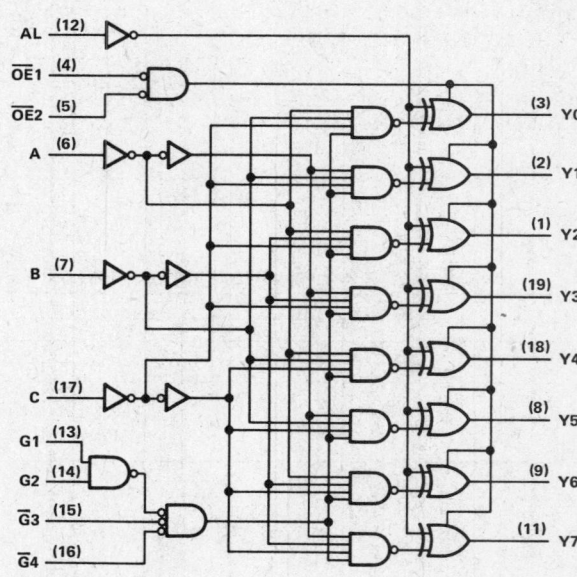

Pin numbers shown are for J and N packages.

absolute maximum ratings over operating free-air temperature range (unless otherwise noted)

Supply voltage, V_{CC} .... 7 V
Input voltage .... 7 V
Voltage applied to a disabled 3-state output .... 5.5 V
Operating free-air temperature range: SN54ALS538 .... −55°C to 125°C
                                       SN74ALS538 .... 0°C to 70°C
Storage temperature range .... −65°C to 150°C

TEXAS INSTRUMENTS
POST OFFICE BOX 225012 • DALLAS, TEXAS 75265

TYPES SN54ALS538, SN74ALS538
3-LINE TO 8-LINE DECODERS/DEMULTIPLEXERS WITH 3-STATE OUTPUTS

recommended operating conditions

		SN54ALS538			SN74ALS538			UNIT
		MIN	NOM	MAX	MIN	NOM	MAX	
V_{CC}	Supply voltage	4.5	5	5.5	4.5	5	5.5	V
V_{IH}	High-level input voltage	2			2			V
V_{IL}	Low-level input voltage			0.8			0.8	V
I_{OH}	High-level output current			−1			−2.6	mA
I_{OL}	Low-level output current			12			24	mA
T_A	Operating free-air temperature	−55		125	0		70	°C

electrical characteristics over recommended operating free-air temperature range (unless otherwise noted)

PARAMETER		TEST CONDITIONS		SN54ALS538			SN74ALS538			UNIT
				MIN	TYP†	MAX	MIN	TYP†	MAX	
V_{IK}		V_{CC} = 4.5 V,	I_I = −18 mA			−1.5			−1.5	V
V_{OH}		V_{CC} = 4.5 V to 5.5 V,	I_{OH} = −0.4 mA	$V_{CC}-2$			$V_{CC}-2$			V
		V_{CC} = 4.5 V,	I_{OH} = −1 mA	2.4	3.3					
		V_{CC} = 4.5 V,	I_{OH} = −2.6 mA				2.4	3.2		
V_{OL}		V_{CC} = 4.5 V,	I_{OL} = 12 mA		0.25	0.4		0.25	0.4	V
		V_{CC} = 4.5 V,	I_{OL} = 24 mA					0.35	0.5	
I_{OZH}		V_{CC} = 5.5 V,	V_O = 2.7 V			20			20	µA
I_{OZL}		V_{CC} = 5.5 V,	V_O = 0.4 V			−20			−20	µA
I_I		V_{CC} = 5.5 V,	V_I = 7 V			0.1			0.1	mA
I_{IH}		V_{CC} = 5.5 V,	V_I = 2.7 V			20			20	µA
I_{IL}	A, B, C or AL	V_{CC} = 5.5 V,	V_I = 0.4 V			−0.2			−0.2	mA
	All other					−0.1			−0.1	
I_O‡		V_{CC} = 5.5 V,	V_O = 2.25 V	−30		−112	−30		−112	mA
I_{CC}		V_{CC} = 5.5 V	Outputs high							mA
			Outputs low							
			Outputs disabled		25			25		

†All typical values are at V_{CC} = 5 V, T_A = 25°C.
‡The output conditions have been chosen to produce a current that closely approximates one half of the true short-circuit output current, I_{OS}.

Additional information on these products can be obtained from the factory as it becomes available.

TEXAS INSTRUMENTS
POST OFFICE BOX 225012 • DALLAS, TEXAS 75265

TYPES SN54ALS538, SN74ALS538
3-LINE TO 8-LINE DECODERS/DEMULTIPLEXERS WITH 3-STATE OUTPUTS

switching characteristics (see Note 1)

PARAMETER	FROM (INPUT)	TO (OUTPUT)	V_{CC} = 4.5 V to 5.5 V, C_L = 50 pF, R1 = 500 Ω, R2 = 500 Ω, T_A = MIN to MAX						UNIT
			SN54ALS538			SN74ALS538			
			MIN	TYP†	MAX	MIN	TYP†	MAX	
t_{PLH}	A, B, C	Any Y		22			22		ns
t_{PHL}				22			22		
t_{PLH}	G1 or G2	Any Y		18			18		ns
t_{PHL}				18			18		
t_{PLH}	$\overline{G3}$ or $\overline{G4}$	Any Y		22			22		ns
t_{PHL}				22			22		
t_{PLH}	AL	Any Y		20			20		ns
t_{PHL}				20			20		
t_{PZH}	$\overline{OE1}$ or $\overline{OE2}$	Any Y		10			10		ns
t_{PZL}				13			13		
t_{PHZ}	$\overline{OE1}$ or $\overline{OE2}$	Any Y		8			8		ns
t_{PLZ}				10			10		

†All typical values are at V_{CC} = 5 V, T_A = 25°C.
NOTE 1: For load circuit and voltage waveforms, see page 1-12.

Additional information on these products can be obtained from the factory as it becomes available.

ALS AND AS CIRCUITS

TEXAS INSTRUMENTS
POST OFFICE BOX 225012 • DALLAS, TEXAS 75265

TYPES SN54ALS539, SN74ALS539
DUAL 2-LINE TO 4-LINE DECODERS/DEMULTIPLEXERS WITH 3-STATE OUTPUTS

D2661, APRIL 1982—REVISED DECEMBER 1983

- 3-State Outputs
- Output Polarity Control
- Data Multiplexing Capability
- Package Options Include Both Plastic and Ceramic Chip Carriers in Addition to Plastic and Ceramic DIPs
- Dependable Texas Instruments Quality and Reliability

SN54ALS539 . . . J PACKAGE
SN74ALS539 . . . N PACKAGE
(TOP VIEW)

```
1Y2  [ 1    20 ] VCC
1Y1  [ 2    19 ] 1Y3
1Y0  [ 3    18 ] 1B
1AL  [ 4    17 ] 1A
1OE  [ 5    16 ] 1G
2A   [ 6    15 ] 2G
2B   [ 7    14 ] 2OE
2Y3  [ 8    13 ] 2AL
2Y2  [ 9    12 ] 2Y0
GND  [ 10   11 ] 2Y1
```

description

The 'ALS539 decoder/demultiplexer contains two independent decoders, each of which accepts two address input signals (A and B) and decodes them to select one of four mutually exclusive outputs. If the polarity-control input (AL) is high, the outputs are active-low; if AL is low, the outputs are active-high. An active-low input enable ($\overline{G}$) is available for data demultiplexing. Data is routed to the selected output in noninverting form in the active-low mode or in inverted form in the active-high mode. A high signal on the output enable ($\overline{OE}$) forces the 3-state outputs to the high-impedance state.

The SN54ALS539 is characterized for operation over the full military temperature range of −55°C to 125°C. The SN74ALS539 is characterized for operation from 0°C to 70°C.

SN54ALS539 . . . FH PACKAGE
SN74ALS539 . . . FN PACKAGE
(TOP VIEW)

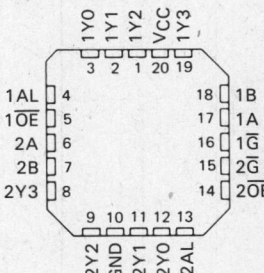

FUNCTION TABLE
(EACH DECODER/DEMULTIPLEXER)

FUNCTION	INPUTS				OUTPUTS			
	$\overline{OE}$	$\overline{G}$	B	A	Y0	Y1	Y2	Y3
High Impedance	H	X	X	X	Z	Z	Z	Z
Disable	L	H	X	X	All outputs same level as AL			
Active-high Output (AL = L)	L	L	L	L	H	L	L	L
	L	L	L	H	L	H	L	L
	L	L	H	L	L	L	H	L
	L	L	H	H	L	L	L	H
Active-low Output (AL = H)	L	L	L	L	L	H	H	H
	L	L	L	H	H	L	H	H
	L	L	H	L	H	H	L	H
	L	L	H	H	H	H	H	L

PRODUCT PREVIEW

This document contains information on a product under development. Texas Instruments reserves the right to change or discontinue this product without notice.

Copyright © 1982 by Texas Instruments Incorporated

TEXAS INSTRUMENTS
POST OFFICE BOX 225012 • DALLAS, TEXAS 75265

ALS AND AS CIRCUITS

TYPES SN54ALS539, SN74ALS539
DUAL 2-LINE TO 4-LINE DECODERS/DEMULTIPLEXERS WITH 3-STATE OUTPUTS

logic symbols (alternatives)

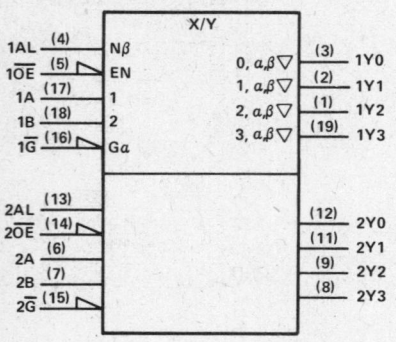
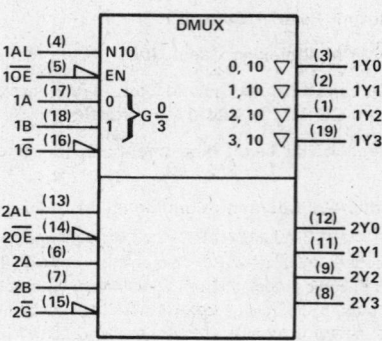

logic diagram (positive logic)

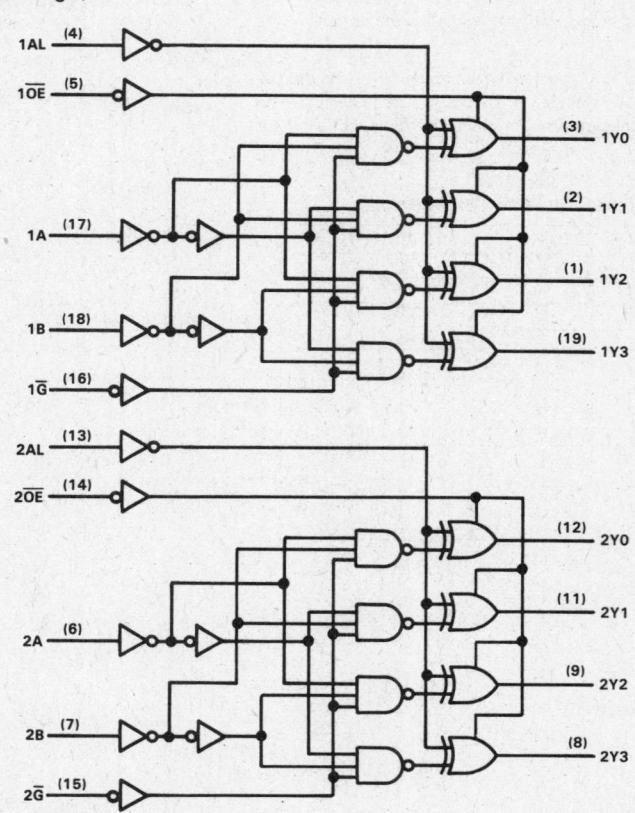

Pin numbers shown are for J and N packages.

TEXAS INSTRUMENTS
POST OFFICE BOX 225012 • DALLAS, TEXAS 75265

TYPES SN54ALS539, SN74ALS539
DUAL 2-LINE TO 4-LINE DECODERS/DEMULTIPLEXERS WITH 3-STATE OUTPUTS

absolute maximum ratings over operating free-air temperature range (unless otherwise noted)

Supply voltage, V_{CC} .... 7 V
Input voltage .... 7 V
Voltage applied to a disabled 3-state output .... 5.5 V
Operating free-air temperature range: SN54ALS539 .... −55 °C to 125 °C
                                          SN74ALS539 .... 0 °C to 70 °C
Storage temperature range .... −65 °C to 150 °C

recommended operating conditions

		SN54ALS539			SN74ALS539			UNIT
		MIN	NOM	MAX	MIN	NOM	MAX	
V_{CC}	Supply voltage	4.5	5	5.5	4.5	5	5.5	V
V_{IH}	High-level input voltage	2			2			V
V_{IL}	Low-level input voltage			0.8			0.8	V
I_{OH}	High-level output current			−1			−2.6	mA
I_{OL}	Low-level output current			12			24	mA
T_A	Operating free-air temperature	−55		125	0		70	°C

electrical characteristics over recommended operating free-air temperature range (unless otherwise noted)

PARAMETER		TEST CONDITIONS		SN54ALS539			SN74ALS539			UNIT
				MIN	TYP†	MAX	MIN	TYP†	MAX	
V_{IK}		V_{CC} = 4.5 V,	I_I = −18 mA			−1.5			−1.5	V
V_{OH}		V_{CC} = 4.5 V to 5.5 V,	I_{OH} = −0.4 mA	V_{CC}−2			V_{CC}−2			V
		V_{CC} = 4.5 V,	I_{OH} = −1 mA	2.4	3.3					
		V_{CC} = 4.5 V,	I_{OH} = −2.6 mA				2.4	3.2		
V_{OL}		V_{CC} = 4.5 V,	I_{OL} = 12 mA		0.25	0.4		0.25	0.4	V
		V_{CC} = 4.5 V,	I_{OL} = 24 mA					0.35	0.5	
I_{OZH}		V_{CC} = 5.5 V,	V_O = 2.7 V			20			20	μA
I_{OZL}		V_{CC} = 5.5 V,	V_O = 0.4 V			−20			−20	μA
I_I		V_{CC} = 5.5 V,	V_I = 7 V			0.1			0.1	mA
I_{IH}		V_{CC} = 5.5 V,	V_I = 2.7 V			20			20	μA
I_{IL}	A, B, C or AL	V_{CC} = 5.5 V,	V_I = 0.4 V			−0.2			−0.2	mA
	All other					−0.1			−0.1	
I_O‡		V_{CC} = 5.5 V,	V_O = 2.25 V	−30		−112	−30		−112	mA
I_{CC}		V_{CC} = 5.5 V	Outputs high							mA
			Outputs low							
			Outputs disabled		24			24		

†All typical values are at V_{CC} = 5 V, T_A = 25 °C.
‡The output conditions have been chosen to produce a current that closely approximates one half of the true short-circuit output current, I_{OS}.

Additional information on these products can be obtained from the factory as it becomes available.

TYPES SN54ALS539, SN74ALS539
DUAL 2-LINE TO 4-LINE DECODERS/DEMULTIPLEXERS WITH 3-STATE OUTPUTS

switching characteristics (see Note 1)

PARAMETER	FROM (INPUT)	TO (OUTPUT)	V_{CC} = 4.5 V to 5.5 V, C_L = 50 pF, R1 = 500 Ω, R2 = 500 Ω, T_A = MIN to MAX						UNIT
			SN54ALS539			SN74ALS539			
			MIN	TYP[†]	MAX	MIN	TYP[†]	MAX	
t_{PLH}	A or B	Y		22			22		ns
t_{PHL}				22			22		
t_{PLH}	$\overline{G}$	Y		18			18		ns
t_{PHL}				18			18		
t_{PLH}	AL	Y		22			22		ns
t_{PHL}				22			22		
t_{PZH}	$\overline{OE}$	Y		10			10		ns
t_{PZL}				13			13		
t_{PHZ}	$\overline{OE}$	Y		8			8		ns
t_{PLZ}				10			10		

[†]All typical values are at V_{CC} = 5 V, T_A = 25°C.
NOTE 2: For load circuit and voltage waveforms, see page 1-12.

Additional information on these products can be obtained from the factory as it becomes available.

ALS AND AS CIRCUITS

TEXAS INSTRUMENTS
POST OFFICE BOX 225012 • DALLAS, TEXAS 75265

TYPES SN54ALS540, SN54ALS541, SN74ALS540, SN74ALS541
OCTAL BUFFERS AND LINE DRIVERS WITH 3-STATE OUTPUTS

D2661, APRIL 1982—REVISED DECEMBER 1983

- 3-State Outputs Drive Bus Lines or Buffer Memory Address Registers
- P-N-P Inputs Reduce D-C Loading
- Data Flow-Thru Pinout (All Inputs on Opposite Side from Outputs)
- Package Options Include Both Plastic and Ceramic Chip Carriers in Addition to Plastic and Ceramic DIPs
- Dependable Texas Instruments Quality and Reliability

SN54ALS540, SN54ALS541 . . . J PACKAGE
SN74ALS540, SN74ALS541 . . . N PACKAGE
(TOP VIEW)

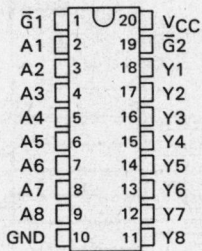

SN54ALS540, SN54ALS541 . . . FH PACKAGE
SN74ALS540, SN74ALS541 . . . FN PACKAGE
(TOP VIEW)

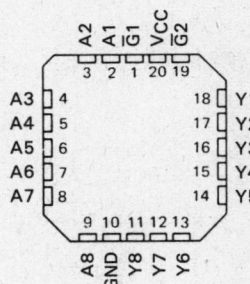

description

These octal buffers and line drivers are designed to have the performance of the popular SN54ALS240/SN74ALS240 series and, at the same time, offer a pinout with inputs and outputs on opposite sides of the package. This arrangement greatly enhances printed circuit board layout.

The three-state control gate is a 2-input NOR such that if either $\overline{G}1$ or $\overline{G}2$ is high, all eight outputs are in the high-impedance state.

The 'ALS540 provides inverted data and the 'ALS541 provides true data at the outputs.

The -1 versions of the SN74ALS540 and SN74ALS541 parts are identical to the standard versions except that the recommended maximum I_{OL} is increased to 48 milliamperes. There are no -1 versions of the SN54ALS540 and SN54ALS541.

The SN54ALS540 and SN54ALS541 are characterized for operation over the full military temperature range of $-55\,°C$ to $125\,°C$. The SN74ALS540 and SN74ALS541 are characterized for operation from $0\,°C$ to $70\,°C$.

ADVANCE INFORMATION
This document contains information on a new product. Specifications are subject to change without notice.

Copyright © 1982 by Texas Instruments Incorporated

TEXAS INSTRUMENTS
POST OFFICE BOX 225012 • DALLAS, TEXAS 75265

TYPES SN54ALS540, SN54ALS541, SN74ALS540, SN74ALS541
OCTAL BUFFERS AND LINE DRIVERS WITH 3-STATE OUTPUTS

logic symbols

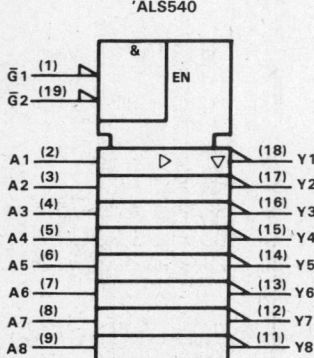

Pin numbers shown are for J and N packages

logic diagrams (positive logic)

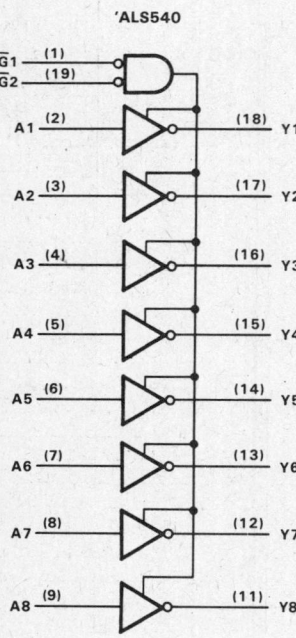

TEXAS INSTRUMENTS
POST OFFICE BOX 225012 • DALLAS, TEXAS 75265

TYPES SN54ALS540, SN54ALS541, SN74ALS540, SN74ALS541
OCTAL BUFFERS AND LINE DRIVERS WITH 3-STATE OUTPUTS

absolute maximum ratings over operating free-air temperature range (unless otherwise noted)

Supply voltage, V_{CC} .... 7 V
Input voltage .... 7 V
Voltage applied to a disabled 3-state output .... 5.5 V
Operating free-air temperature range: SN54ALS540, SN54ALS541 .... −55 °C to 125 °C
                             SN74ALS540, SN74ALS541 .... 0 °C to 70 °C
Storage temperature range .... −65 °C to 150 °C

recommended operating conditions

		SN54ALS540 SN54ALS541			SN74ALS540 SN74ALS541			UNIT
		MIN	NOM	MAX	MIN	NOM	MAX	
V_{CC}	Supply voltage	4.5	5	5.5	4.5	5	5.5	V
V_{IH}	High-level input voltage	2			2			V
V_{IL}	Low-level input voltage			0.8			0.8	V
I_{OH}	High-level output current			−12			−15	mA
I_{OL}	Low-level output current			12			24	mA
							48†	
T_A	Operating free-air temperature	−55		125	0		70	°C

†The extended limit applies only if V_{CC} is maintained between 4.75 V and 5.25 V.
  The 48 mA limit applies for the SN74ALS540-1 and SN74ALS541-1 only.

electrical characteristics over recommended operating free-air temperature range (unless otherwise noted)

PARAMETER		TEST CONDITIONS	SN54ALS540 SN54ALS541			SN74ALS540 SN74ALS541			UNIT
			MIN	TYP‡	MAX	MIN	TYP‡	MAX	
V_{IK}		V_{CC} = 4.5 V, I_I = −18 mA			−1.5			−1.5	V
V_{OH}		V_{CC} = 4.5 V to 5.5 V, I_{OH} = −0.4 mA	V_{CC}−2			V_{CC}−2			V
		V_{CC} = 4.5 V, I_{OH} = −3 mA	2.4	3.2		2.4	3.2		
		V_{CC} = 4.5 V, I_{OH} = −12 mA	2						
		V_{CC} = 4.5 V, I_{OH} = −15 mA				2			
V_{OL}		V_{CC} = 4.5 V, I_{OL} = 12 mA		0.25	0.4		0.25	0.4	V
		V_{CC} = 4.5 V, I_{OL} = 24 mA					0.35	0.5	
I_{OZH}		V_{CC} = 5.5 V, V_O = 2.7 V			20			20	µA
I_{OZL}		V_{CC} = 5.5 V, V_O = 0.4 V			−20			−20	µA
I_I		V_{CC} = 5.5 V, V_I = 7 V			0.1			0.1	mA
I_{IH}		V_{CC} = 5.5 V, V_I = 2.7 V			20			20	µA
I_{IL}		V_{CC} = 5.5 V, V_I = 0.4 V			−0.1			−0.1	mA
I_O §		V_{CC} = 5.5 V, V_O = 2.25 V	−30		−112	−30		−112	mA
I_{CC}	'ALS540	V_{CC} = 5.5 V	Outputs high		15			15		mA
			Outputs low		18			18		
			Outputs disabled		29			29		
	'ALS541	V_{CC} = 5.5 V	Outputs high		15			15		mA
			Outputs low		18			18		
			Outputs disabled		19			19		

‡All typical values are at V_{CC} = 5 V, T_A = 25 °C.
§The output conditions have been chosen to produce a current that closely approximates one half of the true short-circuit output current, I_{OS}.

Additional information on these products can be obtained from the factory as it becomes available.

TYPES SN54ALS540, SN54ALS541, SN74ALS540, SN74ALS541
OCTAL BUFFERS AND LINE DRIVERS WITH 3-STATE OUTPUTS

'ALS540 switching characteristics (see Note 1)

PARAMETER	FROM (INPUT)	TO (OUTPUT)	V_{CC} = 4.5 V to 5.5 V, C_L = 50 pF, $R1$ = 500 Ω, $R2$ = 500 Ω, T_A = MIN to MAX						UNIT
			SN54ALS540			SN74ALS540			
			MIN	TYP†	MAX	MIN	TYP†	MAX	
t_{PLH}	A	Y		6			6		ns
t_{PHL}				6			6		
t_{PZH}	$\overline{G}$	Y		13			13		ns
t_{PZL}				18			18		
t_{PHZ}	$\overline{G}$	Y		7			7		ns
t_{PLZ}				11			11		

'ALS541 switching characteristics (see Note 1)

PARAMETER	FROM (INPUT)	TO (OUTPUT)	V_{CC} = 4.5 V to 5.5 V, C_L = 50 pF, $R1$ = 500 Ω, $R2$ = 500 Ω, T_A = MIN to MAX						UNIT
			SN54ALS541			SN74ALS541			
			MIN	TYP†	MAX	MIN	TYP†	MAX	
t_{PLH}	A	Y		6			6		ns
t_{PHL}				6			6		
t_{PZH}	$\overline{G}$	Y		13			13		ns
t_{PZL}				18			18		
t_{PHZ}	$\overline{G}$	Y		7			7		ns
t_{PLZ}				11			11		

†All typical values are at V_{CC} = 5 V, T_A = 25°C.
NOTE 1: For load circuit and voltage waveforms, see page 1-12.

Texas Instruments
POST OFFICE BOX 225012 • DALLAS, TEXAS 75265

TYPES SN54ALS560A, SN54ALS561A, SN74ALS560A, SN74ALS561A
SYNCHRONOUS 4-BIT COUNTERS WITH 3-STATE OUTPUTS

D2661, DECEMBER 1982—REVISED DECEMBER 1983

- Carry Output for n-Bit Cascading
- Buffer-Type Outputs Drive Bus Lines Directly
- Choice of Asynchronous or Synchronous Load or Clear
- Internal Look-Ahead for Fast Cascading
- Package Options Include Both Plastic and Ceramic Chip Carriers in Addition to Plastic and Ceramic DIPs
- Dependable Texas Instruments Quality and Reliability

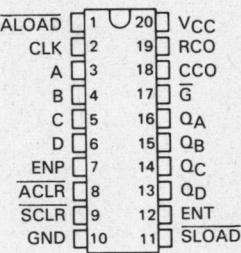

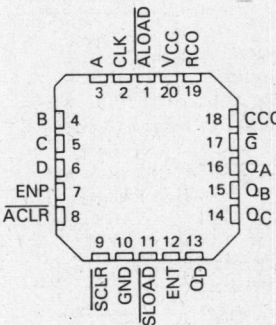

description

The 'ALS560A decade counters and 'ALS561A binary counters are programmable and offer synchronous and asynchronous clearing as well as synchronous and asynchronous loading. All synchronous functions are executed on the positive-going edge of the clock.

The clear function is initiated by applying a low level to either Asynchronous Clear ($\overline{ACLR}$) or Synchronous Clear ($\overline{SCLR}$). Asynchronous (direct) clearing overrides all other functions of the device, while synchronous clearing overrides only the other synchronous functions. Data is loaded from the A, B, C, and D inputs by applying a low level to Asynchronous Load ($\overline{ALOAD}$) or by the combination of a low level at Synchronous Load ($\overline{SLOAD}$) and a positive-going clock transition. The counting function is enabled only when Enable P (ENP), Enable T (ENT), $\overline{ACLR}$, $\overline{ALOAD}$, $\overline{SCLR}$, and $\overline{SLOAD}$ are all high.

A high level at the Output Enable ($\overline{G}$) forces the Q outputs into the high-impedance state, and a low level enables those outputs. Counting is independent of $\overline{G}$. ENT is fed forward to enable the Ripple Carry Output (RCO) to produce a high-level pulse while the count is maximum (9 or 15). The Clocked Carry Output (CCO) produces a high-level pulse for a duration equal to that of the low level of the clock when RCO is high and the counter is enabled (both ENP and ENT are high); otherwise, CCO is low. CCO does not have the glitches commonly associated with a ripple-carry output. Cascading is normally accomplished by connecting RCO or CCO of the first counter to ENT of the next counter. However, for very-high-speed counting, RCO should be used for cascading since CCO does not become active until the clock returns to the low level.

The SN54ALS560A and SN54ALS561A are characterized for operation over the full military temperature range of −55°C to 125°C. The SN74ALS560A and SN74ALS561A are characterized for operation from 0°C to 70°C.

TYPES SN54ALS560A, SN54ALS561A, SN74ALS560A, SN74ALS561A
SYNCHRONOUS 4-BIT COUNTERS WITH 3-STATE OUTPUTS

FUNCTION TABLE

INPUTS								OPERATION
$\overline{G}$	$\overline{ACLR}$	$\overline{ALOAD}$	$\overline{SCLR}$	$\overline{SLOAD}$	ENT	ENP	CLK	
H	X	X	X	X	X	X	X	Q Outputs Disabled
L	L	X	X	X	X	X	X	Asynchronous Clear
L	H	L	X	X	X	X	X	Asynchronous Load
L	H	H	L	X	X	X	↑	Synchronous Clear
L	H	H	H	L	X	X	↑	Synchronous Load
L	H	H	H	H	H	H	↑	Count
L	H	H	H	H	L	X	X	Inhibit Counting
L	H	H	H	H	X	L	X	Inhibit Counting

logic symbols

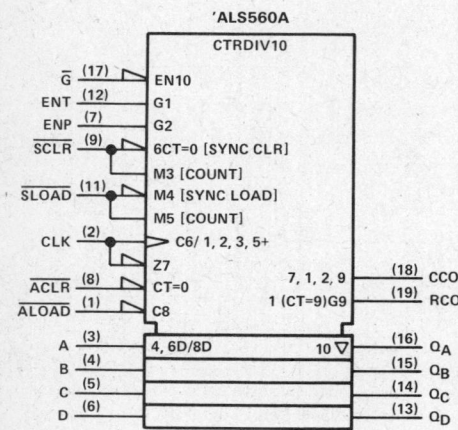
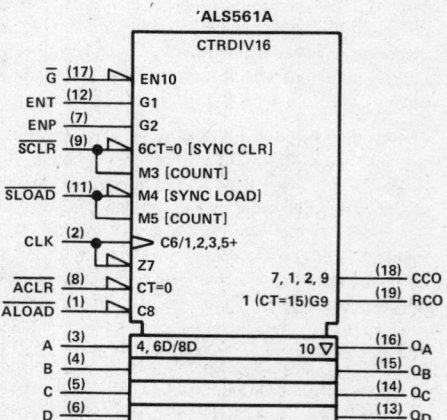

Pin numbers shown are for J and N packages.

TYPES SN54ALS560A, SN74ALS560A
SYNCHRONOUS 4-BIT COUNTERS WITH 3-STATE OUTPUTS

'ALS560A logic diagram (positive logic)

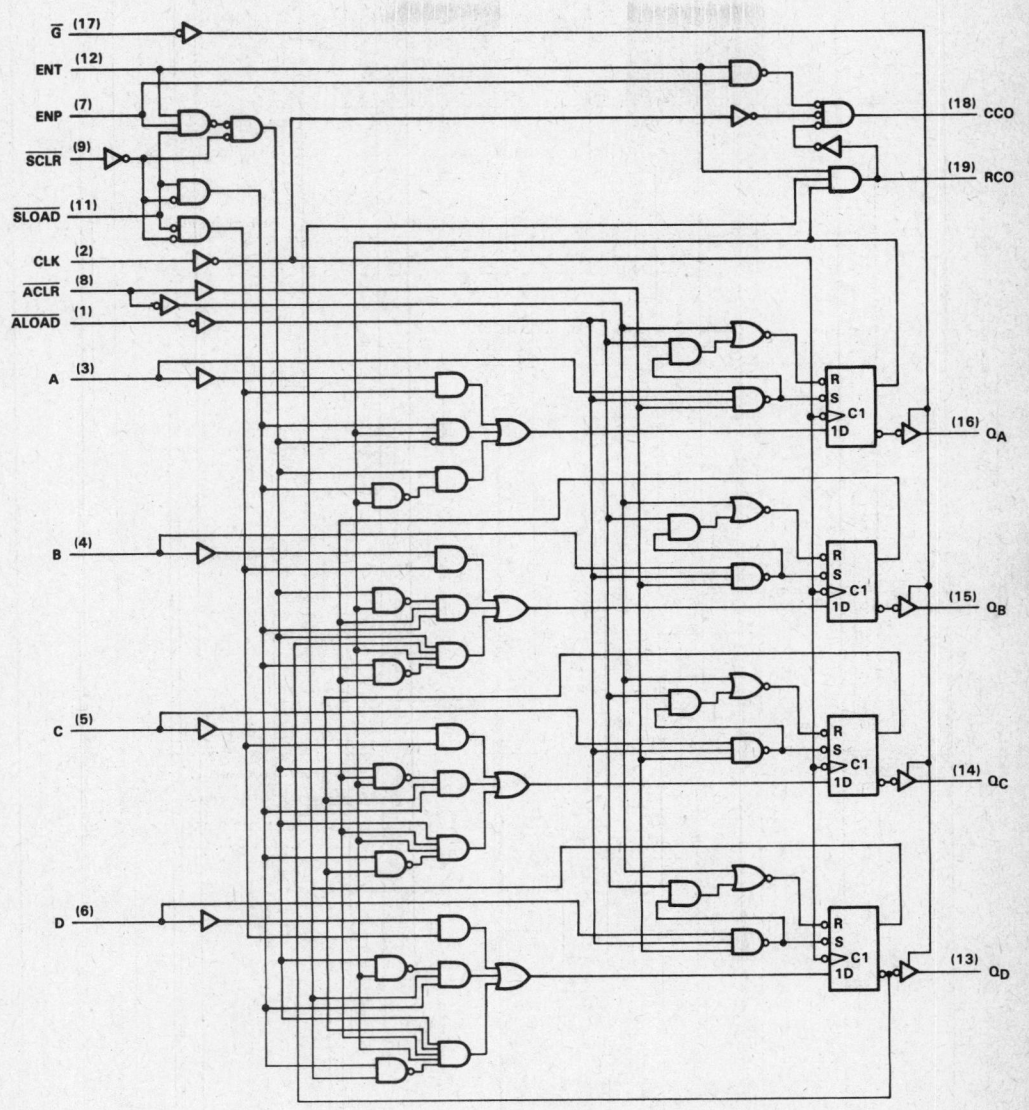

Pin numbers shown are for J and N packages

TYPES SN54ALS561A, SN74ALS561A
SYNTHRONOUS 4-BIT COUNTERS WITH 3-STATE OUTPUTS

'ALS561A logic diagram (positive logic)

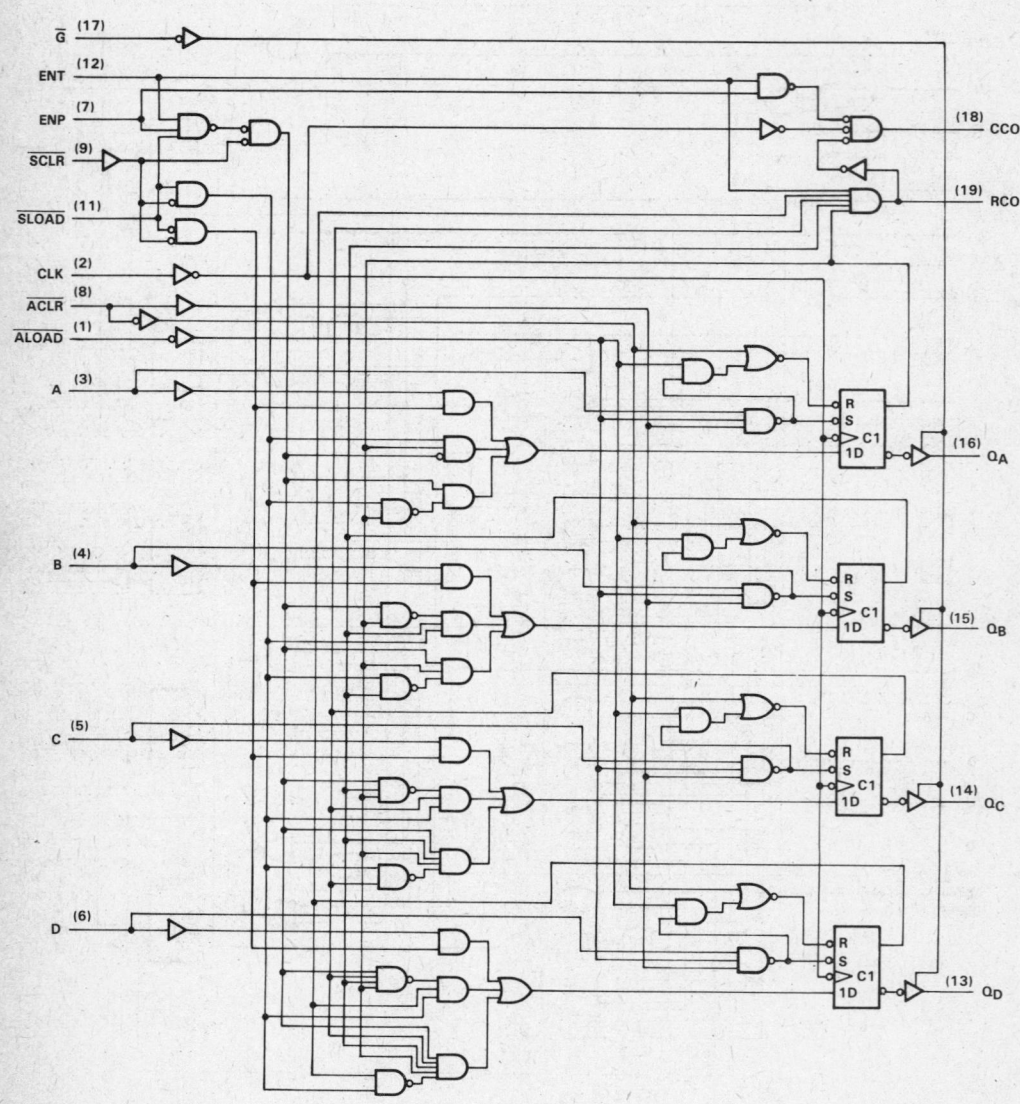

Pin numbers shown are for J and N packages

TYPES SN54ALS560A, SN74ALS560A
SYNCHRONOUS 4-BIT COUNTERS WITH 3-STATE OUTPUTS

'ALS560A typical load, count, and inhibit sequences

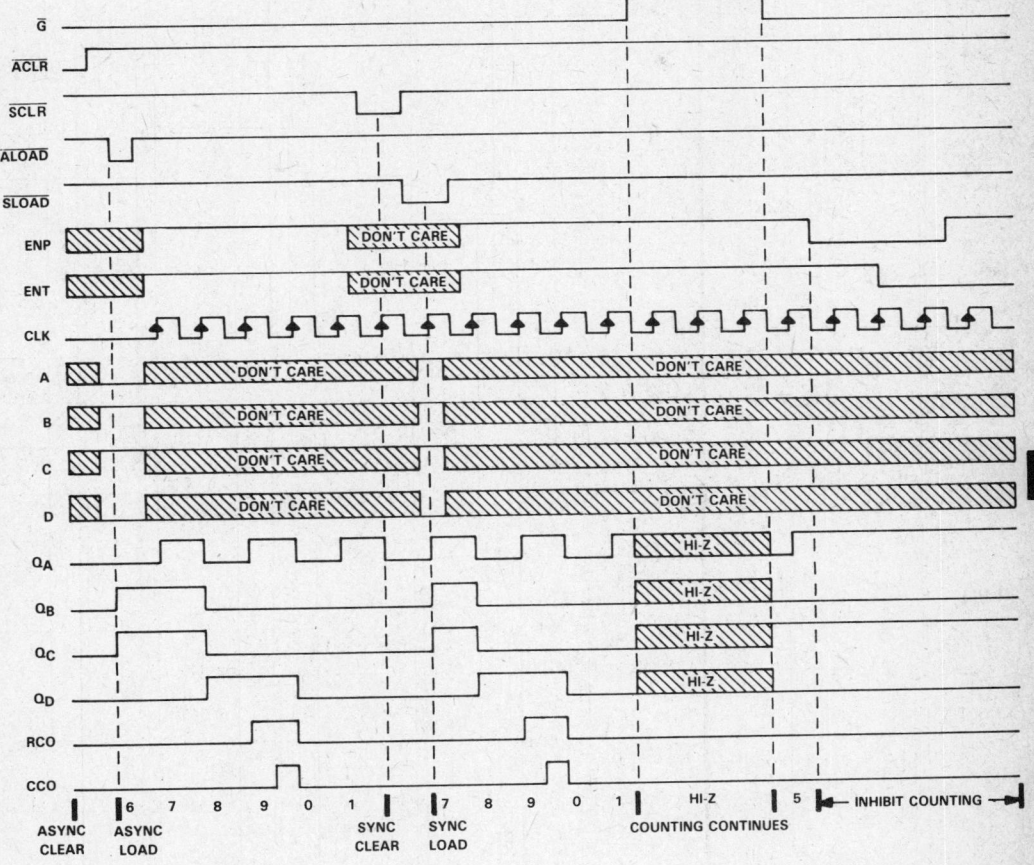

TYPES SN54ALS561A, SN74ALS561A
SYNCHRONOUS 4-BIT COUNTERS WITH 3-STATE OUTPUTS

'ALS561A typical load, count, and inhibit sequences

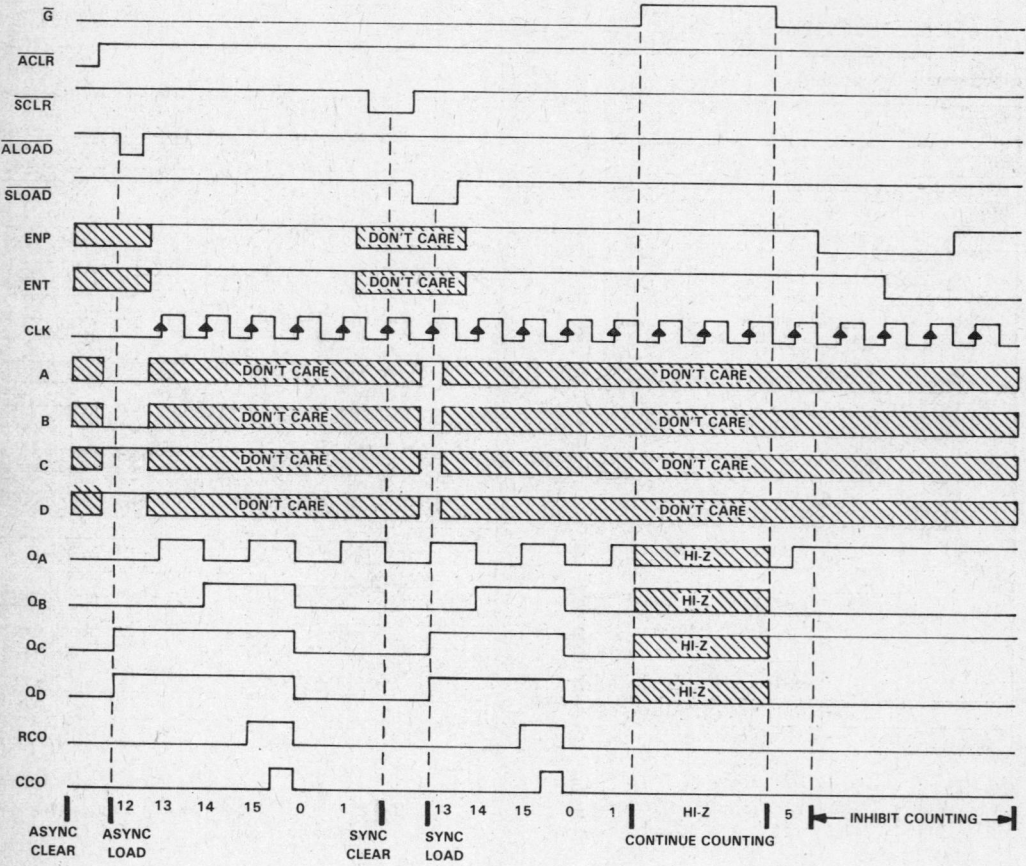

TYPES SN54ALS560A, SN54ALS561A, SN74ALS560A, SN74ALS561A
SYNCHRONOUS 4-BIT COUNTERS WITH 3-STATE OUTPUTS

absolute maximum ratings over operating free-air temperature range (unless otherwise noted)

Supply voltage, V_{CC} .... 7 V
Input voltage .... 7 V
Operating free-air temperature range: SN54ALS560A, SN54ALS561A .... −55°C to 125°C
                                       SN74ALS560A, SN74ALS561A .... 0°C to 70°C
Storage temperature range .... −65°C to 150°C

recommended operating conditions

			SN54ALS560A SN54ALS561A			SN74ALS560A SN74ALS561A			UNIT
			MIN	NOM	MAX	MIN	NOM	MAX	
V_{CC}	Supply voltage		4.5	5	5.5	4.5	5	5.5	V
V_{IH}	High-level input voltage		2			2			V
V_{IL}	Low-level input voltage				0.8			0.8	V
I_{OH}	High-level output current	Q outputs			−1			−2.6	mA
		CCO and RCO			−0.4			−0.4	
I_{OL}	Low-level output current	Q outputs			12			24	mA
		CCO and RCO			4			8	
f_{clock}	Clock frequency	'ALS560A	0		18	0		20	MHz
		'ALS561A	0		25	0		30	
t_w	Pulse duration	$\overline{ACLR}$ or $\overline{ALOAD}$ low	20			15			ns
		'ALS560A CLK high	27.5			25			
		'ALS560A CLK low	27.5			25			
		'ALS561A CLK high	20			16.5			
		'ALS561A CLK low	20			16.5			
t_{su}	Setup time before CLK↑	ENP, ENT High	25			20			ns
		ENP, ENT Low	25			20			
		Data at A, B, C, D	25			20			
		$\overline{SCLR}$ Low	21			15			
		$\overline{SCLR}$ High (inactive)	35			30			
		$\overline{SLOAD}$ Low	20			15			
		$\overline{SLOAD}$ High (inactive)	35			30			
		$\overline{ACLR}$ or $\overline{ALOAD}$ inactive	10			10			
t_h	Hold time after CLK↑ for data, ENP, ENT, $\overline{SCLR}$, or $\overline{SLOAD}$		0			0			ns
T_A	Operating free-air temperature		−55		125	0		70	°C

TYPES SN54ALS560A, SN54ALS561A, SN74ALS560A, SN74ALS561A
SYNCHRONOUS 4-BIT COUNTERS WITH 3-STATE OUTPUTS

electrical characteristics over recommended operating free-air temperature range (unless otherwise noted)

PARAMETER		TEST CONDITIONS		SN54ALS560A SN54ALS561A			SN74ALS560A SN74ALS561A			UNIT
				MIN	TYP†	MAX	MIN	TYP†	MAX	
V_{IK}		$V_{CC} = 4.5$ V,	$I_I = -18$ mA			−1.5			−1.5	V
V_{OH}	All outputs	$V_{CC} = 4.5$ V to 5.5 V,	$I_{OH} = -0.4$ mA	$V_{CC}-2$			$V_{CC}-2$			V
	Q outputs	$V_{CC} = 4.5$ V,	$I_{OH} = -1$ mA	2.4	3.3					
		$V_{CC} = 4.5$ V,	$I_{OH} = -2.6$ mA				2.4	3.2		
V_{OL}	Q outputs	$V_{CC} = 4.5$ V,	$I_{OL} = 12$ mA		0.25	0.4		0.25	0.4	V
		$V_{CC} = 4.5$ V,	$I_{OL} = 24$ mA					0.35	0.5	
	CCO and RCO	$V_{CC} = 4.5$ V,	$I_{OL} = 4$ mA		0.25	0.4		0.25	0.4	
		$V_{CC} = 4.5$ V,	$I_{OL} = 8$ mA					0.35	0.5	
I_{OZH}		$V_{CC} = 5.5$ V,	$V_O = 2.7$ V			20			20	µA
I_{OZL}		$V_{CC} = 5.5$ V,	$V_O = 0.4$ V			−20			−20	µA
I_I	ENT and ENP	$V_{CC} = 5.5$ V,	$V_I = 7$ V			0.2			0.2	mA
	Other inputs					0.1			0.1	
I_{IH}	ENT and ENP	$V_{CC} = 5.5$ V,	$V_I = 2.7$ V			40			40	µA
	Other inputs					20			20	
I_{IL}		$V_{CC} = 5.5$ V,	$V_I = 0.4$ V			−0.2			−0.2	mA
I_O‡	CCO and RCO	$V_{CC} = 5.5$ V,	$V_O = 2.25$ V	−15		−70	−15		−70	mA
	Q			−30		−112	−30		−112	
I_{CC}		$V_{CC} = 5.5$ V	Outputs high		17	27		17	27	mA
			Outputs low		21	33		21	33	
			Outputs disabled		22	36		22	36	

†All typical values are at $V_{CC} = 5$ V, $T_A = 25$°C.
‡The output conditions have been chosen to produce a current that closely approximates one half of the true short-circuit output current, I_{OS}.

TYPES SN54ALS560A, SN54ALS561A, SN74ALS560A, SN74ALS561A
SYNCHRONOUS 4-BIT COUNTERS WITH 3-STATE OUTPUTS

switching characteristics (see Note 1)

PARAMETER	FROM (INPUT)	TO (OUTPUT)	V_{CC} = 4.5 V to 5.5 V, C_L = 50 pF, R1 = 500 Ω, R2 = 500 Ω, T_A = MIN to MAX				UNIT
			SN54ALS560A SN54ALS561A		SN74ALS560A SN74ALS561A		
			MIN	MAX	MIN	MAX	
f_{max}		'ALS560A	18		20		MHz
		'ALS561A	25		30		
t_{PLH}	CLK	Any Q	4	15	4	12	ns
t_{PHL}			5	21	5	18	
t_{PLH}	CLK	RCO	9	35	9	29	ns
t_{PHL}			8	29	8	24	
t_{PLH}	CLK	CCO	8	31	8	26	ns
t_{PHL}			5	20	5	16	
t_{PLH}	$\overline{ALOAD}$	Any Q	10	38	10	35	ns
t_{PHL}			7	27	7	23	
t_{PLH}	$\overline{ALOAD}$	RCO	15	50	15	40	ns
t_{PHL}			12	35	12	30	
t_{PLH}	$\overline{ALOAD}$	CCO	25	65	25	55	ns
t_{PHL}			12	42	12	33	
t_{PLH}	A, B, C, or D	Any Q	8	35	8	30	ns
t_{PHL}			7	27	7	22	
t_{PLH}	ENT	RCO	5	20	5	16	ns
t_{PHL}			4	18	4	14	
t_{PLH}	ENT	CCO	12	35	12	32	ns
t_{PHL}			4	15	4	12	
t_{PLH}	ENP	CCO	5	22	5	18	ns
t_{PHL}			4	14	4	12	
t_{PHL}	$\overline{ACLR}$	Any Q	7	28	7	22	ns
t_{PZH}	$\overline{G}$	Any Q	5	24	5	19	ns
t_{PZL}			8	28	8	23	
t_{PHZ}	$\overline{G}$	Any Q	2	12	2	10	ns
t_{PLZ}			4	20	4	15	

NOTE 1: For load circuit and voltage waveforms, see page 1-12.

2
ALS AND AS CIRCUITS

TYPES SN54ALS563, SN74ALS563
OCTAL D-TYPE TRANSPARENT LATCHES WITH 3-STATE OUTPUTS

D2661, DECEMBER 1982—REVISED DECEMBER 1983

- 3-State Buffer-Type Outputs Drive Bus-Lines Directly
- Bus-Structured Pinout
- Package Options Include Both Plastic and Ceramic Chip Carriers in Addition to Plastic and Ceramic DIPs
- Dependable Texas Instruments Quality and Reliability

SN54ALS563 . . . J PACKAGE
SN74ALS563 . . . N PACKAGE
(TOP VIEW)

```
 OC  [ 1   20 ] VCC
 1D  [ 2   19 ] 1Q̄
 2D  [ 3   18 ] 2Q̄
 3D  [ 4   17 ] 3Q̄
 4D  [ 5   16 ] 4Q̄
 5D  [ 6   15 ] 5Q̄
 6D  [ 7   14 ] 6Q̄
 7D  [ 8   13 ] 7Q̄
 8D  [ 9   12 ] 8Q̄
 GND [ 10  11 ] C
```

description

These 8-bit latches feature three-state outputs designed specifically for driving highly capacitive or relatively low-impedance loads. They are particularly suitable for implementing buffer registers, I/O ports, bidirectional bus drivers, and working registers.

The eight latches are transparent D-type latches. While the enable (C) is high the Q̄ outputs will follow the complements of data (D) inputs. When the enable is taken low the output will be latched at the inverses of the levels that were set up at the D inputs.

A buffered output-control input can be used to place the eight outputs in either a normal logic state (high or low logic levels) or a high-impedance state. In the high-impedance state the outputs neither load nor drive the bus lines significantly. The high-impedance state and increased high-logic level provide the capability to drive the bus lines in a bus-organized system without need for interface or pull-up components.

The output control (OC) does not affect the internal operation of the latches. Old data can be retained or new data can be entered while the outputs are in the high-impedance state.

The SN54ALS563 is characterized for operation over the full military temperature range of −55°C to 125°C. The SN74ALS563 is characterized for operation from 0°C to 70°C.

SN54ALS563 . . . FH PACKAGE
SN74ALS563 . . . FN PACKAGE
(TOP VIEW)

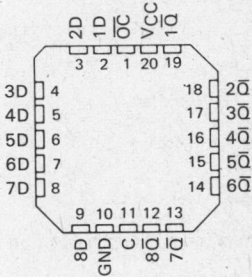

FUNCTION TABLE
(Each Latch)

| INPUTS | | | OUTPUT |
| ENABLE | | | Q̄ |
OC	C	D	
L	H	H	L
L	H	L	H
L	L	X	Q₀
H	X	X	Z

ALS AND AS CIRCUITS

TYPES SN54ALS563, SN74ALS563
OCTAL D-TYPE TRANSPARENT LATCHES WITH 3-STATE OUTPUTS

logic symbol

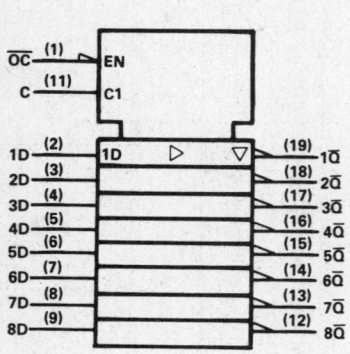

logic diagram (positive logic)

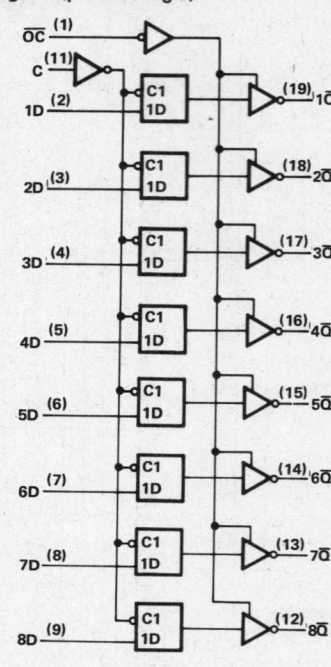

Pin numbers shown are for J and N packages.

absolute maximum ratings over operating free-air temperature range (unless otherwise noted)

Supply voltage, V_{CC} .... 7 V
Input voltage .... 7 V
Voltage applied to a disabled 3-state output .... 5.5 V
Operating free-air temperature range: SN54ALS563 .... −55 °C to 125 °C
                                             SN74ALS563 .... 0 °C to 70 °C
Storage temperature range .... −65 °C to 150 °C

recommended operating conditions

		SN54ALS563			SN74ALS563			UNIT
		MIN	NOM	MAX	MIN	NOM	MAX	
V_{CC}	Supply voltage	4.5	5	5.5	4.5	5	5.5	V
V_{IH}	High-level input voltage	2			2			V
V_{IL}	Low-level input voltage			0.8			0.8	V
I_{OH}	High-level output current			−1			−2.6	mA
I_{OL}	Low-level output current			12			24	mA
t_w	Pulse duration, enable C high	15			15			ns
t_{su}	Setup time, data before enable C↓	10			10			ns
t_h	Hold time, data after enable C↓	10			10			ns
T_A	Operating free-air temperature	−55		125	0		70	°C

TYPES SN54ALS563, SN74ALS563
OCTAL D-TYPE TRANSPARENT LATCHES WITH 3-STATE OUTPUTS

electrical characteristics over recommended operating free-air temperature range (unless otherwise noted)

PARAMETER	TEST CONDITIONS		SN54ALS563			SN74ALS563			UNIT
			MIN	TYP†	MAX	MIN	TYP†	MAX	
V_{IK}	$V_{CC} = 4.5$ V,	$I_I = -18$ mA			-1.5			-1.5	V
V_{OH}	$V_{CC} = 4.5$ V to 5.5 V,	$I_{OH} = -0.4$ mA	$V_{CC}-2$			$V_{CC}-2$			V
	$V_{CC} = 4.5$ V,	$I_{OH} = -1$ mA	2.4	3.3					
	$V_{CC} = 4.5$ V,	$I_{OH} = -2.6$ mA				2.4	3.2		
V_{OL}	$V_{CC} = 4.5$ V,	$I_{OL} = 12$ mA		0.25	0.4		0.25	0.4	V
	$V_{CC} = 4.5$ V,	$I_{OL} = 24$ mA					0.35	0.5	
I_{OZH}	$V_{CC} = 5.5$ V,	$V_O = 2.7$ V			20			20	µA
I_{OZL}	$V_{CC} = 5.5$ V,	$V_I = 0.4$ V			-20			-20	µA
I_I	$V_{CC} = 5.5$ V,	$V_I = 7$ V			0.1			0.1	mA
I_{IH}	$V_{CC} = 5.5$ V,	$V_I = 2.7$ V			20			20	µA
I_{IL}	$V_{CC} = 5.5$ V,	$V_I = 0.4$ V			-0.1			-0.1	mA
I_O‡	$V_{CC} = 5.5$ V,	$V_O = 2.25$ V	-15		-70	-15		-70	mA
I_{CC}	$V_{CC} = 5.5$ V	Outputs high		10	17		10	17	mA
		Outputs low		15	24		15	24	
		Outputs disabled		16	27		16	27	

†All typical values are at $V_{CC} = 5$ V, $T_A = 25°C$.
‡The output conditions have been chosen to produce a current that closely approximates one half of the true short-circuit output current, I_{OS}.

switching characteristics (see Note 1)

PARAMETER	FROM (INPUT)	TO (OUTPUT)	$V_{CC} = 4.5$ V to 5.5 V, $C_L = 50$ pF, R1 = 500Ω, R2 = 500Ω, T_A = MIN to MAX				UNIT
			SN54ALS563		SN74ALS563		
			MIN	MAX	MIN	MAX	
t_{PLH}	D	$\bar{Q}$	3	21	3	18	ns
t_{PHL}			3	15	3	14	
t_{PLH}	C	$\bar{Q}$	8	29	8	22	ns
t_{PHL}			8	22	8	21	
t_{PZH}	$\overline{OC}$	$\bar{Q}$	4	21	4	18	ns
t_{PZL}			4	21	4	18	
t_{PHZ}	$\overline{OC}$	$\bar{Q}$	2	10	2	8	ns
t_{PLZ}			3	15	3	13	

NOTE 1: For load circuit and voltage waveforms, see page 1-12.

ALS AND AS CIRCUITS

TEXAS INSTRUMENTS
POST OFFICE BOX 225012 • DALLAS, TEXAS 75265

2
ALS AND AS CIRCUITS

TYPES SN54ALS564, SN74ALS564
OCTAL D-TYPE EDGE-TRIGGERED FLIP-FLOPS
WITH 3-STATE OUTPUTS

D2661, APRIL 1982—REVISED DECEMBER 1983

- 3-State Buffer-Type Inverting Outputs Drive Bus-Lines Directly
- Bus-Structured Pinout
- Buffered Control Inputs
- Package Options include Both Plastic and Ceramic Chip Carriers in Addition to Plastic and Ceramic DIPs
- Dependable Texas Instruments Quality and Reliability

description

These 8-bit registers feature inverting three-state outputs designed specifically for bus driving. They are particularly suitable for implementing buffer registers, I/O ports, bidirectional bus drivers, and working registers.

The eight-bit edge-triggered D-type flip-flops enter data on the low-to-high transition of the clock.

The output control does not affect the internal operation of the flip-flops. Old data can be retained or new data can be entered while the outputs are in the high-impedance state.

The SN54ALS564 is characterized for operation over the full military temperature range of −55°C to 125°C. The SN74ALS564 is characterized for operation from 0°C to 70°C.

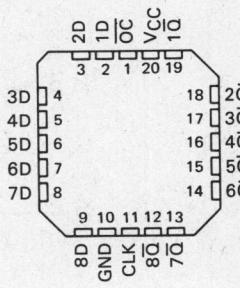

FUNCTION TABLE (EACH FLIP-FLOP)

INPUTS			OUTPUT
$\overline{OC}$	CLK	D	$\overline{Q}$
L	↑	H	L
L	↑	L	H
L	L	X	$\overline{Q}_0$
H	X	X	Z

logic symbol

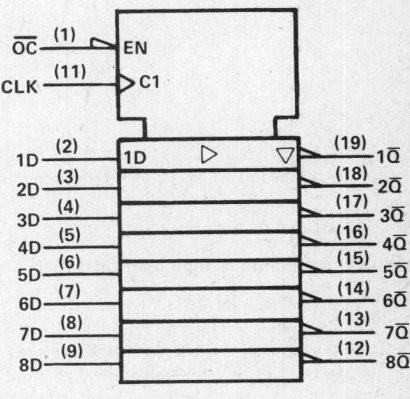

Pin numbers shown are for J and N packages.

Copyright © 1982 by Texas Instruments Incorporated

TYPES SN54ALS564, SN74ALS564
OCTAL D-TYPE EDGE-TRIGGERED FLIP-FLOPS
WITH 3-STATE OUTPUTS

logic diagram (positive logic)

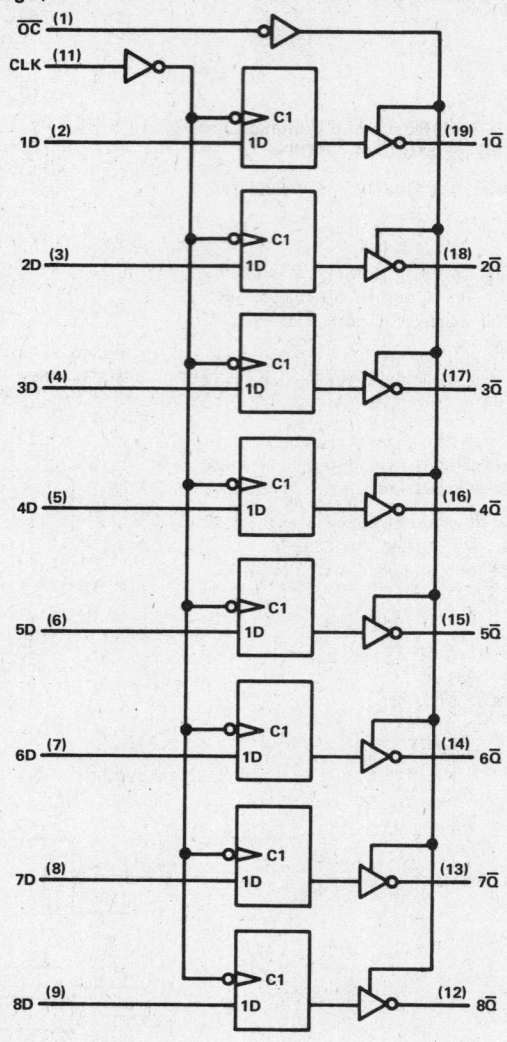

Pin numbers shown are for J and N packages.

absolute maximum ratings over operating free-air temperature range (unless otherwise noted)

Supply voltage, V_{CC} .... 7 V
Input voltage .... 7 V
Voltage applied to a disabled 3-state output .... 5.5 V
Operating free-air temperature range: SN54ALS564 .... −55 °C to 125 °C
                                    SN74ALS564 .... 0 °C to 70 °C
Storage temperature range .... −65 °C to 150 °C

TEXAS INSTRUMENTS
POST OFFICE BOX 225012 • DALLAS, TEXAS 75265

TYPES SN54ALS564, SN74ALS564
OCTAL D-TYPE EDGE-TRIGGERED FLIP-FLOPS
WITH 3-STATE OUTPUTS

recommended operating conditions

		SN54ALS564			SN74ALS564			UNIT
		MIN	NOM	MAX	MIN	NOM	MAX	
V_{CC}	Supply voltage	4.5	5	5.5	4.5	5	5.5	V
V_{IH}	High-level input voltage	2			2			V
V_{IL}	Low-level input voltage			0.8			0.8	V
I_{OH}	High-level output current			−1			−2.6	mA
I_{OL}	Low-level output current			12			24	mA
f_{clock}	Clock frequency	0		30	0		35	MHz
t_w	Pulse duration CLK high	16.5			14			ns
	CLK low	16.5			14			
t_{su}	Setup time, data before CLK↑	15			15			ns
t_h	Hold time, data after CLK↑	4			0			ns
T_A	Operating free-air temperature	−55		125	0		70	°C

electrical characteristics over recommended operating free-air temperature range (unless otherwise noted)

PARAMETER	TEST CONDITIONS		SN54ALS564			SN74ALS564			UNIT
			MIN	TYP†	MAX	MIN	TYP†	MAX	
V_{IK}	$V_{CC} = 4.5$ V,	$I_I = -18$ mA			−1.5			−1.5	V
V_{OH}	$V_{CC} = 4.5$ V to 5.5 V,	$I_{OH} = -0.4$ mA	$V_{CC}-2$			$V_{CC}-2$			V
	$V_{CC} = 4.5$ V,	$I_{OH} = -1$ mA	2.4	3.3					
	$V_{CC} = 4.5$ V,	$I_{OH} = -2.6$ mA				2.4	3.2		
V_{OL}	$V_{CC} = 4.5$ V,	$I_{OL} = 12$ mA		0.25	0.4		0.25	0.4	V
	$V_{CC} = 4.5$ V,	$I_{OL} = 24$ mA					0.35	0.5	
I_{OZH}	$V_{CC} = 5.5$ V,	$V_O = 2.7$ V			20			20	μA
I_{OZL}	$V_{CC} = 5.5$ V,	$V_I = 0.4$ V			−20			−20	μA
I_I	$V_{CC} = 5.5$ V,	$V_I = 7$ V			0.1			0.1	mA
I_{IH}	$V_{CC} = 5.5$ V,	$V_I = 2.7$ V			20			20	μA
I_{IL}	$V_{CC} = 5.5$ V,	$V_I = 0.4$ V			−0.2			−0.2	mA
I_O‡	$V_{CC} = 5.5$ V,	$V_O = 2.25$ V	−15		−70	−15		−70	mA
I_{CC}	$V_{CC} = 5.5$ V	Outputs high		10	17		10	17	mA
		Outputs low		15	24		15	24	
		Outputs disabled		16	27		16	27	

†All typical values are at $V_{CC} = 5$ V, $T_A = 25$°C.
‡The output conditions have been chosen to produce a current that closely approximates one half of the true short-circuit output current, I_{OS}.

switching characteristics (see Note 1)

PARAMETER	FROM (INPUT)	TO (OUTPUT)	$V_{CC} = 4.5$ V to 5.5 V, $C_L = 50$ pF, $R1 = 500$ Ω, $R2 = 500$ Ω, $T_A =$ MIN to MAX				UNIT
			SN54ALS564		SN74ALS564		
			MIN	MAX	MIN	MAX	
f_{max}			30		35		MHz
t_{PLH}	CLK	Q	4	15	4	14	ns
t_{PHL}			4	15	4	14	
t_{PZH}	$\overline{OC}$	Q	4	21	4	18	ns
t_{PZL}			4	21	4	18	
t_{PHZ}	$\overline{OC}$	Q	2	10	2	8	ns
t_{PLZ}			3	15	3	13	

NOTE 1: For load circuit and voltage waveforms, see page 1-12.

Texas Instruments
POST OFFICE BOX 225012 • DALLAS, TEXAS 75265

2
ALS AND AS CIRCUITS

TYPES SN54ALS568A, SN54ALS569A, SN74ALS568A, SN74ALS569A
SYNCHRONOUS 4-BIT UP/DOWN DECADE AND BINARY COUNTERS WITH 3-STATE OUTPUTS

D2661, APRIL 1982—REVISED DECEMBER 1983

- 3-State Q Outputs Drive Bus Lines Directly
- Counter Operation Independent of 3-State Output
- Fully Synchronous Clear, Count, and Load
- Asynchronous Clear Also Provided
- Fully Cascadable
- Package Options Include Both Plastic and Ceramic Chip Carriers in Addition to Plastic and Ceramic DIPs
- Dependable Texas Instruments Quality and Reliability

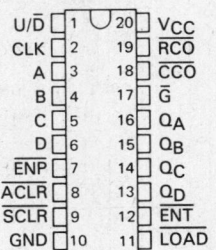

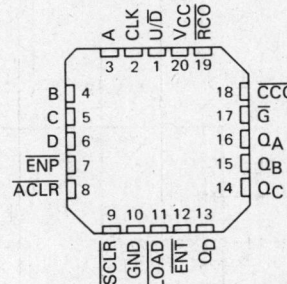

description

The 'ALS568A decade counters and 'ALS569A binary counters are programmable, count up or down, and offer both synchronous and asynchronous clearing. All synchronous functions are executed on the positive-going edge of the clock.

The clear function is initiated by applying a low level to either Asynchronous Clear ($\overline{ACLR}$) or Synchronous Clear ($\overline{SCLR}$). Asynchronous (direct) clearing overrides all other functions of the device, while synchronous clearing overrides only the other synchronous functions. Data is loaded from the A, B, C, and D inputs by holding Load ($\overline{LOAD}$) low during a positive-going clock transition. The counting function is enabled only when Enable P ($\overline{ENP}$) and Enable T ($\overline{ENT}$) are low and $\overline{ACLR}$, $\overline{SCLR}$, and $\overline{LOAD}$ are high. The Up/Down (U/$\overline{D}$) input controls the direction of the count. These counters count up when U/$\overline{D}$ is high and count down when U/$\overline{D}$ is low.

A high level at the Output Enable ($\overline{G}$) forces the Q outputs into the high-impedance state, and a low level enables those outputs. Counting is independent of of $\overline{G}$. $\overline{ENT}$ is fed forward to enable the Ripple Carry Output ($\overline{RCO}$) to produce a low-level pulse while the count is zero (all Q outputs low) when counting down or maximum (9 or 15) when counting up. The Clocked Carry Output ($\overline{CCO}$) produces a low level pulse for a duration equal to that of the low level of the clock when $\overline{RCO}$ is low and the counter is enabled (both $\overline{ENP}$ and $\overline{ENT}$ are low); otherwise, $\overline{CCO}$ is high. $\overline{CCO}$ does not have the glitches commonly associated with a ripple-carry output. Cascading is normally accomplished by connecting $\overline{RCO}$ or $\overline{CCO}$ of the first counter to $\overline{ENT}$ of the next counter. However, for very-high-speed counting, $\overline{RCO}$ should be used for cascading since $\overline{CCO}$ does not become active until the clock returns to the low level.

The SN54ALS568A and SN54ALS569A are characterized for operation over the full military temperature range of −55°C to 125°C. The SN74ALS568A and SN74ALS569A are characterized for operation from 0°C to 70°C.

TYPES SN54ALS568A, SN54ALS569A, SN74ALS568A, SN74ALS569A
SYNCHRONOUS 4-BIT UP/DOWN DECADE AND BINARY COUNTERS WITH 3-STATE OUTPUTS

FUNCTION TABLE

INPUTS								OPERATION
$\overline{G}$	$\overline{ACLR}$	$\overline{SCLR}$	$\overline{LOAD}$	$\overline{ENT}$	$\overline{ENP}$	$U/\overline{D}$	CLK	
H	X	X	X	X	X	X	X	Q Outputs Disabled
L	L	X	X	X	X	X	X	Asynchronous Clear
L	H	L	X	X	X	X	↑	Synchronous Clear
L	H	H	L	X	X	X	↑	Load
L	H	H	H	L	L	H	↑	Count Up
L	H	H	H	L	L	L	↑	Count Down
L	H	H	H	H	X	X	X	Inhibit Count
L	H	H	H	X	H	X	X	Inhibit Count

logic symbols

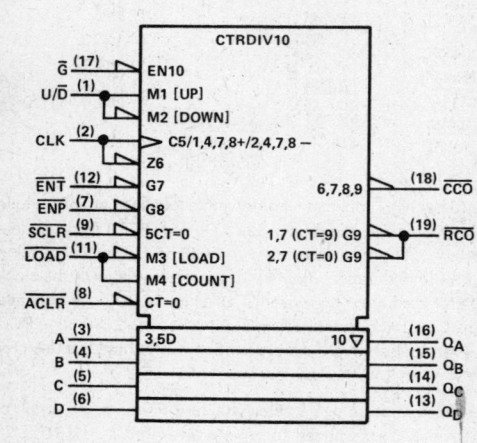
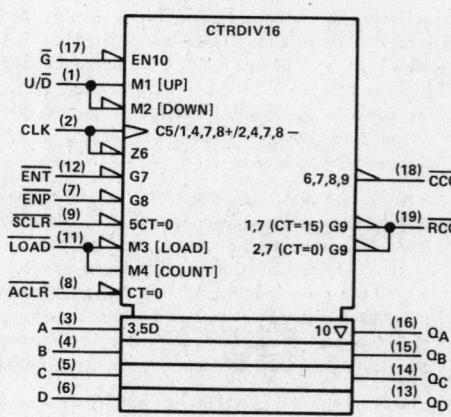

Pin numbers shown are for J and N packages.

TYPES SN54ALS568A, SN74ALS568A
SYNCHRONOUS 4-BIT UP/DOWN DECADE COUNTERS
WITH 3-STATE OUTPUTS

'ALS568A logic diagram (positive logic)

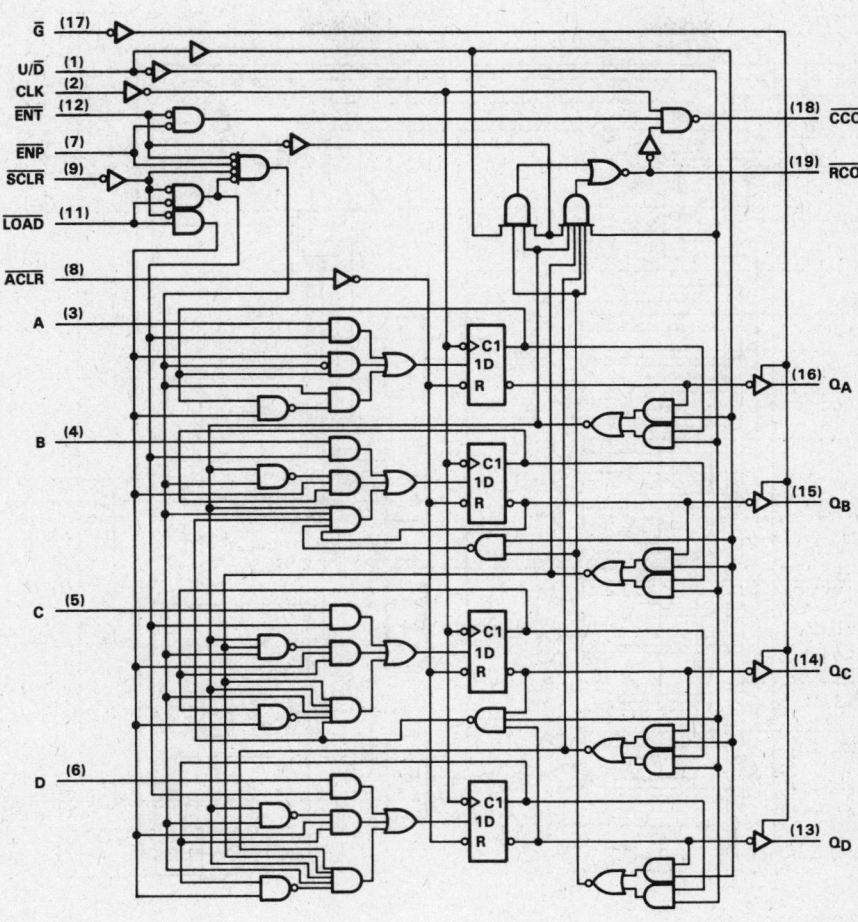

Pin numbers shown are for J and N packages.

TYPES SN54ALS569A, SN74ALS569A
SYNCHRONOUS 4-BIT UP/DOWN BINARY COUNTERS WITH 3-STATE OUTPUTS

'ALS569A logic diagram (positive logic)

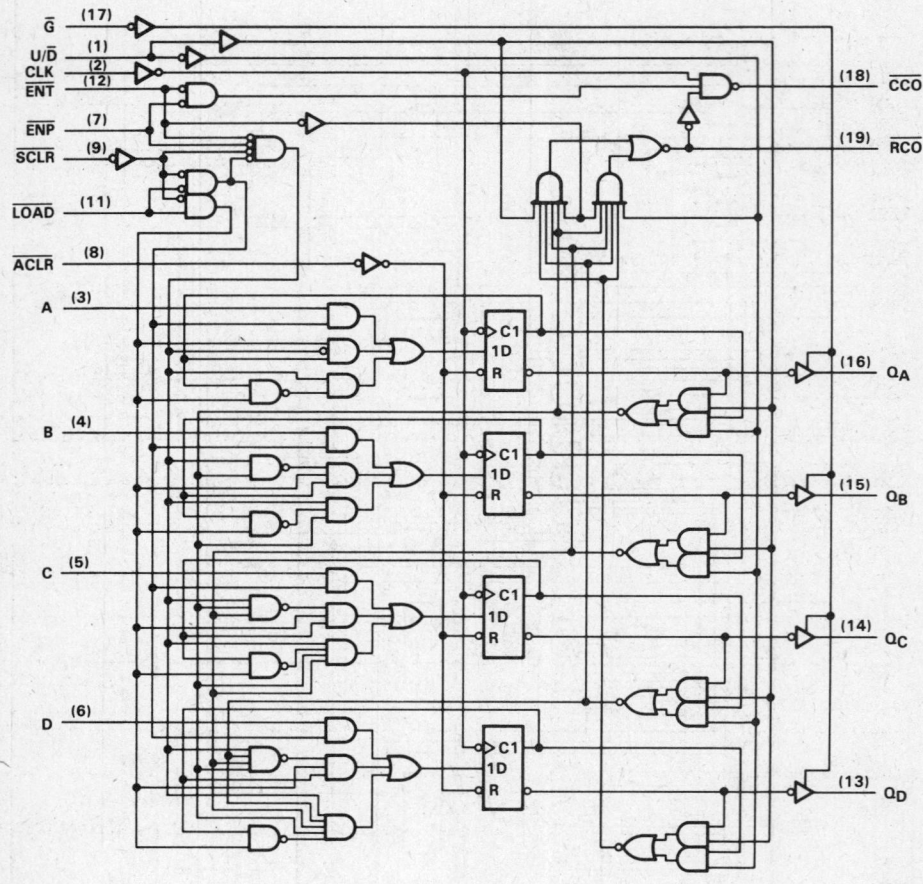

Pin numbers shown are for J and N packages.

TYPES SN54ALS568A, AN74ALS568A
SYNCHRONOUS 4-BIT UP/DOWN DECADE COUNTERS
WITH 3-STATE OUTPUTS

'ALS568A typical load, count, and inhibit sequences

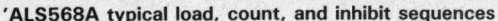

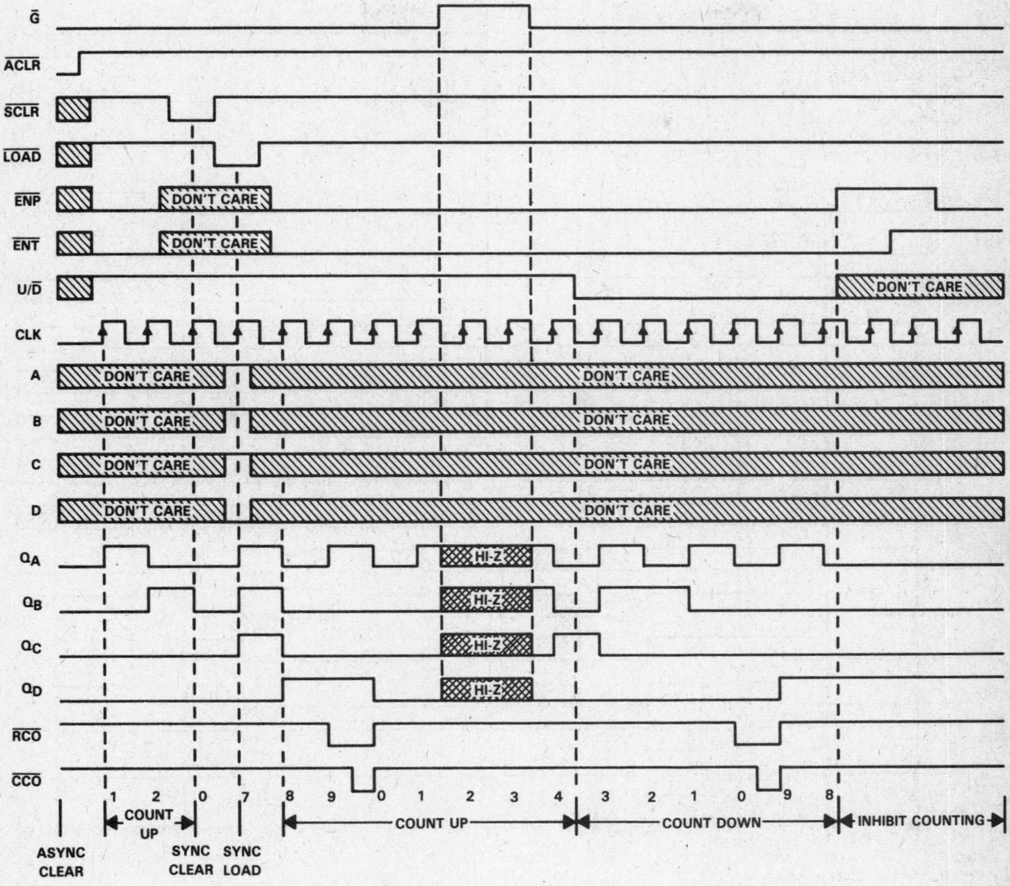

TYPES SN54ALS569A, SN74ALS569A
SYNCHRONOUS 4-BIT UP/DOWN BINARY COUNTERS WITH 3-STATE OUTPUTS

'ALS569A typical load, count, and inhibit sequences

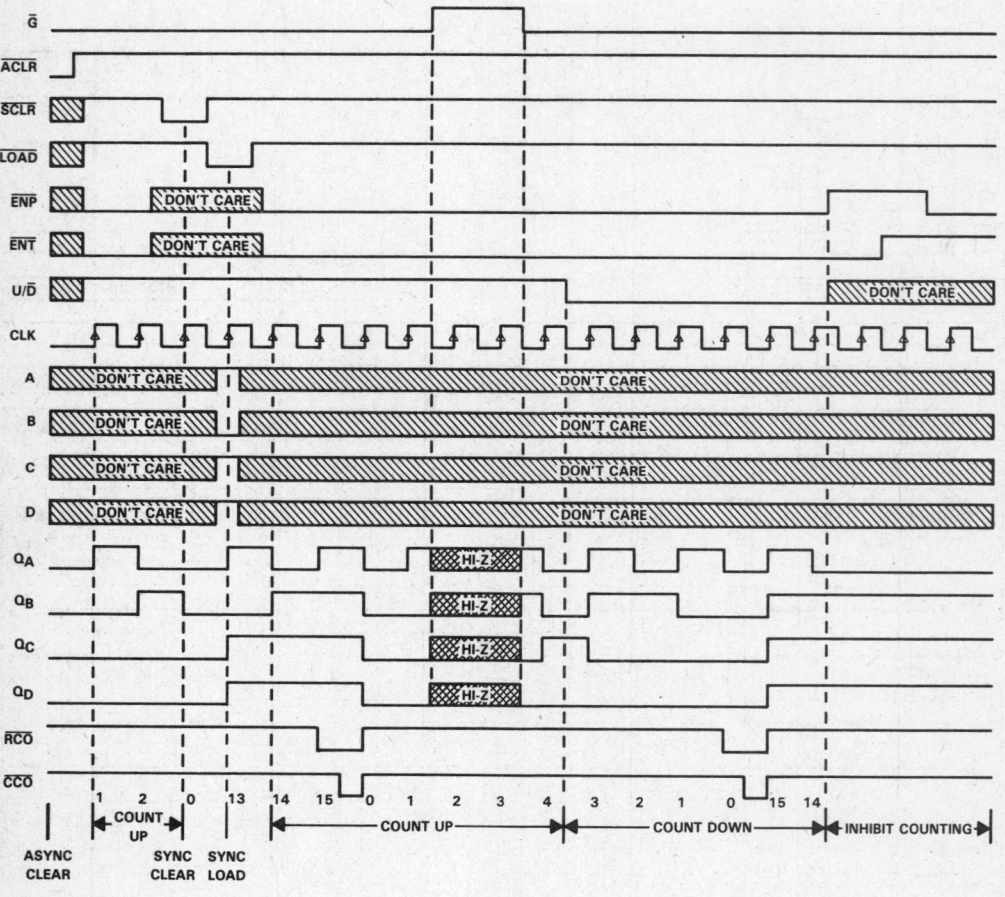

TYPES SN54ALS568A, SN54ALS569A, SN74ALS568A, SN74ALS569A
SYNCHRONOUS 4-BIT UP/DOWN DECADE AND BINARY COUNTERS WITH 3-STATE OUTPUTS

absolute maximum ratings over operating free-air temperature range (unless otherwise noted)

Supply voltage, V_{CC} .... 7 V
Input voltage .... 7 V
Voltage applied to a disabled 3-state output .... 5.5 V
Operating free-air temperature range: SN54ALS568A, SN54ALS569A .... −55 °C to 125 °C
                                      SN74ALS568A, SN74ALS569A .... 0 °C to 70 °C
Storage temperature range .... −65 °C to 150 °C

recommended operating conditions

			SN54ALS568A SN54ALS569A			SN74ALS568A SN74ALS569A			UNIT
			MIN	NOM	MAX	MIN	NOM	MAX	
V_{CC}	Supply voltage		4.5	5	5.5	4.5	5	5.5	V
V_{IH}	High-level input voltage		2			2			V
V_{IL}	Low-level input voltge				0.8			0.8	C
I_{OH}	High-level output current	Q outputs			−1			−2.6	mA
		$\overline{CCO}$ and $\overline{RCO}$			−0.4			−0.4	
I_{OL}	Low-level output current	Q outputs			12			24	mA
		$\overline{CCO}$ and $\overline{RCO}$			4			8	
f_{clock}	Clock frequency	'ALS568A	0		18	0		20	MHz
		'ALS569A	0		25	0		30	
t_w	Pulse duration	$\overline{ALCR}$ or $\overline{LOAD}$ low	20			15			ns
		'ALS568A CLK high	27.5			25			
		'ALS568A CLK low	27.5			25			
		'ALS569A CLK high	20			16.5			
		'ALS569A CLK low	20			16.5			
t_{su}	Setup time before CLK ↑	Data at A, B, C, D	25			20			ns
		$\overline{ENP}$, $\overline{ENT}$ High	35			30			
		$\overline{ENP}$, $\overline{ENT}$ Low	25			20			
		$\overline{SCLR}$ Low	20			15			
		$\overline{SCLR}$ High (inactive)	35			30			
		$\overline{LOAD}$ Low	20			15			
		$\overline{LOAD}$ High (inactive)	35			30			
		$U/\overline{D}$	35			30			
		$\overline{ACLR}$ inactive	10			10			
t_h	Hold time after CLK↑ for any input		0			0			ns
T_A	Operating free-air temperature		−55		125	0		70	°C

TYPES SN54ALS568A, SN54ALS569A, SN74ALS568A, SN74ALS569A
SYNCHRONOUS 4-BIT UP/DOWN DECADE AND BINARY COUNTERS WITH 3-STATE OUTPUTS

electrical characteristics over recommended operating free-air temperature range (unless otherwise noted)

PARAMETER		TEST CONDITIONS		SN54ALS568A SN54ALS569A			SN74ALS568A SN74ALS569A			UNIT
				MIN	TYP†	MAX	MIN	TYP†	MAX	
V_{IK}		V_{CC} = 4.5 V,	I_I = −18 mA			−1.5			−1.5	V
V_{OH}	All outputs	V_{CC} = 4.5 V to 5.5 V,	I_{OH} = −0.4 mA	$V_{CC}-2$			$V_{CC}-2$			V
	Q outputs	V_{CC} = 4.5 V,	I_{OH} = −1 mA	2.4	3.3					
		V_{CC} = 4.5 V,	I_{OH} = −2.6 mA				2.4	3.2		
V_{OL}	Q outputs	V_{CC} = 4.5 V,	I_{OL} = 12 mA		0.25	0.4		0.25	0.4	V
		V_{CC} = 4.5 V,	I_{OL} = 24 mA					0.35	0.5	
	$\overline{CCO}$ and $\overline{RCO}$	V_{CC} = 4.5 V,	I_{OH} = 4 mA		0.25	0.4		0.25	0.4	
		V_{CC} = 4.5 V,	I_{OL} = 8 mA					0.35	0.5	
I_{OZH}		V_{CC} = 5.5 V,	V_O = 2.7 V			20			20	μA
I_{OZL}		V_{CC} = 5.5 V,	V_O = 0.4 V			−20			−20	μA
I_I		V_{CC} = 5.5 V,	V_I = 7 V			0.1			0.1	mA
I_{IH}		V_{CC} = 5.5 V,	V_I = 2.7 V			20			20	μA
I_{IL}		V_{CC} = 5.5 V,	V_I = 0.4 V			−0.2			−0.2	mA
I_O ‡	$\overline{CCO}$ and $\overline{RCO}$	V_{CC} = 5.5 V,	V_O = 2.25 V	−15		−70	−15		−70	mA
	Q outputs			−30		−112	−30		−112	
I_{CC}		V_{CC} = 5.5 V	Outputs high		16	26		16	26	mA
			Outputs low		20	32		20	32	
			Outputs disabled		20	32		20	32	

†All typical values are at V_{CC} = 5 V, T_A = 25°C.
‡The output conditions have been chosen to produce a current that closely approximates one half of the true short-circuit output current, I_{OS}.

TYPES SN54ALS568A, SN54ALS569A, SN74ALS568A, SN74ALS569A
SYNCHRONOUS 4-BIT UP/DOWN DECADE AND BINARY COUNTERS WITH 3-STATE OUTPUTS

switching characteristics (see Note 1)

PARAMETER	FROM (INPUT)	TO (OUTPUT)	V_{CC} = 4.5 V to 5.5 V, C_L = 50 pF, R1 = 500 Ω, R2 = 500 Ω, T_A = MIN to MAX				UNIT
			SN54ALS568A SN54ALS569A		SN74ALS568A SN74ALS569A		
			MIN	MAX	MIN	MAX	
f_{max}		'ALS568A	18		20		MHz
		'ALS569A	25		30		
t_{PLH}	CLK	Any Q	4	17	4	13	ns
t_{PHL}			7	18	7	16	
t_{PLH}	CLK	$\overline{RCO}$	12	31	12	28	ns
t_{PHL}			10	22	10	19	
t_{PLH}	CLK	$\overline{CCO}$	5	15	5	13	ns
t_{PHL}			6	30	6	25	
t_{PLH}	U/$\overline{D}$	$\overline{RCO}$	9	25	9	23	ns
t_{PHL}			9	23	9	19	
t_{PLH}	$\overline{ENT}$	$\overline{RCO}$	6	17	6	15	ns
t_{PHL}			4	17	4	13	
t_{PLH}	$\overline{ENT}$	$\overline{CCO}$	5	15	5	13	ns
t_{PHL}			9	28	9	23	
t_{PLH}	$\overline{ENP}$	$\overline{CCO}$	4	14	4	12	ns
t_{PHL}			5	17	5	14	
t_{PHL}	$\overline{ACLR}$	Any Q	9	22	9	20	ns
t_{PZH}	$\overline{G}$	Any Q	6	21	6	18	ns
t_{PZL}			6	29	6	24	
t_{PHZ}	$\overline{G}$	Any Q	1	12	1	10	ns
t_{PLZ}			3	19	3	13	

NOTE 1: For load circuit and voltage waveforms, see page 1-12.

2
ALS AND AS CIRCUITS

TYPES SN54ALS573, SN54ALS580, SN54AS573, SN54AS580, SN74ALS573, SN74ALS580, SN74AS573, SN74AS580
OCTAL D-TYPE TRANSPARENT LATCHES WITH 3-STATE OUTPUTS

D2661, DECEMBER 1982 — REVISED DECEMBER 1983

- 3-State Buffer-Type Outputs Drive Bus-Lines Directly
- Bus-Structured Pinout
- Choice of True or Inverting Logic

    'ALS573, 'AS573   True Outputs
    'ALS580, 'AS580   Inverting Outputs

- Package Options Include Both Plastic and Ceramic Chip Carriers in Addition to Plastic and Ceramic DIPs
- Dependable Texas Instruments Quality and Reliability

SN54ALS573, SN54AS573 . . . J PACKAGE
SN74ALS573, SN74AS573 . . . N PACKAGE
(TOP VIEW)

```
 OC  [ 1   20 ] VCC
 1D  [ 2   19 ] 1Q
 2D  [ 3   18 ] 2Q
 3D  [ 4   17 ] 3Q
 4D  [ 5   16 ] 4Q
 5D  [ 6   15 ] 5Q
 6D  [ 7   14 ] 6Q
 7D  [ 8   13 ] 7Q
 8D  [ 9   12 ] 8Q
GND  [10   11 ] C
```

description

These 8-bit latches feature three-state outputs designed specifically for driving highly capacitive or relatively low-impedance loads. They are particularly suitable for implementing buffer registers, I/O ports, bidirectional bus drivers, and working registers.

The eight latches are transparent D-type latches. While the enable (C) is high the outputs (Q or $\overline{Q}$) will respond to the data (D) inputs. When the enable is taken low the outputs will be latched to retain the data that was set up.

A buffered output-control input can be used to place the eight outputs in either a normal logic state (high or low logic levels) or a high-impedance state. In the high-impedance state the outputs neither load nor drive the bus lines significantly. The high-impedance state and increased drive provide the capability to drive the bus lines in a bus-organized system without need for interface or pull-up components.

The output control ($\overline{OC}$) does not affect the internal operation of the latches. Old data can be retained or new data can be entered while the outputs are at high impedance.

The SN54ALS573, SN54AS573, SN54ALS580 and SN54AS580 are characterized for operation over the full military temperature range of −55°C to 125°C. The SN74ALS573, SN74AS573, SN74ALS580, and SN74AS580 are characterized for operation from 0°C to 70°C.

SN54ALS573, SN54AS573 . . . FH PACKAGE
SN74ALS573, SN74AS573 . . . FN PACKAGE
(TOP VIEW)

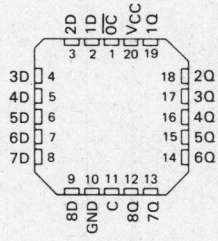

SN54ALS580, SN54AS580 . . . J PACKAGE
SN74ALS580, SN74AS580 . . . N PACKAGE
(TOP VIEW)

```
 OC  [ 1   20 ] VCC
 1D  [ 2   19 ] 1Q̄
 2D  [ 3   18 ] 2Q̄
 3D  [ 4   17 ] 3Q̄
 4D  [ 5   16 ] 4Q̄
 5D  [ 6   15 ] 5Q̄
 6D  [ 7   14 ] 6Q̄
 7D  [ 8   13 ] 7Q̄
 8D  [ 9   12 ] 8Q̄
GND  [10   11 ] C
```

SN54ALS580, SN54AS580 . . . FH PACKAGE
SN74ALS580, SN74AS580 . . . FN PACKAGE
(TOP VIEW)

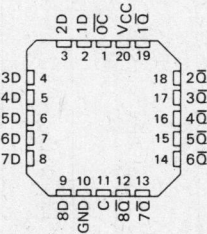

Copyright © 1982 by Texas Instruments Incorporated

Texas Instruments
POST OFFICE BOX 225012 • DALLAS, TEXAS 75265

TYPES SN54ALS573, SN54ALS580, SN54AS573, SN54AS580, SN74ALS573, SN74ALS580, SN74AS573, SN74AS580
OCTAL D-TYPE TRANSPARENT LATCHES WITH 3-STATE OUTPUTS

FUNCTION TABLES

'ALS573, 'AS573
(EACH LATCH)

INPUTS			OUTPUT
ENABLE			Q
$\overline{OC}$	C	D	
L	H	H	H
L	H	L	L
L	L	X	Q_0
H	X	X	Z

'ALS580, 'AS580
(EACH LATCH)

INPUTS			OUTPUT
ENABLE			$\overline{Q}$
$\overline{OC}$	C	D	
L	H	H	L
L	H	L	H
L	L	X	$\overline{Q}_0$
H	X	X	Z

logic symbols

'ALS573, 'AS573

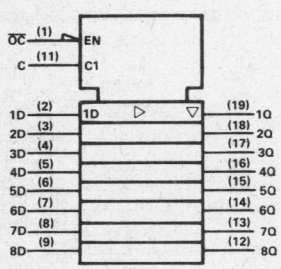

'ALS580, 'AS580

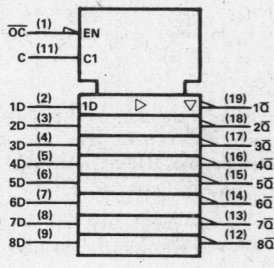

logic diagram (positive logic)

'ALS573, 'AS573

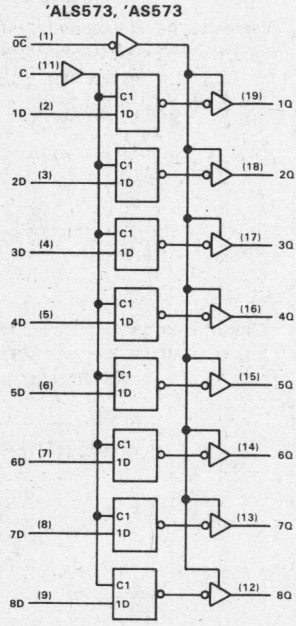

'ALS580, 'AS580

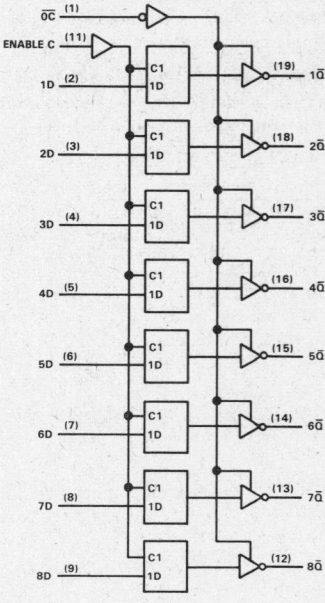

Pin numbers shown are for J and N packages.

Texas Instruments
POST OFFICE BOX 225012 • DALLAS, TEXAS 75265

TYPES SN54ALS573, SN54ALS580, SN74ALS573, SN74ALS580
OCTAL D-TYPE TRANSPARENT LATCHES WITH 3-STATE OUTPUTS

absolute maximum ratings over operating free-air temperature range (unless otherwise noted)

Supply voltage, V_{CC}	7 V
Input voltage	7 V
Voltage applied to a disabled 3-state output	5.5 V
Operating free-air temperature range: SN54ALS573, SN54ALS580	−55 °C to 125 °C
SN74ALS573, SN74ALS580	0 °C to 70 °C
Storage temperature range	−65 °C to 150 °C

recommended operating conditions

			SN54ALS573 SN54ALS580			SN74ALS573 SN74ALS580			UNIT
			MIN	NOM	MAX	MIN	NOM	MAX	
V_{CC}	Supply voltage		4.5	5	5.5	4.5	5	5.5	V
V_{IH}	High-level input voltage		2			2			V
V_{IL}	Low-level input voltage				0.8			0.8	V
I_{OH}	High-level output current				−1			−2.6	mA
I_{OL}	Low-level output current				12			24	mA
t_w	Pulse duration, enable C high	'ALS573	10			10			ns
		'ALS580	15			15			
t_{su}	Setup time, data before enable C↓		10			10			ns
t_h	Hold time, data after enable C↓	'ALS573	7			7			ns
		'ALS580	10			10			
T_A	Operating free-air temperature		−55		125	0		70	°C

electrical characteristics over recommended operating free-air temperature range (unless otherwise noted)

PARAMETER		TEST CONDITIONS		SN54ALS573 SN54ALS580			SN74ALS573 SN74ALS580			UNIT
				MIN	TYP†	MAX	MIN	TYP†	MAX	
V_{IK}		V_{CC} = 4.5 V,	I_I = −18 mA			−1.5			−1.5	V
V_{OH}		V_{CC} = 4.5 V to 5.5 V,	I_{OH} = −0.4 mA	$V_{CC}-2$			$V_{CC}-2$			V
		V_{CC} = 4.5 V,	I_{OH} = −1 mA	2.4	3.3					
		V_{CC} = 4.5 V,	I_{OH} = −2.6 mA				2.4	3.2		
V_{OL}		V_{CC} = 4.5 V,	I_{OL} = 12 mA		0.25	0.4		0.25	0.4	V
		V_{CC} = 4.5 V,	I_{OL} = 24 mA					0.35	0.5	
I_{OZH}		V_{CC} = 5.5 V,	V_O = 2.7 V			20			20	µA
I_{OZL}		V_{CC} = 5.5 V,	V_O = 0.4 V,			−20			−20	µA
I_I		V_{CC} = 5.5 V,	V_I = 7 V			0.1			0.1	mA
I_{IH}		V_{CC} = 5.5 V,	V_I = 2.7 V			20			20	µA
I_{IL}		V_{CC} = 5.5 V,	V_I = 0.4 V			−0.1			−0.1	mA
I_O‡		V_{CC} = 5.5 V,	V_O = 2.25 V	−15		−70	−15		−70	mA
I_{CC}	'ALS573	V_{CC} = 5.5 V	Outputs high		10	17		10	17	mA
			Outputs low		15	24		15	24	
			Outputs disabled		16	27		16	27	
	'ALS580		Outputs high		10	17		10	17	
			Outputs low		15	24		15	24	
			Outputs disabled		16	27		16	27	

†All typical values are at V_{CC} = 5 V, T_A = 25 °C.
‡The output conditions have been chosen to produce a current that closely approximates one half of the true short-circuit output current, I_{OS}.

ALS AND AS CIRCUITS

TEXAS INSTRUMENTS
POST OFFICE BOX 225012 • DALLAS, TEXAS 75265

TYPES SN54ALS573, SN54ALS580, SN74ALS573, SN74ALS580
OCTAL D-TYPE TRANSPARENT LATCHES WITH 3-STATE OUTPUTS

'ALS573 switching characteristics (see Note 1)

PARAMETER	FROM (INPUT)	TO (OUTPUT)	V_{CC} = 4.5 V to 5.5 V, C_L = 50 pF, R1 = 500 Ω, R2 = 500 Ω, T_A = MIN to MAX				UNIT
			SN54ALS573		SN74ALS573		
			MIN	MAX	MIN	MAX	
t_{PLH}	D	Q	2	15	2	14	ns
t_{PHL}			2	15	2	14	
t_{PLH}	C	Q	8	27	8	20	ns
t_{PHL}			8	20	8	19	
t_{PZH}	$\overline{OC}$	Q	4	21	4	18	ns
t_{PZL}			4	21	4	18	
t_{PHZ}	$\overline{OC}$	Q	2	10	2	8	ns
t_{PLZ}			3	15	3	13	

'ALS580 switching characteristics (see Note 1)

PARAMETER	FROM (INPUT)	TO (OUTPUT)	V_{CC} = 4.5 V to 5.5 V, C_L = 50 pF, R1 = 500 Ω, R2 = 500 Ω, T_A = MIN to MAX				UNIT
			SN54ALS580		SN74ALS580		
			MIN	MAX	MIN	MAX	
t_{PLH}	D	$\overline{Q}$	3	21	3	18	ns
t_{PHL}			3	15	3	14	
t_{PLH}	C	$\overline{Q}$	8	29	8	22	ns
t_{PHL}			8	22	8	21	
t_{PZH}	$\overline{OC}$	$\overline{Q}$	4	21	4	18	ns
t_{PZL}			4	21	4	18	
t_{PHZ}	$\overline{OC}$	$\overline{Q}$	2	10	2	8	ns
t_{PLZ}			3	15	3	13	

NOTE 1: For load circuit and voltage waveforms, see page 1-12.

TEXAS INSTRUMENTS
POST OFFICE BOX 225012 • DALLAS, TEXAS 75265

TYPES SN54AS573, SN54AS580, SN74AS573, SN74AS580
OCTAL D-TYPE TRANSPARENT LATCEHS WITH 3-STATE OUTPUTS

absolute maximum ratings over operating free-air temperature range (unless otherwise noted)

Supply voltage, V_{CC} .... 7 V
Input voltage .... 7 V
Voltage applied to a disabled 3-state output .... 5.5 V
Operating free-air temperature range: SN54AS573, SN54AS580 .... −55 °C to 125 °C
                                      SN74AS573, SN74AS580 .... 0 °C to 70 °C
Storage temperature range .... −65 °C to 150 °C

recommended operating conditions

			SN54AS573 SN54AS580			SN74AS573 SN74AS580			UNIT
			MIN	NOM	MAX	MIN	NOM	MAX	
V_{CC}	Supply voltage		4.5	5	5.5	4.5	5	5.5	V
V_{IH}	High-level input voltage		2			2			V
V_{IL}	Low-level input voltage				0.8			0.8	V
I_{OH}	High-level output current				−12			−15	mA
I_{OL}	Low-level output current				32			48	mA
t_w	Pulse duration, enable C high	'AS573	5.5			4.5			ns
		'AS580	3			2			
t_{su}	Setup time, data before enable C↑		2			2			ns
t_h	Hold time, data after enable C↑		3			3			ns
T_A	Operating free-air temperature		−55		125	0		70	°C

electrical characteristics over recommended operating free-air temperature range (unless otherwise noted)

PARAMETER		TEST CONDITIONS		SN54AS573 SN54AS580			SN74AS573 SN74AS580			UNIT
				MIN	TYP†	MAX	MIN	TYP†	MAX	
V_{IK}		V_{CC} = 4.5 V,	I_I = −18 mA			−1.2			−1.2	V
V_{OH}		V_{CC} = 4.5 V to 5.5 V,	I_{OH} = −2 mA	V_{CC}−2			V_{CC}−2			V
		V_{CC} = 4.5 V,	I_{OH} = −12 mA	2.4	3.2					
		V_{CC} = 4.5 V,	I_{OH} = −15 mA				2.4	3.3		
V_{OL}		V_{CC} = 4.5 V,	I_{OL} = 32 mA		0.28	0.5				V
		V_{CC} = 4.5 V,	I_{OL} = 48 mA					0.33	0.5	
I_{OZH}		V_{CC} = 5.5 V,	V_O = 2.7 V			50			50	µA
I_{OZL}		V_{CC} = 5.5 V,	V_O = 0.4 V			−50			−50	µA
I_I		V_{CC} = 5.5 V,	V_I = 7 V			0.1			0.1	mA
I_{IH}		V_{CC} = 5.5 V,	V_I = 2.7 V			20			20	µA
I_{IL}		V_{CC} = 5.5 V,	V_I = 0.4 V			−0.5			−0.5	mA
I_O‡		V_{CC} = 5.5 V,	V_O = 2.25 V	−30		−112	−30		−112	mA
I_{CC}	'AS573	V_{CC} = 5.5 V	Outputs high		56	93		56	93	mA
			Outputs low		55	90		55	90	
			Outputs disabled		65	106		65	106	
	'AS580		Outputs high		62	100		62	100	
			Outputs low		65	106		65	106	
			Outputs disabled		71	115		71	115	

†All typical values are at V_{CC} = 5 V, T_A = 25 °C.
‡The output conditions have been chosen to produce a current that closely approximates one half of the true short-circuit output current, I_{OS}.

TEXAS INSTRUMENTS
POST OFFICE BOX 225012 • DALLAS, TEXAS 75265

TYPES SN54AS573, SN54AS580, SN74AS573, SN74AS580
OCTAL D-TYPE TRANSPARENT LATCHES WITH 3-STATE OUTPUTS

'AS573 switching characteristics (see Note 1)

PARAMETER	FROM (INPUT)	TO (OUTPUT)	V_{CC} = 4.5 V to 5.5 V, C_L = 50 pF, $R1$ = 500 Ω, $R2$ = 500 Ω, T_A = MIN to MAX				UNIT
			SN54AS573		SN74AS573		
			MIN	MAX	MIN	MAX	
t_{PLH}	D	Q	3	9	3	6	ns
t_{PHL}			3	7	3	6	
t_{PLH}	C	Q	6	14	6	11.5	ns
t_{PHL}			4	9	4	7.5	
t_{PZH}	$\overline{OC}$	Q	2	8	2	6.5	ns
t_{PZL}			4	11	4	9.5	
t_{PHZ}	$\overline{OC}$	Q	2	8	2	6.5	ns
t_{PLZ}			2	8	2	7	

'AS580 switching characteristics (see Note 1)

PARAMETER	FROM (INPUT)	TO (OUTPUT)	V_{CC} = 4.5 V to 5.5 V, C_L = 50 pF, $R1$ = 500 Ω, $R2$ = 500 Ω, T_A = MIN to MAX				UNIT
			SN54AS580		SN74AS580		
			MIN	MAX	MIN	MAX	
t_{PLH}	D	$\overline{Q}$	3	10	3	7.5	ns
t_{PHL}			3	7.5	3	7	
t_{PLH}	C	$\overline{Q}$	5	12	5	9	ns
t_{PHL}			4	8.5	4	8	
t_{PZH}	$\overline{OC}$	$\overline{Q}$	2	7.5	2	6.5	ns
t_{PZL}			4	10.5	4	9.5	
t_{PHZ}	$\overline{OC}$	$\overline{Q}$	2	7.5	2	6.5	ns
t_{PLZ}			2	8	2	7	

NOTE 1: For load circuit and voltage waveforms, see page 1-12.

TYPES SN54ALS574, SN54ALS575, SN54AS574, SN54AS575 SN74ALS574, SN74ALS575, SN74AS574, SN74AS575
OCTAL D-TYPE EDGE-TRIGGERED FLIP-FLOPS WITH 3-STATE OUTPUTS

D2661, JUNE 1982—REVISED DECEMBER 1983

- 3-State Buffer-Type Noninverting Outputs Drive Bus-Lines Directly
- Bus-Structured Pinout
- Buffered Control Inputs
- 'ALS575 and 'AS575 Have Synchronous Clear
- Package Options Include Both Plastic and Ceramic Chip Carriers in Addition to Plastic and Ceramic DIPs
- Dependable Texas Instruments Quality and Reliability

description

These 8-bit registers feature three-state outputs designed specifically for bus driving. They are particularly suitable for implementing buffer registers, I/O ports, bidirectional bus drivers, and working registers.

The eight edge-triggered D-type flip-flops enter data on the low-to-high transition of the clock. The 'ALS575 and 'AS575 may be synchronously cleared by taking the CLR input low.

The output-control does not affect the internal operation of the flip-flops. Old data can be retained or new data can be entered while the outputs are in the high-impedance state.

The SN54ALS' and SN54AS' devices are characterized for operation over the full military temperature range of −55°C to 125°C. The SN74ALS' and SN74AS' devices are characterized for operation from 0°C to 70°C.

SN54ALS574, SN54AS574 . . . J PACKAGE
SN74ALS574, SN74AS574 . . . N PACKAGE
(TOP VIEW)

```
 OC  [ 1    20 ] VCC
 1D  [ 2    19 ] 1Q
 2D  [ 3    18 ] 2Q
 3D  [ 4    17 ] 3Q
 4D  [ 5    16 ] 4Q
 5D  [ 6    15 ] 5Q
 6D  [ 7    14 ] 6Q
 7D  [ 8    13 ] 7Q
 8D  [ 9    12 ] 8Q
 GND [10    11 ] CLK
```

SN54ALS574, SN54AS574 . . . FH PACKAGE
SN74ALS574, SN74AS574 . . . FN PACKAGE
(TOP VIEW)

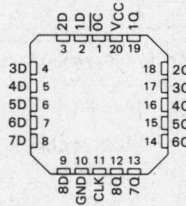

SN54ALS575, SN54AS575 . . . JT PACKAGE
SN74ALS575, SN74AS575 . . . NT PACKAGE
(TOP VIEW)

```
 CLR [ 1    24 ] VCC
 OC  [ 2    23 ] NC
 1D  [ 3    22 ] 1Q
 2D  [ 4    21 ] 2Q
 3D  [ 5    20 ] 3Q
 4D  [ 6    19 ] 4Q
 5D  [ 7    18 ] 5Q
 6D  [ 8    17 ] 6Q
 7D  [ 9    16 ] 7Q
 8D  [10    15 ] 8Q
 NC  [11    14 ] CLK
 GND [12    13 ] NC
```

SN54ALS575, SN54AS575 . . . FH PACKAGE
SN74ALS575, SN74AS575 . . . FN PACKAGE

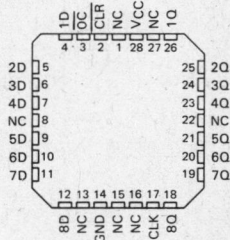

NC — No internal connection

FUNCTION TABLES

'ALS574, 'AS574
(EACH FLIP-FLOP)

INPUTS			OUTPUT
$\overline{OC}$	CLK	D	Q
L	↑	H	H
L	↑	L	L
L	L	X	Q_0
H	X	X	Z

'ALS575, 'AS575
(EACH FLIP-FLOP)

INPUTS				OUTPUT
$\overline{OC}$	$\overline{CLR}$	CLK	D	Q
L	L	↑	X	L
L	H	↑	H	H
L	H	↑	L	L
L	H	L	X	Q_0
H	X	X	X	Z

Copyright © 1982 by Texas Instruments Incorporated

Texas Instruments
POST OFFICE BOX 225012 • DALLAS, TEXAS 75265

TYPES SN54ALS574, SN54ALS575, SN54AS574, SN54AS575 SN74ALS574, SN74ALS575, SN74AS574, SN74AS575 OCTAL D-TYPE EDGE-TRIGGERED FLIP-FLOPS WITH 3-STATE OUTPUTS

logic symbols

'ALS574, 'AS574

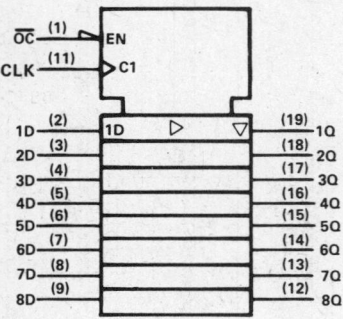

'ALS575, 'AS575

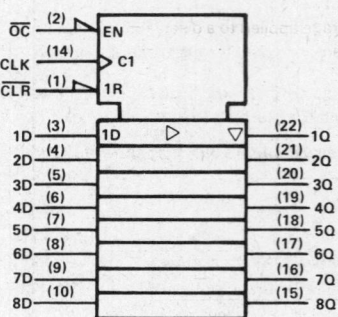

logic diagrams (positive logic)

'ALS574, 'AS574

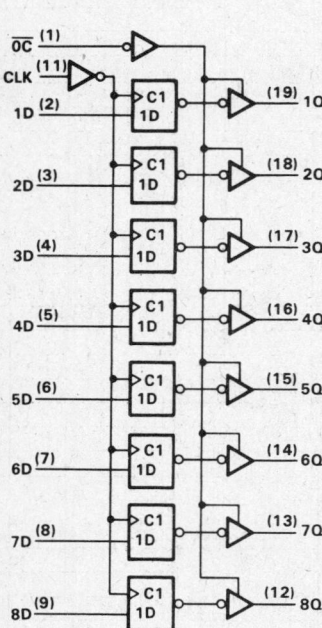

'ALS575, 'AS575

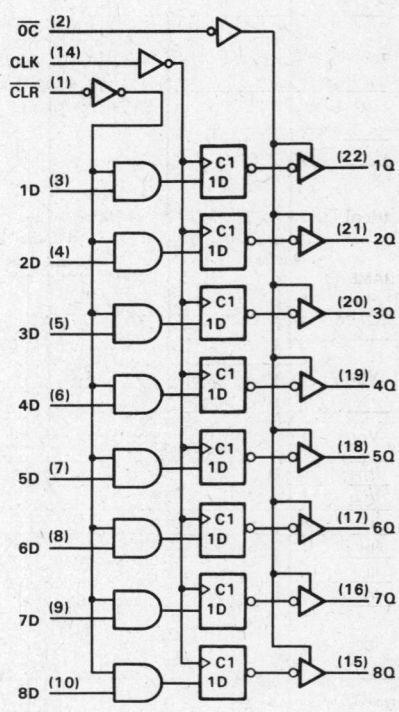

Pin numbers shown are for J and N packages.

Pin numbers shown are for JT and NT packages.

TYPES SN54ALS574, SN54ALS575, SN74ALS574, SN74ALS575
OCTAL D-TYPE EDGE-TRIGGERED FLIP-FLOPS WITH 3-STATE OUTPUTS

absolute maximum ratings over operating free-air temperature range (unless otherwise noted)

Supply voltage, V_{CC} .... 7 V
Input voltage .... 7 V
Voltage applied to a disabled 3-state output .... 5.5 V
Operating free-air temperature range: SN54ALS574, SN54ALS575 .... −55 °C to 125 °C
                                       SN74ALS574, SN74ALS575 .... 0 °C to 70 °C
Storage temperature range .... −65 °C to 150 °C

recommended operating conditions

			SN54ALS574 SN54ALS575			SN74ALS574 SN74ALS575			UNIT
			MIN	NOM	MAX	MIN	NOM	MAX	
V_{CC}	Supply voltage		4.5	5	5.5	4.5	5	5.5	V
V_{IH}	High-level input voltage		2			2			V
V_{IL}	Low-level input voltage				0.8			0.8	V
I_{OH}	High-level output current				−1			−2.6	mA
I_{OL}	Low-level output current				12			24	mA
f_{clock}	Clock frequency	'ALS574	0		30	0		35	MHz
		'ALS575	0		25	0		30	
t_w	Pulse duration	'ALS574 CLK high or low	16.5			14			ns
		'ALS575 CLK high or low	20			16.5			
t_{su}	Setup time before CLK↑	Data	15			15			ns
		'ALS575 CLR high	20			20			
		'ALS575 CLR low	15			15			
t_h	Hold time after CLK↑	Data	4			0			ns
		'ALS575 CLR	0			0			
T_A	Operating free-air temperature		−55		125	0		70	°C

electrical characteristics over recommended operating free-air temperature range (unless otherwise noted)

PARAMETER	TEST CONDITIONS		SN54ALS574 SN54ALS575			SN74ALS574 SN74ALS575			UNIT
			MIN	TYP[†]	MAX	MIN	TYP[†]	MAX	
V_{IK}	V_{CC} = 4.5 V,	I_I = −18 mA			−1.5			−1.5	V
V_{OH}	V_{CC} = 4.5 V to 5.5 V,	I_{OH} = −0.4 mA	V_{CC}−2			V_{CC}−2			V
	V_{CC} = 4.5 V,	I_{OH} = −1 mA	2.4	3.3					
	V_{CC} = 4.5 V,	I_{OH} = −2.6 mA				2.4	3.2		
V_{OL}	V_{CC} = 4.5 V,	I_{OL} = 12 mA		0.25	0.4		0.25	0.4	V
	V_{CC} = 4.5 V,	I_{OL} = 24 mA					0.35	0.5	
I_{OZH}	V_{CC} = 5.5 V,	V_O = 2.7 V			20			20	µA
I_{OZL}	V_{CC} = 5.5 V,	V_O = 0.4 V			−20			−20	µA
I_I	V_{CC} = 5.5 V,	V_I = 7 V			0.1			0.1	mA
I_{IH}	V_{CC} = 5.5 V,	V_I = 2.7 V			20			20	µA
I_{IL}	V_{CC} = 5.5 V,	V_I = 0.4 V			−0.2			−0.2	mA
I_O[‡]	V_{CC} = 5.5 V,	V_O = 2.25 V	−15		−70	−15		−70	mA
I_{CC}	V_{CC} = 5.5 V	Output high		10	17		10	17	mA
		Outputs low		15	24		15	24	
		Outputs disabled		16	27		16	27	

[†] All typical values are at V_{CC} = 5 V, T_A = 25 °C.
[‡] The output conditions have been chosen to produce a current that closely approximates one half of the true short-circuit output current, I_{OS}.

TEXAS INSTRUMENTS
POST OFFICE BOX 225012 • DALLAS, TEXAS 75265

TYPES SN54ALS574, SN54ALS575, SN74ALS574, SN74ALS575
OCTAL D-TYPE EDGE-TRIGGERED FLIP-FLOPS WITH 3-STATE OUTPUTS

switching characteristics (see Note 1)

PARAMETER		FROM (INPUT)	TO (OUTPUT)	V_{CC} = 4.5 V to 5.5 V, C_L = 50 pF, R1 = 500 Ω, R2 = 500 Ω, T_A = MIN to MAX				UNIT
				SN54ALS574 SN54ALS575		SN74ALS574 SN74ALS575		
				MIN	MAX	MIN	MAX	
f_{max}			'ALS574	30		35		MHz
			'ALS575	25		30		
t_{PLH}		CLK	Any Q	4	15	4	14	ns
t_{PHL}				4	15	4	14	
t_{PZH}		OC	Any Q	4	21	4	18	ns
t_{PZL}				4	21	4	18	
t_{PHZ}		$\overline{OC}$	Any Q 'ALS574	2	10	2	8	ns
			Any Q 'ALS575	2	12	2	10	
t_{PLZ}			Any Q	3	15	3	13	

NOTE 1: For load circuit and voltage waveforms, see page 1-12.

D flip-flop signal conventions

It is TI practice to name the outputs and other inputs of a D-type flip-flop and to draw its logic symbol based on the assumption of true data (D) inputs. Then outputs that produce data in phase with the data inputs are called Q and those producing complementary data are called $\overline{Q}$. An input that causes a Q output to go high or a $\overline{Q}$ output to go low is called Preset; an input that causes a $\overline{Q}$ output to go high or a Q output to go low is called Clear. Bars are used over these pin names ($\overline{PRE}$ and $\overline{CLR}$) if they are active low.

In some applications it may be advantageous to redesignate the data input $\overline{D}$. In that case all the other inputs and outputs should be renamed as shown below. Also shown are corresponding changes in the graphical symbol. Arbitrary pin numbers are shown in parentheses.

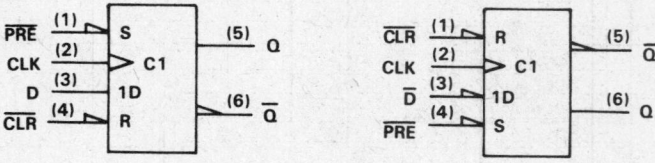

Notice that Q and $\overline{Q}$ exchange names, which causes Preset and Clear to do likewise. Also notice that the polarity indicators (◁) on $\overline{PRE}$ and $\overline{CLR}$ remain since these inputs are still active-low, but that the presence or absence of the polarity indicator changes at $\overline{D}$, Q, and $\overline{Q}$. Of course pin 5 ($\overline{Q}$) is still in phase with the data input $\overline{D}$, but now both are considered active-low.

TYPES SN54AS574, SN54AS575, SN74AS574, SN74AS575
OCTAL D-TYPE EDGE-TRIGGERED FLIP-FLOPS WITH 3-STATE OUTPUTS

absolute maximum ratings over operating free-air temperature range (unless otherwise noted)

Supply voltage, V_{CC} .... 7 V
Input voltage .... 7 V
Voltage applied to a disabled 3-state output .... 5.5 V
Operating free-air temperature range: SN54AS574, SN54AS575 .... −55 °C to 125 °C
                                      SN74AS574, SN74AS575 .... 0 °C to 70 °C
Storage temperature range .... −65 °C to 150 °C

recommended operating conditions

			SN54AS574 SN54AS575			SN74AS574 SN74AS575			UNIT
			MIN	NOM	MAX	MIN	NOM	MAX	
V_{CC}	Supply voltage		4.5	5	5.5	4.5	5	5.5	V
V_{IH}	High-level input voltage		2			2			V
V_{IL}	Low-level input voltage				0.8			0.8	V
I_{OH}	High-level output current				−12			−15	mA
I_{OL}	Low-level output current				32			48	mA
f_{clock}	Clock frequency		0		100	0		125	MHz
t_w	Pulse duration	CLK high	5			4			ns
		CLK low	3			2			
t_{su}	Setup time before CLK↑	Data	3			2			ns
		'AS575 $\overline{CLR}$ high or low	6.5			5.5			
t_h	Hold time after CLK↑	Data	3			2			ns
		'ALS575 $\overline{CLR}$	0			0			
T_A	Operating free-air temperature		−55		125	0		70	°C

electrical characteristics over recommended operating free-air temperature range (unless otherwise noted)

PARAMETER		TEST CONDITIONS		SN54AS574 SN54AS575			SN74AS574 SN74AS575			UNIT
				MIN	TYP†	MAX	MIN	TYP†	MAX	
V_{IK}		$V_{CC} = 4.5$ V,	$I_I = -18$ mA			−1.2			−1.2	V
V_{OH}		$V_{CC} = 4.5$ V to 5.5 V,	$I_{OH} = -2$ mA	$V_{CC}-2$			$V_{CC}-2$			V
		$V_{CC} = 4.5$ V,	$I_{OH} = -12$ mA	2.4	3.2					
		$V_{CC} = 4.5$ V,	$I_{OH} = -15$ mA				2.4	3.3		
V_{OL}		$V_{CC} = 4.5$ V,	$I_{OL} = 32$ mA		0.29	0.5				V
		$V_{CC} = 4.5$ V,	$I_{OL} = 48$ mA					0.34	0.5	
I_{OZH}		$V_{CC} = 5.5$ V,	$V_O = 2.7$ V			50			50	µA
I_{OZL}		$V_{CC} = 5.5$ V,	$V_O = 0.4$ V			−50			−50	µA
I_I		$V_{CC} = 5.5$ V,	$V_I = 7$ V			0.1			0.1	mA
I_{IH}		$V_{CC} = 5.5$ V,	$V_I = 2.7$ V			20			20	µA
I_{IL}	$\overline{OC}$, CLK, $\overline{CLR}$	$V_{CC} = 5.5$ V,	$V_I = 0.4$ V			−0.5			−0.5	mA
	D					−3			−2	
I_O‡		$V_{CC} = 5.5$ V,	$V_O = 2.25$ V	−30		−112	−30		−112	mA
I_{CC}	'AS574	$V_{CC} = 5.5$ V	Outputs high		73	116		73	116	mA
			Outputs low		85	134		85	134	
			Outputs disabled		84	134		84	134	
	'AS575		Outputs high		78	126		78	126	
			Outputs low		88	142		88	142	
			Outputs disabled		88	142		88	142	

†All typical values are at $V_{CC} = 5$ V, $T_A = 25$ °C.
‡The output conditions have been chosen to produce a current that closely approximates one half of the true short-circuit output current, I_{OS}.

TEXAS INSTRUMENTS
POST OFFICE BOX 225012 • DALLAS, TEXAS 75265

TYPES SN54AS574, SN54AS575, SN74AS574, SN74AS575
OCTAL D-TYPE EDGE-TRIGGERED FLIP-FLOPS WITH 3-STATE OUTPUTS

switching characteristics (see Note 1)

PARAMETER	FROM (INPUT)	TO (OUTPUT)	V_{CC} = 4.5 V to 5.5 V, C_L = 50 pF, R1 = 500 Ω, R2 = 500 Ω, T_A = MIN to MAX				UNIT
			SN54AS574 SN54AS575		SN74AS574 SN74AS575		
			MIN	MAX	MIN	MAX	
f_{max}			100		125		MHz
t_{PLH}	CLK	Any Q	3	11	3	8	ns
t_{PHL}			4	11	4	9	
t_{PZH}	$\overline{OC}$	Any Q	2	7	2	6	ns
t_{PZL}			3	11	3	10	
t_{PHZ}	$\overline{OC}$	Any Q	2	7	2	6	ns
t_{PLZ}			2	7	2	6	

NOTE 1: For load circuit and voltage waveforms, see page 1-12.

ALS AND AS CIRCUITS

TEXAS INSTRUMENTS
POST OFFICE BOX 225012 • DALLAS, TEXAS 75265

TYPES SN54ALS576, SN54ALS577, SN54AS576, SN54AS577 SN74ALS576, SN74ALS577, SN74AS576, SN74AS577
OCTAL D-TYPE EDGE-TRIGGERED FLIP-FLOPS WITH 3-STATE OUTPUTS

D2661, DECEMBER 1982 – REVISED DECEMBER 1983

- 3-State Buffer-Type Inverting Outputs Drive Bus-Lines Directly
- Bus-Structured Pinout
- Buffered Control Inputs
- 'ALS577 and 'AS577 Have Synchronous Clear
- Package Options Include Both Plastic and Ceramic Chip Carriers in Addition to Plastic and Ceramic DIPs
- Dependable Texas Instruments Quality and Reliability

description

These 8-bit registers feature three-state outputs designed specifically for bus driving. They are particularly suitable for implementing buffer registers, I/O ports, bidirectional bus drivers, and working registers.

The eight-bit edge-triggered D-type flip-flops enter data on the low-to-high transition of the clock.

The output control does not affect the internal operation of the flip-flops. Old data can be retained or new data can be entered while the outputs are off.

The SN54ALS' and SN54AS' devices are characterized for operation over the full military temperature range of −55°C to 125°C. The SN74ALS' and SN74AS' devices are characterized for operation from 0°C to 70°C.

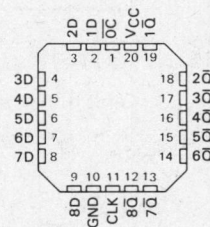

SN54ALS576, SN54AS576 . . . J PACKAGE
SN74ALS576, SN74AS576 . . . N PACKAGE
(TOP VIEW)

SN54ALS576, SN54AS576 . . . FH PACKAGE
SN74ALS576, SN74AS576 . . . FN PACKAGE
(TOP VIEW)

FUNCTION TABLES

ALS576, AS576
(Each Flip-Flop)

INPUTS			OUTPUT
$\overline{OC}$	CLK	D	$\overline{Q}$
L	↑	H	L
L	↑	L	H
L	L	X	$\overline{Q}_0$
H	X	X	Z

ALS577, AS577
(Each Flip-Flop)

INPUTS				OUTPUT
$\overline{OC}$	$\overline{CLR}$	CLK	D	$\overline{Q}$
L	L	↑	X	H
L	H	↑	H	L
L	H	↑	L	H
L	H	L	X	$\overline{Q}_0$
H	X	X	X	Z

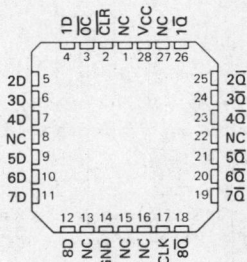

SN54ALS577, SN54AS577 . . . JT PACKAGE
SN74ALS577, SN74AS577 . . . NT PACKAGE
(TOP VIEW)

SN54ALS577, SN54AS577 . . . FH PACKAGE
SN74ALS577, SN74AS577 . . . FN PACKAGE
(TOP VIEW)

NC — No internal connection

Copyright © 1982 by Texas Instruments Incorporated

ALS AND AS CIRCUITS

TYPES SN54ALS576, SN54ALS577, SN54AS576, SN54AS577 SN74ALS576, SN74ALS577, SN74AS576, SN74AS577
OCTAL D-TYPE EDGE-TRIGGERED FLIP-FLOPS WITH 3-STATE OUTPUTS

logic symbols

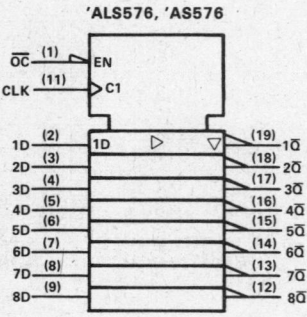
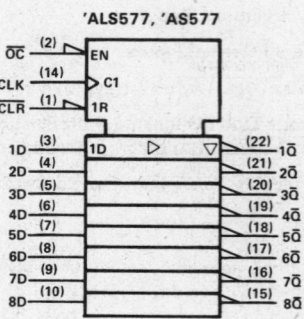

logic diagrams (positive logic)

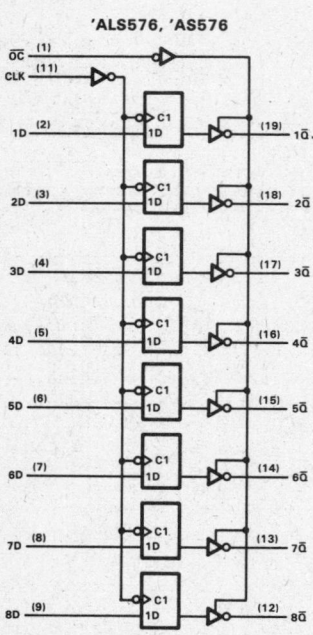

Pin numbers shown are for J and N packages.

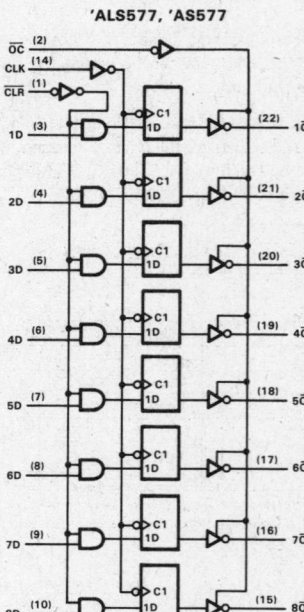

Pin numbers shown are for JT and NT packages.

absolute maximum ratings over operating free-air temperature range (unless otherwise noted)

Supply voltage, V_{CC} .... 7 V
Input voltage .... 7 V
Voltage applied to a disabled 3-state output .... 5.5 V
Operating free-air temperature range: SN54ALS′, SN54AS′ .... −55 °C to 125 °C
                                        SN74ALS′, SN74AS′ .... 0 °C to 70 °C
Storage temperature range .... −65 °C to 150 °C

TEXAS INSTRUMENTS
POST OFFICE BOX 225012 • DALLAS, TEXAS 75265

TYPES SN54ALS576, SN54ALS577, SN74ALS576, SN74ALS577
OCTAL D-TYPE EDGE-TRIGGERED FLIP-FLOPS WITH 3-STATE OUTPUTS

recommended operating conditions

			SN54ALS576 SN54ALS577			SN74ALS576 SN74ALS577			UNIT
			MIN	NOM	MAX	MIN	NOM	MAX	
V_{CC}	Supply voltage		4.5	5	5.5	4.5	5	5.5	V
V_{IH}	High-level input voltage		2			2			V
V_{IL}	Low-level input voltage				0.8			0.8	V
I_{OH}	High-level output current				−1			−2.6	mA
I_{OL}	Low-level output current				12			24	mA
f_{clock}	Clock frequency	'ALS576	0		25	0		30	MHz
		'ALS577	0		25	0		30	
t_w	Pulse duration	CLK high or low 'ALS576	20			16.5			ns
		CLK high or low 'ALS577	20			16.5			
t_{su}	Setup time before CLK↑	Data	15			15			ns
		$\overline{CLR}$ ('ALS577)	15			15			
t_h	Hold time after CLK↑	Data	4			0			ns
		$\overline{CLR}$ ('ALS577)	4			0			
T_A	Operating free-air temperature		−55		125	0		70	°C

electrical characteristics over recommended operating free-air temperature range (unless otherwise noted)

PARAMETER	TEST CONDITIONS		SN54ALS576 SN54ALS577			SN74ALS576 SN74ALS577			UNIT
			MIN	TYP†	MAX	MIN	TYP†	MAX	
V_{IK}	$V_{CC} = 4.5$ V,	$I_I = -18$ mA			−1.5			−1.5	V
V_{OH}	$V_{CC} = 4.5$ V to 5.5 V,	$I_{OH} = -0.4$ mA	$V_{CC}-2$			$V_{CC}-2$			V
	$V_{CC} = 4.5$ V,	$I_{OH} = -1$ mA	2.4	3.3					
	$V_{CC} = 4.5$ V,	$I_{OH} = -2.6$ mA				2.4	3.2		
V_{OL}	$V_{CC} = 4.5$ V,	$I_{OL} = 12$ mA		0.25	0.4				V
	$V_{CC} = 4.5$ V	$I_{OL} = 24$ mA					0.35	0.5	
I_{OZH}	$V_{CC} = 5.5$ V,	$V_O = 2.7$ V			20			20	µA
I_{OZL}	$V_{CC} = 5.5$ V,	$V_O = 0.4$ V			−20			−20	µA
I_I	$V_{CC} = 5.5$ V,	$V_I = 7$ V			0.1			0.1	mA
I_{IH}	$V_{CC} = 5.5$ V,	$V_I = 2.7$ V			20			20	µA
I_{IL}	$V_{CC} = 5.5$ V,	$V_I = 0.4$ V			−0.2			−0.2	mA
I_O‡	$V_{CC} = 5.5$ V,	$V_O = 2.25$ V	−15		−70	−15		−70	mA
I_{CC}	$V_{CC} = 5.5$ V	Outputs high		10	17		10	17	mA
		Outputs low		15	24		15	24	
		Outputs disabled		16	27		16	27	

†All typical values are at $V_{CC} = 5$ V, $T_A = 25$°C.
‡The output conditions have been chosen to produce a current that closely approximates one half of the true short-circuit output current, I_{OS}.

TEXAS INSTRUMENTS
POST OFFICE BOX 225012 • DALLAS, TEXAS 75265

TYPES SN54ALS576, SN54ALS577, SN74ALS576, SN74ALS577
OCTAL D-TYPE EDGE-TRIGGERED FLIP-FLOPS WITH 3-STATE OUTPUTS

switching characteristics (see Note 1)

PARAMETER	FROM (INPUT)	TO (OUTPUT)	V_{CC} = 4.5 V to 5.5 V, C_L = 50 pF, R1 = 500 Ω, R2 = 500 Ω, T_A = MIN to MAX				UNIT
			SN54ALS576 SN54ALS577		SN74ALS576 SN74ALS577		
			MIN	MAX	MIN	MAX	
f_{max}		'ALS576	25		30		MHz
		'ALS577	25		30		
t_{PLH}	CLK	Any $\overline{Q}$	4	15	4	14	ns
t_{PHL}			4	15	4	14	
t_{PZH}	$\overline{OC}$	Any $\overline{Q}$	4	21	4	18	ns
t_{PZL}			4	21	4	18	
t_{PHZ}	$\overline{OC}$	Any $\overline{Q}$ ALS576	2	10	2	8	ns
		Any $\overline{Q}$ ALS577	2	12	2	10	
t_{PLZ}		Any $\overline{Q}$	3	15	3	13	

NOTE 1: For load circuit and voltage waveforms, see page 1-12.

TYPES SN54AS576, SN54AS577, SN74AS576, SN74AS577
OCTAL D-TYPE EDGE-TRIGGERED FLIP-FLOPS WITH 3-STATE OUTPUTS

recommended operating conditions

			SN54AS576 SN54AS577			SN74AS576 SN74AS577			UNIT
			MIN	NOM	MAX	MIN	NOM	MAX	
V_{CC}	Supply voltage		4.5	5	5.5	4.5	5	5.5	V
V_{IH}	High-level input voltage		2			2			V
V_{IL}	Low-level input voltage				0.8			0.8	V
I_{OH}	High-level output current				−12			−15	mA
I_{OL}	Low-level output current				32			48	mA
f_{clock}	Clock frequency		0		100	0		125	MHz
t_w	Pulse duration	CLK high	5			4			ns
		CLK low	3			2			
t_{su}	Setup time before CLK↑	Data	3			2			ns
		$\overline{CLR}$ ('AS577)	6.5			5.5			
t_h	Hold time after CLK↑	Data	3			2			ns
		$\overline{CLR}$ ('AS577)	0			0			
T_A	Operating free-air temperature		−55		125	0		70	°C

electrical characteristics over recommended operating free-air temperature range (unless otherwise noted)

PARAMETER		TEST CONDITIONS		SN54AS576 SN54AS577			SN74AS576 SN74AS577			UNIT
				MIN	TYP†	MAX	MIN	TYP†	MAX	
V_{IK}		V_{CC} = 4.5 V,	I_I = −18 mA			−1.2			−1.2	V
V_{OH}		V_{CC} = 4.5 V to 5.5 V,	I_{OH} = −2 mA	$V_{CC}-2$			$V_{CC}-2$			V
		V_{CC} = 4.5 V,	I_{OH} = −12 mA	2.4	3.2					
		V_{CC} = 4.5 V,	I_{OH} = −15 mA				2.4	3.3		
V_{OL}		V_{CC} = 4.5 V,	I_{OL} = 32 mA		0.29	0.5				V
		V_{CC} = 4.5 V	I_{OL} = 48 mA					0.33	0.5	
I_{OZH}		V_{CC} = 5.5 V,	V_O = 2.7 V			50			50	µA
I_{OZL}		V_{CC} = 5.5 V,	V_O = 0.4 V			−50			−50	µA
I_I		V_{CC} = 5.5 V,	V_I = 7 V			0.1			0.1	mA
I_{IH}		V_{CC} = 5.5 V,	V_I = 2.7 V			20			20	µA
I_{IL}	D	V_{CC} = 5.5 V,	V_I = 0.4 V			−3			−2	mA
	All other					−0.5			−0.5	
I_O‡		V_{CC} = 5.5 V,	V_O = 2.25 V	−30		−112	−30		−112	mA
I_{CC}	'AS576	V_{CC} = 5.5 V	Outputs high		77	125		77	125	mA
			Outputs low		84	135		84	135	
			Outputs disabled		84	135		84	135	
	'AS577		Outputs high		78	126		78	126	
			Outputs low		76	123		76	123	
			Outputs disabled		88	142		88	142	

†All typical values are at V_{CC} = 5 V, T_A = 25°C.
‡The output conditions have been chosen to produce a current that closely approximates one half of the true short-circuit output current, I_{OS}.

ALS AND AS CIRCUITS

TEXAS INSTRUMENTS
POST OFFICE BOX 225012 • DALLAS, TEXAS 75265

TYPES SN54AS576, SN54AS577, SN74AS576, SN74AS577
OCTAL D-TYPE EDGE-TRIGGERED FLIP-FLOPS WITH 3-STATE OUTPUTS

switching characteristics (see Note 1)

PARAMETER	FROM (INPUT)	TO (OUTPUT)	V_{CC} = 4.5 V to 5.5 V, C_L = 50 pF, $R1$ = 500 Ω, $R2$ = 500 Ω, T_A = MIN to MAX				UNIT
			SN54AS576 SN54AS577		SN74AS576 SN74AS577		
			MIN	MAX	MIN	MAX	
f_{max}			100		125		MHz
t_{PLH}	CLK	Any $\overline{Q}$	3	11	3	8	ns
t_{PHL}			4	11	4	9	
t_{PZH}	$\overline{OC}$	Any $\overline{Q}$	2	7	2	6	ns
t_{PZL}			3	11	3	10	
t_{PHZ}	$\overline{OC}$	Any $\overline{Q}$	2	7	2	6	ns
t_{PLZ}			2	7	2	6	

NOTE 1: For load circuit and voltage waveforms, see page 1-12.

ALS AND AS CIRCUITS

TEXAS INSTRUMENTS
POST OFFICE BOX 225012 • DALLAS, TEXAS 75265

TYPES SN54ALS620A THRU SN54ALS623A, SN54AS620 THRU SN54AS623 SN74ALS620A THRU SN74ALS623A, SN74AS620 THRU SN74AS623
OCTAL BUS TRANSCEIVERS

D2661, DECEMBER 1982—REVISED DECEMBER 1983

- Bus Transceivers in High-Density 20-Pin DIP and the New Plastic and Ceramic Chip Carriers
- Local Bus-Latch Capability
- Choice of True or Inverting Logic
- Choice of 3-State or Open-Collector Outputs
- Dependable Texas Instruments Quality and Reliability

SN54ALS', SN54AS' . . . J PACKAGE
SN74ALS', SN74AS' . . . N PACKAGE
(TOP VIEW)

```
GAB  [ 1   U  20 ] VCC
A1   [ 2      19 ] GBA
A2   [ 3      18 ] B1
A3   [ 4      17 ] B2
A4   [ 5      16 ] B3
A5   [ 6      15 ] B4
A6   [ 7      14 ] B5
A7   [ 8      13 ] B6
A8   [ 9      12 ] B7
GND  [10      11 ] B8
```

DEVICE	OUTPUT	LOGIC
'ALS620A, 'AS620	3-State	Inverting
'ALS621A, 'AS621	Open-Collector	True
'ALS622A, 'AS622	Open-Collector	Inverting
'ALS623A, 'AS623	3-State	True

SN54ALS', SN54AS' . . . FH PACKAGE
SN74ALS', SN74AS' . . . FN PACKAGE
(TOP VIEW)

description

These octal bus transceivers are designed for asynchronous two-way communication between data buses. The control function implementation allows for maximum flexibility in timing.

These devices allow data transmission from A bus to the B bus or from the B bus to the A bus depending upon the logic levels at the enable inputs ($\overline{GBA}$ and GAB).

The enable inputs can be used to disable the device so that the buses are effectively isolated.

The dual-enable configuration gives the octal bus transceivers the capability to store data by simultaneous enabling of $\overline{GBA}$ and GAB. Each output reinforces its input in this transceiver configuration. Thus, when both control inputs are enabled and all other data sources to the two sets of bus lines are at high impedance, both sets of bus lines (16 in all) will remain at their last states. The 8-bit codes appearing on the two sets of buses will be identical for the 'ALS621A, 'AS621 and 'ALS623A, 'AS623 or complementary for the 'ALS620A, 'AS620 and 'ALS622A, 'AS622.

The -1 versions of the SN74ALS' parts are identical to their standard versions except that the recommended maximum I_{OL} is increased to 48 mA. There are no -1 versions of the SN54ALS' parts.

The SN54' family is characterized for operation over the full military temperature range of $-55\,°C$ to $125\,°C$. The SN74' family is characterized for operation from $0\,°C$ to $70\,°C$.

FUNCTION TABLE

ENABLE INPUTS		OPERATION	
$\overline{GBA}$	GAB	'ALS620A, 'ALS622A 'AS620, 'AS622	'ALS621A, 'ALS623A 'AS621, 'AS623
L	L	$\overline{B}$ data to A bus	B data to A bus
H	H	$\overline{A}$ data to B bus	A data to B bus
H	L	Isolation	Isolation
L	H	$\overline{B}$ data to A bus, $\overline{A}$ data to B bus	B data to A bus, A data to B bus

Copyright © 1982 by Texas Instruments Incorporated

TEXAS INSTRUMENTS
POST OFFICE BOX 225012 • DALLAS, TEXAS 75265

TYPES SN54ALS620A THRU SN54ALS623A, SN54AS620 THRU SN54AS623 SN74ALS620A THRU SN74ALS623A, SN74AS620 THU SN74AS623 OCTAL BUS TRANSCEIVERS

logic symbols

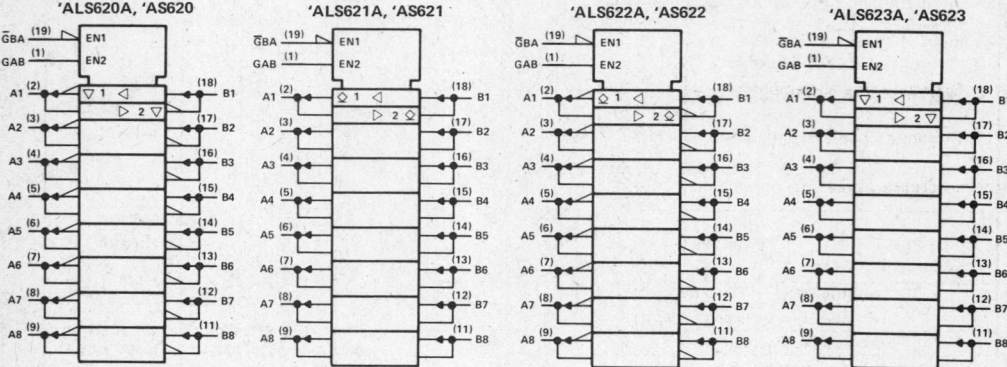

Pin numbers shown are for J and N packages.

logic diagrams (positive logic)

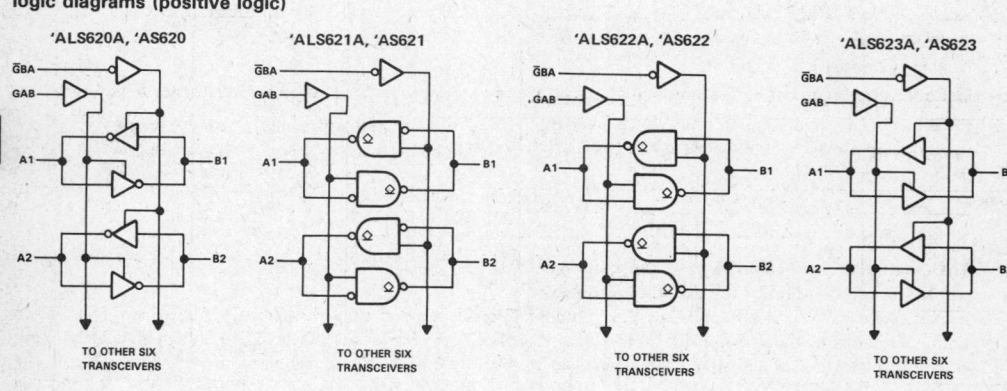

TYPES SN54ALS620A, SN54ALS623A, SN74ALS620A, SN74ALS623A
OCTAL BUS TRANSCEIVERS WITH 3-STATE OUTPUTS

absolute maximum ratings over operating free-air temperature range (unless otherwise noted)

Supply voltage, V_{CC} .... 7 V
Input voltage: All inputs .... 7 V
 I/O ports .... 7 V
Operating free-air temperature range: SN54ALS620A, SN54ALS623A .... −55°C to 125°C
 SN74ALS620A, SN74ALS623A .... 0°C to 70°C
Storage temperature range .... −65°C to 150°C

recommended operating conditions

		SN54ALS620A SN54ALS623A			SN74ALS620A SN74ALS623A			UNIT
		MIN	NOM	MAX	MIN	NOM	MAX	
V_{CC}	Supply voltage	4.5	5	5.5	4.5	5	5.5	V
V_{IH}	High-level input voltage	2			2			V
V_{IL}	Low-level input voltage			0.8			0.8	V
I_{OH}	High-level output current			−12			−15	mA
I_{OL}	Low-level output current			12			24	mA
							48†	
T_A	Operating free-air temperature	−55		125	0		70	°C

†The extended limits apply only if V_{CC} is maintained between 4.75 V and 5.25 V.
 The 48-mA limit applies for the SN74ALS620A-1 and SN74ALS623A-1 only.

electrical characteristics over recommended operating free-air temperature range (unless otherwise noted)

PARAMETER		TEST CONDITIONS		SN54ALS620A SN54ALS623A			SN74ALS620A SN74ALS623A			UNIT
				MIN	TYP‡	MAX	MIN	TYP‡	MAX	
V_{IK}		V_{CC} = 4.5 V,	I_I = −18 mA			−1.5			−1.5	V
V_{OH}		V_{CC} = 4.5 V to 5.5 V,	I_{OH} = −0.4 mA	V_{CC}−2			V_{CC}−2			V
		V_{CC} = 4.5 V,	I_{OH} = −3 mA	2.4	3.2		2.4	3.2		
		V_{CC} = 4.5 V,	I_{OH} = −12 mA	2						
		V_{CC} = 4.5 V,	I_{OH} = −15 mA				2			
V_{OL}		V_{CC} = 4.5 V,	I_{OL} = 12 mA		0.25	0.4		0.25	0.4	V
		V_{CC} = 4.5 V,	I_{OL} = 24 mA					0.35	0.5	
		(I_{OL} = 48 mA for −1 versions)								
I_I	Control inputs	V_{CC} = 5.5 V,	V_I = 7 V			0.1			0.1	mA
	A or B ports	V_{CC} = 5.5 V,	V_I = 5.5 V			0.1			0.1	
I_{IH}	Control inputs	V_{CC} = 5.5 V,	V_I = 2.7 V			20			20	µA
	A or B ports§					20			20	
I_{IL}	Control inputs	V_{CC} = 5.5 V,	V_I = 0.4 V			−0.1			−0.1	mA
	A or B ports§					−0.1			−0.1	
I_O¶		V_{CC} = 5.5 V,	V_O = 2.25 V	−30		−112	−30		−112	mA
I_{CC}	'ALS620A	V_{CC} = 5.5 V	Outputs high		24	39		24	34	mA
			Outputs low		31	49		31	44	
			Outputs disabled		33	52		33	47	
	'ALS623A	V_{CC} = 5.5 V	Outputs high		32	48		32	43	
			Outputs low		39	55		39	50	
			Outputs disabled		42	60		42	55	

‡All typical values are at V_{CC} = 5 V, T_A = 25°C
§For I/O ports, the parameters I_{IH} and I_{IL} include the off-state output current.
¶The output conditions have been chosen to produce a current that closely approximates one half of the true short-circuit output current, I_{OS}.

TEXAS INSTRUMENTS
POST OFFICE BOX 225012 • DALLAS, TEXAS 75265

TYPES SN54ALS620A, SN54ALS623A, SN74ALS620A, SN74ALS623A
OCTAL BUS TRANSCEIVERS WITH 3-STATE OUTPUTS

'ALS620A switching characteristics (see Note 1)

PARAMETER	FROM (INPUT)	TO (OUTPUT)	V_{CC} = 4.5 V to 5.5 V, C_L = 50 pF, $R1$ = 500 Ω, $R2$ = 500 Ω, T_A = MIN to MAX				UNIT
			SN54ALS620A		SN74ALS620A		
			MIN	MAX	MIN	MAX	
t_{PLH}	A	B	2	12	2	10	ns
t_{PHL}			2	12	2	10	
t_{PLH}	B	A	2	12	2	10	ns
t_{PHL}			2	12	2	10	
t_{PZH}	$\overline{GBA}$	A	3	23	3	17	ns
t_{PZL}			5	31	5	25	
t_{PHZ}	$\overline{GBA}$	A	2	14	2	12	ns
t_{PLZ}			3	22	3	18	
t_{PZH}	GAB	B	3	23	3	18	ns
t_{PZL}			5	31	5	25	
t_{PHZ}	GAB	B	2	14	2	12	ns
t_{PLZ}			3	22	3	18	

'ALS623A switching characteristics (see Note 1)

PARAMETER	FROM (INPUT)	TO (OUTPUT)	V_{CC} = 4.5 V to 5.5 V, C_L = 50 pF, $R1$ = 500 Ω, $R2$ = 500 Ω, T_A = MIN to MAX				UNIT
			SN54ALS623A		SN74ALS623A		
			MIN	MAX	MIN	MAX	
t_{PLH}	A	B	2	15	2	13	ns
t_{PHL}			3	13	3	11	
t_{PLH}	B	A	2	15	2	13	ns
t_{PHL}			3	13	3	11	
t_{PZH}	$\overline{GBA}$	A	5	25	5	22	ns
t_{PZL}			5	25	5	22	
t_{PHZ}	$\overline{GBA}$	A	2	19	2	16	ns
t_{PLZ}			2	23	2	19	
t_{PZH}	GAB	B	5	25	5	22	ns
t_{PZL}			5	25	5	22	
t_{PHZ}	GAB	B	2	19	2	16	ns
t_{PLZ}			2	23	2	19	

NOTE 1: For load circuit and voltage waveforms, see page 1-12.

TYPES SN54ALS621A, SN54ALS622A, SN74ALS621A, SN74ALS622A
OCTAL BUS TRANSCEIVERS WITH OPEN-COLLECTOR OUTPUTS

absolute maximum ratings over operating free-air temperature range (unless otherwise noted)

Supply voltage, V_{CC} .... 7 V
Input voltage: All inputs and I/O ports .... 7 V
Operating free-air temperature range: SN54ALS621A, SN54ALS622A .... −55°C to 125°C
                                       SN74ALS621A, SN74ALS622A .... 0°C to 70°C
Storage temperature range .... −65°C to 150°C

recommended operating conditions

		SN54ALS621A SN54ALS622A			SN74ALS621A SN74ALS622A			UNIT
		MIN	NOM	MAX	MIN	NOM	MAX	
V_{CC}	Supply voltage	4.5	5	5.5	4.5	5	5.5	V
V_{IH}	High-level input voltage	2			2			V
V_{IL}	Low-level input voltage			0.8			0.8	V
V_{OH}	High-level output voltage			5.5			5.5	V
I_{OL}	Low-level output current			12			24	mA
							48†	
T_A	Operating free-air temperature	−55		125	0		70	°C

†The extended limits apply only if V_{CC} is maintained between 4.75 V and 5.25 V.
The 48-mA limit applies for the SN74ALS621A-1 and SN74ALS622A-1 only.

electrical characteristics over recommended operating free-air temperature range (unless otherwise noted)

PARAMETER		TEST CONDITIONS		SN54ALS621A SN54ALS622A			SN74ALS621A SN74ALS622A			UNIT
				MIN	TYP‡	MAX	MIN	TYP‡	MAX	
V_{IK}		V_{CC} = 4.5 V,	I_I = −18 mA			−1.5			−1.5	V
I_{OH}		V_{CC} = 4.5 V,	V_{OH} = 5.5 V			0.1			0.1	mA
V_{OL}		V_{CC} = 4.5 V,	I_{OL} = 12 mA		0.25	0.4		0.25	0.4	V
		V_{CC} = 4.5 V,	I_{OL} = 24 mA					0.35	0.5	
		(I_{OL} = 48 mA for -1 versions)								
I_I	Control inputs	V_{CC} = 5.5 V,	V_I = 7 V			0.1			0.1	mA
	A or B ports	V_{CC} = 5.5 V,	V_I = 5.5 V			0.1			0.1	
I_{IH}	Control inputs	V_{CC} = 5.5 V,	V_I = 2.7 V			20			20	µA
	A or B ports §					20			20	
I_{IL}	Control inputs	V_{CC} = 5.5 V,	V_I = 0.4 V			−0.1			−0.1	mA
	A or B ports §					−0.1			−0.1	
I_{CC}	'ALS621A	V_{CC} = 5.5 V	Outputs high		29	45		29	40	mA
			Outputs low		35	53		35	48	
	'ALS622A	V_{CC} = 5.5 V	Outputs high		11	20		11	15	
			Outputs low		20	33		20	28	

‡All typical values are at V_{CC} = 5 V, T_A = 25°C.
§For I/O ports, the parameters I_{IH} and I_{IL} include the off-state output current.

ALS AND AS CIRCUITS

TEXAS INSTRUMENTS
POST OFFICE BOX 225012 • DALLAS, TEXAS 75265

TYPES SN54ALS621A, SN54ALS622A, SN74ALS621A, AN74ALS622A
OCTAL BUS TRANCEIVERS WITH OPEN-COLLECTOR OUTPUTS

'ALS621A switching characteristics (see Note 1)

PARAMETER	FROM (INPUT)	TO (OUTPUT)	V_{CC} = 4.5 V to 5.5 V, C_L = 50 pF, R_L = 680 Ω, T_A = MIN to MAX				UNIT
			SN54ALS621A		SN74ALS621A		
			MIN	MAX	MIN	MAX	
t_{PLH}	A	B	10	45	10	33	ns
t_{PHL}			5	24	5	20	
t_{PLH}	B	A	10	45	10	33	ns
t_{PHL}			5	24	5	20	
t_{PLH}	$\overline{GBA}$	A	10	47	10	39	ns
t_{PHL}			12	40	12	35	
t_{PLH}	GAB	B	10	47	10	39	ns
t_{PHL}			12	40	12	35	

'ALS622A switching characteristics (see Note 1)

PARAMETER	FROM (INPUT)	TO (OUTPUT)	V_{CC} = 4.5 V to 5.5 V, C_L = 50 pF, R_L = 680 Ω, T_A = MIN to MAX				UNIT
			SN54ALS622A		SN74ALS622A		
			MIN	MAX	MIN	MAX	
t_{PLH}	A	B	8	42	8	35	ns
t_{PHL}			5	23	5	19	
t_{PLH}	B	A	8	42	8	35	ns
t_{PHL}			5	23	5	19	
t_{PLH}	$\overline{GBA}$	A	8	45	8	38	ns
t_{PHL}			10	40	10	35	
t_{PLH}	GAB	B	8	45	8	38	ns
t_{PHL}			10	40	10	35	

NOTE 1: For load circuit and voltage waveforms, see page 1-12.

TEXAS INSTRUMENTS
POST OFFICE BOX 225012 • DALLAS, TEXAS 75265

TYPES SN54AS620, SN54AS623, SN74AS620, SN74AS623
OCTAL BUS TRANSCEIVERS WITH 3-STATE OUTPUTS

absolute maximum ratings over operating free-air temperature range (unless otherwise noted)

Supply voltage, V_{CC} .... 7 V
Input voltage: All inputs .... 7 V
              I/O ports .... 5.5 V
Operating free-air temperature range: SN54AS620, SN54AS623 .... −55 °C to 125 °C
                                  SN74AS620, SN74AS623 .... 0 °C to 70 °C
Storage temperature range .... −65 °C to 150 °C

recommended operating conditions

		SN54AS620 SN54AS623			SN74AS620 SN74AS623			UNIT
		MIN	NOM	MAX	MIN	NOM	MAX	
V_{CC}	Supply voltage	4.5	5	5.5	4.5	5	5.5	V
V_{IH}	High-level input voltage	2			2			V
V_{IL}	Low-level input voltage			0.8			0.8	V
I_{OH}	High-level output current			−12			−15	mA
I_{OL}	Low-level output current			48			64	mA
T_A	Operating free-air temperature	−55		125	0		70	°C

electrical characteristics over recommended operating free-air temperature range (unless otherwise noted)

PARAMETER		TEST CONDITIONS		SN54AS620 SN54AS623			SN74AS620 SN74AS623			UNIT
				MIN	TYP†	MAX	MIN	TYP†	MAX	
V_{IK}		V_{CC} = 4.5 V,	I_I = −18 mA			−1.2			−1.2	V
V_{OH}		V_{CC} = 4.5 V to 5.5 V,	I_{OH} = −2 mA	V_{CC}−2			V_{CC}−2			V
		V_{CC} = 4.5 V,	I_{OH} = −3 mA	2.4	3.2		2.4	3.2		
		V_{CC} = 4.5 V,	I_{OH} = −12 mA	2.4						
		V_{CC} = 4.5 V,	I_{OH} = −15 mA				2.4			
V_{OL}		V_{CC} = 4.5 V,	I_{OL} = 48 mA		0.30	0.55				V
		V_{CC} = 4.5 V,	I_{OL} = 64 mA					0.35	0.55	
I_I	Control inputs	V_{CC} = 5.5 V,	V_I = 7 V			0.1			0.1	mA
	A or B ports	V_{CC} = 5.5 V,	V_I = 5.5 V			0.1			0.1	
I_{IH}	Control inputs	V_{CC} = 5.5 V,	V_I = 2.7 V			20			20	µA
	A or B ports‡					50			50	
I_{IL}	Control inputs	V_{CC} = 5.5 V,	V_I = 0.4 V			−0.5			−0.5	mA
	A or B ports‡					−0.75			−0.75	
I_O§		V_{CC} = 5.5 V,	V_O = 2.25 V	−30		−112	−30		−112	mA
I_{CC}	'AS620	V_{CC} = 5.5 V	Outputs high		35	57		35	57	mA
			Outputs low		74	122		74	122	
			Outputs disabled		48	77		48	77	
	'AS623	V_{CC} = 5.5 V	Outputs high		57			57		
			Outputs low		116			116		
			Outputs disabled		71			71		

†All typical values are at V_{CC2} = 5 V, T_A = 25 °C.
‡For I/O ports, the parameters I_{IH} and I_{IL} include the off-state output current.
§The output conditions have been chosen to produce a current that closely approximates one half of the true short-circuit output current, I_{OS}.

PRODUCT PREVIEW

This page contains information on a product under development. Texas Instruments reserves the right to change or discontinue this product without notice.

TEXAS INSTRUMENTS
POST OFFICE BOX 225012 • DALLAS, TEXAS 75265

TYPES SN54AS620, SN54AS623, SN74AS620, SN74AS623
OCTAL BUS TRANSCEIVERS WITH 3-STATE OUTPUTS

'AS620 switching characteristics (see Note 1)

PARAMETER	FROM (INPUT)	TO (OUTPUT)	$V_{CC} = 4.5$ V to 5.5 V, $C_L = 50$ pF, $R1 = 500$ Ω, $R2 = 500$ Ω, T_A = MIN to MAX				UNIT
			SN54AS620		SN74AS620		
			MIN	MAX	MIN	MAX	
t_{PLH}	A	B	1	8	1	7	ns
t_{PHL}			2	7	2	6	
t_{PLH}	B	A	1	8	1	7	ns
t_{PHL}			2	7	2	6	
t_{PZH}	$\overline{GBA}$	A	2	8.5	2	8	ns
t_{PZL}			2	10	2	9	
t_{PHZ}	$\overline{GBA}$	A	1	7.5	1	6	ns
t_{PLZ}			2	15	2	12	
t_{PZH}	GAB	B	2	9	2	8	ns
t_{PZL}			2	10.5	2	9	
t_{PHZ}	GAB	B	1	6.5	1	6	ns
t_{PLZ}			2	16	2	13	

'AS623 switching characteristics (see Note 1)

PARAMETER	FROM (INPUT)	TO (OUTPUT)	$V_{CC} = 4.5$ V to 5.5 V, $C_L = 50$ pF, $R1 = 500$ Ω, $R2 = 500$ Ω, T_A = MIN to MAX						UNIT
			SN54AS623			SN74AS623			
			MIN	TYP†	MAX	MIN	TYP†	MAX	
t_{PLH}	A	B		5			5		ns
t_{PHL}				5			5		
t_{PLH}	B	A		5			5		ns
t_{PHL}				5			5		
t_{PZH}	$\overline{GBA}$	A		4			4		ns
t_{PZL}				6			6		
t_{PHZ}	$\overline{GBA}$	A		4			4		ns
t_{PLZ}				5			5		
t_{PZH}	GAB	B		5			5		ns
t_{PZL}				7			7		
t_{PHZ}	GAB	B		4			4		ns
t_{PLZ}				5			5		

†All typical values are at $V_{CC} = 5$ V, $T_A = 25$ °C.
NOTE 1: For load circuit and voltage waveforms, see page 1-12.

PRODUCT PREVIEW

This page contains information on a product under development. Texas Instruments reserves the right to change or discontinue this product without notice.

Texas Instruments
POST OFFICE BOX 225012 • DALLAS, TEXAS 75265

TYPES SN54AS621, SN54AS622, SN74AS621, SN74AS622
OCTAL BUS TRANSCEIVERS WITH OPEN-COLLECTOR OUTPUTS

absolute maximum ratings over operating free-air temperature range (unless otherwise noted)

Supply voltage, V_{CC} .... 7 V
Input voltage: All inputs and I/O ports .... 7 V
Operating free-air temperature range: SN54AS621, SN54AS622 .... −55°C to 125°C
 SN74AS621, SN74AS622 .... 0°C to 70°C
Storage temperature range .... −65°C to 150°C

recommended operating conditions

		SN54AS621 SN54AS622			SN74AS621 SN74AS622			UNIT
		MIN	NOM	MAX	MIN	NOM	MAX	
V_{CC}	Supply voltage	4.5	5	5.5	4.5	5	5.5	V
V_{IH}	High-level input voltage	2			2			V
V_{IL}	Low-level input voltage			0.8			0.8	V
V_{OH}	High-level output voltage			5.5			5.5	V
I_{OL}	Low-level output current			48			64	mA
T_A	Operating free-air temperature	−55		125	0		70	°C

electrical characteristics over recommended operating free-air temperature range (unless otherwise noted)

PARAMETER		TEST CONDITIONS		SN54AS621 SN54AS622			SN74AS621 SN74AS622			UNIT
				MIN	TYP†	MAX	MIN	TYP†	MAX	
V_{IK}		V_{CC} = 4.5 V,	I_I = −18 mA			−1.2			−1.2	V
I_{OH}		V_{CC} = 4.5 V,	V_{OH} = 5.5 V			0.1			0.1	mA
V_{OL}		V_{CC} = 4.5 V,	I_{OL} = 48 mA		0.25	0.5				V
		V_{CC} = 4.5 V,	I_{OL} = 64 mA					0.35	0.5	
I_I	Control inputs	V_{CC} = 5.5 V,	V_I = 7 V			0.1			0.1	mA
	A or B ports	V_{CC} = 5.5 V,	V_I = 5.5 V			0.1			0.1	
I_{IH}	Control inputs	V_{CC} = 5.5 V,	V_I = 2.7 V			20			20	µA
	A or B ports‡					20			20	
I_{IL}	Control inputs	V_{CC} = 5.5 V,	V_I = 0.4 V			−0.5			−0.5	mA
	A or B ports‡					−0.5			−0.5	
I_{CC}	'AS621	V_{CC} = 5.5 V	Outputs high			48	79	48	79	mA
			Outputs low			116	189	116	189	
	'AS622	V_{CC} = 5.5 V	Outputs high			25		25		
			Outputs low			62		62		

†All typical values are at V_{CC} = 5 V, T_A = 25°C
‡For I/O ports, the parameters I_{IH} and I_{IL} include the off-state output current.

PRODUCT PREVIEW
This page contains information on a product under development. Texas Instruments reserves the right to change or discontinue this product without notice.

TEXAS INSTRUMENTS
POST OFFICE BOX 225012 • DALLAS, TEXAS 75265

TYPES SN54AS621, SN54AS622, SN74AS621, SN74AS622
OCTAL BUS TRANSCEIVERS WITH OPEN-COLLECTOR OUTPUTS

'AS621 switching characteristics (see Note 1)

PARAMETER	FROM (INPUT)	TO (OUTPUT)	V_{CC} = 4.5 V to 5.5 V, C_L = 50 pF, R_L = 680 Ω, T_A = MIN to MAX				UNIT
			SN54AS621		SN74AS621		
			MIN	MAX	MIN	MAX	
t_{PLH}	A	B	5	28.5	5	24	ns
t_{PHL}			1	8.5	1	7.5	
t_{PLH}	B	A	5	23	5	21	ns
t_{PHL}			1	8.5	1	7.5	
t_{PLH}	$\overline{G}BA$	A	5	24	5	21	ns
t_{PHL}			1	10	1	9	
t_{PLH}	GAB	B	5	26	5	22	ns
t_{PHL}			1	11	1	10	

'AS622 switching characteristics (see Note 1)

PARAMETER	FROM (INPUT)	TO (OUTPUT)	V_{CC} = 4.5 V to 5.5 V, C_L = 50 pF, R_L = 680 Ω, T_A = MIN to MAX						UNIT
			SN54AS622			SN74AS622			
			MIN	TYP†	MAX	MIN	TYP†	MAX	
t_{PLH}	A	B		20			20		ns
t_{PHL}				6			6		
t_{PLH}	B	A		20			20		ns
t_{PHL}				6			6		
t_{PLH}	$\overline{G}BA$	A		22			22		ns
t_{PHL}				8			8		
t_{PLH}	GAB	B		23			23		ns
t_{PHL}				9			9		

†All typical values are at V_{CC} = 5 V, T_A = 25°C.
NOTE 1: For load circuit and voltage waveforms, see page 1-12.

PRODUCT PREVIEW

This page contains information on a product under development. Texas Instruments reserves the right to change or discontinue this product without notice.

TEXAS INSTRUMENTS
POST OFFICE BOX 225012 • DALLAS, TEXAS 75265

TYPES SN54ALS632 THRU SN54ALS635, SN74ALS632 THRU SN74ALS635
32-BIT PARALLEL ERROR DETECTION AND CORRECTION CIRCUITS

D2661, DECEMBER 1982 – REVISED DECEMBER 1983

- Detects and Corrects Single-Bit Errors
- Detects and Flags Dual-Bit Errors
- Built-In Diagnostic Capability
- Fast Write and Read Cycle Processing Times
- Byte-Write Capability . . . 'ALS632 and 'ALS633
- Dependable Texas Instruments Quality and Reliability

DEVICE	PACKAGE	BYTE-WRITE	OUTPUT
'ALS632	52-pin	yes	3-State
'ALS633	52-pin	yes	Open-Collector
'ALS634	48-pin	no	3-State
'ALS635	48-pin	no	Open-Collector

'ALS632, 'ALS633, . . . JD PACKAGE
(TOP VIEW)

'ALS634, 'ALS635 . . . . JD PACKAGE
(TOP VIEW)

FOR CHIP CARRIER INFORMATION,
CONTACT THE FACTORY

description

The 'ALS632 through 'ALS635 devices are 32-bit parallel error detection and correction circuits (EDACs) in 52-pin ('ALS632 and 'ALS633) or 48-pin ('ALS634 and 'ALS635), 600-mil packages. The EDACs use a modified Hamming code to generate a 7-bit check word from a 32-bit data word. This check word is stored along with the data word during the memory write cycle. During the memory read cycle, the 39-bit words from memory are processed by the EDACs to determine if errors have occurred in memory.

Single-bit errors in the 32-bit data word are flagged and corrected.

Single-bit errors in the 7-bit check word are flagged, and the CPU sends the EDAC through the correction cycle even though the 32-bit data word is not in error. The correction cycle will simply pass along the original 32-bit data word in this case and produce error syndrome bits to pinpoint the error-generating location.

Dual-bit errors are flagged but not corrected. These errors may occur in any two bits of the 39-bit word from memory (two errors in the 32-bit data word, two errors in the 7-bit check word, or one error in each word). The gross-error condition of all lows or all highs from memory will be detected. Otherwise, errors in three or more bits of the 39-bit word are beyond the capabilities of these devices to detect.

Read-modify-write (byte-control) operations can be performed with the 'ALS632 and 'ALS633 EDACs by using output latch enable, $\overline{\text{LEDBO}}$, and the individual $\overline{\text{OEB0}}$ thru $\overline{\text{OEB3}}$ byte control pins.

Diagnostics are performed on the EDACs by controls and internal paths that allow the user to read the contents of the DB and CB input latches. These will determine if the failure occurred in memory or in the EDAC.

ALS AND AS CIRCUITS

TYPES SN54ALS632 THRU SN54ALS635, SN74ALS632 THRU SN74ALS635
32-BIT PARALLEL ERROR DETECTION AND CORRECTION CIRCUITS

TABLE 1 — WRITE CONTROL FUNCTION

MEMORY CYCLE	EDAC FUNCTION	CONTROL S1 S0	DATA I/O	DB CONTROL $\overline{OEB}_n$ OR $\overline{OEDB}$	DB OUTPUT LATCH ('ALS632,'ALS633) $\overline{LEDBO}$	CHECK I/O	CB CONTROL $\overline{OECB}$	ERROR FLAGS $\overline{ERR}$ $\overline{MERR}$
Write	Generate check word	L L	Input	H	X	Output check bits†	L	H H

†See Table 2 for details on check bit generation.

memory write cycle details

During a memory write cycle, the check bits (CB0 thru CB6) are generated internally in the EDAC by seven 16-input parity generators using the 32-bit data word as defined in Table 2. These seven check bits are stored in memory along with the original 32-bit data word. This 32-bit word will later be used in the memory read cycle for error detection and correction.

TABLE 2 — PARITY ALGORITHM

CHECK WORD BIT	31	30	29	28	27	26	25	24	23	22	21	20	19	18	17	16	15	14	13	12	11	10	9	8	7	6	5	4	3	2	1	0
CB0	X		X	X		X				X	X	X		X		X					X	X	X	X		X						X
CB1				X		X		X		X		X	X	X				X		X		X		X		X				X	X	X
CB2	X	X				X	X		X		X	X			X	X			X	X		X	X			X			X	X		X
CB3			X	X	X				X	X	X			X	X		X	X	X				X	X	X					X	X	
CB4	X	X							X	X	X	X	X	X			X	X					X	X	X	X	X	X			X	X
CB5	X	X	X	X	X	X	X										X	X	X	X	X	X	X	X								
CB6	X	X	X	X	X	X	X	X																	X	X	X	X	X	X	X	X

The seven check bits are parity bits derived from the matrix of data bits as indicated by "X" for each bit.

error detection and correction details

During a memory read cycle, the 7-bit check word is retrieved along with the actual data. In order to be able to determine whether the data from memory is acceptable to use as presented to the bus, the error flags must be tested to determine if they are at the high level.

The first case in Table 3 represents the normal, no-error conditions. The EDAC presents highs on both flags. The next two cases of single-bit errors give a high on $\overline{MERR}$ and a low on $\overline{ERR}$, which is the signal for a correctable error, and the EDAC should be sent through the correction cycle. The last three cases of double-bit errors will cause the EDAC to signal lows on both $\overline{ERR}$ and $\overline{MERR}$, which is the interrupt indication for the CPU.

TABLE 3 — ERROR FUNCTION

Total Number of Errors		Error Flags		Data Correction
32-Bit Data Word	7-Bit Check Word	$\overline{ERR}$	$\overline{MERR}$	
0	0	H	H	Not applicable
1	0	L	H	Correction
0	1	L	H	Correction
1	1	L	L	Interrupt
2	0	L	L	Interrupt
0	2	L	L	Interrupt

TYPES SN54ALS632 THRU SN54ALS635, SN74ALS632 THRU SN74ALS635
32-BIT PARALLEL ERROR DETECTION AND CORRECTION CIRCUITS

Error detection is accomplished as the 7-bit check word and the 32-bit data word from memory are applied to internal parity generators/checkers. If the parity of all seven groupings of data and check bits are correct, it is assumed that no error has occurred and both error flags will be high.

If the parity of one or more of the check groups is incorrect, an error has occurred and the proper error flag or flags will be set low. Any single error in the 32-bit data word will change the state of either three or five bits of the 7-bit check word. Any single error in the 7-bit check word changes the state of only that one bit. In either case, the single error flag ($\overline{ERR}$) will be set low while the dual error flag ($\overline{MERR}$) will remain high.

Any two-bit error will change the state of an even number of check bits. The two-bit error is not correctable since the parity tree can only identify single-bit errors. Both error flags are set low when any two-bit error is detected.

Three or more simultaneous bit errors can cause the EDAC to believe that no error, a correctable error, or an uncorrectable error has occurred and will produce erroneous results in all three cases. It should be noted that the gross-error conditions of all lows and all highs will be detected.

TABLE 4 — READ, FLAG, AND CORRECT FUNCTION

MEMORY CYCLE	EDAC FUNCTION	CONTROL S1 S0		DATA I/O	DB CONTROL $\overline{OEBn}$ OR $\overline{OEDB}$	DB OUTPUT LATCH ('ALS632,'ALS633) $\overline{LEDBO}$	CHECK I/O	CB CONTROL $\overline{OECB}$	ERROR FLAGS	
									$\overline{ERR}$	$\overline{MERR}$
Read	Read & flag	H	L	Input	H	X	Input	H	Enabled†	
Read	Latch input data & check bits	H	H	Latched input data	H	L	Latched input check word	H	Enabled†	
Read	Output corrected data & syndrome bits	H	H	Output corrected data word	L	X	Output syndrome bits‡	L	Enabled†	

†See Table 3 for error description.
‡See Table 5 for error location.

As the corrected word is made available on the data I/O port (DB0 thru DB31), the check word I/O port (CB0 thru CB6) presents a 7-bit syndrome error code. This syndrome error code can be used to locate the bad memory chip. See Table 5 for syndrome decoding.

TYPES SN54ALS632 THRU SN54ALS635, SN74ALS632 THRU SN74ALS635
32-BIT PARALLEL ERROR DETECTION AND CORRECTION CIRCUITS

TABLE 5 — SYNDROME DECODING

SYNDROME BITS 6 5 4 3 2 1 0	ERROR	SYNDROME BITS 6 5 4 3 2 1 0	ERROR	SYNDROME BITS 6 5 4 3 2 1 0	ERROR	SYNDROME BITS 6 5 4 3 2 1 0	ERROR
L L L L L L L	unc	L H L L L L L	2-bit	H L L L L L L	2-bit	H H L L L L L	unc
L L L L L L H	2-bit	L H L L L L H	unc	H L L L L L H	unc	H H L L L L H	2-bit
L L L L L H L	2-bit	L H L L L H L	DB7	H L L L L H L	unc	H H L L L H L	2-bit
L L L L L H H	unc	L H L L L H H	2-bit	H L L L L H H	2-bit	H H L L L H H	DB23
L L L L H L L	2-bit	L H L L H L L	DB6	H L L L H L L	unc	H H L L H L L	2-bit
L L L L H L H	unc	L H L L H L H	2-bit	H L L L H L H	2-bit	H H L L H L H	DB22
L L L L H H L	unc	L H L L H H L	2-bit	H L L L H H L	2-bit	H H L L H H L	DB21
L L L L H H H	2-bit	L H L L H H H	DB5	H L L L H H H	unc	H H L L H H H	2-bit
L L L H L L L	2-bit	L H L H L L L	DB4	H L L H L L L	unc	H H L H L L L	2-bit
L L L H L L H	unc	L H L H L L H	2-bit	H L L H L L H	2-bit	H H L H L L H	DB20
L L L H L H L	DB31	L H L H L H L	2-bit	H L L H L H L	2-bit	H H L H L H L	DB19
L L L H L H H	2-bit	L H L H L H H	DB3	H L L H L H H	DB15	H H L H L H H	2-bit
L L L H H L L	unc	L H L H H L L	2-bit	H L L H H L L	2-bit	H H L H H L L	DB18
L L L H H L H	2-bit	L H L H H L H	DB2	H L L H H L H	unc	H H L H H L H	2-bit
L L L H H H L	2-bit	L H L H H H L	2-bit	H L L H H H L	DB14	H H L H H H L	2-bit
L L L H H H H	DB30	L H L H H H H	2-bit	H L L H H H H	2-bit	H H L H H H H	CB4
L L H L L L L	2-bit	L H H L L L L	DB0	H L H L L L L	unc	H H H L L L L	2-bit
L L H L L L H	unc	L H H L L L H	2-bit	H L H L L L H	2-bit	H H H L L L H	DB16
L L H L L H L	DB29	L H H L L H L	2-bit	H L H L L H L	2-bit	H H H L L H L	unc
L L H L L H H	2-bit	L H H L L H H	unc	H L H L L H H	DB13	H H H L L H H	2-bit
L L H L H L L	DB28	L H H L H L L	2-bit	H L H L H L L	2-bit	H H H L H L L	DB17
L L H L H L H	2-bit	L H H L H L H	DB1	H L H L H L H	DB12	H H H L H L H	2-bit
L L H L H H L	2-bit	L H H L H H L	unc	H L H L H H L	DB11	H H H L H H L	2-bit
L L H L H H H	DB27	L H H L H H H	2-bit	H L H L H H H	2-bit	H H H L H H H	CB3
L L H H L L L	DB26	L H H H L L L	2-bit	H L H H L L L	2-bit	H H H H L L L	unc
L L H H L L H	2-bit	L H H H L L H	unc	H L H H L L H	DB10	H H H H L L H	2-bit
L L H H L H L	2-bit	L H H H L H L	unc	H L H H L H L	DB9	H H H H L H L	2-bit
L L H H L H H	DB25	L H H H L H H	2-bit	H L H H L H H	2-bit	H H H H L H H	CB2
L L H H H L L	2-bit	L H H H H L L	unc	H L H H H L L	DB8	H H H H H L L	2-bit
L L H H H L H	DB24	L H H H H L H	2-bit	H L H H H L H	2-bit	H H H H H L H	CB1
L L H H H H L	unc	L H H H H H L	2-bit	H L H H H H L	2-bit	H H H H H H L	CB0
L L H H H H H	2-bit	L H H H H H H	CB6	H L H H H H H	CB5	H H H H H H H	none

CB X = error in check bit X
DB Y = error in data bit Y
2-bit = double-bit error
unc = uncorrectable multibit error

read-modify-write (byte control) operations

The 'ALS632 and 'ALS633 devices are capable of byte-write operations. The 39-bit word from memory must first be latched into the DB and CB input latches. This is easily accomplished by switching from the read and flag mode (S1 = H, S0 = L) to the latch input mode (S1 = H, S0 = H). The EDAC will then make any corrections, if necessary, to the data word and place it at the input of the output data latch. This data word must then be latched into the output data latch by taking LEDBO from a low to a high.

Byte control can now be employed on the data word through the OEB0 through OEB3 controls. OEB0 controls DB0-DB7 (byte 0), OEB1 controls DB8-DB15 (byte 1), OEB2 controls DB16-DB23 (byte 2), and OEB3 controls DB24-DB31 (byte 3). Placing a high on the byte control will disable the output and the user can modify the byte. If a low is placed on the byte control, then the original byte is allowed to pass onto the data bus unchanged. If the original data word is altered through byte control, a new check word must be generated before it is written back into memory. This is easily accomplished by taking control S1 and S0 low. Table 6 lists the read-modify-write functions.

TYPES SN54ALS632 THRU SN54ALS635, SN74ALS632 THRU SN74ALS635
32-BIT PARALLEL ERROR DETECTION AND CORRECTION CIRCUITS

TABLE 6 — READ-MODIFY-WRITE FUNCTION

MEMORY CYCLE	EDAC FUNCTION	CONTROL S1 S0	BYTEn†	$\overline{OEBn}$†	DB OUTPUT LATCH $\overline{LEDBO}$	CHECK I/O	CB CONTROL	ERROR FLAG $\overline{ERR}$ $\overline{MERR}$
Read	Read & Flag	H L	Input	H	X	Input	H	Enabled
Read	Latch input data & check bits	H H	Latched Input data	H	L	Latched input check word	H	Enabled
Read	Latch corrected data word into output latch	H H	Latched output data word	H	H	Hi-Z / Output Syndrome bits	H / L	Enabled
Modify /write	Modify appropriate byte or bytes & generate new check word	L L	Input modified BYTE0 / Output unchanged BYTE0	H / L	—	Output check word	L	H H

†$\overline{OEB0}$ controls DB0-DB7 (BYTE0), $\overline{OEB1}$ controls DB8-DB15 (BYTE1), $\overline{OEB2}$ controls DB16-DB23 (BYTE2), $\overline{OEB3}$ controls DB24-DB31 (BYTE3).

diagnostic operations

The 'ALS632 thru 'ALS635 are capable of diagnostics that allow the user to determine whether the EDAC or the memory is failing. The diagnostic function tables will help the user to see the possibilities for diagnostic control.

In the diagnostic mode (S1 = L, S0 = H), the checkword is latched into the input latch while the data input latch remains transparent. This lets the user apply various data words against a fixed known checkword. If the user applies a diagnostic data word with an error in any bit location, the $\overline{ERR}$ flag should be low. If a diagnostic data word with two errors in any bit location is applied, the $\overline{MERR}$ flag should be low. After the checkword is latched into the input latch, it can be verified by taking $\overline{OECB}$ low. This outputs the latched checkword. With the 'ALS632 and 'ALS633, the diagnostic data word can be latched into the output data latch and verified. It should be noted that the 'ALS634 and 'ALS635 do not have this pass-through capability because they do not contain an output data latch. By changing from the diagnostic mode (S1 = L, S0 = H) to the correction mode (S1 = H, S0 = H), the user can verify that the EDAC will correct the diagnostic data word. Also, the syndrome bits can be produced to verify that the EDAC pinpoints the error location. Table 7 ('ALS632 and 'ALS633) and Table 8 ('ALS634 and 'ALS635) list the diagnostic functions.

TYPES SN54ALS632 THRU SN54ALS635, SN74ALS632 THRU SN74ALS635
32-BIT PARALLEL ERROR DETECTION AND CORRECTION CIRCUITS

TABLE 7 — 'ALS632, 'ALS633 DIAGNOSTIC FUNCTION

EDAC FUNCTION	CONTROL S1 S0	DATA I/O	DB BYTE CONTROL $\overline{OEBn}$	DB OUTPUT LATCH $\overline{LEDBO}$	CHECK I/O	CB CONTROL $\overline{OECB}$	ERROR FLAGS ERR MERR
Read & flag	H L	Input correct data word	H	X	Input correct check bits	H	H H
Latch input check word while data input latch remains transparent	L H	Input diagnostic data word†	H	L	Latched input check bits	H	Enabled
Latch diagnostic data word into output latch	L H	Input diagnostic data word†	H	H	Output latched check bits / Hi-Z	L / H	Enabled
Latch diagnostic data word into input latch	H H	Latched input diagnostic data word	H	H	Output syndrome bits / Hi-Z	L / H	Enabled
Output diagnostic data word & syndrome bits	H H	Output diagnostic data word	L	H	Output syndrome bits / Hi-Z	L / H	Enabled
Output corrected diagnostic data word & output syndrome bits	H H	Output corrected diagnostic data word	L	L	Output syndrome bits / Hi-Z	L / H	Enabled

†Diagnostic data is a data word with an error in one bit location except when testing the $\overline{MERR}$ error flag. In this case, the diagnostic data word will contain errors in two bit locations.

TABLE 8 — 'ALS634, 'ALS635 DIAGNOSTIC FUNCTION

EDAC FUNCTION	CONTROL S1 S0	DATA I/O	DB CONTROL $\overline{OEDB}$	CHECK I/O	DB CONTROL $\overline{OECB}$	ERROR FLAGS ERR MERR
Read & flag	H L	Input correct data word	H	Input correct check bits	H	H H
Latch input check bits while data input latch remains transparent	L H	Input diagnostic data word†	H	Latched input check bits	H	Enabled
Output input check bits	L H	Input diagnostic data word†	H	Output input check bits	H	Enabled
Latch diagnostic data into input latch	H H	Latched input diagnostic data word	H	Output syndrome bits / Hi-Z	L / H	Enabled
Output corrected diagnostic data word	H H	Output corrected diagnostic data word	L	Output syndrome bits / $\overline{Hi-Z}$	L / H	Enabled

†Diagnostic data is a data word with an error in one bit location except when testing the $\overline{MERR}$ error flag. In this case, the diagnostic data word will contain errors in two bit locations.

TYPES SN54ALS632, SN54ALS633, SN74ALS632, SN74ALS633
32-BIT PARALLEL ERROR DETECTION AND CORRECTION CIRCUITS

'ALS632, 'ALS633 logic diagram (positive logic)

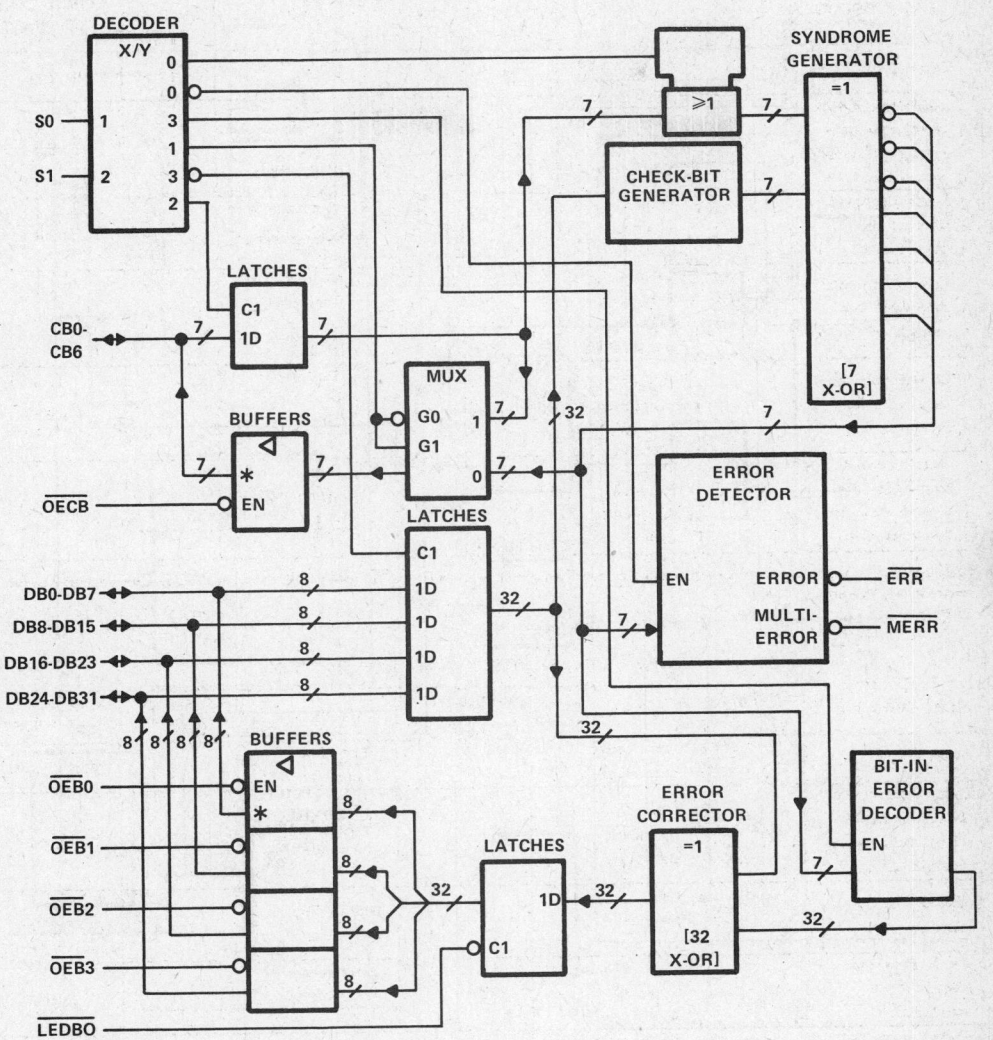

* 'ALS632 has 3-state (▽) check-bit and data outputs.
  'ALS633 has open-collector (◇) check-bit and data outputs.

ALS AND AS CIRCUITS

TEXAS INSTRUMENTS
POST OFFICE BOX 225012 • DALLAS, TEXAS 75265

2-431

TYPES SN54ALS634, SN54ALS635, SN74ALS634, SN74ALS635
32-BIT PARALLEL ERROR DETECTION AND CORRECTION CIRCUITS

'ALS634, 'ALS635 logic diagram (positive logic)

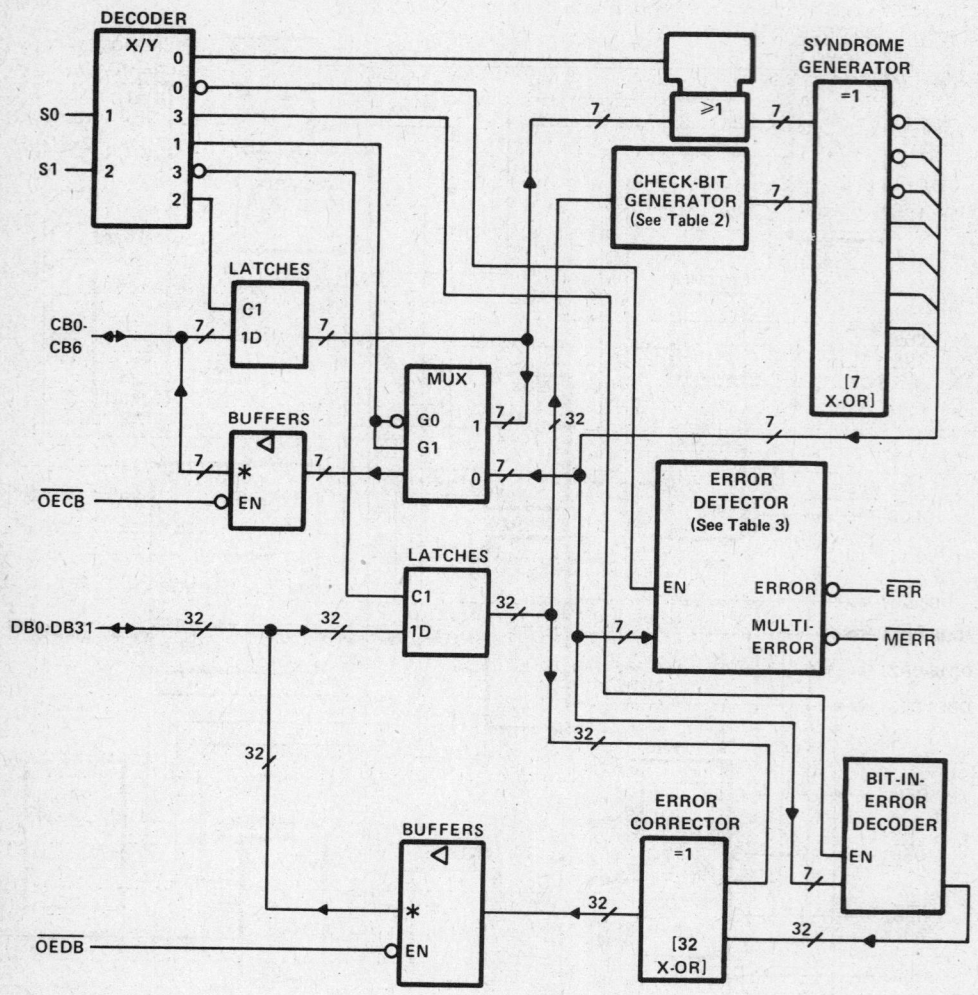

* 'ALS634 has 3-state (▽) check-bit and data outputs.
  'ALS635 has open-collector (◇) check-bit and data outputs.

TYPES SN54ALS632 THRU SN54ALS635, SN74ALS632 THRU SN74ALS635
32-BIT PARALLEL ERROR DETECTION AND CORRECTION CIRCUITS

absolute maximum ratings over operating free-air temperature range (unless otherwise noted)

Supply voltage, V_{CC} (see Note 1) .... 7 V
Input voltage: CB and DB .... 5.5 V
             All others .... 7 V
Operating free-air temperature range: SN54ALS632 thru SN54ALS635 .... −55°C to 125°C
                                 SN74ALS632 thru SN74ALS635 .... 0°C to 70°C
Storage temperature range .... −65°C to 150°C

recommended operating conditions

			SN54ALS632 THRU SN54ALS635			SN74ALS632 THRU SN74ALS635			UNIT
			MIN	NOM	MAX	MIN	NOM	MAX	
V_{CC}	Supply voltage		4.5	5	5.5	4.5	5	5.5	V
V_{IH}	High-level input voltage		2			2			V
V_{IL}	Low-level input voltage				0.8			0.8	V
I_{OH}	High-level output current	ERR or MERR			−0.4			−0.4	mA
		DB or CB 'ALS632, 'ALS634			−1			−2.6	
I_{OL}	Low-level output current	ERR or MERR			4			8	mA
		DB or CB			12			24	
t_w	Pulse duration	LEDBO low	25			25			ns
t_{su}	Setup time	(1) Data and check word before S0↑ (S1 = H)	15			10			ns
		(2) S0 high before LEDBO↑ (S1 = H)†	45			45			
		(3) LEDBO high before the earlier of S0↓ or S1↓†	0			0			
		(4) LEDBO high before S1↑ (S0 = H)	0			0			
		(5) Diagnostic data word before S1↑ (S0 = H)	15			10			
		(6) Diagnostic check word before the later of S1↓ or S0↑	15			10			
		(7) Diagnostic data word before LEDBO↑ (S1 = L and S0 = H)‡	25			20			
t_h	Hold time	(8) Read-mode, S0 low and S1 high	35			30			ns
		(9) Data and check word after S0↑ (S1 = H)	20			15			
		(10) Data word after S1↑ (S0 = H)	20			15			
		(11) Check word after the later of S1↓ or S0↑	20			15			
		(12) Diagnostic data word after LEDBO↑ (S1 = L and S0 = H)‡	0			0			
t_{corr}	Correction time (see Figure 1)		65			58			ns
T_A	Operating free-air temperature		−55		125	0		70	°C

† These times ensure that corrected data is saved in the output data latch.
‡ These times ensure that the diagnostic data word is saved in the output data latch.

TEXAS INSTRUMENTS
POST OFFICE BOX 225012 • DALLAS, TEXAS 75265

TYPES SN54ALS632 THRU SN54ALS635, SN74ALS632 THRU SN74ALS635
32-BIT PARALLEL ERROR DETECTION AND CORRECTION CIRCUITS

'ALS632, 'ALS634 electrical characteristics over recommended operating free-air temperature range (unless otherwise noted)

PARAMETER		TEST CONDITIONS		SN54ALS632 SN54ALS634			SN74ALS632 SN74ALS634			UNIT
				MIN	TYP†	MAX	MIN	TYP†	MAX	
V_{IK}		$V_{CC} = 4.5$ V,	$I_I = -18$ mA			−1.5			−1.5	V
V_{OH}	All outputs	$V_{CC} = 4.5$ V to 5.5 V,	$I_{OH} = -0.4$ mA	$V_{CC}-2$			$V_{CC}-2$			V
	DB or CB	$V_{CC} = 4.5$ V,	$I_{OH} = -1$ mA	2.4	3.3					
		$V_{CC} = 4.5$ V,	$I_{OH} = -2.6$ mA				2.4	3.2		
V_{OL}	ERR or MERR	$V_{CC} = 4.5$ V,	$I_{OL} = 4$ mA		0.25	0.4		0.25	0.4	V
		$V_{CC} = 4.5$ V,	$I_{OL} = 8$ mA					0.35	0.5	
	DB or CB	$V_{CC} = 4.5$ V,	$I_{OL} = 12$ mA		0.25	0.4		0.25	0.4	
		$V_{CC} = 4.5$ V,	$I_{OL} = 24$ mA					0.35	0.5	
I_I	S0 or S1	$V_{CC} = 5.5$ V,	$V_I = 7$ V			0.1			0.1	mA
	DB or CB	$V_{CC} = 5.5$ V,	$V_I = 5.5$ V			0.1			0.1	
I_{IH}	S0 or S1	$V_{CC} = 5.5$ V,	$V_I = 2.7$ V			20			20	µA
	DB or CB‡					20			20	
I_{IL}	S0 or S1	$V_{CC} = 5.5$ V,	$V_I = 0.4$ V			−0.4			−0.4	mA
	DB or CB‡					−0.1			−0.1	
I_{OS}§		$V_{CC} = 5.5$ V,	$V_O = 2.25$ V	−30		−112	−30		−112	mA
I_{CC}		$V_{CC} = 5.5$ V,	See Note 1		150	250		150	250	mA

'ALS633, 'ALS635 electrical characteristics over recommended operating free-air temperature range (unless otherwise noted)

PARAMETER		TEST CONDITIONS		SN54ALS633 SN54ALS635			SN74ALS633 SN74ALS635			UNIT
				MIN	TYP†	MAX	MIN	TYP†	MAX	
V_{IK}		$V_{CC} = 4.5$ V,	$I_I = -18$ mA			−1.5			−1.5	V
V_{OH}	ERR or MERR	$V_{CC} = 4.5$ V to 5.5 V,	$I_{OH} = -0.4$ mA	$V_{CC}-2$			$V_{CC}-2$			V
I_{OH}	DB or CB	$V_{CC} = 5.5$ V,	$V_{OH} = 5.5$ V			0.1			0.1	mA
V_{OL}	ERR or MERR	$V_{CC} = 4.5$ V,	$I_{OL} = 4$ mA		0.25	0.4		0.25	0.4	V
		$V_{CC} = 4.5$ V,	$I_{OL} = 8$ mA					0.35	0.5	
	DB or CB	$V_{CC} = 4.5$ V,	$I_{OL} = 12$ mA		0.25	0.4		0.25	0.4	
		$V_{CC} = 4.5$ V,	$I_{OL} = 24$ mA					0.35	0.5	
I_I	S0 or S1	$V_{CC} = 5.5$ V,	$V_I = 7$ V							mA
	DB or CB	$V_{CC} = 5.5$ V,	$V_I = 5.5$ V							
I_{IH}	S0 or S1	$V_{CC} = 5.5$ V,	$V_I = 2.7$ V							µA
	DB or CB‡									
I_{IL}	S0 or S1	$V_{CC} = 5.5$ V,	$V_I = 0.4$ V							mA
	DB or CB‡									
I_{OS}§	ERR or MERR	$V_{CC} = 5.5$ V,	$V_O = 2.25$ V	−30		−112	−30		−112	mA
I_{CC}		$V_{CC} = 5.5$ V,	See Note 1		150			150		mA

†All typical values are at $V_{CC} = 5$ V, $T_A = 25$°C.
‡For I/O ports (Q_A through Q_H), the parameters I_{IH} and I_{IL} include the off-state output current.
§The output conditions have been chosen to produce a current that closely approximates one half of the true short-circuit output current, I_{OS}.
NOTE 1: I_{CC} is measured with S0 and S1 at 4.5 V and all CB and DB pins grounded.

Additional information on these products can be obtained from the factory as it becomes available.

PRODUCT PREVIEW

This page contains information on a product under development. Texas Instruments reserves the right to change or discontinue this product without notice.

Texas Instruments
POST OFFICE BOX 225012 • DALLAS, TEXAS 75265

TYPES SN54ALS632, SN54ALS633, SN74ALS632, SN74ALS633
32-BIT PARALLEL ERROR DETECTION AND CORRECTION CIRCUITS

'ALS632 switching characteristics, V_{CC} = 4.5 V to 5.5 V, C_L = 50 pF, T_A = MIN to MAX (unless otherwise noted)

PARAMETER	FROM (INPUT)	TO (OUTPUT)	TEST CONDITIONS	SN54ALS632 MIN	SN54ALS632 TYP	SN54ALS632 MAX	SN74ALS632 MIN	SN74ALS632 TYP	SN74ALS632 MAX	UNIT
t_{pd}	DB and CB	$\overline{ERR}$	S1 = H, S0 = L, R_L = 500 Ω	10		43	10		40	ns
t_{pd}	DB	$\overline{ERR}$	S1 = L, S0 = H, R_L = 500 Ω	10		43	10		40	ns
t_{pd}	DB and CB	$\overline{MERR}$	S1 = H, S0 = L, R_L = 500 Ω	15		67	15		60	ns
t_{pd}	DB	$\overline{MERR}$	S1 = L, S0 = H, R_L = 500 Ω	15		67	15		60	ns
t_{pd}	S0↓ and S1↓	CB	R1 = R2 = 500 Ω	10		60	10		54	ns
t_{pd}	DB	CB	S1 = L, S0 = L, R1 = R2 = 500 Ω	10		60	10		54	ns
t_{pd}	$\overline{LEDBO}$↓	DB	S1 = X, S0 = H, R1 = R2 = 500 Ω	7		35	8		30	ns
t_{pd}	S1↑	CB	S0 = H, R1 = R2 = 500 Ω	10		60	10		54	ns
t_{en}	$\overline{OECB}$↓	CB	S0 = H, S1 = X, R1 = R2 = 500 Ω	5		30	7		25	ns
t_{dis}	$\overline{OECB}$↑	CB	S0 = H, S1 = X, R1 = R2 = 500 Ω	5		30	7		25	ns
t_{en}	$\overline{OEB0}$ thru $\overline{OEB3}$↓	DB	S0 = H, S1 = X, R1 = R2 = 500 Ω	5		30	7		25	ns
t_{dis}	$\overline{OEB0}$ thru $\overline{OEB3}$↑	DB	S0 = H, S1 = X, R1 = R2 = 500 Ω	5		30	7		25	ns

'ALS633 switching characteristics, V_{CC} = 4.5 V to 5.5 V, C_L = 50 pF, T_A = MIN to MAX (unless otherwise noted)

PARAMETER	FROM (INPUT)	TO (OUTPUT)	TEST CONDITIONS	SN54ALS633 MIN	SN54ALS633 TYP[†]	SN54ALS633 MAX	SN74ALS633 MIN	SN74ALS633 TYP[†]	SN74ALS633 MAX	UNIT
t_{pd}	DB and CB	$\overline{ERR}$	S1 = H, S0 = L, R_L = 500 Ω		26			26		ns
t_{pd}	DB	$\overline{ERR}$	S1 = L, S0 = H, R_L = 500 Ω		26			26		ns
t_{pd}	DB and CB	$\overline{MERR}$	S1 = H, S0 = L, R_L = 500 Ω		40			40		ns
t_{pd}	DB	$\overline{MERR}$	S1 = L, S0 = H, R_L = 500 Ω		40			40		ns
t_{pd}	S0↓ and S1↓	CB	R_L = 680 Ω		40			40		ns
t_{pd}	DB	CB	S1 = L, S0 = L, R_L = 680 Ω		40			40		ns
t_{pd}	$\overline{LEDBO}$↓	DB	S1 = X, S0 = H, R_L = 680 Ω		26			26		ns
t_{pd}	S1↑	CB	S0 = H, R_L = 680 Ω		40			40		ns
t_{PLH}	$\overline{OECB}$↑	CB	S1 = X, S0 = H, R_L = 680 Ω		24			24		ns
t_{PHL}	$\overline{OECB}$↓	CB	S1 = X, S0 = H, R_L = 680 Ω		24			24		ns
t_{PLH}	$\overline{OEB0}$ thru $\overline{OEB3}$↑	DB	S1 = X, S0 = H, R_L = 680 Ω		24			24		ns
t_{PHL}	$\overline{OEB0}$ thru $\overline{OEB3}$↓	DB	S1 = X, S0 = H, R_L = 680 Ω		24			24		ns

[†]All typical values are at V_{CC} = 5 V, T_A = 25°C.

Additional information on these products can be obtained from the factory as it becomes available.

PRODUCT PREVIEW
This page contains information on a product under development. Texas Instruments reserves the right to change or discontinue this product without notice.

TEXAS INSTRUMENTS
POST OFFICE BOX 225012 • DALLAS, TEXAS 75265

TYPES SN54ALS634, SN54ALS635, SN74ALS634, SN74ALS635
32-BIT PARALLEL ERROR DETECTION AND CORRECTION CIRCUITS

'ALS634 switching characteristics, V_{CC} = 4.5 V to 5.5 V, C_L = 50 pF, T_A = MIN to MAX

PARAMETER	FROM (INPUT)	TO (OUTPUT)	TEST CONDITIONS	SN54ALS634 MIN	SN54ALS634 TYP†	SN54ALS634 MAX	SN74ALS634 MIN	SN74ALS634 TYP†	SN74ALS634 MAX	UNIT
t_{pd}	DB and CB	$\overline{ERR}$	S1=H, S0=L, R_L=500 Ω			26			26	ns
			S1=L, S0=H, R_L=500 Ω			26			26	
t_{pd}	DB and CB	$\overline{MERR}$	S1=H, S0=L, R_L=500 Ω			40			40	ns
			S1=L, S0=H, R_L=500 Ω			40			40	
t_{pd}	S0↓ and S1↓	CB	R1=R2=500 Ω			35			35	ns
t_{pd}	DB	CB	S1=L, S0=L, R1=R2=500 Ω			35			35	ns
t_{pd}	S1↑	CB	S0=H, R1=R2=500 Ω			35			35	ns
t_{en}	$\overline{OECB}$↓	CB	S1=X, S0=H, R1=R2=500 Ω			18			18	ns
t_{dis}	$\overline{OECB}$↑	CB	S1=X, S0=H, R1=R2=500 Ω			18			18	ns
t_{en}	$\overline{OECB}$↓	DB	S1=X, S0=H, R1=R2=500 Ω			18			18	ns
t_{dis}	$\overline{OECB}$↑	DB	S1=X, S0=H, R1=R2=500 Ω			18			18	ns

'ALS635 switching characteristics, V_{CC} = 4.5 V to 5.5 V, C_L = 50 pF, T_A = MIN to MAX

PARAMETER	FROM (INPUT)	TO (OUTPUT)	TEST CONDITIONS	SN54ALS635 MIN	SN54ALS635 TYP†	SN54ALS635 MAX	SN74ALS635 MIN	SN74ALS635 TYP†	SN74ALS635 MAX	UNIT
t_{pd}	DB and CB	$\overline{ERR}$	S1=H, S0=L, R_L=500 Ω			26			26	ns
	DB	$\overline{ERR}$	S1=L, S0=H, R_L=500 Ω			26			26	
t_{pd}	DB and CB	$\overline{MERR}$	S1=H, S0=L, R_L=500 Ω			40			40	ns
			S1=L, S0=H, R_L=500 Ω			40			40	
t_{pd}	S0↓ and S1↓	CB	R_L=680 Ω			40			40	ns
t_{pd}	DB	CB	S1=L, S0=L, R_L=680 Ω			40			40	ns
t_{pd}	S1↑	DB	S0=H, R_L=680 Ω			40			40	ns
t_{PLH}	$\overline{OECB}$↑	CB	S1=X, S0=H, R_L=680 Ω			24			24	ns
t_{PHL}	$\overline{OECB}$↓	CB	S1=X, S0=H, R_L=680 Ω			24			24	ns
t_{PLH}	$\overline{OEDB}$↑	DB	S1=X, S0=H, R_L=680 Ω			24			24	ns
t_{PHL}	$\overline{OEDB}$↓	DB	S1=X, S0=H, R_L=680 Ω			24			24	ns

†All typical values are at V_{CC} = 5 V, T_A = 25°C.

PRODUCT PREVIEW

This page contains information on a product under development. Texas Instruments reserves the right to change or discontinue this product without notice.

TEXAS INSTRUMENTS
POST OFFICE BOX 225012 • DALLAS, TEXAS 75265

TYPES SN54ALS632 THRU SN54ALS635, SN74ALS632 THRU SN74ALS635
32-BIT PARALLEL ERROR DETECTION AND CORRECTION CIRCUITS

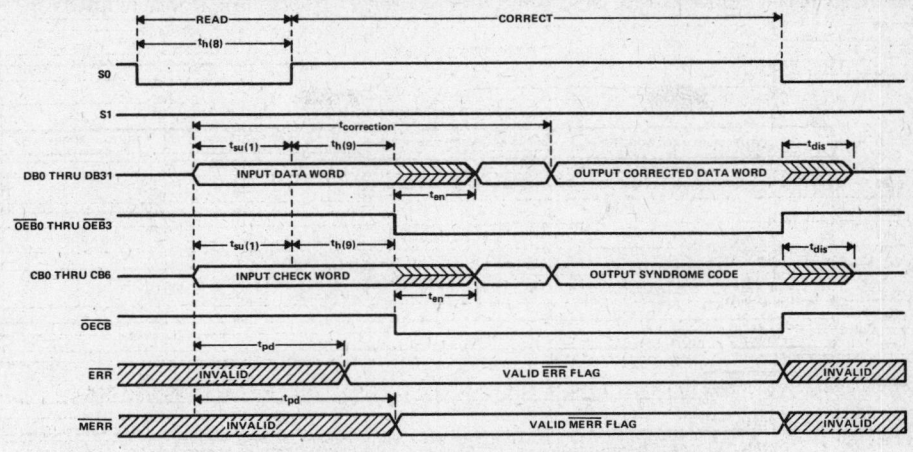

FIGURE 1—READ, FLAG, AND CORRECT MODE SWITCHING WAVEFORMS

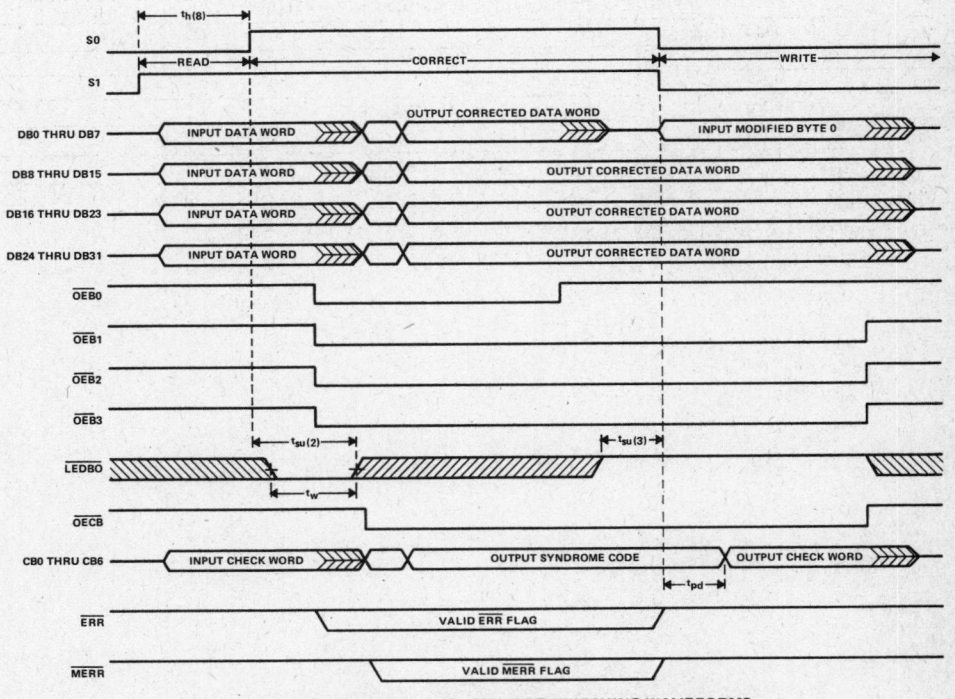

FIGURE 2—READ, CORRECT, MODIFY MODE SWITCHING WAVEFORMS

TEXAS INSTRUMENTS
POST OFFICE BOX 225012 • DALLAS, TEXAS 75265

TYPES SN54ALS632 THRU SN54ALS635, SN74ALS632 THRU SN74ALS635
32-BIT PARALLEL ERROR DETECTION AND CORRECTION CIRCUITS

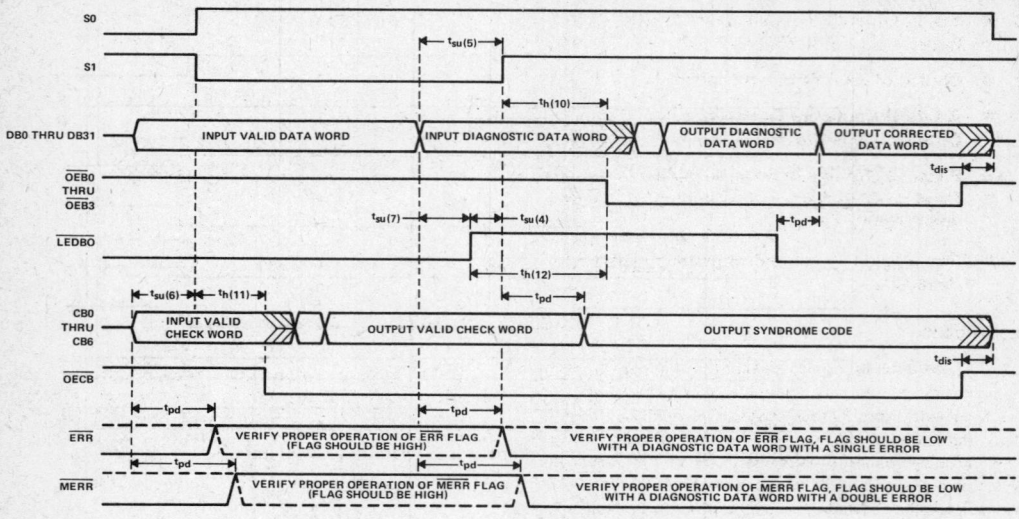

FIGURE 3–DIAGNOSTIC MODE SWITCHING WAVEFORM

TYPES SN54ALS638A, SN54ALS639A, SN54AS638, SN54AS639 SN74ALS638A, SN74ALS639A, SN74AS638, SN74AS639
OCTAL BUS TRANSCEIVERS

D2261, DECEMBER 1983

- Bidirectional Bus Transceivers in High-Density 20-Pin Packages
- Choice of True or Inverting Logic
- A Bus Outputs are Open-Collector; B Bus Outputs are 3-State
- Package Options Include Both Plastic and Ceramic Chip Carriers in Addition to Plastic and Ceramic DIPs
- Dependable Texas Instruments Quality and Reliability

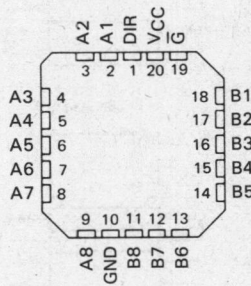

SN54ALS', SN54AS' . . . J PACKAGE
SN74ALS', SN74AS' . . . N PACKAGE
(TOP VIEW)

SN54ALS', SN54AS' . . . FH PACKAGE
SN74ALS', SN74AS' . . . FN PACKAGE
(TOP VIEW)

description

These octal bus transceivers are designed for asynchronous two-way communication between open-collector and 3-state buses. The devices transmit data from the A bus (open-collector) to the B bus (3-state) or from the B bus to the A bus depending upon the level at the direction control (DIR) input. The enable input ($\overline{G}$) can be used to disable the device so the buses are isolated.

DEVICE	A OUTPUT	B OUTPUT	LOGIC
'ALS638A, 'AS638	Open-Collector	3-State	Inverting
'ALS639A, 'AS639	Open-Collector	3-State	True

The −1 versions of the SN74ALS' parts are identical to the standard versions except that recommended maximum of I_{OL} is increased to 48 milliamperes. There are no −1 versions of the SN54ALS' parts.

The SN54' family is characterized for operation over the full military temperature range of −55°C to 125°C. The SN74' family is characterized for operation from 0°C to 70°C.

FUNCTION TABLE

CONTROL INPUTS		OPERATION	
$\overline{G}$	DIR	'ALS638A 'AS638	'ALS639A 'AS639
L	L	$\overline{B}$ data to A bus	B data to A bus
L	H	$\overline{A}$ data to B bus	A data to B bus
H	X	Isolation	Isolation

Copyright © 1982 by Texas Instruments Incorporated

TYPES SN54ALS638A, SN54ALS639A, SN54AS638, SN54AS639 SN74ALS638A, SN74ALS639A, SN74AS638, SN74AS639 OCTAL BUS TRANSCEIVERS

logic symbols

functional block diagrams (positive logic)

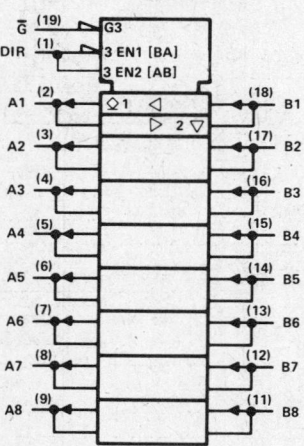

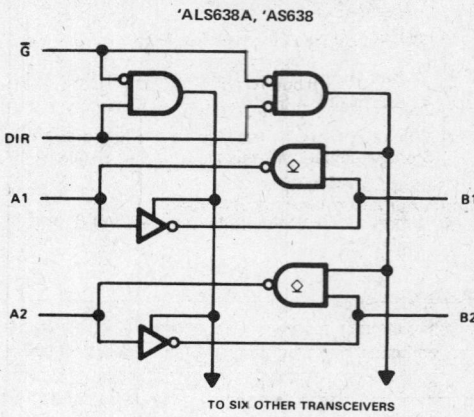

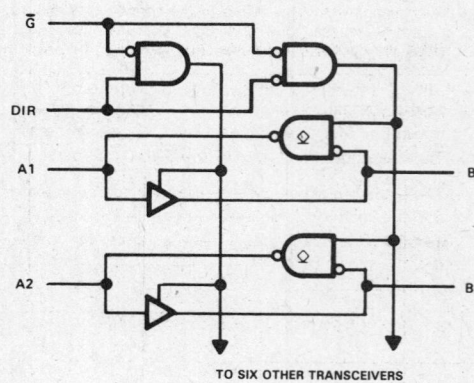

Pin numbers shown are for J and N packages.

TEXAS INSTRUMENTS
POST OFFICE BOX 225012 • DALLAS, TEXAS 75265

TYPES SN54ALS638A, SN54ALS639A, SN74ALS638A, SN74ALS639A
OCTAL BUS TRANSCEIVERS

absolute maximum ratings over operating free-air temperature range (unless otherwise noted)

Supply voltage, V_{CC} .... 7 V
Input voltage: All inputs .... 7 V
                A bus I/O ports .... 7 V
                B bus I/O ports .... 5.5 V
Operating free-air temperature range: SN54ALS638A, SN54ALS639A .... −55°C to 125°C
                        SN74ALS638A, SN74ALS639A .... 0°C to 70°C
Storage temperature range .... −65°C to 150°C

recommended operating conditions

			SN54ALS638A SN54ALS639A			SN74ALS638A SN74ALS639A			UNIT
			MIN	NOM	MAX	MIN	NOM	MAX	
V_{CC}	Supply voltage		4.5	5	5.5	4.5	5	5.5	V
V_{IH}	High-level input voltage		2			2			V
V_{IL}	Low-level input voltage				0.8			0.8	V
V_{OH}	High-level output voltage	A ports			5.5			5.5	V
I_{OH}	High-level output current	B ports			−12			−15	mA
I_{OL}	Low-level output current	A or B ports			12			24	mA
								48†	
T_A	Operating free-air temperature		−55		125	0		70	°C

†The extended limits apply only if V_{CC} is maintained between 4.75 V and 5.25 V.
The 48-mA limit applies for the SN74ALS638A-1 and SN74ALS639A-1 only.

electrical characteristics over recommended operating free-air temperature range (unless otherwise noted)

PARAMETER		TEST CONDITIONS		SN54ALS638A SN54ALS639A			SN74ALS638A SN74ALS639A			UNIT
				MIN	TYP‡	MAX	MIN	TYP‡	MAX	
V_{IK}		V_{CC} = 4.5 V,	I_I = −18 mA			−1.5			−1.5	V
I_{OH}	A ports	V_{CC} = 4.5 V,	V_{OH} = 5.5 V			0.1			0.1	mA
V_{OH}	B ports	V_{CC} = 4.5 V to 5.5 V,	I_{OH} = −0.4 mA	V_{CC}−2			V_{CC}−2			V
		V_{CC} = 4.5 V,	I_{OH} = −3 mA	2.4	3.2		2.4	3.2		
		V_{CC} = 4.5 V,	I_{OH} = −12 mA	2						
		V_{CC} = 4.5 V,	I_{OH} = −15 mA				2			
V_{OL}	A or B ports	V_{CC} = 4.5 V,	I_{OL} = 12 mA		0.25	0.4		0.25	0.4	V
		V_{CC} = 4.5 V,	I_{OL} = 24 mA					0.35	0.5	
		(I_{OL} = 48 mA for −1 versions)								
I_I	Control inputs	V_{CC} = 5.5 V,	V_I = 7 V			0.1			0.1	mA
	A or B ports	V_{CC} = 5.5 V,	V_I = 5.5 V			0.1			0.1	
I_{IH}	Control inputs	V_{CC} = 5.5 V,	V_I = 2.7 V			20			20	μA
	A or B ports§					20			20	
I_{IL}	Control inputs	V_{CC} = 5.5 V,	V_I = 0.4 V			−0.1			−0.1	mA
	A or B ports§					−0.1			−0.1	
I_O¶	B ports	V_{CC} = 5.5 V,	V_O = 2.25 V	−30		−112	−30		−112	mA
I_{CC}	'ALS638A	V_{CC} = 5.5 V	Outputs high		18	36		18	30	mA
			Outputs low		25	48		26	41	
			Outputs disabled		16	35		16	30	
	'ALS639A		Outputs high		25	45		25	40	
			Outputs low		30	55		30	50	
			Outputs disabled		33	60		33	54	

‡All typical values are at V_{CC} = 5 V, T_A = 25°C
§For I/O ports, the parameters I_{IH} and I_{IL} include the off-state output current.
¶The output conditions have been chosen to produce a current that closely approximates one half of the true short-circuit output current, I_{OS}.

TEXAS INSTRUMENTS
POST OFFICE BOX 225012 • DALLAS, TEXAS 75265

TYPES SN54ALS638A, SN54ALS639A, SN74ALS638A, SN74ALS639A
OCTAL BUS TRANSCEIVERS

'ALS638A switching characteristics (see Note 1)

PARAMETER	FROM (INPUT)	TO (OUTPUT)	$V_{CC} = 4.5$ V to 5.5 V, $C_L = 50$ pF, $R_L = 680\ \Omega$ (A outputs), $R1 = R2 = 500\ \Omega$ (B outputs), $T_A = $ MIN to MAX				UNIT
			SN54ALS638A		SN74ALS638A		
			MIN	MAX	MIN	MAX	
t_{PLH}	A	B	2	15	2	12	ns
t_{PHL}			2	15	2	12	
t_{PLH}	B	A	8	30	8	25	ns
t_{PHL}			8	35	8	30	
t_{PLH}	$\overline{G}$	A	5	30	5	25	ns
t_{PHL}			10	50	10	45	
t_{PZH}	$\overline{G}$	B	5	25	5	20	ns
t_{PZL}			5	28	5	22	
t_{PHZ}	$\overline{G}$	B	2	12	2	10	ns
t_{PLZ}			3	18	3	15	

'ALS639A switching characteristics (see Note 1)

PARAMETER	FROM (INPUT)	TO (OUTPUT)	$V_{CC} = 4.5$ V to 5.5 V, $C_L = 50$ pF, $R_L = 680\ \Omega$ (A outputs), $R1 = R2 = 500\ \Omega$ (B outputs), $T_A = $ MIN to MAX				UNIT
			SN54ALS639A		SN74ALS639A		
			MIN	MAX	MIN	MAX	
t_{PLH}	A	B	2	15	2	12	ns
t_{PHL}			2	15	2	12	
t_{PLH}	B	A	10	35	10	30	ns
t_{PHL}			5	28	5	22	
t_{PLH}	$\overline{G}$	A	10	35	10	30	ns
t_{PHL}			10	40	10	35	
t_{PZH}	$\overline{G}$	B	6	28	6	21	ns
t_{PZL}			8	30	8	25	
t_{PHZ}	$\overline{G}$	B	2	12	2	10	ns
t_{PLZ}			3	19	3	16	

NOTE 1: For load circuit and voltage waveforms, see page 1-12.

TYPES SN54AS638, SN54AS639, SN74AS638, SN74AS639
OCTAL BUS TRANSCEIVERS

absolute maximum ratings over operating free-air temperature range (unless otherwise noted)

Supply voltage, V_{CC} .... 7 V
Input voltage: All inputs .... 7 V
               A bus I/O ports .... 7 V
               B bus I/O ports .... 5.5 V
Operating free-air temperature range: SN54AS638, SN54AS639 .... −55 °C to 125 °C
                              SN74AS638, SN74AS639 .... 0 °C to 70 °C
Storage temperature range .... −65 °C to 150 °C

recommended operating conditions

		SN54AS638 SN54AS639			SN74AS638 SN74AS639			UNIT
		MIN	NOM	MAX	MIN	NOM	MAX	
V_{CC}	Supply voltage	4.5	5	5.5	4.5	5	5.5	V
V_{IH}	High-level input voltage	2			2			V
V_{IL}	Low-level input voltage			0.8			0.8	V
V_{OH}	High-level output voltage	A ports		5.5			5.5	V
I_{OH}	High-level output current	B ports		−12			−15	mA
I_{OL}	Low-level output current	A or B ports		48			64	mA
T_A	Operating free-air temperature		−55		125	0		70	°C

electrical characteristics over recommended operating free-air temperature range (unless otherwise noted)

PARAMETER		TEST CONDITIONS		SN54AS638 SN54AS639			SN74AS638 SN74AS639			UNIT
				MIN	TYP†	MAX	MIN	TYP†	MAX	
V_{IK}		$V_{CC} = 4.5$ V,	$I_I = -18$ mA			−1.2			−1.2	V
I_{OH}	A ports	$V_{CC} = 4.5$ V,	$V_{OH} = 5.5$ V			0.1			0.1	mA
V_{OH}	B ports	$V_{CC} = 4.5$ V, to 5.5 V,	$I_{OH} = -2$ mA	$V_{CC}-2$			$V_{CC}-2$			V
		$V_{CC} = 4.5$ V,	$I_{OH} = -3$ mA	2.4	3.2		2.4	3.2		
		$V_{CC} = 4.5$ V,	$I_{OH} = -12$ mA	2.4						
		$V_{CC} = 4.5$ V,	$I_{OH} = -15$ mA				2.4			
V_{OL}	A or B ports	$V_{CC} = 4.5$ V,	$I_{OL} = 48$ mA		0.3	0.55				V
		$V_{CC} = 4.5$ V,	$I_{OL} = 64$ mA					0.35	0.55	
I_I	Control inputs	$V_{CC} = 5.5$ V,	$V_I = 7$ V			0.1			0.1	mA
	A or B ports	$V_{CC} = 5.5$ V,	$V_I = 5.5$ V			0.1			0.1	
I_{IH}	Control inputs	$V_{CC} = 5.5$ V,	$V_I = 2.7$ V			20			20	μA
	A or B ports‡					50			50	
I_{IL}	Control inputs	$V_{CC} = 5.5$ V,	$V_I = 0.4$ V			−0.5			−0.5	mA
	A or B ports‡					−0.75			−0.75	
I_O §		$V_{CC} = 5.5$ V,	$V_O = 2.25$ V	−30		−112	−30		−112	mA
I_{CC}	'AS638	$V_{CC} = 5.5$ V	Outputs high		24	40		24	40	mA
			Outputs low		75	122		75	122	
			Outputs disabled		37	61		37	61	
	'AS639		Outputs high		56	92		56	92	
			Outputs low		95	154		95	154	
			Outputs disabled		62	100		62	100	

†All typical values are at $V_{CC} = 5$ V, $T_A = 25$ °C.
‡For I/O ports, the parameters I_{IH} and I_{IL} include the off-state output current.
§The output conditions have been chosen to produce a current that closely approximates one half of the true short-circuit output current, I_{OS}.

TEXAS INSTRUMENTS
POST OFFICE BOX 225012 • DALLAS, TEXAS 75265

TYPES SN54AS638, SN54AS639, SN74AS638, SN74AS639
OCTAL BUS TRANSCEIVERS

'AS638 switching characteristics (see Note 1)

PARAMETER	FROM (INPUT)	TO (OUTPUT)	V_{CC} = 4.5 V to 5.5 V, C_L = 50 pF, R_L = 680 Ω (A outputs), R1 = R2 = 500 Ω (B outputs), T_A = MIN to MAX				UNIT
			SN54AS638		SN74AS638		
			MIN	MAX	MIN	MAX	
t_{PLH}	A	B	2	8	2	7	ns
t_{PHL}			2	7.5	2	6.5	
t_{PLH}	B	A	5	23	5	20	ns
t_{PHL}			2	8	2	7	
t_{PLH}	$\overline{G}$	A	5	20	5	19	ns
t_{PHL}			2	10	2	9	
t_{PZH}	$\overline{G}$	B	2	10	2	8	ns
t_{PZL}			2	12	2	10	
t_{PHZ}	$\overline{G}$	B	2	8	2	7	ns
t_{PLZ}			2	12	2	10	

'AS639 switching characteristics (see Note 1)

PARAMETER	FROM (INPUT)	TO (OUTPUT)	V_{CC} = 4.5 V to 5.5 V, C_L = 50 pF, R_L = 680 Ω (A outputs), R1 = R2 = 500 Ω (B outputs), T_A = MIN to MAX				UNIT
			SN54AS639		SN74AS639		
			MIN	MAX	MIN	MAX	
t_{PLH}	A	B	2	11	2	9.5	ns
t_{PHL}			2	10.5	2	9	
t_{PLH}	B	A	5	25	5	22	ns
t_{PHL}			2	10	2	9	
t_{PLH}	$\overline{G}$	A	5	23	5	21.5	ns
t_{PHL}			2	12.5	2	11.5	
t_{PZH}	$\overline{G}$	B	2	12	2	10.5	ns
t_{PZL}			2	12	2	10.5	
t_{PHZ}	$\overline{G}$	B	2	7.5	2	7	ns
t_{PLZ}			2	12	2	10.5	

NOTE 1: For load circuit and voltage waveforms, see page 1-12.

TYPES SN54ALS640A THRU SN54ALS645A, SN54AS640 THRU SN54AS645 SN74ALS640A THRU SN74ALS645A, SN74AS640 THRU SN74AS645 OCTAL BUS TRANSCEIVERS

D2661, DECEMBER 1983

- Bidirectional Bus Transceivers in High-Density 20-Pin Packages
- Choice of True or Inverting Logic
- Choice of 3-State or Open-Collector Outputs
- Package Options Include Both Plastic and Ceramic Chip Carriers in Addition to Plastic and Ceramic DIPs
- Dependable Texas Instruments Quality and Reliability

DEVICE	OUTPUT	LOGIC
'ALS640A, 'AS640	3-State	Inverting
'ALS641A, 'AS641	Open-Collector	True
'ALS642A, 'AS642	Open-Collector	Inverting
'ALS643A, 'AS643	3-State	True and Inverting
'ALS644A, 'AS644	Open-Collector	True and Inverting
'ALS645A, 'AS645	3-State	True

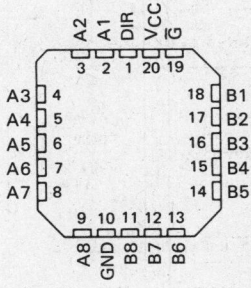

SN54ALS', SN54AS' . . . J PACKAGE
SN74ALS', SN74AS' . . . N PACKAGE
(TOP VIEW)

SN54ALS', SN54AS' . . . FH PACKAGE
SN74ALS', SN74AS' . . . FN PACKAGE
(TOP VIEW)

description

These octal bus transceivers are designed for asynchronous two-way communication between data buses. The devices transmit data from the A bus to the B bus or from the B bus to the A bus depending upon the level at the direction control (DIR) input. The enable input ($\overline{G}$) can be used to disable the device so the buses are effectively isolated.

The −1 versions of the SN74ALS' parts are identical to the standard versions except that the recommended maximum I_{OL} is increased to 48 milliamperes. There are no −1 versions of the SN54ALS' parts.

The SN54' family is characterized for operation over the full military temperature range of −55°C to 125°C. The SN74' family is characterized for operation from 0°C to 70°C.

FUNCTION TABLE

CONTROL INPUTS		OPERATION		
$\overline{G}$	DIR	'ALS640A, 'AS640 'ALS642A, 'AS642	'ALS641A, 'AS641 'ALS645A, 'AS645	'ALS643A, 'AS643 'ALS644A, 'AS644
L	L	$\overline{B}$ data to A bus	B data to A bus	B data to A bus
L	H	$\overline{A}$ data to B bus	A data to B bus	$\overline{A}$ data to B bus
H	X	Isolation	Isolation	Isolation

Copyright © 1982 by Texas Instruments Incorporated

TYPES SN54ALS640A THRU SN54ALS645A, SN54AS640 THRU SN54AS645 SN74ALS640A THRU SN74ALS645A, SN74AS640 THRU SN74AS645 OCTAL BUS TRANSCEIVERS

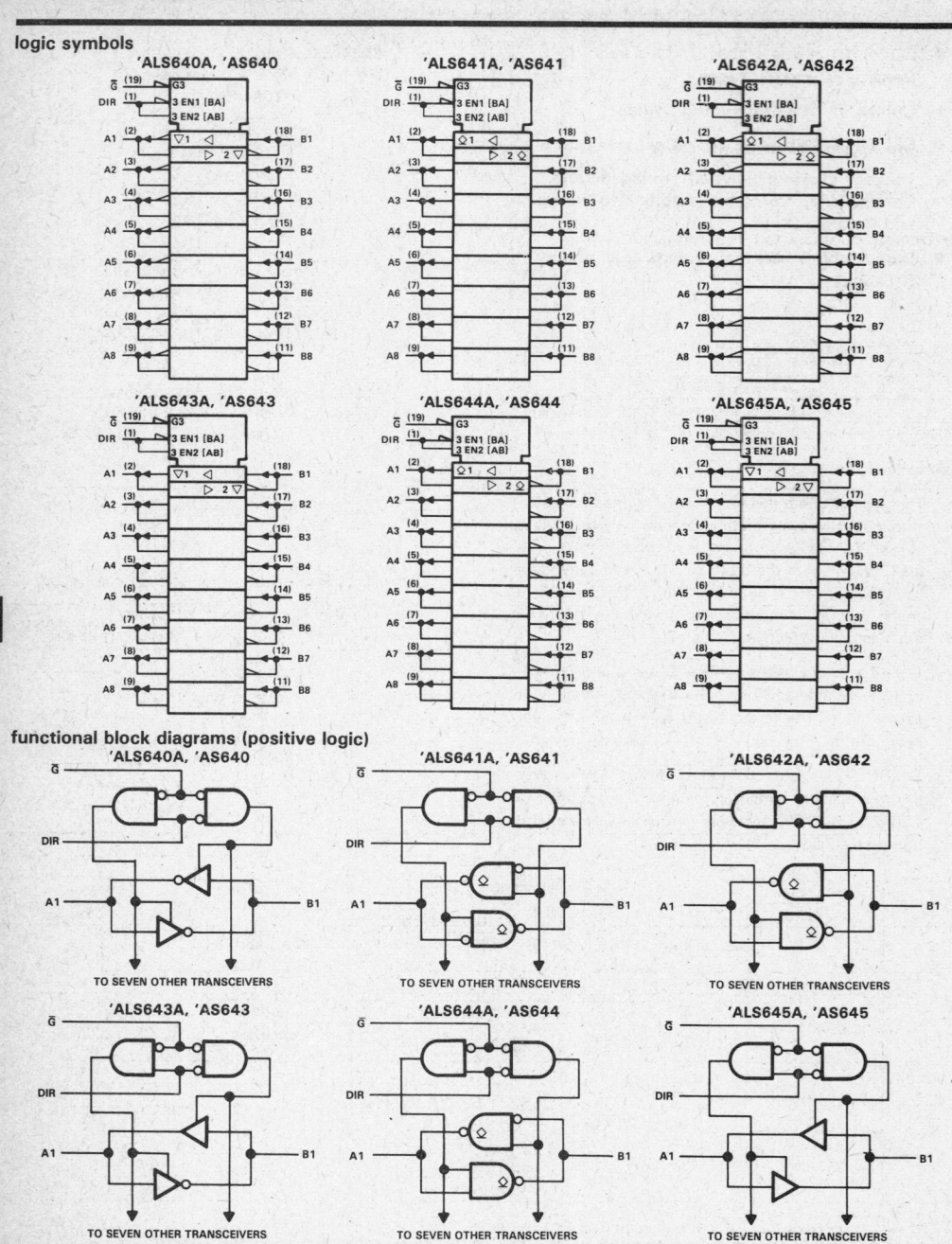

Pin numbers shown are for J and N packages.

**TYPES SN54ALS640A THRU SN54ALS645A
SN74ALS640A THRU SN74ALS645A
OCTAL BUS TRANSCEIVERS**

absolute maximum ratings over operating free-air temperature range (unless otherwise noted)

Supply voltage, V_{CC} .... 7 V
Input voltage: All inputs .... 7 V
               I/O ports .... 5.5 V
Operating free-air temperature range: SN54ALS640A, SN54ALS643A, SN54ALS645A .... −55 °C to 125 °C
                                 SN74ALS640A, SN74ALS643A, SN74ALS645A .... 0 °C to 70 °C
Storage temperature range .... −65 °C to 150 °C

recommended operating conditions

		SN54ALS640A SN54ALS643A SN54ALS645A			SN74ALS640A SN74ALS643A SN74ALS645A			UNIT
		MIN	NOM	MAX	MIN	NOM	MAX	
V_{CC}	Supply voltage	4.5	5	5.5	4.5	5	5.5	V
V_{IH}	High-level input voltage	2			2			V
V_{IL}	Low-level input voltage			0.8			0.8	V
I_{OH}	High-level output current			−12			−15	mA
I_{OL}	Low-level output current			12			24	mA
							48†	
T_A	Operating free-air temperature	−55		125	0		70	°C

† The extended limits apply only if V_{CC} is maintained between 4.75 V and 5.25 V.
The 48-mA limit applies for the SN74ALS640A−1, SN74ALS643A−1, and SN74ALS645A−1 only.

electrical characteristics over recommended operating free-air temperature range (unless otherwise noted)

PARAMETER		TEST CONDITIONS		SN54ALS'			SN74ALS'			UNIT
				MIN	TYP‡	MAX	MIN	TYP‡	MAX	
V_{IK}		V_{CC} = 4.5 V,	I_I = −18 mA			−1.5			−1.5	V
V_{OH}		V_{CC} = 4.5 V to 5.5 V,	I_{OH} = −0.4 mA	V_{CC}−2			V_{CC}−2			V
		V_{CC} = 4.5 V,	I_{OH} = −3 mA	2.4	3.2		2.4	3.2		
		V_{CC} = 4.5 V,	I_{OH} = −12 mA	2						
		V_{CC} = 4.5 V,	I_{OH} = −15 mA				2			
V_{OL}		V_{CC} = 4.5 V,	I_{OL} = 12 mA		0.25	0.4		0.25	0.4	V
		V_{CC} = 4.5 V, (I_{OL} = 48 mA for −1 versions)	I_{OL} = 24 mA					0.35	0.5	
I_I	Control inputs	V_{CC} = 5.5 V,	V_I = 7 V			0.1			0.1	mA
	A or B ports	V_{CC} = 5.5 V,	V_I = 5.5 V			0.1			0.1	
I_{IH}	Control inputs	V_{CC} = 5.5 V,	V_I = 2.7 V			20			20	µA
	A or B ports§					20			20	
I_{IL}	Control inputs	V_{CC} = 5.5 V,	V_I = 0.4 V			−0.1			−0.1	mA
	A or B ports§					−0.1			−0.1	
I_O¶		V_{CC} = 5.5 V,	V_O = 2.25 V	−30		−112	−30		−112	mA
I_{CC}	'ALS640A	V_{CC} = 5.5 V	Outputs high		19	35		19	30	mA
			Outputs low		27	45		27	40	
			Outputs disabled		28	48		28	43	
	'ALS643A		Outputs high		25	37		25	35	
			Outputs low		33	47		33	45	
			Outputs disabled		35	50		35	48	
	'ALS645A		Outputs high		30	48		30	45	
			Outputs low		36	60		36	55	
			Outputs disabled		38	63		38	58	

‡ All typical values are at V_{CC} = 5 V, T_A = 25 °C
§ For I/O ports, the parameters I_{IH} and I_{IL} include the off-state output current.
¶ The output conditions have been chosen to produce a current that closely approximates one half of the true short-circuit output current, I_{OS}.

TEXAS INSTRUMENTS
POST OFFICE BOX 225012 • DALLAS, TEXAS 75265

TYPES SN54ALS640A THRU SN54ALS645A
SN74ALS640A THRU SN74ALS645A
OCTAL BUS TRANSCEIVERS

'ALS640A switching characteristics (see Note 1)

PARAMETER	FROM (INPUT)	TO (OUTPUT)	V_{CC} = 4.5 V to 5.5 V, C_L = 50 pF, R1 = 500 Ω, R2 = 500 Ω, T_A = MIN to MAX				UNIT
			SN54ALS640A		SN74ALS640A		
			MIN	MAX	MIN	MAX	
t_{PLH}	A or B	B or A	2	14	2	11	ns
t_{PHL}			2	13	2	10	
t_{PZH}	$\overline{G}$	A or B	5	25	5	21	ns
t_{PZL}			8	27	8	24	
t_{PHZ}	$\overline{G}$	A or B	2	12	2	10	ns
t_{PLZ}			3	20	3	15	

'ALS643A switching characteristics (see Note 1)

PARAMETER	FROM (INPUT)	TO (OUTPUT)	V_{CC} = 4.5 V to 5.5 V, C_L = 50 pF, R1 = 500 Ω, R2 = 500 Ω, T_A = MIN to MAX				UNIT
			SN54ALS643A		SN74ALS643A		
			MIN	MAX	MIN	MAX	
t_{PLH}	A	B	2	15	2	13	ns
t_{PHL}			2	13	2	11	
t_{PLH}	B	A	2	15	2	13	ns
t_{PHL}			2	13	2	11	
t_{PZH}	$\overline{G}$	A	5	28	5	25	ns
t_{PZL}			5	28	5	25	
t_{PHZ}	$\overline{G}$	A	2	12	2	10	ns
t_{PLZ}			3	22	3	17	
t_{PZH}	$\overline{G}$	B	5	28	5	25	ns
t_{PZL}			5	28	5	25	
t_{PHZ}	$\overline{G}$	B	2	12	2	10	ns
t_{PLZ}			3	22	3	17	

'ALS645A switching characteristics (see Note 1)

PARAMETER	FROM (INPUT)	TO (OUTPUT)	V_{CC} = 4.5 V to 5.5 V, C_L = 50 pF, R1 = 500 Ω, R2 = 500 Ω, T_A = MIN to MAX				UNIT
			SN54ALS645A		SN74ALS645A		
			MIN	MAX	MIN	MAX	
t_{PLH}	A or B	B or A	3	15	3	10	ns
t_{PHL}			3	13	3	10	
t_{PZH}	$\overline{G}$	A or B	5	25	5	20	ns
t_{PZL}			5	25	5	20	
t_{PHZ}	$\overline{G}$	A or B	2	12	2	10	ns
t_{PLZ}			4	18	4	15	

NOTE 1: For load circuit and voltage waveforms, see page 1-12.

TEXAS INSTRUMENTS
POST OFFICE BOX 225012 • DALLAS, TEXAS 75265

TYPES SN54ALS640A THRU SN54ALS645A
SN74ALS640A THRU SN74ALS645A
OCTAL BUS TRANSCEIVERS

absolute maximum ratings over operating free-air temperature range (unless otherwise noted)

Supply voltage, V_{CC} .... 7 V
Input voltage: All inputs and I/O ports .... 7 V
Operating free-air temperature range: SN54ALS641A, SN54ALS642A, SN54ALS644A ...... −55 °C to 125 °C
                                      SN74ALS641A, SN74ALS642A, SN74ALS644A ...... 0 °C to 70 °C
Storage temperature range .... −65 °C to 150 °C

recommended operating conditions

		SN54ALS641A SN54ALS642A SN54ALS644A			SN74ALS641A SN74ALS642A SN74ALS644A			UNIT
		MIN	NOM	MAX	MIN	NOM	MAX	
V_{CC}	Supply voltage	4.5	5	5.5	4.5	5	5.5	V
V_{IH}	High-level input voltage	2			2			V
V_{IL}	Low-level input voltage			0.8			0.8	V
V_{OH}	High-level output voltage			5.5			5.5	V
I_{OL}	Low-level output current			12			24	mA
							48†	
T_A	Operating free-air temperature	−55		125	0		70	°C

†The extended limits apply only if V_{CC} is maintained between 4.75 and 5.25 V.
 The 48-mA limit applies for the SN74ALS641A-1, SN74ALS642A-1, and SN74ALS644A-1 only.

electrical characteristics over recommended operating free-air temperature range (unless otherwise noted)

PARAMETER		TEST CONDITIONS		SN54ALS641A SN54ALS642A SN54ALS644A			SN74ALS641A SN74ALS642A SN74ALS644A			UNIT
				MIN	TYP‡	MAX	MIN	TYP‡	MAX	
V_{IK}		V_{CC} = 4.5 V,	I_I = −18 mA			−1.5			−1.5	V
I_{OH}		V_{CC} = 4.5 V,	V_{OH} = 5.5 V			0.1			0.1	mA
V_{OL}		V_{CC} = 4.5 V,	I_{OL} = 12 mA		0.25	0.4		0.25	0.4	V
		V_{CC} = 4.5 V,	I_{OL} = 24 mA					0.35	0.5	
		(I_{OL} = 48 mA for −1 versions)								
I_I	Control inputs	V_{CC} = 5.5 V,	V_I = 7 V			0.1			0.1	mA
	A or B ports	V_{CC} = 5.5 V,	V_I = 5.5 V			0.1			0.1	
I_{IH}	Control inputs	V_{CC} = 5.5 V,	V_I = 2.7 V			20			20	µA
	A or B ports§					20			20	
I_{IL}	Control inputs	V_{CC} = 5.5 V,	V_I = 0.4 V			−0.1			−0.1	mA
	A or B ports§					−0.1			−0.1	
I_{CC}	'ALS641A	V_{CC} = 5.5 V	Outputs high		25	40		25	37	mA
			Outputs low		33	50		33	47	
	'ALS642A		Outputs high		8	15		8	15	
			Outputs low		18	28		18	28	
	'ALS644A		Outputs high		16	32		16	29	
			Outputs low		25	44		25	40	

‡All typical values are at V_{CC} = 5 V, T_A = 25 °C.
§For I/O ports, the parameters I_{IH} and I_{IL} include the off-state output current.

ALS AND AS CIRCUITS

TEXAS INSTRUMENTS
POST OFFICE BOX 225012 • DALLAS, TEXAS 75265

TYPES SN54ALS640A THRU SN54ALS645A
SN74ALS640A THRU SN74ALS645A
OCTAL BUS TRANSCEIVERS

'ALS641A switching characteristics (see Note 1)

PARAMETER	FROM (INPUT)	TO (OUTPUT)	V_{CC} = 4.5 V to 5.5 V, C_L = 50 pF, R_L = 680 Ω, T_A = MIN to MAX				UNIT
			SN54ALS641A		SN74ALS641A		
			MIN	MAX	MIN	MAX	
t_{PLH}	A or B	B or A	5	30	5	25	ns
t_{PHL}			3	23	3	18	
t_{PLH}	$\overline{G}$	A or B	8	35	8	30	ns
t_{PHL}			8	35	8	30	
t_{PLH}	DIR	A or B	8	37	8	32	ns
t_{PHL}			8	37	8	32	

'ALS642A switching characteristics (see Note 1)

PARAMETER	FROM (INPUT)	TO (OUTPUT)	V_{CC} = 4.5 V to 5.5 V, C_L = 50 pF, R_L = 680 Ω, T_A = MIN to MAX				UNIT
			SN54ALS642A		SN74ALS642A		
			MIN	MAX	MIN	MAX	
t_{PLH}	A or B	B or A	10	35	10	30	ns
t_{PHL}			5	25	5	22	
t_{PLH}	$\overline{G}$ or DIR	A or B	10	35	10	30	ns
t_{PHL}			15	43	15	38	

'ALS644A switching characteristics (see Note 1)

PARAMETER	FROM (INPUT)	TO (OUTPUT)	V_{CC} = 4.5 V to 5.5 V, C_L = 50 pF, R_L = 680 Ω, T_A = MIN to MAX				UNIT
			SN54ALS644A		SN74ALS644A		
			MIN	MAX	MIN	MAX	
t_{PLH}	A	B	10	35	10	30	ns
t_{PHL}			5	25	5	22	
t_{PLH}	B	A	10	35	10	30	ns
t_{PHL}			5	23	5	21	
t_{PLH}	$\overline{G}$	A	8	35	8	30	ns
t_{PHL}			10	38	10	35	
t_{PLH}	$\overline{G}$	B	8	31	8	26	ns
t_{PHL}			15	40	15	35	
t_{PLH}	DIR	A	8	31	8	26	ns
t_{PHL}			10	40	10	35	
t_{PLH}	DIR	B	10	35	10	30	ns
t_{PHL}			15	40	15	35	

NOTE 1: For load circuit and voltage waveforms, see page 1-12.

TYPES SN54AS640 THRU SN54AS645
SN74AS640 THRU SN74AS645
OCTAL BUS TRANSCEIVERS

absolute maximum ratings over operating free-air temperature range (unless otherwise noted)

Supply voltage, V_{CC} .... 7 V
Input voltage: All inputs .... 7 V
               I/O ports .... 5.5 V
Operating free-air temperature range: SN54AS640, SN54AS643, SN54AS645 .... −55 °C to 125 °C
                        SN74AS640, SN74AS643, SN74AS645 .... 0 °C to 70 °C
Storage temperature range .... −65 °C to 150 °C

recommended operating conditions

		SN54AS640 SN54AS643 SN54AS645			SN74AS640 SN74AS643 SN74AS645			UNIT
		MIN	NOM	MAX	MIN	NOM	MAX	
V_{CC}	Supply voltage	4.5	5	5.5	4.5	5	5.5	V
V_{IH}	High-level input voltage	2			2			V
V_{IL}	Low-level input voltage			0.8			0.8	V
I_{OH}	High-level output current			−12			−15	mA
I_{OL}	Low-level output current			48			64	mA
T_A	Operating free-air temperature	−55		125	0		70	°C

electrical characteristics over recommended operating free-air temperature range (unless otherwise noted)

PARAMETER		TEST CONDITIONS		SN54AS′			SN74AS′			UNIT
				MIN	TYP†	MAX	MIN	TYP†	MAX	
V_{IK}		V_{CC} = 4.5 V,	I_I = −18 mA			−1.2			−1.2	V
V_{OH}		V_{CC} = 4.5 V to 5.5 V,	I_{OH} = −2 mA	V_{CC}−2			V_{CC}−2			V
		V_{CC} = 4.5 V,	I_{OH} = −3 mA	2.4	3.2		2.4	3.2		
		V_{CC} = 4.5 V,	I_{OH} = −12 mA	2.4						
		V_{CC} = 4.5 V,	I_{OH} = −15 mA				2.4			
V_{OL}		V_{CC} = 4.5 V,	I_{OL} = 48 mA		0.30	0.55				V
		V_{CC} = 4.5 V,	I_{OL} = 64 mA					0.35	0.55	
I_I	Control inputs	V_{CC} = 5.5 V,	V_I = 7 V			0.1			0.1	mA
	A or B ports	V_{CC} = 5.5 V,	V_I = 5.5 V			0.1			0.1	
I_{IH}	Control inputs	V_{CC} = 5.5 V,	V_I = 2.7 V			20			20	µA
	A or B ports‡					50			50	
I_{IL}	Control inputs	V_{CC} = 5.5 V,	V_I = 0.4 V			−0.5			−0.5	mA
	A or B ports‡					−0.75			−0.75	
I_O§		V_{CC} = 5.5 V,	V_O = 2.25 V	−30		−112	−30		−112	mA
I_{CC}	′AS640	V_{CC} = 5.5 V	Outputs high		37	58		37	58	mA
			Outputs low		78	123		78	123	
			Outputs disabled		51	80		51	80	
	′AS643		Outputs high		48	79		48	79	
			Outputs low		88	143		88	143	
			Outputs disabled		61	100		61	100	
	′AS645		Outputs high		62	97		62	97	
			Outputs low		95	149		95	149	
			Outputs disabled		79	123		79	123	

†All typical values are at V_{CC} = 5 V, T_A = 25 °C
‡For I/O ports, the parameters I_{IH} and I_{IL} include the off-state output current.
§The output conditions have been chosen to produce a current that closely approximates one half of the true short-circuit output current, I_{OS}.

Texas Instruments
POST OFFICE BOX 225012 • DALLAS, TEXAS 75265

TYPES SN54AS640 THRU SN54AS645 SN74AS640 THRU SN74AS645 OCTAL BUS TRANSCEIVERS

'AS640 switching characteristics (see Note 1)

PARAMETER	FROM (INPUT)	TO (OUTPUT)	V_{CC} = 4.5 V to 5.5 V, C_L = 50 pF, R1 = 500 Ω, R2 = 500 Ω, T_A = MIN to MAX				UNIT
			SN54AS640		SN74AS640		
			MIN	MAX	MIN	MAX	
t_{PLH}	A or B	B or A	2	8	2	7	ns
t_{PHL}			2	7	2	6	
t_{PZH}	$\overline{G}$	A or B	2	10	2	8	ns
t_{PZL}			2	12	2	10	
t_{PHZ}	$\overline{G}$	A or B	2	9	2	8	ns
t_{PLZ}			2	16	2	13	

'AS643 switching characteristics (see Note 1)

PARAMETER	FROM (INPUT)	TO (OUTPUT)	V_{CC} = 4.5 V to 5.5 V, C_L = 50 pF, R1 = 500 Ω, R2 = 500 Ω, T_A = MIN to MAX				UNIT
			SN54AS643		SN74AS643		
			MIN	MAX	MIN	MAX	
t_{PLH}	A	B	2	10	2	8	ns
t_{PHL}			2	7.5	2	7	
t_{PLH}	B	A	2	11.5	2	10	ns
t_{PHL}			2	10	2	9	
t_{PZH}	$\overline{G}$	A	2	13	2	11	ns
t_{PZL}			2	13	2	11	
t_{PHZ}	$\overline{G}$	A	2	8.5	2	7.5	ns
t_{PLZ}			2	12	2	10.5	
t_{PZH}	$\overline{G}$	B	2	11.5	2	10	ns
t_{PZL}			2	12	2	10	
t_{PHZ}	$\overline{G}$	B	2	8	2	7	ns
t_{PLZ}			2	12	2	10	

'AS645 switching characteristics (see Note 1)

PARAMETER	FROM (INPUT)	TO (OUTPUT)	V_{CC} = 4.5 V to 5.5 V, C_L = 50 pF, R1 = 500 Ω, R2 = 500 Ω, T_A = MIN to MAX				UNIT
			SN54AS645		SN74AS645		
			MIN	MAX	MIN	MAX	
t_{PLH}	A or B	B or A	2	11	2	9.5	ns
t_{PHL}			2	10.5	2	9	
t_{PZH}	$\overline{G}$	A or B	2	12	2	11	ns
t_{PZL}			2	12	2	10	
t_{PHZ}	$\overline{G}$	A or B	2	8	2	7	ns
t_{PLZ}			2	13	2	12	

NOTE 1: For load circuit and voltage waveforms, see page 1-12.

TEXAS INSTRUMENTS
POST OFFICE BOX 225012 • DALLAS, TEXAS 75265

TYPES SN54AS640 THRU SN54AS645
SN74AS640 THRU SN74AS645
OCTAL BUS TRANSCEIVERS

absolute maximum ratings over operating free-air temperature range (unless otherwise noted)

Supply voltage, V_{CC} .... 7 V
Input voltage: All inputs and I/O ports .... 7 V
Operating free-air temperature range: SN54AS641, SN54AS642, SN54AS644 .... −55 °C to 125 °C
                                    SN74AS641, SN74AS642, SN74AS644 .... 0 °C to 70 °C
Storage temperature range .... −65 °C to 150 °C

recommended operating conditions

		SN54AS641 SN54AS642 SN54AS644			SN74AS641 SN74AS642 SN74AS644			UNIT
		MIN	NOM	MAX	MIN	NOM	MAX	
V_{CC}	Supply voltage	4.5	5	5.5	4.5	5	5.5	V
V_{IH}	High-level input voltage	2			2			V
V_{IL}	Low-level input voltage			0.8			0.8	V
V_{OH}	High-level output voltage			5.5			5.5	V
I_{OL}	Low-level output current			48			64	mA
T_A	Operating free-air temperature	−55		125	0		70	°C

electrical characteristics over recommended operating free-air temperature range (unless otherwise noted)

PARAMETER		TEST CONDITIONS		SN54AS641 SN54AS642 SN54AS644			SN74AS641 SN74AS642 SN74AS644			UNIT
				MIN	TYP†	MAX	MIN	TYP†	MAX	
V_{IK}		V_{CC} = 4.5 V,	I_I = −18 mA			−1.2			−1.2	V
I_{OH}		V_{CC} = 4.5 V,	V_{OH} = 5.5 V			0.1			0.1	mA
V_{OL}		V_{CC} = 4.5 V,	I_{OL} = 48 mA		0.3	0.55				V
		V_{CC} = 4.5 V,	I_{OL} = 64 mA					0.35	0.55	
I_I	Control inputs	V_{CC} = 5.5 V,	V_I = 7 V			0.1			0.1	mA
	A or B ports	V_{CC} = 5.5 V,	V_I = 5.5 V			0.1			0.1	
I_{IH}	Control inputs	V_{CC} = 5.5 V,	V_I = 2.7 V			20			20	µA
	A or B ports‡					50			50	
I_{IL}	Control inputs	V_{CC} = 5.5 V,	V_I = 0.4 V			−0.5			−0.5	mA
	A or B ports‡					−0.75			−0.75	
I_{CC}	'AS641	V_{CC} = 5.5 V	Outputs high		50	82		50	82	mA
			Outputs low		84	136		84	136	
	'AS642		Outputs high		25	42		25	42	
			Outputs low		64	104		64	104	
	'AS644		Outputs high		38	62		38	62	
			Outputs low		76	124		76	124	

†All typical values are at V_{CC} = 5 V, T_A = 25 °C.
‡For I/O ports, the parameters I_{IH} and I_{IL} include the off-state output current.

ALS AND AS CIRCUITS

TEXAS INSTRUMENTS
POST OFFICE BOX 225012 • DALLAS, TEXAS 75265

TYPES SN54AS640 THRU SN54AS645, SN74AS640 THRU SN74AS645 OCTAL BUS TRANSCEIVERS

'AS641 switching characteristics (see Note 1)

PARAMETER	FROM (INPUT)	TO (OUTPUT)	V_{CC} = 4.5 V to 5.5 V, C_L = 50 pF, R_L = 680 Ω, T_A = MIN to MAX				UNIT
			SN54AS641		SN74AS641		
			MIN	MAX	MIN	MAX	
t_{PLH}	A or B	B or A	5	23	5	21	ns
t_{PHL}			1	8.5	1	7.5	
t_{PLH}	$\overline{G}$	A or B	5	24	5	21	ns
t_{PHL}			1	10	1	9	
t_{PLH}	DIR	A or B	5	26	5	22	ns
t_{PHL}			1	11	1	10	

'AS642 switching characteristics (see Note 1)

PARAMETER	FROM (INPUT)	TO (OUTPUT)	V_{CC} = 4.5 V to 5.5 V, C_L = 50 pF, R_L = 680 Ω, T_A = MIN to MAX				UNIT
			SN54AS642		SN74AS642		
			MIN	MAX	MIN	MAX	
t_{PLH}	A or B	B or A	5	28.5	5	24	ns
t_{PHL}			1	8.5	1	7.5	
t_{PLH}	$\overline{G}$	A or B	5	25	5	22	ns
t_{PHL}			1	11	1	10	
t_{PLH}	DIR	A or B	5	26.5	5	23.5	ns
t_{PHL}			1	12.5	1	11.5	

'AS644 switching characteristics (see Note 1)

PARAMETER	FROM (INPUT)	TO (OUTPUT)	V_{CC} = 4.5 V to 5.5 V, C_L = 50 pF, R_L = 680 Ω, T_A = MIN to MAX				UNIT
			SN54AS644		SN74AS644		
			MIN	MAX	MIN	MAX	
t_{PLH}	A	B	5	28.5	5	24	ns
t_{PHL}			1	8.5	1	7.5	
t_{PLH}	B	A	5	23	5	21	ns
t_{PHL}			1	8.5	1	7.5	
t_{PLH}	$\overline{G}$	A or B	5	24	5	21	ns
t_{PHL}			1	10	1	9	
t_{PLH}	DIR	A or B	5	26	5	22	ns
t_{PHL}			1	11	1	10	

NOTE 1: For load circuit and voltage waveforms, see page 1-12.

PRODUCT PREVIEW

This page contains information on a product under development. Texas Instruments reserves the right to change or discontinue this product without notice.

TYPES SN54ALS646 THRU SN54ALS649, SN54AS646, SN54AS648 SN74ALS646 THRU SN74ALS649, SN74AS646, SN74AS648 OCTAL BUS TRANSCEIVERS AND REGISTERS

D2661, DECEMBER 1982—REVISED DECEMBER 1983

- Independent Registers for A and B Buses
- Multiplexed Real-Time and Stored Data
- Choice of True or Inverting Data Paths
- Choice of 3-State or Open-Collector Outputs
- Included Among the Package Options Are Compact 24-pin 300-mil Wide DIPs and Both 28-pin Plastic and Ceramic Chip Carriers
- Dependable Texas Instruments Quality and Reliability

DEVICE	OUTPUT	LOGIC
'ALS646, 'AS646	3-State	True
'ALS647	Open-Collector	True
'ALS648, 'AS648	3-State	Inverting
'ALS649	Open-Collector	Inverting

SN54ALS', SN54AS' . . . JT PACKAGE
SN74ALS', SN74AS' . . . NT PACKAGE
(TOP VIEW)

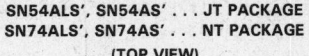

SN54ALS', SN54AS' . . . FH PACKAGE
SN74ALS', SN74AS' . . . FN PACKAGE
(TOP VIEW)

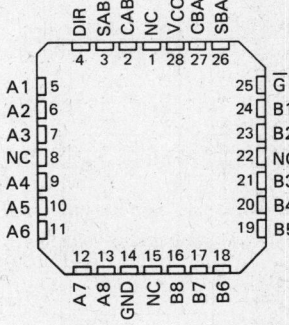

description

These devices consist of bus transceiver circuits with 3-state or open-collector outputs, D-type flip-flops, and control circuitry arranged for multiplexed transmission of data directly from the input bus or from the internal registers. Data on the A or B bus will be clocked into the registers on the low-to-high transition of the appropriate clock pin (CAB or CBA). The following examples demonstrate the four fundamental bus-management functions that can be performed with the octal bus transceivers and registers.

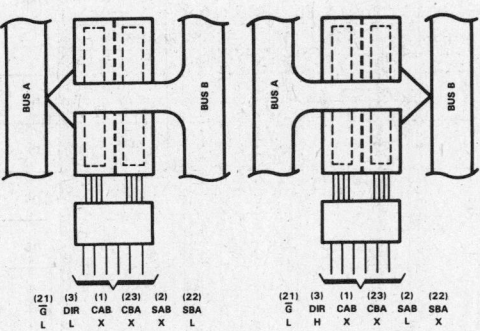

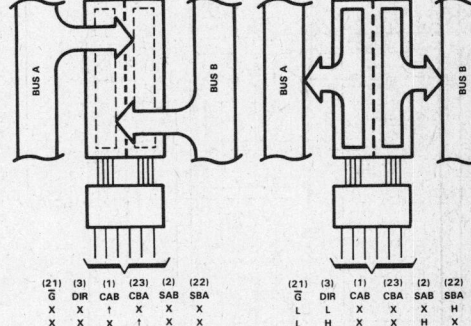

ALS AND AS CIRCUITS

Copyright © 1982 by Texas Instruments Incorporated

TEXAS INSTRUMENTS
POST OFFICE BOX 225012 • DALLAS, TEXAS 75265

TYPES SN54ALS646 THRU SN54ALS649, SN54AS646, SN54AS648, SN74ALS646 THRU SN74ALS649, SN74AS646, SN74AS649 OCTAL BUS TRANSCEIVERS AND REGISTERS

Enable ($\overline{G}$) and direction (DIR) pins are provided to control the transceiver functions. In the transceiver mode, data present at the high-impedance port may be stored in either register or in both. The select controls (SAB and SBA) can multiplex stored and real-time (transparent mode) data. The direction control determines which bus will receive data when enable $\overline{G}$ is active (low). In the isolation mode (control $\overline{G}$ high), A data may be stored in one register and/or B data may be stored in the other register.

When an output function is disabled, the input function is still enabled and may be used to store and transmit data. Only one of the two buses, A or B, may be driven at a time.

The −1 versions of the SN74ALS' parts are identical to the standard versions except that the recommended maximum I_{OL} is increased to 48 milliamperes. There are no −1 versions of the SN54ALS' parts.

The SN54' family is characterized for operation over the full military temperature range of −55°C to 125°C. The SN74' family is characterized for operation from 0° to 70°C.

FUNCTION TABLE

INPUTS						DATA I/O*		OPERATION OR FUNCTION	
								'ALS646, 'ALS647 'AS646	'ALS648, 'ALS649 'AS648
$\overline{G}$	DIR	CAB	CBA	SAB	SBA	A1 THRU A8	B1 THRU B8		
X	X	↑	X	X	X	Input	Not specified	Store A, B unspecified	Store A, B unspecified
X	X	X	↑	X	X	Not specified	Input	Store B, A unspecified	Store B, A unspecified
H	X	↑	↑	X	X	Input	Input	Store A and B Data	Store A and B Data
H	X	H or L	H or L	X	X			Isolation, hold storage	Isolation, hold storage
L	L	X	X	X	L	Output	Input	Real-Time B Data to A Bus	Real-Time $\overline{B}$ Data to A Bus
L	L	X	X	X	H			Stored B Data to A Bus	Stored $\overline{B}$ Data to A Bus
L	H	X	X	L	X	Input	Output	Real-Time A Data to B Bus	Real-Time $\overline{A}$ Data to B Bus
L	H	X	X	H	X			Stored A Data to B Bus	Stored $\overline{A}$ Data to B Bus

*The data output functions may be enabled or disabled by various signals at the $\overline{G}$ and DIR inputs. Data input functions are always enabled, i.e., data at the bus pins will be stored on every low-to-high transition on the clock inputs.

functional block diagrams (positive logic)

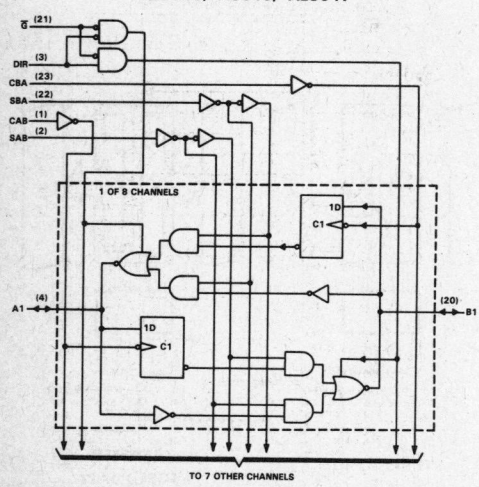

'ALS646, 'AS646, 'ALS647

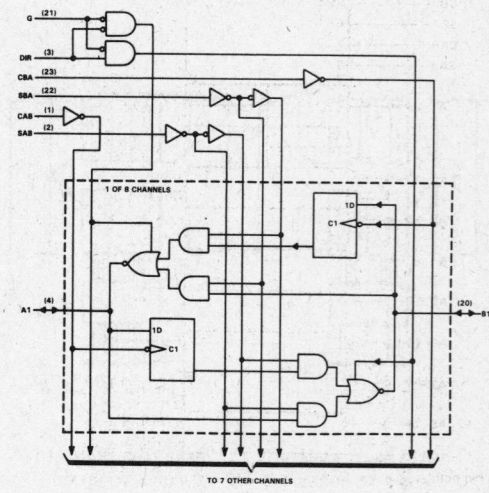

'ALS648, 'AS648, 'ALS649

Pin numbers shown are for JT and NT packages.

TYPES SN54ALS646 THRU SN54ALS649, SN54AS646, SN54AS648
SN74ALS646 THRU SN74ALS649, SN74AS646, SN74AS648
OCTAL BUS TRANSCEIVERS AND REGISTERS

logic symbols

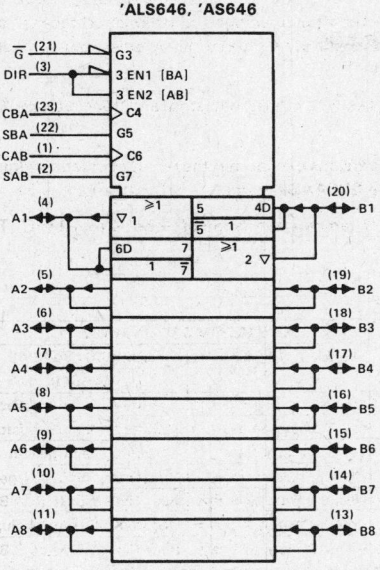

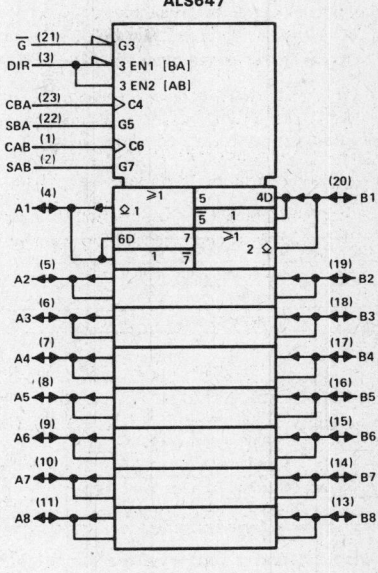

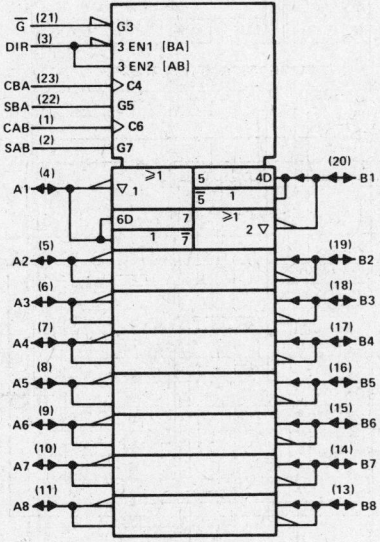

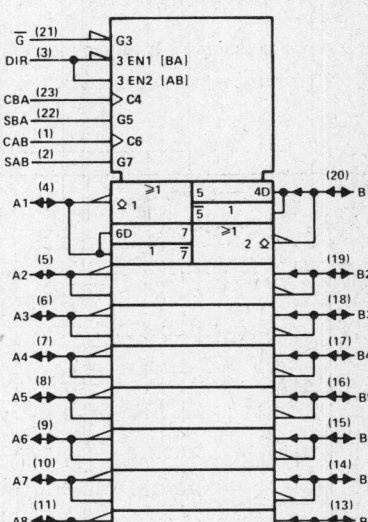

Pin numbers shown are for J and N packages.

TEXAS INSTRUMENTS
POST OFFICE BOX 225012 • DALLAS, TEXAS 75265

TYPES SN54ALS646, SN54ALS648, SN74ALS646, SN74ALS648
OCTAL BUS TRANSCEIVERS AND REGISTERS WITH 3-STATE OUTPUTS

absolute maximum ratings over operating free-air temperature range (unless otherwise noted)

Supply voltage, V_{CC} .... 7 V
Input voltage: Control inputs .... 7 V
          I/O ports .... 5.5 V
Operating free-air temperature range: SN54ALS646, SN54ALS648 .... −55 °C to 125 °C
                                     SN74ALS646, SN74ALS648 .... 0 °C to 70 °C
Storage temperature range .... −65 °C to 150 °C

recommended operating conditions

		SN54ALS646 SN54ALS648			SN74ALS646 SN74ALS648			UNIT
		MIN	NOM	MAX	MIN	NOM	MAX	
V_{CC}	Supply voltage	4.5	5	5.5	4.5	5	5.5	V
V_{IH}	High-level input voltage	2			2			V
V_{IL}	Low-level input voltage			0.8			0.8	V
I_{OH}	High-level output current			−12			−15	mA
I_{OL}	Low-level output current			12			24	mA
							48†	
f_{clock}	Clock frequency							MHz
t_w	Pulse duration, clocks high or low							ns
t_{su}	Setup time, A before CAB↑ or B before CBA↑							ns
t_h	Hold time, A after CAB↑ or B after CBA↑							ns
T_A	Operating free-air temperature	−55		125	0		70	°C

†The extended condition applies if V_{CC} is maintained between 4.75 V and 5.25 V.
 The 48-mA limit applies for the SN74ALS646-1 and SN74ALS648-1 only.

electrical characteristics over recommended operating free-air temperature range (unless otherwise noted)

PARAMETER		TEST CONDITIONS		SN54ALS646 SN54ALS648			SN74ALS646 SN74ALS648			UNIT
				MIN	TYP‡	MAX	MIN	TYP‡	MAX	
V_{IK}		V_{CC} = 4.5 V,	I_I = −18 mA			−1.5			−1.5	V
V_{OH}		V_{CC} = 4.5 V to 5.5 V,	I_{OH} = −0.4 mA	V_{CC}−2			V_{CC}−2			V
		V_{CC} = 4.5 V,	I_{OH} = −3 mA	2.4	3.2		2.4	3.2		
		V_{CC} = 4.5 V,	I_{OH} = −12 mA	2						
		V_{CC} = 4.5 V,	I_{OH} = −15 mA				2			
V_{OL}		V_{CC} = 4.5 V,	I_{OL} = 12 mA		0.25	0.4		0.25	0.4	V
		V_{CC} = 4.5 V,	I_{OL} = 24 mA					0.35	0.5	
		(I_{OL} = 48 mA for −1 versions)								
I_I	Control inputs	V_{CC} = 5.5 V,	V_I = 7 V			0.1			0.1	mA
	A or B ports	V_{CC} = 5.5 V,	V_I = 5.5 V			0.1			0.1	
I_{IH}	Control inputs	V_{CC} = 5.5 V,	V_I = 2.7 V			20			20	µA
	A or B ports§					20			20	
I_{IL}	Control inputs	V_{CC} = 5.5 V,	V_I = 0.4 V			−0.1			−0.1	mA
	A or B ports§					−0.2			−0.2	
I_O¶		V_{CC} = 5.5 V,	V_O = 2.25 V	−30		−112	−30		−112	mA
I_{CC}	'ALS646	V_{CC} = 5.5 V	Outputs high		60			60		mA
			Outputs low		68			68		
			Outputs disabled		68			68		
	'ALS648		Outputs high		52			52		
			Outputs low		57			57		
			Outputs disabled		58			58		

‡All typical values are at V_{CC} = 5 V, T_A = 25 °C.
§For I/O ports, the parameters I_{IH} and I_{IL} include the off-state output current.
¶The output conditions have been chosen to produce a current that closely approximates one half of the true short-circuit output current, I_{OS}.
Additional information on these products can be obtained from the factory as it becomes available.

PRODUCT PREVIEW
This page contains information on a product under development. Texas Instruments reserves the right to change or discontinue this product without notice.

Texas Instruments
POST OFFICE BOX 225012 • DALLAS, TEXAS 75265

TYPES SN54ALS646, SN54ALS648, SN74ALS646, SN74ALS648
OCTAL BUS TRANSCEIVERS AND REGISTERS WITH 3-STATE OUTPUTS

'ALS646 switching characteristics (see Note 1)

PARAMETER	FROM (INPUT)	TO (OUTPUT)	V_{CC} = 4.5 V to 5.5 V, C_L = 50 pF, R1 = 500 Ω, R2 = 500 Ω, T_A = MIN to MAX					UNIT
			SN54ALS646			SN74ALS646		
			MIN	TYP†	MAX	MIN	TYP†	MAX	
f_{max}									MHz
t_{PLH}	CBA or CAB	A or B		11			11		ns
t_{PHL}				13			13		
t_{PLH}	A or B	B or A		8			8		ns
t_{PHL}				8			8		
t_{PLH}	SBA or SAB‡ (with A or B high)	A or B		16			16		ns
t_{PHL}				16			16		
t_{PLH}	SBA or SAB‡ (with A or B low)	A or B		15			15		ns
t_{PHL}				12			12		
t_{PZH}	$\overline{G}$	A or B		17			17		ns
t_{PZL}				20			20		
t_{PHZ}	$\overline{G}$	A or B		10			10		ns
t_{PLZ}				12			12		
t_{PZH}	DIR	A or B		17			17		ns
t_{PZL}				20			20		
t_{PHZ}	DIR	A or B		10			10		ns
t_{PLZ}				12			12		

'ALS648 switching characteristics (see Note 1)

PARAMETER	FROM (INPUT)	TO (OUTPUT)	V_{CC} = 4.5 V to 5.5 V, C_L = 50 pF, R1 = 500 Ω, R2 = 500 Ω, T_A = MIN to MAX					UNIT
			SN54ALS648			SN74ALS648		
			MIN	TYP†	MAX	MIN	TYP†	MAX	
f_{max}									MHz
t_{PLH}	CBA or CAB	A or B		11			11		ns
t_{PHL}				13			13		
t_{PLH}	A or B	B or A		10			10		ns
t_{PHL}				12			12		
t_{PLH}	SBA or SAB‡ (with A or B high)	A or B		16			16		ns
t_{PHL}				16			16		
t_{PLH}	SBA or SAB‡ (with A or B low)	A or B		15			15		ns
t_{PHL}				15			15		
t_{PZH}	$\overline{G}$	A or B		17			17		ns
t_{PZL}				20			20		
t_{PHZ}	$\overline{G}$	A or B		10			10		ns
t_{PLZ}				12			12		
t_{PZH}	DIR	A or B		17			17		ns
t_{PZL}				20			20		
t_{PHZ}	DIR	A or B		10			10		ns
t_{PLZ}				12			12		

†All typical values are at V_{CC} = 5 V, T_A = 25°C.
‡These parameters are measured with the internal output state of the storage register opposite to that of the bus input.
Additional information on these products can be obtained from the factory as it becomes available.

PRODUCT PREVIEW
This page contains information on a product under development. Texas Instruments reserves the right to change or discontinue this product without notice.

TEXAS INSTRUMENTS
POST OFFICE BOX 225012 • DALLAS, TEXAS 75265

ALS AND AS CIRCUITS

TYPES SN54ALS647, SN54ALS649, SN74ALS647, SN74ALS649
OCTAL BUS TRANSCEIVERS AND REGISTERS WITH OPEN-COLLECTOR OUTPUTS

absolute maximum ratings over operating free-air temperature range (unless otherwise noted)

Supply voltage, V_{CC} .... 7 V
Input voltage .... 7 V
Operating free-air temperature range: SN54ALS647, SN54ALS649 .... −55°C to 125°C
                                      SN74ALS647, SN74ALS649 .... 0°C to 70°C
Storage temperature range .... −65°C to 150°C

recommended operating conditions

		SN54ALS647 SN54ALS649			SN74ALS647 SN74ALS649			UNIT
		MIN	NOM	MAX	MIN	NOM	MAX	
V_{CC}	Supply voltage	4.5	5	5.5	4.5	5	5.5	V
V_{IH}	High-level input voltage	2			2			V
V_{IL}	Low-level input voltage			0.8			0.8	V
V_{OH}	High-level output voltage			5.5			5.5	V
I_{OL}	Low-level output current			12			24	mA
							48†	
f_{clock}	Clock frequency							MHz
t_w	Pulse duration, clocks high or low							ns
t_{su}	Setup time, A before CAB↑ or B before CBA↑							ns
t_h	Hold time, A after CAB↑ or B after CBA↑							ns
T_A	Operating free-air temperature	−55		125	0		70	°C

†The extended condition applies if V_{CC} is maintained between 4.75 and 5.25 V.
The 48-mA limit applies for the SN74ALS647−1 and SN74ALS649−1 only.

electrical characteristics over recommended operating free-air temperature range (unless otherwise noted)

PARAMETER		TEST CONDITIONS		SN54ALS647 SN54ALS649			SN74ALS647 SN74ALS649			UNIT
				MIN	TYP‡	MAX	MIN	TYP‡	MAX	
V_{IK}		V_{CC} = 4.5 V,	I_I = −18 mA			−1.5			−1.5	V
I_{OH}		V_{CC} = 4.5 V,	V_{OH} = 5.5 V			0.1			0.1	mA
V_{OL}		V_{CC} = 4.5 V,	I_{OL} = 12 mA		0.25	0.4				V
		V_{CC} = 4.5 V,	I_{OL} = 24 mA					0.35	0.5	
		(I_{OL} = 48 mA for −1 versions)								
I_I	A or B ports	V_{CC} = 5.5 V,	V_I = 5.5 V			0.1			0.1	mA
	Control inputs	V_{CC} = 5.5 V,	V_I = 7 V			0.1			0.1	
I_{IH}	A or B ports§	V_{CC} = 5.5 V,	V_I = 2.7 V			20			20	µA
	Control inputs					20			20	
I_{IL}	Control inputs	V_{CC} = 5.5 V,	V_I = 0.4 V			−0.1			−0.1	mA
	A or B ports§					−0.2			−0.2	
I_{CC}	'ALS647	V_{CC} = 5.5 V	Outputs high		52			52		mA
			Outputs low		62			62		
	'ALS649		Outputs high		50			50		
			Outputs low		60			60		

‡All typical values are at V_{CC} = 5 V, T_A = 25°C.
§For I/O ports, the parameters I_{IH} and I_{IL} include the off-state output current.
Additional information on these products can be obtained from the factory as it becomes available.

PRODUCT PREVIEW

This page contains information on a product under development. Texas Instruments reserves the right to change or discontinue this product without notice.

TEXAS INSTRUMENTS
POST OFFICE BOX 225012 • DALLAS, TEXAS 75265

TYPES SN54ALS647, SN54ALS649, SN74ALS647, SN74ALS649
OCTAL BUS TRANSCEIVERS AND REGISTERS WITH OPEN-COLLECTOR OUTPUTS

'ALS647 switching characteristics (see Note 1)

PARAMETER	FROM (INPUT)	TO (OUTPUT)	$V_{CC} = 4.5$ V to 5.5 V, $C_L = 50$ pF, $R_L = 680\ \Omega$, T_A = MIN to MAX						UNIT
			SN54ALS647			SN74ALS647			
			MIN	TYP†	MAX	MIN	TYP†	MAX	
f_{max}									MHz
t_{PLH}	CBA or CAB	A or B		24			24		ns
t_{PHL}				15			15		
t_{PLH}	A or B	B or A		24			24		ns
t_{PHL}				12			12		
t_{PLH}	SBA or SAB‡ (with A or B high)	A or B		26			26		ns
t_{PHL}				15			15		
t_{PLH}	SBA or SAB‡ (with A or B low)	A or B		26			26		ns
t_{PHL}				15			15		
t_{PLH}	$\overline{G}$	A or B		24			24		ns
t_{PHL}				17			17		
t_{PLH}	DIR	A or B		24			24		ns
t_{PHL}				17			17		

'ALS649 switching characteristics (see Note 1)

PARAMETER	FROM (INPUT)	TO (OUTPUT)	$V_{CC} = 4.5$ V to 5.5 V, $C_L = 50$ pF, $R_L = 680\ \Omega$, T_A = MIN to MAX						UNIT
			SN54ALS649			SN74ALS649			
			MIN	TYP†	MAX	MIN	TYP†	MAX	
f_{max}									MHz
t_{PLH}	CBA or CAB	A or B		24			24		ns
t_{PHL}				15			15		
t_{PLH}	A or B	B or A		24			24		ns
t_{PHL}				10			10		
t_{PLH}	SBA or SAB‡ (with A or B high)	A or B		26			26		ns
t_{PHL}				15			15		
t_{PLH}	SBA or SAB‡ (with A or B low)	A or B		26			26		ns
t_{PHL}				15			15		
t_{PLH}	$\overline{G}$	A or B		24			24		ns
t_{PHL}				17			17		
t_{PLH}	DIR	A or B		24			24		ns
t_{PHL}				17			17		

†All typical values are at $V_{CC} = 5$ V, $T_A = 25°C$.
‡These parameters are measured with the internal output state of the storage register opposite to that of the bus input.
NOTE 1: For load circuit and voltage waveforms, see page 1-12.
Additional information on these products can be obtained from the factory as it becomes available.

ALS AND AS CIRCUITS

TYPES SN54AS646, SN54AS648, SN74AS646, SN74AS648
OCTAL BUS TRANSCEIVERS AND REGISTERS WITH 3-STATE OUTPUTS

absolute maximum ratings over operating free-air temperature range (unless otherwise noted)

Supply voltage, V_{CC} .... 7 V
Input voltage: Control inputs .... 7 V
               I/O ports .... 5.5 V
Operating free-air temperature range: SN54AS646, SN54AS648 .... −55 °C to 125 °C
                                SN74AS646, SN74AS648 .... 0 °C to 70 °C
Storage temperature range .... −65 °C to 150 °C

recommended operating conditions

			SN54AS646 SN54AS648			SN74AS646 SN74AS648			UNIT
			MIN	NOM	MAX	MIN	NOM	MAX	
V_{CC}	Supply voltage		4.5	5	5.5	4.5	5	5.5	V
V_{IH}	High-level input voltage		2			2			V
V_{IL}	Low-level input voltage				0.8			0.8	V
I_{OH}	High-level output current				−12			−15	mA
I_{OL}	Low-level output current				48			64	mA
f_{clock}	Clock frequency		0		75	0		90	MHz
t_w	Pulse duration	Clock high	6			5			ns
		Clock low	7			6			
t_{su}	Setup time, A before CAB↑ or B before CBA↑		7			6			ns
t_h	Hold time, A after CAB↑ or B after CBA↑		0			0			ns
T_A	Operating free-air temperature		−55		125	0		70	°C

electrical characteristics over recommended operating free-air temperature range (unless otherwise noted)

PARAMETER		TEST CONDITIONS		SN54AS646 SN54AS648			SN74AS646 SN74AS648			UNIT
				MIN	TYP†	MAX	MIN	TYP†	MAX	
V_{IK}		V_{CC} = 4.5 V,	I_I = −18 mA			−1.2			−1.2	V
V_{OH}		V_{CC} = 4.5 V to 5.5 V,	I_{OH} = −2 mA	V_{CC}−2			V_{CC}−2			V
		V_{CC} = 4.5 V,	I_{OH} = −3 mA	2.4	3.2		2.4	3.2		
		V_{CC} = 4.5 V,	I_{OH} = −12 mA	2.4						
		V_{CC} = 4.5 V,	I_{OH} = −15 mA				2.4			
V_{OL}		V_{CC} = 4.5 V,	I_{OL} = 48 mA		0.35	0.55				V
		V_{CC} = 4.5 V,	I_{OL} = 64 mA					0.35	0.55	
I_I	Control inputs	V_{CC} = 5.5 V,	V_I = 7 V			0.1			0.1	mA
	A or B ports	V_{CC} = 5.5 V,	V_I = 5.5 V			0.1			0.1	
I_{IH}	Control inputs	V_{CC} = 5.5 V,	V_I = 2.7 V			20			20	µA
	A or B ports‡					50			50	
I_{IL}	Control inputs	V_{CC} = 5.5 V,	V_I = 0.4 V			−0.5			−0.5	mA
	A or B ports‡					−0.5			−0.5	
I_O§		V_{CC} = 5.5 V,	V_O = 2.25 V	−30		−112	−30		−112	mA
I_{CC}	'AS646	V_{CC} = 5.5 V	Outputs high		120	195		120	195	mA
			Outputs low		130	211		130	211	
			Outputs disabled		130	211		130	211	
	'AS648		Outputs high		110	185		110	185	
			Outputs low		120	195		120	195	
			Outputs disabled		120	195		120	195	

†All typical values are at V_{CC} = 5 V, T_A = 25 °C
‡For I/O ports, the parameters I_{IH} and I_{IL} include the off-state output current.
§The output conditions have been chosen to produce a current that closely approximates one half of the true short-circuit output current, I_{OS}.

ALS AND AS CIRCUITS

TEXAS INSTRUMENTS
POST OFFICE BOX 225012 • DALLAS, TEXAS 75265

TYPES SN54AS646, SN54AS648, SN74AS646, SN74AS648
OCTAL BUS TRANSCEIVERS AND REGISTERS WITH 3-STATE OUTPUTS

'AS646 switching characteristics (see Note 1)

PARAMETER	FROM (INPUT)	TO (OUTPUT)	V_{CC} = 4.5 V to 5.5 V, C_L = 50 pF, $R1$ = 500 Ω, $R2$ = 500 Ω, T_A = MIN to MAX				UNIT
			SN54AS646		SN74AS646		
			MIN	MAX	MIN	MAX	
f_{max}			75		90		MHz
t_{PLH}	CBA or CAB	A or B	2	9.5	2	8.5	ns
t_{PHL}			2	10	2	9	
t_{PLH}	A or B	B or A	2	11	2	9	ns
t_{PHL}			1	8	1	7	
t_{PLH}	SBA or SAB†	A or B	2	12	2	11	ns
t_{PHL}	(with A or B high)		2	10	2	9	
t_{PZH}	$\overline{G}$	A or B	2	10	2	9	ns
t_{PZL}			3	15	3	14	
t_{PHZ}	$\overline{G}$	A or B	2	11	2	9	ns
t_{PLZ}			2	11	2	9	
t_{PZH}	DIR	A or B	3	19	3	16	ns
t_{PZL}			3	21	3	18	
t_{PHZ}	DIR	A or B	2	12	2	10	ns
t_{PLZ}			2	12	2	10	

'AS648 switching characteristics (see Note 1)

PARAMETER	FROM (INPUT)	TO (OUTPUT)	V_{CC} = 4.5 V to 5.5 V, C_L = 50 pF, $R1$ = 500 Ω, $R2$ = 500 Ω, T_A = MIN to MAX				UNIT
			SN54AS648		SN74AS648		
			MIN	MAX	MIN	MAX	
f_{max}			75		90		MHz
t_{PLH}	CBA or CAB	A or B	2	9.5	2	8.5	ns
t_{PHL}			2	10	2	9	
t_{PLH}	A or B	B or A	2	9	2	8	ns
t_{PHL}			1	8	1	7	
t_{PLH}	SBA or SAB†	A or B	2	12	2	11	ns
t_{PHL}	(with A or B high)		2	10	2	9	
t_{PZH}	$\overline{G}$	A or B	2	10	2	9	ns
t_{PZL}			3	18	3	15	
t_{PHZ}	$\overline{G}$	A or B	2	11	2	9	ns
t_{PLZ}			2	11	2	9	
t_{PZH}	DIR	A or B	3	19	3	16	ns
t_{PZL}			3	21	3	18	
t_{PHZ}	DIR	A or B	2	12	2	10	ns
t_{PLZ}			2	12	2	10	

†These parameters are measured with the internal output state of the storage register opposite to that of the bus input.

TEXAS INSTRUMENTS
POST OFFICE BOX 225012 • DALLAS, TEXAS 75265

2
ALS AND AS CIRCUITS

TYPES SN54ALS651 THRU SN54ALS654, SN54AS651, SN54AS652
SN74ALS651 THRU SN74ALS654, SN74AS651, SN74AS652
OCTAL BUS TRANSCEIVERS AND REGISTERS

D2661, DECEMBER 1983

- Bus Transceivers/Registers
- Independent Registers and Enables for A and B Buses
- Multiplexed Real-Time and Stored Data
- Choice of True and Inverting Data Paths
- Choice of 3-State or Open-Collector Outputs to A Bus
- Included Among the Package Options Are Compact 24-Pin 300-mil-Wide DIPs and Both 28-Pin Plastic and Ceramic Chip Carriers
- Dependable Texas Instruments Quality and Reliability

SN54ALS', SN54AS' . . . JT PACKAGE
SN74ALS', SN74AS' . . . NT PACKAGE
(TOP VIEW)

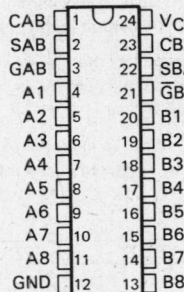

DEVICE	A OUTPUT	B OUTPUT	LOGIC
'ALS651, 'AS651	3-State	3-State	Inverting
'ALS652, 'AS652	3-State	3-State	True
'ALS653	Open-Collector	3-State	Inverting
'ALS654	Open-Collector	3-State	True

SN54ALS', SN54AS' . . . FC PACKAGE
SN74ALS', SN74AS' . . . FN PACKAGE
(TOP VIEW)

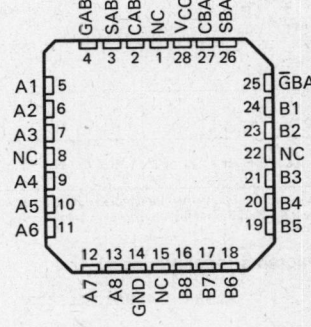

NC — No internal connection

description

These devices consist of bus transceiver circuits, D-type flip-flops, and control circuitry arranged for multiplexed transmission of data directly from the data bus or from the internal storage registers. Enable GAB and GBA are provided to control the transceiver functions. SAB and SBA control pins are provided to select whether real-time or stored data is transferred. A low input level selects real-time data, and a high selects stored data. The following examples demonstrate the four fundamental bus-management functions that can be performed with the octal bus transceivers and registers.

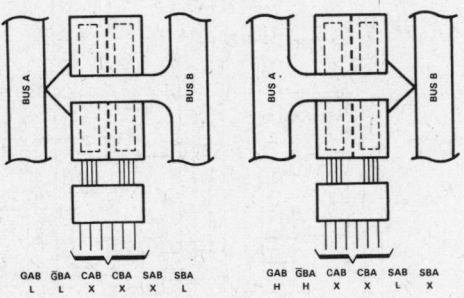

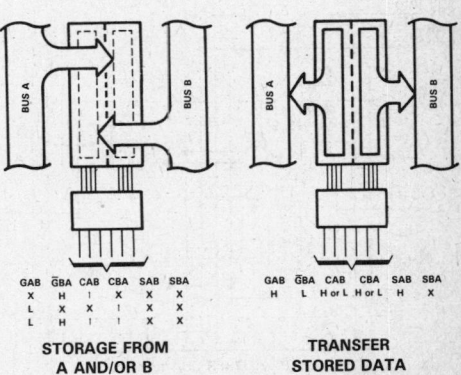

ALS AND AS CIRCUITS

Copyright © 1983 by Texas Instruments Incorporated

TEXAS INSTRUMENTS
POST OFFICE BOX 225012 • DALLAS, TEXAS 75265

TYPES SN54ALS651 THRU SN54ALS654, SN54AS651, SN54AS652 SN74ALS651 THRU SN74ALS654, SN74AS651, SN74AS652 OCTAL BUS TRANSCEIVERS AND REGISTERS

Data on the A or B data bus, or both, can be stored in the internal D flip-flops by low-to-high transitions at the appropriate clock pins (CAB or CBA) regardless of the select or enable control pins. When SAB and SBA are in the real-time transfer mode, it is also possible to store data without using the internal D-type flip-flops by simultaneously enabling GAB and $\overline{GBA}$. In this configuration each output reinforces its input. Thus, when all other data sources to the two sets of bus lines are at high impedance, each set of bus lines will remain at its last state.

The −1 versions of the SN74ALS651 through SN74ALS654 are identical to the standard versions except that the recommended maximum I_{OL} is increased to 48 milliamperes. There are no −1 versions of the SN54ALS651 through SN54ALS654.

The SN54' family is characterized for operation over the full military temperature range of −55 °C to 125 °C. The SN74' family is characterized for operation from 0 °C to 70 °C.

FUNCTION TABLE

INPUTS						DATA I/O*		OPERATION OR FUNCTION	
GAB	$\overline{GBA}$	CAB	CBA	SAB	SBA	A1 THRU A8	B1 THRU B8	'ALS651, 'ALS653 'AS651	'ALS652, 'ALS654 'AS652
L	H	H or L	H or L	X	X	Input	Input	Isolation	Isolation
L	H	↑	↑	X	X	Input	Input	Store A and B Data	Store A and B Data
X	H	↑	H or L	X	X	Input	Not specified	Store A, Hold B	Store A, Hold B
H	H	↑	↑	X	X	Input	Output	Store A in both registers	Store A in both registers
L	X	H or L	↑	X	X	Not specified	Input	Hold A, Store B	Hold A, Store $\overline{B}$
L	L	↑	↑	X	X	Output	Input	Store B in both registers	Store B in both registers
L	L	X	X	X	L	Output	Input	Real-Time $\overline{B}$ Data to A Bus	Real-Time B Data to A Bus
L	L	X	H or L	X	H	Output	Input	Stored $\overline{B}$ Data to A Bus	Stored B Data to A Bus
H	H	X	X	L	X	Input	Output	Real-Time $\overline{A}$ Data to B Bus	Real-Time A Data to B Bus
H	H	H or L	X	H	X	Input	Output	Stored $\overline{A}$ Data to B Bus	Stored A Data to B Bus
H	L	H or L	H or L	H	H	Output	Output	Stored $\overline{A}$ Data to B Bus and Stored $\overline{B}$ Data to A Bus	Stored A Data to B Bus and Stored B Data to A Bus

*The data output functions may be enabled or disabled by various signals at the GAB and $\overline{GBA}$ inputs. Data input functions are always enabled, i.e., data at the bus pins will be stored on every low-to-high transition on the clock inputs.

logic diagrams (positive logic)

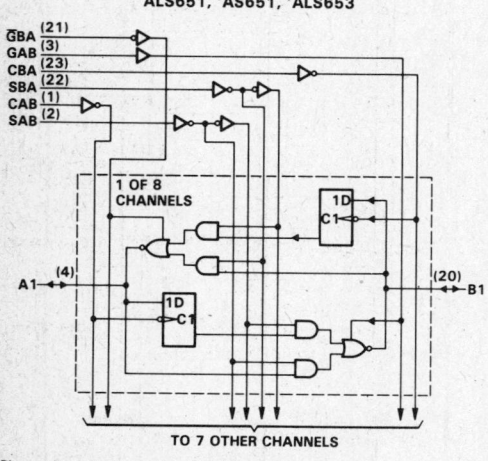

'ALS651, 'AS651, 'ALS653   'ALS652, 'AS652, 'ALS654

Pin numbers shown are for JT and NT packages.

TEXAS INSTRUMENTS
POST OFFICE BOX 225012 • DALLAS, TEXAS 75265

TYPES SN54ALS651 THRU SN54ALS654, SN54AS651, SN54AS652
SN74ALS651 THRU SN74ALS654, SN74AS651, SN74AS652
OCTAL BUS TRANSCEIVERS AND REGISTERS

logic symbols

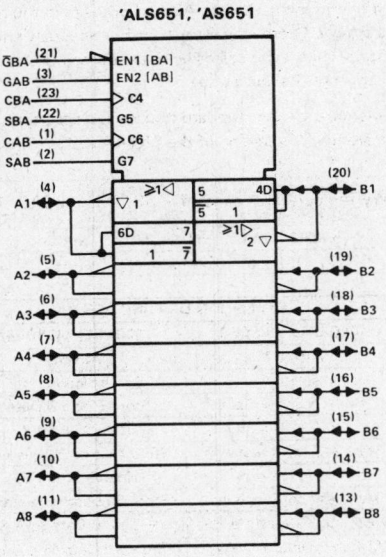

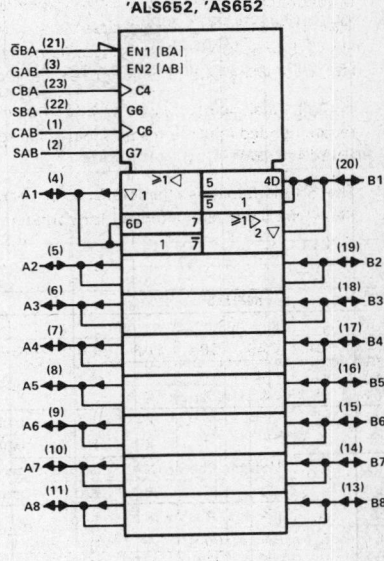

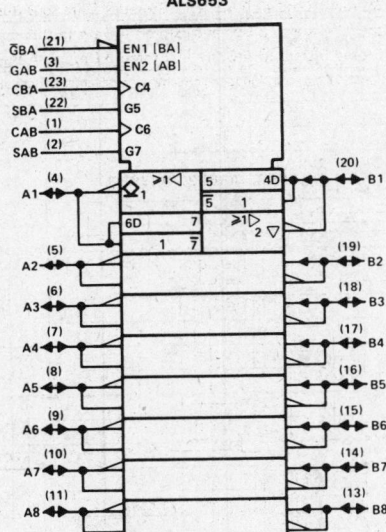

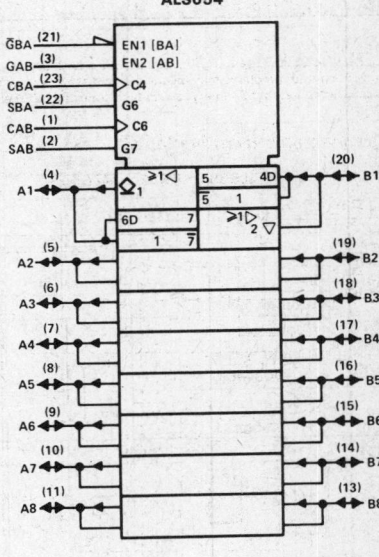

Pin numbers shown are for JT and NT packages.

ALS AND AS CIRCUITS

TYPES SN54ALS651, SN54ALS652, SN74ALS651, SN74ALS652
OCTAL BUS TRANSCEIVERS AND REGISTERS

absolute maximum ratings over operating free-air temperature range (unless otherwise noted)

Supply voltage, V_{CC} .... 7 V
Input voltage: Control inputs .... 7 V
              I/O ports .... 5.5 V
Operating free-air temperature range: SN54ALS651, SN54ALS652 .... −55 °C to 125 °C
                            SN74ALS651, SN74ALS652 .... 0 °C to 70 °C
Storage temperature range .... −65 °C to 150 °C

recommended operating conditions

		SN54ALS651 SN54ALS652			SN74ALS651 SN74ALS652			UNIT
		MIN	NOM	MAX	MIN	NOM	MAX	
V_{CC}	Supply voltage	4.5	5	5.5	4.5	5	5.5	V
V_{IH}	High-level input voltage	2			2			V
V_{IL}	Low-level input voltage			0.8			0.8	V
I_{OH}	High-level output current			−12			−15	mA
I_{OL}	Low-level output current			12			24	mA
							48†	
f_{clock}	Clock frequency							MHz
t_w	Pulse duration	CBA or CAB high						ns
		CBA or CAB low						
t_{su}	Setup time before CAB↑ or CBA↑	A or B						ns
t_h	Hold time after CAB↑ or CBA↑	A or B						ns
T_A	Operating free-air temperature		−55		125	0		70	°C

†The extended condition applies if V_{CC} is maintained between 4.75 V and 5.25 V.
The 48-mA limit applies for the SN74ALS651-1 and SN74ALS652-1 only.

electrical characteristics over recommended operating free-air temperature range (unless otherwise noted)

PARAMETER		TEST CONDITIONS		SN54ALS651 SN54ALS652			SN74ALS651 SN74ALS652			UNIT
				MIN	TYP‡	MAX	MIN	TYP‡	MAX	
V_{IK}		V_{CC} = 4.5 V,	I_I = −18 mA			−1.5			−1.5	V
V_{OH}		V_{CC} = 4.5 V to 5.5 V,	I_{OH} = −0.4 mA	V_{CC}−2			V_{CC}−2			V
		V_{CC} = 4.5 V,	I_{OH} = −3 mA	2.4	3.2		2.4	3.2		
		V_{CC} = 4.5 V,	I_{OH} = −12 mA	2						
		V_{CC} = 4.5 V,	I_{OH} = −15 mA				2			
V_{OL}		V_{CC} = 4.5 V,	I_{OL} = 12 mA		0.25	0.4		0.25	0.4	V
		V_{CC} = 4.5 V,	I_{OL} = 24 mA					0.35	0.5	
		(I_{OL} = 48 mA for −1 versions)								
I_I	Control inputs	V_{CC} = 5.5 V,	V_I = 7 V			0.1			0.1	mA
	A or B ports	V_{CC} = 5.5 V,	V_I = 5.5 V			0.1			0.1	
I_{IH}	Control inputs	V_{CC} = 5.5 V,	V_I = 2.7 V			20			20	μA
	A or B ports§					20			20	
I_{IL}	Control inputs	V_{CC} = 5.5 V,	V_I = 0.4 V			−0.1			−0.1	mA
	A or B ports§					−0.2			−0.2	
I_O¶		V_{CC} = 5.5 V,	V_O = 2.25 V	−30		−112	−30		−112	mA
I_{CC}	'ALS651	V_{CC} = 5.5 V	Outputs high		52			52		mA
			Outputs low		57			57		
			Outputs disabled		58			58		
	'ALS652		Outputs high		60			60		
			Outputs low		68			68		
			Outputs disabled		68			68		

‡All typical values are at V_{CC} = 5 V, T_A = 25 °C.
§For I/O ports, the parameters I_{IH} and I_{IL} include the off-state output current.
¶The output conditions have been chosen to produce a current that closely approximates one half of the true short-circuit output current, I_{OS}.

PRODUCT PREVIEW
This page contains information on a product under development. Texas Instruments reserves the right to change or discontinue this product without notice.

TEXAS INSTRUMENTS
POST OFFICE BOX 225012 • DALLAS, TEXAS 75265

TYPES SN54ALS651, SN54ALS652, SN74ALS651, SN74ALS652
OCTAL BUS TRANSCEIVERS AND REGISTERS

'ALS651 switching characteristics (see Note 1)

PARAMETER	FROM (INPUT)	TO (OUTPUT)	V_{CC} = 4.5 V to 5.5 V, C_L = 50 pF, $R1$ = 500 Ω, $R2$ = 500 Ω, T_A = MIN to MAX						UNIT
			SN54ALS651			SN74ALS651			
			MIN	TYP†	MAX	MIN	TYP†	MAX	
f_{max}									MHz
t_{PLH}	CBA or CAB	A or B		11			11		ns
t_{PHL}				13			13		
t_{PLH}	A or B	B or A		10			10		ns
t_{PHL}				12			12		
t_{PLH}	SBA or SAB‡ (with A or B high)	A or B		16			16		ns
t_{PHL}				16			16		
t_{PLH}	SBA or SAB‡ (with A or B low)	A or B		15			15		ns
t_{PHL}				15			15		
t_{PZH}	$\overline{GBA}$	A		17			17		ns
t_{PZL}				20			20		
t_{PHZ}	$\overline{GBA}$	A		10			10		ns
t_{PLZ}				12			12		
t_{PZH}	GAB	B		19			19		ns
t_{PZL}				22			22		
t_{PHZ}	GAB	B		12			12		ns
t_{PLZ}				14			14		

'ALS652 switching characteristics (see Note 1)

PARAMETER	FROM (INPUT)	TO (OUTPUT)	V_{CC} = 4.5 V to 5.5 V, C_L = 50 pF, $R1$ = 500 Ω, $R2$ = 500 Ω, T_A = MIN to MAX						UNIT
			SN54ALS652			SN74ALS652			
			MIN	TYP†	MAX	MIN	TYP†	MAX	
f_{max}									MHz
t_{PLH}	CBA or CAB	A or B		11			11		ns
t_{PHL}				13			13		
t_{PLH}	A or B	B or A		8			8		ns
t_{PHL}				8			8		
t_{PLH}	SBA or SAB‡ (with A or B high)	A or B		16			16		ns
t_{PHL}				16			16		
t_{PLH}	SBA or SAB‡ (with A or B low)	A or B		15			15		ns
t_{PHL}				12			12		
t_{PZH}	$\overline{GBA}$	A		17			17		ns
t_{PZL}				20			20		
t_{PHZ}	$\overline{GBA}$	A		10			10		ns
t_{PLZ}				12			12		
t_{PZH}	GAB	B		19			19		ns
t_{PZL}				22			22		
t_{PHZ}	GAB	B		12			12		ns
t_{PLZ}				14			14		

†All typical values are at V_{CC} = 5 V, T_A = 25°C.
‡These parameters are measured with the internal output state of the storage register opposite to that of the bus input.
NOTE 1: For load circuit and voltage waveforms, see page 1-12.

PRODUCT PREVIEW

This page contains information on a product under development. Texas Instruments reserves the right to change or discontinue this product without notice.

TEXAS INSTRUMENTS
POST OFFICE BOX 225012 • DALLAS, TEXAS 75265

TYPES SN54ALS653, SN54ALS654, SN74ALS653, SN74ALS654 OCTAL BUS TRANSCEIVERS AND REGISTERS

absolute maximum ratings over operating free-air temperature range (unless otherwise noted)

Supply voltage, V_{CC} .... 7 V
Input voltage: All inputs and A I/O ports .... 7 V
               B I/O ports .... 5.5 V
Operating free-air temperature range: SN54ALS653, SN54ALS654 .... −55°C to 125°C
                               SN74ALS653, SN74ALS654 .... 0°C to 70°C
Storage temperature range .... −65°C to 150°C

recommended operating conditions

			SN54ALS653 SN54ALS654			SN74ALS653 SN74ALS654			UNIT
			MIN	NOM	MAX	MIN	NOM	MAX	
V_{CC}	Supply voltage		4.5	5	5.5	4.5	5	5.5	V
V_{IH}	High-level input voltage		2			2			V
V_{IL}	Low-level input voltage				0.8			0.8	V
V_{OH}	High-level output voltage	A ports			5.5			5.5	V
I_{OH}	High-level output current	B ports			−12			−15	mA
I_{OL}	Low-level output current				12			24	mA
								48†	
f_{clock}	Clock frequency								MHz
t_w	Pulse duration	CBA or CAB high							ns
		CBA or CAB low							
t_{su}	Setup time before CAB↑ or CBA↑	A or B							ns
t_h	Hold time after CAB↑ or CBA↑	A or B							ns
T_A	Operating free-air temperature		−55		125	0		70	°C

†The extended condition applies if V_{CC} is maintained between 4.75 V and 5.25 V.
The 48-mA limit applies for the SN74ALS653−1 and SN74ALS654−1 only.

electrical characteristics over recommended operating free-air temperature range (unless otherwise noted)

PARAMETER		TEST CONDITIONS		SN54ALS653 SN54ALS654			SN74ALS653 SN74ALS654			UNIT
				MIN	TYP‡	MAX	MIN	TYP‡	MAX	
V_{IK}		$V_{CC} = 4.5$ V,	$I_I = -18$ mA			−1.5			−1.5	V
V_{OH}	B ports	$V_{CC} = 4.5$ V to 5.5 V,	$I_{OH} = -0.4$ mA	$V_{CC}-2$			$V_{CC}-2$			V
		$V_{CC} = 4.5$ V,	$I_{OH} = -3$ mA	2.4	3.2		2.4	3.2		
		$V_{CC} = 4.5$ V,	$I_{OH} = -12$ mA	2						
		$V_{CC} = 4.5$ V,	$I_{OH} = -15$ mA				2			
I_{OH}	A ports	$V_{CC} = 4.5$ V,	$V_{OH} = 5.5$ V			0.1			0.1	mA
V_{OL}		$V_{CC} = 4.5$ V,	$I_{OL} = 12$ mA		0.25	0.4		0.25	0.4	V
		$V_{CC} = 4.75$ V,	$I_{OL} = 24$ mA					0.35	0.5	
		($I_{OL} = 48$ mA for −1 versions)								
I_I	Control inputs	$V_{CC} = 5.5$ V,	$V_I = 7$ V			0.1			0.1	mA
	A or B ports	$V_{CC} = 5.5$ V,	$V_I = 5.5$ V			0.1			0.1	
I_{IH}	Control inputs	$V_{CC} = 5.5$ V,	$V_I = 2.7$ V			20			20	μA
	A or B ports §					20			20	
I_{IL}	Control inputs	$V_{CC} = 5.5$ V,	$V_I = 0.4$ V			−0.1			−0.1	mA
	A or B ports §					−0.2			−0.2	
I_O ¶	B ports	$V_{CC} = 5.5$ V,	$V_O = 2.25$ V	−30		−112	−30		−112	mA
I_{CC}	'ALS653	$V_{CC} = 5.5$ V	Outputs high		52			52		mA
			Outputs low		57			57		
			Outputs disabled		58			58		
	'ALS654		Outputs high		60			60		
			Outputs low		68			68		
			Outputs disabled		68			68		

‡All typical values are at $V_{CC} = 5$ V, $T_A = 25°C$.
§For I/O ports, the parameters I_{IH} and I_{IL} include the off-state output current.
¶The output conditions have been chosen to produce a current that closely approximates one half of the true short-circuit output current, I_{OS}.

PRODUCT PREVIEW

This page contains information on a product under development. Texas Instruments reserves the right to change or discontinue this product without notice.

TEXAS INSTRUMENTS
POST OFFICE BOX 225012 • DALLAS, TEXAS 75265

TYPES SN54ALS653, SN74ALS653
OCTAL BUS TRANSCEIVERS AND REGISTERS

'ALS653 switching characteristics (see Note 1)

Test conditions: $V_{CC} = 4.5$ V to 5.5 V, $C_L = 50$ pF, $R_L = 680$ Ω, (A outputs), $R1 = R2 = 500$ Ω, (B outputs), $T_A =$ MIN to MAX

PARAMETER	FROM (INPUT)	TO (OUTPUT)	SN54ALS653 MIN	SN54ALS653 TYP†	SN54ALS653 MAX	SN74ALS653 MIN	SN74ALS653 TYP†	SN74ALS653 MAX	UNIT
f_{max}									MHz
t_{PLH}	CBA	A		24			24		ns
t_{PHL}				15			15		
t_{PLH}	CAB	B		11			11		ns
t_{PHL}				13			13		
t_{PLH}	A	B		10			10		ns
t_{PHL}				12			12		
t_{PLH}	B	A		24			24		ns
t_{PHL}				10			10		
t_{PLH}	SBA‡ (with B high)	A		26			26		ns
t_{PHL}				15			15		
t_{PLH}	SBA‡ (with B low)	A		26			26		ns
t_{PHL}				15			15		
t_{PLH}	SAB‡ (with A high)	B		16			16		ns
t_{PHL}				16			16		
t_{PLH}	SAB‡ (with A low)	B		15			15		ns
t_{PHL}				15			15		
t_{PLH}	$\overline{GBA}$	A		24			24		ns
t_{PHL}				17			17		
t_{PZH}	GAB	B		19			19		ns
t_{PZL}				22			22		
t_{PHZ}	GAB	B		12			12		ns
t_{PLZ}				14			14		

† All typical values are at $V_{CC} = 5$ V, $T_A = 25°C$.
‡ These parameters are measured with the internal output state of the storage register opposite to that of the bus input.
NOTE 1: For load circuit and voltage waveforms, see page 1-12.
Additional information on these products can be obtained from the factory as it becomes available.

PRODUCT PREVIEW
This page contains information on a product under development. Texas Instruments reserves the right to change or discontinue this product without notice.

TEXAS INSTRUMENTS
POST OFFICE BOX 225012 • DALLAS, TEXAS 75265

TYPES SN54ALS654, SN74ALS654
OCTAL BUS TRANSCEIVERS AND REGISTERS

'ALS654 switching characteristics (see Note 1)

PARAMETER	FROM (INPUT)	TO (OUTPUT)	V_{CC} = 4.5 V to 5.5 V, C_L = 50 pF, R_L = 680 Ω, (A outputs) R1 = R2 = 500 Ω, (B outputs) T_A = MIN to MAX						UNIT
			SN54ALS654			SN74ALS654			
			MIN	TYP†	MAX	MIN	TYP†	MAX	
f_{max}									MHz
t_{PLH}	CBA	A		24			24		ns
t_{PHL}				15			15		
t_{PLH}	CAB	B		11			11		ns
t_{PHL}				13			13		
t_{PLH}	A	B		8			8		ns
t_{PHL}				8			8		
t_{PLH}	B	A		24			24		ns
t_{PHL}				10			10		
t_{PLH}	SBA‡ (with B high)	A		26			26		ns
t_{PHL}				15			15		
t_{PLH}	SBA‡ (with B low)	A		26			26		ns
t_{PHL}				15			15		
t_{PLH}	SAB‡ (with A high)	B		16			16		ns
t_{PHL}				16			16		
t_{PLH}	SAB‡ (with A low)	B		15			15		ns
t_{PHL}				12			12		
t_{PLH}	$\overline{GBA}$	A		24			24		ns
t_{PHL}				17			17		
t_{PZH}	GAB	B		19			19		ns
t_{PZL}				22			22		
t_{PHZ}	GAB	B		12			12		ns
t_{PLZ}				14			14		

† All typical values are at V_{CC} = 5 V, T_A = 25°C.
‡ These parameters are measured with the internal output state of the storage register opposite to that of the bus input.
NOTE 1: For load circuit and voltage waveforms, see page 1-12.
Additional information on these products can be obtained from the factory as it becomes available.

PRODUCT PREVIEW
This page contains information on a product under development. Texas Instruments reserves the right to change or discontinue this product without notice.

TYPES SN54AS651, SN54AS652, SN74AS651, SN74AS652
OCTAL BUS TRANSCEIVERS AND REGISTERS

absolute maximum ratings over operating free-air temperature range (unless otherwise noted)

Supply voltage, V_{CC} .... 7 V
Input voltage: Control inputs .... 7 V
               I/O ports .... 5.5 V
Operating free-air temperature range: SN54AS651, SN54AS652 .... −55°C to 125°C
                              SN74AS651, SN74AS652 .... 0°C to 70°C
Storage temperature range .... −65°C to 150°C

recommended operating conditions

			SN54AS651 SN54AS652			SN74AS651 SN74AS652			UNIT
			MIN	NOM	MAX	MIN	NOM	MAX	
V_{CC}	Supply voltage		4.5	5	5.5	4.5	5	5.5	V
V_{IH}	High-level input voltage		2			2			V
V_{IL}	Low-level input voltage				0.8			0.8	V
I_{OH}	High-level output current				−12			−15	mA
I_{OL}	Low-level output current				48			64	mA
f_{clock}			0		75	0		90	MHz
t_w	Pulse duration	CBA or CAB high	6			5			ns
		CBA or CAB low	7			6			
t_{su}	Setup time before CAB↑ or CBA↑	A or B	7			6			ns
t_h	Hold time after CAB↑ or CBA↑	A or B	0			0			ns
T_A	Operating free-air temperature		−55		125	0		70	°C

electrical characteristics over recommended operating free-air temperature range (unless otherwise noted)

PARAMETER		TEST CONDITIONS		SN54AS651 SN54AS652			SN74AS651 SN74AS652			UNIT
				MIN	TYP†	MAX	MIN	TYP†	MAX	
V_{IK}		V_{CC} = 4.5 V,	I_I = −18 mA			−1.2			−1.2	V
V_{OH}		V_{CC} = 4.5 V to 5.5 V,	I_{OH} = −2 mA	$V_{CC}-2$			$V_{CC}-2$			V
		V_{CC} = 4.5 V,	I_{OH} = −3 mA	2.4	3.2		2.4	3.2		
		V_{CC} = 4.5 V,	I_{OH} = −12 mA	2.4						
		V_{CC} = 4.5 V,	I_{OH} = −15 mA				2.4			
V_{OL}		V_{CC} = 4.5 V,	I_{OL} = 48 mA		0.35	0.55				V
		V_{CC} = 4.5 V,	I_{OL} = 64 mA					0.35	0.55	
I_I	Control inputs	V_{CC} = 5.5 V,	V_I = 7 V			0.1			0.1	mA
	A or B ports	V_{CC} = 5.5 V,	V_I = 5.5 V			0.1			0.1	
I_{IH}	Control inputs	V_{CC} = 5.5 V,	V_I = 2.7 V			20			20	µA
	A or B ports‡					50			50	
I_{IL}	Control inputs	V_{CC} = 5.5 V,	V_I = 0.4 V			−0.5			−0.5	mA
	A or B ports‡					−0.5			−0.5	
I_O §		V_{CC} = 5.5 V,	V_O = 2.25 V	−30		−112	−30		−112	mA
I_{CC}	'AS651	V_{CC} = 5.5 V	Outputs high		110	185		110	185	mA
			Outputs low		120	195		120	195	
			Outputs disabled		130	195		130	195	
	'AS652		Outputs high		120	195		120	195	
			Outputs low		130	211		130	211	
			Outputs disabled		130	211		130	211	

†All typical values are at V_{CC} = 5 V, T_A = 25°C
‡For I/O ports, the parameters I_{IH} and I_{IL} include the off-state output current.
§The output conditions have been chosen to produce a current that closely approximates one half of the true short-circuit output current, I_{OS}.

TEXAS INSTRUMENTS
POST OFFICE BOX 225012 • DALLAS, TEXAS 75265

TYPES SN54AS651, SN54AS652, SN74AS651, SN74AS652
OCTAL BUS TRANSCEIVERS AND REGISTERS

'AS651 switching characteristics (see Note 1)

PARAMETER	FROM (INPUT)	TO (OUTPUT)	V_{CC} = 4.5 V to 5.5 V, C_L = 50 pF, R1 = 500 Ω, R2 = 500 Ω, T_A = MIN to MAX				UNIT
			SN54AS651		SN74AS651		
			MIN	MAX	MIN	MAX	
f_{max}			75		90		MHz
t_{PLH}	CBA or CAB	A or B	2	9.5	2	8.5	ns
t_{PHL}			2	10	2	9	
t_{PLH}	A or B	B or A	2	9	2	8	ns
t_{PHL}			1	8	1	7	
t_{PLH}	SBA or SAB†	A or B	2	12	2	11	ns
t_{PHL}			2	10	2	9	
t_{PZH}	$\overline{G}BA$	A	2	11	2	10	ns
t_{PZL}			3	18	3	16	
t_{PHZ}	$\overline{G}BA$	A	2	10	2	9	ns
t_{PLZ}			2	10	2	9	
t_{PZH}	GAB	B	3	12	3	11	ns
t_{PZL}			3	20	3	16	
t_{PHZ}	GAB	B	2	11	2	10	ns
t_{PLZ}			2	12	2	11	

'AS652 switching characteristics (see Note 1)

PARAMETER	FROM (INPUT)	TO (OUTPUT)	V_{CC} = 4.5 V to 5.5 V, C_L = 50 pF, R1 = 500 Ω, R2 = 500 Ω, T_A = MIN to MAX				UNIT
			SN54AS652		SN74AS652		
			MIN	MAX	MIN	MAX	
f_{max}			75		90		MHz
t_{PLH}	CBA or CAB	A or B	2	9.5	2	8.5	ns
t_{PHL}			2	10	2	9	
t_{PLH}	A or B	B or A	2	11	2	9	ns
t_{PHL}			1	8	1	7	
t_{PLH}	SBA or SAB†	A or B	2	12	2	11	ns
t_{PHL}			2	10	2	9	
t_{PZH}	$\overline{G}BA$	A	2	11	2	10	ns
t_{PZL}			3	18	3	16	
t_{PHZ}	$\overline{G}BA$	A	2	10	2	9	ns
t_{PHL}			2	10	2	9	
t_{PZH}	GAB	B	3	12	3	11	ns
t_{PZL}			3	20	3	16	
t_{PHZ}	GAB	B	2	11	2	10	ns
t_{PLZ}			2	12	2	11	

†These parameters are measured with the internal output state of the storage register opposite to that of the bus input.
NOTE 1: For load circuit and voltage waveforms, see page 1-12.

TYPES SN54ALS677, SN54ALS678, SN74ALS677, SN74ALS678 ADDRESS COMPARATORS

D2661, JUNE 1982—REVISED DECEMBER 1983

- 'ALS677 is a 16-bit to 4-Bit Comparator with Enable
- 'ALS678 is a 16-Bit to 4-Bit Comparator with Latch
- Package Options Include Both Plastic and Ceramic Chip Carriers in Addition to Plastic and Ceramic DIPs
- Dependable Texas Instruments Quality and Reliability

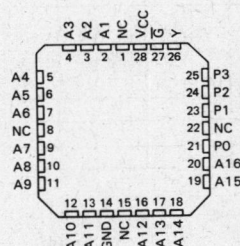

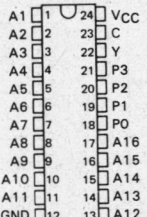

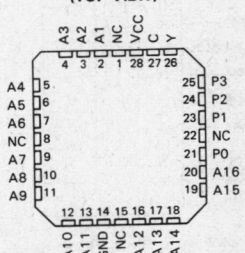

description

The 'ALS677 and 'ALS678 address comparators simplify addressing of memory boards and/or other peripheral devices. The four P inputs are normally hard wired with a preprogrammed address. An internal decoder determines what input information applied to the 16 A inputs must be low or high to cause a low state at the output (Y). For example, a positive-logic bit combination of 0111 (decimal 7) at the P input determines that inputs A1 through A7 must be low and that inputs A8 through A16 must be high to cause the output to go low. Equality of the address applied at the A inputs to the preprogrammed address is indicated by the output being low.

The 'ALS677 features an enable input (G). When G is low, the device is enabled. When G is high, the device is disabled and the output is high regardless of the A and P inputs. The 'ALS678 features a transparent latch and a latch enable input (C). When C is high, the device is in the transparent mode. When C is low, the previous logic state of Y is latched.

The SN54ALS677 and SN54ALS678 are characterized for operation over the full military temperature range of −55°C to 125°C. The SN54ALS677 and SN74ALS678 are characterized for operation from 0°C to 70°C.

PRODUCT PREVIEW

This document contains information on a product under development. Texas Instruments reserves the right to change or discontinue this product without notice.

Copyright © 1982 by Texas Instruments Incorporated

Texas Instruments
POST OFFICE BOX 225012 • DALLAS, TEXAS 75265

TYPES SN54ALS677, SN54ALS678, SN74ALS677, SN74ALS678 ADDRESS COMPARATORS

FUNCTION TABLE

ALS677	ALS678	INPUTS COMMON TO 'ALS677 AND 'ALS678																				OUTPUT
$\overline{G}$	C	P3	P2	P1	P0	A1	A2	A3	A4	A5	A6	A7	A8	A9	A10	A11	A12	A13	A14	A15	A16	Y
L	H	L	L	L	L	H	H	H	H	H	H	H	H	H	H	H	H	H	H	H	H	L
L	H	L	L	L	H	L	H	H	H	H	H	H	H	H	H	H	H	H	H	H	H	L
L	H	L	L	H	L	L	L	H	H	H	H	H	H	H	H	H	H	H	H	H	H	L
L	H	L	L	H	H	L	L	L	H	H	H	H	H	H	H	H	H	H	H	H	H	L
L	H	L	H	L	L	L	L	L	L	H	H	H	H	H	H	H	H	H	H	H	H	L
L	H	L	H	L	H	L	L	L	L	L	H	H	H	H	H	H	H	H	H	H	H	L
L	H	L	H	H	L	L	L	L	L	L	L	H	H	H	H	H	H	H	H	H	H	L
L	H	L	H	H	H	L	L	L	L	L	L	L	H	H	H	H	H	H	H	H	H	L
L	H	H	L	L	L	L	L	L	L	L	L	L	L	H	H	H	H	H	H	H	H	L
L	H	H	L	L	H	L	L	L	L	L	L	L	L	L	H	H	H	H	H	H	H	L
L	H	H	L	H	L	L	L	L	L	L	L	L	L	L	L	H	H	H	H	H	H	L
L	H	H	L	H	H	L	L	L	L	L	L	L	L	L	L	L	H	H	H	H	H	L
L	H	H	H	L	L	L	L	L	L	L	L	L	L	L	L	L	L	H	H	H	H	L
L	H	H	H	L	H	L	L	L	L	L	L	L	L	L	L	L	L	L	H	H	H	L
L	H	H	H	H	L	L	L	L	L	L	L	L	L	L	L	L	L	L	L	H	H	L
L	H	H	H	H	H	L	L	L	L	L	L	L	L	L	L	L	L	L	L	L	H	L
L	H	All other combinations																				H
H		'ALS677: Any combination																				H
	L	'ALS678: Any combination																				Latched

logic symbols

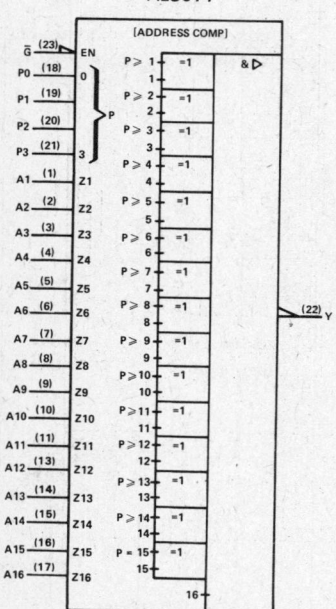

Pin numbers shown are for JT and NT packages.

TEXAS INSTRUMENTS
POST OFFICE BOX 225012 • DALLAS, TEXAS 75265

TYPES SN54ALS677, SN54ALS678, SN74ALS677, SN74ALS678, ADDRESS COMPARATORS

logic diagrams (positive logic)

'ALS677

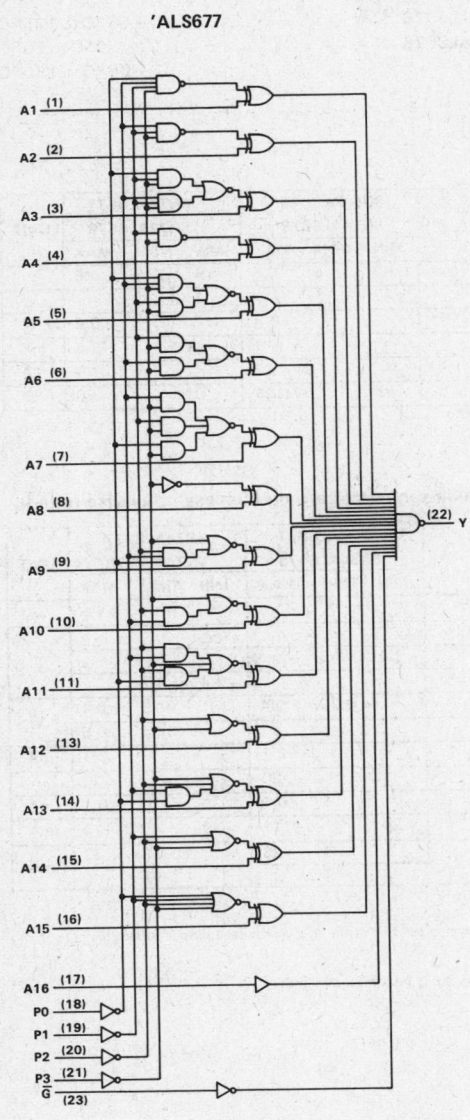

'ALS678

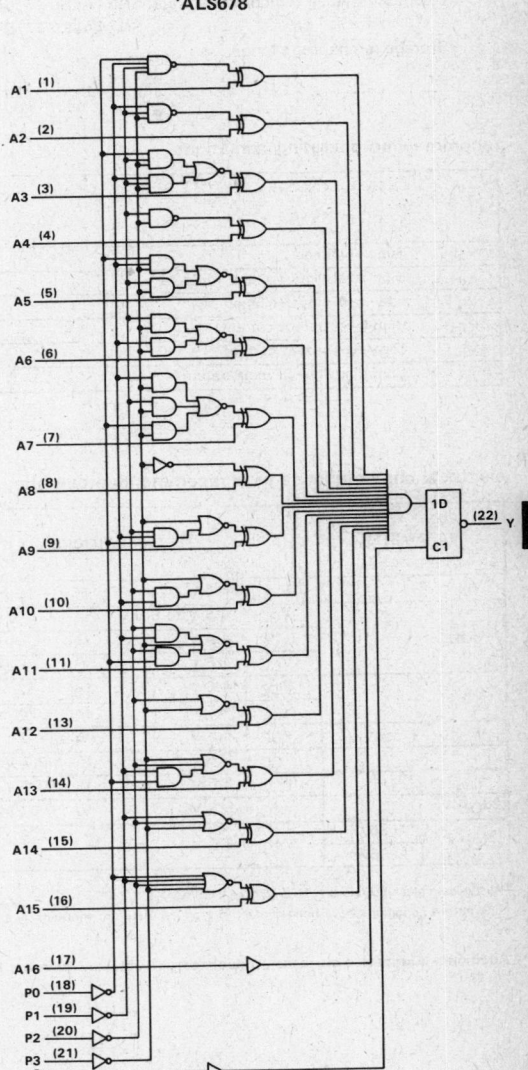

Pin numbers shown are for JT and NT packages.

Texas Instruments
POST OFFICE BOX 225012 • DALLAS, TEXAS 75265

TYPES SN54ALS677, SN54ALS678, SN74ALS677, SN74ALS678 ADDRESS COMPARATORS

absolute maximum ratings over operating free-air temperature range (unless otherwise noted)

Supply voltage, V_{CC} .... 7 V
Input voltage .... 7 V
Operating free-air temperature range: SN54ALS677, SN54ALS678 .... −55 °C to 125 °C
                                    SN74ALS677, SN74ALS678 .... 0 °C to 70 °C
Storage temperature range .... −65 °C to 150 °C

recommended operating conditions

		SN54ALS677 SN54ALS678			SN74ALS677 SN74ALS678			UNIT
		MIN	NOM	MAX	MIN	NOM	MAX	
V_{CC}	Supply voltage	4.5	5	5.5	4.5	5	5.5	V
V_{IH}	High-level input voltage	2			2			V
V_{IL}	Low-level input voltage			0.8			0.8	V
I_{OH}	High-level output current			−1			−2.6	mA
I_{OL}	Low-level output current			12			24	mA
T_A	Operating free-air temperature	−55		125	0		70	°C

electrical characteristics over recommended operating free-air temperature range (unless otherwise noted)

PARAMETER		TEST CONDITIONS		SN54ALS677 SN54ALS678			SN74ALS677 SN74ALS678			UNIT
				MIN	TYP[†]	MAX	MIN	TYP[†]	MAX	
V_{IK}		V_{CC} = 4.5 V,	I_I = −18 mA			−1.5			−1.5	V
V_{OH}		V_{CC} = 4.5 V to 5.5 V,	I_{OH} = −0.4 mA	V_{CC}−2			V_{CC}−2			V
		V_{CC} = 4.5 V,	I_{OH} = −1 mA	2.4	3.3					
		V_{CC} = 4.5 V,	I_{OH} = −2.6 mA				2.4	3.2		
V_{OL}		V_{CC} = 4.5 V,	I_{OL} = 12 mA		0.25	0.4				V
		V_{CC} = 4.5 V,	I_{OL} = 24 mA					0.25 0.35	0.4 0.5	
I_I		V_{CC} = 5.5 V,	V_I = 7 V			0.1			0.1	mA
I_{IH}		V_{CC} = 5.5 V,	V_I = 2.7 V			20			20	μA
I_{IL}		V_{CC} = 5.5 V,	V_I = 0.4 V			−0.1			−0.1	mA
I_O[‡]		V_{CC} = 5.5 V,	V_O = 2.25 V	−30		−112	−30		−112	mA
I_{CC}	'ALS677	V_{CC} = 5.5 V			21	33		21	33	mA
	'ALS678				17			17		

[†] All typical values are at V_{CC} = 5 V, T_A = 25 °C.
[‡] The output conditions have been chosen to produce a current that closely approximates one half of the true short-circuit output current, I_{OS}.

Additional information on these products can be obtained from the factory as it becomes available.

2 ALS AND AS CIRCUITS

TYPES SN54ALS677, SN54ALS678, SN74ALS677, SN74ALS678, ADDRESS COMPARATORS

'ALS677 switching characteristics (see Note 1)

PARAMETER	FROM (INPUT)	TO (OUTPUT)	V_{CC} = 4.5 V to 5.5 V, C_L = 50 pF, R_L = 500 Ω, T_A = MIN to MAX				UNIT
			SN54ALS677		SN74ALS677		
			MIN	MAX	MIN	MAX	
t_{PLH}	Any P	Y	4	28	4	25	ns
t_{PHL}			8	40	8	35	
t_{PLH}	Any A	Y	5	26	5	22	ns
t_{PHL}			5	35	5	30	
t_{PLH}	$\overline{G}$	Y	3	15	3	13	ns
t_{PHL}			5	30	5	25	

'ALS678 switching characteristics (see Note 1)

PARAMETER	FROM (INPUT)	TO (OUTPUT)	V_{CC} = 4.5 V to 5.5 V, C_L = 50 pF, R_L = 500 Ω, T_A = MIN to MAX						UNIT
			SN54ALS678			SN74ALS678			
			MIN	TYP†	MAX	MIN	TYP†	MAX	
t_{PLH}	Any P	Y		18			18		ns
t_{PHL}				18			18		
t_{PLH}	Any A	Y		14			14		ns
t_{PHL}				14			14		
t_{PLH}	C	Y		14			14		ns
t_{PHL}				14			14		

†All typical values are at V_{CC} = 5 V, T_A = 25°C.
NOTE 1: For load circuit and voltage waveforms, see page 1-12.

Additional information on these products can be obtained from the factory as it becomes available.

ALS AND AS CIRCUITS

TEXAS INSTRUMENTS
POST OFFICE BOX 225012 • DALLAS, TEXAS 75265

TYPES SN54ALS677, SN54ALS678, SN74ALS677, SN74ALS678
ADDRESS COMPARATORS

TYPICAL APPLICATION INFORMATION

The 'ALS677 and 'ALS678 can be wired to recognize any one of $2^{16} - 1$ addresses. The number of "lows" in the address determines the input pattern for the P inputs. Then those system address lines that are low in the address to be recognized are connected to the lowest numbered A inputs of the address comparator and the system address lines that are high are connected to the highest numbered A inputs.

For example, assume the comparator is to enable a device when the 16-bit system address is:

A15	A14	A13	A12	A11	A10	A9	A8	A7	A6	A5	A4	A3	A2	A1	A0
H	H	L	L	H	H	L	L	H	H	L	L	H	H	H	H

Since the address contains 6 lows and 10 highs, the following connections are made:

P3 to 0 V, P2 to V_{CC}, P1 to V_{CC}, and P0 to 0 V.

System address lines A13, A12, A9, A8, A5, and A4 to comparator inputs A1 through A6 in any convenient order.

The remaining ten system address lines to comparator inputs A7 through A16 in any convenient order.

The output provides an active-low enabling signal.

The following circuit is a modulo-N synchronous counter. The 'ALS163 is connected to provide a low-level clear signal when $N = FEFF_{16}$.

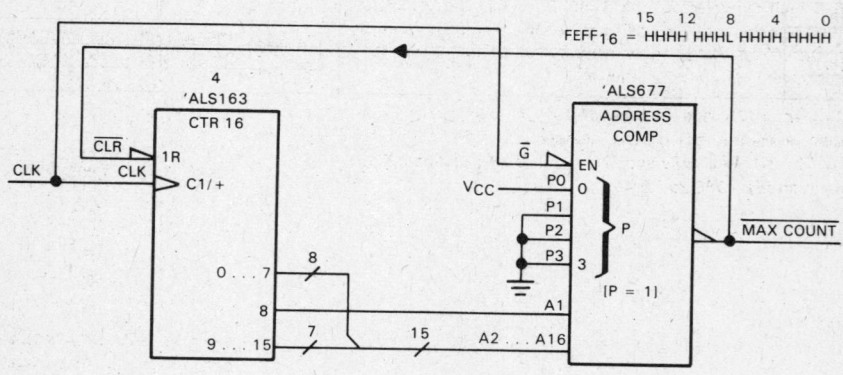

MODULO-N SYNCHRONOUS COUNTER

TYPES SN54ALS679, SN54ALS680, SN74ALS679, SN74ALS680, ADDRESS COMPARATORS

D2661, JUNE 1982—REVISED DECEMBER 1983

- 'ALS679 is a 12-Bit to 4-Bit Comparator With Enable
- 'ALS680 is a 12-Bit to 4-Bit Comparator With Latch
- Package Options Include Both Plastic and Ceramic Chip Carriers in Addition to Plastic and Ceramic DIPs
- Dependable Texas Instruments Quality and Reliability

SN54ALS679 . . . J PACKAGE
SN74ALS679 . . . N PACKAGE
(TOP VIEW)

```
A1  [ 1    20 ] VCC
A2  [ 2    19 ] G
A3  [ 3    18 ] Y
A4  [ 4    17 ] P3
A5  [ 5    16 ] P2
A6  [ 6    15 ] P1
A7  [ 7    14 ] P0
A8  [ 8    13 ] A12
A9  [ 9    12 ] A11
GND [ 10   11 ] A10
```

description

The 'ALS679 and 'ALS680 address comparators simplify addressing of memory boards and/or other peripheral devices. The four P inputs are normally hard wired with a preprogrammed address. An internal decoder determines what input information applied to the 12 A inputs must be low or high to cause a low state at the output (Y). For example, a positive-logic bit combination of 0111 (decimal 7) at the P input determines that inputs A1 through A7 must be low and that inputs A8 through A12 must be high to cause the output to go low. Equality of the address applied at the A inputs to the preprogrammed address is indicated by the output being low.

The 'ALS679 features an enable input ($\overline{G}$). When $\overline{G}$ is low, the device is enabled. When $\overline{G}$ is high, the device is disabled and the output is high regardless of the A and P inputs. The 'ALS680 features a transparent latch and a latch enable input (C). When C is high, the device is in the transparent mode. When C is low, the previous logical state of Y is latched.

The SN54ALS679 and SN54ALS680 are characterized for operation over the full military temperature of −55°C to 125°C. The SN74ALS679 and SN74ALS680 are characterized for operation from 0°C to 70°C.

SN54ALS679 . . . FH PACKAGE
SN74ALS679 . . . FN PACKAGE
(TOP VIEW)

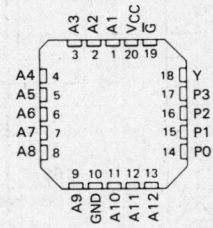

SN54ALS680 . . . J PACKAGE
SN74ALS680 . . . N PACKAGE
(TOP VIEW)

```
A1  [ 1    20 ] VCC
A2  [ 2    19 ] C
A3  [ 3    18 ] Y
A4  [ 4    17 ] P3
A5  [ 5    16 ] P2
A6  [ 6    15 ] P1
A7  [ 7    14 ] P0
A8  [ 8    13 ] A12
A9  [ 9    12 ] A11
GND [ 10   11 ] A10
```

SN54ALS680 . . . FH PACKAGE
SN74ALS680 . . . FN PACKAGE
(TOP VIEW)

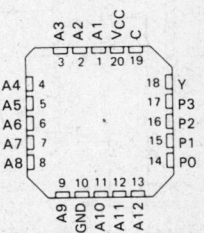

PRODUCT PREVIEW

This document contains information on a product under development. Texas Instruments reserves the right to change or discontinue this product without notice.

Copyright © 1982 by Texas Instruments Incorporated

TEXAS INSTRUMENTS
POST OFFICE BOX 225012 • DALLAS, TEXAS 75265

TYPES SN54ALS679, SN54ALS680, SN74ALS679, SN74ALS680
ADDRESS COMPARATORS

FUNCTION TABLE

'ALS679 $\overline{G}$	'ALS680 C	\multicolumn{17}{c	}{INPUTS COMMON TO 'ALS679 AND 'ALS680}	OUTPUT Y
		P3	P2	P1	P0	A1	A2	A3	A4	A5	A6	A7	A8	A9	A10	A11	A12	
L	H	L	L	L	L	H	H	H	H	H	H	H	H	H	H	H	H	L
L	H	L	L	L	H	L	H	H	H	H	H	H	H	H	H	H	H	L
L	H	L	L	H	L	L	L	H	H	H	H	H	H	H	H	H	H	L
L	H	L	L	H	H	L	L	L	H	H	H	H	H	H	H	H	H	L
L	H	L	H	L	L	L	L	L	L	H	H	H	H	H	H	H	H	L
L	H	L	H	L	H	L	L	L	L	L	H	H	H	H	H	H	H	L
L	H	L	H	H	L	L	L	L	L	L	L	H	H	H	H	H	H	L
L	H	L	H	H	H	L	L	L	L	L	L	L	H	H	H	H	H	L
L	H	H	L	L	L	L	L	L	L	L	L	L	L	H	H	H	H	L
L	H	H	L	L	H	L	L	L	L	L	L	L	L	L	H	H	H	L
L	H	H	L	H	L	L	L	L	L	L	L	L	L	L	L	H	H	L
L	H	H	L	H	H	L	L	L	L	L	L	L	L	L	L	L	H	L
L	H	H	H	L	L	L	L	L	L	L	L	L	L	L	H	H	H	L*
L	H	H	H	L	H	L	L	L	L	L	L	L	L	L	L	H	H	L*
L	H	H	H	H	L	L	L	L	L	L	L	L	L	L	L	L	H	L*
L	H	H	H	H	H	L	L	L	L	L	L	L	L	L	L	L	L	L
L	H	\multicolumn{16}{c	}{All other combinations}	H
H		\multicolumn{16}{c	}{'ALS679: Any combination}	H
	L	\multicolumn{16}{c	}{'ALS680: Any combination}	Latched

logic symbols

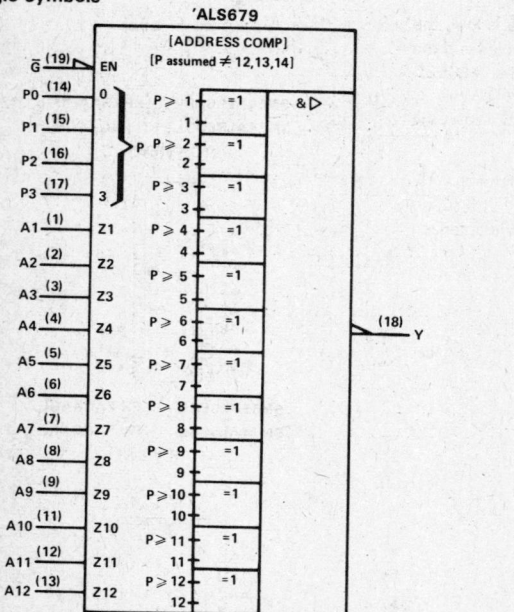

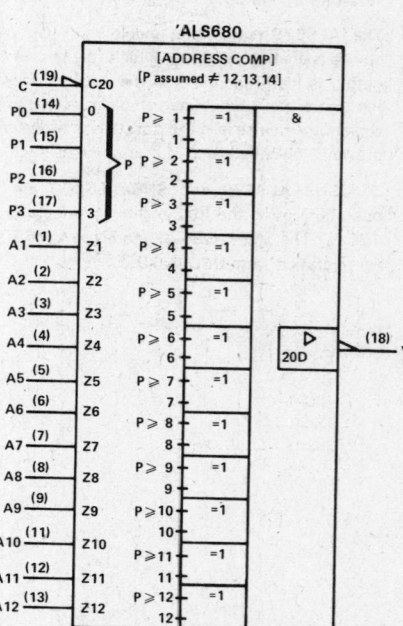

*The three shaded rows of the function table show combinations that would normally not be used in address comparator applications. The logic symbols above are not valid for these combinations in which P = 12, 13, and 14. If symbols valid for all combinations are required, starting with the fourth Exclusive-OR from the bottom, change $P \geq 9$ to $P = 9...11/13...15$, $P \geq 10$ to $P = 10/11/14/15$, and $P \geq 11$ to $P = 11/15$.

Pin numbers shown are for J and N packages.

Texas Instruments
POST OFFICE BOX 225012 • DALLAS, TEXAS 75265

TYPES SN54ALS679, SN54ALS680, SN74ALS679, SN74ALS680, ADDRESS COMPARATORS

logic diagrams (positive logic)

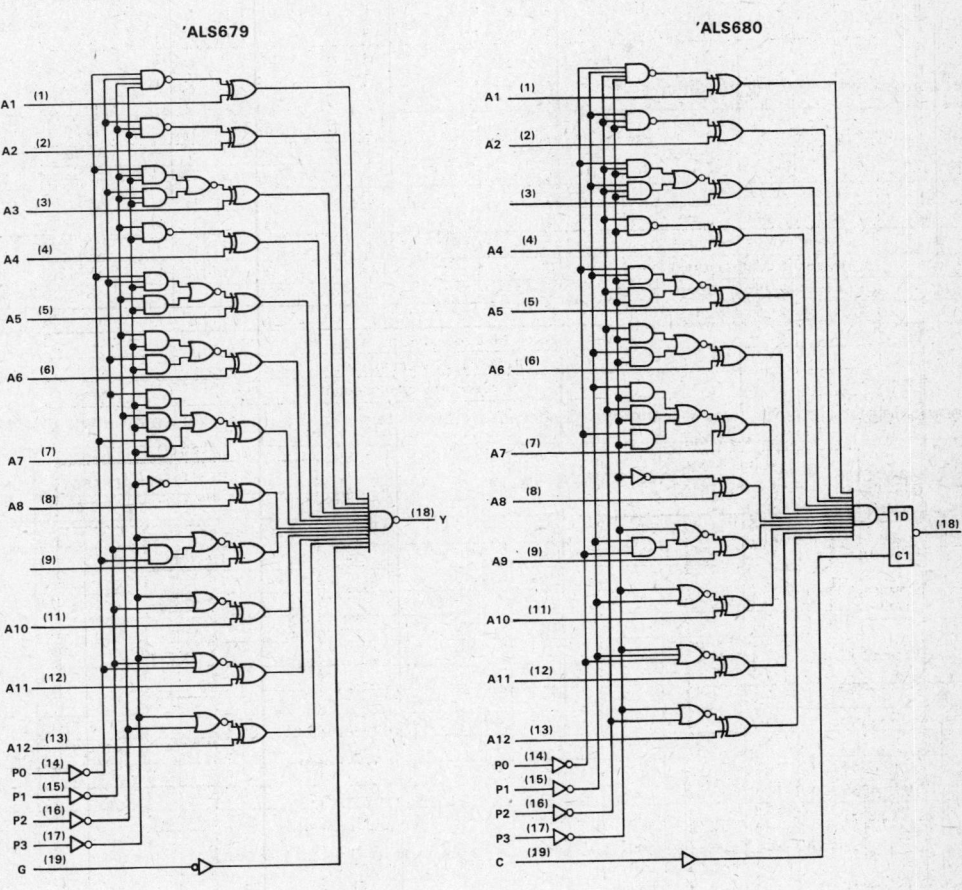

Pin numbers shown are for J and N packages.

TYPES SN54ALS679, SN54ALS680, SN74ALS679, SN74ALS680 ADDRESS COMPARATORS

absolute maximum ratings over operating free-air temperature range (unless otherwise noted)

Supply voltage, V_{CC} .... 7 V
Input voltage .... 7 V
Operating free-air temperature range: SN54ALS679, SN54ALS680 .... −55 °C to 125 °C
 SN74ALS679, SN74ALS680 .... 0 °C to 70 °C
Storage temperature range .... −65 °C to 150 °C

recommended operating conditions

		SN54ALS679 SN54ALS680			SN74ALS679 SN74ALS680			UNIT
		MIN	NOM	MAX	MIN	NOM	MAX	
V_{CC}	Supply voltage	4.5	5	5.5	4.5	5	5.5	V
V_{IH}	High-level input voltage	2			2			V
V_{IL}	Low-level input voltage			0.8			0.8	V
I_{OH}	High-level output current			−1			−2.6	mA
I_{OL}	Low-level output current			12			24	mA
T_A	Operating free-air temperature	−55		125	0		70	°C

electrical characteristics over recommended operating free-air temperature range (unless otherwise noted)

PARAMETER		TEST CONDITIONS		SN54ALS679 SN54ALS680			SN74ALS679 SN74ALS680			UNIT
				MIN	TYP†	MAX	MIN	TYP†	MAX	
V_{IK}		V_{CC} = 4.5 V,	I_I = −18 mA			−1.5			−1.5	V
V_{OH}		V_{CC} = 4.5 V to 5.5 V,	I_{OH} = −0.4 mA	V_{CC} −2			V_{CC} −2			V
		V_{CC} = 4.5 V,	I_{OH} = −1 mA	2.4	3.3					
		V_{CC} = 4.5 V,	I_{OH} = −2.6 mA				2.4	3.2		
V_{OL}		V_{CC} = 4.5 V,	I_{OL} = 12 mA		0.25	0.4		0.25	0.4	V
		V_{CC} = 4.5 V,	I_{OL} = 24 mA					0.35	0.5	
I_I		V_{CC} = 5.5 V,	V_I = 7 V			0.1			0.1	mA
I_{IH}		V_{CC} = 5.5 V,	V_I = 2.7 V			20			20	μA
I_{IL}		V_{CC} = 5.5 V,	V_I = 0.4 V			−0.1			−0.1	mA
I_O‡		V_{CC} = 5.5 V,	V_O = 2.25 V	−30		−112	−30		−112	mA
I_{CC}	'ALS679	V_{CC} = 5.5 V			17	28		17	28	mA
	'ALS680				13.4			13.4		

†All typical values are at V_{CC} = 5 V, T_A = 25 °C.
‡The output conditions have been chosen to produce a current that closely approximates one half of the true short-circuit output current, I_{OS}.

Additional information on these products can be obtained from the factory as it becomes available.

Texas Instruments
POST OFFICE BOX 225012 • DALLAS, TEXAS 75265

TYPES SN54ALS679, SN54ALS680, SN74ALS679, SN74ALS680, ADDRESS COMPARATORS

'ALS679 switching characteristics (see Note 1)

PARAMETER	FROM (INPUT)	TO (OUTPUT)	V_{CC} = 4.5 V to 5.5 V, C_L = 50 pF, R_L = 500 Ω, T_A = MIN to MAX				UNIT
			SN54ALS679		SN74ALS679		
			MIN	MAX	MIN	MAX	
t_{PLH}	Any P	Y	4	28	4	25	ns
t_{PHL}			8	40	8	35	
t_{PLH}	Any A	Y	5	26	5	22	ns
t_{PHL}			5	35	5	30	
t_{PLH}	$\overline{G}$	Y	3	15	3	13	ns
t_{PHL}			5	30	5	25	

'ALS680 switching characteristics (see Note 1)

PARAMETER	FROM (INPUT)	TO (OUTPUT)	V_{CC} = 4.5 V to 5.5 V, C_L = 50 pF, R_L = 500 Ω, T_A = MIN To MAX						UNIT
			SN54ALS680			SN74ALS680			
			MIN	TYP†	MAX	MIN	TYP†	MAX	
t_{PLH}	Any P	Y		18			18		ns
t_{PHL}				18			18		
t_{PLH}	Any A	Y		14			14		ns
t_{PHL}				14			14		
t_{PLH}	C	Y		14			14		ns
t_{PHL}				14			14		

†All typical values are at V_{CC} = 5 V, T_A = 25°C.
NOTE 1: For load circuit and voltage waveforms, see page 1-12.

Additional information on these products can be obtained from the factory as it becomes available.

TEXAS INSTRUMENTS
POST OFFICE BOX 225012 • DALLAS, TEXAS 75265

TYPES SN54ALS679, SN54ALS680, SN74ALS679, SN74ALS680
ADDRESS COMPARATORS

TYPICAL APPLICATION INFORMATION

The 'ALS679 and 'ALS680 can be wired to recognize any one of 2^{12} addresses. The number of "lows" in the address determines the input pattern for the P inputs. Then those system address lines that are low in the address to be recognized are connected to the lowest numbered A inputs of the address comparator and the system address lines that are high are connected to the highest numbered A inputs.

For example, assume the comparator is to enable a device when the 12-bit system address is:

A11	A10	A9	A8	A7	A6	A5	A4	A3	A2	A1	A0
H	H	L	L	H	H	L	L	H	H	H	H

Since the address contains 4 lows and 8 highs, the following connections are made:

P3 to 0 V, P2 to V_{CC}, P1 to 0 V, and P0 to 0 V.

System address lines A9, A8, A5, and A4 to comparator inputs A1 through A4 in any convenient order.

The remaining eight system address lines to comparator inputs A5 through A12 in any convenient order.

The output provides an active-low enabling signal.

The following circuit is a register bank decoder that examines the 14 most significant bits (A0 through A13) of a 20-bit address to select banks corresponding to the hex addresses 10000, 10040, 10080, and 100C0.

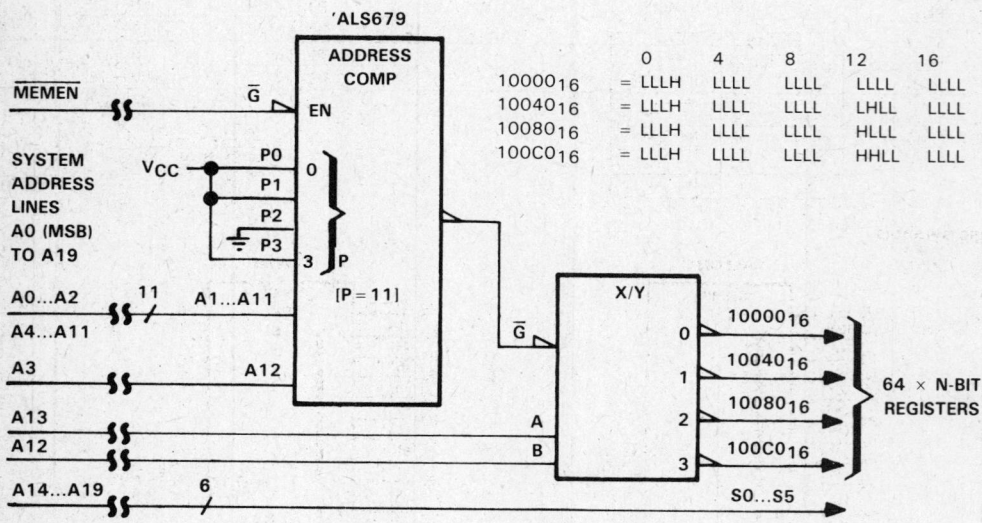

REGISTER BANK DECODER

TYPES SN54ALS688, SN54ALS689, SN74ALS688, SN74ALS689
8-BIT IDENTITY COMPARATORS

D2661, JUNE 1982-REVISED DECEMBER 1983

- Compares Two Eight-Bit Words
- Choice of Totem-Pole or Open-Collector Outputs
- Package Options Include Both Plastic and Ceramic Chip Carriers in Addition to Plastic and Ceramic DIPs
- Dependable Texas Instruments Quality and Reliablility

TYPE	OUTPUT FUNCTION AND CONFIGURATION
'ALS688†	$\overline{P=Q}$ totem-pole
'ALS689	$\overline{P=Q}$ open-collector

†'ALS688 is identical to 'ALS521

SN54ALS688, SN54ALS689 . . . J PACKAGE
SN74ALS688, SN74ALS689 . . . N PACKAGE
(TOP VIEW)

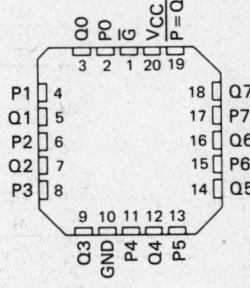

description

These identity comparators perform comparisons of two eight-bit binary or BCD words. The 'ALS688 and 'ALS689 provide $\overline{P=Q}$ outputs. The 'ALS688 has totem-pole outputs, while 'ALS689 has open-collector outputs.

The SN54ALS688 and SN54ALS689 are characterized for operation over the full military temperature range of −55°C to 125°C. The SN74ALS688 and SN74ALS689 are characterized for operation from 0°C to 70°C.

FUNCTION TABLE

INPUTS		OUTPUT
DATA P,Q	ENABLE $\overline{G}$	$\overline{P=Q}$
P=Q	L	L
P>Q	L	H
P<Q	L	H
X	H	H

SN54ALS688, SN54ALS689 . . . FH PACKAGE
SN74ALS688, SN74ALS689 . . . FN PACKAGE
(TOP VIEW)

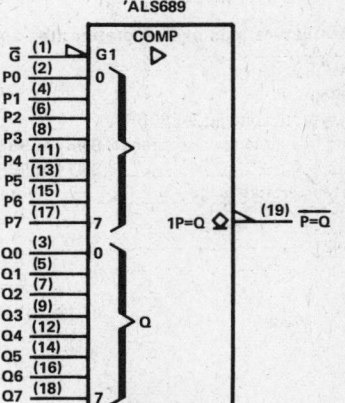

logic symbols

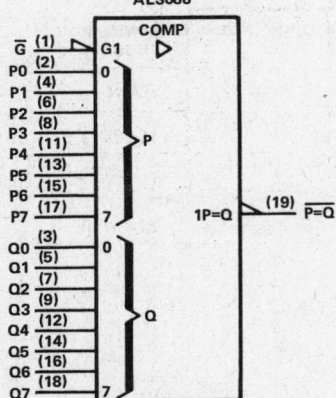

Pin numbers shown are for J and N packages.

Copyright © 1982 by Texas Instruments Incorporated

ALS AND AS CIRCUITS

TEXAS INSTRUMENTS
POST OFFICE BOX 225012 • DALLAS, TEXAS 75265

TYPES SN54ALS688, SN54ALS689, SN74ALS688, SN74ALS689
8-BIT IDENTITY COMPARATORS

logic diagram (positive logic)

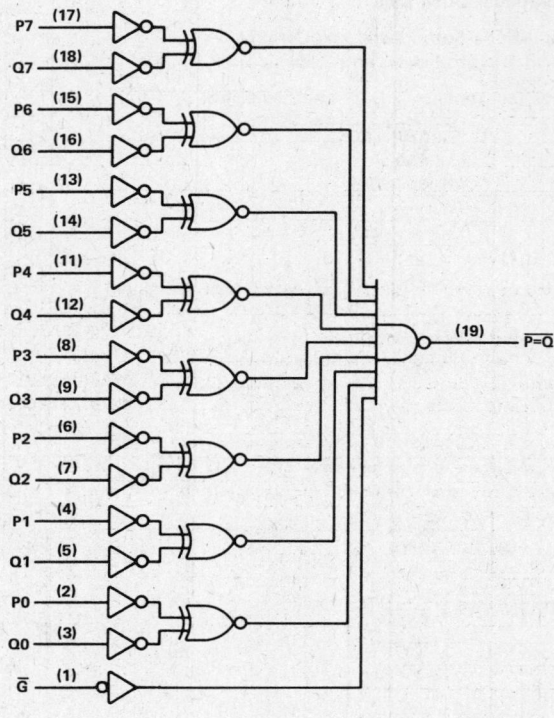

Pin numbers shown are for J and N packages.

absolute maximum ratings over operating free-air temperature range (unless otherwise noted)

Supply voltage, V_{CC} .... 7 V
Input voltage: .... 7 V
Off-state output voltage: 'ALS689 .... 7 V
Operating free-air temperature range: SN54ALS688, SN54AS689 .... −55°C to 125°C
                                         SN74ALS688, SN74AS689 .... 0°C to 70°C
Storage temperature range .... −65°C to 150°C

TYPES SN54ALS688, SN74ALS688
8-BIT IDENTITY COMPARATORS WITH TOTEM-POLE OUTPUTS

recommended operating conditions

		SN54ALS688			SN74ALS688			UNIT
		MIN	NOM	MAX	MIN	NOM	MAX	
V_{CC}	Supply voltage	4.5	5	5.5	4.5	5	5.5	V
V_{IH}	High-level input voltage	2			2			V
V_{IL}	Low-level input voltage			0.8			0.8	V
I_{OH}	High-level output current			−1			−2.6	mA
I_{OL}	Low-level output current			12			24	mA
T_A	Operating free-air temperature	−55		125	0		70	°C

electrical characteristics over recommended operating free-air temperature range (unless otherwise noted)

PARAMETER	TEST CONDITIONS		SN54ALS688			SN74ALS688			UNIT
			MIN	TYP[†]	MAX	MIN	TYP[†]	MAX	
V_{IK}	V_{CC} = 4.5 V,	I_I = −18 mA			−1.5			−1.5	V
V_{OH}	V_{CC} = 4.5 V to 5.5 V,	I_{OH} = −0.4 mA	V_{CC}−2			V_{CC}−2			V
	V_{CC} = 4.5 V,	I_{OH} = −1 mA	2.4	3.3					
	V_{CC} = 4.5 V,	I_{OH} = −2.6 mA				2.4	3.2		
V_{OL}	V_{CC} = 4.5 V,	I_{OL} = 12 mA		0.25	0.4		0.25	0.4	V
	V_{CC} = 4.5 V,	I_{OL} = 24 mA					0.35	0.5	
I_I	V_{CC} = 5.5 V,	V_I = 7 V			0.1			0.1	mA
I_{IH}	V_{CC} = 5.5 V,	V_I = 2.7 V			20			20	µA
I_{IL}	V_{CC} = 5.5 V,	V_I = 0.4 V			−0.1			−0.1	mA
I_O[‡]	V_{CC} = 5.5 V,	V_O = 2.25 V	−30		−112	−30		−112	mA
I_{CC}	V_{CC} = 5.5 V	See Note 1		12	19		12	19	mA

[†] All typical values are at V_{CC} = 5 V, T_A = 25°C.
[‡] The output conditions have been chosen to produce a current that closely approximates one half of the true short-circuit output current, I_{OS}.
NOTE 1: I_{CC} is measured with G grounded, P and Q at 4.5 V.

switching characteristics (see Note 2)

PARAMETER	FROM (INPUT)	TO (OUTPUT)	V_{CC} = 4.5 V to 5.5 V, C_L = 50 pF, R_L = 500 Ω, T_A = MIN to MAX				UNIT
			SN54ALS688		SN74ALS688		
			MIN	MAX	MIN	MAX	
t_{PLH}	P	$\overline{P=Q}$	3	16	3	12	ns
t_{PHL}			5	25	5	20	
t_{PLH}	Q	$\overline{P=Q}$	3	16	3	12	ns
t_{PHL}			5	25	5	20	
t_{PLH}	$\overline{G}$	$\overline{P=Q}$	3	15	3	12	ns
t_{PHL}			5	25	5	22	

NOTE 2: For load circuit and voltage waveforms, see page 1-12.

TEXAS INSTRUMENTS
POST OFFICE BOX 225012 • DALLAS, TEXAS 75265

TYPES SN54ALS689, SN74ALS689
8-BIT IDENTITY COMPARATORS WITH OPEN-COLLECTOR OUTPUTS

recommended operating conditions

		SN54ALS689			SN74ALS689			UNIT
		MIN	NOM	MAX	MIN	NOM	MAX	
V_{CC}	Supply voltage	4.5	5	5.5	4.5	5	5.5	V
V_{IH}	High-level input voltage	2			2			V
V_{IL}	Low-level input voltage			0.8			0.8	V
I_{OH}	High-level output current			5.5			5.5	V
I_{OL}	Low-level output current			12			24	mA
T_A	Operating free-air temperature	−55		125	0		70	°C

electrical characteristics over recommended operating free-air temperature range (unless otherwise noted)

PARAMETER	TEST CONDITIONS		SN54ALS689			SN74ALS689			UNIT
			MIN	TYP†	MAX	MIN	TYP†	MAX	
V_{IK}	$V_{CC} = 4.5$ V,	$I_I = -18$ mA			−1.5			−1.5	V
I_{OH}	$V_{CC} = 4.5$ V,	$V_{OH} = 5.5$ V			0.1			0.1	mA
V_{OL}	$V_{CC} = 4.5$ V,	$I_{OL} = 12$ mA		0.25	0.4		0.25	0.4	V
	$V_{CC} = 4.5$ V,	$I_{OL} = 24$ mA					0.35	0.5	
I_I	$V_{CC} = 5.5$ V,	$V_I = 7$ V			0.1			0.1	mA
I_{IH}	$V_{CC} = 5.5$ V,	$V_I = 2.7$ V			20			20	µA
I_{IL}	$V_{CC} = 5.5$ V,	$V_I = 0.4$ V			−0.1			−0.1	mA
I_{CC}	$V_{CC} = 5.5$ V,	See Note 1		12	19		12	19	mA

†All typical values are at $V_{CC} = 5$ V, $T_A = 25$°C.
NOTE 1: I_{CC} is measured with G grounded, P and Q at 4.5 V.

switching characteristics (see Note 2)

PARAMETER	FROM (INPUT)	TO (OUTPUT)	$V_{CC} = 4.5$ V to 5.5 V, $C_L = 50$ pF, $R_L = 680$ Ω, T_A = MIN to MAX				UNIT
			SN54ALS689		SN74ALS689		
			MIN	MAX	MIN	MAX	
t_{PLH}	P	$\overline{P=Q}$	10	30	10	25	ns
t_{PHL}			5	25	5	23	
t_{PLH}	Q	$\overline{P=Q}$	10	30	10	25	ns
t_{PHL}			5	25	5	23	
t_{PLH}	$\overline{G}$	$\overline{P=Q}$	8	30	8	25	ns
t_{PHL}			8	30	8	25	

NOTE 2: For load circuit and voltage waveforms, see page 1-12.

TEXAS INSTRUMENTS
POST OFFICE BOX 225012 • DALLAS, TEXAS 75265

TYPES SN54AS756, SN54AS757, SN74AS756, SN74AS757
OCTAL BUFFERS AND LINE DRIVERS WITH OPEN-COLLECTOR OUTPUTS

D2261, DECEMBER 1983

- Open-Collector Outputs Drive Bus Lines or Buffer Memory Address Registers
- Eliminates the Need for 3-State Overlap Protection
- P-N-P Inputs Reduce DC Loading
- Dependable Texas Instruments Quality and Reliability
- Open-Collector Versions of 'AS240, 'AS241

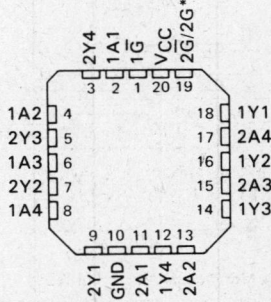

*2$\overline{G}$ for 'AS756 or 2G for 'AS757

description

These octal bus transceivers are designed specifically to improve both the performance and density of three-state memory address drivers, clock drivers, and bus-oriented receivers and transmitters by eliminating the need for three-state overlap protection. The designer has a choice of selected combinations of inverting and noninverting outputs, symmetrical $\overline{G}$ (active-low output control) inputs, and complementary G and $\overline{G}$ inputs. These devices feature high fan-out and improved fan-in.

The SN54' family is characterized for operation over the full military temperature range of $-55\,°C$ to $125\,°C$. The SN74' family is characterized for operation from $0\,°C$ to $70\,°C$.

PRODUCT PREVIEW

This document contains information on a product under development. Texas Instruments reserves the right to change or discontinue this product without notice.

Copyright © 1983 by Texas Instruments Incorporated

TEXAS INSTRUMENTS
POST OFFICE BOX 225012 • DALLAS, TEXAS 75265

TYPES SN54AS756, SN54AS757, SN74AS756, SN74AS757
OCTAL BUFFERS AND LINE DRIVERS WITH OPEN-COLLECTOR OUTPUTS

logic symbols

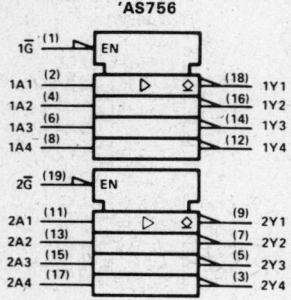

logic diagrams (positive logic)

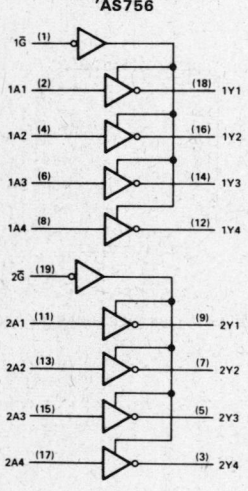

Pin numbers shown are for J and N packages.

TYPES SN54AS756, SN54AS757, SN74AS756, SN74AS757
OCTAL BUFFERS AND LINE DRIVERS WITH OPEN-COLLECTOR OUTPUTS

absolute maximum ratings over operating free-air temperature range (unless otherwise noted)

Supply voltage, V_{CC} .... 7 V
Input voltage .... 7 V
Off-state output voltage .... 7 V
Operating free-air temperature range: SN54AS756, SN54AS757 .... −55 °C to 125 °C
 SN74AS756, SN74AS757 .... 0 °C to 70 °C
Storage temperature range .... −65 °C to 150 °C

recommended operating conditions

		SN54AS756 SN54AS757			SN74AS756 SN74AS757			UNIT
		MIN	NOM	MAX	MIN	NOM	MAX	
V_{CC}	Supply voltage	4.5	5	5.5	4.5	5	5.5	V
V_{IH}	High-level input voltage	2			2			V
V_{IL}	Low-level input voltage			0.8			0.8	V
V_{OH}	High-level output voltage			5.5			5.5	V
I_{OL}	Low-level output current			48			64	mA
T_A	Operating free-air temperature	−55		125	0		70	°C

electrical characteristics over recommended operating free-air temperature range (unless otherwise noted)

PARAMETER		TEST CONDITIONS		SN54AS756 SN54AS757			SN74AS756 SN74AS757			UNIT
				MIN	TYP†	MAX	MIN	TYP†	MAX	
V_{IK}		V_{CC} = 4.5 V,	I_I = −18 mA			−1.2			−1.2	V
I_{OH}		V_{CC} = 4.5 V,	V_{OH} = 5.5 V			0.1			0.1	mA
V_{OL}		V_{CC} = 4.5 V,	I_{OL} = 48 mA			0.55				V
		V_{CC} = 4.5 V,	I_{OL} = 64 mA						0.55	
I_I		V_{CC} = 5.5 V,	V_I = 7 V			0.1			0.1	mA
I_{IH}		V_{CC} = 5.5 V,	V_I = 2.7 V			20			20	µA
I_{IL}		V_{CC} = 5.5 V,	V_I = 0.4 V			−0.3			−0.3	mA
I_{CC}	'AS756	V_{CC} = 5.5 V	Outputs high		16			16		mA
			Outputs low		51			51		
	'AS757		Outputs high		27			27		
			Outputs low		61			61		

† All typical values are at V_{CC} = 5 V, T_A = 25 °C.
Additional information on these products can be obtained from the factory as it becomes available.

ALS AND AS CIRCUITS

TEXAS INSTRUMENTS
POST OFFICE BOX 225012 • DALLAS, TEXAS 75265

TYPES SN54AS756, SN54AS757, SN74AS756, SN74AS757
OCTAL BUFFERS AND LINE DRIVERS WITH OPEN-COLLECTOR OUTPUTS

'AS756 switching characteristics (see Note 1)

PARAMETER	FROM (INPUT)	TO (OUTPUT)	V_{CC} = 4.5 V to 5.5 V, C_L = 50 pF, R_L = 500 Ω, T_A = MIN to MAX						UNIT
			SN54AS756			SN74AS756			
			MIN	TYP†	MAX	MIN	TYP†	MAX	
t_{PLH}	A	Y		20			20		ns
t_{PHL}				6			6		
t_{PLH}	$\overline{G}$	Y		22			22		ns
t_{PHL}				8			8		

'AS757 switching characteristics (see Note 1)

PARAMETER	FROM (INPUT)	TO (OUTPUT)	V_{CC} = 4.5 V to 5.5 V, C_L = 50 pF, R_L = 500 Ω, T_A = MIN to MAX						UNIT
			SN54AS757			SN74AS757			
			MIN	TYP†	MAX	MIN	TYP†	MAX	
t_{PLH}	A	Y		20			20		ns
t_{PHL}				6			6		
t_{PLH}	$1\overline{G}$	Y		22			22		ns
t_{PHL}				8			8		
t_{PLH}	2G	Y		23			23		ns
t_{PHL}				9			9		

†All typical values are at V_{CC} = 5 V, T_A = 25°C.
NOTE 1: For load circuit and voltage waveforms, see page 1-12.
Additional information on these products can be obtained from the factory as it becomes available.

TYPES SN54AS758, SN54AS759, SN74AS758, SN74AS759
QUADRUPLE BUS TRANSCEIVERS WITH OPEN-COLLECTOR OUTPUTS

DECEMBER 1983

- 2-Way Asynchronous Communication Between Data Buses
- P-N-P Inputs Reduce Loading
- Dependable Texas Instruments Quality and Reliability
- Open-Collector Versions of 'AS242, 'AS243

description

These four-data-line transceivers are designed for asynchronous two-way communications between data buses.

The SN54' family is characterized for operation over the full military temperature range of $-55\,°C$ to $125\,°C$. The SN74' family is characterized for operation from $0\,°C$ to $70\,°C$.

SN54' . . . J PACKAGE
SN74' . . . N PACKAGE
(TOP VIEW)

$\overline{GAB}$	1	14	V_{CC}
NC	2	13	GBA
A1	3	12	NC
A2	4	11	B1
A3	5	10	B2
A4	6	9	B3
GND	7	8	B4

SN54' . . . FH PACKAGE
SN74' . . . FN PACKAGE
(TOP VIEW)

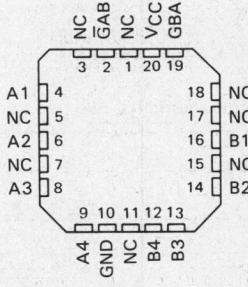

NC—No internal connection

logic symbol

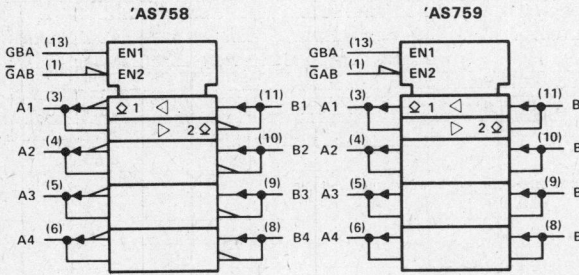

logic diagrams (positive logic)

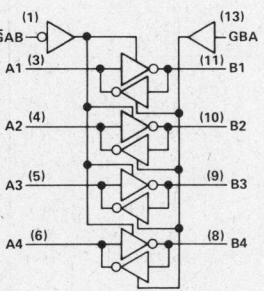
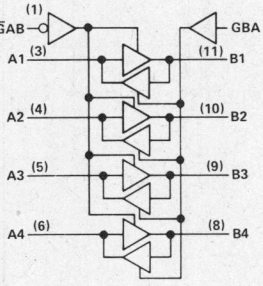

Pin numbers shown are for J and N packages.

FUNCTION TABLE

INPUTS		'AS758	'AS759
$\overline{GAB}$	GBA		
L	L	$\overline{A}$ to B	A to B
H	H	$\overline{B}$ to A	B to A
H	L	Isolation	Isolation
L	H	Latch A and B ($A=\overline{B}$)	Latch A and B ($A=B$)

PRODUCT PREVIEW

This document contains information on a product under development. Texas Instruments reserves the right to change or discontinue this product without notice.

Copyright © 1983 by Texas Instruments Incorporated

TEXAS INSTRUMENTS
POST OFFICE BOX 225012 • DALLAS, TEXAS 75265

TYPES SN54AS758, SN54AS759, SN74AS758, SN74AS759
QUADRUPLE BUS TRANSCEIVERS WITH OPEN-COLLECTOR OUTPUTS

absolute maximum ratings over operating free-air temperature range (unless otherwise noted)

Supply voltage, V_{CC} .... 7 V
Input voltage: All inputs and I/O ports .... 7 V
Operating free-air temperature range: SN54AS758, SN54AS759 .... −55 °C to 125 °C
                                         SN74AS758, SN74AS759 .... 0 °C to 70 °C
Storage temperature range .... −65 °C to 150 °C

recommended operating conditions

		SN54AS758 SN54AS759			SN74AS758 SN74AS759			UNIT
		MIN	NOM	MAX	MIN	NOM	MAX	
V_{CC}	Supply voltage	4.5	5	5.5	4.5	5	5.5	V
V_{IH}	High-level input voltage	2			2			V
V_{IL}	Low-level input voltage			0.8			0.8	V
V_{OH}	High-level output voltage			5.5			5.5	V
I_{OL}	Low-level output current			48			64	mA
T_A	Operating free-air temperature	−55		125	0		70	°C

electrical characteristics over recommended operating free-air temperature range (unless otherwise noted)

PARAMETER		TEST CONDITIONS		SN54AS758 SN54AS759			SN74AS758 SN74AS759			UNIT
				MIN	TYP†	MAX	MIN	TYP†	MAX	
V_{IK}		V_{CC} = 4.5 V,	I_I = −18 mA			−1.2			−1.2	V
I_{OH}		V_{CC} = 4.5 V,	V_{OH} = 5.5 V			0.1			0.1	mA
V_{OL}		V_{CC} = 4.5 V,	I_{OL} = 48 mA			0.55				V
		V_{CC} = 4.5 V,	I_{OL} = 64 mA						0.55	
I_I	Control inputs	V_{CC} = 5.5 V,	V_I = 7 V			0.1			0.1	mA
	A or B ports	V_{CC} = 5.5 V,	V_I = 5.5 V			0.1			0.1	
I_{IH}	Control inputs	V_{CC} = 5.5 V,	V_I = 2.7 V			20			20	µA
	A or B ports‡									
I_{IL}	Control inputs	V_{CC} = 5.5 V,	V_I = 0.4 V			−0.1			−0.1	mA
	A or B ports‡					−0.1			−0.1	
I_{CC}	'AS758	V_{CC} = 5.5 V	Outputs high			18			18	mA
			Outputs low			34			34	
	'AS759		Outputs high			28			28	
			Outputs low			34			34	

†All typical values are at V_{CC} = 5 V, T_A = 25 °C
‡For I/O ports, the parameters I_{IH} and I_{IL} include the off-state output current.
Additional information on these products can be obtained from the factory as it becomes available.

TEXAS INSTRUMENTS
POST OFFICE BOX 225012 • DALLAS, TEXAS 75265

TYPES SN54AS758, SN54AS759, SN74AS758, SN74AS759
QUADRUPLE BUS TRANSCEIVERS WITH OPEN-COLLECTOR OUTPUTS

'AS758 switching characteristics (see Note 1)

PARAMETER	FROM (INPUT)	TO (OUTPUT)	V_{CC} = 4.5 V to 5.5 V, C_L = 50 pF, R_L = 500 Ω, T_A = MIN to MAX						UNIT
			SN54AS758			SN74AS758			
			MIN	TYP†	MAX	MIN	TYP†	MAX	
t_{PLH}	A or B	B or A		20			20		ns
t_{PHL}				6			6		
t_{PLH}	GBA	A		23			23		ns
t_{PHL}				9			9		
t_{PLH}	$\overline{G}$AB	B		22			22		ns
t_{PHL}				8			8		

'AS759 switching characteristics (see Note 1)

PARAMETER	FROM (INPUT)	TO (OUTPUT)	V_{CC} = 4.5 V to 5.5 V, C_L = 50 pF, R_L = 500 Ω, T_A = MIN to MAX						UNIT
			SN54AS759			SN74AS759			
			MIN	TYP†	MAX	MIN	TYP†	MAX	
t_{PLH}	A or B	B or A		20			20		ns
t_{PHL}				6			6		
t_{PLH}	GBA	A		23			23		ns
t_{PHL}				9			9		
t_{PLH}	$\overline{G}$AB	B		22			22		ns
t_{PHL}				8			8		

†All typical values are at V_{CC} = 5 V, T_A = 25°C.
NOTE 1: For load circuit and voltage waveforms, see page 1-12.
Additional information on these products can be obtained from the factory as it becomes available.

TEXAS INSTRUMENTS
POST OFFICE BOX 225012 • DALLAS, TEXAS 75265

2 ALS AND AS CIRCUITS

TYPES SN54AS760, SN74AS760
OCTAL BUFFERS AND LINE DRIVERS WITH OPEN-COLLECTOR OUTPUTS

DECEMBER 1983

- Open-Collector Outputs Drive Bus Lines or Buffer Memory Address Registers
- Eliminates the Need For 3-State Overlap Protection
- P-N-P Inputs Reduce DC Loading
- Package Options Include Both Plastic and Ceramic Chip Carriers in Addition to Plastic and Ceramic DIPs
- Dependable Texas Instruments Quality and Reliability
- Open-Collector Version of 'AS244

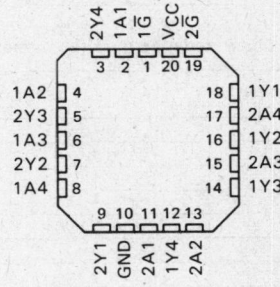

description

These octal buffers and line drivers are designed specifically to improve both the performance and density of three-state memory address drivers, clock drivers, and bus-oriented receivers and transmitters by eliminating the need for 3-state overlap protection. Taken together with the 'AS756 and 'AS757, these devices provide the choice of selected combinations of inverting outputs, symmetrical $\overline{G}$ (active-low input control) inputs, and complementary G and $\overline{G}$ inputs.

The SN54AS760 is characterized for operation over the full military temperature range of −55°C to 125°C. The SN74AS760 is characterized for operation from 0°C to 70°C.

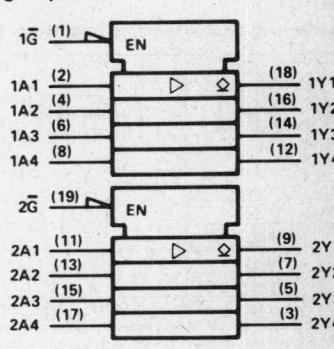

functional block diagram (positive logic)

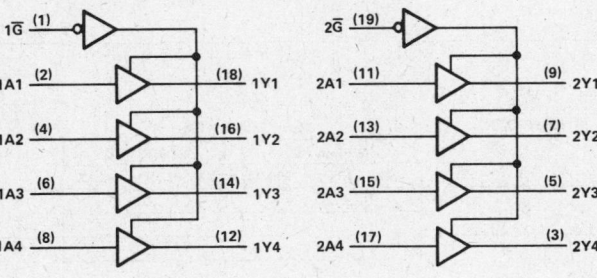

Pin numbers shown are for J and N packages.

logic symbol

ALS AND AS CIRCUITS

PRODUCT PREVIEW
This document contains information on a product under development. Texas Instruments reserves the right to change or discontinue this product without notice.

Copyright © 1983 by Texas Instruments Incorporated

TEXAS INSTRUMENTS
POST OFFICE BOX 225012 • DALLAS, TEXAS 75265

2-499

TYPES SN54AS760, SN74AS760
OCTAL BUFFERS AND LINE DRIVERS WITH OPEN-COLLECTOR OUTPUTS

absolute maximum ratings over operating free-air temperature range (unless otherwise noted)

Supply voltage, V_{CC} .... 7 V
Input voltage .... 7 V
Off-state output voltage .... 7 V
Operating free-air temperature range: SN54AS760 .... $-55\,°C$ to $125\,°C$
$\qquad\qquad\qquad\qquad\qquad\qquad\;\;$ SN74AS760 .... $0\,°C$ to $70\,°C$
Storage temperature range .... $-65\,°C$ to $150\,°C$

recommended operating conditions

		SN54AS760			SN74AS760			UNIT
		MIN	NOM	MAX	MIN	NOM	MAX	
V_{CC}	Supply voltage	4.5	5	5.5	4.5	5	5.5	V
V_{IH}	High-level input voltage	2			2			V
V_{IL}	Low-level input voltage			0.8			0.8	V
V_{OH}	High-level output voltage			5.5			5.5	V
I_{OL}	Low-level output current			48			64	mA
T_A	Operating free-air temperature	-55		125	0		70	°C

electrical characteristics over recommended operating free-air temperature range (unless otherwise noted)

PARAMETER	TEST CONDITIONS		SN54AS760			SN74AS760			UNIT
			MIN	TYP†	MAX	MIN	TYP†	MAX	
V_{IK}	$V_{CC} = 4.5\,V$,	$I_I = -18\,mA$			-1.2			-1.2	V
I_{OH}	$V_{CC} = 4.5\,V$,	$V_{OH} = 5.5\,V$			0.1			0.1	mA
V_{OL}	$V_{CC} = 4.5\,V$,	$I_{OL} = 48\,mA$			0.55				V
	$V_{CC} = 4.75\,V$,	$I_{OL} = 64\,mA$						0.55	
I_I	$V_{CC} = 5.5\,V$,	$V_I = 7\,V$			0.1			0.1	mA
I_{IH}	$V_{CC} = 5.5\,V$,	$V_I = 2.7\,V$			20			20	µA
I_{IL}	$V_{CC} = 5.5\,V$,	$V_I = 0.4\,V$			-0.3			-0.3	mA
I_{CC}	$V_{CC} = 5.5\,V$	Outputs high			22			22	mA
		Outputs low			60			60	

switching characteristics (see Note 1)

PARAMETER	FROM (INPUT)	TO (OUTPUT)	$V_{CC} = 4.5\,V$ to $5.5\,V$, $C_L = 50\,pF$, $R_L = 500\,\Omega$, $T_A = $ MIN to MAX						UNIT
			SN54AS760			SN74AS760			
			MIN	TYP†	MAX	MIN	TYP†	MAX	
t_{PLH}	A	Y		20			20		ns
t_{PHL}				6			6		
t_{PLH}	$\overline{G}$	Y		22			22		ns
t_{PHL}				8			8		

† All typical values are at $V_{CC} = 5\,V$, $T_A = 25\,°C$.
NOTE 1: For load circuit and voltage waveforms, see page 1-12.
Additional information on these products can be obtained from the factory as it becomes available.

TYPES SN54AS762, SN54AS763, SN74AS762, SN74AS763
OCTAL BUFFERS AND LINE DRIVERS WITH OPEN-COLLECTOR OUTPUTS

DECEMBER 1983

- Included Among the Package Options Are 20-Pin DIPs and Both Plastic and Ceramic Chip Carriers
- 'AS762 Has True and Complementary Outputs
- 'AS763 Has Complementary G and $\overline{G}$ Inputs
- Open-Collector Outputs Drive Bus Lines or Buffer Memory Address Registers
- Eliminates the Need for 3-State Overlap Protection
- Current Sinking Capability Up to 64 mA
- Dependable Texas Instruments Quality and Reliability

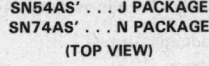

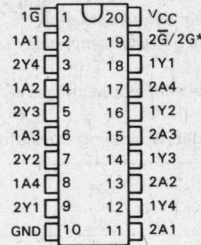

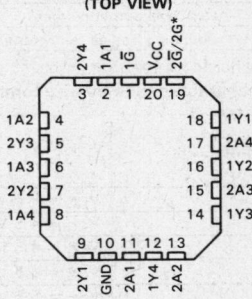

*2$\overline{G}$ for 'AS762 or 2G for 'AS763

description

These octal buffers and line drivers are designed specifically to improve the performance of three-state memory address drivers, clock drivers, and bus-oriented receivers and transmitters by eliminating the need for 3-state overlap protection. The designer has a choice of selected combinations of inverting and noninverting outputs, symmetrical $\overline{G}$ (active-low output control) inputs, and complementary G and $\overline{G}$ inputs.

The SN74AS762 and SN74AS763 can be used to drive terminated lines down to 133 ohms.

The SN54AS762 and SN54AS763 are characterized for operation over the full military temperature range of $-55\,°C$ to $125\,°C$. The SN74AS762 and SN74AS763 are characterized for operation from $0\,°C$ to $70\,°C$.

logic symbols

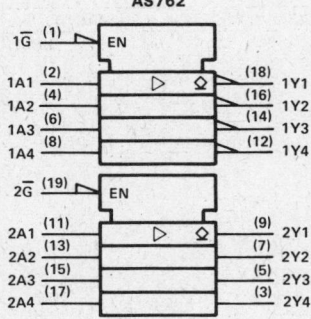

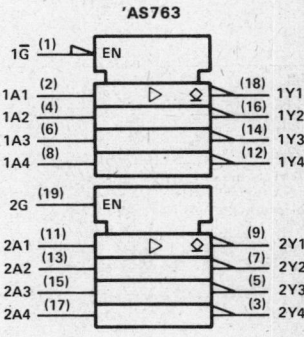

Pin numbers shown are for J and N packages.

PRODUCT PREVIEW

This document contains information on a product under development. Texas Instruments reserves the right to change or discontinue this product without notice.

Copyright © 1983 by Texas Instruments Incorporated

TEXAS INSTRUMENTS
POST OFFICE BOX 225012 • DALLAS, TEXAS 75265

TYPES SN54AS762, SN54AS763, SN74AS762, SN74AS763
OCTAL BUFFERS AND LINE DRIVERS WITH OPEN-COLLECTOR OUTPUTS

absolute maximum ratings over operating free-air temperature range (unless otherwise noted)

Supply voltage, V_{CC} .... 7 V
Input voltage .... 7 V
Off-state output voltage .... 7 V
Operating free-air temperature range: SN54AS762, SN54AS763 .... −55 °C to 125 °C
                                      SN74AS762, SN74AS763 .... 0 °C to 70 °C
Storage temperature range .... −65 °C to 150 °C

recommended operating conditions

		SN54AS762 SN54AS763			SN74AS762 SN74AS763			UNIT
		MIN	NOM	MAX	MIN	NOM	MAX	
V_{CC}	Supply voltage	4.5	5	5.5	4.5	5	5.5	V
V_{IH}	High-level input voltage	2			2			V
V_{IL}	Low-level input voltage			0.8			0.8	V
V_{OH}	High-level output voltage			5.5			5.5	V
I_{OL}	Low-level output current			48			64	mA
T_A	Operating free-air temperature	−55		125	0		70	°C

electrical characteristics over recommended operating free-air temperature range (unless otherwise noted)

PARAMETER		TEST CONDITIONS		SN54AS762 SN54AS763			SN74AS762 SN74AS763			UNIT
				MIN	TYP†	MAX	MIN	TYP†	MAX	
V_{IK}		V_{CC} = 4.5 V,	I_I = −18 mA			−1.2			−1.2	V
I_{OH}		V_{CC} = 4.5 V,	V_{OH} = 5.5 V			0.1			0.1	mA
V_{OL}		V_{CC} = 4.5 V,	I_{OL} = 48 mA			0.55				V
		V_{CC} = 4.5 V,	I_{OL} = 64 mA						0.55	
I_I		V_{CC} = 5.5 V,	V_I = 7 V			0.1			0.1	mA
I_{IH}		V_{CC} = 5.5 V,	V_I = 2.7 V			20			20	μA
I_{IL}		V_{CC} = 5.5 V,	V_I = 0.4 V			−0.5			−0.5	mA
I_{CC}	'AS762	V_{CC} = 5.5 V	Outputs high		21			21		mA
			Outputs low		55			55		
	'AS763	V_{CC} = 5.5 V	Outputs high		17			17		mA
			Outputs low		52			52		

†All typical values are at V_{CC} = 5 V, T_A = 25 °C.
Additional information on these products can be obtained from the factory as it becomes available.

TYPES SN54AS762, SN54AS763, SN74AS762, SN74AS763
OCTAL BUFFERS AND LINE DRIVERS WITH OPEN-COLLECTOR OUTPUTS

'AS762 switching characteristics (see Note 1)

PARAMETER	FROM (INPUT)	TO (OUTPUT)	V_{CC} = 4.5 V to 5.5 V, C_L = 50 pF, R_L = 500 Ω, T_A = MIN to MAX						UNIT
			SN54AS762			SN74AS762			
			MIN	TYP†	MAX	MIN	TYP†	MAX	
t_{PLH}	1A	1Y		20			20		ns
t_{PHL}				6			6		
t_{PLH}	2A	2Y		20			20		ns
t_{PHL}				6			6		
t_{PLH}	$\overline{G}$	1Y		22			22		ns
t_{PHL}				8			8		
t_{PLH}	$\overline{G}$	2Y		22			22		ns
t_{PHL}				8			8		

'AS763 switching characteristics (see Note 1)

PARAMETER	FROM (INPUT)	TO (OUTPUT)	V_{CC} = 4.5 V to 5.5 V, C_L = 50 pF, R_L = 500 Ω, T_A = MIN to MAX						UNIT
			SN54AS763			SN74AS763			
			MIN	TYP†	MAX	MIN	TYP†	MAX	
t_{PLH}	A	Y		20			20		ns
t_{PHL}				6			6		
t_{PLH}	$\overline{G}$	Y		22			22		ns
t_{PHL}				8			8		
t_{PLH}	G	Y		23			23		ns
t_{PHL}				9			9		

†All typical values are at V_{CC} = 5 V, T_A = 25°C.
NOTE 1: For load circuit and voltage waveforms, see page 1-12.
Additional information on these products can be obtained from the factory as it becomes available.

TEXAS INSTRUMENTS
POST OFFICE BOX 225012 • DALLAS, TEXAS 75265

ALS AND AS CIRCUITS

TYPES SN54AS800, SN74AS800
TRIPLE 4-INPUT AND/NAND DRIVERS

D2661, DECEMBER 1982–REVISED DECEMBER 1983

- Less than 0.5 ns Skew between True and Complementary Outputs
- High Capacitive-Drive Capability
- Current Sink/Source Capability Up to 48 mA
- Approximately 35% Improvement in AC Performance over Schottky TTL
- Package Options Include DIPs and Both Plastic and Ceramic Chip Carriers
- Suitable for Use in Applications such as:
  — Differential Line Drivers
  — Complementary Input Circuit for Decoders and Code Converters
  — Symmetrical Complementary Clock Generators
- Dependable Texas Instruments Quality and Reliability

SN54AS800 . . . . J PACKAGE
SN74AS800 . . . . N PACKAGE
(TOP VIEW)

```
1A  [ 1    20 ] VCC
2A  [ 2    19 ] 1D
2B  [ 3    18 ] 1C
2C  [ 4    17 ] 1B
2D  [ 5    16 ] 1Y
3A  [ 6    15 ] 1Z
3B  [ 7    14 ] 2Z
3C  [ 8    13 ] 2Y
3D  [ 9    12 ] 3Y
GND [ 10   11 ] 3Z
```

SN54AS800 . . . . FH PACKAGE
SN74AS800 . . . . FN PACKAGE
(TOP VIEW)

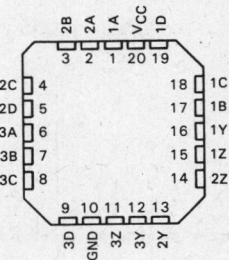

description

The 'AS800 is especially suitable for symmetrical complementary clock-generator applications due to the delay time in either function (AND/NAND) being typically 4 ns with less than 0.5 ns skew between the true and complementary outputs. Elimination of decode spikes in symmetrical decoder and code converter applications, and the high capacitive-drive capability coupled with high current-sinking capability (48 mA), make the device useful for applications such as a decoder or differential line driver.

The SN54AS800 is characterized for operation over the full military temperature range of $-55°C$ to $125°C$. The SN74AS800 is characterized for operation from $0°C$ to $70°C$.

PRODUCT PREVIEW

This document contains information on a product under development. Texas Instruments reserves the right to change or discontinue this product without notice.

Copyright © 1982 by Texas Instruments Incorporated

POST OFFICE BOX 225012 • DALLAS, TEXAS 75265

TYPES SN54AS800, SN74AS800
TRIPLE 4-INPUT AND/NAND DRIVERS

logic symbol

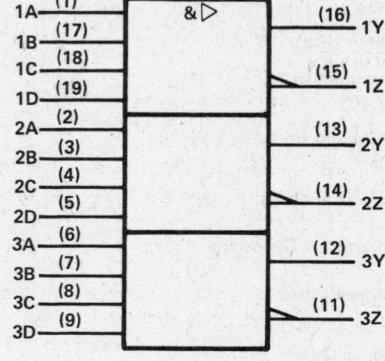

Pin numbers shown are for J and N packages.

positive logic: $Y = ABCD$
$Z = \overline{ABCD}$

absolute maximum ratings over free-air temperature range (unless otherwise noted)

Supply voltage, V_{CC} .... 7 V
Input voltage .... 7 V
Operating free-air temperature range: SN54AS800 .... −55 °C to 125 °C
                                       SN74AS800 .... 0 °C to 70 °C
Storage temperature range .... −65 °C to 150 °C

recommended operating conditions

		SN54AS800			SN74AS800			UNIT
		MIN	NOM	MAX	MIN	NOM	MAX	
V_{CC}	Supply voltage	4.5	5	5.5	4.5	5	5.5	V
V_{IH}	High-level input voltage	2			2			V
V_{IL}	Low-level input voltage			0.8			0.8	V
I_{OH}	High-level output current			−40			−48	mA
I_{OL}	Low-level output current			40			48	mA
T_A	Operating free-air temperature	−55		125	0		70	°C

TYPES SN54AS800, SN74AS800
TRIPLE 4-INPUT AND/NAND DRIVERS

electrical characteristics over recommended operating free-air temperature range (unless otherwise noted)

PARAMETER	TEST CONDITIONS		SN54AS800			SN74AS800			UNIT
			MIN	TYP†	MAX	MIN	TYP†	MAX	
V_{IK}	V_{CC} = 4.5 V,	I_I = −18 mA			−1.2			−1.2	V
V_{OH}	V_{CC} = 4.5 V to 5.5 V,	I_{OH} = −2 mA	V_{CC}−2			V_{CC}−2			V
	V_{CC} = 4.5 V,	I_{OH} = −3 mA	2.4	3.2		2.4	3.2		
	V_{CC} = 4.5 V,	I_{OH} = −40 mA	2						
	V_{CC} = 4.5 V,	I_{OH} = −48 mA				2			
V_{OL}	V_{CC} = 4.5 V,	I_{OL} = 40 mA		0.25	0.5				V
	V_{CC} = 4.5 V,	I_{OL} = 48 mA					0.35	0.5	
I_I	V_{CC} = 5.5 V,	V_I = 7 V			0.1			0.1	mA
I_{IH}	V_{CC} = 5.5 V,	V_I = 2.7 V			20			20	µA
I_{IL}	V_{CC} = 5.5 V,	V_I = 0.4 V			−0.3			−0.3	mA
I_O‡	V_{CC} = 5.5 V,	V_O = 2.25 V	−150			−150			mA
I_{CC}	V_{CC} = 5.5 V,			13			13		mA

†All typical values are at V_{CC} = 5 V, T_A = 25 °C.
‡The output conditions have been chosen to produce a current that closely approximates one half of the true short-circuit output current, I_{OS}.

switching characteristics (see Note 1)

PARAMETER	FROM (INPUT)	TO (OUTPUT)	V_{CC} = 4.5 V to 5.5 V, C_L = 50 pF, R_L = 500 Ω, T_A = MIN to MAX						UNIT
			SN54AS800			SN74AS800			
			MIN	TYP†	MAX	MIN	TYP†	MAX	
t_{PLH}	A,B,C, or D	Z		3.5			3.5		ns
t_{PHL}				3.5			3.5		
t_{PLH}		Y		3			3		ns
t_{PHL}				4			4		

†All typical values are at V_{CC} = 5 V, T_A = 25 °C.
NOTE 1: For load circuit and voltage waveforms, see page 1-12.

2
ALS AND AS CIRCUITS

TYPES SN54AS802, SN74AS802
TRIPLE 4-INPUT OR/NOR LINE DRIVERS

D2662, DECEMBER 1982—REVISED DECEMBER 1983

- **True and Complementary Outputs**
- **Less than 0.5 ns Skew between Outputs**
- **High Capacitive Drive Capability**
- **Approximately 35% Improvement in AC Performance over Schottky TTL**
- **Current Sink/Source Capability Up to 48 mA**
- **Package Options Include Plastic and Ceramic DIPs as well as Both Plastic and Ceramic Chip Carriers**
- **Designed Specifically for Use in Applications such as:**
  — Symmetrical Complementary Clock Generators
  — Complementary Input Circuit for Decoders and Code Converters
  — Differential Line Drivers
- **Dependable Texas Instruments Quality and Reliability**

SN54AS802 .... J PACKAGE
SN74AS802 .... N PACKAGE
(TOP VIEW)

```
 1A [ 1    20 ] Vcc
 2A [ 2    19 ] 1D
 2B [ 3    18 ] 1C
 2C [ 4    17 ] 1B
 2D [ 5    16 ] 1Y
 3A [ 6    15 ] 1Z
 3B [ 7    14 ] 2Z
 3C [ 8    13 ] 2Y
 3D [ 9    12 ] 3Y
GND [10    11 ] 3Z
```

SN54AS802 .... FH PACKAGE
SN74AS802 .... FN PACKAGE
(TOP VIEW)

```
         2B 2A 1A Vcc 1D
          3  2  1  20 19
    2C [ 4            18 ] 1C
    2D [ 5            17 ] 1B
    3A [ 6            16 ] 1Y
    3B [ 7            15 ] 1Z
    3C [ 8            14 ] 2Z
          9 10 11 12 13
         3D GND 3Z 3Y 2Y
```

description

The 'AS802 is uniquely suitable for symmetrical complementary clock-generator applications due to the delay time in either function (OR/NOR) being typically 4 ns with less than 0.5 ns skew between the true and complementary outputs. Elimination of decode spikes in symmetrical decoder and code converter applications, and the high capacitive drive capability coupled with high current-sinking capability (48 mA), make the device useful for applications such as a decoder or differential line driver.

The SN54AS802 is characterized for operation over the full military temperature range of −55°C to 125°C. The SN74AS802 is characterized for operation from 0°C to 70°C.

PRODUCT PREVIEW

This document contains information on a product under development. Texas Instruments reserves the right to change or discontinue this product without notice.

Copyright © 1982 Texas Instruments Incorporated

TEXAS INSTRUMENTS
POST OFFICE BOX 225012 • DALLAS, TEXAS 75265

TYPES SN54AS802, SN74AS802
TRIPLE 4-INPUT OR/NOR LINE DRIVERS

logic symbol

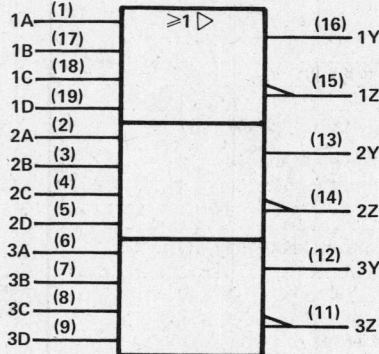

Pin numbers shown are for J and N packages.

positive logic: $Y = A + B + C + D$
$Z = \overline{A + B + C + D}$

absolute maximum ratings over operating free-air temperature range (unless otherwise noted)

Supply voltage, V_{CC} .... 7 V
Input voltage .... 7 V
Operating free-air temperature range: SN54AS802 .... −55 °C to 125 °C
 SN74AS802 .... 0 °C to 70 °C
Storage temperature range .... −65 °C to 150 °C

recommended operating conditions

		SN54AS802			SN74AS802			UNIT
		MIN	NOM	MAX	MIN	NOM	MAX	
V_{CC}	Supply voltage	4.5	5	5.5	4.5	5	5.5	V
V_{IH}	High-level input voltage	2			2			V
V_{IL}	Low-level input voltage			0.8			0.8	V
I_{OH}	High-level output current			−40			−48	mA
I_{OL}	Low-level output current			40			48	mA
T_A	Operating free-air temperature	−55		125	0		70	°C

TEXAS
INSTRUMENTS
POST OFFICE BOX 225012 • DALLAS, TEXAS 75265

TYPES SN54AS802, SN74AS802
TRIPLE 4-INPUT OR/NOR LINE DRIVERS

electrical characteristics over recommended operating free-air temperature range (unless otherwise noted)

PARAMETER	TEST CONDITIONS		SN54AS802			SN74AS802			UNIT
			MIN	TYP†	MAX	MIN	TYP†	MAX	
V_{IK}	V_{CC} = 4.5 V,	I_I = −18 mA			−1.2			−1.2	V
V_{OH}	V_{CC} = 4.5 V to 5.5 V,	I_{OH} = −2 mA	V_{CC}−2			V_{CC}−2			V
	V_{CC} = 4.5 V,	I_{OH} = −3 mA	2.4	3.2		2.4	3.2		
	V_{CC} = 4.5 V,	I_{OH} = −40 mA	2						
	V_{CC} = 4.5 V,	I_{OH} = −48 mA				2			
V_{OL}	V_{CC} = 4.5 V,	I_{OL} = 40 mA		0.25	0.5				V
	V_{CC} = 4.5 V,	I_{OL} = 48 mA					0.35	0.5	
I_I	V_{CC} = 5.5 V,	V_I = 7 V			0.1			0.1	mA
I_{IH}	V_{CC} = 5.5 V,	V_I = 2.7 V			20			20	µA
I_{IL}	V_{CC} = 5.5 V,	V_I = 0.4 V			−0.3			−0.3	mA
I_O‡	V_{CC} = 5.5 V,	V_O = 2.25 V	−150			−150			mA
I_{CC}	V_{CC} = 5.5 V,			20			20		mA

†All typical values are at V_{CC} = 5 V, T_A = 25°C.
‡The output conditions have been chosen to produce a current that closely approximates one half of the true short-circuit output current, I_{OS}.

switching characteristics (see Note 1)

PARAMETER	FROM (INPUT)	TO (OUTPUT)	V_{CC} = 4.5 V to 5.5 V, C_L = 50 pF, R_L = 500 Ω, T_A = MIN to MAX						UNIT
			SN54AS802			SN74AS802			
			MIN	TYP†	MAX	MIN	TYP†	MAX	
t_{PLH}	A, B, C, D	Y		3.5			3.5		ns
t_{PHL}				4.5			4.5		
t_{PLH}		Z		4			4		ns
t_{PHL}				5			5		

†All typical values are at V_{CC} = 5 V, T_A = 25°C.
NOTE 1: For load circuit and voltage waveforms, see page 1-12.

Texas Instruments
POST OFFICE BOX 225012 • DALLAS, TEXAS 75265

2
ALS AND AS CIRCUITS

TYPES SN54ALS804, SN54AS804A, SN74ALS804, SN74AS804A
HEX 2-INPUT NAND DRIVERS

D2661, DECEMBER 1982—REVISED DECEMBER 1983

- **High Capacitive Drive Capability**
- **'ALS804 Has Typical Delay Time of 4 ns (C_L = 50 pF) and Typical Power Dissipation of 3.4 mW per Gate**
- **'AS804A Has Typical Delay Time of 2.6 ns (C_L = 50 pF) and Typical Power Dissipation of Less than 9 mW per Gate**
- **Package Options Include Both Plastic and Ceramic Chip Carriers in Addition to Plastic and Ceramic DIPs**
- **Dependable Texas Instruments Quality and Reliability**

SN54ALS804, SN54AS804A . . . J PACKAGE
SN74ALS804, SN74AS804A . . . N PACKAGE
(TOP VIEW)

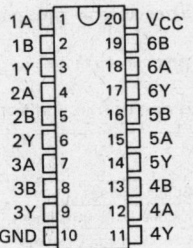

description

These devices contain six independent 2-input NAND drivers. They perform the Boolean functions $Y = \overline{A \cdot B}$ or $Y = \overline{A} + \overline{B}$ in positive logic.

The −1 version of the SN74ALS804 parts is identical to the standard version except that the recommended maximum I_{OL} is increased to 48 milliamperes. There is no −1 version of the SN54ALS804 parts.

The SN54ALS804 and SN54AS804A are characterized for operation over the full military temperature range of −55°C to 125°C. The SN74ALS804 and SN74AS804A are characterized for operation from 0°C to 70°C.

SN54ALS804, SN54AS804A . . . FH PACKAGE
SN74ALS804, SN74AS804A . . . FN PACKAGE
(TOP VIEW)

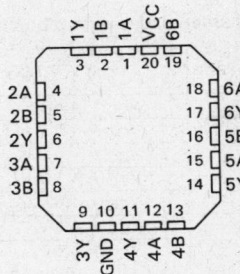

FUNCTION TABLE (each driver)

INPUTS		OUTPUT
A	B	Y
H	H	L
L	X	H
X	L	H

logic symbol

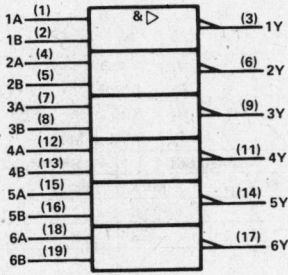

Pin numbers shown are for J and N packages.

Copyright © 1983 by Texas Instruments Incorporated

TEXAS INSTRUMENTS
INCORPORATED
POST OFFICE BOX 225012 • DALLAS, TEXAS 75265

TYPES SN54ALS804, SN74ALS804
HEX 2-INPUT NAND DRIVERS

absolute maximum ratings over operating free-air temperature range (unless otherwise noted)

Supply voltage, V_{CC} .... 7 V
Input voltage .... 7 V
Operating free-air temperature range: SN54ALS804 .... −55 °C to 125 °C
                                       SN74ALS804 .... 0 °C to 70 °C
Storage temperature range .... −65 °C to 150 °C

recommended operating conditions

		SN54ALS804			SN74ALS804			UNIT
		MIN	NOM	MAX	MIN	NOM	MAX	
V_{CC}	Supply voltage	4.5	5	5.5	4.5	5	5.5	V
V_{IH}	High-level input voltage	2			2			V
V_{IL}	Low-level input voltage			0.8			0.8	V
I_{OH}	High-level output current			−12			−15	mA
I_{OL}	Low-level output current			12			24	mA
							48†	
T_A	Operating free-air temperature	−55		125	0		70	°C

† The extended limit applies if V_{CC} is maintained between 4.75 V and 5.25 V.
The 48 mA limit applies for the SN74ALS804−1 only.

electrical characteristics over recommended operating free-air temperature range (unless otherwise noted)

PARAMETER	TEST CONDITIONS		SN54ALS804			SN74ALS804			UNIT
			MIN	TYP‡	MAX	MIN	TYP‡	MAX	
V_{IK}	V_{CC} = 4.5 V,	I_I = −18 mA			−1.5			−1.5	V
	V_{CC} = 4.5 V to 5.5 V,	I_{OH} = −0.4 mA	V_{CC}−2			V_{CC}−2			
V_{OH}	V_{CC} = 4.5 V,	I_{OH} = −3 mA	2.4	3.2		2.4	3.2		V
	V_{CC} = 4.5 V,	I_{OH} = −12 mA	2						
	V_{CC} = 4.5 V,	I_{OH} = −15 mA				2			
V_{OL}	V_{CC} = 4.5 V,	I_{OL} = 12 mA		0.25	0.4		0.25	0.4	V
	V_{CC} = 4.5 V,	I_{OL} = 24 mA					0.35	0.5	
	(I_{OL} = 48 mA for −1 version)								
I_I	V_{CC} = 5.5 V,	V_I = 7 V			0.1			0.1	mA
I_{IH}	V_{CC} = 5.5 V,	V_I = 2.7 V			20			20	μA
I_{IL}	V_{CC} = 5.5 V,	V_I = 0.4 V			−0.1			−0.1	mA
I_O §	V_{CC} = 5.5 V,	V_O = 2.25 V	−30		−112	−30		−112	mA
I_{CCH}	V_{CC} = 5.5 V,	V_I = 0 V		0.9	2.5		0.9	2.5	mA
I_{CCL}	V_{CC} = 5.5 V,	V_I = 4.5 V		7	12		7	12	mA

‡ All typical values are at V_{CC} = 5 V, T_A = 25 °C.
§ The output conditions have been chosen to produce a current that closely approximates one half of the true short-circuit output current, I_{OS}.

switching characteristics (see Note 1)

PARAMETER	FROM (INPUT)	TO (OUTPUT)	V_{CC} = 4.5 V to 5.5 V, C_L = 50 pF, R_L = 500 Ω, T_A = MIN to MAX				UNIT
			SN54ALS804		SN74ALS804		
			MIN	MAX	MIN	MAX	
t_{PLH}	A or B	Y	2	8	2	6	ns
t_{PHL}			2	9	2	7	

NOTE 1: For load circuit and voltage waveforms, see page 1-12.

TEXAS INSTRUMENTS
POST OFFICE BOX 225012 • DALLAS, TEXAS 75265

TYPES SN54AS804A, SN74AS804A
HEX 2-INPUT NAND DRIVERS

absolute maximum ratings over operating free-air temperature range (unless otherwise noted)

Supply voltage, V_{CC} .... 7 V
Input voltage .... 7 V
Operating free-air temperature range: SN54AS804A .... −55°C to 125°C
 SN74AS804A .... 0°C to 70°C
Storage temperature range .... −65°C to 150°C

recommended operating conditions

		SN54AS804A			SN74AS804A			UNIT
		MIN	NOM	MAX	MIN	NOM	MAX	
V_{CC}	Supply voltage	4.5	5	5.5	4.5	5	5.5	V
V_{IH}	High-level input voltage	2			2			V
V_{IL}	Low-level input voltage			0.8			0.8	V
I_{OH}	High-level output current			−40			−48	mA
I_{OL}	Low-level output current			40			48	mA
T_A	Operating free-air temperature	−55		125	0		70	°C

electrical characteristics over recommended operating free-air temperature range (unless otherwise noted)

PARAMETER	TEST CONDITIONS		SN54AS804A			SN74AS804A			UNIT
			MIN	TYP†	MAX	MIN	TYP†	MAX	
V_{IK}	$V_{CC} = 4.5$ V,	$I_I = -18$ mA			−1.2			−1.2	V
V_{OH}	$V_{CC} = 4.5$ V to 5.5 V,	$I_{OH} = -2$ mA	$V_{CC}-2$			$V_{CC}-2$			V
	$V_{CC} = 4.5$ V,	$I_{OH} = -3$ mA	2.4	3.2		2.4	3.2		
	$V_{CC} = 4.5$ V,	$I_{OH} = -40$ mA	2						
	$V_{CC} = 4.5$ V,	$I_{OH} = -48$ mA				2			
V_{OL}	$V_{CC} = 4.5$ V,	$I_{OL} = 40$ mA		0.25	0.5				V
	$V_{CC} = 4.5$ V,	$I_{OL} = 48$ mA					0.35	0.5	
I_I	$V_{CC} = 5.5$ V,	$V_I = 7$ V			0.1			0.1	mA
I_{IH}	$V_{CC} = 5.5$ V,	$V_I = 2.7$ V			20			20	µA
I_{IL}	$V_{CC} = 5.5$ V,	$V_I = 0.4$ V			−0.5			−0.5	mA
I_O‡	$V_{CC} = 5.5$ V,	$V_O = 2.25$ V		−135			−135		mA
I_{CCH}	$V_{CC} = 5.5$ V,	$V_I = 0$ V		2.5	4		2.5	4	mA
I_{CCL}	$V_{CC} = 5.5$ V,	$V_I = 4.5$ V		16	27		16	27	mA

† All typical values are at $V_{CC} = 5$ V, $T_A = 25$°C.
‡ The output conditions have been chosen to produce a current that closely approximates one half of the true short-circuit output current, I_{OS}.

switching characteristics (see Note 1)

PARAMETER	FROM (INPUT)	TO (OUTPUT)	$V_{CC} = 4.5$ V to 5.5 V, $C_L = 50$ pF, $R_L = 500$ Ω, T_A = MIN to MAX				UNIT
			SN54AS804A		SN74AS804A		
			MIN	MAX	MIN	MAX	
t_{PLH}	A or B	Y	2	4.5	2	3.5	ns
t_{PHL}			2	4.5	2	3.5	

NOTE 1: For load circuit and voltage waveforms, see page 1-12.

TEXAS INSTRUMENTS
POST OFFICE BOX 225012 • DALLAS, TEXAS 75265

ALS AND AS CIRCUITS

2

ALS AND AS CIRCUITS

TYPES SN54ALS805, SN54AS805A, SN74ALS805, SN74AS805A
HEX 2-INPUT NOR DRIVERS

D2661, DECEMBER 1982 – REVISED DECEMBER 1983

- High Capacitive Drive Capability
- 'ALS805 Has Typical Delay Time of 4.2 ns (C_L = 50 pF) and Typical Power Dissipation of 4.2 mW per Gate
- 'AS805A Has Typical Delay Time of 2.6 ns (C_L = 50 pF) and Typical Power Dissipation of 12 mW per Gate
- Package Options Include Both Plastic and Ceramic Chip Carriers in Addition to Plastic and Ceramic DIPs
- Dependable Texas Instruments Quality and Reliability

SN54ALS805, SN54AS805A . . . J PACKAGE
SN74ALS805, SN74AS805A . . . N PACKAGE
(TOP VIEW)

```
     1A  [ 1    20 ] VCC
     1B  [ 2    19 ] 6B
     1Y  [ 3    18 ] 6A
     2A  [ 4    17 ] 6Y
     2B  [ 5    16 ] 5B
     2Y  [ 6    15 ] 5A
     3A  [ 7    14 ] 5Y
     3B  [ 8    13 ] 4B
     3Y  [ 9    12 ] 4A
    GND  [10    11 ] 4Y
```

description

These devices contain six independent 2-input NOR drivers. They perform the Boolean functions $Y = \overline{A+B}$ or $Y = \overline{A} \cdot \overline{B}$ in positive logic.

The –1 version of the SN74ALS805 parts is identical to the standard version except that the recommended maximum I_{OL} is increased to 48 milliamperes. There is no –1 version of the SN54ALS805 parts.

The SN54ALS805 and SN54AS805A are characterized for operation over the full military temperature range of –55°C to 125°C. The SN74ALS805 and SN74AS805A are characterized for operation from 0°C to 70°C.

SN54ALS805, SN54AS805A . . . FH PACKAGE
SN74ALS805, SN74AS805A . . . FN PACKAGE
(TOP VIEW)

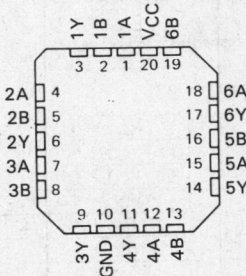

FUNCTION TABLE (each driver)

INPUTS		OUTPUT
A	B	Y
H	X	L
X	H	L
L	L	H

logic symbol

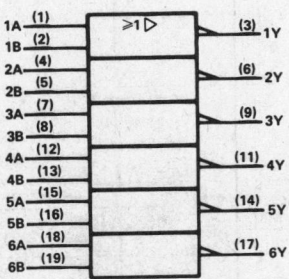

Pin numbers shown are for J and N packages.

Copyright © 1983 by Texas Instruments Incorporated

TEXAS INSTRUMENTS
POST OFFICE BOX 225012 • DALLAS, TEXAS 75265

TYPES SN54ALS805, SN74ALS805
HEX 2-INPUT NOR DRIVERS

absolute maximum ratings over operating free-air temperature range (unless otherwise noted)

Supply voltage, V_{CC} .... 7 V
Input voltage .... 7 V
Operating free-air temperature range: SN54ALS805 .... −55 °C to 125 °C
                                      SN74ALS805 .... 0 °C to 70 °C
Storage temperature range .... −65 °C to 150 °C

recommended operating conditions

		SN54ALS805			SN74ALS805			UNIT
		MIN	NOM	MAX	MIN	NOM	MAX	
V_{CC}	Supply voltage	4.5	5	5.5	4.5	5	5.5	V
V_{IH}	High-level input voltage	2			2			V
V_{IL}	Low-level input voltage			0.8			0.8	V
I_{OH}	High-level output current			−12			−15	mA
I_{OL}	Low-level output current			12			24	mA
							48†	
T_A	Operating free-air temperature	−55		125	0		70	°C

†The extended limit applies if V_{CC} is maintained between 4.75 V and 5.25 V.
The 48 mA limit applies for the SN74ALS805-1 only.

electrical characteristics over recommended operating free-air temperature range (unless otherwise noted)

PARAMETER	TEST CONDITIONS		SN54ALS805			SN74ALS805			UNIT
			MIN	TYP‡	MAX	MIN	TYP‡	MAX	
V_{IK}	V_{CC} = 4.5 V,	I_I = −18 mA			−1.5			−1.5	V
V_{OH}	V_{CC} = 4.5 V to 5.5 V,	I_{OH} = −0.4 mA	V_{CC}−2			V_{CC}−2			V
	V_{CC} = 4.5 V,	I_{OH} = −3 mA	2.4	3.2		2.4	3.2		
	V_{CC} = 4.5 V,	I_{OH} = −12 mA	2						
	V_{CC} = 4.5 V,	I_{OH} = −15 mA				2			
V_{OL}	V_{CC} = 4.5 V,	I_{OL} = 12 mA		0.25	0.4		0.25	0.4	V
	V_{CC} = 4.5 V,	I_{OL} = 24 mA					0.35	0.5	
	(I_{OL} = 48 mA for −1 version)								
I_I	V_{CC} = 5.5 V,	V_I = 7 V			0.1			0.1	mA
I_{IH}	V_{CC} = 5.5 V,	V_I = 2.7 V			20			20	µA
I_{IL}	V_{CC} = 5.5 V,	V_I = 0.4 V			−0.1			−0.1	mA
I_O §	V_{CC} = 5.5 V,	V_O = 2.25 V	−30		−112	−30		−112	mA
I_{CCH}	V_{CC} = 5.5 V,	V_I = 0 V		2	4		2	4	mA
I_{CCL}	V_{CC} = 5.5 V,	V_I = 4.5 V		8	14		8	14	mA

‡All typical values are at V_{CC} = 5 V, T_A = 25 °C.
§The output conditions have been chosen to produce a current that closely approximates one half of the true short-circuit output current, I_{OS}.

switching characteristics (see Note 1)

PARAMETER	FROM (INPUT)	TO (OUTPUT)	V_{CC} = 4.5 V to 5.5 V, C_L = 50 pF, R_L = 500 Ω, T_A = MIN to MAX				UNIT
			SN54ALS805		SN74ALS805		
			MIN	MAX	MIN	MAX	
t_{PLH}	A or B	Y	2	8	2	6	ns
t_{PHL}			2	9	2	7	

NOTE 1: For load circuit and voltage waveforms, see page 1-12.

Texas Instruments
POST OFFICE BOX 225012 • DALLAS, TEXAS 75265

TYPES SN54AS805A, SN74AS805A
HEX 2-INPUT NOR DRIVERS

absolute maximum ratings over operating free-air temperature range (unless otherwise noted)

Supply voltage, V_{CC} .... 7 V
Input voltage .... 7 V
Operating free-air temperature range: SN54AS805A .... −55 °C to 125 °C
 SN74AS805A .... 0 °C to 70 °C
Storage temperature range .... −65 °C to 150 °C

recommended operating conditions

		SN54AS805A			SN74AS805A			UNIT
		MIN	NOM	MAX	MIN	NOM	MAX	
V_{CC}	Supply voltage	4.5	5	5.5	4.5	5	5.5	V
V_{IH}	High-level input voltage	2			2			V
V_{IL}	Low-level input voltage			0.8			0.8	V
I_{OH}	High-level output current			−40			−48	mA
I_{OL}	Low-level output current			40			48	mA
T_A	Operating free-air temperature	−55		125	0		70	°C

electrical characteristics over recommended operating free-air temperature range (unless otherwise noted)

PARAMETER	TEST CONDITIONS		SN54AS805A			SN74AS805A			UNIT
			MIN	TYP†	MAX	MIN	TYP†	MAX	
V_{IK}	V_{CC} = 4.5 V,	I_I = −18 mA			−1.2			−1.2	V
V_{OH}	V_{CC} = 4.5 V to 5.5 V,	I_{OH} = −2 mA	$V_{CC}-2$			$V_{CC}-2$			V
	V_{CC} = 4.5 V,	I_{OH} = −3 mA	2.4	3.2		2.4	3.2		
	V_{CC} = 4.5 V,	I_{OH} = −40 mA	2						
	V_{CC} = 4.5 V,	I_{OH} = −48 mA				2			
V_{OL}	V_{CC} = 4.5 V,	I_{OL} = 40 mA		0.25	0.5				V
	V_{CC} = 4.5 V,	I_{OL} = 48 mA					0.35	0.5	
I_I	V_{CC} = 5.5 V,	V_I = 7 V			0.1			0.1	mA
I_{IH}	V_{CC} = 5.5 V,	V_I = 2.7 V			20			20	µA
I_{IL}	V_{CC} = 5.5 V,	V_I = 0.4 V			−0.5			−0.5	mA
I_O‡	V_{CC} = 5.5 V,	V_O = 2.25 V	−135			−135			mA
I_{CCH}	V_{CC} = 5.5 V,	V_I = 0 V		5	9		5	9	mA
I_{CCL}	V_{CC} = 5.5 V,	V_I = 4.5 V		18	32		18	32	mA

† All typical values are at V_{CC} = 5 V, T_A = 25 °C.
‡ The output conditions have been chosen to produce a current that closely approximates one half of the true short-circuit output current, I_{OS}.

switching characteristics (see Note 1)

PARAMETER	FROM (INPUT)	TO (OUTPUT)	V_{CC} = 4.5 V to 5.5 V, C_L = 50 pF, R_L = 500 Ω, T_A = MIN to MAX				UNIT
			SN54AS805A		SN74AS805A		
			MIN	MAX	MIN	MAX	
t_{PLH}	A or B	Y	1	4.5	1	4	ns
t_{PHL}			1	4.5	1	4	

NOTE 1: For load circuit and voltage waveforms, see page 1-12.

TEXAS INSTRUMENTS
POST OFFICE BOX 225012 • DALLAS, TEXAS 75265

ALS AND AS CIRCUITS

TYPES SN54ALS808, SN54AS808A, SN74ALS808, SN74AS808A
HEX 2-INPUT AND DRIVERS

D2661, DECEMBER 1982–REVISED DECEMBER 1983

- **High Capacitive Drive Capability**
- **'ALS808 Has Typical Delay Time of 4.8 ns (C_L = 50 pF) and Typical Power Dissipation of 4.5 mW per Gate**
- **'AS808A Has Typical Delay Time of 3.2 ns (C_L = 50 pF) and Typical Power Dissipation of Less than 13 mW per Gate**
- **Package Options Include Both Plastic and Ceramic Chip Carriers in Addition to Plastic and Ceramic DIPs**
- **Dependable Texas Instruments Quality and Reliability**

SN54ALS808, SN54AS808A . . . J PACKAGE
SN74ALS808, SN74AS808A . . . N PACKAGE
(TOP VIEW)

```
1A  [ 1   U  20 ] VCC
1B  [ 2      19 ] 6B
1Y  [ 3      18 ] 6A
2A  [ 4      17 ] 6Y
2B  [ 5      16 ] 5B
2Y  [ 6      15 ] 5A
3A  [ 7      14 ] 5Y
3B  [ 8      13 ] 4B
3Y  [ 9      12 ] 4A
GND [ 10     11 ] 4Y
```

description

These devices contain six independent 2-input AND drivers. They perform the Boolean functions $Y = A \cdot B$ or $Y = \overline{A} + \overline{B}$ in positive logic.

The –1 version of the SN74ALS808 parts is identical to the standard version except that the recommended maximum I_{OL} is increased to 48 milliamperes. There is no –1 version of the SN54ALS808 parts.

The SN54ALS808 and SN54AS808A are characterized for operation over the full military temperature range of –55°C to 125°C. The SN74ALS808 and SN74AS808A are characterized for operation from 0°C to 70°C.

SN54ALS808, SN54AS808A . . . FH PACKAGE
SN74ALS808, SN74AS808A . . . FN PACKAGE
(TOP VIEW)

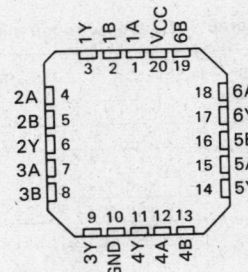

FUNCTION TABLE (each driver)

INPUTS		OUTPUT
A	B	Y
H	H	H
L	X	L
X	L	L

logic symbol

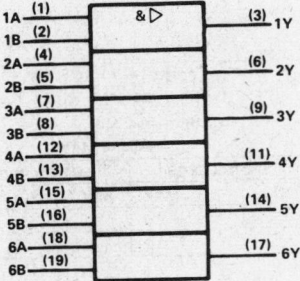

Pin numbers shown are for J and N packages.

ALS AND AS CIRCUITS

Copyright © 1983 by Texas Instruments Incorporated

TEXAS INSTRUMENTS
POST OFFICE BOX 225012 • DALLAS, TEXAS 75265

TYPES SN54ALS808, SN74ALS808
HEX 2-INPUT AND DRIVERS

absolute maximum ratings over operating free-air temperature range (unless otherwise noted)

Supply voltage, V_{CC} .... 7 V
Input voltage .... 7 V
Operating free-air temperature range: SN54ALS808 .... −55 °C to 125 °C
                                       SN74ALS808 .... 0 °C to 70 °C
Storage temperature range .... −65 °C to 150 °C

recommended operating conditions

		SN54ALS808			SN74ALS808			UNIT
		MIN	NOM	MAX	MIN	NOM	MAX	
V_{CC}	Supply voltage	4.5	5	5.5	4.5	5	5.5	V
V_{IH}	High-level input voltage	2			2			V
V_{IL}	Low-level input voltage			0.8			0.8	V
I_{OH}	High-level output current			−12			−15	mA
I_{OL}	Low-level output current			12			24	mA
							48†	
T_A	Operating free-air temperature	−55		125	0		70	°C

† The extended limit applies if V_{CC} is maintained between 4.75 V and 5.25 V.
  The 48 mA limit applies for the SN74ALS808−1 only.

electrical characteristics over recommended operating free-air temperature range (unless otherwise noted)

PARAMETER	TEST CONDITIONS		SN54ALS808			SN74ALS808			UNIT
			MIN	TYP‡	MAX	MIN	TYP‡	MAX	
V_{IK}	V_{CC} = 4.5 V,	I_I = −18 mA			−1.5			−1.5	V
V_{OH}	V_{CC} = 4.5 V to 5.5 V,	I_{OH} = −0.4 mA	V_{CC}−2			V_{CC}−2			V
	V_{CC} = 4.5 V,	I_{OH} = −3 mA	2.4	3.2		2.4	3.2		
	V_{CC} = 4.5 V,	I_{OH} = −12 mA	2						
	V_{CC} = 4.5 V,	I_{OH} = −15 mA				2			
V_{OL}	V_{CC} = 4.5 V,	I_{OL} = 12 mA		0.25	0.4		0.25	0.4	V
	V_{CC} = 4.5 V,	I_{OL} = 24 mA					0.35	0.5	
	(I_{OL} = 48 mA for −1 version)								
I_I	V_{CC} = 5.5 V,	V_I = 7 V			0.1			0.1	mA
I_{IH}	V_{CC} = 5.5 V,	V_I = 2.7 V			20			20	µA
I_{IL}	V_{CC} = 5.5 V,	V_I = 0.4 V			−0.1			−0.1	mA
I_O §	V_{CC} = 5.5 V,	V_O = 2.25 V	−30		−112	−30		−112	mA
I_{CCH}	V_{CC} = 5.5 V,	V_I = 4.5 V		3	6		3	6	mA
I_{CCL}	V_{CC} = 5.5 V,	V_I = 0 V		8	16		8	16	mA

‡ All typical values are at V_{CC} = 5 V, T_A = 25 °C.
§ The output conditions have been chosen to produce a current that closely approximates one half of the true short-circuit output current, I_{OS}.

switching characteristics (see Note 1)

PARAMETER	FROM (INPUT)	TO (OUTPUT)	V_{CC} = 4.5 V to 5.5 V, C_L = 50 pF, R_L = 500 Ω, T_A = MIN to MAX				UNIT
			SN54ALS808		SN74ALS808		
			MIN	MAX	MIN	MAX	
t_{PLH}	A or B	Y	2	10	2	8	ns
t_{PHL}			2	10	2	8	

NOTE 1: For load circuit and voltage waveforms, see page 1-12.

TEXAS INSTRUMENTS
POST OFFICE BOX 225012 • DALLAS, TEXAS 75265

TYPES SN54AS808A, SN74AS808A
HEX 2-INPUT AND DRIVERS

absolute maximum ratings over operating free-air temperature range (unless otherwise noted)

Supply voltage, V_{CC} .... 7 V
Input voltage .... 7 V
Operating free-air temperature range: SN54AS808A .... −55 °C to 125 °C
                                      SN74AS808A .... 0 °C to 70 °C
Storage temperature range .... −65 °C to 150 °C

recommended operating conditions

		SN54AS808A			SN74AS808A			UNIT
		MIN	NOM	MAX	MIN	NOM	MAX	
V_{CC}	Supply voltage	4.5	5	5.5	4.5	5	5.5	V
V_{IH}	High-level input voltage	2			2			V
V_{IL}	Low-level input voltage			0.8			0.8	V
I_{OH}	High-level output current			−40			−48	mA
I_{OL}	Low-level output current			40			48	mA
T_A	Operating free-air temperature	−55		125	0		70	°C

electrical characteristics over recommended operating free-air temperature range (unless otherwise noted)

PARAMETER	TEST CONDITIONS		SN54AS808A			SN74AS808A			UNIT
			MIN	TYP†	MAX	MIN	TYP†	MAX	
V_{IK}	V_{CC} = 4.5 V,	I_I = −18 mA			−1.2			−1.2	V
V_{OH}	V_{CC} = 4.5 V to 5.5 V,	I_{OH} = −2 mA	V_{CC}−2			V_{CC}−2			V
	V_{CC} = 4.5 V,	I_{OH} = −3 mA	2.4	3.2		2.4	3.2		
	V_{CC} = 4.5 V,	I_{OH} = −40 mA	2						
	V_{CC} = 4.5 V,	I_{OH} = −48 mA				2			
V_{OL}	V_{CC} = 4.5 V,	I_{OL} = 40 mA		0.25	0.5				V
	V_{CC} = 4.5 V,	I_{OL} = 48 mA					0.35	0.5	
I_I	V_{CC} = 5.5 V,	V_I = 7 V			0.1			0.1	mA
I_{IH}	V_{CC} = 5.5 V,	V_I = 2.7 V			20			20	µA
I_{IL}	V_{CC} = 5.5 V,	V_I = 0.4 V			−0.5			−0.5	mA
I_O‡	V_{CC} = 5.5 V,	V_O = 2.25 V	−135			−135			mA
I_{CCH}	V_{CC} = 5.5 V,	V_I = 4.5 V		6.5	11		6.5	11	mA
I_{CCL}	V_{CC} = 5.5 V,	V_I = 0 V		19	32		19	32	mA

† All typical values are at V_{CC} = 5 V, T_A = 25 °C.
‡ The output conditions have been chosen to produce a current that closely approximates one half of the true short-circuit output current, I_{OS}.

switching characteristics (see Note 1)

PARAMETER	FROM (INPUT)	TO (OUTPUT)	V_{CC} = 4.5 V to 5.5 V, C_L = 50 pF, R_L = 500 Ω, T_A = MIN to MAX				UNIT
			SN54AS808A		SN74AS808A		
			MIN	MAX	MIN	MAX	
t_{PLH}	A or B	Y	1	6	1	5	ns
t_{PHL}	A or B	Y	1	6	1	5	ns

NOTE 1: For load circuit and voltage waveforms, see page 1-12.

TEXAS INSTRUMENTS
POST OFFICE BOX 225012 • DALLAS, TEXAS 75265

2 ALS AND AS CIRCUITS

TYPES SN54AS821, SN54AS822, SN74AS821, SN74AS822
10-BIT BUS INTERFACE FLIP-FLOPS WITH 3-STATE OUTPUTS

D2825, DECEMBER 1983

- 10-Bit Versions of 'AS574 and 'AS576 with Improved I_{OH} Specifications
- Ideal for Data Synchronization of Wider Data Paths
- Provides Extra Data Width Necessary for Wider Address/Data Paths or Buses with Parity
- Outputs Have Undershoot Protection Circuitry
- Power-Up High-Impedance State
- Package Options Include Both Plastic and Ceramic Carriers in Addition to Plasstic and Ceramic DIPs
- Buffered Control Inputs to Reduce DC Loading Effects
- Dependable Texas Instruments Quality and Reliability

SN54AS821 . . . JT PACKAGE
SN74AS821 . . . NT PACKAGE
(TOP VIEW)

```
 OC  [ 1  U 24 ] VCC
 1D  [ 2    23 ] 1Q
 2D  [ 3    22 ] 2Q
 3D  [ 4    21 ] 3Q
 4D  [ 5    20 ] 4Q
 5D  [ 6    19 ] 5Q
 6D  [ 7    18 ] 6Q
 7D  [ 8    17 ] 7Q
 8D  [ 9    16 ] 8Q
 9D  [ 10   15 ] 9Q
10D  [ 11   14 ] 10Q
GND  [ 12   13 ] CLK
```

SN54AS821 . . . FH PACKAGE
SN74AS821 . . . FN PACKAGE
(TOP VIEW)

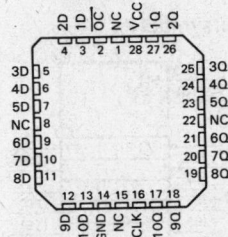

description

These 10-bit flip-flops feature three-state outputs designed specifically for driving highly-capacitive or relatively low-impedance loads. They are particularly suitable for implementing wider buffer registers, I/O ports, bidirectional bus drivers with parity, and working registers.

The ten flip-flops are edge-triggered D-type flip-flops. On the positive transition of the clock the Q outputs on the 'AS821 will be true, and on the 'AS822 will be complementary, to the data input.

A buffered output-control input can be used to place the ten outputs in either a normal logic state (high or low levels) or a high-impedance state. In the high-impedance state the outputs neither load nor drive the bus lines significantly. The high-impedance state and increased drive provide the capability to drive the bus lines in a bus-organized system without need for interface or pull-up components.

SN54AS822 . . . JT PACKAGE
SN74AS822 . . . NT PACKAGE
(TOP VIEW)

```
 OC   [ 1  U 24 ] VCC
 1D̄   [ 2    23 ] 1Q
 2D̄   [ 3    22 ] 2Q
 3D̄   [ 4    21 ] 3Q
 4D̄   [ 5    20 ] 4Q
 5D̄   [ 6    19 ] 5Q
 6D̄   [ 7    18 ] 6Q
 7D̄   [ 8    17 ] 7Q
 8D̄   [ 9    16 ] 8Q
 9D̄   [ 10   15 ] 9Q
10D̄   [ 11   14 ] 10Q
GND   [ 12   13 ] CLK
```

SN54AS822 . . . FH PACKAGE
SN74AS822 . . . FN PACKAGE
(TOP VIEW)

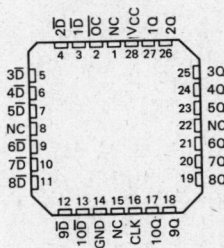

NC — No internal connection

ADVANCE INFORMATION
This document contains information on a new product. Specifications are subject to change without notice.

Copyright © 1983 by Texas Instruments Incorporated

TEXAS INSTRUMENTS
POST OFFICE BOX 225012 • DALLAS, TEXAS 75265

TYPES SN54AS821, SN54AS822, SN74AS821, SN74AS822
10-BIT BUS INTERFACE FLIP-FLOPS WITH 3-STATE OUTPUTS

The output control ($\overline{OC}$) does not affect the internal operation of the flip-flops. Old data can be retained or new data can be entered while the outputs are in the high-impedance state.

The SN54AS821 and SN54AS822 are characterized for operation over the full military temperature range of −55°C to 125°C. The SN74AS821 and SN74AS822 are characterized for operation from 0°C to 70°C.

'AS821 FUNCTION TABLE (EACH FLIP-FLOP)

INPUTS			OUTPUT
$\overline{OC}$	CLK	D	Q
L	↑	H	H
L	↑	L	L
L	L	X	Q_0
H	X	X	Z

'AS821 logic diagram (positive logic)

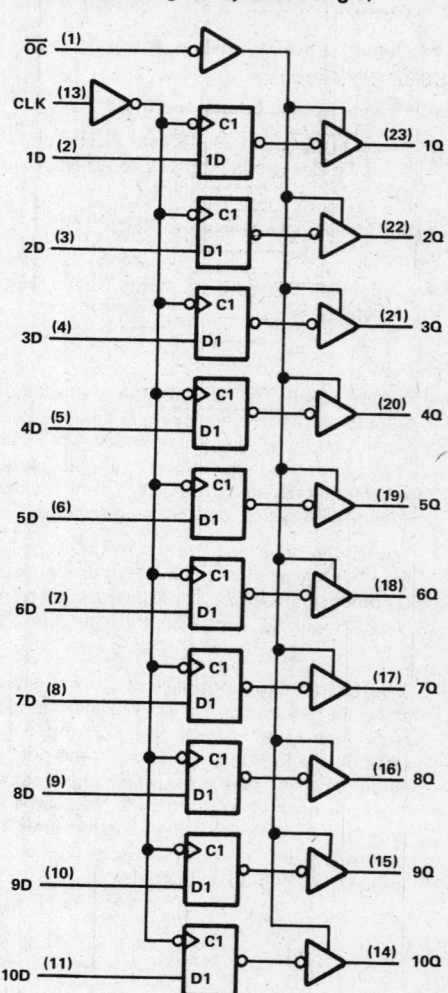

'AS821 logic symbol

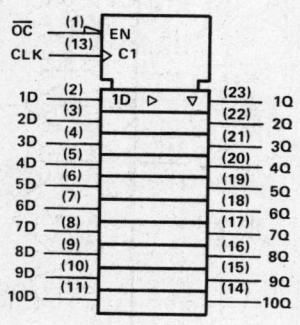

Pin numbers shown are for JT and NT packages.

TYPES SN54AS822, SN74AS822
10-BIT BUS INTERFACE FLIP-FLOPS WITH 3-STATE OUTPUTS

'AS822 FUNCTION TABLE (EACH FLIP-FLOP)

INPUTS			OUTPUT
$\overline{OC}$	CLK	$\overline{D}$	Q
L	↑	H	L
L	↑	L	H
L	L	X	Q_0
H	X	X	Z

'AS822 logic symbol

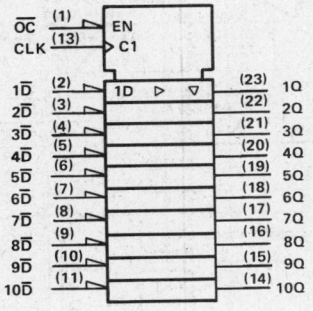

'AS822 logic diagram (positive logic)

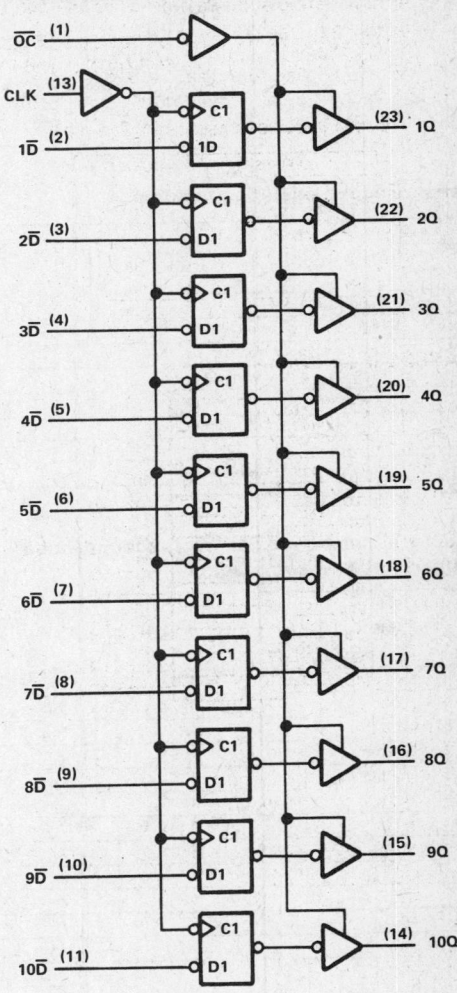

Pin numbers shown are for JT and NT packages

TEXAS INSTRUMENTS
POST OFFICE BOX 225012 • DALLAS, TEXAS 75265

TYPES SN54AS821, SN54AS822, SN74AS821, SN74AS822
10-BIT BUS INTERFACE FLIP-FLOPS WITH 3-STATE OUTPUTS

absolute maximum ratings over operating free-air temperature range (unless otherwise noted)

Supply voltage, V_{CC} .... 7 V
Input voltage .... 7 V
Voltage applied to a disabled 3-state output .... 5.5 V
Operating free-air temperature range: SN54AS821, SN54AS822 .... −55°C to 125°C
                                        SN74AS821, SN74AS822 .... 0°C to 70°C
Storage temperature range .... −65°C to 150°C

recommended operating conditions

		SN54AS821 SN54AS822			SN74AS821 SN74AS822			UNIT
		MIN	NOM	MAX	MIN	NOM	MAX	
V_{CC}	Supply voltage	4.5	5	5.5	4.5	5	5.5	V
V_{IH}	High-level input voltage	2			2			V
V_{IL}	Low-level input voltage			0.8			0.8	V
I_{OH}	High-level output current			−24			−24	mA
I_{OL}	Low-level output current			32			48	mA
t_w	Pulse duration, CLK high or low	9			8			ns
t_{su}	Setup time, data before CLK↑	7			6			ns
t_h	Hold time, data after CLK↑	0			0			ns
T_A	Operating free-air temperature	−55		−125	0		70	°C

electrical characteristics over recommended operating free-air temperature range (unless otherwise noted)

PARAMETER		TEST CONDITIONS		SN54AS821 SN54AS822			SN74AS821 SN74AS822			UNIT
				MIN	TYP†	MAX	MIN	TYP†	MAX	
V_{IK}		$V_{CC} = 4.5$ V,	$I_I = -18$ mA			−1.2			−1.2	V
V_{OH}		$V_{CC} = 4.5$ V to 5.5 V,	$I_{OH} = -2$ mA	$V_{CC}-2$			$V_{CC}-2$			V
		$V_{CC} = 4.5$ V,	$I_{OH} = -15$ mA	2.4	3.2		2.4	3.2		
		$V_{CC} = 4.5$ V,	$I_{OH} = -24$ mA	2			2			
V_{OL}		$V_{CC} = 4.5$ V,	$I_{OL} = 32$ mA		0.25	0.5				V
		$V_{CC} = 4.5$ V,	$I_{OL} = 48$ mA					0.35	0.5	
I_{OZH}		$V_{CC} = 5.5$ V,	$V_O = 2.7$ V			50			50	µA
I_{OZL}		$V_{CC} = 5.5$ V,	$V_O = 0.4$ V			−50			−50	µA
I_I		$V_{CC} = 5.5$ V,	$V_I = 7$ V			0.1			0.1	mA
I_{IH}		$V_{CC} = 5.5$ V,	$V_I = 2.7$ V			20			20	µA
I_{IL}		$V_{CC} = 5.5$ V,	$V_I = 0.4$ V			−0.5			−0.5	mA
I_O‡		$V_{CC} = 5.5$ V,	$V_O = 2.25$ V	−30		−112	−30		−112	mA
I_{CC}	'AS821	$V_{CC} = 5.5$ V	Outputs high		55	88		55	88	mA
			Outputs low		68	109		68	109	
			Outputs disabled		70	113		70	113	
	'AS822		Outputs high		55	88		55	88	
			Outputs low		68	109		68	109	
			Outputs disabled		70	113		70	113	

†All typical values are at $V_{CC} = 5$ V, $T_A = 25$°C.
‡The output conditions have been chosen to produce a current that closely approximates one half of the true short-circuit output current, I_{OS}.

ALS AND AS CIRCUITS

TEXAS INSTRUMENTS
POST OFFICE BOX 225012 • DALLAS, TEXAS 75265

TYPES SN54AS821, SN54AS822, SN74AS821, SN74AS822
10-BIT BUS INTERFACE FLIP-FLOPS WITH 3-STATE OUTPUTS

switching characteristics (see Note 1)

PARAMETER	FROM (INPUT)	TO (OUTPUT)	V_{CC} = 4.5 V to 5.5 V, C_L = 50 pF, R1 = 500 Ω, R2 = 500 Ω, T_A = MIN to MAX				UNIT
			SN54AS821 SN54AS822		SN74AS821 SN74AS822		
			MIN	MAX	MIN	MAX	
t_{PLH}	CLK	Any Q	3.5	9	3.5	7.5	ns
t_{PHL}			3.5	11.5	3.5	10.5	
t_{PZH}	$\overline{OC}$	Any Q	4	12	4	11	ns
t_{PZL}			4	13	4	12	
t_{PHZ}	$\overline{OC}$	Any Q	2	10	2	8	ns
t_{PZL}			2	10	2	8	

NOTE 1: For load circuit and voltage waveforms, see page 1-12.

ALS AND AS CIRCUITS

TYPES SN54AS823, SN54AS824, SN74AS823, SN74AS824
9-BIT BUS INTERFACE FLIP-FLOPS WITH 3-STATE OUTPUTS

D2825, DECEMBER 1983

- Similar to 'AS574 and 'AS576 with Clock Enable and Clear and Improved I_{OH} Specifications
- Ideal for Data Synchronization of Wider Data Paths
- Provides Extra Data Width Necessary for Wider Address/Data Paths or Buses with Parity
- Outputs Have Undershoot Protection Circuitry
- Power-Up High-Impedance State
- Buffered Control Inputs to Reduce DC Loading Effects
- Package Options Include both Plastic and Ceramic Carriers in Addition to Plastic and Ceramic DIPs
- Dependable Texas Instruments Quality and Reliability

description

These 9-bit flip-flops feature three-state outputs designed specifically for driving highly-capacitive or relatively low-impedance loads. They are particularly suitable for implementing wider buffer registers, I/O ports, bidirectional bus drivers, parity bus interfacing and working registers.

With the clock enable ($\overline{CLKEN}$) low, the nine D-type edge-triggered flip-flops enter data on the low-to-high transitions of the clock. Taking $\overline{CLKEN}$ high will disable the clock buffer, thus latching the outputs. The 'AS823 has noninverting D inputs and the 'AS824 has inverting $\overline{D}$ inputs. Taking the $\overline{CLR}$ input low causes the nine Q outputs to go low independently of the clock.

A buffered output-control input can be used to place the nine outputs in either a normal logic state (high or low levels) or a high-impedance state. In the high-impedance state the outputs neither load nor drive the bus lines significantly. The high-impedance state and increased drive provide the capability to drive the bus lines in a bus-organized system without need for interface or pull-up components.

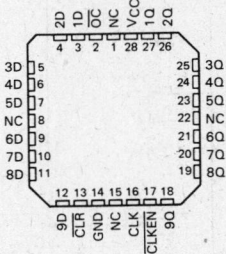

PRODUCT PREVIEW

This document contains information on a product under development. Texas Instruments reserves the right to change or discontinue this product without notice.

Copyright © 1983 by Texas Instruments Incorporated

TEXAS INSTRUMENTS
POST OFFICE BOX 225012 • DALLAS, TEXAS 75265

ALS AND AS CIRCUITS

2-531

TYPES SN54AS823, SN54AS824, SN74AS823, SN74AS824
9-BIT INTERFACE FLIP-FLOPS WITH 3-STATE OUTPUTS

The output control ($\overline{OC}$) does not affect the internal operation of the flip-flops. Old data can be retained or new data can be entered while the outputs are in the high-impedance state.

The SN54AS823 and SN54AS824 are characterized for operation over the full military temperature range of −55 °C to 125 °C. The SN74AS823 and SN74AS824 are characterized for operation from 0 °C to 70 °C.

FUNCTION TABLES

'AS823

INPUTS					OUTPUT
$\overline{OC}$	$\overline{CLR}$	$\overline{CLKEN}$	CLK	D	Q
L	L	X	X	X	L
L	H	L	↑	H	H
L	H	L	↑	L	L
L	H	H	X	X	Q_0
H	X	X	X	X	Z

'AS824

INPUTS					OUTPUT
$\overline{OC}$	$\overline{CLR}$	$\overline{CLKEN}$	CLK	$\overline{D}$	Q
L	L	X	X	X	L
L	H	L	↑	H	L
L	H	L	↑	L	H
L	H	H	X	X	Q_0
H	X	X	X	X	Z

'AS823 logic symbol

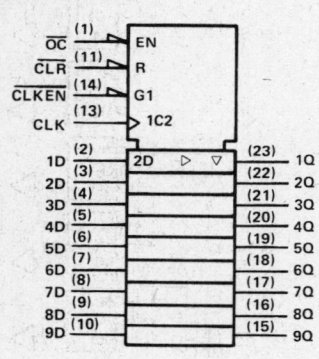

'AS823 logic diagram (positive logic)

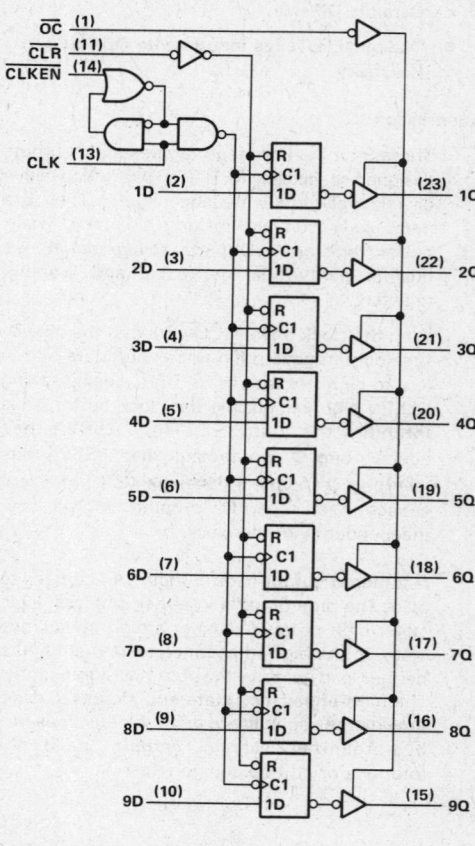

Pin numbers shown are for JT and NT packages.

TYPES SN54AS823, SN54AS824, SN74AS823, SN74AS824
9-BIT INTERFACE FLIP-FLOPS WITH 3-STATE OUTPUTS

'AS824 logic symbol

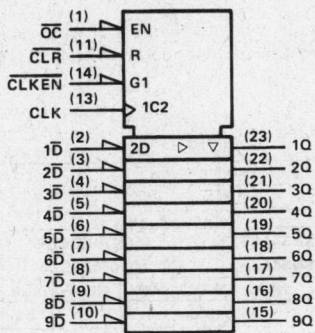

'AS824 logic diagram (positive logic)

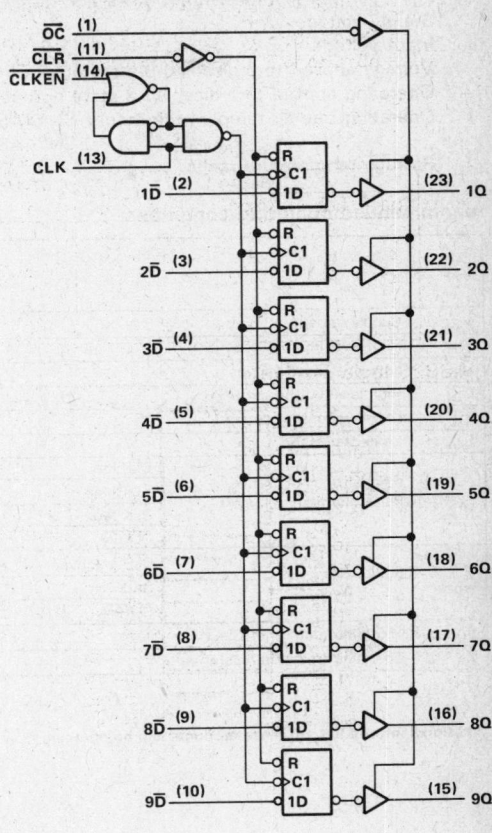

Pin numbers shown are for JT and NT packages.

TYPES SN54AS823, SN54AS824, SN74AS823, SN74AS824
9-BIT INTERFACE FLIP-FLOPS WITH 3-STATE OUTPUTS

absolute maximum ratings over operating free-air temperature range (unless otherwise noted)

Supply voltage, V_{CC} .... 7 V
Input voltage .... 7 V
Voltage applied to a disabled 3-state output .... 5.5 V
Operating applied to a disabled 3-state output .... 5.5 V
Operating free-air temperature range: SN54AS823, SN54AS824 .... −55°C to 125°C
                                   SN74AS823, SN74AS824 .... 0°C to 70°C
Storage temperature range .... −65°C to 150°C

recommended operating conditions

		SN54AS823 SN54AS824			SN74AS823 SN74AS824			UNIT
		MIN	NOM	MAX	MIN	NOM	MAX	
V_{CC}	Supply voltage	4.5	5	5.5	4.5	5	5.5	V
V_{IH}	High-level input voltage	2			2			V
V_{IL}	Low-level input voltage			0.8			0.8	V
I_{OH}	High-level output current			−24			−24	mA
I_{OL}	Low-level output current			32			48	mA
f_{clock}	Clock frequency							MHz
t_w	Pulse duration — $\overline{CLR}$ low							ns
	CLK high							
	CLK low							
	$\overline{CLKEN}$							
t_{su}	Setup time before CLK↑ — $\overline{CLR}$ inactive							ns
	Data							
	$\overline{CLKEN}$							
t_h	Hold time, data after CLK↑							ns
T_A	Operating free-air temperature	−55		125	0		70	°C

Additional information on these products can be obtained from the factory as it becomes available.

TYPES SN54AS823, SN54AS824, SN74AS823, SN74AS824
9-BIT INTERFACE FLIP-FLOPS WITH 3-STATE OUTPUTS

electrical characteristics over recommended operating free-air temperature range (unless otherwise noted)

PARAMETER		TEST CONDITIONS		SN54AS823 SN54AS824			SN74AS823 SN74AS824			UNIT
				MIN	TYP†	MAX	MIN	TYP†	MAX	
V_{IK}		V_{CC} = 4.5 V,	I_I = −18 mA			−1.2			−1.2	V
V_{OH}		V_{CC} = 4.5 V to 5.5 V,	I_{OH} = −2 mA	V_{CC}−2			V_{CC}−2			V
		V_{CC} = 4.5 V,	I_{OH} = −15 mA	2.4	3.2		2.4	3.2		
		V_{CC} = 4.5 V,	I_{OH} = −24 mA	2			2			
V_{OL}		V_{CC} = 4.5 V,	I_{OL} = 32 mA		0.25	0.5				V
		V_{CC} = 4.5 V,	I_{OL} = 48 mA					0.35	0.5	
I_{OZH}		V_{CC} = 5.5 V,	V_O = 2.7 V			50			50	µA
I_{OZL}		V_{CC} = 5.5 V,	V_O = 0.4 V			−50			−50	µA
I_I		V_{CC} = 5.5 V,	V_I = 7 V							mA
I_{IH}		V_{CC} = 5.5 V,	V_I = 2.7 V							µA
I_{IL}		V_{CC} = 5.5 V,	V_I = 0.4 V							mA
I_O‡		V_{CC} = 5.5 V,	V_O = 2.25 V	−30		−112	−30		−112	mA
I_{CC}	'AS823	V_{CC} = 5.5 V	Outputs high							mA
			Outputs low							
			Outputs disabled		58			58		
	'AS824	V_{CC} = 5.5 V	Outputs high							mA
			Outputs low							
			Outputs disabled		58			58		

†All typical values are at V_{CC} = 5 V, T_A = 25°C.
‡The output conditions have been chosen to produce a current that closely approximates one half of the true short-circuit output current, I_{OS}.

Additional information on these products can be obtained from the factory as it becomes available.

TEXAS INSTRUMENTS
POST OFFICE BOX 225012 • DALLAS, TEXAS 75265

TYPES SN54AS823, SN54AS824, SN74AS823, SN74AS824
9-BIT INTERFACE FLIP-FLOPS WITH 3-STATE OUTPUTS

switching characteristics (see Note 1)

PARAMETER	FROM (INPUT)	TO (OUTPUT)	V_{CC} = 4.5 V to 5.5 V, C_L = 50 pF, $R1$ = 500 Ω, $R2$ = 500 Ω, T_A = MIN to MAX						UNIT
			SN54AS823 SN54AS824			SN74AS823 SN74AS824			
			MIN	TYP†	MAX	MIN	TYP†	MAX	
f_{max}									MHz
t_{PLH}	CLK	Any Q		7.5			7.5		ns
t_{PHL}				9.5			9.5		
t_{PHL}	$\overline{CLR}$	Any Q		11			11		ns
t_{PZH}	$\overline{OC}$	Any Q		6			6		ns
t_{PZL}				7			7		
t_{PHZ}	$\overline{OC}$	Any Q		6			6		ns
t_{PLZ}				7			7		

†All typical values are at V_{CC} = 5 V, T_A = 25°C.
NOTE 1: For load circuit and voltage waveforms, see page 1-12.

Additional information on these products can be obtained from the factory as it becomes available.

D flip-flop signal conventions

It is normal TI practice to name the outputs and other inputs of a D-type flip-flop and to draw its logic symbol based on the assumption of true data (D) inputs. Then outputs that produce data in phase with the data inputs are called Q and those producing complementary data are called $\overline{Q}$. An input that causes a Q output to go high or a $\overline{Q}$ output to go low is called Preset; an input that causes a $\overline{Q}$ output to go high or a Q output to go low is called Clear. Bars are used over these pin names ($\overline{PRE}$ and $\overline{CLR}$) if they are active-low.

The devices on this data sheet are second-source designs and the pin-name convention used by the original manufacturer has been retained. That makes it necessary to designate the inputs and outputs of the inverting circuit $\overline{D}$ and Q. In some applications it may be advantageous to redesignate the inputs and outputs as D and $\overline{Q}$. In that case, outputs should be renamed as shown below. Also shown are corresponding changes in the graphical symbol. Arbitrary pin numbers are shown in parentheses.

Notice that Q and $\overline{Q}$ exchange names, which causes Preset and Clear to do likewise. Also notice that the polarity indicators (◁) on $\overline{PRE}$ and $\overline{CLR}$ remain since these inputs are still active-low, but that the presence or absence of the polarity indicator changes at $\overline{D}$, Q, and $\overline{Q}$. Of course pin 5 (Q) is still in phase with the data input D, but now both are considered active high.

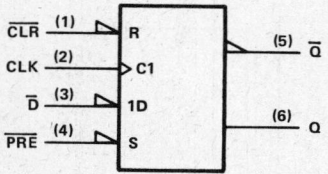 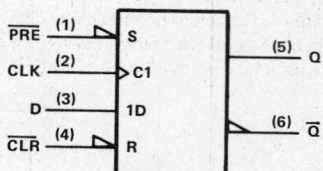

TYPES SN54AS825, SN54AS826, SN74AS825, SN74AS826
8-BIT BUS INTERFACE FLIP-FLOPS WITH 3-STATE OUTPUTS

D2825, DECEMBER 1983

- Similar to 'AS574 and 'AS576 with Clock Enable, Clear, and Multiple Output Controls
- Improved I_{OH} Specifications
- Multiple Output Enables Allow Multiuser Control of the Interface
- Outputs Have Undershoot Protection Circuitry
- Power-Up High-Impedance State
- Package Options Include Both Plastic and Ceramic Chip Carriers in Addition to Plastic and Ceramic DIPs
- Buffered Control Inputs to Reduce DC Loading Effect
- Dependable Texas Instruments Quality and Reliability

description

These 8-bit flip-flops feature three-state outputs designed specifically for driving highly-capacitive or relatively low-impedance loads. They are particularly suitable for implementing multiuser registers, I/O ports, bidirectional bus drivers, and working registers.

With the clock enable ($\overline{CLKEN}$) low, the eight D-type edge-triggered flip-flops enter data on the low-to-high transitions of the clock. Taking $\overline{CLKEN}$ high will disable the clock buffer, thus latching the outputs. The 'AS825 has noninverting D inputs and the 'AS826 has inverting $\overline{D}$ inputs. Taking the $\overline{CLR}$ input low causes the eight Q outputs to go low independently of the clock.

A multiuser buffered output-control input can be used to place the eight outputs in either a normal logic state (high or low levels) or a high-impedance state. In the high-impedance state the outputs neither load nor drive the bus lines significantly. The high-impedance state and increased drive provide the capability to drive the bus lines in a bus-organized system without need for interface or pull-up components.

SN54AS825 . . . JT PACKAGE
SN74AS825 . . . NT PACKAGE
(TOP VIEW)

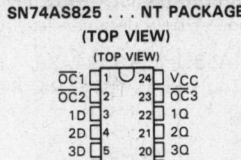

SN54AS825 . . . FH PACKAGE
SN74AS825 . . . FN PACKAGE
(TOP VIEW)

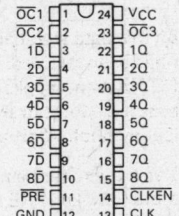

SN54AS826 . . . JT PACKAGE
SN74AS826 . . . NT PACKAGE
(TOP VIEW)

SN54AS826 . . . FH PACKAGE
SN74AS826 . . . FN PACKAGE
(TOP VIEW)

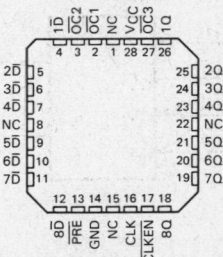

NC—No internal connection

PRODUCT PREVIEW

This document contains information on a product under development. Texas Instruments reserves the right to change or discontinue this product without notice.

ALS AND AS CIRCUITS

Copyright © 1983 by Texas Instruments Incorporated

TEXAS INSTRUMENTS
POST OFFICE BOX 225012 • DALLAS, TEXAS 75265

TYPES SN54AS825, SN54AS826, SN74AS825, SN74AS826
8-BIT BUS INTERFACE FLIP-FLOPS WITH 3-STATE OUTPUTS

The output controls ($\overline{OC}1$, $\overline{OC}2$, and $\overline{OC}3$) do not affect the internal operation of the flip-flops. Old data can be retained or new data can be entered while the outputs are in the high-impedance state.

The SN54AS825 and SN54AS826 are characterized for operation over the full military temperature range of $-55\,°C$ to $125\,°C$. The SN74AS825 and SN74AS826 are characterized for operation from $0\,°C$ to $70\,°C$.

FUNCTION TABLES

'AS825

INPUTS					OUTPUT
$\overline{OC}$*	$\overline{CLR}$	$\overline{CLKEN}$	CLK	D	Q
L	L	X	X	X	L
L	H	L	↑	H	H
L	H	L	↑	L	H
L	H	H	X	X	Q_0
H	X	X	X	X	Z

$\overline{OC}$* = $\overline{OC}1 \cdot \overline{OC}2 \cdot \overline{OC}3$

'AS826

INPUTS					OUTPUT
$\overline{OC}$*	$\overline{CLR}$	$\overline{CLKEN}$	CLK	$\overline{D}$	Q
L	L	X	X	X	L
L	H	L	↑	H	L
L	H	L	↑	L	H
L	H	H	X	X	Q_0
H	X	X	X	X	Z

'AS825 logic symbol

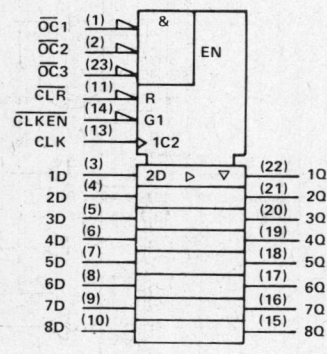

'AS825 logic diagram (positive logic)

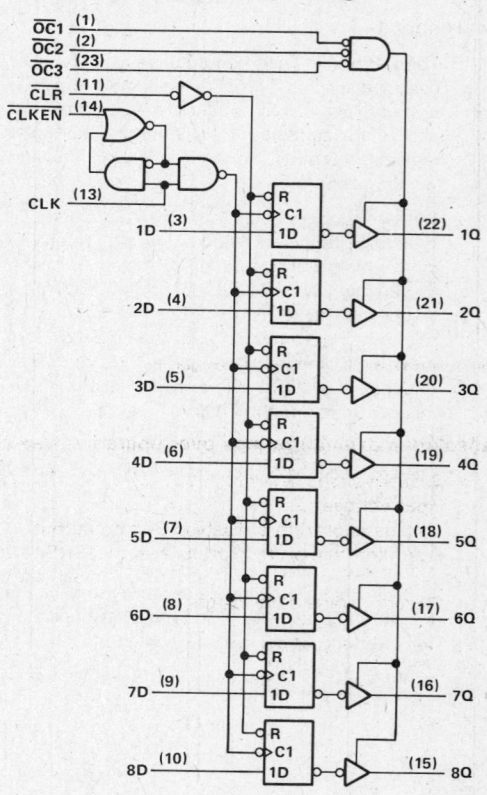

Pin numbers shown are for JT and NT packages.

TYPES SN54AS825, SN54AS826, SN74AS825, SN74AS826
8-BIT BUS INTERFACE FLIP-FLOPS WITH 3-STATE OUTPUTS

'AS826 logic symbol

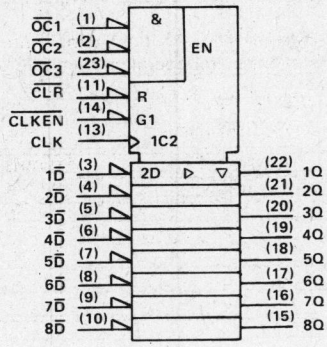

'AS826 logic diagram (positive logic)

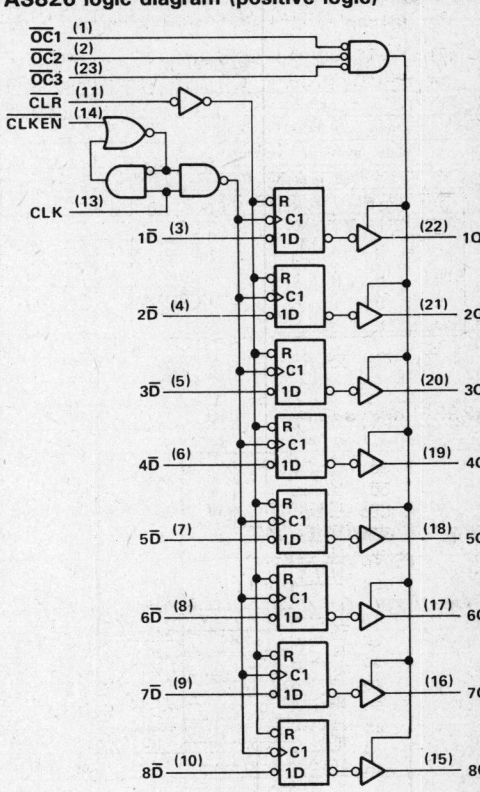

Pin numbers shown are for JT and NT packages.

absolute maximum ratings over operating free-air temperature range (unless otherwise noted)

Supply voltage, V_{CC} . . . . 7 V
Input voltage . . . . 7 V
Voltage applied to a disabled 3-state output . . . . 5.5 V
Operating free-air temperature range: SN54AS825, SN54AS826 . . . . −55°C to 125°C
             SN74AS825, SN74AS826 . . . . 0°C to 70°C
Storage temperature range . . . . −65 to 150°C

ALS AND AS CIRCUITS

TEXAS INSTRUMENTS
POST OFFICE BOX 225012 • DALLAS, TEXAS 75265

TYPES SN54AS825, SN54AS826, SN74AS825, SN74AS826
8-BIT BUS INTERFACE FLIP-FLOPS WITH 3-STATE OUTPUTS

recommended operating conditions

			SN54AS825 SN54AS826			SN74AS825 SN74AS826			UNIT
			MIN	NOM	MAX	MIN	NOM	MAX	
V_{CC}	Supply voltage		4.5	5	5.5	4.5	5	5.5	V
V_{IH}	High-level input voltage		2			2			V
V_{IL}	Low-level input voltage				0.8			0.8	V
I_{OH}	High-level output current				−24			−24	mA
I_{OL}	Low-level output current				32			48	mA
f_{clock}	Clock frequency								MHz
t_w	Pulse duration	$\overline{CLR}$ low							ns
		CLK high							
		CLK low							
		$\overline{CLKEN}$							
t_{su}	Setup time before CLK ↑	$\overline{CLR}$ inactive							ns
		Data							
		$\overline{CLKEN}$							
t_h	Hold time, data after CLK ↑								ns
T_A	Operating free-air temperature		−55		125	0		70	°C

electrical characteristics over recommended operating free-air temperature range (unless otherwise noted)

PARAMETER		TEST CONDITIONS		SN54AS825 SN54AS826			SN74AS825 SN74AS826			UNIT
				MIN	TYP†	MAX	MIN	TYP†	MAX	
V_{IK}		V_{CC} = 4.5 V,	I_I = −18 mA			−1.2			−1.2	V
V_{OH}		V_{CC} = 4.5 V to 5.5 V,	I_{OH} = −2 mA	V_{CC}−2			V_{CC}−2			V
		V_{CC} = 4.5 V,	I_{OH} = −15 mA	2.4	3.2		2.4	3.2		
		V_{CC} = 4.5 V,	I_{OH} = −24 mA	2			2			
V_{OL}		V_{CC} = 4.5 V,	I_{OL} = 32 mA		0.25	0.5				V
		V_{CC} = 4.5 V,	I_{OL} = 48 mA					0.25	0.5	
I_{OZH}		V_{CC} = 5.5 V,	V_O = 2.7 V			50			50	µA
I_{OZL}		V_{CC} = 5.5 V,	V_O = 0.4 V			−50			−50	µA
I_I		V_{CC} = 5.5 V,	V_I = 7 V							mA
I_{IH}		V_{CC} = 5.5 V,	V_I = 2.7 V							µA
I_{IL}		V_{CC} = 5.5 V,	V_I = 0.4 V							mA
I_O‡		V_{CC} = 5.5 V,	V_O = 2.25 V	−30		−112	−30		−112	mA
I_{CC}	'AS825	V_{CC} = 5.5 V	Outputs high							mA
			Outputs low							
			Outputs disabled			58			58	
	'AS826	V_{CC} = 5.5 V	Outputs high							mA
			Outputs low							
			Outputs disabled			58			58	

†All typical values are at V_{CC} = 5 V, T_A = 25°C.
‡The output conditions have been chosen to produce a current that closely approximates one half of the true short-circuit output current, I_{OS}.

Additional information on these products can be obtained from the factory as it becomes available.

ALS AND AS CIRCUITS

Texas Instruments
POST OFFICE BOX 225012 • DALLAS, TEXAS 75265

TYPES SN54AS825, SN54AS826, SN74AS825, SN74AS826
8-BIT BUS INTERFACE FLIP-FLOPS WITH 3-STATE OUTPUTS

switching characteristics (see Note 1)

PARAMETER	FROM (INPUT)	TO (OUTPUT)	V_{CC} = 4.5 V to 5.5 V, C_L = 50 pF, R1 = 500 Ω, R2 = 500 Ω, T_A = MIN to MAX						UNIT
			SN54AS825 SN54AS826			SN74AS825 SN74AS826			
			MIN	TYP†	MAX	MIN	TYP†	MAX	
f_{max}									MHz
t_{PLH}	CLK	Any Q		7.5			7.5		ns
t_{PHL}				9.5			9.5		
t_{PHL}	$\overline{CLR}$	Any Q		11			11		ns
t_{PZH}	$\overline{OC}$	Any Q		6			6		ns
t_{PZL}				7			7		
t_{PHZ}	$\overline{OC}$	Any Q		6			6		ns
t_{PLZ}				7			7		

†All typical values are at V_{CC} = 5 V, T_A = 25°C.
NOTE 1: For load circuit and voltage waveforms, see page 1-12.

D flip-flop signal conventions

It is normal TI practice to name the outputs and other inputs of a D-type flip-flop and to draw its logic symbol based on the assumption of true data (D) inputs. Then outputs that produce data in phase with the data inputs are called Q and those producing complementary data are called $\overline{Q}$. An input that causes a Q output to go high or a $\overline{Q}$ output to go low is called Preset; an input that causes a $\overline{Q}$ output to go high or a Q output to go low is called Clear. Bars are used over these pin names ($\overline{PRE}$ and $\overline{CLR}$) if they are active-low.

The devices on this data sheet are second-source designs and the pin-name convention used by the original manufacturer has been retained. That makes it necessary to designate the inputs and outputs of the inverting circuit $\overline{D}$ and Q. In some applications it may be advantageous to redesignate the inputs and outputs as D and $\overline{Q}$. In that case, outputs should be renamed as shown below. Also shown are corresponding changes in the graphical symbol. Arbitrary pin numbers are shown in parentheses.

Notice that Q and $\overline{Q}$ exchange names, which causes Preset and Clear to do likewise. Also notice that the polarity indicators (⊳) on $\overline{PRE}$ and $\overline{CLR}$ remain since these inputs are still active-low, but that the presence or absence of the polarity indicator changes at $\overline{D}$, Q, and $\overline{Q}$. Of course pin 5 (Q) is still in phase with the data input D, but now both are considered active high.

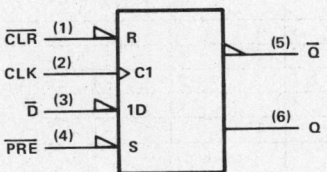 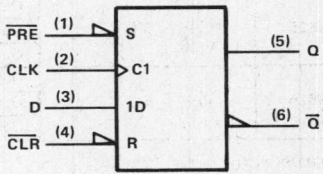

TEXAS INSTRUMENTS
POST OFFICE BOX 225012 • DALLAS, TEXAS 75265

2
ALS AND AS CIRCUITS

TYPES SN54ALS832, SN54AS832A, SN74ALS832, SN74AS832A
HEX 2-INPUT OR DRIVERS

D2661, DECEMBER 1982—REVISED DECEMBER 1983

- High Capacitive Drive Capability
- 'ALS832 Has Typical Delay Time of 5 ns (C_L = 50 pF) and Typical Power Dissipation of 5.3 mW per Gate
- 'AS832A Has Typical Delay Time of 3.9 ns (C_L = 50 pF) and Typical Power Dissipation of Less than 17 mW per Gate
- Package Options Include Both Plastic and Ceramic Chip Carriers in Addition to Plastic and Ceramic DIPs
- Dependable Texas Instruments Quality and Reliability

SN54ALS832, SN54AS832A . . . J PACKAGE
SN74ALS832, SN74AS832A . . . N PACKAGE
(TOP VIEW)

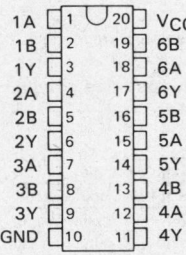

description

These devices contain six independent 2-input OR drivers. They perform the Boolean functions $Y = A + B$ or $Y = \overline{\overline{A} \cdot \overline{B}}$ in positive logic.

The –1 version of the SN74ALS832 parts is identical to the standard version except that the recommended maximum I_{OL} is increased to 48 milliamperes. There is no –1 version of the SN54ALS832 parts.

The SN54ALS832 and SN54AS832A are characterized for operation over the full military temperature range of –55 °C to 125 °C. The SN74ALS832 and SN74AS832A are characterized for operation from 0 °C to 70 °C.

SN54ALS832, SN54AS832A . . . FH PACKAGE
SN74ALS832, SN74AS832A . . . FN PACKAGE
(TOP VIEW)

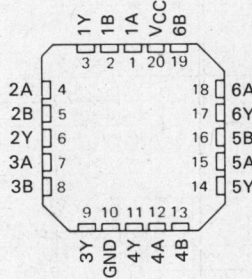

FUNCTION TABLE (each driver)

INPUTS		OUTPUT
A	B	Y
H	X	H
X	H	H
L	L	L

logic symbol

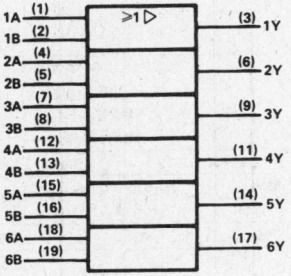

Pin numbers shown are for J and N packages.

Copyright © 1983 by Texas Instruments Incorporated

TYPES SN54ALS832, SN74ALS832
HEX 2-INPUT OR DRIVERS

absolute maximum ratings over operating free-air temperature range (unless otherwise noted)

Supply voltage, V_{CC} .... 7 V
Input voltage .... 7 V
Operating free-air temperature range: SN54ALS832 .... −55 °C to 125 °C
                                          SN74ALS832 .... 0 °C to 70 °C
Storage temperature range .... −65 °C to 150 °C

recommended operating conditions

		SN54ALS832			SN74ALS832			UNIT
		MIN	NOM	MAX	MIN	NOM	MAX	
V_{CC}	Supply voltage	4.5	5	5.5	4.5	5	5.5	V
V_{IH}	High-level input voltage	2			2			V
V_{IL}	Low-level input voltage			0.8			0.8	V
I_{OH}	High-level output current			−12			−15	mA
I_{OL}	Low-level output current			12			24	mA
							48†	
T_A	Operating free-air temperature	−55		125	0		70	°C

† The extended limit applies if V_{CC} is maintained between 4.75 V and 5.25 V.
  The 48 mA limit applies for the SN74ALS832−1 only.

electrical characteristics over recommended operating free-air temperature range (unless otherwise noted)

PARAMETER	TEST CONDITIONS		SN54ALS832			SN74ALS832			UNIT
			MIN	TYP‡	MAX	MIN	TYP‡	MAX	
V_{IK}	V_{CC} = 4.5 V,	I_I = −18 mA			−1.5			−1.5	V
V_{OH}	V_{CC} = 4.5 V to 5.5 V,	I_{OH} = −0.4 mA	$V_{CC}-2$			$V_{CC}-2$			V
	V_{CC} = 4.5 V,	I_{OH} = −3 mA	2.4	3.2		2.4	3.2		
	V_{CC} = 4.5 V,	I_{OH} = −12 mA	2						
	V_{CC} = 4.5 V,	I_{OH} = −15 mA				2			
V_{OL}	V_{CC} = 4.5 V,	I_{OL} = 12 mA		0.25	0.4		0.25	0.4	V
	V_{CC} = 4.5 V,	I_{OL} = 24 mA					0.35	0.5	
	(I_{OL} = 48 mA for −1 version)								
I_I	V_{CC} = 5.5 V,	V_I = 7 V			0.1			0.1	mA
I_{IH}	V_{CC} = 5.5 V,	V_I = 2.7 V			20			20	µA
I_{IL}	V_{CC} = 5.5 V,	V_I = 0.4 V			−0.1			−0.1	mA
I_O §	V_{CC} = 5.5 V,	V_O = 2.25 V	−30		−112	−30		−112	mA
I_{CCH}	V_{CC} = 5.5 V,	V_I = 4.5 V		4	8		4	8	mA
I_{CCL}	V_{CC} = 5.5 V,	V_I = 0 V		9.5	16		9.5	16	mA

‡ All typical values are at V_{CC} = 5 V, T_A = 25 °C.
§ The output conditions have been chosen to produce a current that closely approximates one half of the true short-circuit output current, I_{OS}.

switching characteristics (see Note 1)

PARAMETER	FROM (INPUT)	TO (OUTPUT)	V_{CC} = 4.5 V to 5.5 V, C_L = 50 pF, R_L = 500 Ω, T_A = MIN to MAX				UNIT
			SN54ALS832		SN74ALS832		
			MIN	MAX	MIN	MAX	
t_{PLH}	A or B	Y	2	10	2	8	ns
t_{PHL}			2	10	2	8	

NOTE 1: For load circuit and voltage waveforms, see page 1-12.

TEXAS INSTRUMENTS
POST OFFICE BOX 225012 • DALLAS, TEXAS 75265

TYPES SN54AS832A, SN74AS832A
HEX 2-INPUT OR DRIVERS

absolute maximum ratings over operating free-air temperature range (unless otherwise noted)

Supply voltage, V_{CC}	7 V
Input voltage	7 V
Operating free-air temperature range: SN54AS832A	$-55\,°C$ to $125\,°C$
SN74AS832A	$0\,°C$ to $70\,°C$
Storage temperature range	$-65\,°C$ to $150\,°C$

recommended operating conditions

		SN54AS832A			SN74AS832A			UNIT
		MIN	NOM	MAX	MIN	NOM	MAX	
V_{CC}	Supply voltage	4.5	5	5.5	4.5	5	5.5	V
V_{IH}	High-level input voltage	2			2			V
V_{IL}	Low-level input voltage			0.8			0.8	V
I_{OH}	High-level output current			-40			-48	mA
I_{OL}	Low-level output current			40			48	mA
T_A	Operating free-air temperature	-55		125	0		70	°C

electrical characteristics over recommended operating free-air temperature range (unless otherwise noted)

PARAMETER	TEST CONDITIONS		SN54AS832A			SN74AS832A			UNIT
			MIN	TYP†	MAX	MIN	TYP†	MAX	
V_{IK}	$V_{CC} = 4.5$ V,	$I_I = -18$ mA			-1.2			-1.2	V
V_{OH}	$V_{CC} = 4.5$ V to 5.5 V,	$I_{OH} = -2$ mA	$V_{CC}-2$			$V_{CC}-2$			V
	$V_{CC} = 4.5$ V,	$I_{OH} = -3$ mA	2.4	3.2		2.4	3.2		
	$V_{CC} = 4.5$ V,	$I_{OH} = -40$ mA	2						
	$V_{CC} = 4.5$ V,	$I_{OH} = -48$ mA				2			
V_{OL}	$V_{CC} = 4.5$ V,	$I_{OL} = 40$ mA		0.25	0.5				V
	$V_{CC} = 4.5$ V,	$I_{OL} = 48$ mA					0.35	0.5	
I_I	$V_{CC} = 5.5$ V,	$V_I = 7$ V			0.1			0.1	mA
I_{IH}	$V_{CC} = 5.5$ V,	$V_I = 2.7$ V			20			20	µA
I_{IL}	$V_{CC} = 5.5$ V,	$V_I = 0.4$ V			-0.5			-0.5	mA
I_O‡	$V_{CC} = 5.5$ V,	$V_O = 2.25$ V	-135			-135			mA
I_{CCH}	$V_{CC} = 5.5$ V,	$V_I = 4.5$ V		9	15		9	15	mA
I_{CCL}	$V_{CC} = 5.5$ V,	$V_I = 0$ V		22	36		22	36	mA

†All typical values are at $V_{CC} = 5$ V, $T_A = 25\,°C$.
‡The output conditions have been chosen to produce a current that closely approximates one half of the true short-circuit output current, I_{OS}.

switching characteristics (see Note 1)

PARAMETER	FROM (INPUT)	TO (OUTPUT)	$V_{CC} = 4.5$ V to 5.5 V, $C_L = 50$ pF, $R_L = 500\,\Omega$, $T_A = $ MIN to MAX				UNIT
			SN54AS832A		SN74AS832A		
			MIN	MAX	MIN	MAX	
t_{PLH}	A or B	Y	1	7	1	5.5	ns
t_{PHL}			1	6.5	1	5.5	

NOTE 1: For load circuit and voltage waveforms, see page 1-12.

TEXAS INSTRUMENTS
POST OFFICE BOX 225012 • DALLAS, TEXAS 75265

2 ALS AND AS CIRCUITS

TYPES SN54ALS841, SN54AS841, SN54ALS842, SN54AS842, SN74ALS841, SN74AS841, SN74ALS842, SN74AS842
10-BIT BUS INTERFACE D-TYPE LATCHES WITH 3-STATE OUTPUTS

D2825, DECEMBER 1983

- 3-State Buffer-Type Outputs Drive Bus-Lines Directly
- Bus-Structured Pinout
- Provide Extra Bus Driving Latches Necessary for Wider Address/Data Paths or Buses with Parity
- Buffered Control Inputs to Reduce DC Loading
- Power-Up High-Impedance State
- Package Options Include Both Plastic and Ceramic Chip Carriers in Addition to Plastic and Ceramic DIPs
- Dependable Texas Instruments Quality and Reliability

SN54ALS841, SN54AS841 . . . JT PACKAGE
SN74ALS841, SN74AS841 . . . NT PACKAGE
(TOP VIEW)

```
 OC  [ 1    24 ] VCC
 1D  [ 2    23 ] 1Q
 2D  [ 3    22 ] 2Q
 3D  [ 4    21 ] 3Q
 4D  [ 5    20 ] 4Q
 5D  [ 6    19 ] 5Q
 6D  [ 7    18 ] 6Q
 7D  [ 8    17 ] 7Q
 8D  [ 9    16 ] 8Q
 9D  [ 10   15 ] 9Q
10D  [ 11   14 ] 10Q
GND  [ 12   13 ] C
```

SN54ALS841, SN54AS841 . . . FH PACKAGE
SN74ALS841, SN74AS841 . . . FN PACKAGE
(TOP VIEW)

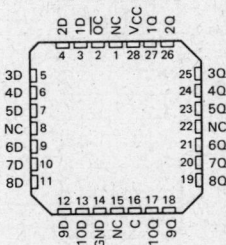

description

These 10-bit latches feature three-state outputs designed specifically for driving highly capacitive or relatively low-impedance loads. They are particularly suitable for implementing buffer registers, I/O ports, bidirectional bus drivers, and working registers.

The ten latches are transparent D-type. The 'ALS841 and 'AS841 have noninverting data (D) inputs. The 'ALS842 and 'AS842 have inverting $\overline{D}$ inputs.

A buffered output control ($\overline{OC}$) input can be used to place the ten outputs in either a normal logic state (high or low levels) or a high-impedance state. In the high-impedance state, the outputs neither load nor drive the bus lines significantly. The high-impedance state and increased drive provide the capability to drive the bus lines in a bus-organized system without need for interface or pull-up components.

The output control does not affect the internal operation of the latches. Old data can be retained or new data can be entered while the outputs are off.

The SN54ALS841, SN54AS841, SN54ALS842, and SN54AS842 are characterized for operation over the full military temperature range of −55°C to 125°C. The SN74ALS841, SN74AS841, SN74ALS842, and SN74AS842 are characterized for operation from 0°C to 70°C.

SN54ALS842, SN54AS842 . . . JT PACKAGE
SN74ALS842, SN74AS842 . . . NT PACKAGE
(TOP VIEW)

```
 OC  [ 1    24 ] VCC
 1D  [ 2    23 ] 1Q
 2D  [ 3    22 ] 2Q
 3D  [ 4    21 ] 3Q
 4D  [ 5    20 ] 4Q
 5D  [ 6    19 ] 5Q
 6D  [ 7    18 ] 6Q
 7D  [ 8    17 ] 7Q
 8D  [ 9    16 ] 8Q
 9D  [ 10   15 ] 9Q
10D  [ 11   14 ] 10Q
GND  [ 12   13 ] C
```

SN54ALS842, SN54AS842 . . . FH PACKAGE
SN74ALS842, SN74AS842 . . . FN PACKAGE
(TOP VIEW)

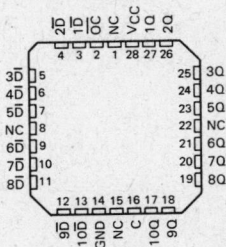

NC — No internal connection

PRODUCT PREVIEW

This document contains information on a product under development. Texas Instruments reserves the right to change or discontinue this product without notice.

Copyright © 1983 by Texas Instruments Incorporated

TEXAS INSTRUMENTS
POST OFFICE BOX 225012 • DALLAS, TEXAS 75265

TYPES SN54ALS841, SN54AS841, SN54ALS842, SN54AS842, SN74ALS841, SN74AS841, SN74ALS842, SN74AS842
10-BIT BUS INTERFACE D-TYPE LATCHES WITH 3-STATE OUTPUTS

FUNCTION TABLES

'ALS841, 'AS841

INPUTS			OUTPUT
$\overline{OC}$	C	D	Q
L	H	H	H
L	H	L	L
L	L	X	Q_0
H	X	X	Z

'ALS842, 'AS842

INPUTS			OUTPUT
$\overline{OC}$	C	$\overline{D}$	Q
L	H	H	L
L	H	L	H
L	L	X	Q_0
H	X	X	Z

'ALS841, 'AS841 logic symbol

'ALS841, 'AS841 logic diagram (positive logic)

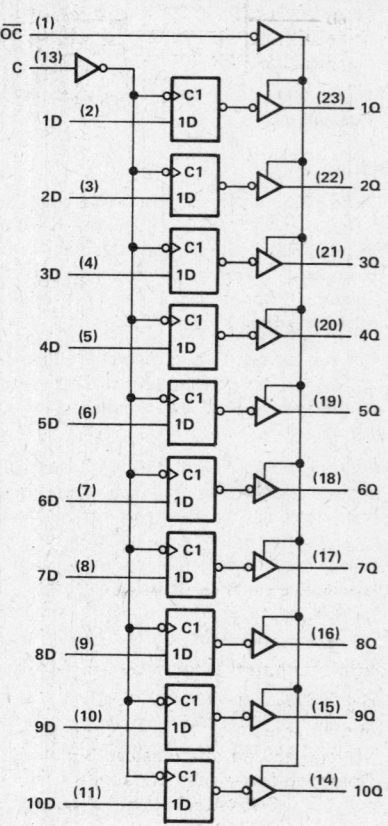

Pin numbers shown are for JT and NT packages.

TEXAS INSTRUMENTS
POST OFFICE BOX 225012 • DALLAS, TEXAS 75265

TYPES SN54ALS841, SN54AS841, SN54ALS842, SN54AS842, SN74ALS841, SN74AS841, SN74ALS842, SN74AS842
10-BIT BUS INTERFACE D-TYPE LATCHES WITH 3-STATE OUTPUTS

'ALS842, 'AS842 logic symbol

'ALS842, 'AS842 logic diagram (positive logic)

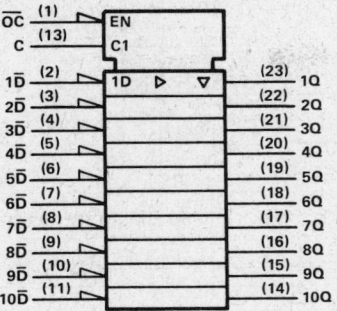
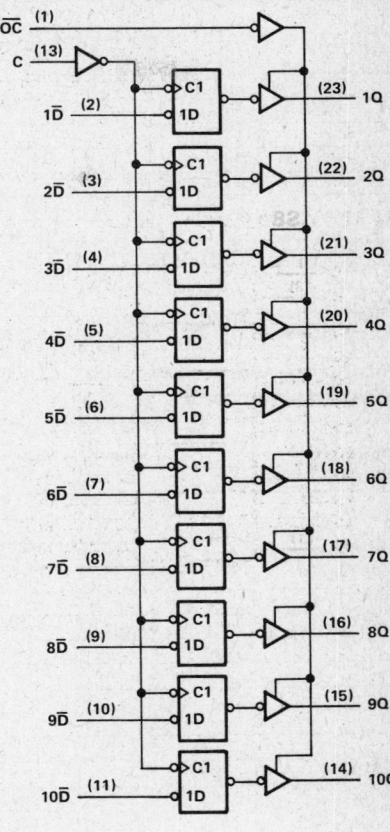

Pin numbers shown are for JT and NT packages.

absolute maximum ratings over operating free-air temperature range (unless otherwise noted)

Supply voltage, V_{CC} . . . . 7 V
Input voltage . . . . 7 V
Voltage applied to a disabled 3-state output . . . . 5.5 V
Operating free-air temperature range:
    SN54ALS841, SN54AS841, SN54ALS842, SN54AS842 . . . . −55°C to 125°C
    SN74ALS841, SB74AS841, SB74ALS842, SN74AS842 . . . . 0°C to 70°C
Storage temperature range . . . . −65°C to 150°C

TEXAS INSTRUMENTS
POST OFFICE BOX 225012 • DALLAS, TEXAS 75265

TYPES SN54ALS841, SN54ALS842, SN74ALS841, SN74ALS842
10-BIT BUS INTERFACE D-TYPE LATCHES WITH 3-STATE OUTPUTS

recommended operating conditions

		SN54ALS841 SN54ALS842			SN74ALS841 SN74ALS842			UNIT
		MIN	NOM	MAX	MIN	NOM	MAX	
V_{CC}	Supply voltage	4.5	5	5.5	4.5	5	5.5	V
V_{IH}	High-level input voltage	2			2			V
V_{IL}	Low-level input voltage			0.8			0.8	V
I_{OH}	High-level output current			−1			−2.6	mA
I_{OL}	Low-level output current			12			24	mA
t_w	Pulse duration, enable C high 'ALS841							ns
	'ALS842							
t_{su}	Setup time, data before enable C↓							ns
t_h	Hold time, data after enable C↓ 'ALS841							ns
	'ALS842							
T_A	Operating free-air temperature	−55		125	0		70	°C

electrical characteristics over recommended operating free-air temperature range (unless otherwise noted)

PARAMETER		TEST CONDITIONS		SN54ALS841 SN54ALS842			SN74ALS841 SN74ALS842			UNIT
				MIN	TYP†	MAX	MIN	TYP†	MAX	
V_{IK}		$V_{CC} = 4.5$ V,	$I_I = -18$ mA			−1.5			−1.5	V
V_{OH}		$V_{CC} = 4.5$ V to 5.5 V,	$I_{OH} = -0.4$ mA	$V_{CC}-2$			$V_{CC}-2$			V
		$V_{CC} = 4.5$ V,	$I_{OH} = -1$ mA	2.4	3.3					
		$V_{CC} = 4.5$ V,	$I_{OH} = -2.6$ mA				2.4	3.2		
V_{OL}		$V_{CC} = 4.5$ V,	$I_{OL} = 12$ mA		0.25	0.4		0.25	0.4	V
		$V_{CC} = 4.5$ V,	$I_{OL} = 24$ mA					0.35	0.5	
I_{OZH}		$V_{CC} = 5.5$ V,	$V_O = 2.7$ V			20			20	µA
I_{OZL}		$V_{CC} = 5.5$ V,	$V_O = 0.4$ V			−20			−20	µA
I_I		$V_{CC} = 5.5$ V,	$V_I = 7$ V			0.1			0.1	mA
I_{IH}		$V_{CC} = 5.5$ V,	$V_I = 2.7$ V			20			20	µA
I_{IL}		$V_{CC} = 5.5$ V,	$V_I = 0.4$ V			−0.1			−0.1	mA
I_O‡		$V_{CC} = 5.5$ V,	$V_O = 2.25$ V	−15		−70	−15		−70	mA
I_{CC}	'ALS841	$V_{CC} = 5.5$ V	Outputs high							mA
			Outputs low							
			Outputs disabled		25			25		
	'ALS842		Outputs high							
			Outputs low							
			Outputs disabled		28			28		

†All typical values are at $V_{CC} = 5$ V, $T_A = 25°C$.
‡The output conditions have been chosen to produce a current that closely approximates one half of the true short-circuit output current, I_{OS}.

Additional information on these products can be obtained from the factory as it becomes available.

TYPES SN54ALS841, SN54ALS842, SN74ALS841, SN74ALS842
10-BIT BUS INTERFACE D-TYPE LATCHES WITH 3-STATE OUTPUTS

'ALS841 switching characteristics (see Note 1)

PARAMETER	FROM (INPUT)	TO (OUTPUT)	V_{CC} = 4.5 V to 5.5 V, C_L = 50 pF, R1 = 500 Ω, R2 = 500 Ω, T_A = MIN to MAX						UNIT
			SN54ALS841			SN74ALS841			
			MIN	TYP†	MAX	MIN	TYP†	MAX	
t_{PLH}	D	Q		7			7		ns
t_{PHL}				9			9		
t_{PLH}	C	Q							ns
t_{PHL}									
t_{PZH}	$\overline{OC}$	Q							ns
t_{PZL}									
t_{PHZ}	$\overline{OC}$	Q							ns
t_{PLZ}									

'ALS842 switching characteristics (see Note 1)

PARAMETER	FROM (INPUT)	TO (OUTPUT)	V_{CC} = 4.5 V to 5.5 V, C_L = 50 pF, R1 = 500 Ω, R2 = 500 Ω, T_A = MIN to MAX						UNIT
			SN54ALS842			SN74ALS842			
			MIN	TYP†	MAX	MIN	TYP†	MAX	
t_{PLH}	$\overline{D}$	Q		11			11		ns
t_{PHL}				9			9		
t_{PLH}	C	Q							ns
t_{PHL}									
t_{PZH}	$\overline{OC}$	Q							ns
t_{PZL}									
t_{PHZ}	$\overline{OC}$	Q							ns
t_{PLZ}									

†All typical values are at T_A = 25 °C.
NOTE 1: For load circuit and voltage waveforms, see page 1-12.

Additional information on these products can be obtained from the factory as it becomes available.

Texas Instruments
POST OFFICE BOX 225012 • DALLAS, TEXAS 75265

TYPES SN54AS841, SN54AS842, SN74AS841, SN74AS842
10-BIT BUS INTERFACE D-TYPE LATCHES WITH 3-STATE OUTPUTS

recommended operating conditions

		SN54AS841 SN54AS842			SN74AS841 SN74AS842			UNIT
		MIN	NOM	MAX	MIN	NOM	MAX	
V_{CC}	Supply voltage	4.5	5	5.5	4.5	5	5.5	V
V_{IH}	High-level input voltage	2			2			V
V_{IL}	Low-level input voltage			0.8			0.8	V
I_{OH}	High-level output current			-24			-24	mA
I_{OL}	Low-level output current			32			48	mA
t_w	Pulse duration, enable C high							ns
t_{su}	Setup time, data before enable C↓							ns
t_h	Hold time, data after enable C↓							ns
T_A	Operating free-air temperature	-55		125	0		70	°C

electrical characteristics over recommended operating free-air temperature range (unless otherwise noted)

PARAMETER		TEST CONDITIONS		SN54AS841 SN54AS842			SN74AS841 SN74AS842			UNIT
				MIN	TYP†	MAX	MIN	TYP†	MAX	
V_{IK}		$V_{CC} = 4.5$ V,	$I_I = -18$ mA			-1.2			-1.2	V
V_{OH}		$V_{CC} = 4.5$ V to 5.5 V,	$I_{OH} = -2$ mA	$V_{CC}-2$			$V_{CC}-2$			V
		$V_{CC} = 4.5$ V,	$I_{OH} = -15$ mA	2.4	3.2		2.4	3.2		
		$V_{CC} = 4.5$ V,	$I_{OH} = -24$ mA	2			2			
V_{OL}		$V_{CC} = 4.5$ V,	$I_{OL} = 32$ mA		0.25	0.5				V
		$V_{CC} = 4.5$ V,	$I_{OL} = 48$ mA					0.25	0.5	
I_{OZH}		$V_{CC} = 5.5$ V,	$V_O = 2.7$ V			50			50	µA
I_{OZL}		$V_{CC} = 5.5$ V,	$V_O = 0.4$ V			-50			-50	µA
I_I		$V_{CC} = 5.5$ V,	$V_I = 7$ V							mA
I_{IH}		$V_{CC} = 5.5$ V,	$V_I = 2.7$ V							µA
I_{IL}		$V_{CC} = 5.5$ V,	$V_I = 0.4$ V							mA
I_O‡		$V_{CC} = 5.5$ V,	$V_O = 2.25$ V	-30		-112	-30		-112	mA
I_{CC}	'AS841	$V_{CC} = 5.5$ V	Outputs high		36			36		mA
			Outputs low		58			58		
			Outputs disabled		56			56		
	'AS842		Outputs high		38			38		
			Outputs low		60			60		
			Outputs disabled		58			58		

†All typical values are at $V_{CC} = 5$ V, $T_A = 25$°C.
‡The output conditions have been chosen to produce a current that closely approximates one half of the true short-circuit output current, I_{OS}.

Additional information on these products can be obtained from the factory as it becomes available.

TYPES SN54AS841, SN54AS842, SN74AS841, SN74AS842
10-BIT BUS INTERFACE D-TYPE LATCHES WITH 3-STATE OUTPUTS

'AS841 switching characteristics (see Note 1)

PARAMETER	FROM (INPUT)	TO (OUTPUT)	$V_{CC} = 4.5$ V to 5.5 V, $C_L = 50$ pF, R1 = 500 Ω, R2 = 500 Ω, T_A = MIN to MAX						UNIT
			SN54AS841			SN74AS841			
			MIN	TYP†	MAX	MIN	TYP†	MAX	
t_{PLH}	D	Q		4			4		ns
t_{PHL}				4.5			4.5		
t_{PLH}	C	Q							ns
t_{PHL}									
t_{PZH}	$\overline{OC}$	Q		6			6		ns
t_{PZL}				6			6		
t_{PHZ}	$\overline{OC}$	Q		4			4		ns
t_{PLZ}				5			5		

'AS842 switching characteristics (see Note 1)

PARAMETER	FROM (INPUT)	TO (OUTPUT)	$V_{CC} = 4.5$ V to 5.5 V, $C_L = 50$ pF, R1 = 500 Ω, R2 = 500 Ω, T_A = MIN to MAX						UNIT
			SN54AS842			SN74AS842			
			MIN	TYP†	MAX	MIN	TYP†	MAX	
t_{PLH}	$\overline{D}$	Q		4			4		ns
t_{PHL}				4.5			4.5		
t_{PLH}	C	Q							ns
t_{PHL}									
t_{PZH}	$\overline{OC}$	Q		6			6		ns
t_{PZL}				6			6		
t_{PHZ}	$\overline{OC}$	Q		4			4		ns
t_{PLZ}				5			5		

†All typical values are at $T_A = 25°C$.
NOTE 1: For load circuits and voltage waveforms, see page 1-12.

Additional information on these products can be obtained from the factory as it becomes available.

ALS AND AS CIRCUITS

TEXAS INSTRUMENTS
POST OFFICE BOX 225012 • DALLAS, TEXAS 75265

2
ALS AND AS CIRCUITS

TYPES SN54ALS843, SN54AS843, SN54ALS844, SN54AS844, SN74ALS843, SN74AS843, SN74ALS844, SN74AS844
9-BIT BUS INTERFACE D-TYPE LATCHES WITH 3-STATE OUTPUTS

DECEMBER 1983

- 3-State Buffer-Type Outputs Drive Bus-Lines Directly
- Bus-Structured Pinout
- Provide Extra Bus Driving Latches Necessary for Wider Address/Data Paths or Buses with Parity
- Buffered Control Inputs to Reduce DC Loading
- Power-Up High Impedance
- Package Options Include Both Plastic and Ceramic Chip Carriers in Addition to Plastic and Ceramic DIPs
- Dependable Texas Instruments Quality and Reliability

description

These 9-bit latches feature three-state outputs designed specifically for driving highly capacitive or relatively low-impedance loads. They are particularly suitable for implementing buffer registers, I/O ports, bidirectional bus drivers, and working registers.

The nine latches are transparent D-type. The 'ALS843 and 'AS843 have noninverting data (D) inputs. The 'ALS844 and 'AS844 have inverting $\overline{D}$ inputs.

A buffered output control ($\overline{OC}$) input can be used to place the ten outputs in either a normal logic state (high or low levels) or a high-impedance state. In the high-impedance state, the outputs neither load nor drive the bus lines significantly. The high-impedance state and increased drive provide the capability to drive the bus lines in a bus-organized system without need for interface or pull-up components.

The output control ($\overline{OC}$) does not affect the internal operation of the latches. Old data can be retained or new data can be entered while the outputs are off.

The SN54ALS843, SN54AS843, SN54ALS844, and SN54AS844 are characterized for operation over the full military temperature range of −55°C to 125°C. The SN74ALS843, SN74AS843, SN74ALS844, and SN74AS844 are characterized for operation from 0°C to 70°C.

SN54ALS843, SN54AS843 . . . JT PACKAGE
SN74ALS843, SN74AS843 . . . NT PACKAGE
(TOP VIEW)

```
 OC  [ 1    24]  Vcc
 1D  [ 2    23]  1Q
 2D  [ 3    22]  2Q
 3D  [ 4    21]  3Q
 4D  [ 5    20]  4Q
 5D  [ 6    19]  5Q
 6D  [ 7    18]  6Q
 7D  [ 8    17]  7Q
 8D  [ 9    16]  8Q
 9D  [10    15]  9Q
CLR  [11    14]  PRE
GND  [12    13]  C
```

SN54ALS843, SN54AS843 . . . FH PACKAGE
SN74ALS843, SN74AS843 . . . FN PACKAGE
(TOP VIEW)

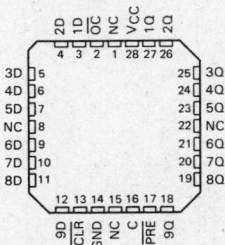

SN54ALS844, SN54AS844 . . . JT PACKAGE
SN74ALS844, SN74AS844 . . . NT PACKAGE
(TOP VIEW)

```
 OC  [ 1    24]  Vcc
 1D  [ 2    23]  1Q
 2D  [ 3    22]  2Q
 3D  [ 4    21]  3Q
 4D  [ 5    20]  4Q
 5D  [ 6    19]  5Q
 6D  [ 7    18]  6Q
 7D  [ 8    17]  7Q
 8D  [ 9    16]  8Q
 9D  [10    15]  9Q
CLR  [11    14]  PRE
GND  [12    13]  C
```

SN54ALS844, SN54AS844 . . . FH PACKAGE
SN74ALS844, SN74AS844 . . . FN PACKAGE
(TOP VIEW)

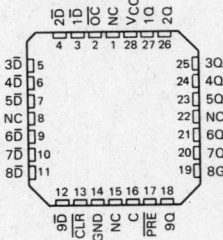

NC—No internal connection

PRODUCT PREVIEW

This document contains information on a product under development. Texas Instruments reserves the right to change or discontinue this product without notice.

Copyright © 1983 by Texas Instruments Incorporated

ALS AND AS CIRCUITS

TYPES SN54ALS843, SN54AS843, SN54ALS844, SN54AS844, SN74ALS843, SN74AS843, SN74ALS844, SN74AS844
9-BIT BUS INTERFACE D-TYPE LATCHES WITH 3-STATE OUTPUTS

FUNCTION TABLES

'ALS843, 'AS843

INPUTS					OUTPUT
PRE	CLR	OC	C	D	Q
L	H	L	X	X	H
H	L	L	X	X	L
L	L	L	X	X	H
H	H	L	H	L	L
H	H	L	H	H	H
H	H	L	L	X	Q_0
X	X	H	X	X	Z

'ALS844, 'AS844

INPUTS					OUTPUT
PRE	CLR	OC	C	$\overline{D}$	Q
L	H	L	X	X	H
H	L	L	X	X	L
L	L	L	X	X	H
H	H	L	H	L	H
H	H	L	H	H	L
H	H	L	L	X	Q_0
X	X	H	X	X	Z

logic symbol

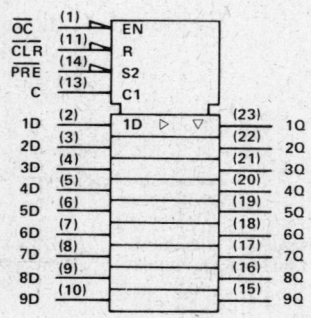

'ALS843, 'AS843 logic diagram (positive logic)

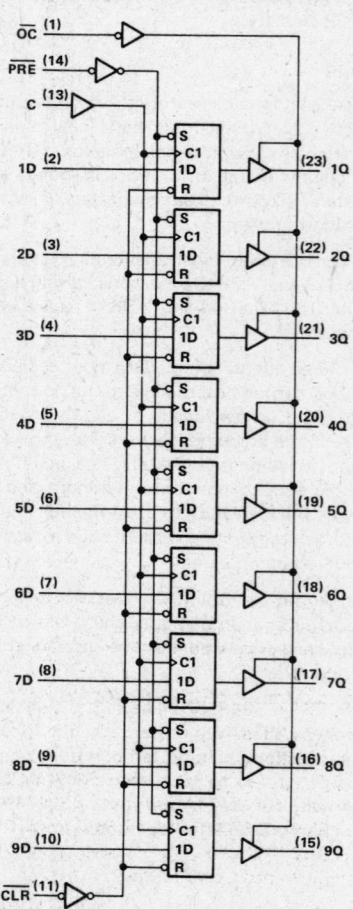

Pin numbers shown are for JT and NT packages.
This symbol is in accordance with IEEE Std 9 and recent decisions of IEEE.

TYPES SN54ALS843, SN54AS843, SN54ALS844, SN54AS844 SN74ALS843, SN74AS843, SN74ALS844, SN74AS844
9-BIT BUS INTERFACE D-TYPE LATCHES WITH 3-STATE OUTPUTS

logic symbol

'ALS844, 'AS844 logic diagram (positive logic)

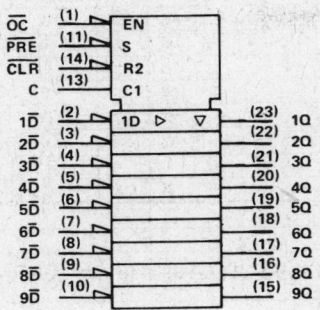

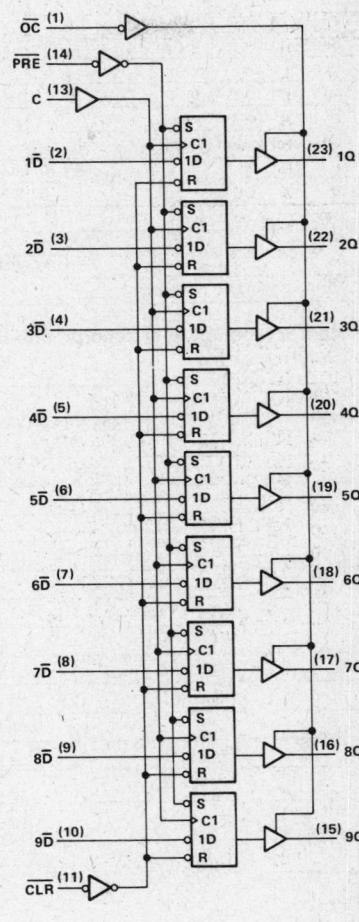

Pin numbers shown are for JT and NT packages.
This symbol is in accordance with IEEE Std 9 and recent decisions of IEEE.

absolute maximum ratings over operating free-air temperature range (unless otherwise noted)

Supply voltage, V_{CC} . . . . 7 V
Input voltage . . . . 7 V
Voltage applied to a disabled 3-state output . . . . 5.5 V
Operating free-air temperature range:
   SN54ALS843, SN54AS843, SN54ALS844, SN54AS844 . . . . −55°C
   SN74ALS843, SN74AS843, SN74ALS844, SN74AS844 . . . . −0°C to 70°C
Storage temperature range . . . . −65°C to 150°C

TEXAS INSTRUMENTS
POST OFFICE BOX 225012 • DALLAS, TEXAS 75265

TYPES SN54ALS843, SN54ALS844, SN74ALS843, SN74ALS844
9-BIT BUS INTERFACE D-TYPE LATCHES WITH 3-STATE OUTPUTS

recommended operating conditions

			SN54ALS843 SN54ALS844			SN74ALS843 SN74ALS844			UNIT
			MIN	NOM	MAX	MIN	NOM	MAX	
V_{CC}	Supply voltage		4.5	5	5.5	4.5	5	5.5	V
V_{IH}	High-level input voltage		2			2			V
V_{IL}	Low-level input voltage				0.8			0.8	V
I_{OH}	High-level output current				−1			−2.6	mA
I_{OL}	Low-level output current				12			24	mA
t_w	Pulse duration, enable C high	'ALS843							ns
		'ALS844							
t_{su}	Setup time, data before enable C↓								ns
t_h	Hold time, data after enable C↓	'ALS843							ns
		'ALS844							
T_A	Operating free-air temperature		−55		125	0		70	°C

electrical characteristics over recommended operating free-air temperature range (unless otherwise noted)

PARAMETER		TEST CONDITIONS		SN54ALS843 SN54ALS844			SN74ALS843 SN74ALS844			UNIT
				MIN	TYP†	MAX	MIN	TYP†	MAX	
V_{IK}		V_{CC} = 4.5 V,	I_I = −18 mA			−1.5			−1.5	V
V_{OH}		V_{CC} = 4.5 to 5.5 V,	I_{OH} = −0.4 mA	V_{CC}−2			V_{CC}−2			V
		V_{CC} = 4.5 V,	I_{OH} = −1 mA	2.4	3.3					
		V_{CC} = 4.5 V,	I_{OH} = −2.6 mA				2.4	3.2		
V_{OL}		V_{CC} = 4.5 V,	I_{OL} = 12 mA		0.25	0.4		0.25	0.4	V
		V_{CC} = 4.5 V,	I_{OL} = 24 mA					0.35	0.5	
I_{OZH}		V_{CC} = 5.5 V,	V_O = 2.7 V			20			20	μA
I_{OZL}		V_{CC} = 5.5 V,	V_O = 0.4 V			−20			−20	μA
I_I		V_{CC} = 5.5 V,	V_I = 7 V			0.1			0.1	mA
I_{IH}		V_{CC} = 5.5 V,	V_I = 2.7 V			20			20	μA
I_{IL}		V_{CC} = 5.5 V,	V_I = 0.4 V			−0.1			−0.1	mA
I_O‡		V_{CC} = 5.5 V,	V_O = 2.25 V	−15		−70	−15		−70	mA
I_{CC}	'ALS843	V_{CC} = 5.5 V	Outputs high							mA
			Outputs low							
			Outputs disabled		25			25		
	'ALS844		Outputs high							
			Outputs low							
			Outputs disabled		28			28		

†All typical values are at V_{CC} = 5 V, T_A = 25°C.
‡The output conditions have been chosen to produce a current that closely approximates one half of the true short-circuit output current, I_{OS}.

Additional information on these products can be obtained from the factory as it becomes available.

TYPES SN54ALS843, SN54ALS844, SN74ALS843, SN74ALS844
9-BIT BUS INTERFACE D-TYPE LATCHES WITH 3-STATE OUTPUTS

ALS843 switching characteristics (see Note 1)

PARAMETER	FROM (INPUT)	TO (OUTPUT)	$V_{CC} = 4.5$ V to 5.5 V, $C_L = 50$ pF, $R1 = 500\ \Omega$, $R2 = 500\ \Omega$, $T_A = $ MIN to MAX						UNIT
			SN54ALS843			SN74ALS843			
			MIN	TYP†	MAX	MIN	TYP†	MAX	
t_{PLH}	D	Q		7			7		ns
t_{PHL}				9			9		
t_{PLH}	C	Q							ns
t_{PHL}									
t_{PLH}	$\overline{PRE}$	Q							ns
t_{PHL}	$\overline{CLR}$	Q							ns
t_{PZH}	$\overline{OC}$	O							ns
t_{PZL}									
t_{PHZ}	$\overline{OC}$	O							ns
t_{PLZ}									

ALS844 switching characteristics (see Note 1)

PARAMETER	FROM (INPUT)	TO (OUTPUT)	$V_{CC} = 4.5$ V to 5.5 V, $C_L = 50$ pF, $R1 = 500\ \Omega$, $R2 = 500\ \Omega$, $T_A = $ MIN to MAX						UNIT
			SN54ALS844			SN74ALS844			
			MIN	TYP†	MAX	MIN	TYP†	MAX	
t_{PLH}	$\overline{D}$	Q		7			7		ns
t_{PHL}				9			9		
t_{PLH}	C	Q							ns
t_{PHL}									
t_{PLH}	$\overline{PRE}$	Q							ns
t_{PHL}	$\overline{CLR}$	Q							ns
t_{PZH}	$\overline{OC}$	O							ns
t_{PZL}									
t_{PHZ}	$\overline{OC}$	O							ns
t_{PLZ}									

†All typical values are at $T_A = 25°C$.
NOTE 1: For load circuit and voltage waveforms, see page 1-12.

Additional information on these products can be obtained from the factory as it becomes available.

TYPES SN54AS843, SN54AS844, SN74AS843, SN74AS844
9-BIT BUS INTERFACE D-TYPE LATCHES WITH 3-STATE OUTPUTS

recommended operating conditions

			SN54AS843 SN54AS844			SN74AS843 SN74AS844			UNIT
			MIN	NOM	MAX	MIN	NOM	MAX	
V_{CC}	Supply voltage		4.5	5	5.5	4.5	5	5.5	V
V_{IH}	High-level input voltage		2			2			V
V_{IL}	Low-level input voltage				0.8			0.8	V
I_{OH}	High-level output current				−24			−24	mA
I_{OL}	Low-level output current				32			48	mA
t_w	Pulse duration, enable C high	CLR or PRE low							ns
		C high							
t_{su}	Setup time, data before enable C↓								ns
t_h	Hold time, data after enable C↓								ns
T_A	Operating free-air temperature		−55		125	0		70	°C

electrical characteristics over recommended operating free-air temperature range (unless otherwise noted)

PARAMETER		TEST CONDITIONS		SN54AS843 SN54AS844			SN74AS843 SN74AS844			UNIT
				MIN	TYP†	MAX	MIN	TYP†	MAX	
V_{IK}		$V_{CC} = 4.5$ V,	$I_I = -18$ mA			−1.2			−1.2	V
V_{OH}		$V_{CC} = 4.5$ V,	$I_{OH} = -2$ mA	$V_{CC}-2$			$V_{CC}-2$			V
		$V_{CC} = 4.5$ V,	$I_{OH} = -15$ mA	2.4	3.2		2.4	3.2		
		$V_{CC} = 4.5$ V,	$I_{OH} = -24$ mA	2			2			
V_{OL}		$V_{CC} = 4.5$ V,	$I_{OL} = 32$ mA		0.25	0.5				V
		$V_{CC} = 4.5$ V,	$I_{OL} = 48$ mA					0.25	0.5	
I_{OZH}		$V_{CC} = 5.5$ V,	$V_O = 2.7$ V			50			50	µA
I_{OZL}		$V_{CC} = 5.5$ V,	$V_O = 0.4$ V			−50			−50	µA
I_I		$V_{CC} = 5.5$ V,	$V_I = 7$ V							mA
I_{IH}		$V_{CC} = 5.5$ V,	$V_I = 2.7$ V							µA
I_{IL}		$V_{CC} = 5.5$ V,	$V_I = 0.4$ V							mA
I_O‡		$V_{CC} = 5.5$ V,	$V_O = 2.25$ V	−30		−112	−30		−112	mA
I_{CC}	'AS843	$V_{CC} = 5.5$ V	Outputs high		38			38		mA
			Outputs low		57			57		
			Outputs disabled		56			56		
	'AS844		Outputs high		39			39		
			Outputs low		58			58		
			Outputs disabled		58			58		

†All typical values are at $V_{CC} = 5$ V, $T_A = 25°C$.
‡The output conditions have been chosen to produce a current that closely approximates one half of the true short-circuit output current, I_{OS}.

Additional information on these products can be obtained from the factory as it becomes available.

TEXAS INSTRUMENTS
POST OFFICE BOX 225012 • DALLAS, TEXAS 75265

TYPES SN54AS843, SN54AS844, SN74AS843, SN74AS844
9-BIT BUS INTERFACE D-TYPE LATCHES WITH 3-STATE OUTPUTS

'AS843 switching chaacteristics (see Note 1)

PARAMETER	FROM (INPUT)	TO (OUTPUT)	V_{CC} = 4.5 V to 5.5 V, C_L = 50 pF, R1 = 500 Ω, R2 = 500 Ω, T_A = MIN to MAX						UNIT
			SN54AS843			SN74AS843			
			MIN	TYP†	MAX	MIN	TYP†	MAX	
t_{PLH}	D	Q		4			4		ns
t_{PHL}				4.5			4.5		
t_{PLH}	C	O							ns
t_{PHL}									
t_{PLH}	$\overline{PRE}$	Q		5			5		ns
t_{PHL}	$\overline{CLR}$	Q		5.5			5.5		ns
t_{PZH}	$\overline{OC}$	O		6			6		ns
t_{PZL}				6			6		
t_{PHZ}	$\overline{OC}$	O		4			4		ns
t_{PLZ}				5			5		

'AS844 switching characteristics (see Note 1)

PARAMETER	FROM (INPUT)	TO (OUTPUT)	V_{CC} = 4.5 V to 5.5 V, C_L = 50 pF, R1 = 500 Ω, R2 = 500 Ω, T_A = MIN to MAX						UNIT
			SN54AS844			SN74AS844			
			MIN	TYP†	MAX	MIN	TYP†	MAX	
t_{PLH}	$\overline{D}$	Q		4			4		ns
t_{PHL}				4.5			4.5		
t_{PLH}	C	Q							ns
t_{PHL}									
t_{PLH}	$\overline{PRE}$	Q		5			5		ns
t_{PHL}	$\overline{CLR}$	Q		5.5			5.5		ns
t_{PZH}	$\overline{OC}$	Q		6			6		ns
t_{PZL}				6			6		
t_{PHZ}	$\overline{OC}$	Q		4			4		ns
t_{PLZ}				5			5		

†All typical values are at T_A = 25°C.
NOTE 1: For load circuit and voltage waveforms, see page 1-12.

Additional information on these products can be obtained from the factory as it becomes available.

TEXAS INSTRUMENTS
POST OFFICE BOX 225012 • DALLAS, TEXAS 75265

TYPES SN54ALS843, SN54AS843, SN54ALS844, SN54AS844, SN74ALS843, SN74AS843, SN74ALS844, SN74AS844
9-BIT BUS INTERFACE D-TYPE LATCHES WITH 3-STATE OUTPUTS

D flip-flop signal conventions

It is normal TI practice to name the outputs and other inputs of a D-type flip-flop and to draw its logic symbol based on the assumption of true data (D) inputs. Then outputs that produce data in phase with the data inputs are called $\overline{Q}$ and those producing complementary data are called Q. An input that causes a Q output to go high or a $\overline{Q}$ output to go low is called Preset; an input that causes a $\overline{Q}$ output to go high or a Q output to go low is called Clear. Bars are used over these pin names ($\overline{PRE}$ and $\overline{CLR}$) if they are active-low.

The devices on this data sheet are second-source designs and the pin-name convention used by the original manufacturer has been retained. That makes it necessary to designate the inputs and outputs of the inverting circuit $\overline{D}$ and Q. In some applications it may be advantageous to redesignate the inputs and outputs as D and $\overline{Q}$. In that case, outputs should be renamed as shown below. Also shown are corresponding changes in the graphical symbol. Arbitrary pin numbers are shown in parentheses.

Notice that Q and $\overline{Q}$ exchange names, which causes Preset and Clear to do likewise. Also notice that the polarity indicators ( ) on $\overline{PRE}$ and $\overline{CLR}$ remain since these inputs are still active-low, but that the presence or absence of the polarity changes at $\overline{D}$, Q, and $\overline{Q}$. Of course pin 5 (Q) is still in phase with the data input D, but now both are considered active high.

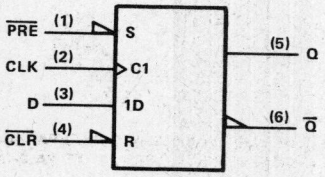 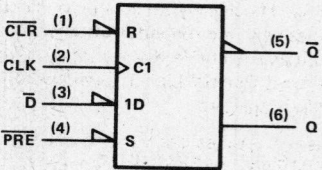

TYPES SN54ALS845, SN54AS845, SN54ALS846, SN54AS846, SN74ALS845, SN74AS845, SN74ALS846, SN74AS846
8-BIT BUS INTERFACE D-TYPE LATCHES WITH 3-STATE OUTPUTS

D2825, DECEMBER 1983

- 3-State Buffer-Type Outputs Drive Bus-Lines Directly
- Bus-Structured Pinout
- Provides Extra Bus Driving Latches Necessary for Wider Address/Data Paths or Buses with Parity
- Buffered Control Inputs to Reduce DC Loading
- Power-Up High-Impedance State
- Package Options Include Both Plastic and Ceramic Chip Carriers in Addition to Plastic and Ceramic DIPs
- Dependable Texas Instruments Quality and Reliability

SN54ALS845, SN54AS845 . . . JT PACKAGE
SN74ALS845, SN74AS845 . . . NT PACKAGE
(TOP VIEW)

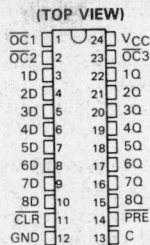

SN54ALS845, SN54AS845 . . . FH PACKAGE
SN74ALS845, SN74AS845 . . . FN PACKAGE
(TOP VIEW)

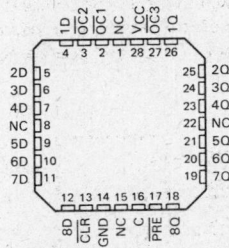

description

These 8-bit latches feature three-state outputs designed specifically for driving highly capacitive or relatively low-impedance loads. They are particularly suitable for implementing buffer registers, I/O ports, bidirectional bus drivers, and working registers.

The eight latches are transparent D-type. The 'ALS845 and 'AS845 have noninverting data (D) inputs. The 'ALS846 and 'AS846 have inverting $\overline{D}$ inputs. Since $\overline{CLR}$ and $\overline{PRE}$ are independent of the clock, taking the $\overline{CLR}$ input low will cause the eight Q outputs to go low. Taking the $\overline{PRE}$ input low will cause the eight Q outputs to go high. When both $\overline{PRE}$ and $\overline{CLR}$ are taken low, the outputs will follow the preset condition.

A buffered output control ($\overline{OC}$) input can be used to place the eight outputs in either a normal logic state (high or low levels) or a high-impedance state. In the high-impedance state, the outputs neither load nor drive the bus lines significantly. The high-impedance state and increased drive provide the capability to drive the bus lines in a bus-organized system without need for interface or pull-up components.

SN54ALS846, SN54AS846 . . . JT PACKAGE
SN74ALS846, SN74AS846 . . . NT PACKAGE
(TOP VIEW)

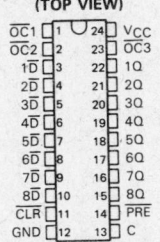

SN54ALS846, SN54AS846 . . . FH PACKAGE
SN74ALS846, SN74AS846 . . . FN PACKAGE
(TOP VIEW)

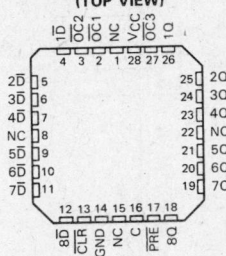

NC—No internal connection

PRODUCT PREVIEW
This document contains information on a product under development. Texas Instruments reserves the right to change or discontinue this product without notice.

Copyright © 1983 by Texas Instruments Incorporated

TEXAS INSTRUMENTS
POST OFFICE BOX 225012 • DALLAS, TEXAS 75265

ALS AND AS CIRCUITS

TYPES SN54ALS845, SN54AS845, SN54ALS846, SN54AS846, SN74ALS845, SN74AS845, SN74ALS846, SN74AS846
8-BIT BUS INTERFACE D-TYPE LATCHES WITH 3-STATE OUTPUTS

The output controls ($\overline{OC1}$, $\overline{OC2}$, $\overline{OC3}$) do not affect the internal operation of the latches. Old data can be retained or new data can be entered while the outputs are in the high-impedance state.

The SN54ALS845, SN54AS845, SN54ALS846, and SN54AS846 are characterized for operation over the full military temperature range of −55°C to 125°C. The SN74ALS845, SN74AS845, SN74ALS846, and SN74AS846 are characterized for operation from 0°C to 70°C.

FUNCTION TABLES

'ALS845, 'AS845

INPUTS							OUTPUT
PRE	CLR	$\overline{OC1}$	$\overline{OC2}$	$\overline{OC3}$	C	D	Q
L	H	L	L	L	X	X	H
H	L	L	L	L	X	X	L
L	L	L	L	L	X	X	H
H	H	L	L	L	H	L	L
H	H	L	L	L	H	H	H
H	H	L	L	L	L	X	Q_0
X	X	L	L	H	X	X	Z
X	X	L	H	L	X	X	Z
X	X	L	H	H	X	X	Z
X	X	H	L	L	X	X	Z
X	X	H	L	H	X	X	Z
X	X	H	H	L	X	X	Z
X	X	H	H	H	X	X	Z

'ALS846, 'AS846

INPUTS							OUTPUT
PRE	CLR	$\overline{OC1}$	$\overline{OC2}$	$\overline{OC3}$	C	$\overline{D}$	Q
L	H	L	L	L	X	X	H
H	L	L	L	L	X	X	L
L	L	L	L	L	X	X	H
H	H	L	L	L	H	L	H
H	H	L	L	L	H	H	L
H	H	L	L	L	L	X	Q_0
X	X	L	L	H	X	X	Z
X	X	L	H	L	X	X	Z
X	X	L	H	H	X	X	Z
X	X	H	L	L	X	X	Z
X	X	H	L	H	X	X	Z
X	X	H	H	L	X	X	Z
X	X	H	H	H	X	X	Z

logic symbols

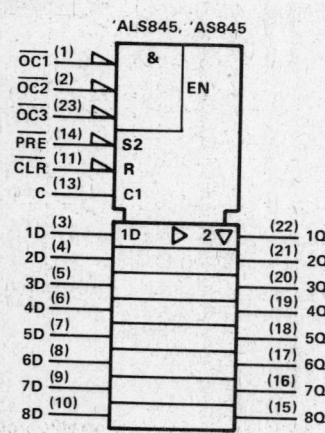

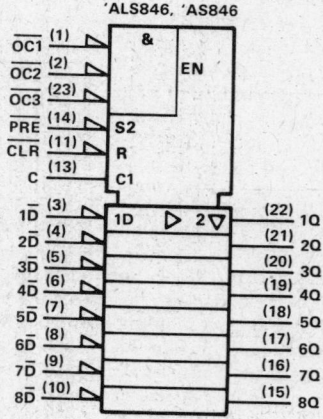

Pin numbers shown are for JT and NT packages.
These symbols are in accordance with IEEE Std 9 and recent decisions of IEEE.

TEXAS INSTRUMENTS
POST OFFICE BOX 225012 • DALLAS, TEXAS 75265

TYPES SN54ALS845, SN54AS845, SN54ALS846, SN54AS846 SN74ALS845, SN74AS845, SN74ALS846, SN74AS846
8-BIT BUS INTERFACE D-TYPE LATCHES WITH 3-STATE OUTPUTS

logic diagrams (positive logic)

'ALS845, 'AS845

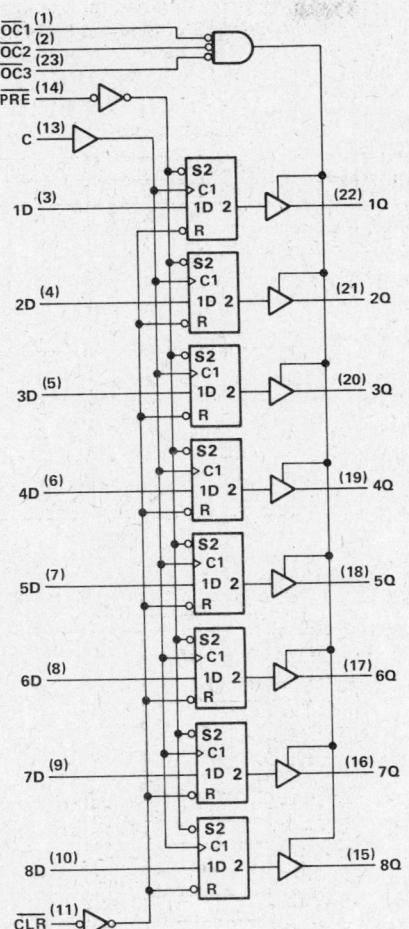

'ALS846, 'AS846

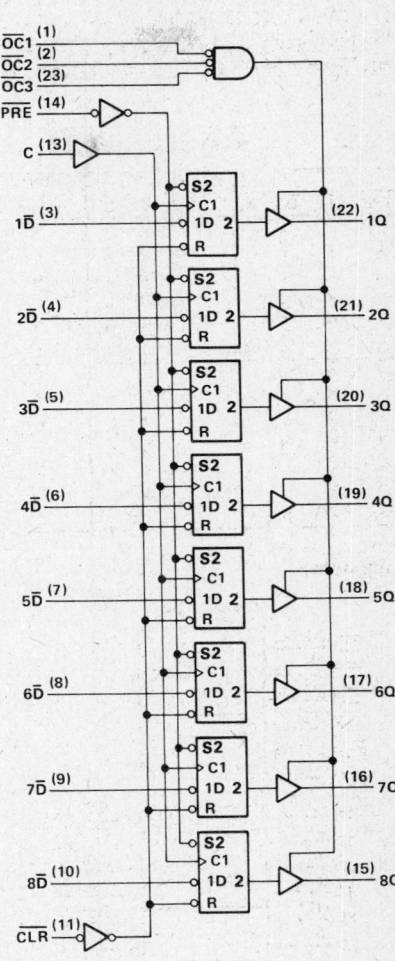

absolute maximum ratings over operating free-air temperature range (unless otherwise noted)

Supply voltage, V_{CC} .... 7 V
Input voltage .... 7 V
Voltage applied to a disabled 3-state output .... 5.5 V
Operating free-air temperature range:
    SN54ALS845, SN54AS845, SN54ALS846, SN54AS846 .... −55°C to 125°C
    SN74ALS845, SN74AS845, SN74ALS846, SN74AS846 .... −0°C to 70°C
Storage temperature range .... −65°C to 150°C

TYPES SN54ALS845, SN54ALS846, SN74ALS845, SN74ALS846
8-BIT BUS INTERFACE D-TYPE LATCHES WITH 3-STATE OUTPUTS

recommended operating conditions

		SN54ALS845 SN54ALS846			SN74ALS845 SN74ALS846			UNIT
		MIN	NOM	MAX	MIN	NOM	MAX	
V_{CC}	Supply voltage	4.5	5	5.5	4.5	5	5.5	V
V_{IH}	High-level input voltage	2			2			V
V_{IL}	Low-level input voltage			0.8			0.8	V
I_{OH}	High-level output current			−1			−2.6	mA
I_{OL}	Low-level output current			12			24	mA
t_w	Pulse duration	CLR or PRE low						ns
		C high						
t_{su}	Setup time, data before enable C↓							ns
t_h	Hold time, data after enable C↓	'ALS845						ns
		'ALS846						
T_A	Operating free-air temperature	−55		125	0		70	°C

electrical characteristics over recommended operating free-air temperature range (unless otherwise noted)

PARAMETER		TEST CONDITIONS	SN54ALS845 SN54ALS846			SN74ALS845 SN74ALS846			UNIT
			MIN	TYP†	MAX	MIN	TYP†	MAX	
V_{IK}		$V_{CC} = 4.5$ V, $I_I = -18$ mA			−1.5			−1.5	V
V_{OH}		$V_{CC} = 4.5$ V to 5.5 V, $I_{OH} = -0.4$ mA	$V_{CC}-2$			$V_{CC}-2$			V
		$V_{CC} = 4.5$ V, $I_{OH} = -1$ mA	2.4	3.3					
		$V_{CC} = 4.5$ V, $I_{OH} = -2.6$ mA				2.4	3.2		
V_{OL}		$V_{CC} = 4.5$ V, $I_{OL} = 12$ mA		0.25	0.4		0.25	0.4	V
		$V_{CC} = 4.5$ V, $I_{OL} = 24$ mA					0.35	0.5	
I_{OZH}		$V_{CC} = 5.5$ V, $V_O = 2.7$ V			20			20	µA
I_{OZL}		$V_{CC} = 5.5$ V, $V_O = 0.4$ V			−20			−20	µA
I_I		$V_{CC} = 5.5$ V, $V_I = 7$ V			0.1			0.1	mA
I_{IH}		$V_{CC} = 5.5$ V, $V_I = 2.7$ V			20			20	µA
I_{IL}		$V_{CC} = 5.5$ V, $V_I = 0.4$ V			−0.1			−0.1	mA
I_O‡		$V_{CC} = 5.5$ V, $V_O = 2.25$ V	−15		−70	−15		−70	mA
I_{CC}	'ALS845	$V_{CC} = 5.5$ V	Outputs high						mA
			Outputs low						
			Outputs disabled		25			25	
	'ALS846		Outputs high						
			Outputs low						
			Outputs disabled		28			28	

†All typical values are at $V_{CC} = 5$ V, $T_A = 25°C$.
‡The output conditions have been chosen to produce a current that closely approximates one half of the true short-circuit output current, I_{OS}.

Additional information on these products can be obtained from the factory as it becomes available.

TYPES SN54ALS845, SN54ALS846, SN74ALS845, SN74ALS846
8-BIT BUS INTERFACE D-TYPE LATCHES WITH 3-STATE OUTPUTS

'ALS845 switching characteristics (see Note 1)

PARAMETER	FROM (INPUT)	TO (OUTPUT)	V_{CC} = 4.5 V to 5.5 V, C_L = 50 pF, R1 = 500 Ω, R2 = 500 Ω, T_A = MIN to MAX						UNIT
			SN54ALS845			SN74ALS845			
			MIN	TYP†	MAX	MIN	TYP†	MAX	
t_{PLH}	D	Q		7			7		ns
t_{PHL}				9			9		
t_{PLH}	C	Q							ns
t_{PHL}									ns
t_{PLH}	$\overline{PRE}$	Q							ns
t_{PHL}	$\overline{CLR}$	Q							ns
t_{PZH}	$\overline{OC}$	Q							ns
t_{PZL}									
t_{PHZ}	$\overline{OC}$	Q							ns
t_{PLZ}									

'ALS846 switching characteristics (see Note 1)

PARAMETER	FROM (INPUT)	TO (OUTPUT)	V_{CC} = 4.5 V to 5.5 V, C_L = 50 pF, R1 = 500 Ω, R2 = 500 Ω, T_A = MIN to MAX						UNIT
			SN54ALS846			SN74ALS846			
			MIN	TYP†	MAX	MIN	TYP†	MAX	
t_{PLH}	$\overline{D}$	Q		7			7		ns
t_{PHL}				9			9		
t_{PLH}	C	Q							ns
t_{PHL}									ns
t_{PLH}	$\overline{PRE}$	Q							ns
t_{PHL}	$\overline{CLR}$	Q							ns
t_{PZH}	$\overline{OC}$	Q							ns
t_{PZL}									
t_{PHZ}	$\overline{OC}$	Q							ns
t_{PLZ}									

†All typical values are at T_A = 25 °C.
NOTE 1: For load circuit and voltage waveforms, see page 1-12.

Additional information on these products can be obtained from the factory as it becomes available.

ALS AND AS CIRCUITS

TEXAS INSTRUMENTS
POST OFFICE BOX 225012 • DALLAS, TEXAS 75265

TYPES SN54AS845, SN54AS846, SN74AS845, SN74AS846
8-BIT BUS INTERFACE D-TYPE LATCHES WITH 3-STATE OUTPUTS

recommended operating conditions

			SN54AS845 SN54AS846			SN74AS845 SN74AS846			UNIT
			MIN	NOM	MAX	MIN	NOM	MAX	
V_{CC}	Supply voltage		4.5	5	5.5	4.5	5	5.5	V
V_{IH}	High-level input voltage		2			2			V
V_{IL}	Low-level input voltage				0.8			0.8	V
I_{OH}	High-level output current				−24			−24	mA
I_{OL}	Low-level output current				32			48	mA
t_w	Pulse duration	CLR or PRE low							ns
		C high							
t_{su}	Setup time, data before enable C↓								ns
t_h	Hold time, data after enable C↓								ns
T_A	Operating free-air temperature		−55		125	0		70	°C

electrical characteristics over recommended operating free-air temperature range (unless otherwise noted)

PARAMETER		TEST CONDITIONS		SN54AS845 SN54AS846			SN74AS845 SN74AS846			UNIT
				MIN	TYP†	MAX	MIN	TYP†	MAX	
V_{IK}		V_{CC} = 4.5 V,	I_I = −18 mA			−1.2			−1.2	V
V_{OH}		V_{CC} = 4.5 to 5.5 V,	I_{OH} = −2 mA	V_{CC}−2			V_{CC}−2			V
		V_{CC} = 4.5 V,	I_{OH} = −15 mA	2.4	3.2		2.4	3.2		
		V_{CC} = 4.5 V,	I_{OH} = −24 mA	2			2			
V_{OL}		V_{CC} = 4.5 V,	I_{OL} = 32 mA		0.25	0.5				V
		V_{CC} = 4.5 V,	I_{OL} = 48 mA					0.25	0.5	
I_{OZH}		V_{CC} = 5.5 V,	V_O = 2.7 V			50			50	µA
I_{OZL}		V_{CC} = 5.5 V,	V_O = 0.4 V			−50			−50	µA
I_I		V_{CC} = 5.5 V,	V_I = 7 V							mA
I_{IH}		V_{CC} = 5.5 V,	V_I = 2.7 V							µA
I_{IL}		V_{CC} = 5.5 V,	V_I = 0.4 V							mA
I_O‡		V_{CC} = 5.5 V,	V_O = 2.25 V	−30		−112	−30		−112	mA
I_{CC}	'AS845	V_{CC} = 5.5 V	Outputs high		35			35		mA
			Outputs low		52			52		
			Outputs disabled		52			52		
	'AS846		Outputs high		36			36		
			Outputs low		53			53		
			Outputs disabled		53			53		

†All typical values are at V_{CC} = 5 V, T_A = 25°C.
‡The output conditions have been chosen to produce a current that closely approximates one half of the true short-circuit output current, I_{OS}.

Additional information on these products can be obtained from the factory as it becomes available.

TYPES SN54AS845, SN54AS846, SN74AS845, SN74AS846
8-BIT BUS INTERFACE D-TYPE LATCHES WITH 3-STATE OUTPUTS

'AS845 switching characteristics (see Note 1)

PARAMETER	FROM (INPUT)	TO (OUTPUT)	$V_{CC} = 4.5$ V to 5.5 V, $C_L = 50$ pF, $R1 = 500\ \Omega$, $R2 = 500\ \Omega$, T_A = MIN to MAX						UNIT
			SN54AS845			SN74AS845			
			MIN	TYP†	MAX	MIN	TYP†	MAX	
t_{PLH}	D	Q		4			4		ns
t_{PHL}				4.5			4.5		
t_{PLH}	C	Q							ns
t_{PHL}									
t_{PLH}	$\overline{PRE}$	Q		5			5		ns
t_{PHL}	$\overline{CLR}$	Q		5.5			5.5		ns
t_{PZH}	$\overline{OC}$	Q		6			6		ns
t_{PZL}				6			6		
t_{PHZ}	$\overline{OC}$	Q		4			4		ns
t_{PLZ}				5			5		

'AS846 switching characteristics (see Note 1)

PARAMETER	FROM (INPUT)	TO (OUTPUT)	$V_{CC} = 4.5$ V to 5.5 V, $C_L = 50$ pF, $R1 = 500\ \Omega$, $R2 = 500\ \Omega$, T_A = MIN to MAX						UNIT
			SN54AS846			SN74AS846			
			MIN	TYP†	MAX	MIN	TYP†	MAX	
t_{PLH}	$\overline{D}$	Q		4			4		ns
t_{PHL}				4.5			4.5		
t_{PLH}	C	Q							ns
t_{PHL}									
t_{PLH}	$\overline{PRE}$	Q		5			5		ns
t_{PHL}	$\overline{CLR}$	Q		5.5			5.5		ns
t_{PZH}	$\overline{OC}$	Q		6			6		ns
t_{PZL}				6			6		
t_{PHZ}	$\overline{OC}$	Q		4			4		ns
t_{PLZ}				5			5		

†All typical values are at $T_A = 25°C$.
NOTE 1: For load circuit and voltage waveforms, see page 1-12.

Additional information on these products can be obtained from the factory as it becomes available.

ALS AND AS CIRCUITS

TYPES SN54ALS845, SN54AS845, SN54ALS846, SN54AS846 SN74ALS845, SN74AS845, SN74ALS846, SN74AS846
8-BIT BUS INTERFACE D-TYPE LATCHES WITH 3-STATE OUTPUTS

D flip-flop signal conventions

It is normal TI practice to name the outputs and other inputs of a D-type latch and to draw its logic symbol based on the assumption of true data (D) inputs. Then outputs that produce data in phase with the data inputs are called Q and those producing complementary data are called $\overline{Q}$. An input that causes a Q output to go high or a $\overline{Q}$ output to go low is called Preset; an input that causes a $\overline{Q}$ output to go high or a Q output to go low is called Clear. Bars are used over these pin names ($\overline{PRE}$ and $\overline{CLR}$) if they are active-low.

The devices on this data sheet are second-source designs and the pin name convention used by the original manufacturer has been retained. That makes it necessary to designate the inputs and outputs of the inverting circuit $\overline{D}$ and Q. In some applications it may be advantageous to redesignate the inputs and outputs as D and $\overline{Q}$. In that case, outputs should be renamed as shown below. Also shown are corresponding changes in the graphical symbol. Arbitrary pin numbers are shown in parentheses.

Notice that Q and $\overline{Q}$ exchange names, which causes Preset and Clear to do likewise. Also notice that the polarity indicators (⊳) on $\overline{PRE}$ and $\overline{CLR}$ remain since these inputs are still active-low, but that the presence or absence of the polarity indicator changes at $\overline{D}$, Q, and $\overline{Q}$. Of course pin 5 (Q) is still in phase with the data input D, but now both are considered active high.

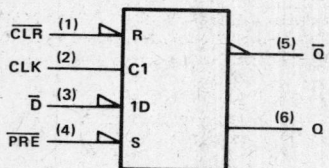 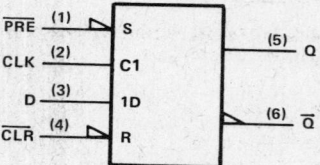

TYPES SN54AS850, SN54AS851, SN74AS850, SN74AS851
1 OF 16 DATA SELECTORS/MULTIPLEXERS WITH 3-STATE OUTPUTS

DECEMBER 1983

- **4-Line to 1-Line Data Selectors/Multiplexers That Can Select 1 of 16 Data Inputs.**
  Typical Applications:
    Boolean Function Generators
    Parallel-to-Serial Converters
    Data Source Selectors
- **Cascadable to n-Bits**
- **3-State Bus Driver Outputs**
- **'AS850 Offers Clocked Selects; 'AS851 Offers Enable-Controlled Selects**
- **Has a Master Output Control ($\overline{G}$) for Cascading and Individual Output Controls ($\overline{GY}$, GW) for Each Output**
- **Package Options Include both Plastic and Ceramic Carriers in Addition to Plastic and Ceramic DIPs**
- **Dependable Texas Instruments Quality and Reliability**

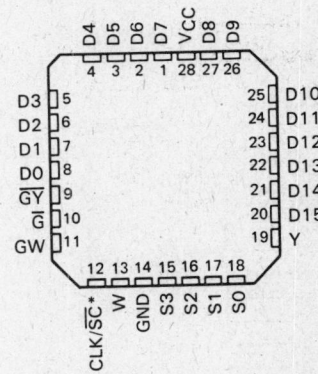

*CLK for 'AS850 or $\overline{SC}$ for 'AS851

description

These four-line to one-line data selectors/multiplexers provide full binary decoding to select one-of-sixteen data sources with complementary Y and W outputs. The 'AS850 has a clock-controlled select register allowing for a symmetrical presentation of the select inputs to the decoder while the 'AS851 has an enable-controlled select register allowing the user to select and hold one particular data line.

A buffered group of output controls ($\overline{G}$, $\overline{GY}$, $\overline{GW}$) can be used to place the two outputs in either a normal logic (high or low logic level) or a high-impedance state. In the high-impedance state the outputs neither load nor drive the bus lines significantly. The high-impedance state and increased drive provide the capability to drive the bus lines in a bus-organized system without the need for interface or pull-up components.

The output controls do not affect the internal operations of the data selector/multiplexer. New data can be up while the outputs are in the high-impedance state.

The SN54AS850 and SN54AS851 are characterized for operation over the full military temperature range from −55°C to 125°C. The SN74AS850 and SN74AS851 are characterized for operation from 0°C to 70°C.

PRODUCT PREVIEW

This document contains information on a product under development. Texas Instruments reserves the right to change or discontinue this product without notice.

Copyright © 1983 by Texas Instruments Incorporated

TEXAS INSTRUMENTS
POST OFFICE BOX 225012 • DALLAS, TEXAS 75265

TYPES SN54AS850, SN54AS851, SN74AS850, SN74AS851
1 OF 16 DATA SELECTORS/MULTIPLEXERS WITH 3-STATE OUTPUTS

INPUT SELECTION TABLE

SELECT INPUTS				'AS850	'AS851	INPUT
S3	S2	S1	S0	CLK	$\overline{SC}$	SELECTED
L	L	L	L	↑	L	D0
L	L	L	H	↑	L	D1
L	L	H	L	↑	L	D2
L	L	H	H	↑	L	D3
L	H	L	L	↑	L	D4
L	H	L	H	↑	L	D5
L	H	H	L	↑	L	D6
L	H	H	H	↑	L	D7
H	L	L	L	↑	L	D8
H	L	L	H	↑	L	D9
H	L	H	L	↑	L	D10
H	L	H	H	↑	L	D11
H	H	L	L	↑	L	D12
H	H	L	H	↑	L	D13
H	H	H	L	↑	L	D14
H	H	H	H	↑	L	D15
X	X	X	X	H or L	H	Dn

Dn = the input selected before the most-recent low-to-high transition of CLK or SC.

OUTPUT FUNCTION TABLE

$\overline{G}$	$\overline{GY}$	GW	OUTPUTS	
			Y	W
H	X	X	Z	Z
L	H	L	Z	Z
L	L	L	D	Z
L	H	H	Z	$\overline{D}$
L	L	H	D	$\overline{D}$

D = level of selected input D0–D15

logic symbols

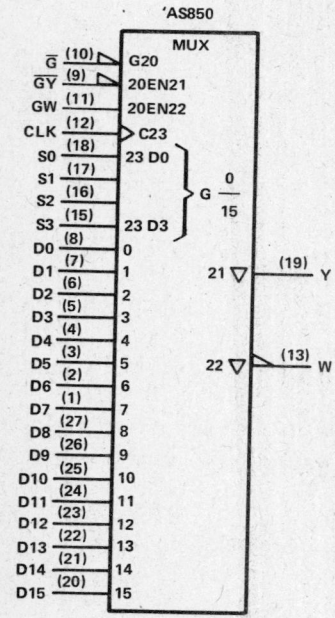

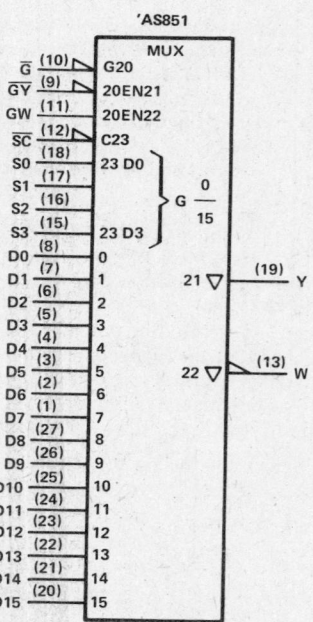

TYPES SN54AS850, SN54AS851, SN74AS850, SN74AS851
1 OF 16 DATA SELECTORS/MULTIPLEXERS WITH 3-STATE OUTPUTS

'AS850 logic diagram (positive logic) (see inset for 'AS851)

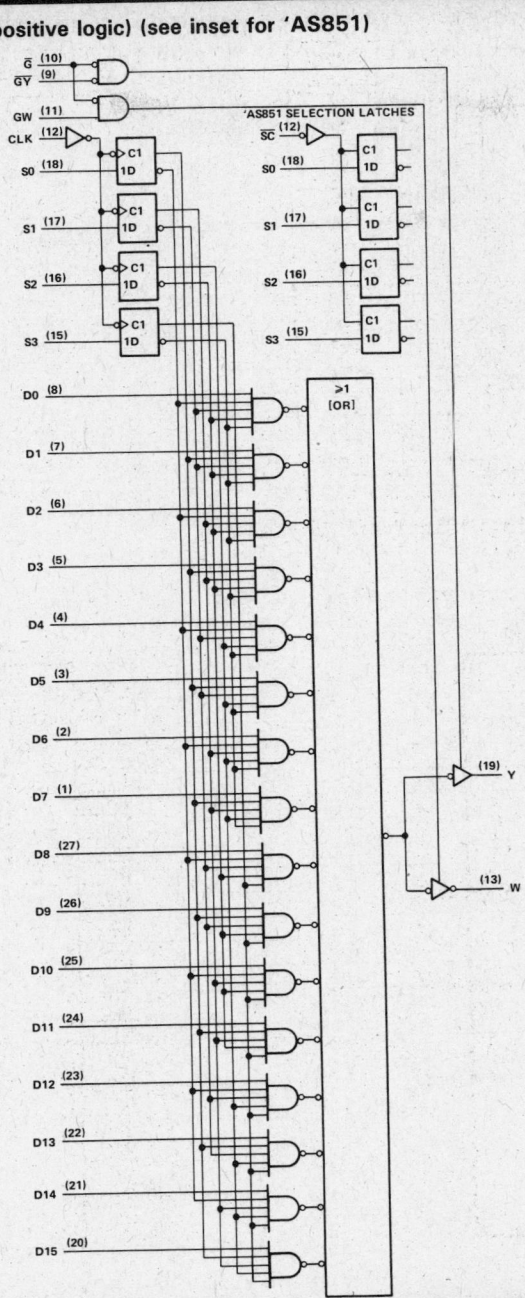

TYPES SN54AS850, SN54AS851, SN74AS850, SN74AS851
1 OF 16 DATA SELECTORS/MULTIPLEXERS WITH 3-STATE OUTPUTS

absolute maximum ratings over operating free-air temperature range (unless otherwise noted)

Supply voltage, V_{CC} .... 7 V
Input voltage .... 7 V
Operating free-air temperature range: SN54AS850, SN54AS851 .... −55°C to 125°C
                                                SN74AS850, SN74AS851 .... 0°C to 70°C
Storage temperature range .... −65°C to 150°C

recommended operating conditions

		SN54AS850			SN74AS850			UNIT
		MIN	NOM	MAX	MIN	NOM	MAX	
V_{CC}	Supply voltage	4.5	5	5.5	4.5	5	5.5	V
V_{IH}	High-level input voltage	2			2			V
V_{IL}	Low-level input voltage			0.8			0.8	V
I_{OH}	High-level output current			−12			−15	mA
I_{OL}	Low-level output current			32			48	mA
f_{clock}	Clock frequency							MHz
t_w	Pulse duration CLK high							ns
	CLK low							
t_{su}	Setup time, select inputs before CLK↑							ns
t_h	Hold time, select inputs after CLK↑							ns
T_A	Operating free-air temperature	−55		125	0		70	°C

electrical characteristics over recommended operating free-air temperature range (unless otherwise noted)

PARAMETER	TEST CONDITIONS		SN54AS850			SN74AS850			UNIT
			MIN	TYP†	MAX	MIN	TYP†	MAX	
V_{IK}	V_{CC} = 4.5 V,	I_I = −18 mA			−1.2			−1.2	V
V_{OH}	V_{CC} = 4.5 V to 5.5 V,	I_{OH} = −2 mA	V_{CC}−2			V_{CC}−2			V
	V_{CC} = 4.5 V,	I_{OH} = −12 mA	2.4	3.2					
	V_{CC} = 4.5 V,	I_{OH} = −15 mA				2.4	3.3		
V_{OL}	V_{CC} = 4.5 V,	I_{OL} = 32 mA		0.25	0.5				V
	V_{CC} = 4.5 V,	I_{OL} = 48 mA					0.35	0.5	
I_{OZH}	V_{CC} = 5.5 V,	V_O = 2.7 V			50			50	μA
I_{OZL}	V_{CC} = 5.5 V,	V_O = 0.4 V			−50			−50	μA
I_I	V_{CC} = 5.5 V,	V_I = 7 V			0.1			0.1	mA
I_{IH}	V_{CC} = 5.5 V,	V_I = 2.7 V			20			20	μA
I_{IL}	V_{CC} = 5.5 V,	V_I = 0.4 V							mA
I_O‡	V_{CC} = 5.5 V,	V_O = 2.25 V	−30		−112	−30		−112	mA
I_{CC}	V_{CC} = 5.5 V	Outputs active							mA
		Outputs disabled							

†All typical values are at V_{CC} = 5 V, T_A = 25°C.
‡The output conditions have been chosen to produce a current that closely approximates one-half of the true short-circuit current, I_{OS}.

Additional information on these products can be obtained from the factory as it becomes available.

TYPES SN54AS850, SN74AS850
1 OF 16 DATA SELECTORS/MULTIPLEXERS WITH 3-STATE OUTPUTS

switching characteristics (see Note 1)

PARAMETER	FROM (INPUT)	TO (OUTPUT)	V_{CC} = 4.5 V to 5.5 V, C_L = 50 pF, R1 = 500 Ω, R2 = 500 Ω, T_A = MIN to MAX						UNIT
			SN54AS850			SN74AS850			
			MIN	TYP†	MAX	MIN	TYP†	MAX	
f_{max}									MHz
t_{PLH}	Any D	Y		4.7			4.7		ns
t_{PHL}				5			5		
t_{PLH}	Any D	W		5.5			5.5		ns
t_{PHL}				6.2			6.2		
t_{PLH}	CLK	Y		10.2			10.2		ns
t_{PHL}				8.3			8.3		
t_{PLH}	CLK	W		8.8			8.8		ns
t_{PHL}				11.6			11.6		
t_{PZH}	$\overline{G}$	Y		4			4		ns
t_{PZL}				4.9			4.9		
t_{PHZ}	$\overline{G}$	Y		3.1			3.1		ns
t_{PLZ}				3.9			3.9		
t_{PZH}	$\overline{G}$	W		4			4		ns
t_{PZL}				4.9			4.9		
t_{PHZ}	$\overline{G}$	W		3.1			3.1		ns
t_{PLZ}				3.9			3.9		
t_{PZH}	$\overline{GY}$	Y		4			4		ns
t_{PZL}				4.9			4.9		
t_{PHZ}	$\overline{GY}$	Y		3.1			3.1		ns
t_{PLZ}				3.9			3.9		
t_{PZH}	GW	W		6.8			6.8		ns
t_{PZL}				6.6			6.6		
t_{PHZ}	GW	W		4.3			4.3		ns
t_{PLZ}				5.6			5.6		

†All typical values are at V_{CC} = 5 V, T_A = 25°C.
NOTE 1: For load circuit and voltage waveforms, see page 1-12.

ALS AND AS CIRCUITS

TEXAS INSTRUMENTS
POST OFFICE BOX 225012 • DALLAS, TEXAS 75265

TYPES SN54AS851, SN74AS851
1 OF 16 DATA SELECTROS/MULTIPLEXERS WITH 3-STATE OUTPUTS

recommended operating conditions

			SN54AS851			SN74AS851			UNIT
			MIN	NOM	MAX	MIN	NOM	MAX	
V_{CC}	Supply voltage		4.5	5	5.5	4.5	5	5.5	V
V_{IH}	High-level input voltage		2			2			V
V_{IL}	Low-level input voltage				0.8			0.8	V
I_{OH}	High-level output current				−12			−15	mA
I_{OL}	Low-level output current				32			48	mA
t_w	Pulse duration	C high							ns
		C low							
t_{su}	Setup time, select inputs before $\overline{CS}\uparrow$								ns
t_h	Hold time, select inputs after $\overline{CS}\uparrow$								ns
T_A	Operating free-air temperature		−55		125	0		70	°C

electrical characteristics over recommended operating free-air temperature range (unless otherwise noted)

PARAMETER	TEST CONDITIONS		SN54AS851			SN74AS851			UNIT
			MIN	TYP†	MAX	MIN	TYP†	MAX	
V_{IK}	$V_{CC} = 4.5$ V,	$I_I = -18$ mA			−1.2			−1.2	V
V_{OH}	$V_{CC} = 4.5$ V to 5.5 V,	$I_{OH} = -2$ mA	$V_{CC}-2$			$V_{CC}-2$			V
	$V_{CC} = 4.5$ V,	$I_{OH} = -12$ mA	2.4	3.2					
	$V_{CC} = 4.5$ V,	$I_{OH} = -15$ mA				2.4	3.3		
V_{OL}	$V_{CC} = 4.5$ V,	$I_{OL} = 32$ mA		0.25	0.5				V
	$V_{CC} = 4.5$ V,	$I_{OL} = 48$ mA					0.35	0.5	
I_{OZH}	$V_{CC} = 5.5$ V,	$V_O = 2.7$ V			50			50	μA
I_{OZL}	$V_{CC} = 5.5$ V,	$V_O = 0.4$ V			−50			−50	μA
I_I	$V_{CC} = 5.5$ V,	$V_I = 7$ V			0.1			0.1	mA
I_{IH}	$V_{CC} = 5.5$ V,	$V_I = 2.7$ V			20			20	μA
I_{IL}	$V_{CC} = 5.5$ V,	$V_I = 0.4$ V							mA
I_O‡	$V_{CC} = 5.5$ V,	$V_O = 2.25$ V	−30		−112	−30		−112	mA
I_{CC}	$V_{CC} = 5.5$ V	Outputs active							mA
		Outputs disabled							

†All typical values are at $V_{CC} = 5$ V, $T_A = 25$°C.
‡The output conditions have been chosen to produce a current that closely approximates one-half of the true short-circuit current, I_{OS}.

Additional information on these products can be obtained from the factory as it becomes available.

ALS AND AS CIRCUITS

Texas Instruments
POST OFFICE BOX 225012 • DALLAS, TEXAS 75265

TYPES SN54AS851, SN74AS851
1 OF 16 DATA SELECTORS/MULTIPLEXERS WITH 3-STATE OUTPUTS

switching characteristics (see Note 1)

PARAMETER	FROM (INPUT)	TO (OUTPUT)	V_{CC} = 4.5 V to 5.5 V, C_L = 50 pF, R1 = 500 Ω, R2 = 500 Ω, T_A = MIN to MAX						UNIT
			SN54AS851			SN74AS851			
			MIN	TYP†	MAX	MIN	TYP†	MAX	
t_{PLH}	Any D	Y		4.7			4.7		ns
t_{PHL}				5			5		
t_{PLH}	Any D	W		5.5			5.5		ns
t_{PHL}				6.2			6.2		
t_{PLH}	S0, S1, S2, S3	Y		7.9			7.9		ns
t_{PHL}				9.6			9.6		
t_{PLH}	S0, S1, S2, S3	W		10.1			10.1		ns
t_{PHL}				11.1			11.1		
t_{PLH}	$\overline{SC}$	Y		12			12		ns
t_{PHL}				12.5			12.5		
t_{PLH}	$\overline{SC}$	W		12			12		ns
t_{PHL}				13			13		
t_{PZH}	$\overline{G}$	Y		4			4		ns
t_{PZL}				4.9			4.9		
t_{PHZ}	$\overline{G}$	Y		3.1			3.1		ns
t_{PLZ}				3.9			3.9		
t_{PZH}	$\overline{G}$	W		4			4		ns
t_{PZL}				4.9			4.9		
t_{PHZ}	$\overline{G}$	W		3.1			3.1		ns
t_{PLZ}				3.9			3.9		
t_{PZH}	$\overline{GY}$	Y		4			4		ns
t_{PZL}				4.9			4.9		
t_{PHZ}	$\overline{GY}$	Y		3.1			3.1		ns
t_{PZL}				3.9			3.9		
t_{PZH}	$\overline{GW}$	W		6.8			6.8		ns
t_{PZL}				6.6			6.6		
t_{PHZ}	$\overline{GW}$	W		4.3			4.3		ns
t_{PLZ}				5.6			5.6		

†All typical values are at V_{CC} = 5 V, T_A = 25°C.
NOTE 1: For load circuit and voltage waveforms, see page 1-12.

Texas Instruments
POST OFFICE BOX 225012 • DALLAS, TEXAS 75265

TYPES SN54AS850, SN54AS851, SN74AS850, SN74AS851
1 OF 16 DATA SELECTORS/MULTIPLEXERS WITH 3-STATE OUTPUTS

TYPICAL APPLICATION DATA

The 'AS850 or 'AS851 can be used as a 1-of-16 Boolean function generator. Figure 1 shows the 'AS850 in one example.

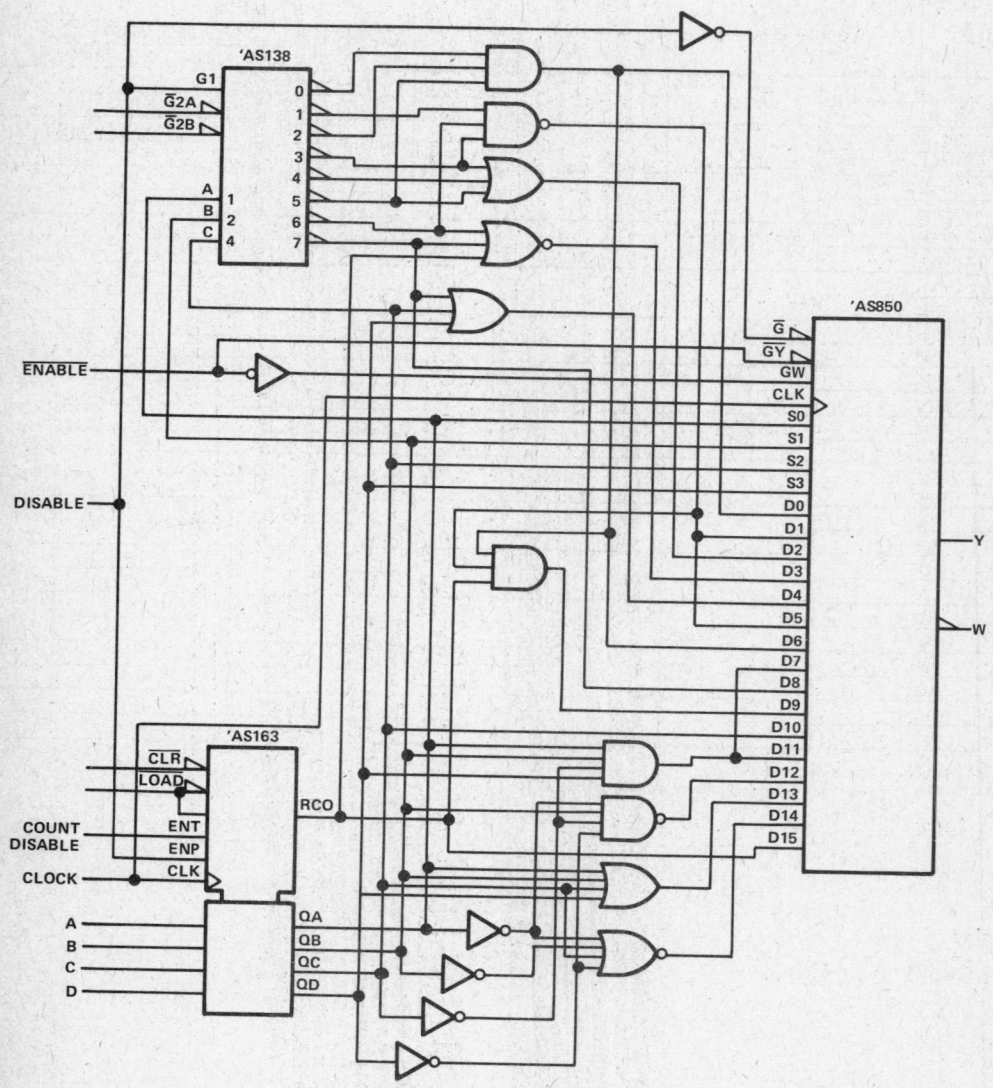

FIGURE 1—1-OF-16 BOOLEAN FUNCTION GENERATOR

TYPES SN54AS850, SN54AS851, SN74AS850, SN74AS851
1 OF 16 DATA SELECTORS/MULTIPLEXERS WITH 3-STATE OUTPUTS

TYPICAL APPLICATION DATA

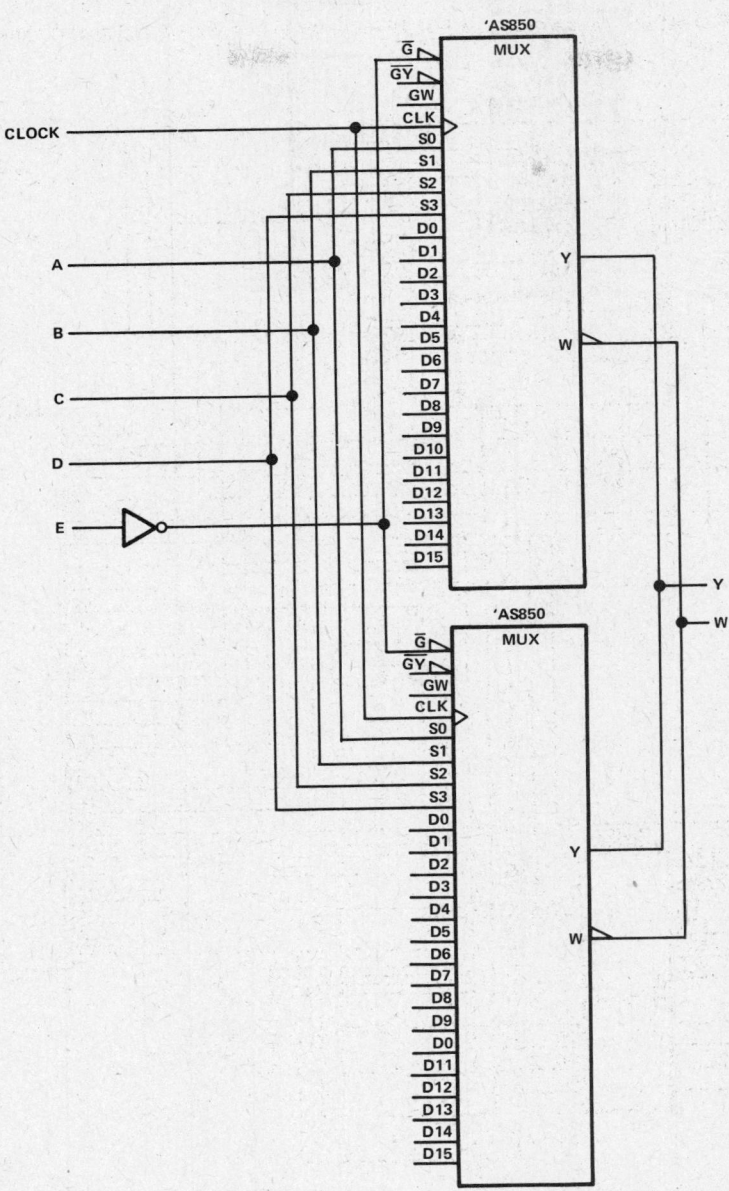

FIGURE 2–1-OF-32 DATA/SELECTOR/MULTIPLEXER

TYPES SN54AS850, SN54AS851, SN74AS850, SN74AS851
1 OF 16 DATA SELECTORS/MULTIPLEXERS WITH 3-STATE OUTPUTS

TYPICAL APPLICATION DATA

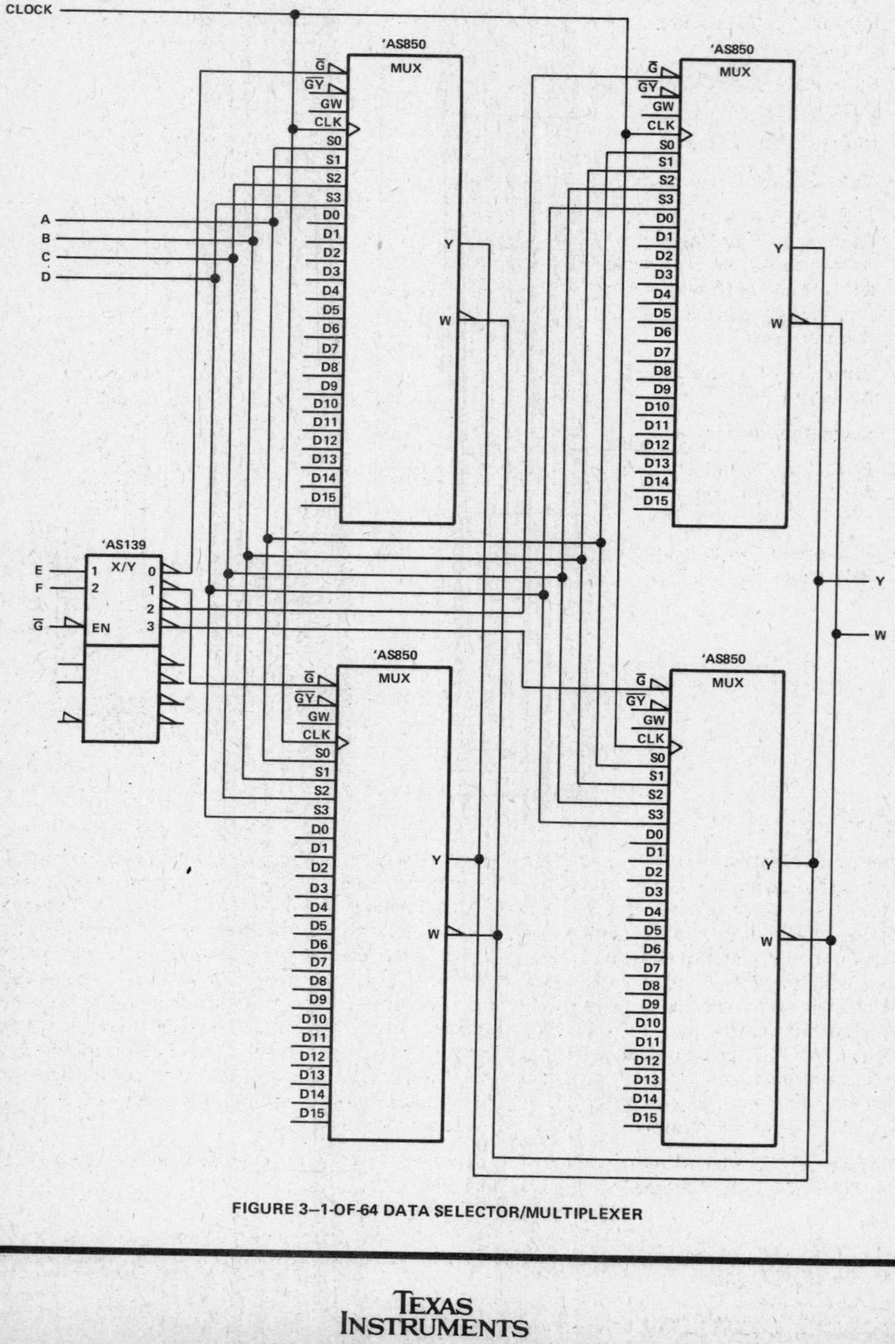

FIGURE 3—1-OF-64 DATA SELECTOR/MULTIPLEXER

TYPES SN54AS852, SN74AS852
8-BIT UNIVERSAL TRANSCEIVER PORT CONTROLLERS

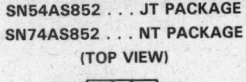

D2810, DECEMBER 1983

- Included Among the Package Options are Compact, 24-Pin, 300-mil-Wide DIPs and Both 28-Pin Plastic and Ceramic Chip Carriers
- Buffered 3-State Outputs Drive Bus Lines Directly
- Cascadable to n-Bits
- Eight Selectable Transceiver/Port Functions:

  A to B or B to A
  Register to A or Register to B
  Shifted to A from B or Shifted to B from A
  Off-Line Shifts (A and B Ports Transceiving or in High-Impedance State)
  Register Clear

- Particularly Suitable for Use in Diagnostics Circuitry
- Serial Register Provides:

  Parallel Storage of Either A or B Input Data
  Serial Transmission of Data from Either A or B Port

- Dependable Texas Instruments Quality and Reliability

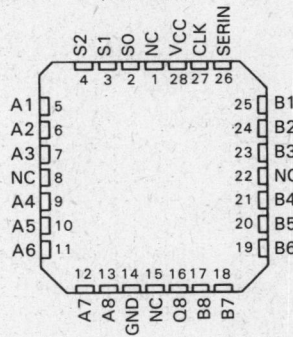

NC — No internal connection

description

The 'AS852 features two 8-bit I/O ports (A1-A8 and B1-B8), and 8-bit parallel-load, serial-in, parallel-out shift register, and control logic. With these features, this device is capable of performing eight selectable transceiver or port functions, depending on the state of the three select lines S0, S1, and S2. These functions include: transferring data from port A to port B or vice versa (i.e., the transceiver function), transferring data from the register to either port, serial shifting data to either port from the opposite port, performing off-line shifts (with A and B ports in high-impedance state), and clearing the register. The 'AS852 can simultaneously transfer data from A to B or B to A and perform an off-line serial shift of data in the register. Synchronous parallel loading of the internal register can be accomplished from either port on the positive transition of the clock while serially shifting data in via the SERIN input. The 'AS852 is ideally suited for applications implementing diagnostic circuitry to enhance system verification and/or fault analysis. All serial data is shifted right. All outputs are buffer-type outputs designed specifically to drive bus lines directly and all are 3-state except for Q8, which is a totem-pole output.

The SN54AS852 is characterized for operation over the full military temperature range of −55°C to 125°C. The SN74AS852 is characterized for operation from 0°C to 70°C.

PRODUCT PREVIEW

This document contains information on a product under development. Texas Instruments reserves the right to change or discontinue this product without notice.

Copyright © 1983 by Texas Instruments Incorporated

TEXAS INSTRUMENTS
POST OFFICE BOX 225012 • DALLAS, TEXAS 75265

TYPES SN54AS852, SN74AS852
8-BIT UNIVERSAL TRANSCEIVER PORT CONTROLLERS

logic diagram (positive logic)

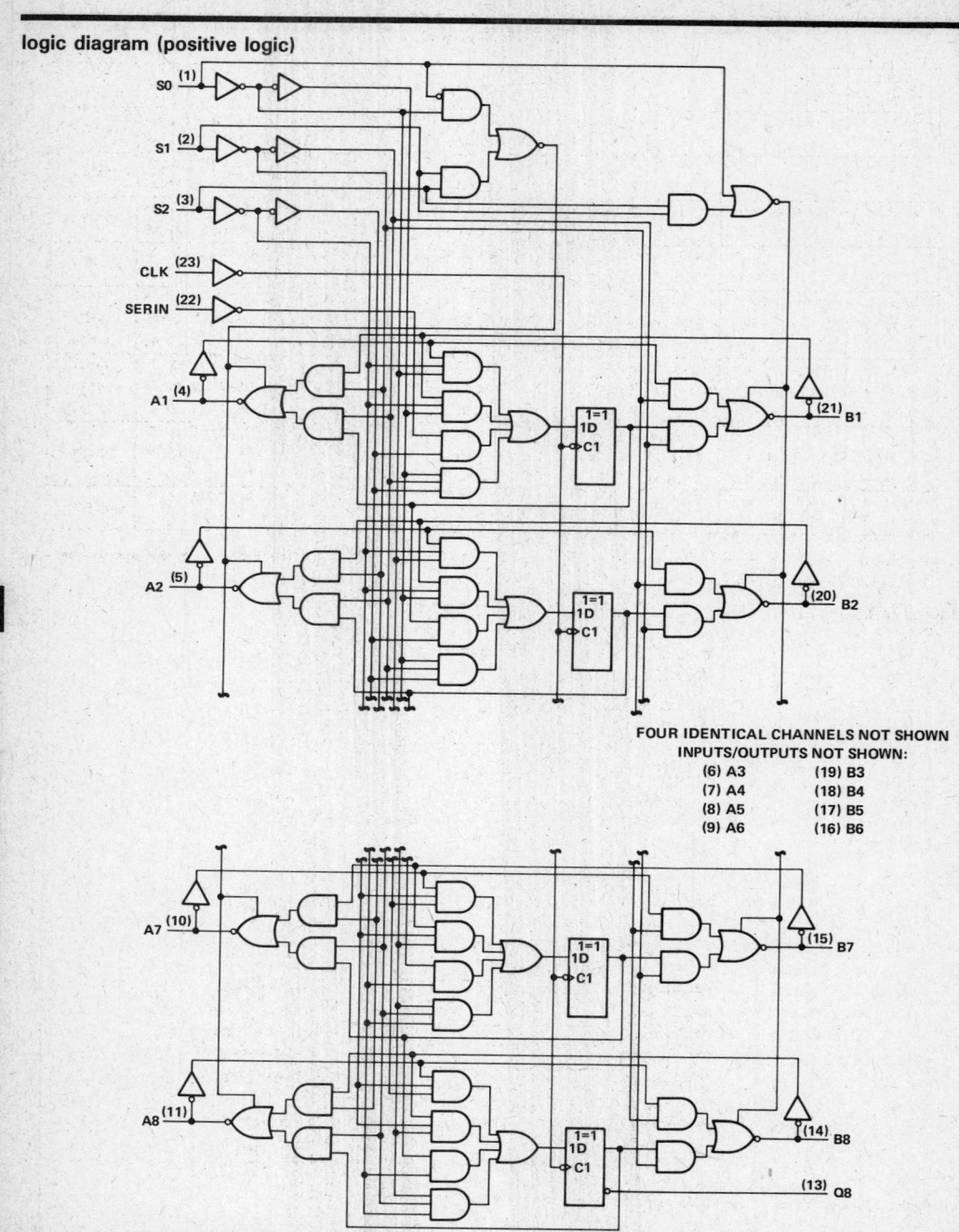

FOUR IDENTICAL CHANNELS NOT SHOWN
INPUTS/OUTPUTS NOT SHOWN:
(6) A3   (19) B3
(7) A4   (18) B4
(8) A5   (17) B5
(9) A6   (16) B6

TYPES SN54AS852, SN74AS852
8-BIT UNIVERSAL TRANSCEIVER PORT CONTROLLERS

FUNCTION TABLE

MODE S2 S1 S0	CLOCK	SERIN	A1 Q1 B1	A2 Q2 B2	A3 Q3 B3	A4 Q4 B4	A5 Q5 B5	A6 Q6 B6	A7 Q7 B7	A8 Q8 B8	PORT FUNCTION
L L L	H or L	X	Z Q_n A1	Z Q_n A2	Z Q_n A3	Z Q_n A4	Z Q_n A5	Z Q_n A6	Z Q_n A7	Z Q_n A8	A TO B
L L L	↑	X	Z A1 A1	Z A2 A2	A A3 A3	Z A4 A4	Z A5 A5	Z A6 A6	Z A7 A7	Z A8 A8	
L L H	H or L	X	B1 Q_n Z	B2 Q_n Z	B3 Q_n Z	B4 Q_n Z	B5 Q_n Z	B6 Q_n Z	B7 Q_n Z	B8 Q_n Z	B TO A
L L H	↑	X	B1 B1 Z	B2 B2 Z	B3 B3 Z	B4 B4 Z	B5 B5 Z	B6 B6 Z	B7 B7 Z	B8 B8 Z	
L H L	H or L	X	X Q_n Q1	X Q_n Q2	X Q_n Q3	X Q_n Q4	X Q_n Q5	X Q_n Q6	X Q_n Q7	X Q_n Q8	Q_N TO B_N
L H L	↑	X	Z A1 A1	Z A2 A2	Z A3 A3	Z A4 A4	Z A5 A5	Z A6 A6	Z A7 A7	Z A8 A8	
L H H	H or L	X	Q1 Q_n X	Q2 Q_n X	Q3 Q_n X	Q4 Q_n X	Q5 Q_n X	Q6 Q_n X	Q7 Q_n X	Q8 Q_n X	Q_N TO A_N
L H H	↑	X	B1 B1 Z	B2 B2 Z	B3 B3 Z	B4 B4 Z	B5 B5 Z	B6 B6 Z	B7 B7 Z	B8 B8 Z	
H L L	H or L	X	Z Q_n A1	Z Q_n A2	Z Q_n A3	Z Q_n A4	Z Q_n A5	Z Q_n A6	Z Q_n A7	Z Q_n A8	SHIFT
H L L	↑	H	Z H A1	Z Q1 A2	Z Q2 A3	Z Q3 A4	Z Q4 A5	Z Q5 A6	Z Q6 A7	Z Q7 A8	AND
H L L	↑	L	Z L A1	Z Q1 A2	Z Q2 A3	Z Q3 A4	Z Q4 A5	Z Q5 A6	Z Q6 A7	Z Q7 A8	A TO B
H L H	H or L	X	B1 Q_n Z	B2 Q_n Z	B3 Q_n Z	B4 Q_n Z	B5 Q_n Z	B6 Q_n Z	B7 Q_n Z	B8 Q_n Z	SHIFT
H L H	↑	H	B1 H Z	B2 Q1 Z	B3 Q2 Z	B4 Q3 Z	B5 Q4 Z	B6 Q5 Z	B7 Q6 Z	B8 Q7 Z	AND
H L H	↑	L	B1 L Z	B2 Q1 Z	B3 Q2 Z	B4 Q3 Z	B5 Q4 Z	B6 Q5 Z	B7 Q6 Z	B8 Q7 Z	B TO A
H H L	H or L	X	Z Q_n Z	Z Q_n Z	Z Q_n Z	Z Q_n Z	Z Q_n Z	Z Q_n Z	Z Q_n Z	Z Q_n Z	
H H H	↑	H	Z H Z	Z Q1 Z	Z Q2 Z	Z Q3 Z	Z Q4 Z	Z Q5 Z	Z Q6 Z	Z Q7 Z	SHIFT
H H H	↑	L	Z L Z	Z Q1 Z	Z Q2 Z	Z Q3 Z	Z Q4 Z	Z Q5 Z	Z Q6 Z	Z Q7 Z	
H H H	H or L	X	Z Q_n Z	Z Q_n	Z Q_n Z	Z Q_n Z	Z Q_n Z	Z Q_n Z	Z Q_n Z	Z Q_n Z	CLEAR
H H H	↑	X	Z L Z	Z L Z	Z L Z	Z L Z	Z L Z	Z L Z	Z L Z	Z L Z	

n = level of Q_n (n = 1, 2, . . . 8) established on most recent ↑ transition of CLK. Q1 through Q8 are the shift register outputs; only Q8 is available externally.

The double inversions that take place as data travels from port to port are ignored in this table.

logic symbol

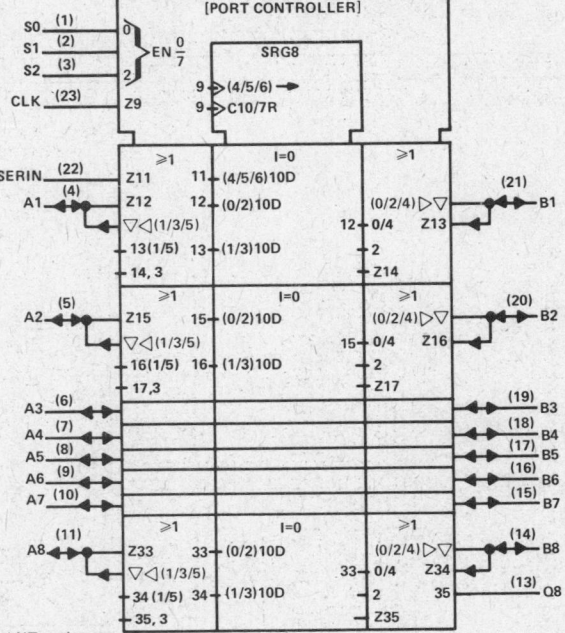

Pin numbers shown are for JT and NT packages.

Texas Instruments
POST OFFICE BOX 225012 • DALLAS, TEXAS 75265

TYPES SN54AS852, SN74AS852
8-BIT UNIVERSAL TRANSCEIVER PORT CONTROLLERS

absolute maximum ratings over free-air temperature range

Supply voltage, V_{CC} .... 7 V
Input voltage: All inputs .... 7 V
                 I/O ports .... 5.5 V
Voltage aplied to a disabled 3-state output .... 5.5 V
Operating free-air temperature range: SN54AS852 .... −55 °C to 125 °C
                                      SN74AS852 .... 0 °C to 70 °C
Storage temperature range .... −65 °C to 150 °C

recommended operating conditions

			SN54AS852			SN74AS852			UNIT
			MIN	NOM	MAX	MIN	NOM	MAX	
V_{CC}	Supply voltage		4.5	5	5.5	4.5	5	5.5	V
V_{IH}	High-level input voltage		2			2			V
V_{IL}	Low-level input voltage				0.8			0.8	V
I_{OH}	High-level output current	A1-A8, B1-B8			−12			−15	mA
		Q8			−2			−2	
I_{OL}	Low-level output current	A1-A8, B1-B8			32			48	mA
		Q8			20			20	
f_{clock}	Clock frequency								MHz
t_w	Duration of clock pulse								ns
t_{su}	Setup time before CLK↑	A1-A8, B1-B8 SERIN							ns
		S0, S1, S2							
t_h	Hold time, data after CLK↑	A1-A8, B1-B8 SERIN	0			0			ns
		S0, S1, S2	0			0			
T_A	Operating free-air temperature		−55		125	0		70	°C

Additional information on these products can be obtained from the factory as it becomes available.

ALS AND AS CIRCUITS

TEXAS INSTRUMENTS
POST OFFICE BOX 225012 • DALLAS, TEXAS 75265

TYPES SN54AS852, SN74AS852
8-BIT UNIVERSAL TRANSCEIVER PORT CONTROLLERS

electrical characteristics over recommended operating free-air temperature range (unless otherwise noted)

PARAMETER		TEST CONDITIONS		SN54AS852 MIN	SN54AS852 TYP†	SN54AS852 MAX	SN74AS852 MIN	SN74AS852 TYP†	SN74AS852 MAX	UNIT
V_{IK}		$V_{CC} = 4.5$ V,	$I_I = -18$ mA			-1.2			-1.2	V
V_{OH}	A1-A8	$V_{CC} = 4.5$ V,	$I_{OH} = -12$ mA	2.4	3.2					V
	B1-B8	$V_{CC} = 4.5$ V,	$I_{OH} = -15$ mA				2.4	3.3		
	All outputs	$V_{CC} = 4.5$ V to 5.5 V,	$I_{OH} = -2$ mA	$V_{CC}-2$			$V_{CC}-2$			
V_{OL}	All outputs except Q8	$V_{CC} = 4.5$ V,	$I_{OL} = 32$ mA		0.25	0.5				V
		$V_{CC} = 4.5$ V,	$I_{OL} = 48$ mA					0.35	0.5	
	Q8	$V_{CC} = 4.5$ V,	$I_{OL} = 20$ mA		0.25	0.5		0.25	0.5	
I_I	S0, S1, S2	$V_{CC} = 5.5$ V,	$V_I = 7$ V			0.3			0.3	mA
	CLK and SERIN					0.1			0.1	
	A1-A8, B1-B8	$V_{CC} = 5.5$ V,	$V_I = 5.5$ V			0.2			0.2	
I_{IH}	S0, S1, S2	$V_{CC} = 5.5$ V,	$V_I = 2.7$ V			60			60	μA
	CLK and SERIN					20			20	
	A1-A8, B1-B8‡					70			70	
I_{IL}	S0, S1, S2	$V_{CC} = 5.5$ V,	$V_I = 0.4$ V			-2			-2	mA
	CLK and SERIN					-0.3			-0.3	
	A1-A8, B1-B8‡					-0.35			-0.35	
I_{OS}§	Except Q8	$V_{CC} = 5.5$ V,	$V_O = 2.25$ V	-30		-112	-30		-112	mA
	Q8			-20		-112	-20		-112	
I_{CC}		$V_{CC} = 5.5$ V			122			122		mA

†All typical values are at $V_{CC} = 5$ V, $T_A = 25$°C.
‡For I/O ports, the parameters I_{IH} and I_{IL} include the output currents I_{OZH} and I_{OZL}, respectively.
§The output conditions have been chosen to produce a current that closely approximates one half of the true short-circuit output current, I_{OS}.

switching characteristics (see Note 1)

PARAMETER	FROM (INPUT)	TO (OUTPUT)	$V_{CC} = 4.5$ V to 5.5 V, $C_L = 50$ pF, $R1 = 500$ Ω, $R2 = 500$ Ω, T_A = MIN to MAX						UNIT
			SN54AS852 MIN	SN54AS852 TYP†	SN54AS852 MAX	SN74AS852 MIN	SN74AS852 TYP†	SN74AS852 MAX	
f_{max}				75			75		MHz
t_{PLH}	Any A port	Any B port		9.5			9.5		ns
t_{PHL}				8			8		
t_{PLH}	Any B port	Any A port		9.5			9.5		ns
t_{PHL}				8			8		
t_{PLH}	S0, S1, S2	Any A or B port		12			12		ns
t_{PHL}				12			12		
t_{PLH}	CLK	Any A or B port		6.5			6.5		ns
t_{PHL}				12.5			12.5		
t_{PLH}	CLK	Q8		9			9		ns
t_{PHL}				9			9		
t_{PHZ}	S0, S1, S2	Any A or B port		6			6		ns
t_{PLZ}				6			6		
t_{PZH}				10			10		ns
t_{PZL}				10			10		

†All typical values are at $V_{CC} = 5$ V, $T_A = 25$°C.
NOTE 1: For load circuit and voltage waveforms, see page 1-12.

ALS AND AS CIRCUITS

TEXAS INSTRUMENTS
POST OFFICE BOX 225012 • DALLAS, TEXAS 75265

TYPES SN54AS852, SN74AS852
8-BIT UNIVERSAL TRANSCEIVER PORT CONTROLLERS

TYPICAL APPLICATION DATA

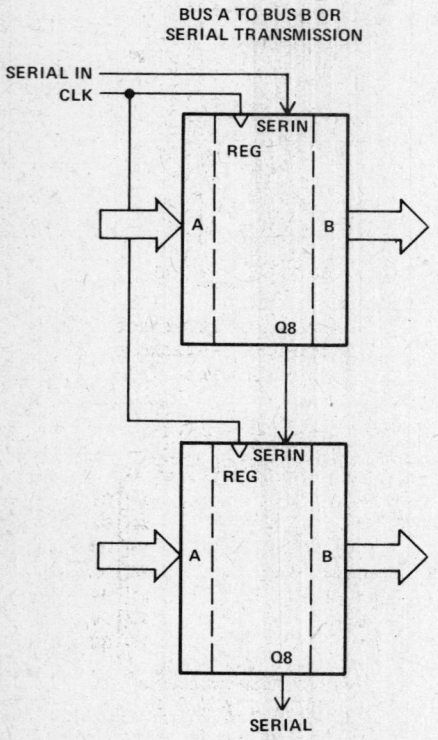

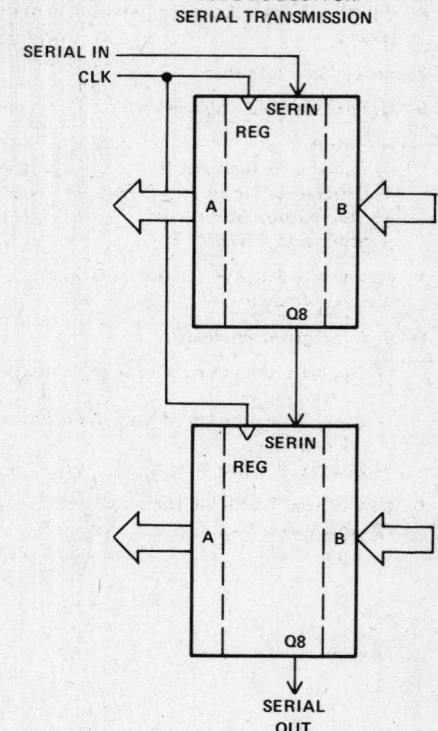

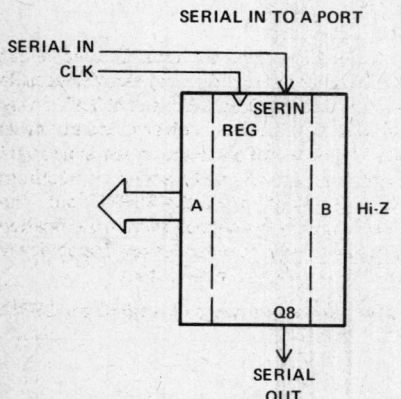

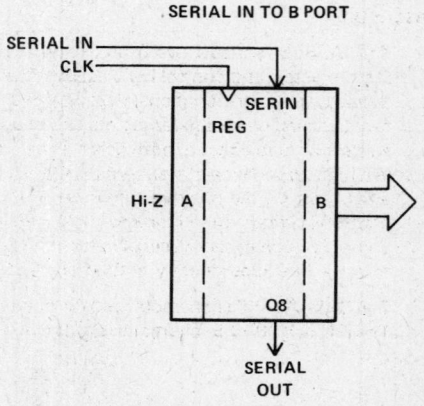

TYPES SN54AS856, SN74AS856
8-BIT UNIVERSAL TRANSCEIVER PORT CONTROLLERS

D2814, DECEMBER 1983

- Included Among the Package Options are Compact, 24-Pin, 300-mil-Wide DIPs and Both 28-Pin Plastic and Ceramic Chip Carriers
- Buffered 3-State Outputs Drive Bus Lines Directly
- Cascadable to n-Bits
- Eight Selectable Transceiver/Port Functions:
  — B to A
  — Register to A and/or B
  — Off-Line Shifts (A and B Ports in High-Impedance State)
  — Shifted to A and/or B
- Particularly Suitable for Use in Diagnostics Analysis Circuitry
- Serial Register Provides:
  — Parallel Storage of Either A or B Input Data
  — Serial Transmission of Data from Either A or B Port
  — Readback Mode B to A
- Dependable Texas Instruments Quality and Reliability

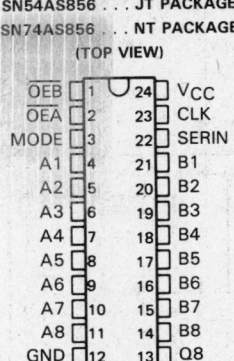

SN54AS856 . . . JT PACKAGE
SN74AS856 . . . NT PACKAGE
(TOP VIEW)

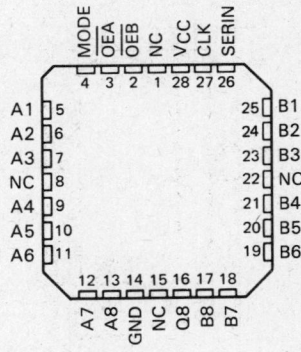

SN54AS856 . . . FH PACKAGE
SN74AS856 . . . FN PACKAGE
(TOP VIEW)

NC—No internal connection

description

The 'AS856 features two 8-bit I/O ports (A1-A8 and B1-B8), an 8-bit parallel-load, serial-in, parallel-out shift register, and control logic. With these features, this device is capable of performing eight selectable transceiver or port functions, depending on the state of the three control lines $\overline{OEA}$, $\overline{OEB}$, and MODE. These functions include: transferring data from port A to port B or vice versa (i.e., the transceiver function), serial shifting data to either or both ports, and performing off-line shifts (with A and B ports active as transceivers in a high-impedance state). Synchronous parallel loading of the internal register can be accomplished from either port on the positive transition of the clock while serially shifting data in via the SERIN input. The 'AS856 is ideally suited for applications needing signature-analysis circuitry to enhance system verification and/or fault analysis. All serial data is shifted right. All outputs are buffer-type outputs designed specifically to drive bus lines directly and all are 3-state except for Q8, which is a totem-pole output.

The SN54AS856 is characterized for operation over the full military temperaure range of $-55\,°C$ to $125\,°C$. The SN74AS856 is characterized for operation from $0\,°C$ to $70\,°C$.

PRODUCT PREVIEW

This document contains information on a product under development. Texas Instruments reserves the right to change or discontinue this product without notice.

Copyright © 1983 by Texas Instruments Incorporated

TEXAS INSTRUMENTS
POST OFFICE BOX 225012 • DALLAS, TEXAS 75265

TYPES SN54AS856, SN74AS856
8-BIT UNIVERSAL TRANSCEIVER PORT CONTROLLERS

logic diagram (positive logic)

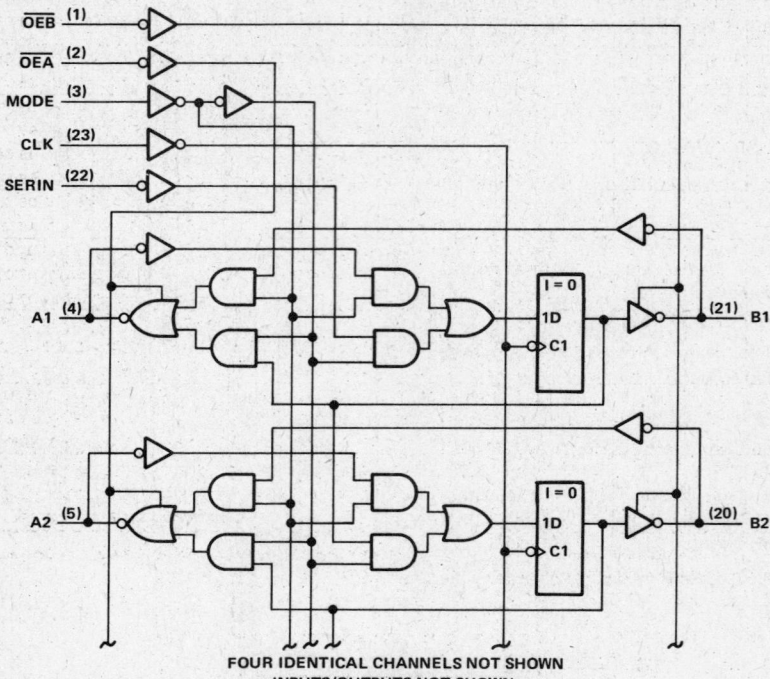

FOUR IDENTICAL CHANNELS NOT SHOWN
INPUTS/OUTPUTS NOT SHOWN:

(6) A3 (19) B3
(7) A4 (18) B4
(8) A5 (17) B5
(9) A6 (16) B6

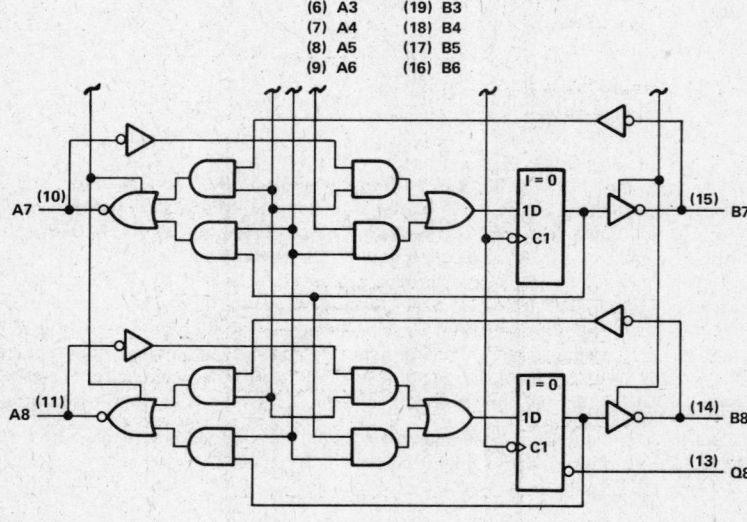

TYPES SN54AS856, SN74AS856
8-BIT UNIVERSAL TRANSCEIVER PORT CONTROLLERS

FUNCTION TABLE

MODE	OEA	OEB	CLOCK	SERIN	A1 Q1 B1	A2 Q2 B2	A3 Q3 B3	A4 Q4 B4	A5 Q5 B5	A6 Q6 B6	A7 Q7 B7	A8 Q8 B8	FUNCTION
L	L	L	H or L	X	Q1 Q1 Q1	Q2 Q2 Q2	Q3 Q3 Q3	Q4 Q4 Q4	Q5 Q5 Q5	Q6 Q6 Q6	Q7 Q7 Q7	Q8 Q8 Q8	FEEDBACK
L	L	L	↑	X	Q1 Q1 Q1	Q2 Q2 Q2	Q3 Q3 Q3	Q4 Q4 Q4	Q5 Q5 Q5	Q6 Q6 Q6	Q7 Q7 Q7	Q8 Q8 Q8	
L	L	H	H or L	X	B1 Q1 Z	B2 Q2 Z	B3 Q3 Z	B4 Q4 Z	B5 Q5 Z	B6 Q6 Z	B7 Q7 Z	B8 Q8 Z	B to A
L	L	H	↑	X	B1 B1 Z	B2 B2 Z	B3 B3 Z	B4 B4 Z	B5 B5 Z	B6 B6 Z	B7 B7 Z	B8 B8 Z	A to Q
L	H	L	H or L	X	Z Q1 Q1	Z Q2 Q2	Z Q3 Q3	Z Q4 Q4	Z Q5 Q5	Z Q6 Q6	Z Q7 Q7	Z Q8 Q8	A to Q
L	H	L	↑	X	Z A1 A1	Z A2 A2	Z A3 A3	Z A4 A4	Z A5 A5	Z A6 A6	Z A7 A7	Z A8 A8	Q to B
L	H	H	H or L	X	Z Q1 Z	Z Q2 Z	Z Q3 Z	Z Q4 Z	Z Q5 Z	Z Q6 Z	Z Q7 Z	Z Q8 Z	A to Q
L	H	H	↑	X	Z A1 Z	Z A2 Z	Z A3 Z	Z A4 Z	Z A5 Z	Z A6 Z	Z A7 Z	Z A8 Z	
H	L	L	H or L	X	Q1 Q_n Q1	Q2 Q_n Q2	Q3 Q_n Q3	Q4 Q_n Q4	Q5 Q_n Q5	Q6 Q_n Q6	Q7 Q_n Q7	Q8 Q_n Q8	SHIFT TO A and B
H	L	L	↑	H	H H H	Q1 Q1 Q1	Q2 Q2 Q2	Q3 Q3 Q3	Q4 Q4 Q4	Q5 Q5 Q5	Q6 Q6 Q6	Q7 Q7 Q7	
H	L	L	↑	L	L L L	Q1 Q1 Q1	Q2 Q2 Q2	Q3 Q3 Q3	Q4 Q4 Q4	Q5 Q5 Q5	Q6 Q6 Q6	Q7 Q7 Q7	
H	L	H	H or L	X	Q1 Q_n Z	Q2 Q_n Z	Q3 Q_n Z	Q4 Q_n Z	Q5 Q_n Z	Q6 Q_n Z	Q7 Q_n Z	Q8 Q_n Z	SHIFT TO A
H	L	H	↑	H	H H Z	Q1 Q1 Z	Q2 Q2 Z	Q3 Q3 Z	Q4 Q4 Z	Q5 Q5 Z	Q6 Q6 Z	Q7 Q7 Z	
H	L	H	↑	L	L L Z	Q1 Q1 Z	Q2 Q2 Z	Q3 Q3 Z	Q4 Q4 Z	Q5 Q5 Z	Q6 Q6 Z	Q7 Q7 Z	
H	H	L	H or L	X	Z Q_n Q1	Z Q_n Q2	Z Q_n Q3	Z Q_n Q4	Z Q_n Q5	Z Q_n Q6	Z Q_n Q7	Z Q_n Q8	SHIFT TO B
H	H	L	↑	H	Z H H	Z Q1 Q1	Z Q2 Q2	Z Q3 Q3	Z Q4 Q4	Z Q5 Q5	Z Q6 Q6	Z Q7 Q7	
H	H	L	↑	L	Z L L	Z Q1 Q1	Z Q2 Q2	Z Q3 Q3	Z Q4 Q4	Z Q5 Q5	Z Q6 Q6	Z Q7 Q7	
H	H	H	H or L	X	Z Q_n Z	Z Q_n Z	Z Q_n Z	Z Q_n Z	Z Q_n Z	Z Q_n Z	Z Q_n Z	Z Q_n Z	SHIFT
H	H	H	↑	H	Z H Z	Z Q1 Z	Z Q2 Z	Z Q3 Z	Z Q4 Z	Z Q5 Z	Z Q6 Z	Z Q7 Z	
H	H	H	↑	L	Z L H	Z Q1 Z	Z Q2 Z	Z Q3 Z	Z Q4 Z	Z Q5 Z	Z Q6 Z	Z Q7 Z	

n = level of Q_n (n = 1, 2 . . . 8) established on most recent ↑ transition of CLK. Q1 through Q8 are the shift register outputs; only Q8 is available externally. The double inversions that take place as data travels from port to port are ignored in this table.

logic symbol

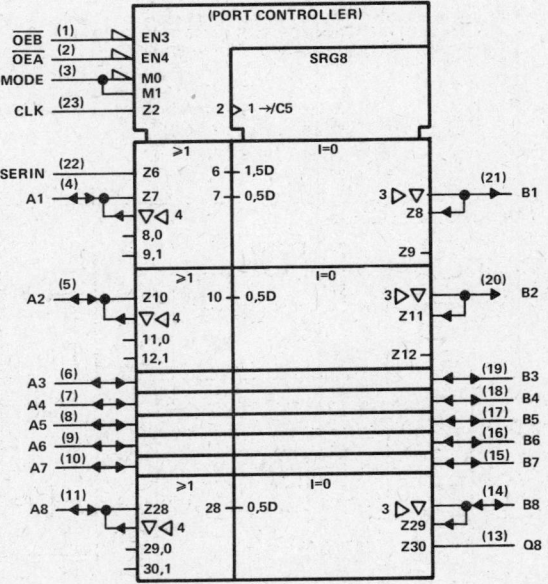

Pin numbers shown are for JT and NT packages.

TEXAS INSTRUMENTS
POST OFFICE BOX 225012 • DALLAS, TEXAS 75265

TYPES SN54AS856, SN74AS856
8-BIT UNIVERSAL TRANSCEIVER PORT CONTROLLERS

absolute maximum ratings over free-air temperature range

Supply voltage, V_{CC} .... 7 V
Input voltage: All inputs .... 7 V
              I/O ports .... 5.5 V
Voltage applied to a disabled 3-state output .... 5.5 V
Operating free-air temperature range: SN54AS856 .... −55°C to 125°C
                                    SN74AS856 .... 0°C to 70°C
Storage temperature range .... −65°C to 150°C

recommended operating conditions

			SN54AS856			SN74AS856			UNIT
			MIN	NOM	MAX	MIN	NOM	MAX	
V_{CC}	Supply voltage		4.5	5	5.5	4.5	5	5.5	V
V_{IH}	High-level input voltage		2			2			V
V_{IL}	Low-level input voltage				0.8			0.8	V
I_{OH}	High-level output current	A1-A8, B1-B8			−12			−15	mA
		Q8			−2			−2	
I_{OL}	Low-level output current	A1-A8, B1-B8			32			48	mA
		Q8			20			20	
f_{clock}	Clock frequency								MHz
t_w	Duration of clock pulse								ns
t_{su}	Setup time before CLK↑	A1-A8, B1-B8 SERIN							ns
		$\overline{OEB}$, $\overline{OEA}$, MODE							
t_h	Hold-time, data after CLK↑	A1-A8, B1-B8 SERIN	0			0			ns
		$\overline{OEB}$, $\overline{OEA}$, MODE	0			0			
T_A	Operating free-air temperature		−55		125	0		70	°C

Additional information on these products can be obtained from the factory as it becomes available.

ALS AND AS CIRCUITS

TEXAS INSTRUMENTS
POST OFFICE BOX 225012 • DALLAS, TEXAS 75265

TYPES SN54AS856, SN74AS856
8-BIT UNIVERSAL TRANSCEIVER PORT CONTROLLERS

electrical characteristics over recommended operating free-air temperature range (unless otherwise noted)

PARAMETER		TEST CONDITIONS		SN54AS856 MIN	SN54AS856 TYP[†]	SN54AS856 MAX	SN74AS856 MIN	SN74AS856 TYP[†]	SN74AS856 MAX	UNIT
V_{IK}		$V_{CC} = 4.5$ V,	$I_I = -18$ mA			-1.2			-1.2	V
V_{OH}	A1-A8	$V_{CC} = 4.5$ V,	$I_{OH} = -12$ mA	2.4	3.2					V
	B1-B8	$V_{CC} = 4.5$ V,	$I_{OH} = -15$ mA				2.4	3.3		
	All outputs	$V_{CC} = 4.5$ V to 5.5 V,	$I_{OH} = -2$ mA	$V_{CC}-2$			$V_{CC}-2$			
V_{OL}	All outputs except Q8	$V_{CC} = 4.5$ V,	$I_{OL} = 32$ mA		0.25	0.5				V
		$V_{CC} = 4.5$ V,	$I_{OL} = 48$ mA					0.35	0.5	
	Q8	$V_{CC} = 4.5$ V,	$I_{OL} = 20$ mA			0.5			0.5	
I_I	$\overline{OEB}$, $\overline{OEA}$, MODE	$V_{CC} = 5.5$ V,	$V_I = 7$ V			1			1	mA
	CLK and SERIN					0.1			0.1	
	A1-A8, B1-B8	$V_{CC} = 5.5$ V,	$V_I = 5.5$ V			0.2			0.2	
I_{IH}	$\overline{OEB}$, $\overline{OEA}$, MODE	$V_{CC} = 5.5$ V,	$V_I = 2.7$ V			800			400	µA
	CLK and SERIN					20			20	
	A1-A8, B1-B8[‡]					70			70	
I_{IL}	$\overline{OEB}$, $\overline{OEA}$, MODE	$V_{CC} = 5.5$ V,	$V_I = 0.4$ V			-2			-2	mA
	CLK and SERIN					-0.3			-0.3	
	A1-A8, B1-B8[‡]					-0.35			-0.35	
I_O[§]	Except Q8	$V_{CC} = 5.5$ V,	$V_O = 2.25$ V	-30		-112	-30		-112	mA
	Q8			-20		-112	-20		-112	
I_{CC}		$V_{CC} = 5.5$ V				118			118	mA

[†] All typical values are at $V_{CC} = 5$ V, $T_A = 25°C$.
[‡] For I/O ports, the parameters I_{IH} and I_{IL} include the output currents I_{OZH} and I_{OZL}, respectively.
[§] The output conditions have been chosen to produce a current that closely approximates one half of the true short-circuit output current, I_{OS}.

switching characteristics (see Note 1)

PARAMETER	FROM (INPUT)	TO (OUTPUT)	$V_{CC} = 4.5$ V to 5.5 V, $C_L = 50$ pF, $R1 = 500$ Ω, $R2 = 500$ Ω, T_A = MIN to MAX						UNIT
			SN54AS856 MIN	SN54AS856 TYP[†]	SN54AS856 MAX	SN74AS856 MIN	SN74AS856 TYP[†]	SN74AS856 MAX	
f_{max}				75			75		MHz
t_{PLH}	Any A port	Any B port		9.5			9.5		ns
t_{PHL}				8			8		
t_{PLH}	Any B port	Any A port		9.5			9.5		ns
t_{PHL}				8			8		
t_{PLH}	$\overline{OEB}$, $\overline{OEA}$, MODE	Any A or B port		12			12		ns
t_{PHL}				12			12		
t_{PLH}	CLK	Any A or B port		6.5			6.5		ns
t_{PHL}				12.5			12.5		
t_{PLH}	CLK	Q8		9			9		ns
t_{PHL}				9			9		
t_{PHZ}	$\overline{OEB}$, $\overline{OEA}$, MODE	Any A or B port		6			6		ns
t_{PLZ}				6			6		
t_{PZH}				10			10		ns
t_{PZL}				10			10		

[†] All typical values are at $V_{CC} = 5$ V, $T_A = 25°C$.
NOTE 1: For load circuit and voltage waveforms, see page 1-12.

2
ALS AND AS CIRCUITS

TYPES SN54ALS857, SN54AS857, SN74ALS857, SN74AS857
HEX 2-TO-1 UNIVERSAL MULTIPLEXERS

D2661, DECEMBER 1982 – REVISED DECEMBER 1983

- Selects True or Complementary Data
- Performs AND/NAND (masking) of A or B Operand
- Cascadable to Expand Number of Operands
- Detects Zeros on A or B Operands
- 3-State Outputs Interface Directly with System Bus
- Included Among the Package Options Are 24-Pin, 300-Mil-Wide DIPs and Both 28-Pin Plastic and Ceramic Chip Carriers
- Dependable Texas Instruments Quality and Reliability

SN54ALS857, SN54AS857 . . . JT PACKAGE
SN74ALS857, SN74AS857 . . . NT PACKAGE
(TOP VIEW)

```
    S0  [ 1   24 ] VCC
    1A  [ 2   23 ] S1
    1B  [ 3   22 ] 6A
    1Y  [ 4   21 ] 6B
    2A  [ 5   20 ] 6Y
    2B  [ 6   19 ] 5A
    2Y  [ 7   18 ] 5B
    3A  [ 8   17 ] 5Y
    3B  [ 9   16 ] 4A
    3Y  [10   15 ] 4B
 OPER=0 [11   14 ] 4Y
    GND [12   13 ] COMP
```

SN54ALS857, SN54AS857 . . . FH PACKAGE
SN74ALS857, SN74AS857 . . . FN PACKAGE

description

The 'ALS857 and 'AS857 are hextuple 2-line to 1-line multiplexers with three-state outputs. The devices can provide either true (COMP low) or inverted (COMP high) data at the Y outputs. In addition, the 'ALS857 and 'AS857 perform the logical AND function (A·B) and the clear function as well. The four modes of operation are:

Select A data inputs,
Select B data inputs,
AND A inputs with B inputs,
Clear

In either of the first two modes, OPER=0 is high if all the selected A or B inputs are low.

The six Y outputs and the OPER=0 output are all three-state and rated at 12 mA and 24 mA I_{OL} for the SN54ALS857 and SN74ALS857, respectively, and at 32 mA and 48 mA I_{OL} for the SN54AS857 and SN74AS857, respectively. All outputs can be placed into the high-impedance state by applying a high level to the COMP, S0, and S1 inputs simultaneously. The complete function table is shown below.

The SN54ALS857 and SN54AS857 are characterized for operation over the full military temperature range of −55°C to 125°C. The SN74ALS857 and SN74AS857 are characterized for operation from 0°C to 70°C.

For chip carrier information contact factory

logic symbol

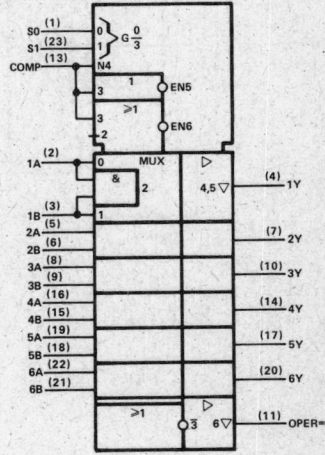

Pin numbers shown are for JT and NT packages.

FUNCTION TABLE

COMP	S1	S0	Y OUTPUTS	OPER = ZERO
L	L	L	A	H = all A inputs L
L	L	H	B	H = all B inputs L
L	H	L	A·B	Z
L	H	H	L	L
H	L	L	$\overline{A}$	H = all A inputs L
H	L	H	$\overline{B}$	H = all B inputs L
H	H	L	$\overline{A·B}$	Z
H	H	H	Z	Z

Copyright © 1982 by Texas Instruments Incorporated

TEXAS INSTRUMENTS
POST OFFICE BOX 225012 • DALLAS, TEXAS 75265

TYPES SN54ALS857, SN74ALS857
HEX 2-TO-1 UNIVERSAL MULTIPLEXERS

'ALS857 logic diagram (positive logic)

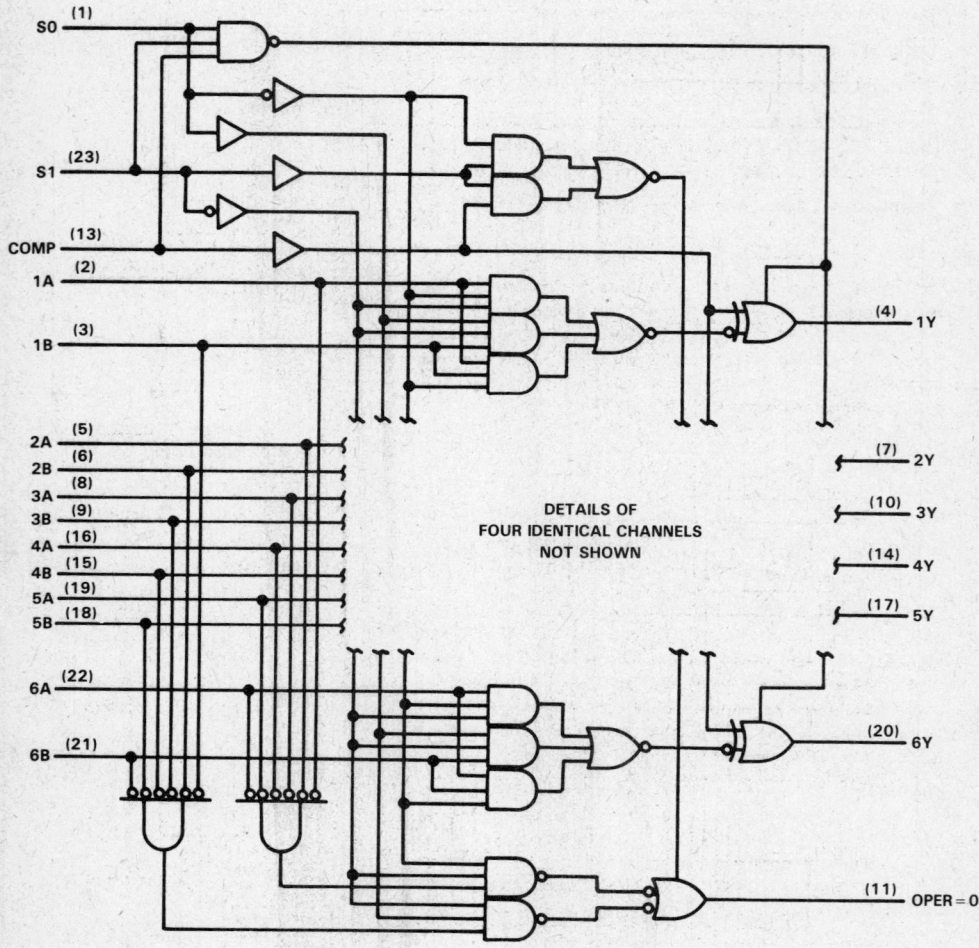

Pin numbers shown are for JT and NT packages.

TYPES SN54AS857, SN74AS857
HEX 2-TO-1 UNIVERSAL MULTIPLEXERS

'AS857 logic diagram (positive logic)

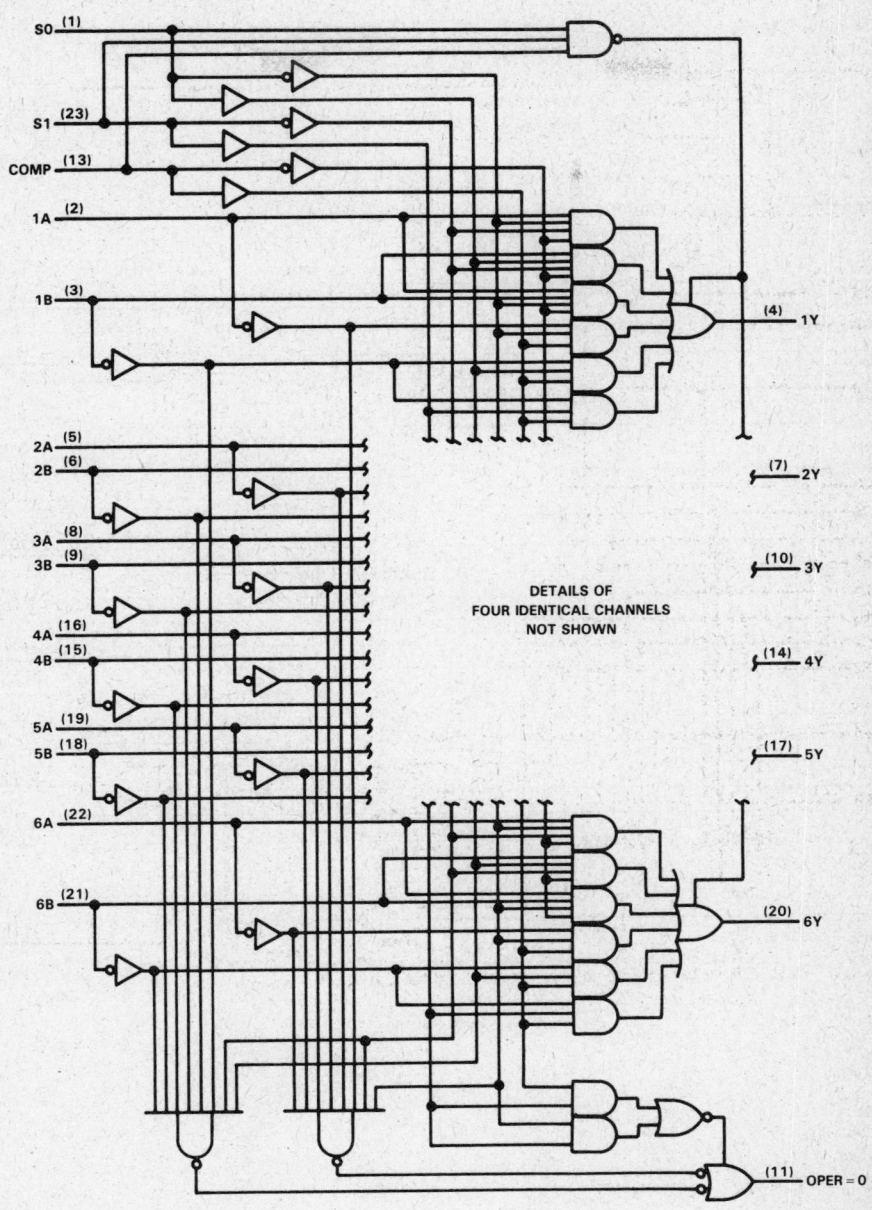

Pin numbers shown are for JT and NT packages.

TYPES SN54ALS857, SN74ALS857
HEX 2-TO-1 UNIVERSAL MULTIPLEXERS

absolute maximum ratings over operating free-air temperature range (unless otherwise noted)

Supply voltage, V_{CC} .... 7 V
Input voltage .... 7 V
Voltage applied to a disabled 3-state output .... 5.5 V
Operating free-air temperature range: SN54ALS857 .... −55 °C to 125 °C
                                      SN74ALS857 .... 0 °C to 70 °C
Storage temperature range .... −65 °C to 150 °C

recommended operating conditions

		SN54ALS857			SN74ALS857			UNIT
		MIN	NOM	MAX	MIN	NOM	MAX	
V_{CC}	Supply voltage	4.5	5	5.5	4.5	5	5.5	V
V_{IH}	High-level input voltage	2			2			V
V_{IL}	Low-level input voltage			0.8			0.8	V
I_{OH}	High-level output current			−1			−2.6	mA
I_{OL}	Low-level output current			12			−24	mA
T_A	Operating free-air temperature	−55		125	0		70	°C

electrical characteristics over recommended operating free-air temperature range (unless otherwise noted)

PARAMETER	TEST CONDITIONS		SN54ALS857			SN74ALS857			UNIT
			MIN	TYP†	MAX	MIN	TYP†	MAX	
V_{IK}	V_{CC} = 4.5 V,	I_I = −18 mA			−1.5			−1.5	V
V_{OH}	V_{CC} = 4.5 V to 5.5 V,	I_{OH} = −0.4 mA	$V_{CC}-2$			$V_{CC}-2$			V
	V_{CC} = 4.5 V,	I_{OH} = −1 mA	2.4	3.3					
	V_{CC} = 4.5 V,	I_{OH} = −2.6 mA				2.4	3.2		
V_{OL}	V_{CC} = 4.5 V,	I_{OL} = 12 mA		0.25	0.4		0.25	0.4	V
	V_{CC} = 4.5 V,	I_{OL} = 24 mA					0.35	0.5	
I_{OZH}	V_{CC} = 5.5 V,	V_O = 2.7 V			20			20	µA
I_{OZL}	V_{CC} = 5.5 V,	V_O = 0.4 V			−20			−20	µA
I_I	V_{CC} = 5.5 V,	V_I = 7 V			0.1			0.1	mA
I_{IH}	V_{CC} = 5.5 V,	V_I = 2.7 V			20			20	µA
I_{IL}	V_{CC} = 5.5 V,	V_I = 0.4 V			−0.2			−0.2	mA
I_O ‡	V_{CC} = 5.5 V,	V_O = 2.25 V	−15		−70	−15		−70	mA
I_{CC}	V_{CC} = 5.5 V, See Note 1	Outputs high		11	24		11	24	mA
		Outputs low		16	33		16	33	
		Outputs disabled		18	36		18	36	

†All typical values are at V_{CC} = 5 V, T_A = 25 °C.
‡The output conditions have been chosen to produce a current that closely approximates one half of the true short-circuit output current, I_{OS}.
NOTE 1: I_{CC} is measured with all possible inputs grounded while achieving the stated output conditions.

TYPES SN54ALS857, SN74ALS857
HEX 2-TO-1 UNIVERSAL MULTIPLEXERS

switching charactersitcs (see Note 1)

PARAMETER	FROM (INPUT)	TO (OUTPUT)	V_{CC} = 4.5 V to 5.5 V, C_L = 50 pF, R1 = 500 Ω, R2 = 500 Ω, T_A = MIN to MAX				UNIT
			SN54ALS857		SN74ALS857		
			MIN	MAX	MIN	MAX	
t_{pd}	A or B (COMP high)	Y (Inverting)	4	18	4	14	ns
t_{pd}	A or B (COMP low)	Y (Noninverting)	4	18	4	14	ns
t_{pd}	S0 or S1	Y	7	37	7	33	ns
t_{pd}	COMP	Y	6	22	6	18	
t_{pd}	A or B	OPER = 0	5	45	5	37	
t_{pd}	S0 or S1	OPER = 0	5	30	5	23	
t_{en}	S0 or S1	Y	7	38	7	35	ns
t_{dis}			2	29	2	23	
t_{en}	COMP	Y	8	27	8	24	ns
t_{dis}			6	27	6	21	
t_{en}	S0	OPER = 0	6	24	6	20	ns
t_{dis}			11	34	11	27	
t_{en}	S1	OPER = 0	6	28	6	25	ns
t_{dis}			3	23	3	19	
t_{en}	COMP	OPER = 0	9	30	9	25	ns
t_{dis}			6	24	6	20	

t_{pd} = t_{PLH} or t_{PHL}
t_{en} = t_{PZH} or t_{PZL}
t_{dis} = t_{PHZ} or t_{PLZ}
NOTE 1: For load circuit and voltage waveforms, see page 1-12.

ALS AND AS CIRCUITS

TEXAS INSTRUMENTS
POST OFFICE BOX 225012 • DALLAS, TEXAS 75265

TYPES SN54AS857, SN74AS857
HEX 2-TO-1 UNIVERSAL MULTIPLEXERS

absolute maximum ratings over operating free-air temperature range (unless otherwise noted)

Supply voltage, V_{CC} .... 7 V
Input voltage .... 7 V
Voltage applied to a disabled 3-state output .... 5.5 V
Operating free-air temperature range: SN54AS857 .... −55 °C to 125 °C
                                   SN74AS857 .... 0 °C to 70 °C
Storage temperature range .... −65 °C to 150 °C

recommended operating conditions

			SN54AS857			SN74AS857			UNIT
			MIN	NOM	MAX	MIN	NOM	MAX	
V_{CC}	Supply voltage		4.5	5	5.5	4.5	5	5.5	V
V_{IH}	High-level input voltage		2			2			V
V_{IL}	Low-level input voltage				0.8			0.8	V
I_{OH}	High-level output current	Y Outputs			−12			−15	mA
		OPER = 0			−2			−2	
I_{OL}	Low-level output current	Y Outputs			32			48	mA
		OPER = 0			20			20	
T_A	Operating free-air temperature		−55		125	0		70	°C

electrical characteristics over recommended operating free-air temperature range (unless otherwise noted)

PARAMETER		TEST CONDITIONS		SN54AS857			SN74AS857			UNIT
				MIN	TYP[†]	MAX	MIN	TYP[†]	MAX	
V_{IK}		V_{CC} = 4.5 V,	I_I = −18 mA			−1.2			−1.2	V
V_{OH}	Y Outputs	V_{CC} = 4.5 V,	I_{OH} = −12 mA	2.4	3.2					V
		V_{CC} = 4.5 V,	I_{OH} = −15 mA				2.4	3.3		
	All Outputs	V_{CC} = 4.5 V to 5.5 V,	I_{OH} = −2 mA	V_{CC}−2			V_{CC}−2			
V_{OL}	Y Outputs	V_{CC} = 4.5 V,	I_{OL} = 32 mA		0.35	0.5				V
		V_{CC} = 4.5 V,	I_{OL} = 48 mA					0.35	0.5	
	OPER = 0	V_{CC} = 4.5 V,	I_{OL} = 20 mA		0.25	0.5		0.25	0.5	
I_{OZH}		V_{CC} = 5.5 V,	V_O = 2.7 V			50			50	µA
I_{OZL}		V_{CC} = 5.5 V,	V_O = 0.4 V			−50			−50	µA
I_I		V_{CC} = 5.5 V,	V_I = 7 V			0.1			0.1	mA
I_{IH}		V_{CC} = 5.5 V,	V_I = 2.7 V			20			20	µA
I_{IL}		V_{CC} = 5.5 V,	V_I = 0.4 V			−2			−2	mA
I_O[‡]		V_{CC} = 5.5 V,	V_O = 2.25 V	−30		−112	−30		−112	mA
I_{CC}		V_{CC} = 5.5 V, See Note 1	Outputs high		97	140		97	140	mA
			Outputs low		127	175		127	175	
			Outputs disabled		92	135		92	135	

[†] All typical values are at V_{CC} = 5 V, T_A = 25 °C.
[‡] The output conditions have been chosen to produce a current that closely approximates one half of the true short-circuit output current, I_{OS}.
NOTE 1: I_{CC} is measured with all possible inputs grounded while achieving the stated output conditions.

TYPES SN54AS857, SN74AS857
HEX 2-TO-1 UNIVERSAL MULTIPLEXERS

switching characteristics (see Note 1)

PARAMETER	FROM (INPUT)	TO (OUTPUT)	V_{CC} = 4.5 V, to 5.5 V, C_L = 50 pF, R1 = 500 Ω, R2 = 500 Ω, T_A = MIN to MAX				UNIT
			SN54AS857		SN74AS857		
			MIN	MAX	MIN	MAX	
t_{pd}	A or B (COMP high)	Y (Inverting)	2	15	2	12	ns
t_{pd}	A or B (COMP low)	Y (Noninverting)	2	12	2	10	ns
t_{pd}	S0 or S1	Y	2	15	2	13	ns
t_{pd}	COMP	Y	2	15	2	13	
t_{pd}	A or B	OPER = 0	2	16	2	14	
t_{pd}	S0 to S1	OPER = 0	2	20	2	18	
t_{en}	S0 to S1	Y	2	14	2	12	ns
t_{dis}			2	13	2	11	
t_{en}	COMP	Y	2	14	2	12	ns
t_{dis}			2	10	2	9	
t_{en}	S0	OPER = 0	2	14	2	12	ns
t_{dis}			2	10	2	9	
t_{en}	S1	OPER = 0	2	14	2	12	ns
t_{dis}			2	10	2	9	
t_{en}	COMP	OPER = 0	2	10	2	13	ns
t_{dis}			2	10	2	9	

t_{pd} = t_{PLH} or t_{PHL}
t_{en} = t_{PZH} or t_{PAL}
t_{dis} = t_{PHZ} or t_{PLZ}
NOTE 1: For load circuit and voltage waveforms, see page 1-12.

2 ALS AND AS CIRCUITS

TYPES SN54AS866, SN74AS866
8-BIT MAGNITUDE COMPARATORS

D2661, DECEMBER 1982 – REVISED DECEMBER 1983

- Included among the Package Options Are 28-Pin DIPs and Both Plastic and Ceramic Chip Carriers
- Input and Output Latches with Active-High Enables
- Fast Compare to Zero
- Arithmetic and Logical Comparison
- Open-Collector P = Q Output
- Dependable Texas Instruments Quality and Reliability

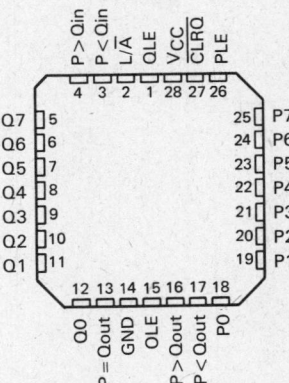

SN54AS866 ... JD PACKAGE
SN74AS866 ... N PACKAGE
(TOP VIEW)

SN54AS866 ... FH PACKAGE
SN74AS866 ... FN PACKAGE
(TOP VIEW)

description

These Advanced Schottky devices are capable of performing high-speed arithmetic or logical comparisons on two 8-bit binary or two's complement words. Three fully decoded decisions about words P and Q are externally available at the outputs. These devices are fully expandable to any word length by connecting the totem pole P > Q and P < Q outputs of each stage to the P > Q and P < Q inputs of the next higher-order stage. The cascading paths are implemented with only a two-gate-level delay to reduce overall comparison times for long words. The open-collector P = Q outputs may be wire-ANDed together.

Both input words P and Q plus all three outputs (P > Q, P < Q, and P = Q) are equipped with latches to provide the designer with temporary data storage for avoiding race conditions. The enable circuitry is implemented with minimal delay times to enhance performance when the devices are cascaded for longer word lengths. Each latch is transparent when the appropriate latch enable, PLE, QLE, or OLE is high.

The enable inputs PLE and QLE and data inputs P and Q utilize p-n-p input transistors to reduce the low-level input current requirement to typically −0.25 mA, which minimizes loading effects.

The Q register may be cleared to zero for a fast comparison of the P word to zero.

The SN54AS866 is characterized for operation over the full military temperature range of −55 °C to 125 °C. The SN74AS866 is characterized for operation from 0 °C to 70 °C.

Copyright © 1982 by Texas Instruments Incorporated

TEXAS INSTRUMENTS
POST OFFICE BOX 225012 • DALLAS, TEXAS 75265

TYPES SN54AS866, SN74AS866
8-BIT MAGNITUDE COMPARATORS

logic symbol

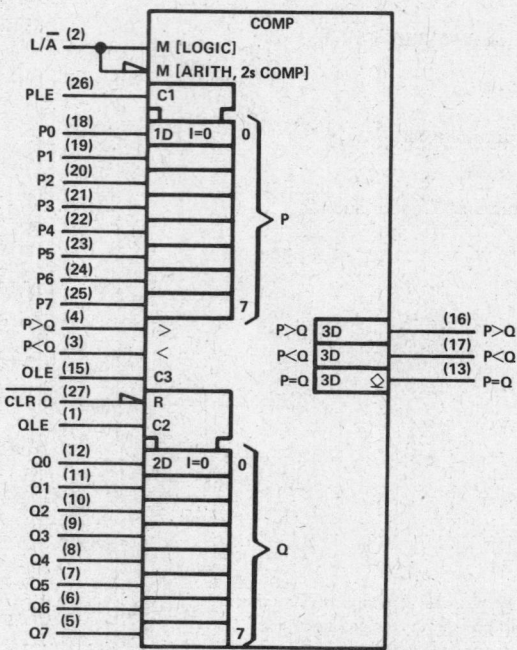

Pin numbers shown are for JD and N packages.

TYPES SN54AS866, SN74AS866
8-BIT MAGNITUDE COMPARATORS

logic diagram (positive logic)

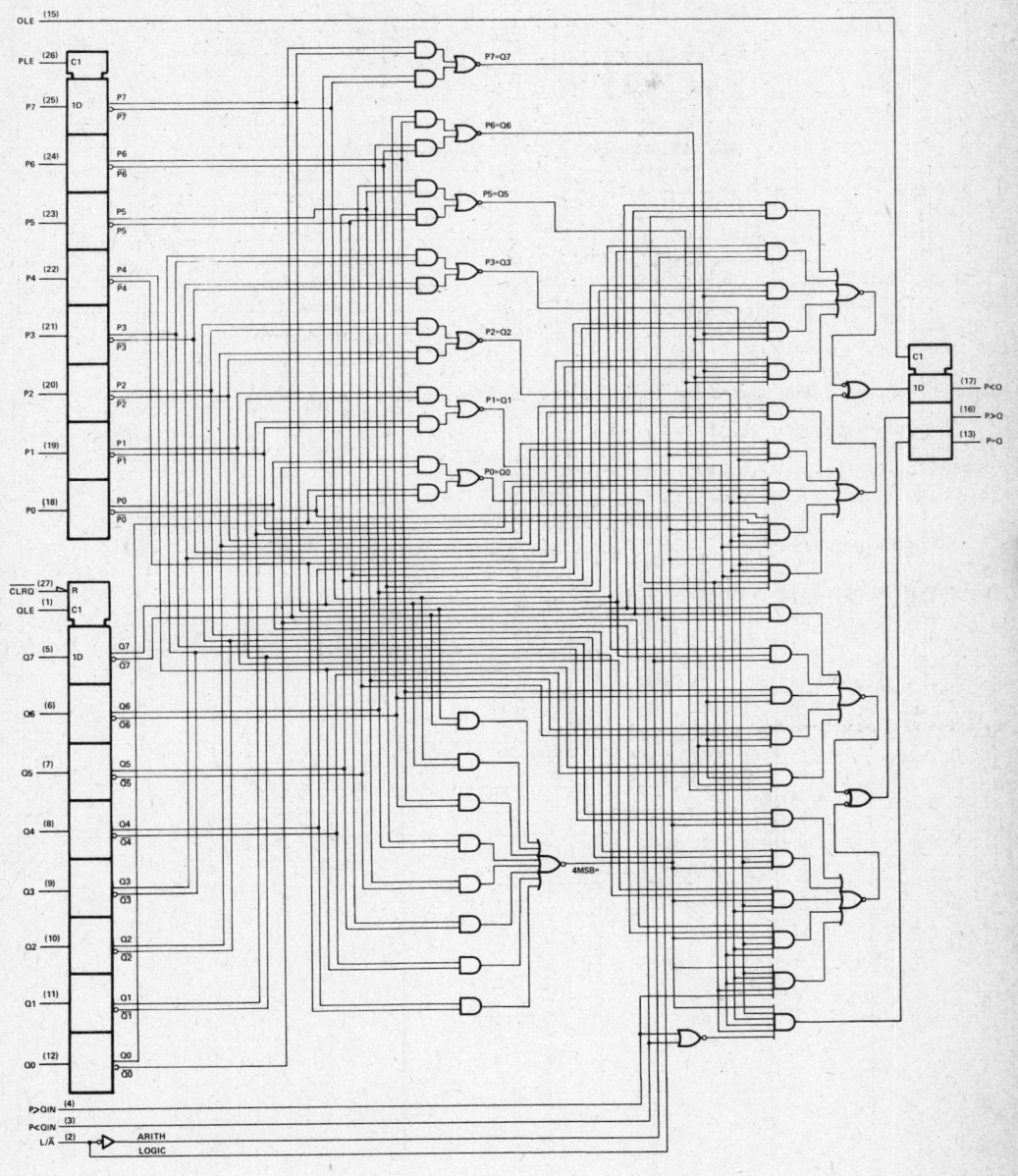

TYPES SN54AS866, SN74AS866
8-BIT MAGNITUDE COMPARATORS

FUNCTION TABLE

COMPARISON	L/$\overline{A}$	DATA INPUTS P0-P7, Q0-Q7	INPUTS P>Q	INPUTS P<Q	OUTPUTS P>Q	OUTPUTS P<Q	OUTPUTS P=Q
Logical	H	P > Q	X	X	H	L	L
Logical	H	P < Q	X	X	L	H	L
Logical	H	P = Q	L	L	L	L	H
Logical	H	P = Q	L	H	L	H	L
Logical	H	P = Q	H	L	H	L	L
Logical	H	P = Q	H	H	H	H	L
Arithmetic	L	P AG Q	X	X	H	L	L
Arithmetic	L	Q AG P	X	X	L	H	L
Arithmetic	L	P = Q	L	L	L	L	H
Arithmetic	L	P = Q	L	H	L	H	L
Arithmetic	L	P = Q	H	L	H	L	L
Arithmetic	L	P = Q	H	H	H	H	L

AG = arithmetically greater than

absolute maximum ratings over operating free-air temperature range (unless otherwise noted)

Supply voltage, V_{CC} .... 7 V
Input voltage .... 7 V
Off-state output voltage, P = Q output .... 7 V
Operating free-air temperature range: SN54AS866 .... −55 °C to 125 °C
                                         SN74AS866 .... 0 °C to 70 °C
Storage temperature range .... −65 °C to 150 °C

recommended operating conditions

	PARAMETER	SN54AS866 MIN	SN54AS866 NOM	SN54AS866 MAX	SN74AS866 MIN	SN74AS866 NOM	SN74AS866 MAX	UNIT
V_{CC}	Supply voltage	4.5	5	5.5	4.5	5	5.5	V
V_{IH}	High-level input voltage	2			2			V
V_{IL}	Low-level input voltage			0.8			0.8	V
I_{OH}	High-level output current, all outputs except P = Q			−2			−2	mA
V_{OH}	High-level output voltage, P = Q output			5.5			5.5	V
I_{OL}	Low-level output current			20			20	mA
t_{su}	Setup time to PLE, QLE, OLEi	2			2			ns
t_h	Hold time after PLE, QLE, OLEi	4			4			ns
T_A	Operating free-air temperature	−55		125	0		70	°C

TEXAS INSTRUMENTS
POST OFFICE BOX 225012 • DALLAS, TEXAS 75265

TYPES SN54AS866, SN74AS866
8-BIT MAGNITUDE COMPARATORS

electrical characteristics over recommended operating free-air temperature range (unless otherwise noted)

PARAMETER		TEST CONDITIONS		SN54AS866 MIN	TYP†	MAX	SN74AS866 MIN	TYP†	MAX	UNIT
V_{IK}		$V_{CC} = 4.5$ V,	$I_I = -18$ mA			-1.2			-1.2	V
V_{OH}	P > Q, P < Q	$V_{CC} = 4.5$ V to 5.5 V,	$I_{OH} = -2$ mA	$V_{CC}-2$			$V_{CC}-2$			
I_{OH}	P = Q only	$V_{CC} = 4.5$ V,	$V_{OH} = 5.5$ V			250			250	μA
V_{OL}		$V_{CC} = 4.5$ V,	$I_{OL} = 20$ mA		0.35	0.5		0.35	0.5	V
I_I		$V_{CC} = 5.5$ V,	$V_I = 7$ V			0.1			0.1	mA
I_{IH}	L/$\overline{A}$, OLE	$V_{CC} = 5.5$ V,	$V_I = 2.7$ V			40			40	μA
	Others					20			20	
I_{IL}	L/$\overline{A}$, OLE, P > Qin, P < Qin	$V_{CC} = 5.5$ V,	$V_I = 0.4$ V			-4			-4	mA
	$\overline{CLRQ}$					-2			-2	
	P, Q, PLE, QLE				-0.25	-1		-0.25	-1	
I_O‡		$V_{CC} = 5.5$ V,	$V_O = 2.25$ V	-20		-112	-20		-112	mA
I_{CC}		$V_{CC} = 5.5$ V,	See Note 1		160	240		160	240	mA

†All typical values are at $V_{CC} = 5$ V, $T_A = 25°C$.
‡The output conditions have been chosen to produce a current that closely approximates one-half of the true short-circuit, I_{OS}.
NOTE 1: I_{CC} is measured with all inputs high except L/$\overline{A}$, which is low.

switching characteristics (see Note 2)

PARAMETER	FROM (INPUT)	TO (OUTPUT)	$V_{CC} = 4.5$ V to 5.5 V, $C_L = 50$ pF, $R_L = 500$ Ω, $T_A = $ MIN to MAX						UNIT
			SN54AS866 MIN	TYP†	MAX	SN74AS866 MIN	TYP†	MAX	
t_{PLH}	L/$\overline{A}$	P < Q, P > Q	1	8.5	14	1	8.5	13	ns
t_{PHL}			1	7.5	14	1	7.5	13	
t_{PLH}	P < Q, P > Q		1	5	10	1	5	8	ns
t_{PLH}			1	5.5	10	1	5.5	8	
t_{PHL}	Any P or Q Data Input		1	13.5	21	1	13.5	17.5	ns
t_{PHL}			1	10	17	1	10	15	
t_{PLH}	$\overline{CLRQ}$		1	16	21	1	16	20	ns
t_{PHL}			1	12	17	1	12	16	

PARAMETER	FROM (INPUT)	TO (OUTPUT)	$V_{CC} = 4.5$ V to 5.5 V, $C_L = 50$ pF, $T_A = 280$ Ω, $T_A = $ MIN to MAX						UNIT
			SN54AS866 MIN	TYP†	MAX	SN74AS866 MIN	TYP†	MAX	
t_{PLH}	P < Q, P > Q	P = Q	1	6.5	12	1	6.5	11	ns
t_{PHL}			1	8	14	1	8	13	
t_{PLH}	Any P or Q Data Input	P = Q	1	10	15	1	10	14	ns
t_{PHL}			1	9	14	1	9	13	
t_{PLH}	$\overline{CLRQ}$	P = Q	1	12	17	1	12	16	ns
t_{PHL}			1	13	18	1	13	17	

†All typical vallues ree at $V_{CC} = 5$ V, $T_A = 25°C$.
NOTE 2: For load circuit and voltage waveforms, see page 1-12.

ALS AND AS CIRCUITS

TEXAS INSTRUMENTS
POST OFFICE BOX 225012 • DALLAS, TEXAS 75265

TYPES SN54AS866, SN74AS866
8-BIT MAGNITUDE COMPARATORS

TYPICAL APPLICATION DATA

This sequence of comparisons illustrates how the $\overline{\text{CLRQ}}$ function can be used to perform dual comparisons of the varying P terms (P0, P1, etc). When $\overline{\text{CLRQ}}$ is high, the P term is compared to the Q term. When $\overline{\text{CLRQ}}$ is taken low, the P term is compared to zero. This or similar sequences can enhance performance and reduce package count to perform value range checks.

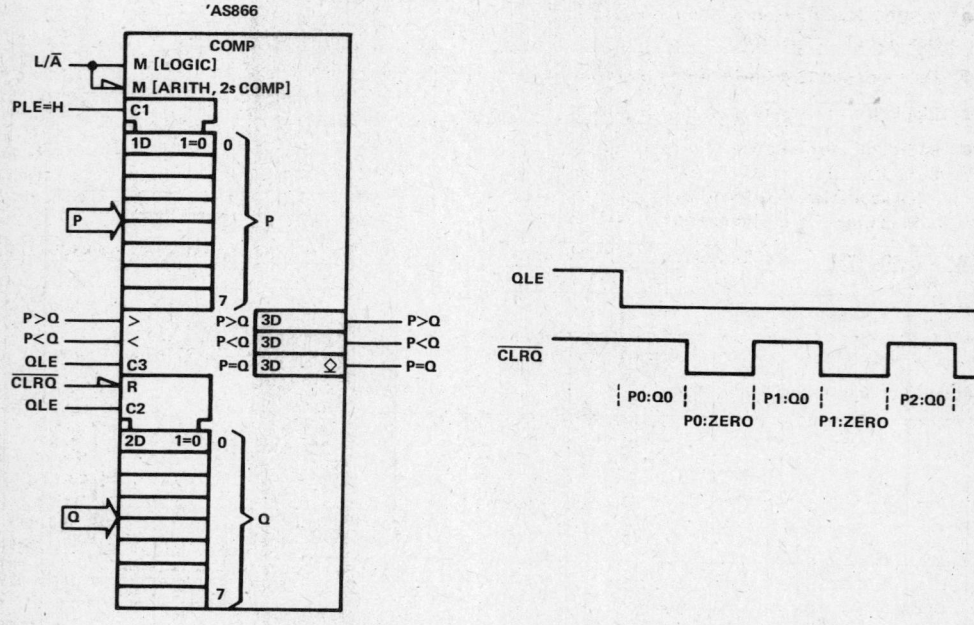

FIGURE 1—MAGNITUDE COMPARISONS COMBINED WITH QUICK COMPARISONS TO ZERO (RANGE VERIFICATIONS)

TYPES SN54AS867, SN54AS869, SN74AS867, SN74AS869
SYNCHRONOUS 8-BIT UP/DOWN COUNTERS

D2661, DECEMBER 1982—REVISED DECEMBER 1983

- Included among the Package Options are Compact, 24-Pin, 300-mil-Wide Dips and 28-Pin Ceramic Chip Carriers
- Fully Programmable with Synchronous Counting and Loading
- 'AS867 Has Asynchronous Clear, 'AS869 Has Synchronous Clear
- Fully Independent Clock circuit Simplifies Use
- Ripple Carry Output for n-Bit Cascading
- Improved Performance Compared to Schottky TTL:
  Typical Power Reduced by 38%
  Maximum Count Frequency is 25% Higher
- Dependable Texas Instruments Quality and Reliability

SN54AS867, SN54AS869 .... JT PACKAGE
SN74AS867, SN74AS869 .... NT PACKAGE
(TOP VIEW)

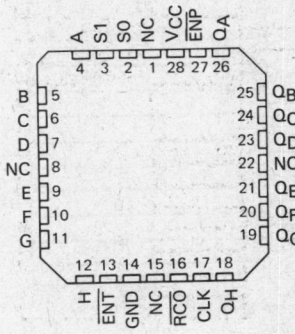

SN54AS867, SN54AS869 .... FH PACKAGE
(TOP VIEW)

NC — No internal connection

description

These synchronous presettable counters feature an internal carry look-ahead for cascading in high-speed counting applications. Synchronous operation is provided by having all flip-flops clocked simultaneously so that the outputs change coincident with each other when so instructed by the count-enable inputs and internal gating. This mode of operation helps eliminate the output counting spikes that are normally associated with asynchronous (ripple-clock) counters. A buffered clock input triggers the eight flip-flops on the rising (positive-going) edge of the clock waveform.

These counters are fully programmable; that is, the outputs may each be preset to either level. The load mode circuitry allows parallel loading of the cascaded counters. As loading is synchronous, selecting the load mode disables the counter and causes the outputs to agree with the data inputs after the next clock pulse.

The carry look-ahead circuitry provides for cascading counters for n-bit synchronous applications without additional gating. Instrumental in accomplishing this function are two count-enable inputs and a carry output. Both count enable inputs ($\overline{ENP}$ and $\overline{ENT}$) must be low to count. The direction of the count is determined by the levels of the select inputs (see Function Table). Input $\overline{ENT}$ is fed forward to enable the carry output. The ripple carry output thus enabled will produce a low-level pulse while the count is zero (all outputs low) counting down or 255 counting up (all outputs high). This low-level overflow carry pulse can be used to enable successive cascaded stages. Transitions at the enable $\overline{ENP}$ and $\overline{ENT}$ inputs are allowed regardless of the level of the clock input. All inputs are diode-clamped to minimize transmission-line effects, thereby simplifying system design.

Copyright © 1983 by Texas Instruments Incorporated

TYPES SN54AS867, SN54AS869, SN74AS867, SN74AS869
SYNCHRONOUS 8-BIT UP/DOWN COUNTERS

These counters feature a fully independent clock circuit. With the exception of thee asynchronous clear on the 'AS867, changes at control inputs (S0, S1) that will modify the operating mode have no effect on the Q outputs until clocking occurs. Anytime the $\overline{ENP}$ and/or $\overline{ENT}$ is taken high, $\overline{RCO}$ will either go or remain high. The function of the counter (whether enabled, disabled, loading, or counting) will be dictated solely by the conditions meeting the stable setup and hold times.

The SN54AS867 and SN54AS869 are characterized for operation over the full military temperature range of $-55°C$ to $125°C$. The SN74AS867 and SN74AS869 are characterized for operation from $0°C$ to $70°C$.

logic symbols

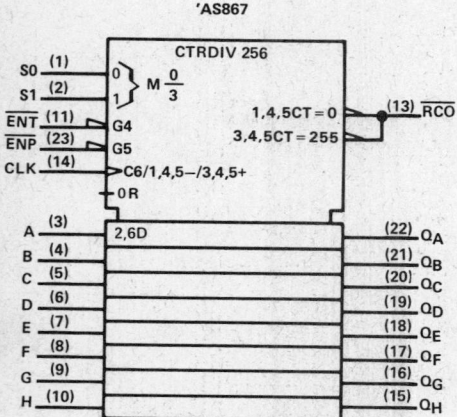

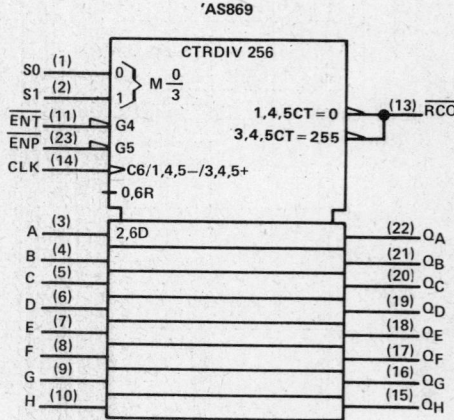

Pin numbers shown are for JT and NT packages.

FUNCTION TABLE

S1	S0	FUNCTION
L	L	Clear
L	H	Count down
H	L	Load
H	H	Count up

absolute maximum ratings over operating free-air temperature range (unless otherwise noted)

Supply voltage, V_{CC} .... 7 V
Input voltage .... 7 V
Operating free-air temperature: SN54AS867, SN54AS869 .... $-55°C$ to $125°C$
                              SN74AS867, SN74AS869 .... $0°C$ to $70°C$
Storage temperature range .... $-65°C$ to $150°C$

TYPES SN54AS867, SN54AS869, SN74AS867, SN74AS869
SYNCHRONOUS 8-BIT UP/DOWN COUNTERS

logic diagram (positive logic)

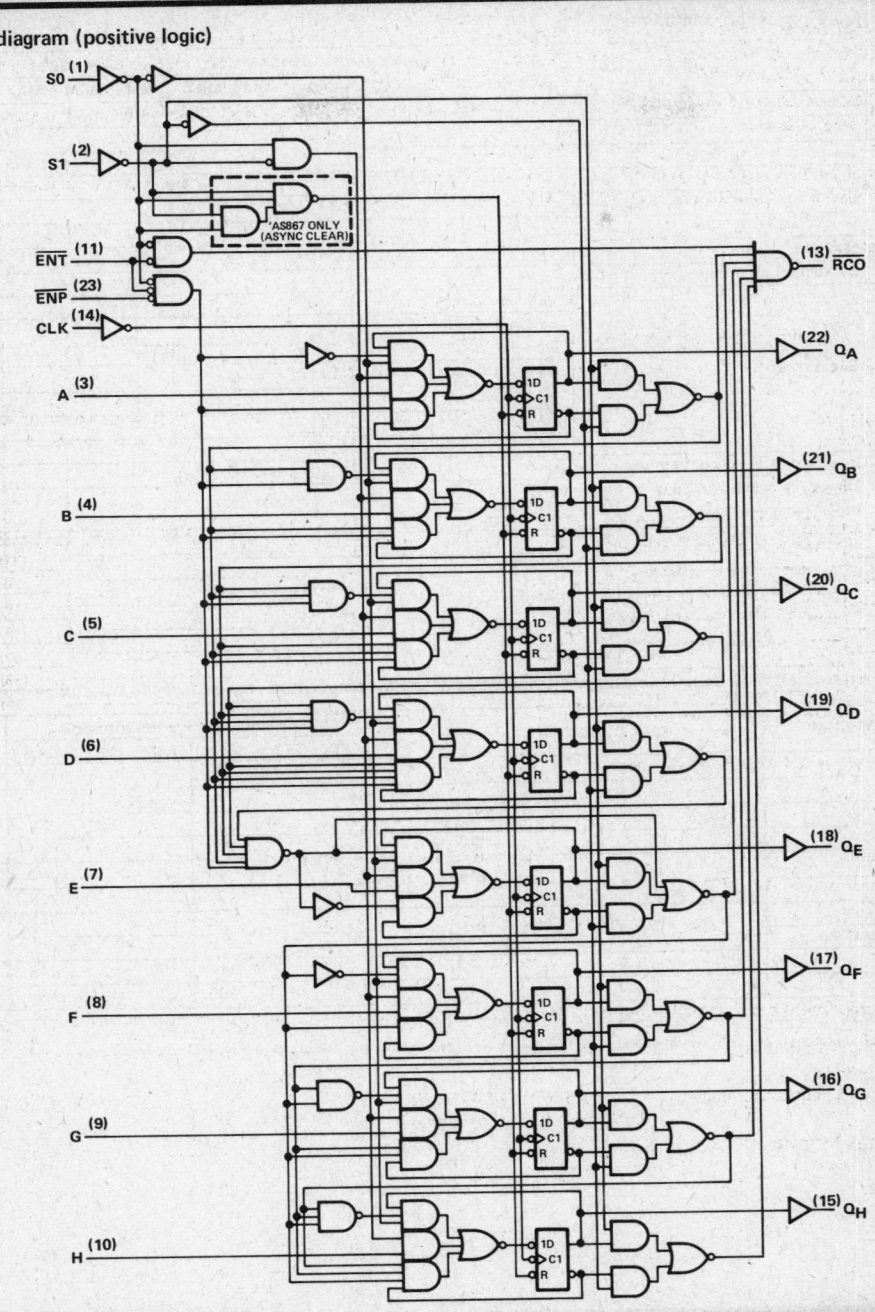

TYPES SN54AS867, SN74AS867,
SYNCHRONOUS 8-BIT UP/DOWN COUNTERS WITH ASYNCHRONOUS CLEAR

recommended operating conditions

			SN54AS867			SN74AS867			UNIT
			MIN	NOM	MAX	MIN	NOM	MAX	
V_{CC}	Supply voltage		4.5	5	5.5	4.5	5	5.5	V
V_{IH}	High-level input voltage		2			2			V
V_{IL}	Low-level input voltage				0.8			0.8	V
I_{OH}	High-level output current				-2			-2	mA
I_{OL}	Low-level output current				20			20	mA
f_{clock}	Clock frequency		0		40	0		50	MHz
$t_{w(clock)}$	Duration		12.5			10			ns
$t_{w(clear)}$	Duration of clear pulse (S0 and S1 low)		12.5			10			ns
t_{su}	Setup time†	Data inputs A-H	5			4			ns
		Enable P ($\overline{ENP}$) or Enable T ($\overline{ENT}$)	9			8			ns
		S0 or S1 (load)	11			10			ns
		S0 or S1 (clear)	11			10			ns
		S0 or S1 (count down)	42			40			ns
		S0 or S1 (count up)	42			40			ns
t_h	Hold time at any input with respect to clock↑		0			0			ns
t_{skew}	Skew time between S0 and S1 (maximum to avoid inadvertent clear)		8			7			ns
T_A	Operating free-air temperature		-55		125	0		70	°C

†This setup time is required to ensure stable data.

electrical characteristics over recommended operating free-air temperature range (unless otherwise noted)

PARAMETER		TEST CONDITIONS		SN54AS867			SN74AS867			UNIT
				MIN	TYP‡	MAX	MIN	TYP‡	MAX	
V_{IK}		$V_{CC} = 4.5$ V,	$I_I = 18$ mA			-1.2			-1.2	V
V_{OH}		$V_{CC} = 4.5$ V to 5.5 V,	$I_{OH} = -2$ mA	$V_{CC}-2$			$V_{CC}-2$			V
V_{OL}		$V_{CC} = 4.5$ V,	$I_{OL} = 20$ mA		0.34	0.5		0.34	0.5	V
I_I		$V_{CC} = 5.5$ V,	$V_I = 7$ V			0.1			0.1	mA
I_{IH}	ENT	$V_{CC} = 5.5$ V,	$V_I = 2.7$ V			40			40	μA
	Other inputs					20			20	
I_{IL}	ENT	$V_{CC} = 5.5$ V,	$V_I = 0.4$ V			-4			-4	mA
	Other inputs					-2			-2	
I_O §		$V_{CC} = 5.5$ V,	$V_O = 2.25$ V	-30		-112	-30		-112	mA
I_{CC}		$V_{CC} = 5.5$ V			134	195		134	195	mA

‡All typical values are at $V_{CC} = 5$ V, $T_A = 25$°C.
§The output conditions have been chosen to produce a current that closely approximates one half of the true short-circuit output current, I_{OS}.

TEXAS INSTRUMENTS
POST OFFICE BOX 225012 • DALLAS, TEXAS 75265

TYPES SN54AS869, SN74AS869
SYNCHRONOUS 8-BIT UP/DOWN COUNTERS WITH SYNCHRONOUS CLEAR

recommended operating conditions

			SN54AS869			SN74AS869			UNIT
			MIN	NOM	MAX	MIN	NOM	MAX	
V_{CC}	Supply voltage		4.5	5	5.5	4.5	5	5.5	V
V_{IH}	High-level input voltage		2			2			V
V_{IL}	Low-level input voltage				0.8			0.8	V
I_{OH}	High-level output current				−2			−2	mA
I_{OL}	Low-level output current				20			20	mA
f_{clock}	Clock frequency		0		40	0		45	MHz
$t_{w(clock)}$	Duration		12.5			11			ns
t_{su}	Setup time†	Data inputs A-H	6			5			ns
		Enable P (ENP) or Enable T (ENT)	10			9			ns
		S0 or S1 (load)	13			11			ns
		S0 or S1 (clear)	13			11			ns
		S0 or S1 (count down)	52			50			ns
		S0 or S1 (count up)	52			50			ns
t_h	Hold time at any input with respect to clock↑		0			0			ns
T_A	Operating free-air temperature		−55		125	0		70	°C

†This setup time is required to ensure stable data.

electrical characteristics over recommended operating free-air temperature range (unless otherwise noted)

PARAMETER		TEST CONDITIONS		SN54AS869			SN74AS869			UNIT
				MIN	TYP‡	MAX	MIN	TYP‡	MAX	
V_{IK}		V_{CC} = 4.5 V,	I_I = 18 mA			−1.2			−1.2	V
V_{OH}		V_{CC} = 4.5 V to 5.5 V,	I_{OH} = −2 mA	V_{CC}−2			V_{CC}−2			V
V_{OL}		V_{CC} = 4.5 V,	I_{OL} = 20 mA		0.34	0.5		0.34	0.5	V
I_I		V_{CC} = 5.5 V,	V_I = 7 V			0.1			0.1	mA
I_{IH}	ENT	V_{CC} = 5.5 V,	V_I = 2.7 V			40			40	µA
	Other inputs					20			20	
I_{IL}	ENT	V_{CC} = 5.5 V,	V_I = 0.4 V			−4			−4	mA
	Other inputs					−2			−2	
I_O§		V_{CC} = 5.5 V,	V_O = 2.25 V	−30		−112	−30		−112	mA
I_{CC}		V_{CC} = 5.5 V			125	180		125	180	mA

‡All typical values are at V_{CC} = 5 V, T_A = 25°C.
§The output conditions have been chosen to produce a current that closely approximates one half of the true short-circuit output current, I_{OS}.

ALS AND AS CIRCUITS

TEXAS INSTRUMENTS
POST OFFICE BOX 225012 • DALLAS, TEXAS 75265

TYPES SN54AS867, SN54AS869, SN74AS867, SN74AS869
SYNCHRONOUS 8-BIT UP/DOWN COUNTERS

'AS867 switching characteristics (see Note 1)

PARAMETER	FROM (INPUT)	TO (OUTPUT)	V_{CC} = 4.5 V to 5.5 V, C_L = 50 pF, R_L = 500 Ω, T_A = MIN to MAX				UNIT
			SN54AS867		SN74AS867		
			MIN	MAX	MIN	MAX	
f_{max}			40		50		MHz
t_{PLH}	CLK	$\overline{RCO}$	5	31	5	22	ns
t_{PHL}			6	19	6	16	
t_{PLH}	CLK	Any Q	3	12	3	11	ns
t_{PHL}			4	16	4	15	
t_{PLH}	$\overline{ENT}$	$\overline{RCO}$	3	19	3	10	ns
t_{PHL}			5	21	5	17	
t_{PLH}	$\overline{ENP}$	$\overline{RCO}$	5	14	5	14	ns
t_{PHL}			5	21	5	17	
t_{PHL}	Clear (S0, S1 low)	Any Q	7	23	7	21	ns

'AS869 switching characteristics (see Note 1)

PARAMETER	FROM (INPUT)	TO (OUTPUT)	V_{CC} = 4.5 V to 5.5 V, C_L = 50 pF, R_L = 500 Ω, T_A = MIN to MAX				UNIT
			SN54AS869		SN74AS869		
			MIN	MAX	MIN	MAX	
f_{max}			40		45		MHz
t_{PLH}	CLK	$\overline{RCO}$	6	35	6	35	ns
t_{PHL}			6	20	6	18	
t_{PLH}	CLK	Any Q	3	12	3	11	ns
t_{PHL}			4	16	4	15	
t_{PLH}	$\overline{ENT}$	$\overline{RCO}$	3	25	3	15	ns
t_{PHL}			6	21	6	17	
t_{PLH}	$\overline{ENP}$	$\overline{RCO}$	5	27	5	19	ns
t_{PHL}			6	21	6	18	

NOTE 1: For load circuit and voltage waveforms, see page 1-12.

TEXAS INSTRUMENTS
POST OFFICE BOX 225012 • DALLAS, TEXAS 75265

TYPES SN54AS870, SN54AS871, SN74AS870, SN74AS871
DUAL 16-BY-4 REGISTER FILES

D2661, DECEMBER 1982–REVISED DECEMBER 1983

- 'AS870 in Compact 24-Pin, 300-mil DIP and Both Plastic and Ceramic 28-Pin Chip Carriers
- 'AS871 in 28-Pin 600-mil DIP and Both Plastic and Ceramic Chip Carriers
- 3-State Buffer-Type Outputs Drive Bus Lines Directly
- Typical Access Time Is 11 ns
- Each Register File Has Individual Write Enable Controls and Address Lines
- Designed Specifically for Multibus Architecture and Overlapping File Operations
- Prioritized B Input Port Prevents Write Conflicts During Dual Input Mode
- Dependable Texas Instruments Quality and Reliability

SN54AS870 . . . . JT PACKAGE
SN74AS870 . . . . NT PACKAGE
(TOP VIEW)

```
S0   [1   24] VCC
1A0  [2   23] S1
1A1  [3   22] 2A3
1A2  [4   21] 2A2
1A3  [5   20] 2A1
1W   [6   19] 2A0
S2   [7   18] 2W
DQA1 [8   17] S3
DQA2 [9   16] DQB4
DQA3 [10  15] DQB3
DQA4 [11  14] DQB2
GND  [12  13] DQB1
```

SN54AS871 . . . . JD PACKAGE
SN74AS871 . . . . N PACKAGE
(TOP VIEW)

```
DA1  [1   28] VCC
DA2  [2   27] DA4
S0   [3   26] DA3
1A0  [4   25] S1
1A1  [5   24] 2A3
1A2  [6   23] 2A2
1A3  [7   22] 2A1
1W   [8   21] 2A0
S2   [9   20] 2W
QA1  [10  19] S3
QA2  [11  18] DQB4
QA3  [12  17] DQB3
QA4  [13  16] DQB2
GND  [14  15] DQB1
```

description

These devices feature two 16-word by 4-bit register files. Each register file has individual write-enable controls and address lines. The 'AS870 has two 4-bit data I/O ports (DQA1-DQA4 and DQB1-DQB4). The 'AS871 has one 4-bit data I/O port (DQB1-DQB4) with the other data port having individual data inputs (DA1-DA4) and data outputs (QA1-QA4). The data I/O ports can output to Bus A and Bus B, receive input from Bus A and Bus B, receive input from Bus A and output to Bus B, or output to Bus A and receive input from Bus B. To prevent writing conflicts in the dual-input mode, the B input port takes priority. Two select lines, S0 and S1, control which port has access to which register. S2 determines whether the A ports are in the input or the output modes and S3 does likewise for the B ports. The address lines (1A0-1A3 or 2A0-2A3) are decoded by an internal 1-of-16 decoder to select which register word is to be accessed. All outputs are 3-state buffer-type outputs designed specifically to drive bus lines directly.

The SN54AS870 and SN54AS871 are characterized for operation over the full military temperature range of −55°C to 125°C. The SN74AS870 and SN74AS871 are characterized for operation from 0°C to 70°C.

SN54AS870 . . . . FH PACKAGE
SN74AS870 . . . . FN PACKAGE
(TOP VIEW)

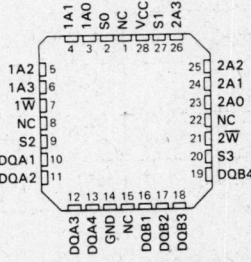

SN54AS871 . . . . FH PACKAGE
SN74AS871 . . . . FN PACKAGE
(TOP VIEW)

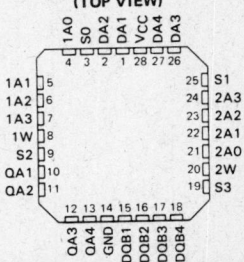

NC — No internal connection

Copyright © 1982 by Texas Instruments Incorporated

Texas Instruments
POST OFFICE BOX 225012 • DALLAS, TEXAS 75265

TYPES SN54AS870, SN54AS871, SN74AS870, SN74AS871
DUAL 16-BY-4 REGISTER FILES

logic symbols

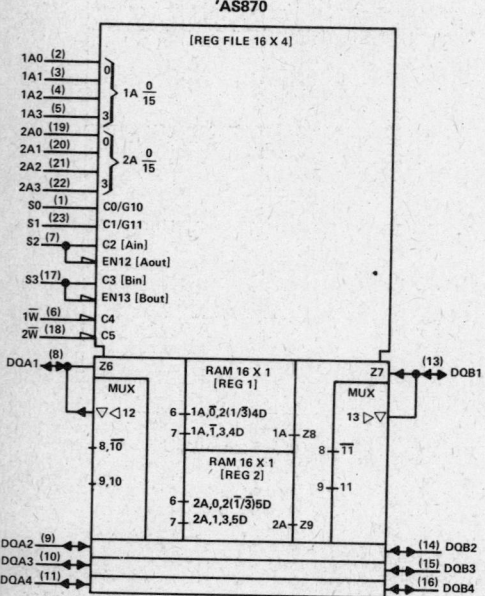
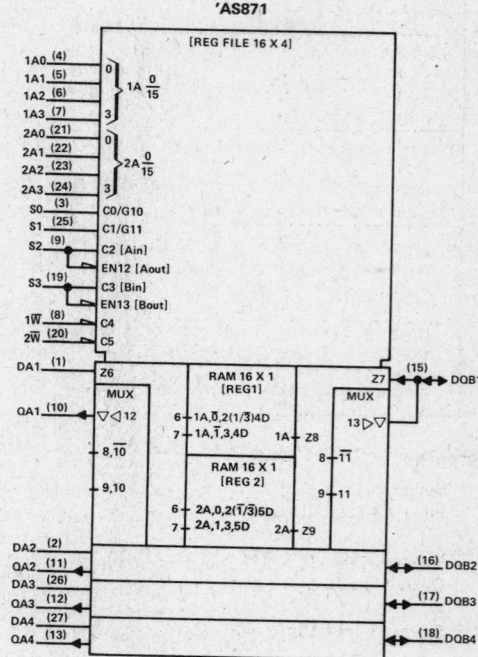

Pin numbers shown are for JD, JT, N, and NT packages.

TYPES SN54AS870, SN54AS871, SN74AS870, SN74AS871
DUAL 16-BY-4 REGISTER FILES

FUNCTION TABLE

FILE SELECT			INPUT/OUTPUT		
S0	S1	FILE SEL	S2	S3	I/O SEL
L	L	1R TO A, 1R TO B	L	L	A OUT, B OUT
H	L	2R TO A, 1R TO B			
L	H	1R TO A, 2R TO B			
H	H	2R TO A, 2R TO B			
L	L	A TO 1R, 1R TO B	H	L	A IN, B OUT
H	L	A TO 2R, 1R TO B			
L	H	A TO 1R, 2R TO B			
H	H	A TO 2R, 2R TO B			
L	L	1R TO A, B TO 1R	L	H	A OUT, B IN
H	L	2R TO A, B TO 1R			
L	H	1R TO A, B TO 2R			
H	H	2R TO A, B TO 2R			
L	L	B TO 1R	H	H	A IN, B IN
H	L	A TO 2R, B TO 1R			
L	H	A TO 1R, B TO 2R			
H	H	B TO 2R			

absolute maximum ratings over operating free-air temperature range (unless otherwise noted)

Supply voltage, V_{CC} .... 7 V
Input voltage: All inputs .... 7 V
      I/O ports .... 5.5 V
Voltage applied to a disabled 3-state output .... 5.5 V
Operating free-air temperature range: SN54AS870, SN54AS871 .... −55°C to 125°C
                SN74AS870, SN74AS871 .... 0°C to 70°C
Storage temperature range .... −65°C to 150°C

recommended operating conditions

			SN54AS870 SN54AS871			SN74AS870 SN74AS871			UNIT
			MIN	NOM	MAX	MIN	NOM	MAX	
V_{CC}	Supply voltage		4.5	5	5.5	4.5	5	5.5	V
V_{IH}	High-level input voltage		2			2			V
V_{IL}	Low-level input voltage				0.8			0.8	V
I_{OH}	High-level output current				−12			−15	mA
I_{OL}	Low-level output current				32			48	mA
t_w	Duration of write pulse		12			12			ns
t_{su}	Setup times	Address before write ↓	5			5			ns
		Data before write ↑	15			15			
		Select before write ↓	12			12			
t_h	Hold times	Address after write ↑	0			0			ns
		Data after write ↑	0			0			
		Select after write ↑	12			12			
T_A	Operating free-air temperature		−55		125	0		70	°C

TYPES SN54AS870, SN54AS871, SN74AS870, SN74AS871
DUAL 16-BY-4 REGISTER FILES

'AS870 electrical characteristics over recommended operating free-air temperature range (unless otherwise noted)

PARAMETER		TEST CONDITIONS		SN54AS870 MIN	SN54AS870 TYP†	SN54AS870 MAX	SN74AS870 MIN	SN74AS870 TYP†	SN74AS870 MAX	UNIT
V_{IK}		$V_{CC} = 4.5$ V,	$I_I = -18$ mA			-1.2			-1.2	V
V_{OH}		$V_{CC} = 4.5$ V to 5.5 V,	$I_{OH} = -2$ mA	$V_{CC}-2$			$V_{CC}-2$			V
		$V_{CC} = 4.5$ V,	$I_{OH} = -12$ mA	2.4	3.2					
		$V_{CC} = 4.5$ V,	$I_{OH} = -15$ mA				2.4	3.2		
V_{OL}		$V_{CC} = 4.5$ V,	$I_{OL} = 32$ mA		0.25	0.5				V
		$V_{CC} = 4.5$ V,	$I_{OL} = 48$ mA					0.35	0.5	
I_I	Control inputs	$V_{CC} = 5.5$ V,	$V_I = 7$ V			0.1			0.1	mA
	DQA and DQB ports	$V_{CC} = 5.5$ V,	$V_I = 5.5$ V			0.2			0.2	
I_{IH}	W1 and W2	$V_{CC} = 5.5$ V,	$V_I = 2.7$ V			20			20	μA
	Other control inputs					40			40	
	DQA and DQB ports‡					50			50	
I_{IL}	Control inputs	$V_{CC} = 5.5$ V,	$V_I = 0.4$ V			-2			-2	mA
	DQA and DQB ports‡					-2			-2	
I_O §		$V_{CC} = 5.5$ V,	$V_O = 2.25$ V	-30		-112	-30		-112	mA
I_{CC}		$V_{CC} = 5.5$ V			120	190		120	190	mA

'AS871 electrical characteristics over recommended operating free-air temperature range (unless otherwise noted)

PARAMETER		TEST CONDITIONS		SN54AS871 MIN	SN54AS871 TYP†	SN54AS871 MAX	SN74AS871 MIN	SN74AS871 TYP†	SN74AS871 MAX	UNIT
V_{IK}		$V_{CC} = 4.5$ V,	$I_I = -18$ mA			-1.2			-1.2	V
V_{OH}		$V_{CC} = 4.5$ V to 5.5 V,	$I_{OH} = -2$ mA	$V_{CC}-2$			$V_{CC}-2$			V
		$V_{CC} = 4.5$ V,	$I_{OH} = -12$ mA	2.4	3.2					
		$V_{CC} = 4.5$ V,	$I_{OH} = -15$ mA				2.4	3.2		
V_{OL}		$V_{CC} = 4.5$ V,	$I_{OL} = 32$ mA		0.25	0.5				V
		$V_{CC} = 4.5$ V,	$I_{OL} = 48$ mA					0.35	0.5	
I_{OZH}	QA outputs	$V_{CC} = 5.5$ V,	$V_O = 2.7$ V			50			50	μA
I_{OZL}	QA outputs	$V_{CC} = 5.5$ V,	$V_O = 0.4$ V			-50			-50	μA
I_I	Control and DA inputs	$V_{CC} = 5.5$ V,	$V_I = 7$ V			0.1			0.1	mA
	DQB ports	$V_{CC} = 5.5$ V,	$V_I = 5.5$ V			0.2			0.2	
I_{IH}	W1, W2 and DA inputs	$V_{CC} = 5.5$ V,	$V_I = 2.7$ V			20			20	μA
	Other control inputs					40			40	
	DQB ports‡					50			50	
I_{IL}	Control and DA inputs	$V_{CC} = 5.5$ V,	$V_I = 0.4$ V			-2			-2	mA
	DQB ports‡					-2			-2	
I_O §		$V_{CC} = 5.5$ V,	$V_O = 2.25$ V	-30		-112	-30		-112	mA
I_{CC}		$V_{CC} = 5.5$ V			120	190		120	190	mA

†All typical values are at $V_{CC} = 5$ V, $T_A = 25°C$.
‡For I/O ports, the parameters I_{IH} and I_{IL} include the output currents I_{OZH} and I_{OZL}, respectively.
§The output conditions have been chosen to produce a current that closely approximates one-half of the true short-circuit current, I_{OS}.

ALS AND AS CIRCUITS

TEXAS INSTRUMENTS
POST OFFICE BOX 225012 • DALLAS, TEXAS 75265

TYPES SN54AS870, SN54AS871, SN74AS870, SN74AS871
DUAL 16-BY-4 REGISTER FILES

'AS870 switching characteristics (see Note 1)

PARAMETER	FROM (INPUT)	TO (OUTPUT)	V_{CC} = 4.5 V to 5.5 V, C_L = 50 pF, R_L = 500 Ω, T_A = MIN to MAX				UNIT
			SN54AS870		SN74AS870		
			MIN	MAX	MIN	MAX	
$t_{a(A)}$	Any A	Any DQ	5	20	5	15	ns
$t_{a(S)}$	S0	Any DQA	3	15	3	13	ns
	S1	Any DQB	3	15	3	13	
t_{dis}	S2	Any DQA	3	12	3	11	ns
	S3	Any DQB	3	12	3	11	
t_{en}	S2	Any DQA	3	15	3	12	ns
	S3	Any DQB	3	15	3	12	
t_{pd}	$\overline{W}$	Any DQ	5	23	5	19	ns
	DQA	DQB	5	25	5	22	
	DQB	DQA	5	25	5	22	

'AS871 switching characteristics (see Note 1)

PARAMETER	FROM (INPUT)	TO (OUTPUT)	V_{CC} = 4.5 V to 5.5 V, C_L = 50 pF, R_L = 500 Ω, T_A = MIN to MAX				UNIT
			SN54AS871		SN74AS871		
			MIN	MAX	MIN	MAX	
$t_{a(A)}$	Any A	Any QA or DQB	5	20	5	16	ns
$t_{a(S)}$	S0	Any QA	3	15	3	13	ns
	S1	Any DQB	3	15	3	13	
t_{dis}	S2	Any QA	3	12	3	11	ns
	S3	Any DQB	3	12	3	11	
t_{en}	S2	Any QA	3	15	3	12	ns
	S3	Any DQB	3	15	3	12	
t_{pd}	$\overline{W}$	Any QA or DQB	5	23	5	19	ns
	DA	DQB	5	26	5	23	
	DQB	QA	5	26	5	23	

NOTE 1: For load circuit and voltage waveforms, see page 1-12.

Texas Instruments
POST OFFICE BOX 225012 • DALLAS, TEXAS 75265

2
ALS AND AS CIRCUITS

TYPES SN54ALS873, SN54AS873, SN74ALS873, SN74AS873
DUAL 4-BIT D-TYPE LATCHES WITH 3-STATE OUTPUTS

D2661, APRIL 1982—REVISED DECEMBER 1983

- 3-State Buffer-Type Outputs Drive Bus-Lines Directly
- Bus-Structured Pinout
- 'ALS880 and 'AS880 Are Alternative Versions with Inverting Outputs
- Package Options Include Both Plastic and Ceramic Chip Carriers in Addition to Plastic and Ceramic DIPs
- Dependable Texas Instruments Quality and Reliability

SN54ALS873, SN54AS873 . . . JT PACKAGE
SN74ALS873, SN74AS873 . . . NT PACKAGE
(TOP VIEW)

```
1CLR  [ 1    24 ] VCC
1OC   [ 2    23 ] 1C
1D1   [ 3    22 ] 1Q1
1D2   [ 4    21 ] 1Q2
1D3   [ 5    20 ] 1Q3
1D4   [ 6    19 ] 1Q4
2D1   [ 7    18 ] 2Q1
2D2   [ 8    17 ] 2Q2
2D3   [ 9    16 ] 2Q3
2D4   [ 10   15 ] 2Q4
2OC   [ 11   14 ] 2C
GND   [ 12   13 ] 2CLR
```

description

These dual 4-bit registers feature three-state outputs designed specifically for bus driving. This makes these devices particularly suitable for implementing buffer registers, I/O ports, bidirectional bus drivers, and working registers.

The dual 4-bit latches are transparent D-type. While the latch enable input (1C or 2C) is high, the Q outputs will follow the data (D) inputs in true form, according to the function table. When the latch enable input is taken low, the outputs will be latched. When $\overline{CLR}$ goes low, the Q outputs go low independently of enable C. The outputs are in a high-impedance state when $\overline{OC}$ (output control) is at a high logic level.

The SN54ALS873 and SN54AS873 are characterized for operation over the full military temperature range of −55 °C to 125 °C. The SN74ALS873 and SN74AS873 are characterized for operation from 0 °C to 70 °C.

SN54ALS873, SN54AS873 . . . FH PACKAGE
SN74ALS873, SN74AS873 . . . FN PACKAGE
(TOP VIEW)

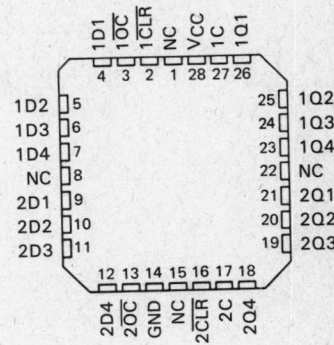

NC — No internal connection

FUNCTION TABLE (EACH LATCH)

INPUTS				OUTPUT
$\overline{OC}$	$\overline{CLR}$	ENABLE C	D	Q
L	L	X	X	L
L	H	H	H	H
L	H	H	L	L
L	H	L	X	Q_0
H	X	X	X	Z

Copyright © 1982 by Texas Instruments Incorporated

TEXAS INSTRUMENTS
POST OFFICE BOX 225012 • DALLAS, TEXAS 75265

TYPES SN54ALS873, SN54AS873, SN74ALS873, SN74AS873
DUAL 4-BIT D-TYPE LATCHES WITH 3-STATE OUTPUTS

logic symbol

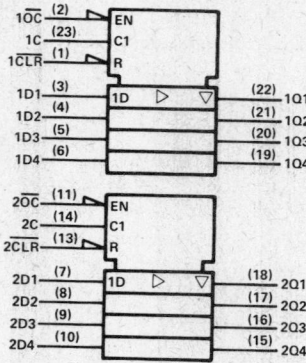

functional block diagram (each quad latch, positive logic)

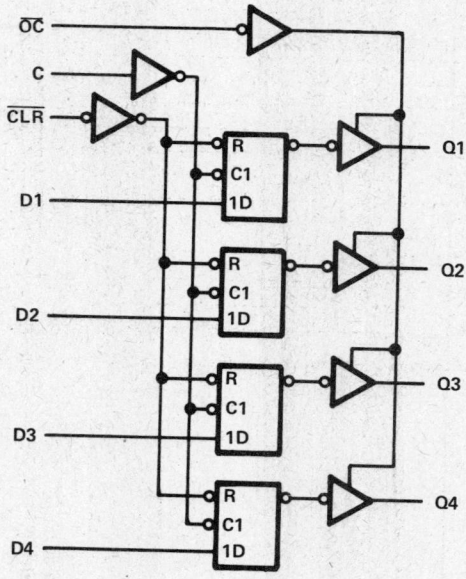

Pin numbers shown are for JT and NT packages.

absolute maximum ratings over operating free-air temperature range (unless otherwise noted)

Supply voltage, V_{CC} .... 7 V
Input voltage .... 7 V
Voltage applied to a disabled 3-state output .... 5.5 V
Operating free-air temperature range: SN54ALS873, SN54AS873 .... −55°C to 125°C
             SN74ALS873, SN74AS873 .... 0°C to 70°C
Storage temperature range .... −65°C to 150°C

TEXAS INSTRUMENTS
POST OFFICE BOX 225012 • DALLAS, TEXAS 75265

TYPES SN54ALS873, SN74ALS873
DUAL 4-BIT D-TYPE LATCHES WITH 3-STATE OUTPUTS

recommended operating conditions

			SN54ALS873			SN74ALS873			UNIT
			MIN	NOM	MAX	MIN	NOM	MAX	
V_{CC}	Supply voltage		4.5	5	5.5	4.5	5	5.5	V
V_{IH}	High-level input voltage		2			2			V
V_{IL}	Low-level input voltage				0.8			0.8	V
I_{OH}	High-level output current				−1			−2.6	mA
I_{OL}	Low-level output current				12			24	mA
t_w	Pulse duration	CLR low	15			15			ns
		Enable C high	10			10			
t_{su}	Setup time, data before enable C↓		10			10			ns
t_h	Hold time, data after enable C↓		7			7			ns
T_A	Operating free-air temperature		−55		125	0		70	°C

electrical characteristics over recommended operating free-air temperature range (unless otherwise noted)

PARAMETER	TEST CONDITIONS		SN54ALS873			SN74ALS873			UNIT
			MIN	TYP†	MAX	MIN	TYP†	MAX	
V_{IK}	V_{CC} = 4.5 V,	I_I = −18 mA			−1.5			−1.5	V
V_{OH}	V_{CC} = 4.5 V to 5.5 V,	I_{OH} = −0.4 mA	V_{CC}−2			V_{CC}−2			V
	V_{CC} = 4.5 V,	I_{OH} = −1 mA	2.4	3.3					
	V_{CC} = 4.5 V,	I_{OH} = −2.6 mA				2.4	3.2		
V_{OL}	V_{CC} = 4.5 V,	I_{OL} = 12 mA		0.25	0.4		0.25	0.4	V
	V_{CC} = 4.5 V	I_{OL} = 24 mA					0.35	0.5	
I_{OZH}	V_{CC} = 5.5 V,	V_O = 2.7 V			20			20	µA
I_{OZL}	V_{CC} = 5.5 V,	V_O = 0.4 V			−20			−20	µA
I_I	V_{CC} = 5.5 V,	V_I = 7 V			0.1			0.1	mA
I_{IH}	V_{CC} = 5.5 V,	V_I = 2.7 V			20			20	µA
I_{IL}	V_{CC} = 5.5 V,	V_I = 0.4 V			−0.1			−0.1	mA
I_O‡	V_{CC} = 5.5 V,	V_O = 2.25 V	−15		−70	−15		−70	mA
I_{CC}	V_{CC} = 5.5 V	Outputs high		10	21		10	21	mA
		Outputs low		15	29		15	29	
		Outputs disabled		16	31		16	31	

†All typical values are at V_{CC} = 5 V, T_A = 25°C.
‡The output conditions have been chosen to produce a current that closely approximates one half of the true short-circuit output current, I_{OS}.

Texas Instruments
POST OFFICE BOX 225012 • DALLAS, TEXAS 75265

TYPES SN54ALS873, SN74ALS873
DUAL 4-BIT D-TYPE LATCHES WITH 3-STATE OUTPUTS

switching characteristics (see Note 1)

PARAMETER	FROM (INPUT)	TO (OUTPUT)	V_{CC} = 4.5 V to 5.5 V, C_L = 50 pF, R1 = 500 Ω, R2 = 500 Ω, T_A = MIN to MAX				UNIT
			SN54ALS873		SN74ALS873		
			MIN	MAX	MIN	MAX	
t_{PLH}	D	Q	2	15	2	14	ns
t_{PHL}			2	15	2	14	
t_{PLH}	C	Q	8	29	8	22	ns
t_{PHL}			8	22	8	21	
t_{PHL}	$\overline{CLR}$	Q	6	24	6	24	ns
t_{PZH}	$\overline{OC}$	Q	4	21	4	18	ns
t_{PZL}			4	21	4	18	
t_{PHZ}	$\overline{OC}$	Q	2	10	2	8	ns
t_{PLZ}			2	15	2	13	

NOTE 1: For load circuit and voltage waveforms, see page 1-12.

D latch signal conventions

It is TI practice to name the outputs and other inputs of a D-type latch and to draw its logic symbol based on the assumption of true data (D) inputs. Then outputs that produce data in phase with the data inputs are called Q and those producing complementary data are called $\overline{Q}$. An input that causes a Q output to go high or a $\overline{Q}$ output to go low is called Preset; an input that causes a $\overline{Q}$ output to go high or a Q output to go low is called Clear. Bars are used over these pin names ($\overline{PRE}$ and $\overline{CLR}$) if they are active low.

In some applications it may be advantageous to redesignate the data input $\overline{D}$. In that case all the other inputs and outputs should be renamed as shown below. Also shown are corresponding changes in the graphical symbol. Arbitrary pin numbers are shown in parentheses.

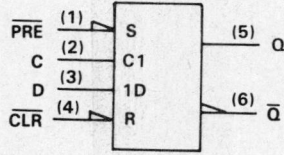 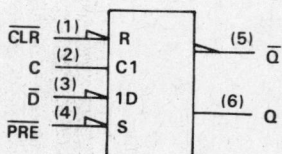

Notice that Q and $\overline{Q}$ exchange names, which causes Preset and Clear to do likewise. Also notice that the polarity indicators (◺) on $\overline{PRE}$ and $\overline{CLR}$ remain since these inputs are still active-low, but that the presence or absence of the polarity indicator changes at $\overline{D}$, Q, and $\overline{Q}$. Of course pin 5 ($\overline{Q}$) is still in phase with the data input $\overline{D}$, but now both are considered active-low.

TYPES SN54AS873, SN74AS873
DUAL 4-BIT D-TYPE LATCHES WITH 3-STATE OUTPUTS

recommended operating conditions

		SN54AS873			SN74AS873			UNIT
		MIN	NOM	MAX	MIN	NOM	MAX	
V_{CC}	Supply voltage	4.5	5	5.5	4.5	5	5.5	V
V_{IH}	High-level input voltage	2			2			V
V_{IL}	Low-level input voltage			0.8			0.8	V
I_{OH}	High-level output current			−12			−15	mA
I_{OL}	Low-level output current			32			48	mA
t_w	Pulse duration CLR low	4.5			3.5			ns
	Enable C high	5.5			4.5			
t_{su}	Setup time, data before enable C↓	2			2			ns
t_h	Hold time, data after enable C↓	3			3			ns
T_A	Operating free-air temperature	−55		125	0		70	°C

electrical characteristics over recommended operating free-air temperature range (unless otherwise noted)

PARAMETER	TEST CONDITIONS		SN54AS873			SN74AS873			UNIT
			MIN	TYP†	MAX	MIN	TYP†	MAX	
V_{IK}	V_{CC} = 4.5 V,	I_I = −18 mA			−1.2			−1.2	V
V_{OH}	V_{CC} = 4.5 V to 5.5 V,	I_{OH} = −2 mA	V_{CC}−2			V_{CC}−2			V
	V_{CC} = 4.5 V,	I_{OH} = −12 mA	2.4	3.2					
	V_{CC} = 4.5 V,	I_{OH} = −15 mA				2.4	3.3		
V_{OL}	V_{CC} = 4.5 V,	I_{OL} = 32 mA		0.25	0.5				V
	V_{CC} = 4.5 V,	I_{OL} = 48 mA					0.35	0.5	
I_{OZH}	V_{CC} = 5.5 V,	V_O = 2.7 V			50			50	µA
I_{OZL}	V_{CC} = 5.5 V,	V_O = 0.4 V			−50			−50	µA
I_I	V_{CC} = 5.5 V,	V_I = 7 V			0.1			0.1	mA
I_{IH}	V_{CC} = 5.5 V,	V_I = 2.7 V			20			20	µA
I_{IL}	V_{CC} = 5.5 V,	V_I = 0.4 V			−0.5			−0.5	mA
I_O‡	V_{CC} = 5.5 V,	V_O = 2.25 V	−30		−112	−30		−112	mA
I_{CC}	V_{CC} = 5.5 V	Outputs high		68	110		68	110	mA
		Outputs low		67	109		67	109	
		Outputs disabled		80	129		80	129	

†All typical values are at V_{CC} = 5 V, T_A = 25°C.
‡The output conditions have been chosen to produce a current that closely approximates one half of the true short-circuit output current, I_{OS}.

ALS AND AS CIRCUITS

TEXAS INSTRUMENTS
POST OFFICE BOX 225012 • DALLAS, TEXAS 75265

TYPES SN54AS873, SN74AS873
DUAL 4-BIT D-TYPE LATCHES WITH 3-STATE OUTPUTS

switching characteristics (see Note 1)

PARAMETER	FROM (INPUT)	TO (OUTPUT)	V_{CC} = 4.5 V to 5.5 V, C_L = 50 pF, R1 = 500 Ω, R2 = 500 Ω, T_A = MIN to MAX				UNIT
			SN54AS873		SN74AS873		
			MIN	MAX	MIN	MAX	
t_{PLH}	D	Q	3	9	3	6	ns
t_{PHL}			3	7	3	6	
t_{PLH}	C	Q	6	14	6	11.5	ns
t_{PHL}			4	9	4	7.5	
t_{PHL}	$\overline{CLR}$	Q	3	8.5	3	7.5	ns
t_{PZH}	$\overline{OC}$	Q	2	8	2	6.5	ns
t_{PZL}			4	11	4	9.5	
t_{PHZ}	$\overline{OC}$	Q	2	8	2	6.5	ns
t_{PLZ}			2	8.5	2	7.5	

NOTE 1: For load circuit and voltage waveforms, see page 1-12.

TYPES SN54ALS874, SN54ALS876, SN54AS874, SN54AS876, SN74ALS874, SN74ALS876, SN74AS874, SN74AS876
DUAL 4-BIT D-TYPE EDGE-TRIGGERED FLIP-FLOPS

D2661, APRIL 1982—REVISED DECEMBER 1983

- 3-State Buffer-Type Outputs Drive Bus-Lines Directly
- Bus-Structured Pinout
- Choice of True or Inverting Logic

  'ALS874, 'AS874   True Outputs
  'ALS876, 'AS876   Inverting Outputs

- Asynchronous Clear
- Package Options Include Both Plastic and Ceramic Chip Carriers in Addition to Plastic and Ceramic DIPs
- Dependable Texas Instruments Quality and Reliability

description

These dual four-bit registers feature three-state outputs designed specifically for bus driving. This makes these devices particularly suitable for implementing buffer registers, I/O ports, bidirectional bus drivers, and working registers.

The edge-triggered flip-flops enter data on the low-to-high transition of the clock. The 'ALS874 and 'AS874 have $\overline{CLR}$ inputs and noninverting Q outputs; the 'ALS876 and 'AS876 have $\overline{PRE}$ inputs and inverting $\overline{Q}$ outputs. In each case, taking this input low causes the four Q or $\overline{Q}$ outputs to go low independently of the clock.

The SN54ALS874, SN54AS874, SN54ALS876 and SN54AS876 are characterized for operation over the full military temperature range of −55°C to 125°C. The SN74ALS874, SN74AS874, SN74ALS876, and SN74AS876 are characterized for operation from 0°C to 70°C.

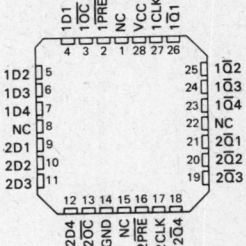

NC — No internal connection

Copyright © 1982 by Texas Instruments Incorporated

ALS AND AS CIRCUITS

TEXAS INSTRUMENTS
POST OFFICE BOX 225012 • DALLAS, TEXAS 75265

2-625

TYPES SN54ALS874, SN54ALS876, SN54AS874, SN54AS876, SN74ALS874, SN74ALS876, SN74AS874, SN74AS876
DUAL 4-BIT D-TYPE EDGE-TRIGGERED FLIP-FLOPS

FUNCTION TABLES

'ALS874, 'AS874 (EACH FLIP-FLOP)

INPUTS				OUTPUT
$\overline{OC}$	CLR	CLK	D	Q
L	L	X	X	L
L	H	↑	H	H
L	H	↑	L	L
L	H	L	X	Q_0
H	X	X	X	Z

'ALS876, 'AS876 (EACH FLIP-FLOP)

INPUTS				OUTPUT
$\overline{OC}$	$\overline{PRE}$	CLK	D	$\overline{Q}$
L	L	X	X	L
L	H	↑	H	L
L	H	↑	L	H
L	H	L	X	$\overline{Q}_0$
H	X	X	X	Z

logic symbols

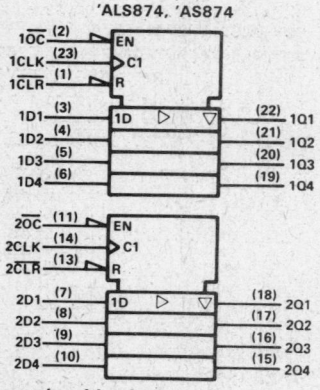

logic diagrams (positive logic)

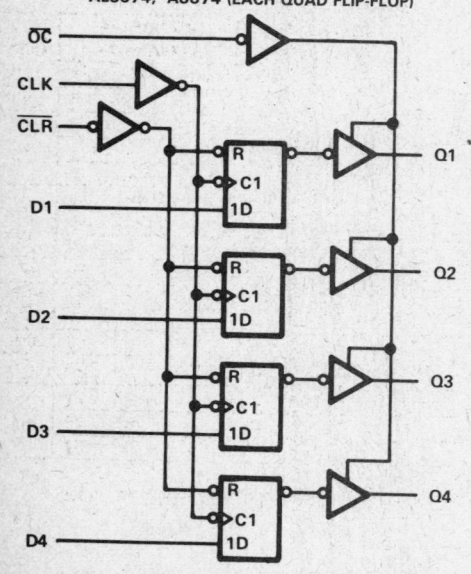

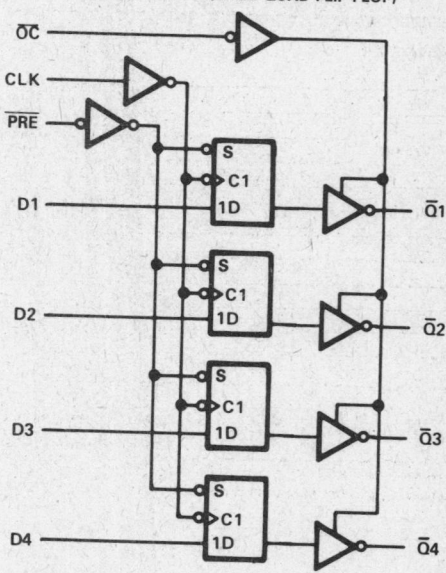

Pin numbers shown are for JT and NT packages.

TYPES SN54ALS874, SN54ALS876, SN74ALS874, SN74ALS876
DUAL 4-BIT D-TYPE EDGE-TRIGGERD FLIP-FLOPS

absolute maximum ratings over operating free-air temperature range (unless otherwise noted)

Supply voltage, V_{CC} .... 7 V
Input voltage .... 7 V
Voltage applied to a disabled 3-state output .... 5.5 V
Operating free-air temperature range: SN54ALS874, SN54ALS876 .... −55 °C to 125 °C
                                  SN74ALS874, SN74ALS876 .... 0 °C to 70 °C
Storage temperature range .... −65 °C to 150 °C

recommended operating conditions

		SN54ALS874 SN54ALS876			SN74ALS874 SN74ALS876			UNIT
		MIN	NOM	MAX	MIN	NOM	MAX	
V_{CC}	Supply voltage	4.5	5	5.5	4.5	5	5.5	V
V_{IH}	High-level input voltage	2			2			V
V_{IL}	Low-level input voltage			0.8			0.8	V
I_{OH}	High-level output current			−1			−2.6	mA
I_{OL}	Low-level output current			12			24	mA
f_{clock}	Clock frequency	0		25	0		30	MHz
t_w	Pulse duration — PRE or CLR low	10			10			ns
	CLK high	20			16.5			
	CLK low	20			16.5			
t_{su}	Setup time before CLK↑ — Data	15			15			ns
	PRE or CLR inactive	10			10			
t_h	Hold time, data after CLK↑	4			0			ns
T_A	Operating free-air temperature	−55		125	0		70	°C

electrical characteristics over recommended operating free-air temperature range (unless otherwise noted)

PARAMETER	TEST CONDITIONS		SN54ALS874 SN54ALS876			SN74ALS874 SN74ALS876			UNIT
			MIN	TYP†	MAX	MIN	TYP†	MAX	
V_{IK}	V_{CC} = 4.5 V,	I_I = −18 mA			−1.5			−1.5	V
V_{OH}	V_{CC} = 4.5 V to 5.5 V,	I_{OH} = −0.4 mA	V_{CC}−2			V_{CC}−2			V
	V_{CC} = 4.5 V,	I_{OH} = −1 mA	2.4	3.3					
	V_{CC} = 4.5 V,	I_{OH} = −2.6 mA				2.4	3.2		
V_{OL}	V_{CC} = 4.5 V,	I_{OL} = 12 mA		0.25	0.4		0.25	0.4	V
	V_{CC} = 4.5 V,	I_{OL} = 24 mA					0.35	0.5	
I_{OZH}	V_{CC} = 5.5 V,	V_O = 2.7 V			20			20	µA
I_{OZL}	V_{CC} = 5.5 V,	V_O = 0.4 V			−20			−20	µA
I_I	V_{CC} = 5.5 V,	V_I = 7 V			0.1			0.1	mA
I_{IH}	V_{CC} = 5.5 V,	V_I = 2.7 V			20			20	µA
I_{IL}	V_{CC} = 5.5 V,	V_I = 0.4 V			−0.2			−0.2	mA
I_O‡	V_{CC} = 5.5 V,	V_O = 2.25 V	−15		−70	−15		−70	mA
I_{CC}	V_{CC} = 5.5 V	Outputs high		14	21		14	21	mA
		Outputs low		18	29		18	29	
		Outputs disabled		20	31		20	31	

†All typical values are at V_{CC} = 5 V, T_A = 25 °C.
‡The output conditions have been chosen to produce a current that closely approximates one half of the true short-circuit output current, I_{OS}.

ALS AND AS CIRCUITS

TEXAS INSTRUMENTS
POST OFFICE BOX 225012 • DALLAS, TEXAS 75265

TYPES SN54ALS874, SN54ALS876, SN74ALS874, SN74ALS876
DUAL 4-BIT D-TYPE EDGE-TRIGGERED FLIP-FLOPS

'ALS874 switching characteristics (see Note 1)

PARAMETER	FROM (INPUT)	TO (OUTPUT)	V_{CC} = 4.5 V to 5.5 V, C_L = 50 pF, R1 = 500 Ω, R2 = 500 Ω, T_A = MIN to MAX				UNIT
			SN54ALS874		SN74ALS874		
			MIN	MAX	MIN	MAX	
f_{max}			25		30		MHz
t_{PLH}	CLK	Any Q	4	15	4	14	ns
t_{PHL}			4	15	4	14	
t_{PHL}	$\overline{CLR}$	Any Q	6	22	6	19	ns
t_{PZH}	$\overline{OC}$	Any Q	4	21	4	18	ns
t_{PZL}			4	21	4	18	
t_{PHZ}	$\overline{OC}$	Any Q	2	10	2	8	ns
t_{PLZ}			3	15	3	13	

'ALS876 switching characteristics (see Note 1)

PARAMETER	FROM (INPUT)	TO (OUTPUT)	V_{CC} = 4.5 V to 5.5 V, C_L = 50 pF, R1 = 500 Ω, R2 = 500 Ω, T_A = MIN to MAX				UNIT
			SN54ALS876		SN74ALS876		
			MIN	MAX	MIN	MAX	
f_{max}			25		30		MHz
t_{PLH}	CLK	Any $\overline{Q}$	4	15	4	14	ns
t_{PHL}			4	15	4	14	
t_{PHL}	$\overline{PRE}$	Any $\overline{Q}$	6	22	6	19	ns
t_{PZH}	$\overline{OC}$	Any $\overline{Q}$	4	21	4	18	ns
t_{PZL}			4	21	4	18	
t_{PHZ}	$\overline{OC}$	Any $\overline{Q}$	2	10	2	8	ns
t_{PLZ}			3	15	3	13	

NOTE 1: For load circuit and voltage waveforms, see page 1-12.

Texas Instruments
POST OFFICE BOX 225012 • DALLAS, TEXAS 75265

TYPES SN54AS874, SN54AS876, SN74AS874, SN74AS876
DUAL 4-BIT D-TYPE EDGE-TRIGGERED FLIP-FLOPS

absolute maximum ratings over operating free-air temperature range (unless otherwise noted)

Supply voltage, V_{CC} .... 7 V
Input voltage .... 7 V
Operating free-air temperature range: SN54AS874, SN54AS876 .... −55 °C to 125 °C
                                      SN74AS874, SN74AS876 .... 0 °C to 70 °C
Storage temperature range .... −65 °C to 150 °C

recommended operating conditions

		SN54AS874 SN54AS876			SN74AS874 SN74AS876			UNIT
		MIN	NOM	MAX	MIN	NOM	MAX	
V_{CC}	Supply voltage	4.5	5	5.5	4.5	5	5.5	V
V_{IH}	High-level input voltage	2			2			V
V_{IL}	Low-level input voltage			0.8			0.8	V
I_{OH}	High-level output current			−12			−15	mA
I_{OL}	Low-level output current			32			48	mA
f_{clock}	Clock frequency	0		100	0		125	MHz
t_w	Pulse duration — PRE or CLR low	3			2			ns
	Pulse duration — CLK high	4			3			ns
	Pulse duration — CLK low	5			4			ns
t_{su}	Setup time before CLK↑ — Data	2.5			2			ns
	Setup time before CLK↑ — PRE or CLR inactive	5			4			ns
t_h	Hold time, data after CLK↑	1			1			ns
T_A	Operating free-air temperature	−55		125	0		70	°C

electrical characteristics over recommended operating free-air temperature range (unless otherwise noted)

PARAMETER		TEST CONDITIONS		SN54AS874 SN54AS876			SN74AS874 SN74AS876			UNIT
				MIN	TYP†	MAX	MIN	TYP†	MAX	
V_{IK}		V_{CC} = 4.5 V,	I_I = −18 mA			−1.2			−1.2	V
V_{OH}		V_{CC} = 4.5 V to 5.5 V,	I_{OH} = −2 mA	$V_{CC}-2$			$V_{CC}-2$			V
		V_{CC} = 4.5 V,	I_{OH} = −12 mA	2.4	3.2					
		V_{CC} = 4.5 V,	I_{OH} = −15 mA				2.4	3.3		
V_{OL}		V_{CC} = 4.5 V,	I_{OL} = 32 mA		0.25	0.5				V
		V_{CC} = 4.5 V,	I_{OL} = 48 mA					0.35	0.5	
I_{OZH}		V_{CC} = 5.5 V,	V_O = 2.7 V			50			50	µA
I_{OZL}		V_{CC} = 5.5 V,	V_O = 0.4 V			−50			−50	µA
I_I		V_{CC} = 5.5 V,	V_I = 7 V			0.1			0.1	mA
I_{IH}		V_{CC} = 5.5 V,	V_I = 2.7 V			20			20	µA
I_{IL}	D	V_{CC} = 5.5 V,	V_I = 0.4 V			−3			−2	mA
	All other					−0.5			−0.5	
I_O‡		V_{CC} = 5.5 V,	V_O = 2.25 V	−30		−112	−30		−112	mA
I_{CC}	'AS874	V_{CC} = 5.5 V	Outputs high		82	133		82	133	mA
			Outputs low		92	149		92	149	
			Outputs disabled		100	160		100	160	
	'AS876		Outputs high		88	142		88	142	
			Outputs low		94	150		94	150	
			Outputs disabled		100	160		100	160	

†All typical values are at V_{CC} = 5 V, T_A = 25 °C.
‡The output conditions have been chosen to produce a current that closely approximates one half of the true short-circuit output current, I_{OS}.

TEXAS INSTRUMENTS
POST OFFICE BOX 225012 • DALLAS, TEXAS 75265

TYPES SN54AS874, SN54AS876, SN74AS874, SN74AS876
DUAL 4-BIT D-TYPE EDGE-TRIGGERED FLIP-FLOPS

'AS874 switching characteristics (see Note 1)

PARAMETER	FROM (INPUT)	TO (OUTPUT)	V_{CC} = 4.5 V to 5.5 V, C_L = 50 pF, R1 = 500 Ω, R2 = 500 Ω, T_A = MIN to MAX				UNIT
			SN54AS874		SN74AS874		
			MIN	MAX	MIN	MAX	
f_{max}			100		125		MHz
t_{PLH}	CLK	Any Q	3	11.5	3	8.5	ns
t_{PHL}			4	12.5	4	10.5	
t_{PHL}	$\overline{CLR}$	Any Q	4	11	4	9.5	ns
t_{PZH}	$\overline{OC}$	Any Q	2	8	2	7	ns
t_{PZL}			3	11.5	3	10.5	
t_{PHZ}	$\overline{OC}$	Any Q	2	7	2	6	ns
t_{PLZ}			2	8.5	2	7.5	

'AS876 switching characteristics (see Note 1)

PARAMETER	FROM (INPUT)	TO (OUTPUT)	V_{CC} = 4.5 V to 5.5 V, C_L = 50 pF, R1 = 500 Ω, R2 = 500 Ω, T_A = MIN to MAX				UNIT
			SN54AS876		SN74AS876		
			MIN	MAX	MIN	MAX	
f_{max}			100		125		MHz
t_{PLH}	CLK	Any $\overline{Q}$	3	11.5	3	8.5	ns
t_{PHL}			4	12.5	4	10.5	
t_{PHL}	$\overline{PRE}$	Any $\overline{Q}$	4	11	4	9.5	ns
t_{PZH}	$\overline{OC}$	Any $\overline{Q}$	2	8	2	7	ns
t_{PZL}			3	11.5	3	10.5	
t_{PHZ}	$\overline{OC}$	Any $\overline{Q}$	2	7	2	6	ns
t_{PLZ}			2	7	2	6	

NOTE 1: For load circuit and voltage waveforms, see page 1-12.

TYPES SN54AS877, SN74AS877
8-BIT UNIVERSAL TRANSCEIVER PORT CONTROLLERS

D2661, DECEMBER 1982—REVISED DECEMBER 1983

- Included among the Package Options Are Compact, 24-Pin, 300-mil-Wide Dips and Both 28-Pin Plastic and Ceramic Chip Carriers
- Buffered 3-State Outputs Drive Bus Lines Directly
- Cascadable to n-Bits
- Eight Selectable Transceiver/Port Functions:
  - A to B or B to A
  - Register to A or Register to B
  - Shifted to A or Shifted to B
  - Off-Line Shifts (A and B Ports in High-Impedance State)
  - Register Clear
- Particularly Suitable for Use in Signature-Analysis Circuitry
- Serial Register Provides:
  - Parallel Storage of Either A or B Input Data
  - Serial Transmission of Data from Either A or B Port
- Dependable Texas Instruments Quality and Reliability

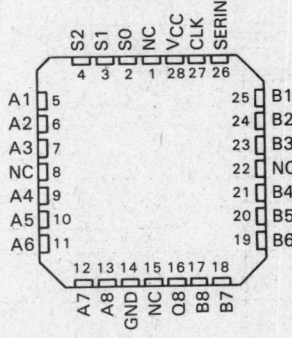

NC — No internal connection

description

The 'AS877 features two 8-bit I/O ports (A1-A8 and B1-B8), an 8-bit parallel-load, serial-in, parallel-out shift register, and control logic. With these features, this device is capable of performing eight selectable transceiver or port functions, depending on the state of the three select lines S0, S1, and S2. These functions include: transferring data from port A to port B or vice versa (i.e., the transceiver function), transferring data from the register to either port, serial shifting data to either port, performing off-line shifts (with A and B ports in high-impedance state), and clearing the register. Synchronous parallel loading of the internal register can be accomplished from either port on the positive transition of the clock while serially shifting data in via the SERIN input. The 'AS877 is ideally suited for applications needing signature-analysis circuitry to enhance system verification and/or fault analysis. All serial data is shifted right. All outputs are buffer-type outputs designed specifically to drive bus lines directly and all are 3-state except for Q8, which is a totem-pole output.

The SN54AS877 is characterized for operation over the full military temperature range of $-55°C$ to $125°C$. The SN74AS877 is characterized for operation from $0°C$ to $70°C$.

TYPES SN54AS877, SN74AS877
8-BIT UNIVERSAL TRANSCEIVER PORT CONTROLLERS

FUNCTION TABLE

MODE S2 S1 S0	CLOCK	SERIN	A1 Q1 B1	A2 Q2 B2	A3 Q3 B3	A4 Q4 B4	A5 Q5 B5	A6 Q6 B6	A7 Q7 B7	A8 Q8 B8	PORT FUNCTION
L L L	H or L	X	Z Q_n A1	Z Q_n A2	Z Q_n A2	Z Q_n A4	Z Q_n A5	Z Q_n A6	Z Q_n A7	Z Q_n A8	A TO B
L L L	↑	X	Z A1 A1	Z A2 A2	Z A3 A3	Z A4 A4	Z A5 A5	Z A6 A6	Z A7 A7	Z A8 A8	
L L H	H or L	X	B1 Q_n Z	B2 Q_n Z	B3 Q_n Z	B4 Q_n Z	B5 Q_n Z	B6 Q_n Z	B7 Q_n Z	B8 Q_n Z	B TO A
L L H	↑	X	B1 B1 Z	B2 B2 Z	B3 B3 Z	B4 B4 Z	B5 B5 Z	B6 B6 Z	B7 B7 Z	B8 B8 Z	
L H L	H or L	X	X Q_n Q1	X Q_n Q2	X Q_n Q3	X Q_n Q4	X Q_n Q5	X Q_n Q6	X Q_n Q7	X Q_n Q8	Q_N TO B_N
L H L	↑	X	Z A1 A1	Z A2 A2	Z A3 A3	Z A4 A4	Z A5 A5	Z A6 A6	Z A7 A7	Z A8 A8	
L H H	H or L	X	Q1 Q_n X	Q2 Q_n X	Q3 Q_n X	Q4 Q_n X	Q5 Q_n X	Q6 Q_n X	Q7 Q_n X	Q8 Q_n X	Q_N TO A_N
L H H	↑	X	B1 B1 Z	B2 B2 Z	B3 B3 Z	B4 B4 Z	B5 B5 Z	B6 B6 Z	B7 B7 Z	B8 B8 Z	
H L L	H or L	X	Z Q_n Q1	Z Q_n Q2	Z Q_n Q3	Z Q_n Q4	Z Q_n Q5	Z Q_n Q6	Z Q_n Q7	Z Q_n Q8	SHIFT TO B
H L L	↑	H	Z H H	Z Q1 Q1	Z Q2 Q2	Z Q3 Q3	Z Q4 Q4	Z Q5 Q5	Z Q6 Q6	Z Q7 Q7	
H L L	↑	L	Z L L	Z Q1 Q1	Z Q2 Q2	Z Q3 Q3	Z Q4 Q4	Z Q5 Q5	Z Q6 Q6	Z Q7 Q7	
H L H	H or L	X	Q1 Q_n Z	Q2 Q_n Z	Q3 Q_n Z	Q4 Q_n Z	Q5 Q_n Z	Q6 Q_n Z	Q7 Q_n Z	Q8 Q_n Z	SHIFT TO A
H L H	↑	H	H H Z	Q1 Q1 Z	Q2 Q2 Z	Q3 Q3 Z	Q4 Q4 Z	Q5 Q5 Z	Q6 Q6 Z	Q7 Q7 Z	
H L H	↑	L	L L Z	Q1 Q1 Z	Q2 Q2 Z	Q3 Q3 Z	Q4 Q4 Z	Q5 Q5 Z	Q6 Q6 Z	Q7 Q7 Z	
H H L	H or L	X	Z Q_n Z	Z Q_n Z	Z Q_n Z	Z Q_n Z	Z Q_n Z	Z Q_n Z	Z Q_n Z	Z Q_n Z	SHIFT
H H L	↑	H	Z H Z	Z Q1 Z	Z Q2 Z	Z Q3 Z	Z Q4 Z	Z Q5 Z	Z Q6 Z	Z Q7 Z	
H H L	↑	L	Z L Z	Z Q1 Z	Z Q2 Z	Z Q3 Z	Z Q4 Z	Z Q5 Z	Z Q6 Z	Z Q7 Z	
H H H	H or L	X	Z Q_n Z	Z Q_n Z	Z Q_n Z	Z Q_n Z	Z Q_n Z	Z Q_n Z	Z Q_n Z	Z Q_n Z	CLEAR
H H H	↑	X	Z L Z	Z L Z	Z L Z	Z L Z	Z L Z	Z L Z	Z L Z	Z L Z	

n = level of Q_n (n = 1, 2 ... 8) established on most recent ↑ transition of CLK. Q1 thru Q8 are the shift register outputs; only Q8 is available externally. The double inversions that take place as data travels from port to port are ignored in this table.

logic symbol

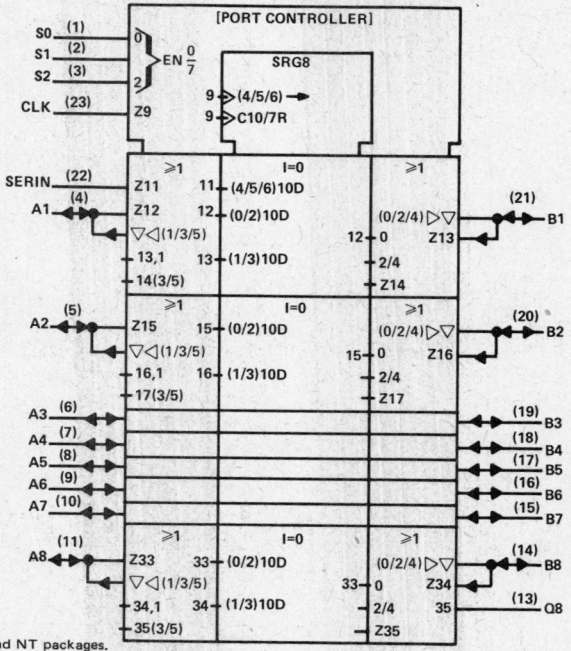

Pin numbers shown are for JT and NT packages.

TYPES SN54AS877, SN74AS877
8-BIT UNIVERSAL TRANSCEIVER PORT CONTROLLERS

logic diagram (positive logic)

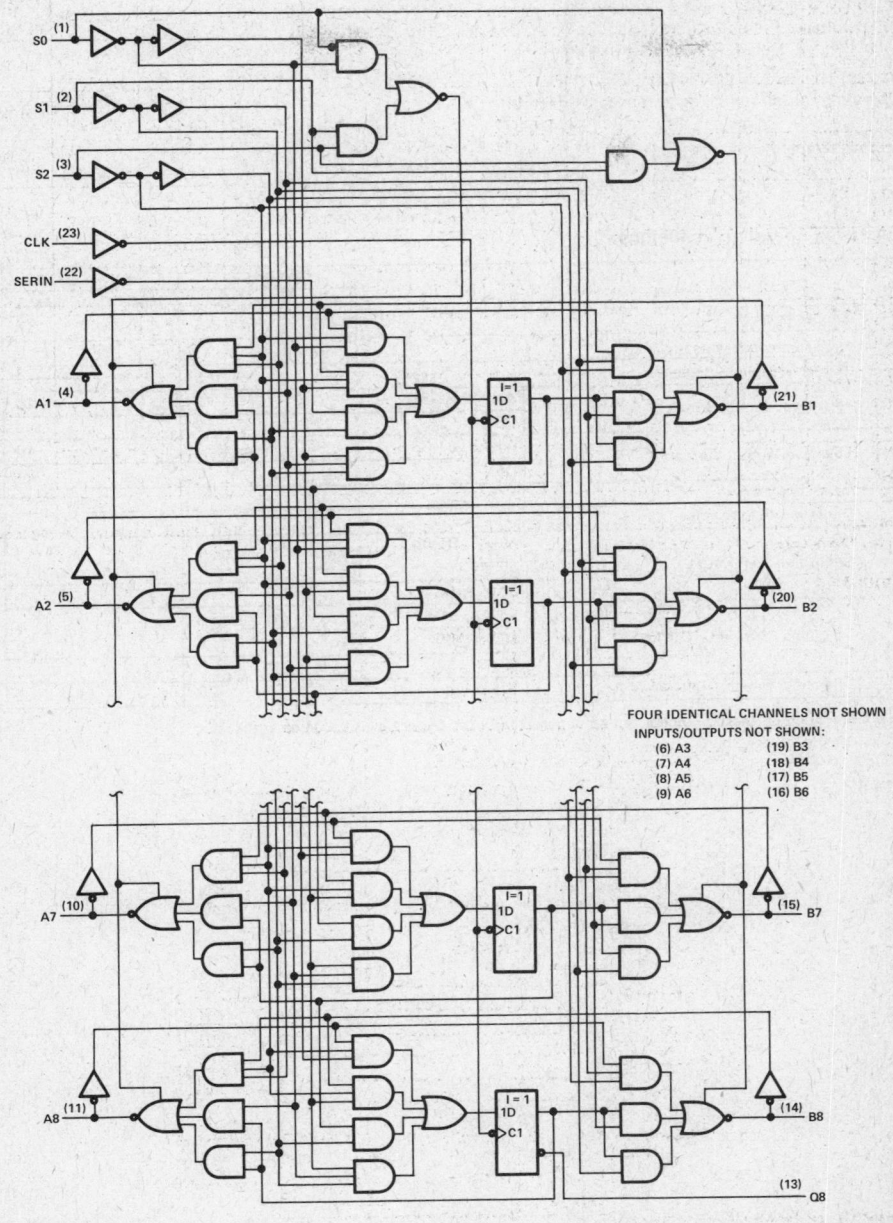

FOUR IDENTICAL CHANNELS NOT SHOWN
INPUTS/OUTPUTS NOT SHOWN:
(6) A3   (19) B3
(7) A4   (18) B4
(8) A5   (17) B5
(9) A6   (16) B6

TEXAS INSTRUMENTS
POST OFFICE BOX 225012 • DALLAS, TEXAS 75265

TYPES SN54AS877, SN74AS877
8-BIT UNIVERSAL TRANSCEIVER PORT CONTROLLERS

absolute maximum ratings over free-air temperature range

Supply voltage, V_{CC} .... 7 V
Input voltage: All inputs .... 7 V
                I/O ports .... 5.5 V
Voltage applied to a disabled 3-state output .... 5.5 V
Operating free-air temperature range: SN54AS877 .... −55°C to 125°C
                                          SN74AS877 .... 0°C to 70°C
Storage temperature range .... −65°C to 150°C

recommended operating conditions

			SN54AS877			SN74AS877			UNIT
			MIN	NOM	MAX	MIN	NOM	MAX	
V_{CC}	Supply voltage		4.5	5	5.5	4.5	5	5.5	V
V_{IH}	High-level input voltage		2			2			V
V_{IL}	Low-level input voltage				0.8			0.8	V
I_{OH}	High-level output current	A1-A8, B1-B8			−12			−15	mA
		Q8			−2			−2	
I_{OL}	Low-level output current	A1-A8, B1-B8			32			48	mA
		Q8			20			20	
f_{clock}	Clock frequency								MHz
t_w	Duration of clock pulse								ns
t_{su}	Setup time before CLK↑	A1-A8, B1-B8 SERIN							ns
		S0, S1, S2							
t_h	Hold time, data after CLK↑	A1-A8, B1-B8 SERIN	0			0			ns
		S0, S1, S2	0			0			
T_A	Operating free-air temperature		−55		125	0		70	°C

Additional information on these products can be obtained from the factory as it becomes available.

TYPES SN54AS877, SN74AS877
8-BIT UNIVERSAL TRANSCEIVER PORT CONTROLLERS

electrical characteristics over recommended operating free-air temperature range (unless otherwise noted)

PARAMETER		TEST CONDITIONS		SN54AS877			SN74AS877			UNIT
				MIN	TYP†	MAX	MIN	TYP†	MAX	
V_{IK}		$V_{CC} = 4.5$ V,	$I_I = -18$ mA			-1.2			-1.2	V
V_{OH}	A1-A8	$V_{CC} = 4.5$ V,	$I_{OH} = -12$ mA	2.4	3.2					V
	B1-B8	$V_{CC} = 4.5$ V,	$I_{OH} = -15$ mA				2.4	3.3		
	All outputs	$V_{CC} = 4.5$ V to 5.5 V,	$I_{OH} = -2$ mA	$V_{CC}-2$			$V_{CC}-2$			
V_{OL}	All outputs except Q8	$V_{CC} = 4.5$ V,	$I_{OL} = 32$ mA		0.25	0.5				V
		$V_{CC} = 4.5$ V,	$I_{OL} = 48$ mA					0.35	0.5	
	Q8	$V_{CC} = 4.5$ V,	$I_{OL} = 20$ mA		0.25	0.5		0.25	0.5	
I_I	S0, S1, S2	$V_{CC} = 5.5$ V,	$V_I = 7$ V			0.3			0.3	mA
	CLK and SERIN					0.1			0.1	
	A1-A8, B1-B8	$V_{CC} = 5.5$ V,	$V_I = 5.5$ V			0.2			0.2	
I_{IH}	S0, S1, S2	$V_{CC} = 5.5$ V,	$V_I = 2.7$ V			60			60	µA
	CLK and SERIN					20			20	
	A1-A8, B1-B8‡					70			70	
I_{IL}	S0, S1, S2	$V_{CC} = 5.5$ V,	$V_I = 0.4$ V			-2			-2	mA
	CLK and SERIN					-0.3			-0.3	
	A1-A8, B1-B8‡					-0.35			-0.35	
I_O§	Except Q8	$V_{CC} = 5.5$ V,	$V_O = 2.25$ V	-30		-112	-30		-112	mA
	Q8			-20		-112	-20		-112	
I_{CC}		$V_{CC} = 5.5$ V			136			136		mA

†All typical values are at $V_{CC} = 5$ V, $T_A = 25°C$.
‡For I/O ports, the parameters I_{IH} and I_{IL} include the output currents I_{OZH} and I_{OZL}, respectively.
§The output conditions have been chosen to produce a current that closely approximates one half of the true short-circuit output current, I_{OS}.

switching characteristics (see Note 1)

PARAMETER	FROM (INPUT)	TO (OUTPUT)	$V_{CC} = 4.5$ V to 5.5 V, $C_L = 50$ pF, R1 = 500 Ω, R2 = 500 Ω, T_A = MIN to MAX						UNIT
			SN54AS877			SN74AS877			
			MIN	TYP†	MAX	MIN	TYP†	MAX	
f_{max}				75			75		MHz
t_{PLH}	Any A port	Any B port		9.5			9.5		ns
t_{PHL}				8			8		
t_{PLH}	Any B port	Any A port		9.5			9.5		ns
t_{PHL}				8			8		
t_{PLH}	S0, S1, S2	Any A or B port		12			12		ns
t_{PHL}				12			12		
t_{PLH}	CLK	Any A or B port		6.5			6.5		ns
t_{PHL}				12.5			12.5		
t_{PLH}	CLK	Q8		9			9		ns
t_{PHL}				9			9		
t_{PHZ}	S0, S1, S2	Any A or B port		6			6		ns
t_{PLZ}				6			6		
t_{PZH}				10			10		ns
t_{PZL}				10			10		

†All typical values are at $V_{CC} = 5$ V, $T_A = 25°C$.
NOTE 1: For load circuit and voltage waveforms, see page 1-12.

ALS AND AS CIRCUITS

TEXAS INSTRUMENTS
POST OFFICE BOX 225012 • DALLAS, TEXAS 75265

TYPES SN54AS877, SN74AS877
8-BIT UNIVERSAL TRANSCEIVER PORT CONTROLLERS

TYPICAL APPLICATION DATA

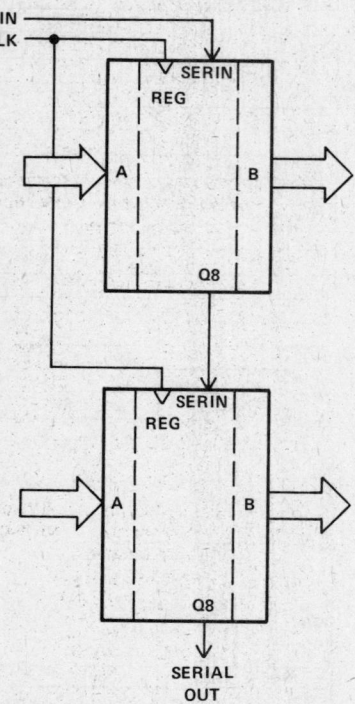

BUS A TO BUS B OR SERIAL TRANSMISSION

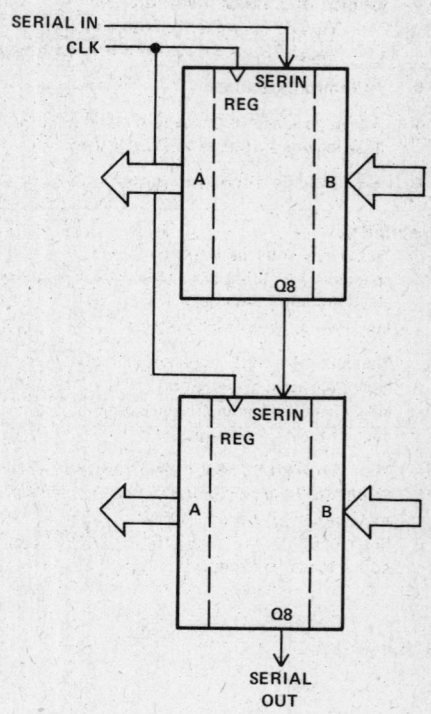

BUS B TO BUS A OR SERIAL TRANSMISSION

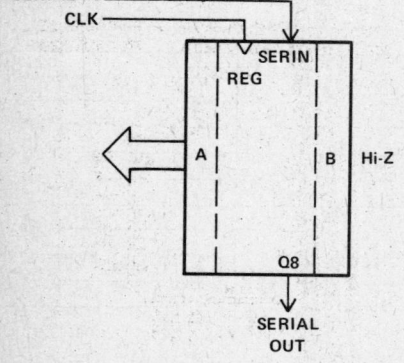

SERIAL IN TO A PORT

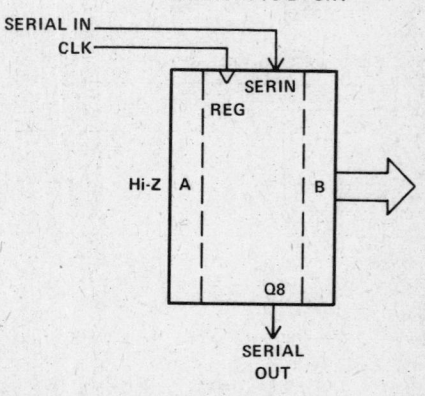

SERIAL IN TO B PORT

TYPES SN54ALS878, SN54ALS879, SN54AS878, SN54AS879, SN74ALS878, SN74ALS879, SN74AS878, SN74AS879
DUAL 4-BIT D-TYPE EDGE-TRIGGERED FLIP-FLOPS WITH 3-STATE OUTPUTS

D2661, APRIL 1982—REVISED DECEMBER 1983

- 3-State Bus-Driving Outputs
- Full Parallel-Access for Loading
- Buffered Control Inputs
- Choice of True or Inverting Logic
  'ALS878, 'AS878 True Outputs
  'ALS879, 'AS879 Inverting Outputs
- Synchronous Clear
- Package Options Include Both Plastic and Ceramic Chip Carriers in Addition to Plastic and Ceramic DIPs
- Dependable Texas Instruments Quality and Reliability

SN54ALS878, SN54AS878 . . . JT PACKAGE
SN74ALS878, SN74AS878 . . . NT PACKAGE
(TOP VIEW)

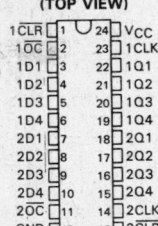

SN54ALS878, SN54AS878 . . . FH PACKAGE
SN74ALS878, SN74AS878 . . . FN PACKAGE
(TOP VIEW)

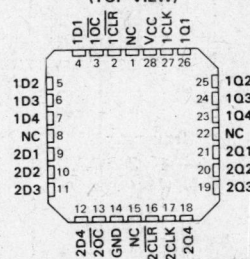

description

These dual 4-bit registers feature three-state outputs designed specifically for bus driving. This makes these devices particularly suitable for implementing buffer registers, I/O ports, bidirectional bus drivers, and working registers.

The dual 4-bit edge-triggered flip-flops enter data on the low-to-high transition of the clock (1CLK and 2CLK). All types have individual synchronous clear inputs and output control pins for each group of 4-bit registers.

The SN54ALS878, SN54ALS879, SN54AS878, and SN54AS879 are characterized for operation over the full military temperature range of −55 °C to 125 °C. The SN74ALS878, SN74ALS879, SN74AS878, and SN74AS879 are characterized for operation from 0 °C to 70 °C.

SN54ALS879, SN54AS879 . . . JT PACKAGE
SN74ALS879, SN74AS879 . . . NT PACKAGE
(TOP VIEW)

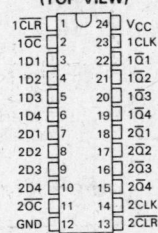

SN54ALS879, SN54AS879 . . . FH PACKAGE
SN74ALS879, SN74AS879 . . . FN PACKAGE
(TOP VIEW)

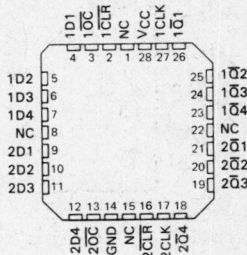

NC—No internal connection

Copyright © 1982 by Texas Instruments Incorporated

TEXAS INSTRUMENTS
POST OFFICE BOX 225012 • DALLAS, TEXAS 75265

TYPES SN54ALS878, SN54ALS879, SN54AS878, SN54AS879, SN74ALS878, SN74ALS879, SN74AS878, SN74AS879
DUAL 4-BIT D-TYPE EDGE-TRIGGERED FLIP-FLOPS WITH 3-STATE OUTPUTS

FUNCTION TABLES

'ALS878, 'AS878 (EACH FLIP-FLOP)

INPUTS				OUTPUT
$\overline{OC}$	$\overline{CLR}$	CLK	D	Q
L	L	↑	X	L
L	H	↑	H	H
L	H	↑	L	L
L	H	L	X	Q_0
H	X	X	X	Z

'ALS879, 'AS879 (EACH FLIP-FLOP)

INPUTS				OUTPUT
$\overline{OC}$	$\overline{CLR}$	CLK	D	$\overline{Q}$
L	L	↑	X	H
L	H	↑	H	L
L	H	↑	L	H
L	H	L	X	Q_0
H	X	X	X	Z

logic symbols

'ALS878, 'AS878

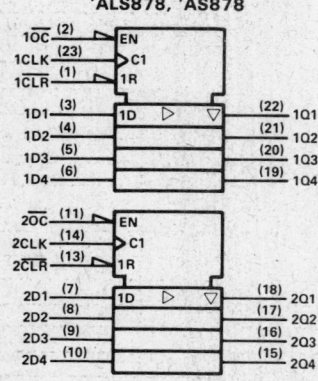

'ALS879, 'AS879

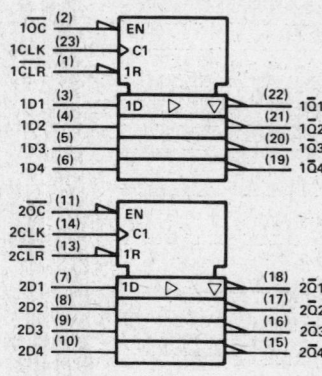

logic diagrams (positive logic)

'ALS878, 'AS878 (EACH QUAD FLIP-FLOP)

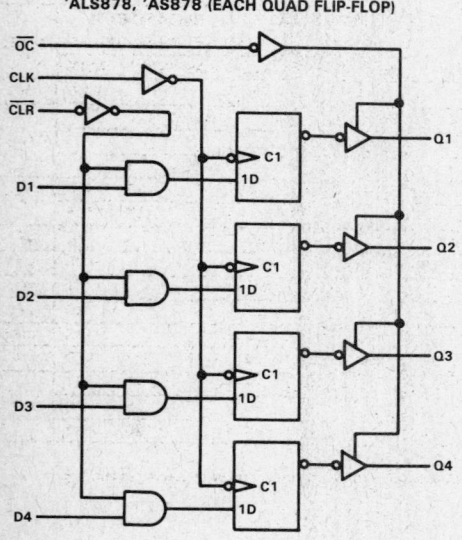

'ALS879, 'AS879 (EACH QUAD FLIP-FLOP)

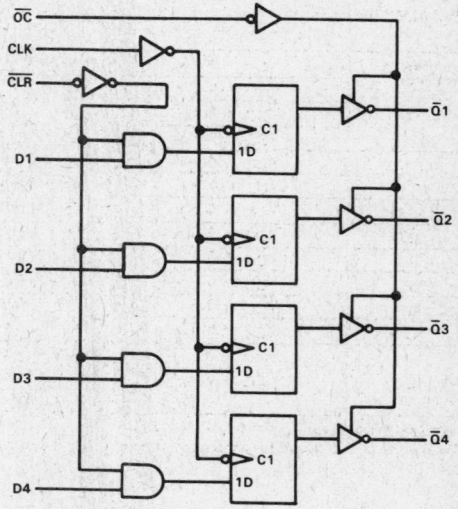

Pin numbers shown are for JT and NT packages.

TEXAS INSTRUMENTS
POST OFFICE BOX 225012 • DALLAS, TEXAS 75265

TYPES SN54ALS878, SN54ALS879, SN74ALS878, SN74ALS879
DUAL 4-BIT D-TYPE EDGE-TRIGGERED FLIP-FLOPS WITH 3-STATE OUTPUTS

absolute maximum ratings over operating free-air temperature range (unless otherwise noted)

Supply voltage, V_{CC} .... 7 V
Input voltage .... 7 V
Voltage applied to a disabled 3-state output .... 5.5 V
Operating free-air temperature range: SN54ALS878, SN54ALS879 .... −55°C to 125°C
　　　　　　　　　　　　　　　　　　　SN74ALS878, SN74ALS879 .... 0°C to 70°C
Storage temperature range .... −65°C to 150°C

recommended operating conditions

			SN54ALS878 SN54ALS879			SN74ALS878 SN74ALS879			UNIT
			MIN	NOM	MAX	MIN	NOM	MAX	
V_{CC}	Supply voltage		4.5	5	5.5	4.5	5	5.5	V
V_{IH}	High-level input voltage		2			2			V
V_{IL}	Low-level input voltage				0.8			0.8	V
I_{OH}	High-level output current				−1			−2.6	mA
I_{OL}	Low-level output current				12			24	mA
f_{clock}	Clock frequency	'ALS878	0		25	0		30	MHz
		'ALS879	0		20	0		25	
t_w	Pulse duration	'ALS878 CLK high or low	20			16.5			ns
		'ALS879 CLK high or low	25			20			
t_{su}	Setup time before CLK↑	Data	15			15			ns
		$\overline{CLR}$	20			20			
t_h	Hold time after CLK↑	Data	4			4			ns
		$\overline{CLR}$	0			0			
T_A	Operating free-air temperature		−55		125	0		70	°C

electrical characteristics over recommended operating free-air temperature range (unless otherwise noted)

PARAMETER	TEST CONDITIONS		SN54ALS878 SN54ALS879			SN74ALS878 SN74ALS879			UNIT
			MIN	TYP[†]	MAX	MIN	TYP[†]	MAX	
V_{IK}	V_{CC} = 4.5 V,	I_I = −18 mA			−1.5			−1.5	V
V_{OH}	V_{CC} = 4.5 V to 5.5 V,	I_{OH} = −0.4 mA	$V_{CC}-2$			$V_{CC}-2$			V
	V_{CC} = 4.5 V,	I_{OH} = −1 mA	2.4	3.3					
	V_{CC} = 4.5 V,	I_{OH} = −2.6 mA				2.4	3.2		
V_{OL}	V_{CC} = 4.5 V,	I_{OL} = 12 mA		0.25	0.4		0.25	0.4	V
	V_{CC} = 4.5 V,	I_{OL} = 24 mA					0.35	0.5	
I_{OZH}	V_{CC} = 5.5 V,	V_O = 2.7 V			20			20	µA
I_{OZL}	V_{CC} = 5.5 V,	V_O = 0.4 V			−20			−20	µA
I_I	V_{CC} = 5.5 V,	V_I = 7 V			0.1			0.1	mA
I_{IH}	V_{CC} = 5.5 V,	V_I = 2.7 V			20			20	µA
I_{IL}	V_{CC} = 5.5 V,	V_I = 0.4 V			−0.2			−0.2	mA
I_O[‡]	V_{CC} = 5.5 V,	V_O = 2.25 V	−15		−70	−15		−70	mA
I_{CC}	V_{CC} = 5.5 V	Outputs high		14	21		14	21	mA
		Outputs low		18	29		18	29	
		Outputs disabled		20	31		20	31	

[†] All typical values are at V_{CC} = 5 V, T_A = 25°C.
[‡] The output conditions have been chosen to produce a current that closely approximates one half of the true short-circuit output current, I_{OS}.

TEXAS INSTRUMENTS
POST OFFICE BOX 225012 • DALLAS, TEXAS 75265

ALS AND AS CIRCUITS

TYPES SN54ALS878, SN54ALS879, SN74ALS878, SN74ALS879
DUAL 4-BIT D-TYPE EDGE-TRIGGERED FLIP-FLOPS WITH 3-STATE OUTPUTS

switching characteristics (see Note 1)

PARAMETER	FROM (INPUT)	TO (OUTPUT)	V_{CC} = 4.5 V to 5.5 V, C_L = 50 pF, R1 = 500 Ω, R2 = 500 Ω, T_A = MIN to MAX				UNIT
			SN54ALS878 SN54ALS879		SN74ALS878 SN74ALS879		
			MIN	MAX	MIN	MAX	
f_{max}	'ALS878		25		30		MHz
	'ALS879		20		25		
t_{PLH}	CLK	Q ('ALS878) or $\overline{Q}$ ('ALS879)	4	15	4	14	ns
t_{PHL}			4	17	4	16	
t_{PZH}	$\overline{OC}$	Q ('ALS878) or $\overline{Q}$ ('ALS879)	4	22	4	20	ns
t_{PZL}			4	22	4	20	
t_{PHZ}	$\overline{OC}$	Q ('ALS878) or $\overline{Q}$ ('ALS879)	2	12	2	10	ns
t_{PLZ}			3	16	3	13	

NOTE 1: For load circuit and voltage waveforms, see page 1-12.

D flip-flop signal conventions

It is TI practice to name the outputs and other inputs of a D-type flip-flop and to draw its logic symbol based on the assumption of true data (D) inputs. Then outputs that produce data in phase with the data inputs are called Q and those producing complementary data are called $\overline{Q}$. An input that causes a Q output to go high or a $\overline{Q}$ output to go low is called Preset; an input that causes a $\overline{Q}$ output to go high or a Q output to go low is called Clear. Bars are used over these pin names ($\overline{PRE}$ and $\overline{CLR}$) if they are active low.

In some applications it may be advantageous to redesignate the data input $\overline{D}$. In that case all the other inputs and outputs should be renamed as shown below. Also shown are corresponding changes in the graphical symbol. Arbitrary pin numbers are shown in parentheses.

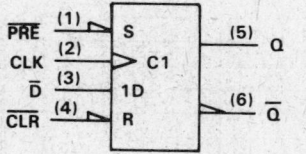 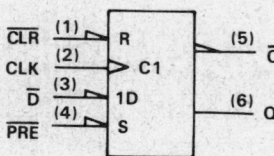

Notice that Q and $\overline{Q}$ exchange names, which causes Preset and Clear to do likewise. Also notice that the polarity indicators ($\triangleright$) on $\overline{PRE}$ and $\overline{CLR}$ remain since these inputs are still active-low, but that the presence or absence of the polarity indicator changes at $\overline{D}$, Q, and $\overline{Q}$. Of course pin 5 ($\overline{Q}$) is still in phase with the data input $\overline{D}$, but now both are considered active-low.

TEXAS INSTRUMENTS
POST OFFICE BOX 225012 • DALLAS, TEXAS 75265

TYPES SN54AS878, SN54AS879, SN74AS878, SN74AS879
DUAL 4-BIT D-TYPE EDGE-TRIGGERED FLIP-FLOPS WITH 3-STATE OUTPUTS

absolute maximum ratings over operating free-air temperature range (unless otherwise noted)

Supply voltage, V_{CC} .... 7 V
Input voltage .... 7 V
Voltage applied to a disabled 3-state output .... 5.5 V
Operating free-air temperature range: SN54AS878, SN54AS879 .... −55 °C to 125 °C
                                    SN74AS878, SN74AS879 .... 0 °C to 70 °C
Storage temperature range .... −65 °C to 150 °C

recommended operating conditions

			SN54AS878 SN54AS879			SN74AS878 SN74AS879			UNIT
			MIN	NOM	MAX	MIN	NOM	MAX	
V_{CC}	Supply voltage		4.5	5	5.5	4.5	5	5.5	V
V_{IH}	High-level input voltage		2			2			V
V_{IL}	Low-level input voltage				0.8			0.8	V
I_{OH}	High-level output current				−12			−15	mA
I_{OL}	Low-level output current				32			48	mA
f_{clock}	Clock frequency		0		100	0		125	MHz
t_w	Pulse duration	CLK low	3			2			ns
		CLK high	5			4			
t_{su}	Setup time before CLK↑	Data	3			2			ns
		$\overline{CLR}$	6.5			5.5			
t_h	Hold time after CLK↑	Data	3			2			ns
		$\overline{CLR}$	0			0			
T_A	Operating free-air temperature		−55		125	0		70	°C

electrical characteristics over recommended operating free-air temperature range (unless otherwise noted)

PARAMETER		TEST CONDITIONS		SN54AS878 SN54AS879			SN74AS878 SN74AS879			UNIT
				MIN	TYP†	MAX	MIN	TYP†	MAX	
V_{IK}		V_{CC} = 4.5 V,	I_I = −18 mA			−1.2			−1.2	V
V_{OH}		V_{CC} = 4.5 V to 5.5 V,	I_{OH} = −2 mA	V_{CC}−2			V_{CC}−2			V
		V_{CC} = 4.5 V,	I_{OH} = −12 mA	2.4	3.2					
		V_{CC} = 4.5 V,	I_{OH} = −15 mA				2.4	3.3		
V_{OL}		V_{CC} = 4.5 V,	I_{OL} = 32 mA		0.29	0.5				V
		V_{CC} = 4.5 V	I_{OL} = 48 mA					0.33	0.5	
I_{OZH}		V_{CC} = 5.5 V,	V_O = 2.7 V			50			50	µA
I_{OZL}		V_{CC} = 5.5 V,	V_O = 0.4 V			−50			−50	µA
I_I		V_{CC} = 5.5 V,	V_I = 7 V			0.1			0.1	mA
I_{IH}		V_{CC} = 5.5 V,	V_I = 2.7 V			20			20	µA
I_{IL}	D	V_{CC} = 5.5 V,	V_I = 0.4 V			−3			−2	mA
	All other					−0.5			−0.5	
I_O‡		V_{CC} = 5.5 V,	V_O = 2.25 V	−30		−112	−30		−112	mA
I_{CC}	'AS878	V_{CC} = 5.5 V, See Note 1	Outputs high		82	132		82	132	mA
			Outputs low		96	155		96	155	
			Outputs disabled		100	160		100	160	
	'AS879		Outputs high		88	142		88	142	
			Outputs low		94	150		94	150	
			Outputs disabled		100	160		100	160	

†All typical values are at V_{CC} = 5 V, T_A = 25 °C.
‡The output conditions have been chosen to produce a current that closely approximates one half of the true short-circuit output current, I_{OS}.
NOTE 1: I_{CC} is measured with CLR and all D inputs grounded, and CLK and OC at 4.5 V.

ALS AND AS CIRCUITS

TEXAS INSTRUMENTS
POST OFFICE BOX 225012 • DALLAS, TEXAS 75265

TYPES SN54AS878, SN54AS879, SN74AS878, SN74AS879
DUAL 4-BIT D-TYPE EDGE-TRIGGERED FLIP-FLOPS WITH 3-STATE OUTPUTS

switching characteristics (see Note 1)

PARAMETER	FROM (INPUT)	TO (OUTPUT)	V_{CC} = 4.5 V to 5.5 V, C_L = 50 pF, R1 = 500 Ω, R2 = 500 Ω, T_A = MIN to MAX				UNIT
			SN54AS878 SN54AS879		SN74AS878 SN74AS879		
			MIN	MAX	MIN	MAX	
f_{max}			100		125		MHz
t_{PLH}	CLK	Q ('AS878) or $\overline{Q}$ ('AS879)	3	11.5	3	8.5	ns
t_{PHL}			4	12.5	4	10.5	
t_{PZH}	$\overline{OC}$	Q ('AS878) or $\overline{Q}$ ('AS879)	2	8	2	7	ns
t_{PZL}			3	11.5	3	10.5	
t_{PHZ}	$\overline{OC}$	Q ('AS878) or $\overline{Q}$ ('AS879)	2	7	2	6	ns
t_{PLZ}			2	7	2	6	

NOTE 1: For load circuit and voltage waveforms, see page 1-12.

ALS AND AS CIRCUITS

TEXAS INSTRUMENTS
POST OFFICE BOX 225012 • DALLAS, TEXAS 75265

TYPES SN54ALS880, SN54AS880, SN74ALS880, SN74AS880
DUAL 4-BIT D-TYPE LATCHES WITH 3-STATE OUTPUTS

D2661, DECEMBER 1982—REVISED DECEMBER 1983

- 3-State Buffer-Type Outputs Drive Bus-Lines Directly
- Bus-Structured Pinout
- 'ALS873 Is Alternative Version with Noninverting Outputs
- Package Options Include Both Plastic and Ceramic Chip Carriers in Addition to Plastic and Ceramic DIPs
- Dependable Texas Instruments Quality and Reliability

SN54ALS880, SN54AS880 . . . JT PACKAGE
SN74ALS880, SN74AS880 . . . NT PACKAGE
(TOP VIEW)

description

These dual 4-bit registers feature three-state outputs designed specifically for bus driving. This makes these devices particularly suitable for implementing buffer registers, I/O ports, bidirectional bus drivers, and working registers.

The dual 4-bit latches are transparent D-type. While the latch enable input (1C or 2C) is high, the $\overline{Q}$ outputs will follow the data (D) inputs in inverted form, according to the function table. When the latch enable input is taken low, the outputs will be latched. When $\overline{PRE}$ goes low, the $\overline{Q}$ outputs go low independently of the clock. The outputs are in a high-impedance state when $\overline{OC}$ (output control) is at a high logic level.

The SN54ALS880 and SN54AS880 are characterized for operation over the full military temperature range of −55 °C to 125 °C. The SN74ALS880 and SN74AS880 are characterized for operation from 0 °C to 70 °C.

```
 1PRE [ 1    U  24] VCC
  1OC [ 2       23] 1C
  1D1 [ 3       22] 1Q1
  1D2 [ 4       21] 1Q2
  1D3 [ 5       20] 1Q3
  1D4 [ 6       19] 1Q4
  2D1 [ 7       18] 2Q1
  2D2 [ 8       17] 2Q2
  2D3 [ 9       16] 2Q3
  2D4 [10       15] 2Q4
  2OC [11       14] 2C
  GND [12       13] 2PRE
```

SN54ALS880, SN54AS880 . . . FH PACKAGE
SN74ALS880, SN74AS880 . . . FN PACKAGE
(TOP VIEW)

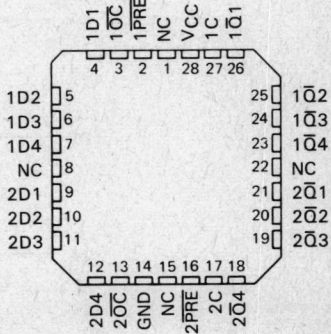

NC — No internal connection

FUNCTION TABLES (EACH LATCH)

INPUTS				OUTPUT
$\overline{OC}$	$\overline{PRE}$	ENABLE C	D	$\overline{Q}$
L	L	X	X	L
L	H	H	H	L
L	H	H	L	H
L	H	L	X	$\overline{Q}_0$
H	X	X	X	Z

ALS AND AS CIRCUITS

Copyright © 1982 by Texas Instruments Incorporated

Texas Instruments
POST OFFICE BOX 225012 • DALLAS, TEXAS 75265

2-643

TYPES SN54ALS880, SN54AS880, SN74ALS880, SN74AS880
DUAL 4-BIT D-TYPE LATCHES WITH 3-STATE OUTPUTS

logic symbol

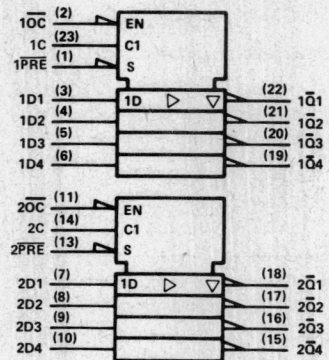

logic diagram (each quad latch, positive logic)

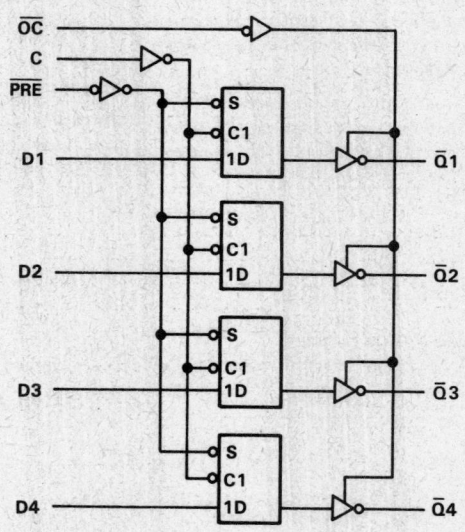

Pin numbers shown are for JT and NT packages.

absolute maximum ratings over operating free-air temperature range (unless otherwise noted)

Supply voltage, V_{CC} .... 7 V
Input voltage .... 7 V
Voltage applied to a disabled 3-state output .... 5.5 V
Operating free-air temperature range: SN54ALS880, SN54AS880 .... −55°C to 125°C
                                       SN74ALS880, SN74AS880 .... 0°C to 70°C
Storage temperature range .... −65°C to 150°C

TYPES SN54ALS880, SN74ALS880
DUAL 4-BIT D-TYPE LATCHES WITH 3-STATE OUTPUTS

recommended operating conditions

		SN54ALS880 MIN	SN54ALS880 NOM	SN54ALS880 MAX	SN74ALS880 MIN	SN74ALS880 NOM	SN74ALS880 MAX	UNIT
V_{CC}	Supply voltage	4.5	5	5.5	4.5	5	5.5	V
V_{IH}	High-level input voltage	2			2			V
V_{IL}	Low-level input voltage			0.8			0.8	V
I_{OH}	High-level output current			−1			−2.6	mA
I_{OL}	Low-level output current			12			24	mA
t_w	Pulse duration PRE low	15			15			ns
	Pulse duration Enable C high	15			15			ns
t_{su}	Setup time, data before enable C↓	10			10			ns
t_h	Hold time, data after enable C↓	10			10			ns
T_A	Operating free-air temperature	−55		125	0		70	°C

electrical characteristics over recommended operating free-air temperature range (unless otherwise noted)

PARAMETER	TEST CONDITIONS		SN54ALS880 MIN	SN54ALS880 TYP†	SN54ALS880 MAX	SN74ALS880 MIN	SN74ALS880 TYP†	SN74ALS880 MAX	UNIT
V_{IK}	$V_{CC} = 4.5$ V,	$I_I = -18$ mA			−1.5			−1.5	V
V_{OH}	$V_{CC} = 4.5$ V to 5.5 V,	$I_{OH} = -0.4$ mA	$V_{CC}-2$			$V_{CC}-2$			V
	$V_{CC} = 4.5$ V,	$I_{OH} = -1$ mA	2.4	3.3					
	$V_{CC} = 4.5$ V,	$I_{OH} = -2.6$ mA				2.4	3.2		
V_{OL}	$V_{CC} = 4.5$ V,	$I_{OL} = 12$ mA		0.25	0.4		0.25	0.4	V
	$V_{CC} = 4.5$ V,	$I_{OL} = 24$ mA					0.35	0.5	
I_{OZH}	$V_{CC} = 5.5$ V,	$V_O = 2.7$ V			20			20	µA
I_{OZL}	$V_{CC} = 5.5$ V,	$V_O = 0.4$ V			−20			−20	µA
I_I	$V_{CC} = 5.5$ V,	$V_I = 7$ V			0.1			0.1	mA
I_{IH}	$V_{CC} = 5.5$ V,	$V_I = 2.7$ V			20			20	µA
I_{IL}	$V_{CC} = 5.5$ V,	$V_I = 0.4$ V			−0.1			−0.1	mA
I_O‡	$V_{CC} = 5.5$ V,	$V_O = 2.25$ V	−15		−70	−15		−70	mA
I_{CC}	$V_{CC} = 5.5$ V	Outputs high		14	21		14	21	mA
		Outputs low		19	29		19	29	
		Outputs disabled		20	31		20	31	

†All typical values are at $V_{CC} = 5$ V, $T_A = 25°C$.
‡The output conditions have been chosen to produce a current that closely approximates one half of the true short-circuit output current, I_{OS}.

ALS AND AS CIRCUITS

Texas Instruments
POST OFFICE BOX 225012 • DALLAS, TEXAS 75265

TYPES SN54ALS880, SN74ALS880
DUAL 4-BIT D-TYPE LATCHES WITH 3-STATE OUTPUTS

switching characteristics (see Note 1)

PARAMETER	FROM (INPUT)	TO (OUTPUT)	V_{CC} = 4.5 V to 5.5 V, C_L = 50 pF, R_1 = 500 Ω, R_2 = 500 Ω, T_A = MIN to MAX				UNIT
			SN54ALS880		SN74ALS880		
			MIN	MAX	MIN	MAX	
t_{PLH}	D	$\bar{Q}$	3	23	3	20	ns
t_{PHL}			3	15	3	14	
t_{PLH}	C	$\bar{Q}$	8	31	8	24	ns
t_{PHL}			8	22	8	21	
t_{PHL}	$\overline{PRE}$	$\bar{Q}$	6	24	6	21	ns
t_{PZH}	$\overline{OC}$	$\bar{Q}$	4	21	5	18	ns
t_{PZL}			4	21	5	18	
t_{PHZ}	$\overline{OC}$	$\bar{Q}$	2	10	2	8	ns
t_{PLZ}			3	15	3	13	

NOTE 1: For load circuit and voltage waveforms, see page 1-12.

D latch signal conventions

It is TI practice to name the outputs and other inputs of a D-type latch and to draw its logic symbol based on the assumption of true data (D) inputs. Then outputs that produce data in phase with the data inputs are called Q and those producing complementary data are called $\bar{Q}$. An input that causes a Q output to go high or a $\bar{Q}$ output to go low is called Preset; an input that causes a $\bar{Q}$ output to go high or a Q output to go low is called Clear. Bars are used over these pin names ($\overline{PRE}$ and $\overline{CLR}$) if they are active low.

In some applications it may be advantageous to redesignate the data input $\bar{D}$. In that case all the other inputs and outputs should be renamed as shown below. Also shown are corresponding changes in the graphical symbol. Arbitrary pin numbers are shown in parentheses.

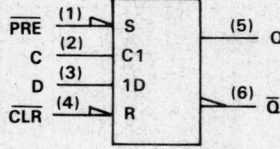 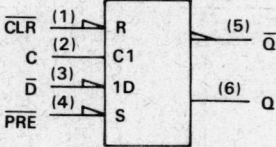

Notice that Q and $\bar{Q}$ exchange names, which causes Preset and Clear to do likewise. Also notice that the polarity indicators (◁) on $\overline{PRE}$ and $\overline{CLR}$ remain since these inputs are still active-low, but that the presence or absence of the polarity indicator changes at $\bar{D}$, Q, and $\bar{Q}$. Of course pin 5 ($\bar{Q}$) is still in phase with the data input $\bar{D}$, but now both are considered active-low.

TYPES SN54AS880, SN74AS880
DUAL 4-BIT D-TYPE LATCHES WITH 3-STATE OUTPUTS

recommended operating conditions

		SN54AS880			SN74AS880			UNIT
		MIN	NOM	MAX	MIN	NOM	MAX	
V_{CC}	Supply voltage	4.5	5	5.5	4.5	5	5.5	V
V_{IH}	High-level input voltage	2			2			V
V_{IL}	Low-level input voltage			0.8			0.8	V
I_{OH}	High-level output current			−12			−15	mA
I_{OL}	Low-level output current			32			48	mA
t_w	Pulse duration — PRE low	4.5			3.5			ns
	Pulse duration — Enable C high	3.5			2.5			
t_{su}	Setup time, data before enable C↓	2			2			ns
t_h	Hold time, data after enable C↓	1			1			ns
T_A	Operating free-air temperature	−55		125	0		70	°C

electrical characteristics over recommended operating free-air temperature range (unless otherwise noted)

PARAMETER	TEST CONDITIONS		SN54AS880			SN74AS880			UNIT
			MIN	TYP†	MAX	MIN	TYP†	MAX	
V_{IK}	V_{CC} = 4.5 V,	I_I = −18 mA			−1.2			−1.2	V
V_{OH}	V_{CC} = 4.5 V to 5.5 V,	I_{OH} = −2 mA	V_{CC}−2			V_{CC}−2			V
	V_{CC} = 4.5 V,	I_{OH} = −12 mA	2.4	3.2					
	V_{CC} = 4.5 V,	I_{OH} = −15 mA				2.4	3.3		
V_{OL}	V_{CC} = 4.5 V,	I_{OL} = 32 mA		0.30	0.5				V
	V_{CC} = 4.5 V,	I_{OL} = 48 mA					0.35	0.5	
I_{OZH}	V_{CC} = 5.5 V,	V_O = 2.7 V			50			50	µA
I_{OZL}	V_{CC} = 5.5 V,	V_O = 0.4 V			−50			−50	µA
I_I	V_{CC} = 5.5 V,	V_I = 7 V			0.1			0.1	mA
I_{IH}	V_{CC} = 5.5 V,	V_I = 2.7 V			20			20	µA
I_{IL}	V_{CC} = 5.5 V,	V_I = 0.4 V			−0.5			−0.5	mA
I_O‡	V_{CC} = 5.5 V,	V_O = 2.25 V	−30		−112	−30		−112	mA
I_{CC}	V_{CC} = 5.5 V	Outputs high		73	118		73	118	mA
		Outputs low		76	122		76	122	
		Outputs disabled		86	137		86	137	

†All typical values are at V_{CC} = 5 V, T_A = 25°C.
‡The output conditions have been chosen to produce a current that closely approximates one half of the true short-circuit output current, I_{OS}.

ALS AND AS CIRCUITS

Texas Instruments
POST OFFICE BOX 225012 • DALLAS, TEXAS 75265

TYPES SN54AS880, SN74AS880
DUAL 4-BIT D-TYPE LATCHES WITH 3-STATE OUTPUTS

'ALS465 switching characteristics (see Note 1)

PARAMETER	FROM (INPUT)	TO (OUTPUT)	V_{CC} = 4.5 V to 5.5 V, C_L = 50 pF, R1 = 500 Ω, R2 = 500 Ω, T_A = MIN to MAX				UNIT
			SN54AS880		SN74AS880		
			MIN	MAX	MIN	MAX	
t_{PLH}	D	$\overline{Q}$	4	11	4	9.5	ns
t_{PHL}			4	9	4	8.5	
t_{PLH}	C	$\overline{Q}$	6	14	6	11.5	ns
t_{PHL}			4	10	4	8	
t_{PHL}	$\overline{PRE}$	$\overline{Q}$	4	11.5	4	10	ns
t_{PZH}	$\overline{OC}$	$\overline{Q}$	2	8	2	7.5	ns
t_{PZL}			4	11	4	10	
t_{PHZ}	$\overline{OC}$	$\overline{Q}$	2	8	2	6.5	ns
t_{PLZ}			2	9	2	8	

NOTE 1: For load circuit and voltage waveforms, see page 1-12.

TYPES SN54AS881A, SN74AS881A
ARITHMETIC LOGIC UNITS/FUNCTION GENERATORS

D2661, DECEMBER 1982—REVISED DECEMBER 1983

- The 'AS881A is Offered in 300-mil DIPS and is Available in Both Plastic and Ceramic Chip Carriers
- Full Look-Ahead for High-Speed Operations on Long Words
- Arithmetic Operating Modes:
  Addition
  Subtraction
  Shift Operand A One Position
  Magnitude Comparison
  Plus Twelve Other Arithmetic Operations
- Logic Function Modes
  Exclusive-OR
  Comparator
  AND, NAND, OR, NOR
  'AS881A Provides Status Register Checks
  Plus Ten Other Logic Operations
- Dependable Texas Instruments Quality and Reliability

SN54AS181A . . . J OR JT PACKAGE
SN54AS881A . . . JT PACKAGE
SN74AS181A . . . N OR NT PACKAGE
SN74AS881A . . . NT PACKAGE
(TOP VIEW)

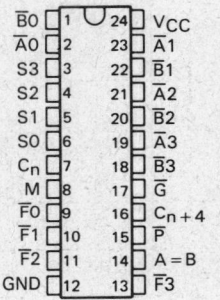

SN54AS181A, SN54AS881A . . . FH PACKAGE
SN74AS181A, SN74AS881A . . . FN PACKAGE
'AS181A, 'AS881A
(TOP VIEW)

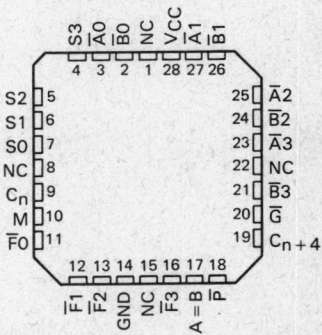

NC—No internal connection

logic symbol

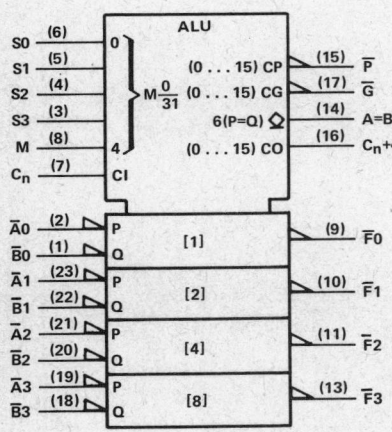

Pin numbers shown are J, JT, N and NT packages.

For complete information on the SN54AS881A and the SN74AS881A, see page   .

ADVANCE INFORMATION
This document contains information on a new product.
Specifications are subject to change without notice.

Copyright © 1982 by Texas Instruments Incorporated

TEXAS INSTRUMENTS
POST OFFICE BOX 225012 • DALLAS, TEXAS 75265

2
ALS AND AS CIRCUITS

TYPES SN54AS882, SN74AS882
32-BIT LOOK-AHEAD CARRY GENERATORS

D2661, DECEMBER 1982 – REVISED DECEMBER 1983

- Directly Compatible with the New 'AS181A and 'AS881A ALU's
- Included among the Package Options Are Compact, 24-Pin, 300-mil-Wide DIPs and Both 28-Pin Plastic and Ceramic Chip Carriers
- Capable of Anticipating the Carry Across a Group of Eight 4-Bit Binary Adders
- Cascadable to Perform Look-Ahead Across n-Bit Adders
- Typical Carry Time, C_n to Any C_{n+i}, Is Less Than 6 ns (C_L = 15 pF)
- Dependable Texas Instruments Quality and Reliability

SN54AS882 . . . JT PACKAGE
SN74AS882 . . . NT PACKAGE
(TOP VIEW)

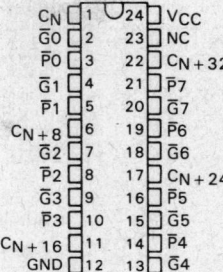

SN54AS882 . . . FH PACKAGE
SN74AS882 . . . FN PACKAGE
(TOP VIEW)

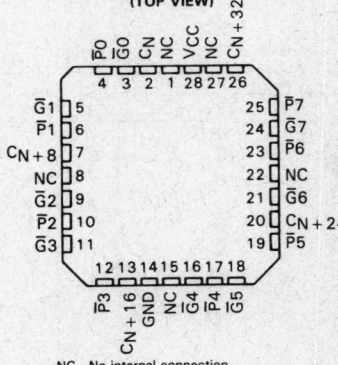

NC – No internal connection

description

The 'AS882 is a high-speed look-ahead carry generator capable of anticipating the carry across a group of eight 4-bit adders permitting the designer to implement look-ahead for a 32-bit ALU with a single package or, by cascading 'AS882's, full look-ahead is possible across n-bit adders.

The SN54AS882 is characterized for operation over the full military temperature range of −55 °C to 125 °C. The SN74AS882 is characterized for operation from 0 °C to 70 °C.

'AS882 LOGIC EQUATIONS

C_{n+8} = G1 + P1G0 + P1P0C_n

C_{n+16} = G3 + P3G2 + P3P2G1 + P3P2P1G0
  + P3P2P1P0C_n

C_{n+24} = G5 + P5G4 + P5P4G3 + P5P4P3G2
  + P5P4P3P2G1 + P5P4P3P2P1G0
  + P5P4P3P2P1P0C_n

C_{n+32} = G7 + P7G6 + P7P6G5 + P7P6P5G4
  + P7P6P5P4G3 + P7P6P5P4P3G2
  + P7P6P5P4P3P2G1 + P7P6P5P4P3P2P1G0
  + P7P6P5P4P3P2P1P0C_n

logic symbol

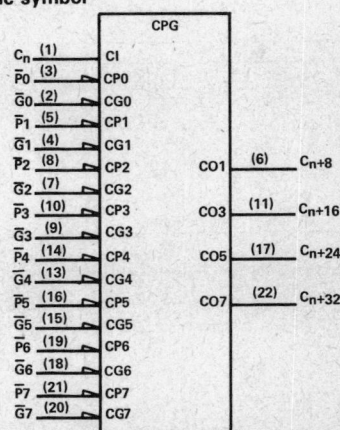

Pin numbers shown are for JT and NT packages.

ADVANCE INFORMATION
This document contains information on a new product. Specifications are subject to change without notice.

Copyright © 1982 Texas Instruments Incorporated

TEXAS INSTRUMENTS
POST OFFICE BOX 225012 • DALLAS, TEXAS 75265

TYPES SN54AS882, SN74AS882
32-BIT LOOK-AHEAD CARRY GENERATORS

FUNCTION TABLE FOR C_{n+32} OUTPUT

INPUTS																OUTPUT
$\overline{G}7$	$\overline{G}6$	$\overline{G}5$	$\overline{G}4$	$\overline{G}3$	$\overline{G}2$	$\overline{G}1$	$\overline{G}0$	$\overline{P}7$	$\overline{P}6$	$\overline{P}5$	$\overline{P}4$	$\overline{P}3$	$\overline{P}2$	$\overline{P}1$	$\overline{P}0$	C_n	C_{n+32}
L	X	X	X	X	X	X	X	X	X	X	X	X	X	X	X	X	H
X	L	X	X	X	X	X	X	L	X	X	X	X	X	X	X	X	H
X	X	L	X	X	X	X	X	L	L	X	X	X	X	X	X	X	H
X	X	X	L	X	X	X	X	L	L	L	X	X	X	X	X	X	H
X	X	X	X	L	X	X	X	L	L	L	L	X	X	X	X	X	H
X	X	X	X	X	L	X	X	L	L	L	L	L	X	X	X	X	H
X	X	X	X	X	X	L	X	L	L	L	L	L	L	X	X	X	H
X	X	X	X	X	X	X	L	L	L	L	L	L	L	L	X	X	H
X	X	X	X	X	X	X	X	L	L	L	L	L	L	L	L	H	H
All other combinations																	L

FUNCTION TABLE FOR C_{n+24} OUTPUT

INPUTS													OUTPUT
$\overline{G}5$	$\overline{G}4$	$\overline{G}3$	$\overline{G}2$	$\overline{G}1$	$\overline{G}0$	$\overline{P}5$	$\overline{P}4$	$\overline{P}3$	$\overline{P}2$	$\overline{P}1$	$\overline{P}0$	C_n	C_{n+24}
L	X	X	X	X	X	X	X	X	X	X	X	X	H
X	L	X	X	X	X	L	X	X	X	X	X	X	H
X	X	L	X	X	X	L	L	X	X	X	X	X	H
X	X	X	L	X	X	L	L	L	X	X	X	X	H
X	X	X	X	L	X	L	L	L	L	X	X	X	H
X	X	X	X	X	L	L	L	L	L	L	X	X	H
X	X	X	X	X	X	L	L	L	L	L	L	H	H
All other combinations													L

FUNCTION TABLE FOR C_{n+16} OUTPUT

INPUTS								OUTPUT
$\overline{G}3$	$\overline{G}2$	$\overline{G}1$	$\overline{G}0$	$\overline{P}3$	$\overline{P}2$	$\overline{P}1$	$\overline{P}0$	C_n	C_{n+16}
L	X	X	X	X	X	X	X	X	H
X	L	X	X	L	X	X	X	X	H
X	X	L	X	L	L	X	X	X	H
X	X	X	L	L	L	L	X	X	H
X	X	X	X	L	L	L	L	H	H
All other combinations									L

FUNCTION TABLE FOR C_{n+8} OUTPUT

INPUTS					OUTPUT
$\overline{G}1$	$\overline{G}0$	$\overline{P}1$	$\overline{P}0$	C_n	C_{n+8}
L	X	X	X	X	H
X	L	L	X	X	H
X	X	L	L	H	H
All other combinations					L

Any inputs not shown in a given table are irrelevant with respect to that output.

TEXAS INSTRUMENTS
POST OFFICE BOX 225012 • DALLAS, TEXAS 75265

TYPES SN54AS882, SN74AS882
32-BIT LOOK-AHEAD CARRY GENERATORS

logic diagram (positive logic)

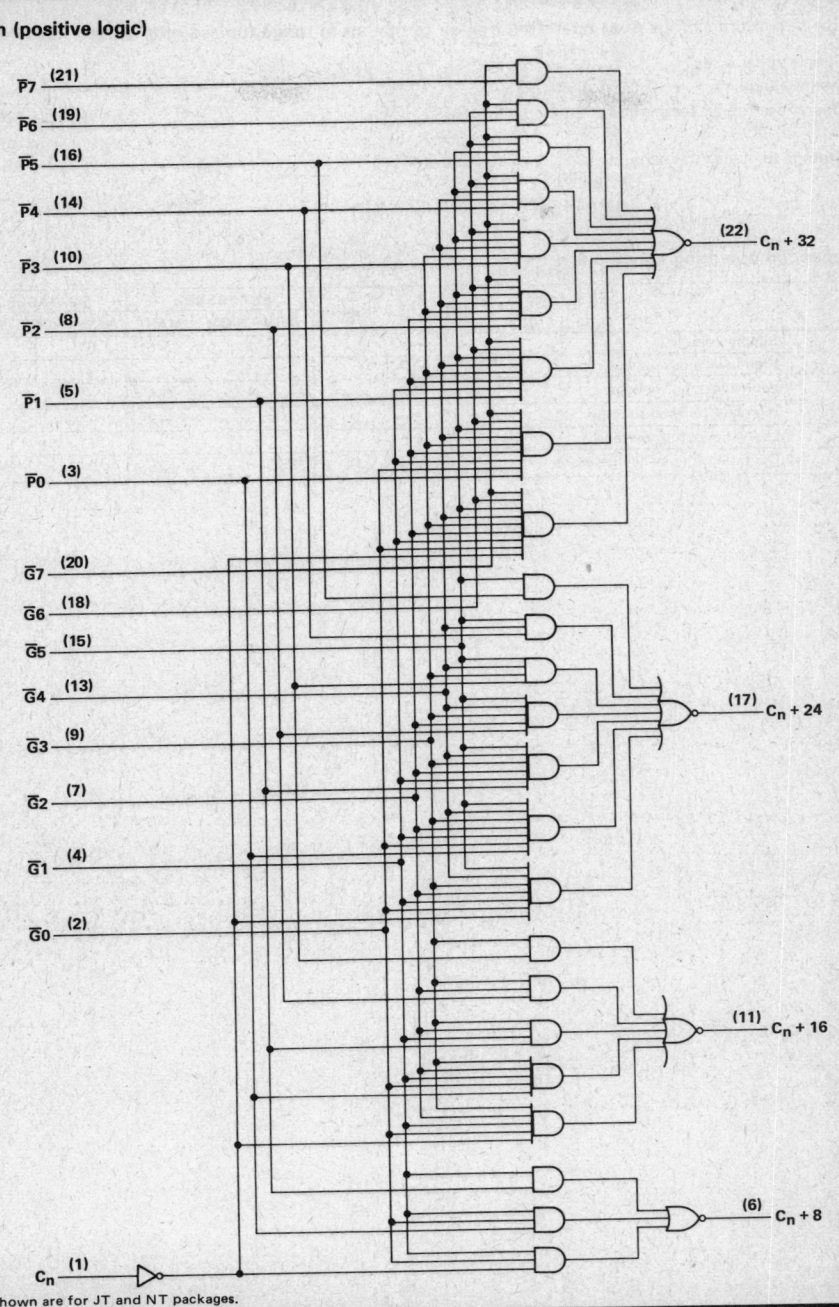

Pin numbers shown are for JT and NT packages.

TYPES SN54AS882, SN74AS882
32-BIT LOOK-AHEAD CARRY GENERATORS

absolute maximum ratings over operating free-air temperature range (unless otherwise noted)

Supply voltage, V_{CC} .... 7 V
Input voltage .... 7 V
Operating free-air temperature range: SN54AS882 .... −55 °C to 125 °C
                                       SN74AS882 .... 0 °C to 70 °C
Storage temperature range .... −65 °C to 150 °C

recommended operating conditions

		SN54AS882			SN74AS882			UNIT
		MIN	NOM	MAX	MIN	NOM	MAX	
V_{CC}	Supply voltage	4.5	5	5.5	4.5	5	5.5	V
V_{IH}	High-level input voltage	2			2			V
V_{IL}	Low-level input voltage			0.8			0.8	V
I_{OH}	High-level output current			−2			−2	mA
I_{OL}	Low-level output current			20			20	mA
T_A	Operating free-air temperature	−55		125	0		70	°C

TEXAS INSTRUMENTS
POST OFFICE BOX 225012 • DALLAS, TEXAS 75265

TYPES SN54AS882, SN74AS882
32-BIT LOOK-AHEAD CARRY GENERATORS

electrical characteristics over recommended operating free-air temperature range (unless otherwise noted)

PARAMETER		TEST CONDITIONS		SN54AS882 MIN	SN54AS882 TYP†	SN54AS882 MAX	SN74AS882 MIN	SN74AS882 TYP†	SN74AS882 MAX	UNIT
V_{IK}		$V_{CC} = 4.5$ V,	$I_I = -18$ mA			−1.2			−1.2	V
V_{OH}		$V_{CC} = 4.5$ V, to 5.5 V,	$I_{OH} = -2$ mA	$V_{CC}-2$			$V_{CC}-2$			V
V_{OL}		$V_{CC} = 4.5$ V,	$I_{OL} = 20$ mA		0.3	0.5		0.3	0.5	V
I_I	C_n	$V_{CC} = 5.5$ V,	$V_I = 7$ V			0.5			0.5	mA
	$\overline{G}0, \overline{G}6$					4			4	
	$\overline{G}1, \overline{G}2, \overline{G}4$					6			6	
	$\overline{G}3$					7.5			7.5	
	$\overline{G}5$					7			7	
	$\overline{G}7$					4.5			4.5	
	$\overline{P}0, \overline{P}1$					2			2	
	$\overline{P}2, \overline{P}3$					1.5			1.5	
	$\overline{P}4, \overline{P}5$					1			1	
	$\overline{P}6, \overline{P}7$					0.5			0.5	
I_{IH}	C_n	$V_{CC} = 5.5$ V,	$V_I = 2.7$ V			0.2			0.2	mA
	$\overline{G}0, \overline{G}6$					1.6			1.6	
	$\overline{G}1, \overline{G}2, \overline{G}4$					2.4			2.4	
	$\overline{G}3$					3			3	
	$\overline{G}5$					2.8			2.8	
	$\overline{G}7$					1.8			1.8	
	$\overline{P}0, \overline{P}1$					0.8			0.8	
	$\overline{P}2, \overline{P}3$					0.6			0.6	
	$\overline{P}4, \overline{P}5$					0.4			0.4	
	$\overline{P}6, \overline{P}7$					0.2			0.2	
I_{IL}	C_n	$V_{CC} = 5.5$ V,	$V_I = 0.4$ V			−3			−3	mA
	$\overline{G}0$					−26			−26	
	$\overline{G}1$					−37			−37	
	$\overline{G}2$					−40			−40	
	$\overline{G}3$					−47			−47	
	$\overline{G}4$					−41			−41	
	$\overline{G}5$					−44			−44	
	$\overline{G}6, \overline{G}7$					−28			−28	
	$\overline{P}0$					−13			−13	
	$\overline{P}1$					−11			−11	
	$\overline{P}2$					−9			−9	
	$\overline{P}3$					−8			−8	
	$\overline{P}4$					−5			−5	
	$\overline{P}5$					−4			−4	
	$\overline{P}6, \overline{P}7$					−2			−2	
I_O‡		$V_{CC} = 5.5$ V,	$V_O = 2.25$ V	−150			−150			mA
I_{CC}		$V_{CC} = 5.5$ V			72	105		72	105	mA

†All typical values are at $V_{CC} = 5$ V, $T_A = 25°C$.
‡The output conditions have been chosen to produce a current that closely approximates one-half of the true short-circuit, I_{OS}.

ALS AND AS CIRCUITS

TEXAS INSTRUMENTS
POST OFFICE BOX 225012 • DALLAS, TEXAS 75265

TYPES SN54AS882, SN74AS882
32-BIT LOOK-AHEAD CARRY GENERATORS

switching characteristics (see Note 1)

PARAMETER	FROM (INPUT)	TO (OUTPUT)	V_{CC} = 4.5 V to 5.5 V, C_L = 50 pF, R_L = 500 Ω, T_A = MIN to MAX				UNIT
			SN54AS822		SN74AS882		
			MIN	MAX	MIN	MAX	
t_{pd}	C_n	Any output	4	15	4	14	
t_{pd}	$\overline{P}$ or $\overline{G}$	$C_n + 8$	2	9	2	8	
t_{pd}	$\overline{P}$ or $\overline{G}$	$C_n + 16$	2	9	2	8	ns
t_{pd}	$\overline{P}$ or $\overline{G}$	$C_n + 24$	2	11	2	10	
t_{pd}	$\overline{P}$ or $\overline{G}$	$C_n + 32$	2	13	2	12	

t_{pd} = t_{PHL} or t_{PLH}.
NOTE 1: For load circuit and voltage waveforms, see page 1-12.

TYPICAL APPLICATION DATA

The application given in figure 1 illustrates how the 'AS882 can implement look-ahead carry for a 32-bit ALU (in this case, the popular 'AS881) with a single package. Typical carry times shown are derived using the standard Advanced Schottky load circuit with C_L = 15 pF.

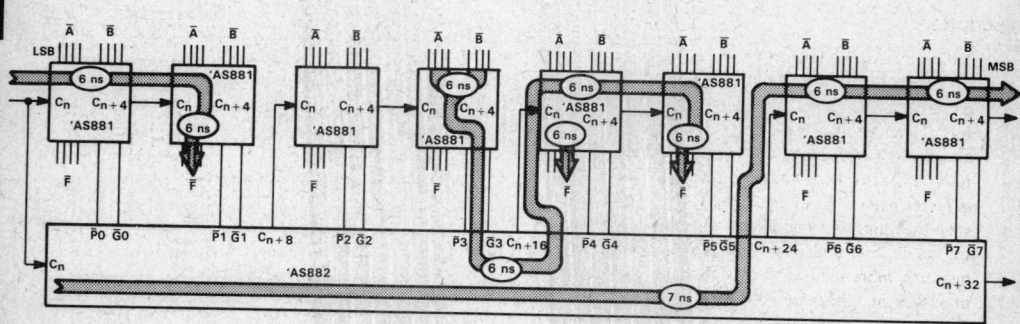

TYPES SN54AS885, SN74AS885
8-BIT MAGNITUDE COMPARATORS

D2661, DECEMBER 1982–REVISED DECEMBER 1983

- Included among the Package Options Are Compact, 24-Pin, 300-mil DIPs and Both 28-Pin Ceramic and Plastic Chip Carriers
- Latchable P Input Ports with Power-Up Clear
- Choice of Logical or Arithmetic (2's Complement) Comparison
- Data and PLE Inputs Utilize P-N-P Input Transistors to Reduce DC Loading Effects
- Approximately 35% Improvement in AC Performance Over Schottky TTL while Performing More Functions
- Cascadable to n-Bits while Maintaining High Performance
- 10% Less Power than STTL for an 8-Bit Comparison
- Dependable Texas Instruments Quality and Reliability

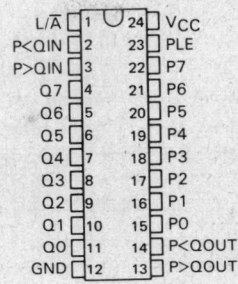

SN54AS885 . . . . JT PACKAGE
SN74AS885 . . . . NT PACKAGE
(TOP VIEW)

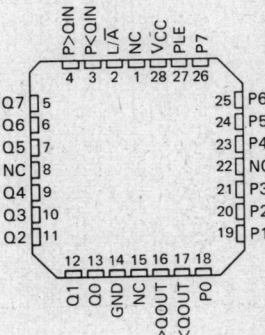

SN54AS885 . . . . FH PACKAGE
SN74AS885 . . . . FN PACKAGE
(TOP VIEW)

description

These advanced Schottky devices are capable of performing high-speed arithmetic or logic comparisons on two 8-bit binary or two's complement words. Two fully decoded decisions about words P and Q are externally available at two outputs. These devices are fully expandable to any number of bits without external gates. The P > Q and P < Q outputs of a stage handling less-significant bits may be connected to the P > Q and P < Q inputs of the next stage handling more-significant bits to obtain comparisons of words of longer lengths. The cascading paths are implemented with only a two-gate-level delay to reduce overall comparison times for long words. Two alternative methods of cascading are shown in the typical application data.

The latch is transparent when P Latch Enable (PLE) is high; the P input port is latched when PLE is low. This provides the designer with temporary storage for the P data word. The enable circuitry is implemented with minimal delay times to enhance performance when cascaded for longer words. The PLE and P and Q data inputs utilize p-n-p input transistors to reduce the low-level current input requirement to typically −0.25 mA, which minimizes dc loading effects.

The SN54AS885 is characterized for operation over the full military temperature range of −55°C to 125°C. The SN74AS885 is characterized for operation from 0°C to 70°C.

Copyright © 1982 by Texas Instruments Incorporated

TEXAS INSTRUMENTS
POST OFFICE BOX 225012 • DALLAS, TEXAS 75265

TYPES SN54AS885, SN74AS885
8-BIT MAGNITUDE COMPARATORS

logic diagram (positive logic)

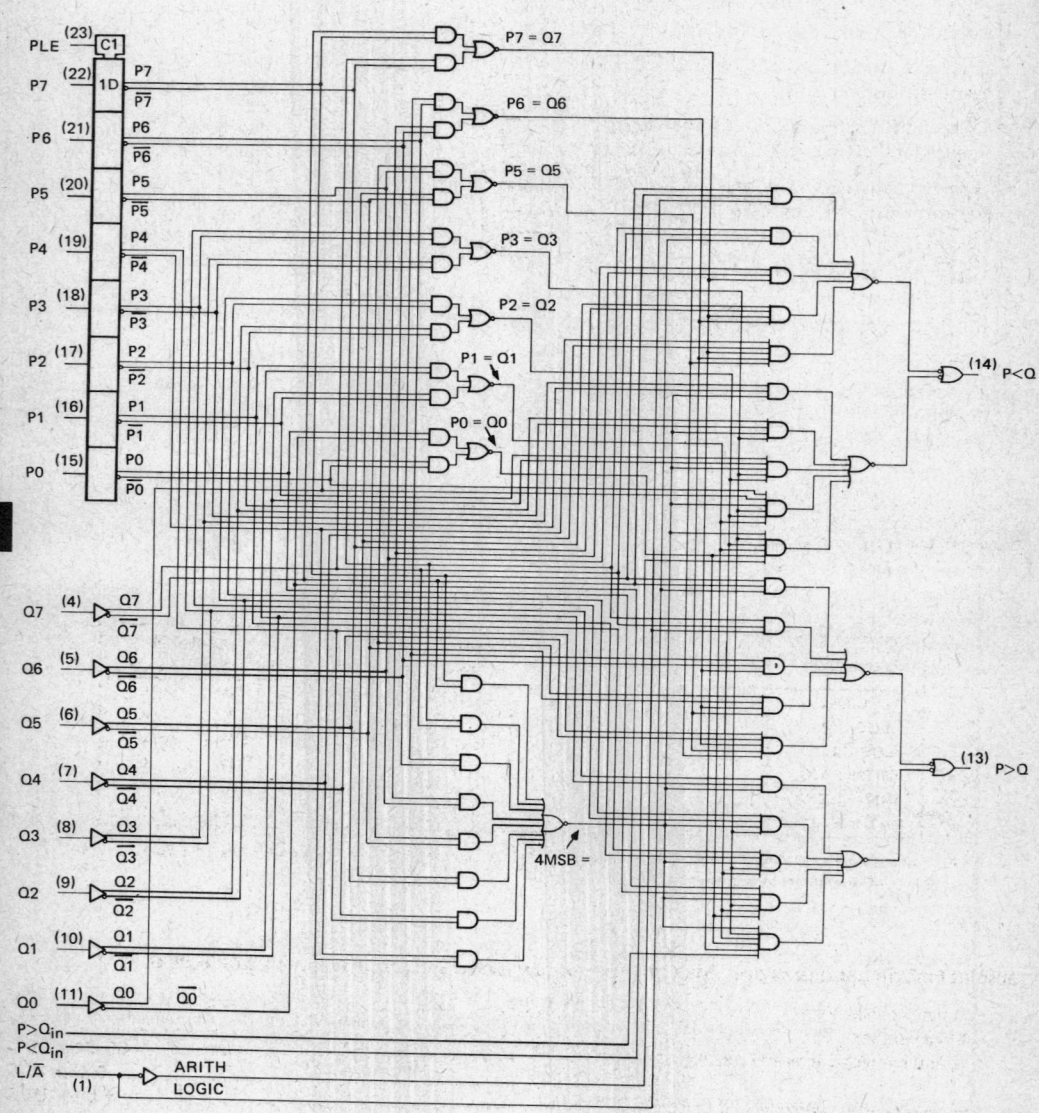

Pin numbers shown are for JT and NT packages.

TYPES SN54AS885, SN74AS885
8-BIT MAGNITUDE COMPARATORS

logic symbol

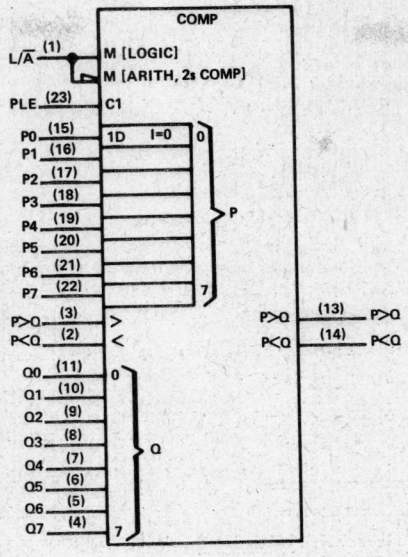

Pin numbers shown are for JT and NT packages.

FUNCTION TABLE

COMPARISON	L/$\overline{A}$	DATA INPUTS P0-P7, Q0-Q7	INPUT P>Q	INPUT P<Q	OUTPUTS P>Q	OUTPUTS P<Q
LOGICAL	H	P>Q	X	X	H	L
LOGICAL	H	P<Q	X	X	L	H
LOGICAL*	H	P=Q	H OR L	H OR L	H OR L	H OR L
ARITHMETIC	L	P AG Q	X	X	H	L
ARITHMETIC	L	Q AG P	X	X	L	H
ARITHMETIC*	L	P=Q	H OR L	H OR L	H OR L	H OR L

*In these cases the P>Q output will follow the P>Q input, and the P<Q output will follow the P<Q input.
AG — arithmetically greater than

absolute maximum ratings over operating free-air temperature range (unless otherwise noted)

Supply voltage, V_{CC} .... 7 V
Input voltage .... 7 V
Operating free-air temperature range: SN54AS885 .... −55°C to 125°C
                                       SN74AS885 .... 0°C to 70°C
Storage temperature range .... −65°C to 150°C

TYPES SN54AS885, SN74AS885
8-BIT MAGNITUDE COMPARATORS

recommended operating conditions

	PARAMETER	SN54AS885			SN74AS885			UNIT
		MIN	NOM	MAX	MIN	NOM	MAX	
V_{CC}	Supply voltage	4.5	5	5.5	4.5	5	5.5	V
V_{IH}	High-level input voltage	2			2			V
V_{IL}	Low-level input voltage			0.8			0.8	V
I_{OH}	High-level output current			−2			−2	mA
I_{OL}	Low-level output current			20			20	mA
t_{su}	Setup time to PLE ↓	2			2			ns
t_h	Hold time after PLE ↓	4			4			ns
T_A	Operating free-air temperature	−55		125	0		70	°C

electrical characteristics over recommended operating free-air temperature range (unless otherwise noted)

PARAMETER		TEST CONDITIONS		SN54AS885			SN74AS885			UNIT
				MIN	TYP[†]	MAX	MIN	TYP[†]	MAX	
V_{IK}		V_{CC} = 4.5 V,	I_I = −18 mA			−1.2			−1.2	V
V_{OH}		V_{CC} = 4.5 V to 5.5 V,	I_{OH} = −2 mA	V_{CC}−2			V_{CC}−2			V
V_{OL}		V_{CC} = 4.5 V,	I_{OL} = 20 mA		0.35	0.5		0.35	0.5	V
I_I		V_{CC} = 5.5 V,	V_I = 7 V			0.1			0.1	µA
I_{IH}	L/$\overline{A}$	V_{CC} = 5.5 V,	V_I = 2.7 V			40			40	µA
	Others					20			20	
I_{IL}	L/$\overline{A}$	V_{CC} = 5.5 V,	V_I = 0.4 V			−4			−4	mA
	P > Q_{in}					−2			−2	
	P < Q_{in}									
	P, Q, PLE					−1			−1	
I_O[‡]		V_{CC} = 5.5 V,	V_O = 2.25 V	−20		−112	−20		−112	mA
I_{CC}		V_{CC} = 5.5 V	See Note 1		130	210		130	210	mA

[†] All typical values are at V_{CC} = 5 V, T_A = 25°C.
[‡] The output conditions have been chosen to produce a current that closely approximates one half of the true short-circuit output current, I_{OS}.
NOTE 1: I_{CC} is measured with all inputs high except L/$\overline{A}$, which is low.

switching characteristics (see Note 2)

PARAMETER	FROM (INPUT)	TO (OUTPUT)	V_{CC} = 4.5 V to 5.5 V, C_L = 50 pF, R_L = 500 Ω, T_A = MIN to MAX						UNIT
			SN54AS885			SN74AS885			
			MIN	TYP[†]	MAX	MIN	TYP[†]	MAX	
t_{PLH}	L/$\overline{A}$			8.5	14		8.5	13	ns
t_{PHL}				7.5	14		7.5	13	
t_{PLH}	P < Q_{in}	P < Q,		5	10		5	8	ns
t_{PHL}	P > Q_{in}	P > Q		5.5	10		5.5	8	
t_{PLH}	Any P or Q			13.5	21		13.5	17.5	ns
t_{PHL}	Data Input			10	17		10	15	

[†] All typical values are at V_{CC} = 5 V, T_A = 25°C.
NOTE 1: For load circuit and voltage waveforms, see page 1-12.

TYPES SN54AS885, SN74AS885
8-BIT MAGNITUDE COMPARATORS

TYPICAL APPLICATION DATA

The 'AS885 can be cascaded to compare words longer than 8-bits. Figure 1 shows the comparison of two 32-bit words; however, the design is expandable to n-bits. Figure 1 shows the optimum cascading arrangement for comparing words of 32 bits or greater. Typical delay times shown are at $V_{CC} = 5$ V, $T_A = 25°C$, and use the standard Advanced Schottky load of $R_L = 500\ \Omega$, $C_L = 50$ pF.

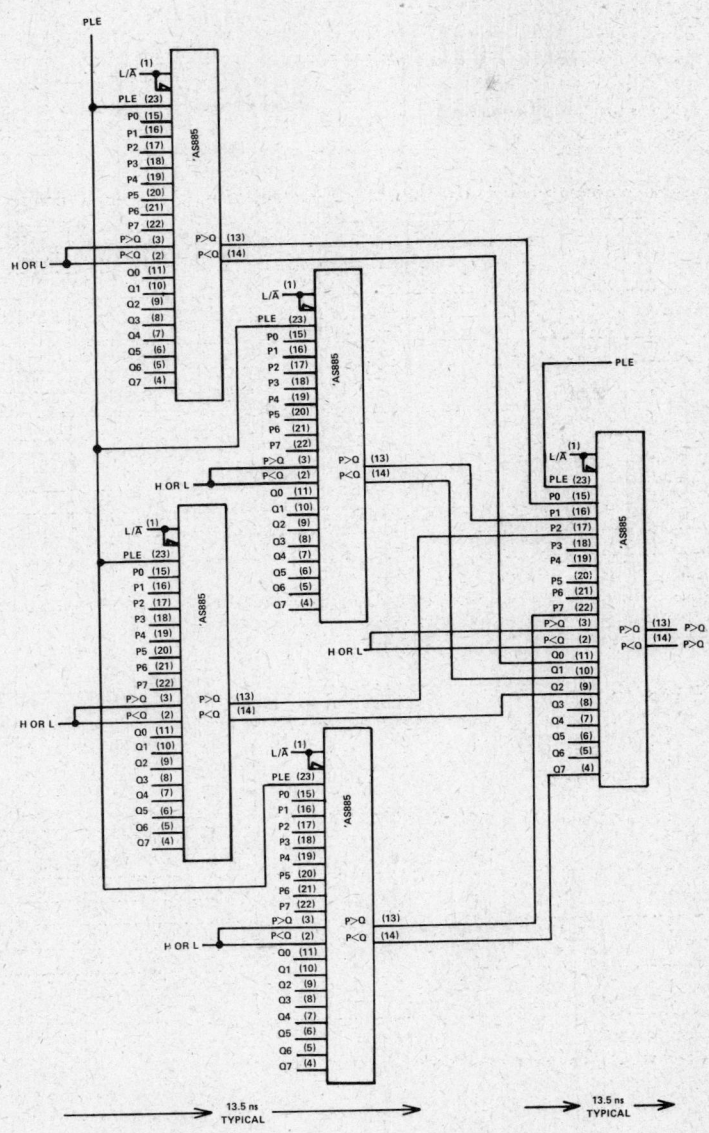

FIGURE 1 – 32-BIT TO 72 (N)-BIT MAGNITUDE COMPARATOR

TYPES SN54AS885, SN74AS885
8-BIT MAGNITUDE COMPARATORS

TYPICAL APPLICATION DATA

The method shown in Figure 2 is the fastest cascading arrangement for comparing 16-bit or 24-bit words. Typical delay times shown are at $V_{CC} = 5$ V, $T_A = 25°C$, and use the standard Advanced Schottky load of $R_L = 500\,\Omega$, $C_L = 50$ pF.

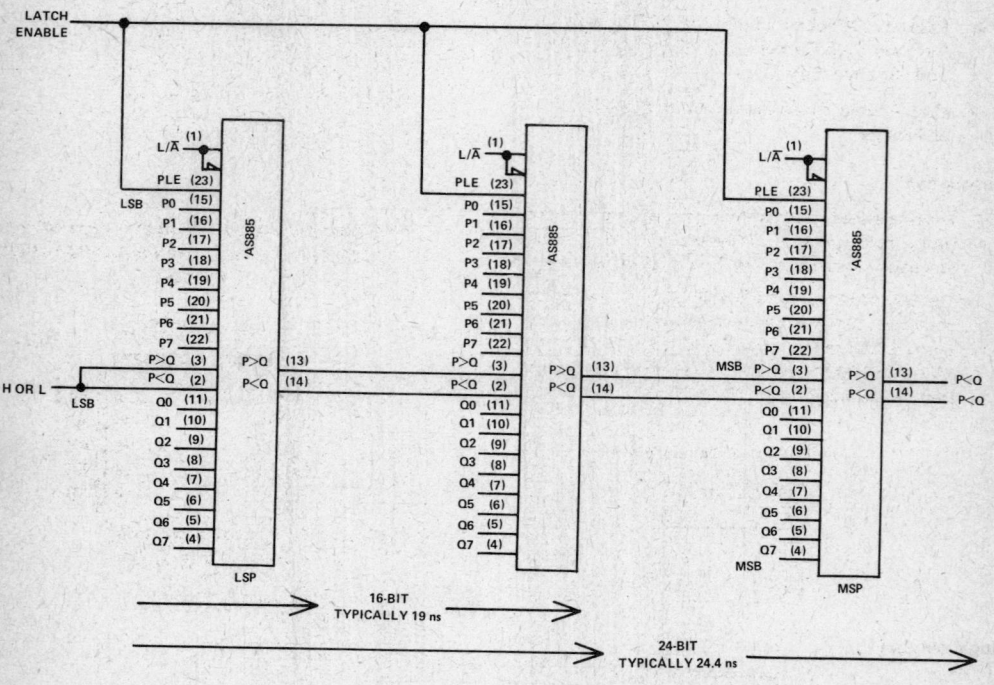

FIGURE 2

TYPES SN54ALS1000A, SN54AS1000, SN74ALS1000A, SN74AS1000
QUADRUPLE 2-INPUT POSITIVE-NAND BUFFERS/DRIVERS

D2661, APRIL 1982 – REVISED DECEMBER 1983

- 'ALS1000A is a Buffer Version of 'ALS00A
- 'AS1000 is a Driver Version of 'AS00
- 'AS1000 Offers High Capacitive Drive Capability
- Package Options Include Both Plastic and Ceramic Chip Carriers in Addition to Plastic and Ceramic DIPs
- Dependable Texas Instruments Quality and Reliability

SN54ALS1000A, SN54AS1000 . . . J PACKAGE
SN74ALS1000A, SN74AS1000 . . . N PACKAGE
(TOP VIEW)

```
1A  [ 1   14 ] VCC
1B  [ 2   13 ] 4B
1Y  [ 3   12 ] 4A
2A  [ 4   11 ] 4Y
2B  [ 5   10 ] 3B
2Y  [ 6    9 ] 3A
GND [ 7    8 ] 3Y
```

description

These devices contain four independent 2-input NAND buffers/drivers. They perform the Boolean functions $Y = \overline{A \cdot B}$ or $Y = \overline{A} + \overline{B}$ in positive logic.

The SN54ALS1000A and SN54AS1000 are characterized for operation over the full military temperature range of −55°C to 125°C. The SN74ALS1000A and SN74AS1000 are characterized for operation from 0°C to 70°C.

SN54ALS1000A, SN54AS1000 . . . FH PACKAGE
SN74ALS1000A, SN74AS1000 . . . FN PACKAGE
(TOP VIEW)

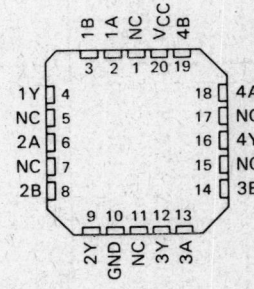

NC – No internal connection

FUNCTION TABLE (each gate)

INPUTS		OUTPUT
A	B	Y
H	H	L
L	X	H
X	L	H

logic symbol

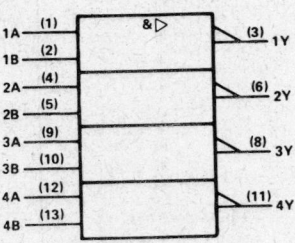

Pin numbers shown are for J and N packages.

Copyright © 1982 by Texas Instruments Incorporated

TEXAS INSTRUMENTS
POST OFFICE BOX 225012 • DALLAS, TEXAS 75265

TYPES SN54ALS1000A, SN74ALS1000A
QUADRUPLE 2-INPUT POSITIVE-NAND BUFFERS

absolute maximum ratings over operating free-air temperature range (unless otherwise noted)

Supply voltage, V_{CC} .... 7 V
Input voltage .... 7 V
Operating free-air temperature range: SN54ALS1000A .... −55 °C to 125 °C
                                         SN74ALS1000A .... 0 °C to 70 °C
Storage temperature range .... −65 °C to 150 °C

recommended operating conditions

		SN54ALS1000A			SN74ALS1000A			UNIT
		MIN	NOM	MAX	MIN	NOM	MAX	
V_{CC}	Supply voltage	4.5	5	5.5	4.5	5	5.5	V
V_{IH}	High-level input voltage	2			2			V
V_{IL}	Low-level input voltage			0.8			0.8	V
I_{OH}	High-level output current			−1			−2.6	mA
I_{OL}	Low-level output current			12			24	mA
T_A	Operating free-air temperature	−55		125	0		70	°C

electrical characteristics over recommended operating free-air temperature range (unless otherwise noted)

PARAMETER	TEST CONDITIONS		SN54ALS1000A			SN74ALS1000A			UNIT
			MIN	TYP†	MAX	MIN	TYP†	MAX	
V_{IK}	V_{CC} = 4.5 V,	I_I = −18 mA			−1.5			−1.5	V
V_{OH}	V_{CC} = 4.5 V to 5.5 V,	I_{OH} = −0.4 mA	V_{CC}−2			V_{CC}−2			V
	V_{CC} = 4.5 V,	I_{OH} = −1 mA	2.4	3.3					
	V_{CC} = 4.5 V,	I_{OH} = −2.6 mA				2.4	3.2		
V_{OL}	V_{CC} = 4.5 V,	I_{OL} = 12 mA		0.25	0.4		0.25	0.4	V
	V_{CC} = 4.5 V,	I_{OL} = 24 mA					0.35	0.5	
I_I	V_{CC} = 5.5 V,	V_I = 7 V			0.1			0.1	mA
I_{IH}	V_{CC} = 5.5 V,	V_I = 2.7 V			20			20	µA
I_{IL}	V_{CC} = 5.5 V,	V_I = 0.4 V			−0.1			−0.1	mA
I_O‡	V_{CC} = 5.5 V,	V_O = 2.25 V	−30		−112	−30		−112	mA
I_{CCH}	V_{CC} = 5.5 V,	V_I = 0 V		0.86	1.6		0.86	1.6	mA
I_{CCL}	V_{CC} = 5.5 V,	V_I = 4.5 V		4.8	7.8		4.8	7.8	mA

†All typical values are at V_{CC} = 5 V, T_A = 25 °C.
‡The output conditions have been chosen to produce a current that closely approximates one half of the true short-circuit output current, I_{OS}.

switching characteristics (see Note 1)

PARAMETER	FROM (INPUT)	TO (OUTPUT)	V_{CC} = 4.5 V to 5.5 V, C_L = 50 pF, R_L = 500 Ω, T_A = MIN to MAX				UNIT
			SN54ALS1000A		SN74ALS1000A		
			MIN	MAX	MIN	MAX	
t_{PLH}	A or B	Y	2	10	2	8	ns
t_{PHL}			2	10	2	7	

NOTE 1: For load circuit and voltage waveforms, see page 1-12.

TEXAS INSTRUMENTS
POST OFFICE BOX 225012 • DALLAS, TEXAS 75265

TYPES SN54AS1000, SN74AS1000
QUADRUPLE 2-INPUT POSITIVE-NAND DRIVERS

absolute maximum ratings over operating free-air temperature range (unless otherwise noted)

Supply voltage, V_{CC} .... 7 V
Input voltage .... 7 V
Operating free-air temperature range: SN54AS1000 .... −55 °C to 125 °C
                                   SN74AS1000 .... 0 °C to 70 °C
Storage temperature range .... −65 °C to 150 °C

recommended operating conditions

		SN54AS1000			SN74AS1000			UNIT
		MIN	NOM	MAX	MIN	NOM	MAX	
V_{CC}	Supply voltage	4.5	5	5.5	4.5	5	5.5	V
V_{IH}	High-level input voltage	2			2			V
V_{IL}	Low-level input voltage			0.8			0.8	V
I_{OH}	High-level output current			−40			−48	mA
I_{OL}	Low-level output current			40			48	mA
T_A	Operating free-air temperature	−55		125	0		70	°C

electrical characteristics over recommended operating free-air temperature range (unless otherwise noted)

PARAMETER	TEST CONDITIONS		SN54AS1000			SN74AS1000			UNIT
			MIN	TYP†	MAX	MIN	TYP†	MAX	
V_{IK}	V_{CC} = 4.5 V,	I_I = −18 mA			−1.2			−1.2	V
V_{OH}	V_{CC} = 4.5 V to 5.5 V,	I_{OH} = −2 mA	V_{CC}−2			V_{CC}−2			V
	V_{CC} = 4.5 V,	I_{OH} = −3 mA	2.4	3.2		2.4	3.2		
	V_{CC} = 4.5 V,	I_{OH} = −40 mA	2						
	V_{CC} = 4.5 V,	I_{OH} = −48 mA				2			
V_{OL}	V_{CC} = 4.5 V,	I_{OL} = 40 mA		0.25	0.5				V
	V_{CC} = 4.5 V,	I_{OL} = 48 mA					0.35	0.5	
I_I	V_{CC} = 5.5 V,	V_I = 7 V			0.1			0.1	mA
I_{IH}	V_{CC} = 5.5 V,	V_I = 2.7 V			20			20	µA
I_{IL}	V_{CC} = 5.5 V,	V_I = 0.4 V			−0.5			−0.5	mA
I_O‡	V_{CC} = 5.5 V,	V_O = 2.25 V			−135			−135	mA
I_{CCH}	V_{CC} = 5.5 V,	V_I = 0 V		2.1	3.5		2.1	3.5	mA
I_{CCL}	V_{CC} = 5.5 V,	V_I = 4.5 V		11.5	19		11.5	19	mA

† All typical values are at V_{CC} = 5 V, T_A = 25 °C.
‡ The output conditions have been chosen to produce a current that closely approximates one half of the true short-circuit output current, I_{OS}.

switching characteristics (see Note 1)

PARAMETER	FROM (INPUT)	TO (OUTPUT)	V_{CC} = 4.5 V to 5.5 V, C_L = 50 pF, R_L = 500 Ω, T_A = MIN to MAX				UNIT
			SN54AS1000		SN74AS1000		
			MIN	MAX	MIN	MAX	
t_{PLH}	A or B	Y	1	4.5	1	3.5	ns
t_{PHL}			1	4.5	1	3.5	

NOTE 1: For load circuit and voltage waveforms, see page 1-12.

ALS AND AS CIRCUITS

TYPES SN54ALS1002A, SN74ALS1002A
QUADRUPLE 2-INPUT POSITIVE-NOR BUFFERS

D2661, DECEMBER 1983 – REVISED DECEMBER 1983

- Quad Versions of 'ALS805
- Buffer Version of 'ALS02
- Package Options Include Both Plastic and Ceramic Chip Carriers in Addition to Plastic and Ceramic DIPs
- Dependable Texas Instruments Quality and Reliability

SN54ALS1002A . . . J PACKAGE
SN74ALS1002A . . . N PACKAGE
(TOP VIEW)

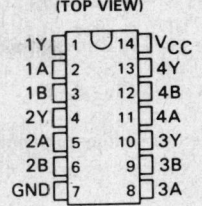

description

These devices contain four independent 2-input NOR buffers. They perform the Boolean functions $Y = \overline{A + B}$ or $Y = \overline{A} \cdot \overline{B}$ in positive logic.

The SN54ALS1002A is characterized for operation over the full military temperature range of $-55\,°C$ to $125\,°C$. The SN74ALS1002A is characterized for operation from $0\,°C$ to $70\,°C$.

SN54ALS1002A . . . FH PACKAGE
SN74ALS1002A . . . FN PACKAGE
(TOP VIEW)

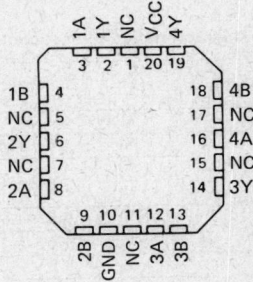

NC – No internal connection

FUNCTION TABLE (each gate)

INPUTS		OUTPUT
A	B	Y
H	X	L
X	H	L
L	L	H

logic symbol

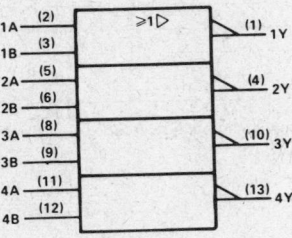

Pin numbers shown are for J and N packages.

Copyright © 1982 by Texas Instruments Incorporated

TEXAS INSTRUMENTS
POST OFFICE BOX 225012 • DALLAS, TEXAS 75265

TYPES SN54ALS1002A, SN74ALS1002A
QUADRUPLE 2-INPUT POSITIVE-NOR BUFFERS

absolute maximum ratings over operating free-air temperature range (unless otherwise noted)

Supply voltage, V_{CC} .... 7 V
Input voltage .... 7 V
Operating free-air temperature range: SN54ALS1002A .... −55 °C to 125 °C
             SN74ALS1002A .... 0 °C to 70 °C
Storage temperature range .... −65 °C to 150 °C

recommended operating conditions

		SN54ALS1002A			SN74ALS1002A			UNIT
		MIN	NOM	MAX	MIN	NOM	MAX	
V_{CC}	Supply voltage	4.5	5	5.5	4.5	5	5.5	V
V_{IH}	High-level input voltage	2			2			V
V_{IL}	Low-level input voltage			0.8			0.8	V
I_{OH}	High-level output current			−1			−2.6	mA
I_{OL}	Low-level output current			12			24	mA
T_A	Operating free-air temperature	−55		125	0		70	°C

electrical characteristics over recommended operating free-air temperature range (unless otherwise noted)

PARAMETER	TEST CONDITIONS		SN54ALS1002A			SN74ALS1002A			UNIT
			MIN	TYP†	MAX	MIN	TYP†	MAX	
V_{IK}	$V_{CC} = 4.5$ V,	$I_I = -18$ mA			−1.5			−1.5	V
V_{OH}	$V_{CC} = 4.5$ V to 5.5 V,	$I_{OH} = -0.4$ mA	$V_{CC}-2$			$V_{CC}-2$			
	$V_{CC} = 4.5$ V,	$I_{OH} = -1$ mA	2.4	3.3					V
	$V_{CC} = 4.5$ V,	$I_{OH} = -2.6$ mA				2.4	3.2		
V_{OL}	$V_{CC} = 4.5$ V,	$I_{OL} = 12$ mA		0.25	0.4		0.25	0.4	V
	$V_{CC} = 4.5$ V,	$I_{OL} = 24$ mA					0.35	0.5	
I_I	$V_{CC} = 5.5$ V,	$V_I = 7$ V			0.1			0.1	mA
I_{IH}	$V_{CC} = 5.5$ V,	$V_I = 2.7$ V			20			20	µA
I_{IL}	$V_{CC} = 5.5$ V,	$V_I = 0.4$ V			−0.1			−0.1	mA
I_O‡	$V_{CC} = 5.5$ V,	$V_O = 2.25$ V	−30		−112	−30		−112	mA
I_{CCH}	$V_{CC} = 5.5$ V,	$V_I = 0$ V		1.7	2.8		1.7	2.8	mA
I_{CCL}	$V_{CC} = 5.5$ V,	$V_I = 4.5$ V		5.6	9		5.6	9	mA

†All typical values are at $V_{CC} = 5$ V, $T_A = 25$ °C.
‡The output conditions have been chosen to produce a current that closely approximates one half of the true short-circuit output current, I_{OS}.

switching characteristics (see Note 1)

PARAMETER	FROM (INPUT)	TO (OUTPUT)	$V_{CC} = 4.5$ V to 5.5 V, $C_L = 50$ pF, $R_L = 500$ Ω, $T_A = $ MIN to MAX				UNIT
			SN54ALS1002A		SN74ALS1002A		
			MIN	MAX	MIN	MAX	
t_{PLH}	A or B	Y	2	10	2	8	ns
t_{PHL}			2	10	2	7	

NOTE 1: For load circuit and voltage waveforms, see page 1-12.

TYPES SN54ALS1003A, SN74ALS1003A
QUADRUPLE 2-INPUT POSITIVE-NAND BUFFERS WITH OPEN-COLLECTOR OUTPUTS

D2661, APRIL 1982 — REVISED DECEMBER 1983

- Buffer Version of 'ALS03A
- Package Options Include Both Plastic and Ceramic Chip Carriers in Addition to Plastic and Ceramic DIPs
- Dependable Texas Instruments Quality and Reliability

description

These devices contain four independent 2-input NAND buffers. They perform the Boolean functions $Y = \overline{A \cdot B}$ or $Y = \overline{A} + \overline{B}$ in positive logic. The open-collector outputs require pull-up resistors to perform correctly. They may be connected to other open-collector outputs to implement active-low wired-OR or active-high wired-AND functions. Open-collector devices are often used to generate higher V_{OH} levels.

The SN54ALS1003A is characterized for operation over the full military temperature range of −55°C to 125°C. The SN74ALS1003A is characterized for operation from 0°C to 70°C.

SN54ALS1003A . . . J PACKAGE
SN74ALS1003A . . . N PACKAGE
(TOP VIEW)

```
1A  [1   14] VCC
1B  [2   13] 4B
1Y  [3   12] 4A
2A  [4   11] 4Y
2B  [5   10] 3B
2Y  [6    9] 3A
GND [7    8] 3Y
```

SN54ALS1003A . . . FH PACKAGE
SN74ALS1003A . . . FN PACKAGE
(TOP VIEW)

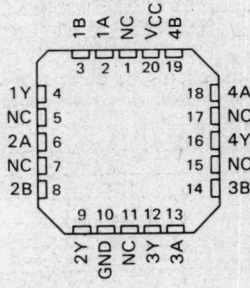

NC — No internal connection

FUNCTION TABLE (each gate)

INPUTS		OUTPUT
A	B	Y
H	H	L
L	X	H
X	L	H

logic symbol

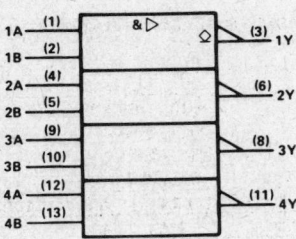

Pin numbers shown are for J and N packages.

Copyright © 1982 by Texas Instruments Incorporated

TYPES SN54ALS1003A, SN74ALS1003A
QUADRUPLE 2-INPUT POSITIVE-NAND BUFFERS WITH OPEN-COLLECTOR OUTPUTS

absolute maximum ratings over operating free-air temperature range (unless otherwise noted)

Supply voltage, V_{CC} .... 7 V
Input voltage .... 7 V
Off-state output voltage .... 7 V
Operating free-air temperature range: SN54ALS1003A .... −55 °C to 125 °C
                                     SN74ALS1003A .... 0 °C to 70 °C
Storage temperature range .... −65 °C to 150 °C

recommended operating conditions

		SN54ALS1003A			SN74ALS1003A			UNIT
		MIN	NOM	MAX	MIN	NOM	MAX	
V_{CC}	Supply voltage	4.5	5	5.5	4.5	5	5.5	V
V_{IH}	High-level input voltage	2			2			V
V_{IL}	Low-level input voltage			0.8			0.8	V
V_{OH}	High-level output voltage			5.5			5.5	V
I_{OL}	Low-level output current			12			24	mA
T_A	Operating free-air temperature	−55		125	0		70	°C

electrical characteristics over recommended operating free-air temperature range (unless otherwise noted)

PARAMETER	TEST CONDITIONS		SN54ALS1003A			SN74ALS1003A			UNIT
			MIN	TYP†	MAX	MIN	TYP†	MAX	
V_{IK}	V_{CC} = 4.5 V,	I_I = −18 mA			−1.5			−1.5	V
I_{OH}	V_{CC} = 4.5 V,	V_{OH} = 5.5 V			0.1			0.1	mA
V_{OL}	V_{CC} = 4.5 V,	I_{OL} = 12 mA		0.25	0.4		0.25	0.4	V
	V_{CC} = 4.5 V,	I_{OL} = 24 mA					0.35	0.5	
I_I	V_{CC} = 5.5 V,	V_I = 7 V			0.1			0.1	mA
I_{IH}	V_{CC} = 5.5 V,	V_I = 2.7 V			20			20	µA
I_{IL}	V_{CC} = 5.5 V,	V_I = 0.4 V			−0.1			−0.1	mA
I_{CCH}	V_{CC} = 5.5 V,	V_I = 0 V		0.86	1.6		0.86	1.6	mA
I_{CCL}	V_{CC} = 5.5 V,	V_I = 4.5 V		4.8	7.8		4.8	7.8	mA

† All typical values are at V_{CC} = 5 V, T_A = 25 °C.

switching characteristics (see Note 1)

PARAMETER	FROM (INPUT)	TO (OUTPUT)	V_{CC} = 4.5 V to 5.5 V, C_L = 50 pF, R_L = 680 Ω, T_A = MIN to MAX				UNIT
			SN54ALS1003A		SN74ALS1003A		
			MIN	MAX	MIN	MAX	
t_{PLH}	A or B	Y	10	40	10	33	ns
t_{PHL}			2	18	2	12	

NOTE 1: For load circuit and voltage waveforms, see page 1-12.

TYPES SN54ALS1004, SN54AS1004, SN74ALS1004, SN74AS1004
HEX INVERTING DRIVERS

D2661, APRIL 1982 – REVISED DECEMBER 1983

- 'AS1004 Offers High Capacitive-Drive Capability
- Driver Version of 'ALS04 and 'AS04
- Package Options Include Both Plastic and Ceramic Chip Carriers in Addition to Plastic and Ceramic DIPs
- Dependable Texas Instruments Quality and Reliability

SN54ALS1004, SN54AS1004 . . . J PACKAGE
SN74ALS1004, SN74AS1004 . . . N PACKAGE
(TOP VIEW)

```
 1A [ 1  U 14 ] VCC
 1Y [ 2    13 ] 6A
 2A [ 3    12 ] 6Y
 2Y [ 4    11 ] 5A
 3A [ 5    10 ] 5Y
 3Y [ 6     9 ] 4A
GND [ 7     8 ] 4Y
```

description

These devices contain six independent inverting drivers. They perform the Boolean function $Y = \overline{A}$.

The SN54ALS1004 and SN54AS1004 are characterized for operation over the full military temperature range of $-55\,°C$ to $125\,°C$. The SN74ALS1004 and SN74AS1004 are characterized for operation from $0\,°C$ to $70\,°C$.

SN54ALS1004, SN54AS1004 . . . FH PACKAGE
SN74ALS1004, SN74AS1004 . . . FN PACKAGE
(TOP VIEW)

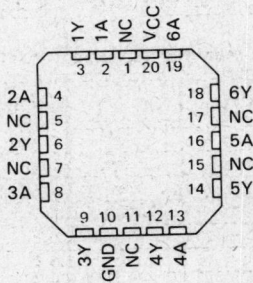

NC – No internal connection

FUNCTION TABLE
(each inverter)

INPUT A	OUTPUT Y
H	L
L	H

logic symbol

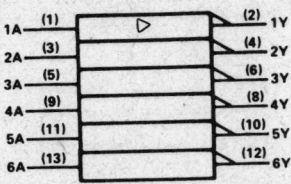

Pin numbers shown are for J and N packages.

Copyright © 1983 by Texas Instruments Incorporated

ALS AND AS CIRCUITS

TEXAS INSTRUMENTS
POST OFFICE BOX 225012 • DALLAS, TEXAS 75265

TYPES SN54ALS1004, SN74ALS1004
HEX INVERTING DRIVERS

absolute maximum ratings over operating free-air temperature range (unless otherwise noted)

Supply voltage, V_{CC} .... 7 V
Input voltage .... 7 V
Operating free-air temperature range: SN54ALS1004 .... −55°C to 125°C
             SN74ALS1004 .... 0°C to 70°C
Storage temperature range .... −65°C to 150°C

recommended operating conditions

		SN54ALS1004			SN74ALS1004			UNIT
		MIN	NOM	MAX	MIN	NOM	MAX	
V_{CC}	Supply voltage	4.5	5	5.5	4.5	5	5.5	V
V_{IH}	High-level input voltage	2			2			V
V_{IL}	Low-level input voltage			0.8			0.8	V
I_{OH}	High-level output current			−12			−15	mA
I_{OL}	Low-level output current			12			24	mA
T_A	Operating free-air temperature	−55		125	0		70	°C

electrical characteristics over recommended operating free-air temperature range (unless otherwise noted)

PARAMETER	TEST CONDITIONS		SN54ALS1004			SN74ALS1004			UNIT
			MIN	TYP†	MAX	MIN	TYP†	MAX	
V_{IK}	V_{CC} = 4.5 V,	I_I = −18 mA			−1.5			−1.5	V
V_{OH}	V_{CC} = 4.5 V to 5.5 V,	I_{OH} = −0.4 mA	V_{CC}−2			V_{CC}−2			V
	V_{CC} = 4.5 V,	I_{OH} = −3 mA	2.4	3.2		2.4	3.2		
	V_{CC} = 4.5 V,	I_{OH} = −12 mA	2						
	V_{CC} = 4.5 V,	I_{OH} = −15 mA				2			
V_{OL}	V_{CC} = 4.5 V,	I_{OL} = 12 mA		0.25	0.4		0.25	0.4	V
	V_{CC} = 4.5 V,	I_{OL} = 24 mA					0.35	0.5	
I_I	V_{CC} = 5.5 V,	V_I = 7 V			0.1			0.1	mA
I_{IH}	V_{CC} = 5.5 V,	V_I = 2.7 V			20			20	µA
I_{IL}	V_{CC} = 5.5 V,	V_I = 0.4 V			−0.1			−0.1	mA
I_O‡	V_{CC} = 5.5 V,	V_O = 2.25 V	−30		−112	−30		−112	mA
I_{CCH}	V_{CC} = 5.5 V,	V_I = 0 V		0.84	3		0.84	3	mA
I_{CCL}	V_{CC} = 5.5 V,	V_I = 4.5 V		7	12		7	12	mA

†All typical values are at V_{CC} = 5 V, T_A = 25°C.
‡The output conditions have been chosen to produce a current that closely approximates one half of the true short-circuit output current, I_{OS}.

switching characteristics (see Note 1)

PARAMETER	FROM (INPUT)	TO (OUTPUT)	V_{CC} = 4.5 V to 5.5 V, C_L = 50 pF, R_L = 500 Ω, T_A = MIN to MAX				UNIT
			SN54ALS1004		SN74ALS1004		
			MIN	MAX	MIN	MAX	
t_{PLH}	A	Y	1	9	1	7	ns
t_{PHL}			1	8	1	6	

NOTE 1: For load circuit and voltage waveforms, see page 1-12.

ALS AND AS CIRCUITS

TYPES SN54AS1004, SN74AS1004
HEX INVERTING DRIVERS

absolute maximum ratings over operating free-air temperature range (unless otherwise noted)

Supply voltage, V_{CC} .... 7 V
Input voltage .... 7 V
Operating free-air temperature range: SN54AS1004 .... −55 °C to 125 °C
                                       SN74AS1004 .... 0 °C to 70 °C
Storage temperature range .... −65 °C to 150 °C

recommended operating conditions

		SN54AS1004			SN74AS1004			UNIT
		MIN	NOM	MAX	MIN	NOM	MAX	
V_{CC}	Supply voltage	4.5	5	5.5	4.5	5	5.5	V
V_{IH}	High-level input voltage	2			2			V
V_{IL}	Low-level input voltage			0.8			0.8	V
I_{OH}	High-level output current			−40			−48	mA
I_{OL}	Low-level output current			40			48	mA
T_A	Operating free-air temperature	−55		125	0		70	°C

electrical characteristics over recommended operating free-air temperature range (unless otherwise noted)

PARAMETER	TEST CONDITIONS		SN54AS1004			SN74AS1004			UNIT
			MIN	TYP†	MAX	MIN	TYP†	MAX	
V_{IK}	$V_{CC} = 4.5$ V,	$I_I = -18$ mA			−1.2			−1.2	V
V_{OH}	$V_{CC} = 4.5$ V to 5.5 V,	$I_{OH} = -2$ mA	$V_{CC}-2$			$V_{CC}-2$			V
	$V_{CC} = 4.5$ V,	$I_{OH} = -3$ mA	2.4	3.2		2.4	3.2		
	$V_{CC} = 4.5$ V,	$I_{OH} = -40$ mA	2						
	$V_{CC} = 4.5$ V,	$I_{OH} = -48$ mA				2			
V_{OL}	$V_{CC} = 4.5$ V,	$I_{OL} = 40$ mA		0.25	0.5				V
	$V_{CC} = 4.5$ V,	$I_{OL} = 48$ mA					0.35	0.5	
I_I	$V_{CC} = 5.5$ V,	$V_I = 7$ V			0.1			0.1	mA
I_{IH}	$V_{CC} = 5.5$ V,	$V_I = 2.7$ V			20			20	µA
I_{IL}	$V_{CC} = 5.5$ V,	$V_I = 0.4$ V			−0.5			−0.5	mA
I_O‡	$V_{CC} = 5.5$ V,	$V_O = 2.25$ V	−135			−135			mA
I_{CCH}	$V_{CC} = 5.5$ V,	$V_I = 0$ V		3.2	5		3.2	5	mA
I_{CCL}	$V_{CC} = 5.5$ V,	$V_I = 4.5$ V		17.2	28		17.2	28	mA

†All typical values are at $V_{CC} = 5$ V, $T_A = 25$ °C.
‡The output conditions have been chosen to produce a current that closely approximates one half of the true short-circuit output current, I_{OS}.

switching characteristics (see Note 1)

PARAMETER	FROM (INPUT)	TO (OUTPUT)	$V_{CC} = 4.5$ V to 5.5 V, $C_L = 50$ pF, $R_L = 500$ Ω, $T_A = $ MIN to MAX				UNIT
			SN54AS1004		SN74AS1004		
			MIN	MAX	MIN	MAX	
t_{PLH}	A or B	Y	1	4.5	1	3.5	ns
t_{PHL}			1	4.5	1	3.5	

NOTE 1: For load circuit and voltage waveforms, see page 1-12.

ALS AND AS CIRCUITS

TEXAS INSTRUMENTS
POST OFFICE BOX 225012 • DALLAS, TEXAS 75265

2
ALS AND AS CIRCUITS

TYPES SN54ALS1005, SN74ALS1005
HEX INVERTING BUFFERS WITH OPEN-COLLECTOR OUTPUTS

D2661, APRIL 1982—REVISED DECEMBER 1983

- Buffer Version of 'ALS05
- Package Options Include Both Plastic and Ceramic Chip Carriers in Addition to Plastic and Ceramic DIPs
- Dependable Texas Instruments Quality and Reliability

description

These devices contain six independent inverting buffers. They perform the Boolean function $Y = \overline{A}$. The open-collector outputs require pull-up resistors to perform correctly. They may be connected to other open-collector outputs to implement active-low wired-OR or active-high wired-AND functions. Open-collector devices are often used to generate higher V_{OH} levels.

The SN54ALS1005 is characterized for operation over the full military temperature range of −55°C to 125°C. The SN74ALS1005 is characterized for operation from 0°C to 70°C.

SN54ALS1005 . . . J PACKAGE
SN74ALS1005 . . . N PACKAGE
(TOP VIEW)

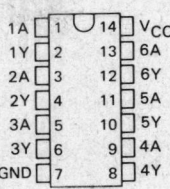

SN54ALS1005 . . . FH PACKAGE
SN74ALS1005 . . . FN PACKAGE
(TOP VIEW)

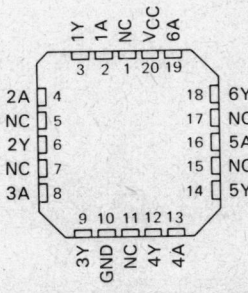

NC—No internal connection

FUNCTION TABLE (each inverter)

INPUT A	OUTPUT Y
H	L
L	H

logic symbol

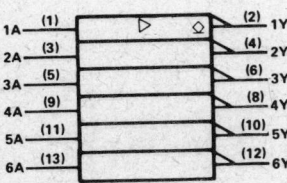

Pin numbers shown are for J and N packages.

ALS AND AS CIRCUITS 2

Copyright © 1983 by Texas Instruments Incorporated

TYPES SN54ALS1005, SN74ALS1005
HEX INVERTING BUFFERS WITH OPEN-COLLECTOR OUTPUTS

absolute maximum ratings over operating free-air temperature range (unless otherwise noted)

Supply voltage, V_{CC} .... 7 V
Input voltage .... 7 V
Off-state output voltage .... 7 V
Operating free-air temperature range: SN54ALS1005 .... −55 °C to 125 °C
                                      SN74ALS1005 .... 0 °C to 70 °C
Storage temperature range .... −65 °C to 150 °C

recommended operating conditions

		SN54ALS1005			SN74ALS1005			UNIT
		MIN	NOM	MAX	MIN	NOM	MAX	
V_{CC}	Supply voltage	4.5	5	5.5	4.5	5	5.5	V
V_{IH}	High-level input voltage	2			2			V
V_{IL}	Low-level input voltage			0.8			0.8	V
V_{OH}	High-level output voltage			5.5			5.5	V
I_{OL}	Low-level output current			12			24	mA
T_A	Operating free-air temperature	−55		125	0		70	°C

electrical characteristics over recommended operating free-air temperature range (unless otherwise noted)

PARAMETER	TEST CONDITIONS		SN54ALS1005			SN74ALS1005			UNIT
			MIN	TYP†	MAX	MIN	TYP†	MAX	
V_{IK}	V_{CC} = 4.5 V,	I_I = −18 mA			−1.5			−1.5	V
I_{OH}	V_{CC} = 4.5 V,	V_{OH} = 5.5 V			0.1			0.1	mA
V_{OL}	V_{CC} = 4.5 V,	I_{OL} = 12 mA		0.25	0.4		0.25	0.4	V
	V_{CC} = 4.5 V,	I_{OL} = 24 mA					0.35	0.5	
I_I	V_{CC} = 5.5 V,	V_I = 7 V			0.1			0.1	mA
I_{IH}	V_{CC} = 5.5 V,	V_I = 2.7 V			20			20	µA
I_{IL}	V_{CC} = 5.5 V,	V_I = 0.4 V			−0.1			−0.1	mA
I_{CCH}	V_{CC} = 5.5 V,	V_I = 0 V		0.9	3		0.9	3	mA
I_{CCL}	V_{CC} = 5.5 V,	V_I = 4.5 V		7	12		7	12	mA

†All typical values are at V_{CC} = 5 V, T_A = 25 °C.

switching characteristics (see Note 1)

PARAMETER	FROM (INPUT)	TO (OUTPUT)	V_{CC} = 4.5 V to 5.5 V, C_L = 50 pF, R_L = 680 Ω, T_A = MIN to MAX				UNIT
			SN54ALS1005		SN74ALS1005		
			MIN	MAX	MIN	MAX	
t_{PLH}	A	Y	5	35	5	30	ns
t_{PHL}			2	12	2	10	

NOTE 1: For load circuit and voltage waveforms, see page 1-12.

TEXAS INSTRUMENTS
POST OFFICE BOX 225012 • DALLAS, TEXAS 75265

TYPES SN54ALS1008A, SN54AS1008, SN74ALS1008A, SN74AS1008
QUADRUPLE 2-INPUT POSITIVE-AND BUFFERS/DRIVERS

D2661, DECEMBER 1982 — REVISED DECEMBER 1983

- 'ALS1008A is a Buffer Version of 'ALS08
- 'AS1008 is a Driver Version of 'AS08
- 'AS1008 Offers High Capacitive Drive Capability
- Package Options Include Both Plastic and Ceramic Chip Carriers in Addition to Plastic and Ceramic DIPs
- Dependable Texas Instruments Quality and Reliability

SN54ALS1008A, SN54AS1008 . . . J PACKAGE
SN74ALS1008A, SN74AS1008 . . . N PACKAGE
(TOP VIEW)

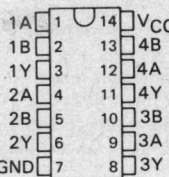

description

These devices contain four independent 2-input AND buffers/drivers. They perform the Boolean functions $Y = A \cdot B$ or $Y = \overline{A} + \overline{B}$ in positive logic.

The SN54ALS1008A and SN54AS1008 are characterized for operation over the full military temperature range of $-55\,°C$ to $125\,°C$. The SN74ALS1008A and SN74AS1008 are characterized for operation from $0\,°C$ to $70\,°C$.

SN54ALS1008A, SN54AS1008 . . . FH PACKAGE
SN74ALS1008A, SN74AS1008 . . . FN PACKAGE
(TOP VIEW)

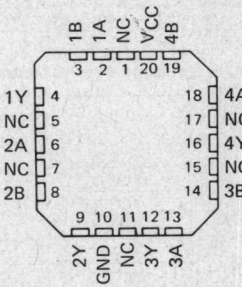

NC — No internal connection

FUNCTION TABLE
(each gate)

INPUTS		OUTPUT
A	B	Y
H	H	H
L	X	L
X	L	L

logic symbol

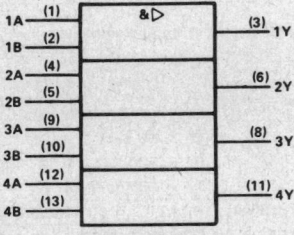

Pin numbers shown are for J and N packages.

Copyright © 1982 by Texas Instruments Incorporated

Texas Instruments
POST OFFICE BOX 225012 • DALLAS, TEXAS 75265

TYPES SN54ALS1008A, SN74ALS1008A
QUADRUPLE 2-INPUT POSITIVE-AND BUFFERS

absolute maximum ratings over operating free-air temperature range (unless otherwise noted)

Supply voltage, V_{CC} .... 7 V
Input voltage .... 7 V
Operating free-air temperature range: SN54ALS1008A .... −55 °C to 125 °C
                                     SN74ALS1008A .... 0 °C to 70 °C
Storage temperature range .... −65 °C to 150 °C

recommended operating conditions

		SN54ALS1008A			SN74ALS1008A			UNIT
		MIN	NOM	MAX	MIN	NOM	MAX	
V_{CC}	Supply voltage	4.5	5	5.5	4.5	5	5.5	V
V_{IH}	High-level input voltage	2			2			V
V_{IL}	Low-level input voltage			0.8			0.8	V
I_{OH}	High-level output current			−1			−2.6	mA
I_{OL}	Low-level output current			12			24	mA
T_A	Operating free-air temperature	−55		125	0		70	°C

electrical characteristics over recommended operating free-air temperature range (unless otherwise noted)

PARAMETER	TEST CONDITIONS		SN54ALS1008A			SN74ALS1008A			UNIT
			MIN	TYP†	MAX	MIN	TYP†	MAX	
V_{IK}	$V_{CC} = 4.5$ V,	$I_I = -18$ mA			−1.5			−1.5	V
V_{OH}	$V_{CC} = 4.5$ V to 5.5 V,	$I_{OH} = -0.4$ mA	$V_{CC}-2$			$V_{CC}-2$			V
	$V_{CC} = 4.5$ V,	$I_{OH} = -1$ mA	2.4	3.3					
	$V_{CC} = 4.5$ V,	$I_{OH} = -2.6$ mA				2.4	3.2		
V_{OL}	$V_{CC} = 4.5$ V,	$I_{OL} = 12$ mA		0.25	0.4				V
	$V_{CC} = 4.5$ V,	$I_{OL} = 24$ mA					0.35	0.5	
I_I	$V_{CC} = 5.5$ V,	$V_I = 7$ V			0.1			0.1	mA
I_{IH}	$V_{CC} = 5.5$ V,	$V_I = 2.7$ V			20			20	μA
I_{IL}	$V_{CC} = 5.5$ V,	$V_I = 0.4$ V			−0.1			−0.1	mA
I_O‡	$V_{CC} = 5.5$ V,	$V_O = 2.25$ V	−30		−112	−30		−112	mA
I_{CCH}	$V_{CC} = 5.5$ V,	$V_I = 4.5$ V		1.8	3		1.8	3	mA
I_{CCL}	$V_{CC} = 5.5$ V,	$V_I = 0$ V		5.7	9.3		5.7	9.3	mA

† All typical values are at $V_{CC} = 5$ V, $T_A = 25$ °C.
‡ The output conditions have been chosen to produce a current that closely approximates one half of the true short-circuit output current, I_{OS}.

switching characteristics (see Note 1)

PARAMETER	FROM (INPUT)	TO (OUTPUT)	$V_{CC} = 4.5$ V to 5.5 V, $C_L = 50$ pF, $R_L = 500$ Ω, $T_A =$ MIN to MAX				UNIT
			SN54ALS1008A		SN74ALS1008A		
			MIN	MAX	MIN	MAX	
t_{PLH}	A or B	Y	2	11	2	9	ns
t_{PHL}			3	11	3	9	

NOTE 1: For load circuit and voltage waveforms, see page 1-12.

Texas Instruments
POST OFFICE BOX 225012 • DALLAS, TEXAS 75265

TYPES SN54AS1008, SN74AS1008
QUADRUPLE 2-INPUT POSITIVE-AND DRIVERS

absolute maximum ratings over operating free-air temperature range (unless otherwise noted)

Supply voltage, V_{CC} .... 7 V
Input voltage .... 7 V
Operating free-air temperature range: SN54AS1008 .... −55 °C to 125 °C
    SN74AS1008 .... 0 °C to 70 °C
Storage temperature range .... −65 °C to 150 °C

recommended operating conditions

		SN54AS1008			SN74AS1008			UNIT
		MIN	NOM	MAX	MIN	NOM	MAX	
V_{CC}	Supply voltage	4.5	5	5.5	4.5	5	5.5	V
V_{IH}	High-level input voltage	2			2			V
V_{IL}	Low-level input voltage			0.8			0.8	V
I_{OH}	High-level output current			−40			−48	mA
I_{OL}	Low-level output current			40			48	mA
T_A	Operating free-air temperature	−55		125	0		70	°C

electrical characteristics over recommended operating free-air temperature range (unless otherwise noted)

PARAMETER	TEST CONDITIONS		SN54AS1008			SN74AS1008			UNIT
			MIN	TYP†	MAX	MIN	TYP†	MAX	
V_{IK}	V_{CC} = 4.5 V,	I_I = −18 mA			−1.2			−1.2	V
V_{OH}	V_{CC} = 4.5 V to 5.5 V,	I_{OH} = −2 mA	V_{CC}−2			V_{CC}−2			V
	V_{CC} = 4.5 V,	I_{OH} = −3 mA	2.4	3.2		2.4	3.2		
	V_{CC} = 4.5 V,	I_{OH} = −40 mA	2						
	V_{CC} = 4.5 V,	I_{OH} = −48 mA				2			
V_{OL}	V_{CC} = 4.5 V,	I_{OL} = 40 mA		0.25	0.5				V
	V_{CC} = 4.5 V,	I_{OL} = 48 mA					0.35	0.5	
I_I	V_{CC} = 5.5 V,	V_I = 7 V			0.1			0.1	mA
I_{IH}	V_{CC} = 5.5 V,	V_I = 2.7 V			20			20	µA
I_{IL}	V_{CC} = 5.5 V,	V_I = 0.4 V			−0.5			−0.5	mA
I_O‡	V_{CC} = 5.5 V,	V_O = 2.25 V			−135			−135	mA
I_{CCH}	V_{CC} = 5.5 V,	V_I = 4.5 V		5.6	9.5		5.6	9.5	mA
I_{CCL}	V_{CC} = 5.5 V,	V_I = 0 V		13.5	22		13.5	22	mA

† All typical values are at V_{CC} = 5 V, T_A = 25 °C.
‡ The output conditions have been chosen to produce a current that closely approximates one half of the true short-circuit output current, I_{OS}.

switching characteristics (see Note 1)

PARAMETER	FROM (INPUT)	TO (OUTPUT)	V_{CC} = 4.5 V to 5.5 V, C_L = 50 pF, R_L = 500 Ω, T_A = MIN to MAX				UNIT
			SN54AS1008		SN74AS1008		
			MIN	MAX	MIN	MAX	
t_{PLH}	A or B	Y	1	6	1	5	ns
t_{PHL}			1	6	1	5	

NOTE 1: For load circuit and voltage waveforms, see page 1-12.

ALS AND AS CIRCUITS

TEXAS INSTRUMENTS
POST OFFICE BOX 225012 • DALLAS, TEXAS 75265

2
ALS AND AS CIRCUITS

TYPES SN54ALS1010A, SN74ALS1010A
TRIPLE 3-INPUT POSITIVE-NAND BUFFERS

D2661, APRIL 1982 – REVISED DECEMBER 1983

- Buffer Version of 'ALS10
- Package Options Include Both Plastic and Ceramic Chip Carriers in Addition to Plastic and Ceramic DIPs
- Dependable Texas Instruments Quality and Reliability

SN54ALS1010A . . . J PACKAGE
SN74ALS1010A . . . N PACKAGE
(TOP VIEW)

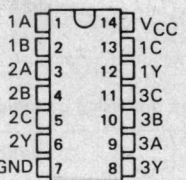

description

These devices contain three independent 3-input NAND buffers. They perform the Boolean functions $Y = \overline{A \cdot B \cdot C}$ or $Y = \overline{A} + \overline{B} + \overline{C}$ in positive logic.

The SN54ALS1010A is characterized for operation over the full military temperature range of $-55\,°C$ to $125\,°C$. The SN74ALS1010A is characterized for operation from $0\,°C$ to $70\,°C$.

SN54ALS1010A . . . FH PACKAGE
SN74ALS1010A . . . FN PACKAGE
(TOP VIEW)

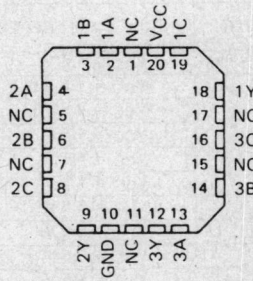

NC – No internal connection

FUNCTION TABLE (each gate)

INPUTS			OUTPUT
A	B	C	Y
H	H	H	L
L	X	X	H
X	L	X	H
X	X	L	H

logic symbol

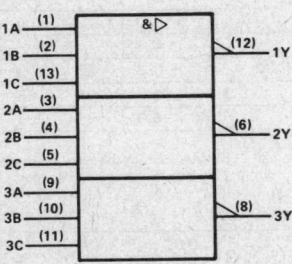

Pin numbers shown are for J and N packages.

Copyright © 1982 by Texas Instruments Incorporated

TEXAS INSTRUMENTS
POST OFFICE BOX 225012 • DALLAS, TEXAS 75265

TYPES SN54ALS1010A, SN74ALS1010A
TRIPLE 3-INPUT POSITIVE-NAND BUFFERS

absolute maximum ratings over operating free-air temperature range (unless otherwise noted)

Supply voltage, V_{CC} .... 7 V
Input voltage .... 7 V
Operating free-air temperature range: SN54ALS1010A .... −55 °C to 125 °C
                                          SN74ALS1010A .... 0 °C to 70 °C
Storage temperature range .... −65 °C to 150 °C

recommended operating conditions

		SN54ALS1010A			SN74ALS1010A			UNIT
		MIN	NOM	MAX	MIN	NOM	MAX	
V_{CC}	Supply voltage	4.5	5	5.5	4.5	5	5.5	V
V_{IH}	High-level input voltage	2			2			V
V_{IL}	Low-level input voltage			0.8			0.8	V
I_{OH}	High-level output current			−1			−2.6	mA
I_{OL}	Low-level output current			12			24	mA
T_A	Operating free-air temperature	−55		125	0		70	°C

electrical characteristics over recommended operating free-air temperature range (unless otherwise noted)

PARAMETER	TEST CONDITIONS		SN54ALS1010A			SN74ALS1010A			UNIT
			MIN	TYP†	MAX	MIN	TYP†	MAX	
V_{IK}	V_{CC} = 4.5 V,	I_I = −18 mA			−1.5			−1.5	V
V_{OH}	V_{CC} = 4.5 V to 5.5 V,	I_{OH} = −0.4 mA	V_{CC}−2			V_{CC}−2			V
	V_{CC} = 4.5 V,	I_{OH} = −1 mA	2.4	3.3					
	V_{CC} = 4.5 V,	I_{OH} = −2.6 mA				2.4	3.2		
V_{OL}	V_{CC} = 4.5 V,	I_{OL} = 12 mA		0.25	0.4		0.25	0.4	V
	V_{CC} = 4.5 V,	I_{OL} = 24 mA					0.35	0.5	
I_I	V_{CC} = 5.5 V,	V_I = 7 V			0.1			0.1	mA
I_{IH}	V_{CC} = 5.5 V,	V_I = 2.7 V			20			20	µA
I_{IL}	V_{CC} = 5.5 V,	V_I = 0.4 V			−0.1			−0.1	mA
I_O‡	V_{CC} = 5.5 V,	V_O = 2.25 V	−30		−112	−30		−112	mA
I_{CCH}	V_{CC} = 5.5 V,	V_I = 0 V		0.65	1.2		0.65	1.2	mA
I_{CCL}	V_{CC} = 5.5 V,	V_I = 4.5 V		3.6	5.8		3.6	5.8	mA

†All typical values are at V_{CC} = 5 V, T_A = 25°C.
‡The output conditions have been chosen to produce a current that closely approximates one half of the true short-circuit output current, I_{OS}.

switching characteristics (see Note 1)

PARAMETER	FROM (INPUT)	TO (OUTPUT)	V_{CC} = 4.5 V to 5.5 V, C_L = 50 pF, R_L = 500 Ω, T_A = MIN to MAX				UNIT
			SN54ALS1010A		SN74ALS1010A		
			MIN	MAX	MIN	MAX	
t_{PLH}	Any	Y	2	10	2	8	ns
t_{PHL}			2	10	2	7	

NOTE 1: For load circuit and voltage waveforms, see page 1-12.

TEXAS INSTRUMENTS
POST OFFICE BOX 225012 • DALLAS, TEXAS 75265

TYPES SN54ALS1011A, SN74ALS1011A
TRIPLE 3-INPUT POSITIVE-AND BUFFERS

D2661, APRIL 1982—REVISED DECEMBER 1983

- Buffer Version of 'ALS11
- Package Options Include Both Plastic and Ceramic Chip Carriers in Addition to Plastic and Ceramic DIPs
- Dependable Texas Instruments Quality and Reliability

description

These devices contain three independent 3-input AND buffers. They perform the Boolean functions $Y = A \cdot B \cdot C$ or $Y = \overline{A} + \overline{B} + \overline{C}$ in positive logic.

The SN54ALS1011A is characterized for operation over the full military temperature range of $-55\,°C$ to $125\,°C$. The SN74ALS1011A is characterized for operation from $0\,°C$ to $70\,°C$.

SN54ALS1011A ... J PACKAGE
SN74ALS1011A ... N PACKAGE
(TOP VIEW)

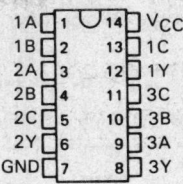

SN54ALS1011A ... FH PACKAGE
SN74ALS1011A ... FN PACKAGE
(TOP VIEW)

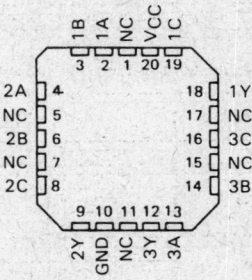

NC—No internal connection

FUNCTION TABLE (each gate)

INPUTS			OUTPUT
A	B	C	Y
H	H	H	H
L	X	X	L
X	L	X	L
X	X	L	L

logic symbol

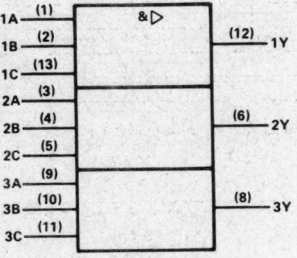

Pin numbers shown are for J and N packages.

ALS AND AS CIRCUITS

Copyright © 1982 by Texas Instruments Incorporated

TEXAS INSTRUMENTS
POST OFFICE BOX 225012 • DALLAS, TEXAS 75265

TYPES SN54ALS1011A, SN74ALS1011A
TRIPLE 3-INPUT POSITIVE-AND BUFFERS

absolute maximum ratings over operating free-air temperature range (unless otherwise noted)

Supply voltage, V_{CC} .... 7 V
Input voltage .... 7 V
Operating free-air temperature range: SN54ALS1011A .... −55 °C to 125 °C
                                      SN74ALS1011A .... 0 °C to 70 °C
Storage temperature range .... −65 °C to 150 °C

recommended operating conditions

		SN54ALS1011A			SN74ALS1011A			UNIT
		MIN	NOM	MAX	MIN	NOM	MAX	
V_{CC}	Supply voltage	4.5	5	5.5	4.5	5	5.5	V
V_{IH}	High-level input voltage	2			2			V
V_{IL}	Low-level input voltage			0.8			0.8	V
I_{OH}	High-level output current			−1			−2.6	mA
I_{OL}	Low-level output current			12			24	mA
T_A	Operating free-air temperature	−55		125	0		70	°C

electrical characteristics over recommended operating free-air temperature range (unless otherwise noted)

PARAMETER	TEST CONDITIONS		SN54ALS1011A			SN74ALS1011A			UNIT
			MIN	TYP†	MAX	MIN	TYP†	MAX	
V_{IK}	V_{CC} = 4.5 V,	I_I = −18 mA			−1.5			−1.5	V
V_{OH}	V_{CC} = 4.5 V to 5.5 V,	I_{OH} = −0.4 mA	V_{CC}−2			V_{CC}−2			V
	V_{CC} = 4.5 V,	I_{OH} = −1 mA	2.4	3.3					
	V_{CC} = 4.5 V,	I_{OH} = −2.6 mA				2.4	3.2		
V_{OL}	V_{CC} = 4.5 V,	I_{OL} = 12 mA		0.25	0.4		0.25	0.4	V
	V_{CC} = 4.5 V,	I_{OL} = 24 mA					0.35	0.5	
I_I	V_{CC} = 5.5 V,	V_I = 7 V			0.1			0.1	mA
I_{IH}	V_{CC} = 5.5 V,	V_I = 2.7 V			20			20	µA
I_{IL}	V_{CC} = 5.5 V,	V_I = 0.4 V			−0.1			−0.1	mA
I_O‡	V_{CC} = 5.5 V,	V_O = 2.25 V	−30		−112	−30		−112	mA
I_{CCH}	V_{CC} = 5.5 V,	V_I = 4.5 V		1.4	2.3		1.4	2.3	mA
I_{CCL}	V_{CC} = 5.5 V,	V_I = 0 V		4.3	7		4.3	7	mA

†All typical values are at V_{CC} = 5 V, T_A = 25 °C.
‡The output conditions have been chosen to produce a current that closely approximates one half of the true short-circuit output current, I_{OS}.

switching characteristics (see Note 1)

PARAMETER	FROM (INPUT)	TO (OUTPUT)	V_{CC} = 4.5 V to 5.5 V, C_L = 50 pF, R_L = 500 Ω, T_A = MIN to MAX				UNIT
			SN54ALS1011A		SN74ALS1011A		
			MIN	MAX	MIN	MAX	
t_{PLH}	Any	Y	2	12	2	10	ns
t_{PHL}			3	11	3	9	

NOTE 1: For load circuit and voltage waveforms, see page 1-12.

TEXAS INSTRUMENTS
POST OFFICE BOX 225012 • DALLAS, TEXAS 75265

TYPES SN54ALS1020A, SN74ALS1020A
DUAL 4-INPUT POSITIVE-NAND BUFFERS

D2661, APRIL 1982 – REVISED DECEMBER 1983

- Buffer Version of 'ALS20A
- Package Options Include Both Plastic and Ceramic Chip Carriers in Addition to Plastic and Ceramic DIPs
- Dependable Texas Instruments Quality and Reliability

description

These devices contain two independent 4-input NAND buffers. They perform the Boolean functions $Y = \overline{A \cdot B \cdot C \cdot D}$ or $Y = \overline{A} + \overline{B} + \overline{C} + \overline{D}$ in positive logic.

The SN54ALS1020A is characterized for operation over the full military temperature range of −55 °C to 125 °C. The SN74ALS1020A is characterized for operation from 0 °C to 70 °C.

SN54ALS1020A . . . J PACKAGE
SN74ALS1020A . . . N PACKAGE
(TOP VIEW)

```
1A  [ 1   14 ] VCC
1B  [ 2   13 ] 2D
NC  [ 3   12 ] 2C
1C  [ 4   11 ] NC
1D  [ 5   10 ] 2B
1Y  [ 6    9 ] 2A
GND [ 7    8 ] 2Y
```

SN54ALS1020A . . . FH PACKAGE
SN74ALS1020A . . . FN PACKAGE
(TOP VIEW)

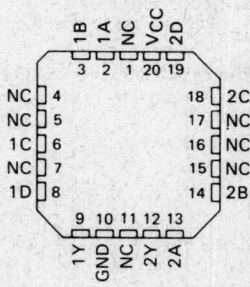

NC – No internal connection

FUNCTION TABLE (each gate)

INPUTS				OUTPUT
A	B	C	D	Y
H	H	H	H	L
L	X	X	X	H
X	L	X	X	H
X	X	L	X	H
X	X	X	L	H

logic symbol

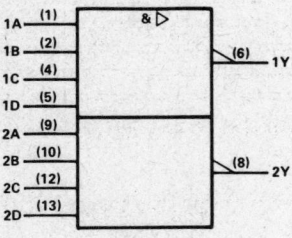

Pin numbers shown are for J and N packages.

Copyright © 1982 by Texas Instruments Incorporated

TYPES SN54ALS1020A, SN74ALS1020A
DUAL 4-INPUT POSITIVE-NAND BUFFERS

absolute maximum ratings over operating free-air temperature range (unless otherwise noted)

Supply voltage, V_{CC} .... 7 V
Input voltage .... 7 V
Operating free-air temperature range: SN54ALS1020A .... −55°C to 125°C
SN74ALS1020A .... 0°C to 70°C
Storage temperature range .... −65°C to 150°C

recommended operating conditions

		SN54ALS1020A			SN74ALS1020A			UNIT
		MIN	NOM	MAX	MIN	NOM	MAX	
V_{CC}	Supply voltage	4.5	5	5.5	4.5	5	5.5	V
V_{IH}	High-level input voltage	2			2			V
V_{IL}	Low-level input voltage			0.8			0.8	V
I_{OH}	High-level output current			−1			−2.6	mA
I_{OL}	Low-level output current			12			24	mA
T_A	Operating free-air temperature	−55		125	0		70	°C

electrical characteristics over recommended operating free-air temperature range (unless otherwise noted)

PARAMETER	TEST CONDITIONS		SN54ALS1020A			SN74ALS1020A			UNIT
			MIN	TYP†	MAX	MIN	TYP†	MAX	
V_{IK}	V_{CC} = 4.5 V,	I_I = −18 mA			−1.5			−1.5	V
V_{OH}	V_{CC} = 4.5 V to 5.5 V,	I_{OH} = −0.4 mA	V_{CC}−2			V_{CC}−2			V
	V_{CC} = 4.5 V,	I_{OH} = −1 mA	2.4	3.3					
	V_{CC} = 4.5 V,	I_{OH} = −2.6 mA				2.4	3.2		
V_{OL}	V_{CC} = 4.5 V,	I_{OL} = 12 mA		0.25	0.4		0.25	0.4	V
	V_{CC} = 4.5 V,	I_{OL} = 24 mA					0.35	0.5	
I_I	V_{CC} = 5.5 V,	V_I = 7 V			0.1			0.1	mA
I_{IH}	V_{CC} = 5.5 V,	V_I = 2.7 V			20			20	μA
I_{IL}	V_{CC} = 5.5 V,	V_I = 0.4 V			−0.1			−0.1	mA
I_O‡	V_{CC} = 5.5 V,	V_O = 2.25 V	−30		−112	−30		−112	mA
I_{CCH}	V_{CC} = 5.5 V,	V_I = 0 V		0.5	0.8		0.5	0.8	mA
I_{CCL}	V_{CC} = 5.5 V,	V_I = 4.5 V		2.4	3.9		2.4	3.9	mA

†All typical values are at V_{CC} = 5 V, T_A = 25°C.
‡The output conditions have been chosen to produce a current that closely approximates one half of the true short-circuit output current, I_{OS}.

switching characteristics (see Note 1)

PARAMETER	FROM (INPUT)	TO (OUTPUT)	V_{CC} = 4.5 V to 5.5 V, C_L = 50 pF, R_L = 500 Ω, T_A = MIN to MAX				UNIT
			SN54ALS1020A		SN74ALS1020A		
			MIN	MAX	MIN	MAX	
t_{PLH}	Any	Y	2	10	2	8	ns
t_{PHL}			2	10	2	7	

NOTE 1: For load circuit and voltage waveforms, see page 1-12.

TYPES SN54ALS1032A, SN54AS1032, SN74ALS1032A, SN74AS1032
QUADRUPLE 2-INPUT POSITIVE-OR BUFFERS/DRIVERS

D2661, DECEMBER 1982 – REVISED DECEMBER 1983

- 'ALS1032A is a Buffer Version of 'ALS32
- 'AS1032 is a Driver Version of 'AS32
- 'AS1032 Offers High Capacitive Drive Capability
- Package Options Include Both Plastic and Ceramic Chip Carriers in Addition to Plastic and Ceramic DIPs
- Dependable Texas Instruments Quality and Reliability

SN54ALS1032A, SN54AS1032 . . . J PACKAGE
SN74ALS1032A, SN74AS1032 . . . N PACKAGE
(TOP VIEW)

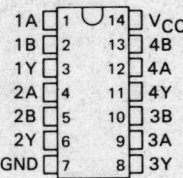

description

These devices contain four independent 2-input OR buffers/drivers. They perform the Boolean functions $Y = A + B$ or $Y = \overline{\overline{A} \cdot \overline{B}}$ in positive logic.

The SN54ALS1032A and SN54AS1032 are characterized for operation over the full military temperature range of $-55\,°C$ to $125\,°C$. The SN74ALS1032A and SN74AS1032 are characterized for operation from $0\,°C$ to $70\,°C$.

SN54ALS1032A, SN54AS1032 . . . FH PACKAGE
SN74ALS1032A, SN74AS1032 . . . FN PACKAGE
(TOP VIEW)

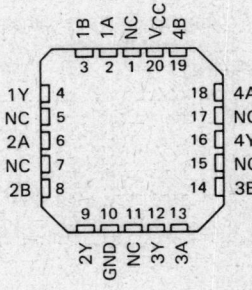

NC – No internal connection

FUNCTION TABLE
(each gate)

| INPUTS | | OUTPUT |
A	B	Y
H	X	H
X	H	H
L	L	L

logic symbol

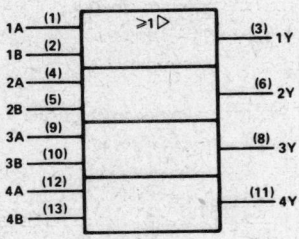

Pin numbers shown are for J and N packages.

Copyright © 1982 by Texas Instruments Incorporated

TEXAS INSTRUMENTS
POST OFFICE BOX 225012 • DALLAS, TEXAS 75265

TYPES SN54ALS1032A, SN74ALS1032A
QUADRUPLE 2-INPUT POSITIVE-OR BUFFERS

absolute maximum ratings over operating free-air temperature range (unless otherwise noted)

Supply voltage, V_{CC} .... 7 V
Input voltage .... 7 V
Operating free-air temperature range: SN54ALS1032A .... −55 °C to 125 °C
                                     SN74ALS1032A .... 0 °C to 70 °C
Storage temperature range .... −65 °C to 150 °C

recommended operating conditions

		SN54ALS1032A			SN74ALS1032A			UNIT
		MIN	NOM	MAX	MIN	NOM	MAX	
V_{CC}	Supply voltage	4.5	5	5.5	4.5	5	5.5	V
V_{IH}	High-level input voltage	2			2			V
V_{IL}	Low-level input voltage			0.8			0.8	V
I_{OH}	High-level output current			−1			−2.6	mA
I_{OL}	Low-level output current			12			24	mA
T_A	Operating free-air temperature	−55		125	0		70	°C

electrical characteristics over recommended operating free-air temperature range (unless otherwise noted)

PARAMETER	TEST CONDITIONS		SN54ALS1032A			SN74ALS1032A			UNIT
			MIN	TYP†	MAX	MIN	TYP†	MAX	
V_{IK}	$V_{CC} = 4.5$ V,	$I_I = -18$ mA			−1.5			−1.5	V
V_{OH}	$V_{CC} = 4.5$ V to 5.5 V,	$I_{OH} = -0.4$ mA	$V_{CC}-2$			$V_{CC}-2$			V
	$V_{CC} = 4.5$ V,	$I_{OH} = -1$ mA	2.4	3.3					
	$V_{CC} = 4.5$ V,	$I_{OH} = -2.6$ mA				2.4	3.2		
V_{OL}	$V_{CC} = 4.5$ V,	$I_{OL} = 12$ mA		0.25	0.4		0.25	0.4	V
	$V_{CC} = 4.5$ V,	$I_{OL} = 24$ mA					0.35	0.5	
I_I	$V_{CC} = 5.5$ V,	$V_I = 7$ V			0.1			0.1	mA
I_{IH}	$V_{CC} = 5.5$ V,	$V_I = 2.7$ V			20			20	µA
I_{IL}	$V_{CC} = 5.5$ V,	$V_I = 0.4$ V			−0.1			−0.1	mA
I_O‡	$V_{CC} = 5.5$ V,	$V_O = 2.25$ V	−30		−112	−30		−112	mA
I_{CCH}	$V_{CC} = 5.5$ V,	$V_I = 4.5$ V		2.5	5		2.5	5	mA
I_{CCL}	$V_{CC} = 5.5$ V,	$V_I = 0$ V		6.6	10.6		6.6	10.6	mA

†All typical values are at $V_{CC} = 5$ V, $T_A = 25$ °C.
‡The output conditions have been chosen to produce a current that closely approximates one half of the true short-circuit output current, I_{OS}.

switching characteristics (see Note 1)

PARAMETER	FROM (INPUT)	TO (OUTPUT)	$V_{CC} = 4.5$ V to 5.5 V, $C_L = 50$ pF, $R_L = 500$ Ω, T_A = MIN to MAX				UNIT
			SN54ALS1032A		SN74ALS1032A		
			MIN	MAX	MIN	MAX	
t_{PLH}	A or B	Y	2	12	2	9	ns
t_{PHL}			3	15	3	12	

NOTE 1: For load circuit and voltage waveforms, see page 1-12.

TEXAS INSTRUMENTS
POST OFFICE BOX 225012 • DALLAS, TEXAS 75265

TYPES SN54AS1032, SN74AS1032
QUADRUPLE 2-INPUT POSITIVE-OR DRIVERS

absolute maximum ratings over operating free-air temperature range (unless otherwise noted)

Supply voltage, V_{CC} .... 7 V
Input voltage .... 7 V
Operating free-air temperature range: SN54AS1032 .... −55 °C to 125 °C
                                     SN74AS1032 .... 0 °C to 70 °C
Storage temperature range .... −65 °C to 150 °C

recommended operating conditions

		SN54AS1032			SN74AS1032			UNIT
		MIN	NOM	MAX	MIN	NOM	MAX	
V_{CC}	Supply voltage	4.5	5	5.5	4.5	5	5.5	V
V_{IH}	High-level input voltage	2			2			V
V_{IL}	Low-level input voltage			0.8			0.8	V
I_{OH}	High-level output current			−40			−48	mA
I_{OL}	Low-level output current			40			48	mA
T_A	Operating free-air temperature	−55		125	0		70	°C

electrical characteristics over recommended operating free-air temperature range (unless otherwise noted)

PARAMETER	TEST CONDITIONS		SN54AS1032			SN74AS1032			UNIT
			MIN	TYP†	MAX	MIN	TYP†	MAX	
V_{IK}	V_{CC} = 4.5 V,	I_I = −18 mA			−1.2			−1.2	V
	V_{CC} = 4.5 V to 5.5 V,	I_{OH} = −2 mA	V_{CC}−2			V_{CC}−2			
V_{OH}	V_{CC} = 4.5 V,	I_{OH} = −3 mA	2.4	3.2		2.4	3.2		V
	V_{CC} = 4.5 V,	I_{OH} = −40 mA	2						
	V_{CC} = 4.5 V,	I_{OH} = −48 mA				2			
V_{OL}	V_{CC} = 4.5 V,	I_{OL} = 40 mA		0.25	0.5				V
	V_{CC} = 4.5 V,	I_{OL} = 48 mA					0.35	0.5	
I_I	V_{CC} = 5.5 V,	V_I = 7 V			0.1			0.1	mA
I_{IH}	V_{CC} = 5.5 V,	V_I = 2.7 V			20			20	μA
I_{IL}	V_{CC} = 5.5 V,	V_I = 0.9 V			−0.5			−0.5	mA
I_O‡	V_{CC} = 5.5 V,	V_O = 2.25 V	−135			−135			mA
I_{CCH}	V_{CC} = 5.5 V,	V_I = 4.5 V		7.7	11.5		7.7	11.5	mA
I_{CCL}	V_{CC} = 5.5 V,	V_I = 0 V		14.7	24		14.7	24	mA

†All typical values are at V_{CC} = 5 V, T_A = 25 °C.
‡The output conditions have been chosen to produce a current that closely approximates one half of the true short-circuit output current, I_{OS}.

switching characteristics (see Note 1)

PARAMETER	FROM (INPUT)	TO (OUTPUT)	V_{CC} = 4.5 V to 5.5 V, C_L = 50 pF, R_L = 500 Ω, T_A = MIN to MAX				UNIT
			SN54AS1032		SN74AS1032		
			MIN	MAX	MIN	MAX	
t_{PLH}	A or B	Y	1	7	1	5.5	ns
t_{PHL}			1	6.5	1	5.5	

NOTE 1: For load circuit and voltage waveforms, see page 1-12.

TEXAS INSTRUMENTS
POST OFFICE BOX 225012 • DALLAS, TEXAS 75265

2
ALS AND AS CIRCUITS

TYPES SN54ALS1034, SN54AS1034, SN74ALS1034, SN74AS1034 HEX DRIVERS

D2661, APRIL 1982—REVISED DECEMBER 1983

- **'AS1034 Offers High Capacitive-Drive Capability**
- **Noninverting Drivers**
- **Package Options Include Both Plastic and Ceramic Chip Carriers in Addition to Plastic and Ceramic DIPs**
- **Dependable Texas Instruments Quality and Reliability**

SN54ALS1034, SN54AS1034 . . . J PACKAGE
SN74ALS1034, SN74AS1034 . . . N PACKAGE
(TOP VIEW)

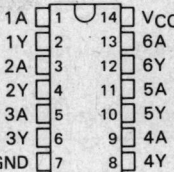

description

These devices contain six independent noninverting drivers. They perform the Boolean function Y = A.

The SN54ALS1034 and SN54AS1034 are characterized for operation over the full military temperature range of −55°C to 125°C. The SN74ALS1034 and SN74AS1034 are characterized for operation from 0°C to 70°C.

SN54ALS1034, SN54AS1034 . . . FH PACKAGE
SN74ALS1034, SN74AS1034 . . . FN PACKAGE
(TOP VIEW)

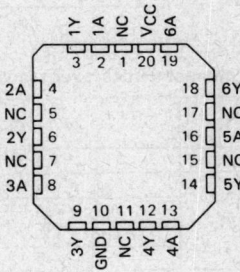

NC—No internal connection

FUNCTION TABLE (each buffer)

INPUT A	OUTPUT Y
H	H
L	L

logic symbol

```
1A ──(1)──▷──(2)── 1Y
2A ──(3)──────(4)── 2Y
3A ──(5)──────(6)── 3Y
4A ──(9)──────(8)── 4Y
5A ──(11)─────(10)── 5Y
6A ──(13)─────(12)── 6Y
```

Pin numbers shown are for J and N packages.

ALS AND AS CIRCUITS

Copyright © 1982 by Texas Instruments Incorporated

TEXAS INSTRUMENTS
POST OFFICE BOX 225012 • DALLAS, TEXAS 75265

TYPES SN54ALS1034, SN74ALS1034
HEX DRIVERS

absolute maximum ratings over operating free-air temperature range (unless otherwise noted)

Supply voltage, V_{CC} .... 7 V
Input voltage .... 7 V
Operating free-air temperature range: SN54ALS1034 .... −55 °C to 125 °C
SN74ALS1034 .... 0 °C to 70 °C
Storage temperature range .... −65 °C to 150 °C

recommended operating conditions

		SN54ALS1034			SN74ALS1034			UNIT
		MIN	NOM	MAX	MIN	NOM	MAX	
V_{CC}	Supply voltage	4.5	5	5.5	4.5	5	5.5	V
V_{IH}	High-level input voltage	2			2			V
V_{IL}	Low-level input voltage			0.8			0.8	V
I_{OH}	High-level output current			−12			−15	mA
I_{OL}	Low-level output current			12			24	mA
T_A	Operating free-air temperature	−55		125	0		70	°C

electrical characteristics over recommended operating free-air temperature range (unless otherwise noted)

PARAMETER	TEST CONDITIONS		SN54ALS1034			SN74ALS1034			UNIT
			MIN	TYP†	MAX	MIN	TYP†	MAX	
V_{IK}	V_{CC} = 4.5 V,	I_I = −18 mA			−1.5			−1.5	V
V_{OH}	V_{CC} = 4.5 V to 5.5 V,	I_{OH} = −0.4 mA	V_{CC}−2			V_{CC}−2			V
	V_{CC} = 4.5 V,	I_{OH} = −3 mA	2.4	3.2		2.4	3.2		
	V_{CC} = 4.5 V,	I_{OH} = −12 mA	2						
	V_{CC} = 4.5 V,	I_{OH} = −15 mA				2			
V_{OL}	V_{CC} = 4.5 V,	I_{OL} = 12 mA		0.25	0.4				V
	V_{CC} = 4.5 V,	I_{OL} = 24 mA					0.35	0.5	
I_I	V_{CC} = 5.5 V,	V_I = 7 V			0.1			0.1	mA
I_{IH}	V_{CC} = 5.5 V,	V_I = 2.7 V			20			20	µA
I_{IL}	V_{CC} = 5.5 V,	V_I = 0.4 V			−0.1			−0.1	mA
I_O‡	V_{CC} = 5.5 V,	V_O = 2.25 V	−30		−112	−30		−112	mA
I_{CCH}	V_{CC} = 5.5 V,	V_I = 4.5 V		3	6		3	6	mA
I_{CCL}	V_{CC} = 5.5 V,	V_I = 0 V		8	14		8	14	mA

†All typical values are at V_{CC} = 5 V, T_A = 25 °C.
‡The output conditions have been chosen to produce a current that closely approximates one half of the true short-circuit output current, I_{OS}.

switching characteristics (see Note 1)

PARAMETER	FROM (INPUT)	TO (OUTPUT)	V_{CC} = 4.5 V to 5.5 V, C_L = 50 pF, R_L = 500 Ω, T_A = MIN to MAX				UNIT
			SN54ALS1034		SN74ALS1034		
			MIN	MAX	MIN	MAX	
t_{PLH}	A	Y	1	10	1	8	ns
t_{PHL}			1	10	1	8	

NOTE 1: For load circuit and voltage waveforms, see page 1-12.

TEXAS INSTRUMENTS
POST OFFICE BOX 225012 • DALLAS, TEXAS 75265

TYPES SN54AS1034, SN74AS1034
HEX DRIVERS

absolute maximum ratings over operating free-air temperature range (unless otherwise noted)

Supply voltage, V_{CC} .... 7 V
Input voltage .... 7 V
Operating free-air temperature range: SN54AS1034 .... −55 °C to 125 °C
                                      SN74AS1034 .... 0 °C to 70 °C
Storage temperature range .... −65 °C to 150 °C

recommended operating conditions

		SN54AS1034			SN74AS1034			UNIT
		MIN	NOM	MAX	MIN	NOM	MAX	
V_{CC}	Supply voltage	4.5	5	5.5	4.5	5	5.5	V
V_{IH}	High-level input voltage	2			2			V
V_{IL}	Low-level input voltage			0.8			0.8	V
I_{OH}	High-level output current			−40			−48	mA
I_{OL}	Low-level output current			40			48	mA
T_A	Operating free-air temperature	−55		125	0		70	°C

electrical characteristics over recommended operating free-air temperature range (unless otherwise noted)

PARAMETER	TEST CONDITIONS		SN54AS1034			SN74AS1034			UNIT
			MIN	TYP†	MAX	MIN	TYP†	MAX	
V_{IK}	V_{CC} = 4.5 V,	I_I = −18 mA			−1.2			−1.2	V
V_{OH}	V_{CC} = 4.5 V to 5.5 V,	I_{OH} = −2 mA	V_{CC}−2			V_{CC}−2			V
	V_{CC} = 4.5 V,	I_{OH} = −3 mA	2.4	3.2		2.4	3.2		
	V_{CC} = 4.5 V,	I_{OH} = −40 mA	2						
	V_{CC} = 4.5 V,	I_{OH} = −48 mA				2			
V_{OL}	V_{CC} = 4.5 V,	I_{OL} = 40 mA		0.25	0.5				V
	V_{CC} = 4.5 V,	I_{OL} = 48 mA					0.35	0.5	
I_I	V_{CC} = 5.5 V,	V_I = 7 V			0.1			0.1	mA
I_{IH}	V_{CC} = 5.5 V,	V_I = 2.7 V			20			20	µA
I_{IL}	V_{CC} = 5.5 V,	V_I = 0.4 V			−0.5			−0.5	mA
I_O‡	V_{CC} = 5.5 V,	V_O = 2.25 V			−135			−135	mA
I_{CCH}	V_{CC} = 5.5 V,	V_I = 0 V		8.5	14		8.5	14	mA
I_{CCL}	V_{CC} = 5.5 V,	V_I = 4.5 V		20	33		20	33	mA

†All typical values are at V_{CC} = 5 V, T_A = 25 °C.
‡The output conditions have been chosen to produce a current that closely approximates one half of the true short-circuit output current, I_{OS}.

switching characteristics (see Note 1)

PARAMETER	FROM (INPUT)	TO (OUTPUT)	V_{CC} = 4.5 V to 5.5 V, C_L = 50 pF, R_L = 500 Ω, T_A = MIN to MAX				UNIT
			SN54AS1034		SN74AS1034		
			MIN	MAX	MIN	MAX	
t_{PLH}	A	Y	1	6	1	5	ns
t_{PHL}			1	6	1	5	

NOTE 1: For load circuit and voltage waveforms, see page 1-12.

TEXAS INSTRUMENTS
POST OFFICE BOX 225012 • DALLAS, TEXAS 75265

TYPES SN54ALS1035, SN74ALS1035
HEX NONINVERTING BUFFERS WITH OPEN-COLLECTOR OUTPUTS

D2661, APRIL 1982 – REVISED DECEMBER 1983

- Noninverting Buffers with Open-Collector Outputs
- Package Options Include Both Plastic and Ceramic Chip Carriers in Addition to Plastic and Ceramic DIPs
- Dependable Texas Instruments Quality and Reliability

description

These devices contain six independent noninverting buffers. They perform the boolean functions Y = A. The open-collector outputs require pull-up resistors to perform correctly. They may be connected to other open-collector outputs to implement active-low wired-OR or active-high wired-AND functions. Open-collector devices are often used to generate higher V_{OH} levels.

The SN54ALS1035 is characterized for operation over the full military temperature range of −55°C to 125°C. The SN74ALS1035 is characterized for operation from 0°C to 70°C.

SN54ALS1035 . . . J PACKAGE
SN74ALS1035 . . . N PACKAGE
(TOP VIEW)

```
  1A [ 1    14 ] VCC
  1Y [ 2    13 ] 6A
  2A [ 3    12 ] 6Y
  2Y [ 4    11 ] 5A
  3A [ 5    10 ] 5Y
  3Y [ 6     9 ] 4A
 GND [ 7     8 ] 4Y
```

FUNCTION TABLE (each buffer)

INPUT A	OUTPUT Y
H	H
L	L

SN54ALS1035 . . . FH PACKAGE
SN74ALS1035 . . . FN PACKAGE
(TOP VIEW)

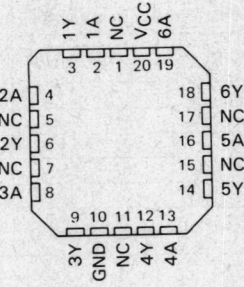

NC—No internal connection

logic symbol

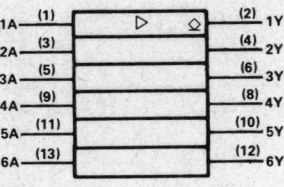

Pin numbers shown are for J and N packages.

Copyright © 1982 by Texas Instruments Incorporated

TYPES SN54ALS1035, SN74ALS1035
HEX NONINVERTING BUFFERS WITH OPEN-COLLECTOR OUTPUTS

absolute maximum ratings over operating free-air temperature range (unless otherwise noted)

Supply voltage, V_{CC} .... 7 V
Input voltage .... 7 V
Off-state output voltage .... 7 V
Operating free-air temperature range: SN54ALS1035 .... −55 °C to 125 °C
 SN74ALS1035 .... 0 °C to 70 °C
Storage temperature range .... −65 °C to 150 °C

recommended operating conditions

		SN54ALS1035			SN74ALS1035			UNIT
		MIN	NOM	MAX	MIN	NOM	MAX	
V_{CC}	Supply voltage	4.5	5	5.5	4.5	5	5.5	V
V_{IH}	High-level input voltage	2			2			V
V_{IL}	Low-level input voltage			0.8			0.8	V
V_{OH}	High-level output voltage			5.5			5.5	V
I_{OL}	Low-level output current			12			24	mA
T_A	Operating free-air temperature	−55		125	0		70	°C

electrical characteristics over recommended operating free-air temperature range (unless otherwise noted)

PARAMETER	TEST CONDITIONS		SN54ALS1035			SN74ALS1035			UNIT
			MIN	TYP[†]	MAX	MIN	TYP[†]	MAX	
V_{IK}	$V_{CC} = 4.5$ V,	$I_I = -18$ mA			−1.5			−1.5	V
I_{OH}	$V_{CC} = 4.5$ V,	$V_{OH} = 5.5$ V			0.1			0.1	mA
V_{OL}	$V_{CC} = 4.5$ V,	$I_{OL} = 12$ mA		0.25	0.4		0.25	0.4	V
	$V_{CC} = 4.5$ V,	$I_{OL} = 24$ mA					0.35	0.5	
I_I	$V_{CC} = 5.5$ V,	$V_I = 7$ V			0.1			0.1	mA
I_{IH}	$V_{CC} = 5.5$ V,	$V_I = 2.7$ V			20			20	µA
I_{IL}	$V_{CC} = 5.5$ V,	$V_I = 0.4$ V			−0.1			−0.1	mA
I_{CCH}	$V_{CC} = 5.5$ V,	$V_I = 4.5$ V	3		6	3		6	mA
I_{CCL}	$V_{CC} = 5.5$ V,	$V_I = 0$ V	8		14	8		14	mA

[†] All typical values are at $V_{CC} = 5$ V, $T_A = 25$ °C.

switching characteristics (see Note 1)

PARAMETER	FROM (INPUT)	TO (OUTPUT)	$V_{CC} = 4.5$ V to 5.5 V, $C_L = 50$ pF, $R_L = 680$ Ω, T_A = MIN to MAX				UNIT
			SN54ALS1035		SN74ALS1035		
			MIN	MAX	MIN	MAX	
t_{PLH}	A	Y	5	35	5	30	ns
t_{PHL}			2	14	2	12	

NOTE 1: For load circuit and voltage waveforms, see page 1-12.

Texas Instruments
POST OFFICE BOX 225012 • DALLAS, TEXAS 75265

TYPES SN54AS1036, SN74AS1036
QUADRUPLE 2-INPUT POSITIVE-NOR DRIVERS

D2661, DECEMBER 1983

- Quad Versions of AS805A
- Offers High Capacitive-Drive Capability
- Package Options Include Both Plastic and Ceramic Chip Carriers in Addition to Plastic and Ceramic DIPs
- Dependable Texas Instruments Quality and Reliability

description

These devices contain four independent 2-input NOR drivers. They perform the Boolean functions $Y = \overline{A+B}$ or $Y = \overline{A} \cdot \overline{B}$ in positive logic.

The SN54AS1036 is characterized for operation over the full military temperature range of $-55\,°C$ to $125\,°C$. The SN74AS1036 is characterized for operation from $0\,°C$ to $70\,°C$.

SN54AS1036 . . . J PACKAGE
SN74AS1036 . . . N PACKAGE
(TOP VIEW)

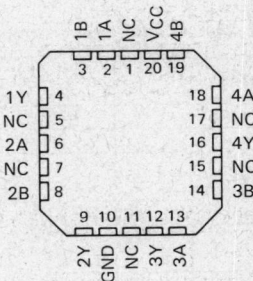

SN54AS1036 . . . FH PACKAGE
SN74AS1036 . . . FN PACKAGE
(TOP VIEW)

NC — No internal connection

FUNCTION TABLE (each gate)

INPUTS		OUTPUT
A	B	Y
H	X	L
X	H	L
L	L	H

logic symbol

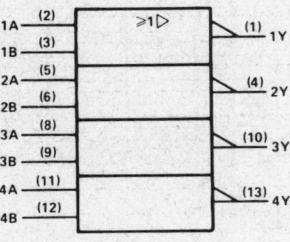

Pin numbers shown are for J and N packages.

Copyright © 1983 by Texas Instruments Incorporated

TEXAS INSTRUMENTS
POST OFFICE BOX 225012 • DALLAS, TEXAS 75265

TYPES SN54AS1036, SN74AS1036
QUADRUPLE 2-INPUT POSITIVE-NOR DRIVERS

absolute maximum ratings over operating free-air temperature range (unless otherwise noted)

Supply voltage, V_{CC} .... 7 V
Input voltage .... 7 V
Operating free-air temperature range: SN54AS1036 .... −55 °C to 125 °C
                                            SN74AS1036 .... 0 °C to 70 °C
Storage temperature range .... −65 °C to 150 °C

recommended operating conditions

		SN54AS1036			SN74AS1036			UNIT
		MIN	NOM	MAX	MIN	NOM	MAX	
V_{CC}	Supply voltage	4.5	5	5.5	4.5	5	5.5	V
V_{IH}	High-level input voltage	2			2			V
V_{IL}	Low-level input voltage			0.8			0.8	V
I_{OH}	High-level output current			−40			−48	mA
I_{OL}	Low-level output current			40			48	mA
T_A	Operating free-air temperature	−55		125	0		70	°C

electrical characteristics over recommended operating free-air temperature range (unless otherwise noted)

PARAMETER	TEST CONDITIONS		SN54AS1036			SN74AS1036			UNIT
			MIN	TYP†	MAX	MIN	TYP†	MAX	
V_{IK}	V_{CC} = 4.5 V,	I_I = −18 mA			−1.2			−1.2	V
V_{OH}	V_{CC} = 4.5 V to 5.5 V,	I_{OH} = −2 mA	V_{CC}−2			V_{CC}−2			V
	V_{CC} = 4.5 V,	I_{OH} = −3 mA	2.4	3.2		2.4	3.2		
	V_{CC} = 4.5 V,	I_{OH} = −40 mA	2						
	V_{CC} = 4.5 V,	I_{OH} = −48 mA				2			
V_{OL}	V_{CC} = 4.5 V,	I_{OL} = 40 mA		0.25	0.5				V
	V_{CC} = 4.5 V,	I_{OL} = 48 mA					0.35	0.5	
I_I	V_{CC} = 5.5 V,	V_I = 7 V			0.1			0.1	mA
I_{IH}	V_{CC} = 5.5 V,	V_I = 2.7 V			20			20	µA
I_{IL}	V_{CC} = 5.5 V,	V_I = 0.9 V			−0.5			−0.5	mA
I_O‡	V_{CC} = 5.5 V,	V_O = 2.25 V		−135			−135		mA
I_{CCH}	V_{CC} = 5.5 V,	V_I = 0 V		4.3	7		4.3	7	mA
I_{CCL}	V_{CC} = 5.5 V,	V_I = 4.5 V		14	23		14	23	mA

†All typical values are at V_{CC} = 5 V, T_A = 25 °C.
‡The output conditions have been chosen to produce a current that closely approximates one half of the true short-circuit output current, I_{OS}.

switching characteristics (see Note 1)

PARAMETER	FROM (INPUT)	TO (OUTPUT)	V_{CC} = 4.5 V to 5.5 V, C_L = 50 pF, R_L = 500 Ω, T_A = MIN to MAX				UNIT
			SN54AS1036		SN74AS1036		
			MIN	MAX	MIN	MAX	
t_{PLH}	A or B	Y	1	4.5	1	4	ns
t_{PHL}			1	4.5	1	4	

NOTE 1: For load circuit and voltage waveforms, see page 1-12.

TYPES SN54ALS1240, SN54ALS1241, SN74ALS1240, SN74ALS1241
OCTAL BUFFERS AND LINE DRIVERS WITH 3-STATE OUTPUTS

D2661, DECEMBER 1982–REVISED DECEMBER 1983

- Low-Power Version of 'ALS240 and 'ALS241
- 3-State Outputs Drive Bus Lines or Buffer Memory Address Registers
- P-N-P Inputs Reduce DC Loading
- Dependable Texas Instruments Quality and Reliability

description

These octal buffers and line drivers are designed specifically to improve both the performance and density of three-state memory address drivers, clock drivers, and bus-oriented receivers and transmitters. The designer has a choice of selected combinations of inverting and non-inverting outputs, symmetrical $\overline{G}$ (active-low output control) inputs, and complementary G and $\overline{G}$ inputs. These devices feature high fan-out and improved fan-in.

The -1 versions of the SN64ALS' parts are identical to the standard versions except that the recommended maximum I_{OL} is increased to 24 milliamperes. There are no -1 versions of the SN54ALS' parts.

The SN54ALS1240 and SN54ALS1241 are characterized for operation over the full military temperature range of −55°C to 125°C. The SN74ALS1240 and SN74ALS1241 are characterized for operation from 0°C to 70°C.

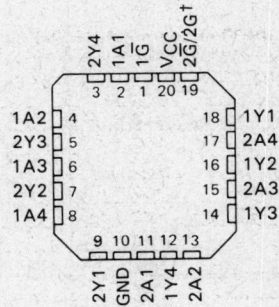

†$2\overline{G}$ for 'ALS1240 or 2G for 'ALS1241

TYPES SN54ALS1240, SN54ALS1241, SN74ALS1240, SN74ALS1241
OCTAL BUFFERS AND LINE DRIVERS WITH 3-STATE OUTPUTS

logic symbols

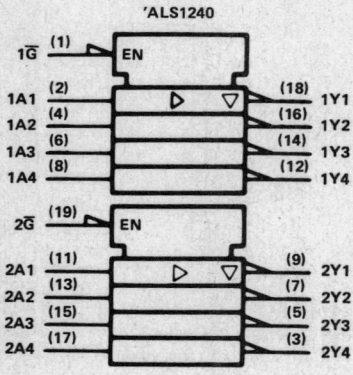 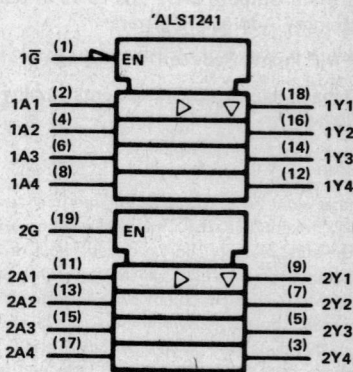

Pin numbers shown are for J and N packages.

functional block diagrams (positive logic)

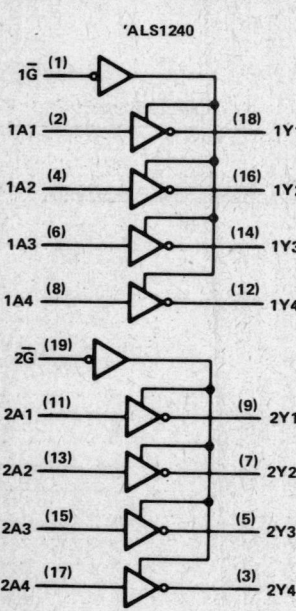 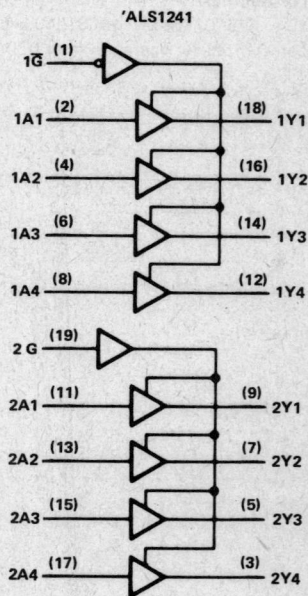

Texas Instruments
POST OFFICE BOX 225012 • DALLAS, TEXAS 75265

TYPES SN54ALS1240, SN54ALS1241, SN74ALS1240, SN74ALS1241
OCTAL BUFFERS AND LINE DRIVERS WITH 3-STATE OUTPUTS

absolute maximum ratings over operating free-air temperature range (unless otherwise noted)

Supply voltage, V_{CC} .... 7 V
Input voltage .... 7 V
Voltage applied to a disabled 3-state output .... 5.5 V
Operating free-air temperature range: SN54ALS1240, SN54ALS1241 .... −55 °C to 125 °C
                                             SN74ALS1240, SN74ALS1241 .... 0 °C to 70 °C
Storage temperature range .... −65 °C to 150 °C

recommended operating conditions

		SN54ALS1240 SN54ALS1241			SN74ALS1240 SN74ALS1241			UNIT
		MIN	NOM	MAX	MIN	NOM	MAX	
V_{CC}	Supply voltage	4.5	5	5.5	4.5	5	5.5	V
V_{IH}	High-level input voltage	2			2			V
V_{IL}	Low-level input voltage			0.8			0.8	V
I_{OH}	High-level output current			−12			−15	mA
I_{OL}	Low-level output current			8			16	mA
							24†	
T_A	Operating free-air temperature	−55		125	0		70	°C

†The extended limits apply only if V_{CC} is maintained between 4.75 V and 5.25 V.
 The 24-mA limit applies for the SN74ALS1240-1 and SN74ALS1241-1 only.

electrical characteristics over recommended operating free-air temperature range (unless otherwise noted)

PARAMETER	TEST CONDITIONS		SN54ALS1240 SN54ALS1241			SN74ALS1240 SN74ALS1241			UNIT
			MIN	TYP‡	MAX	MIN	TYP‡	MAX	
V_{IK}	V_{CC} = 4.5 V,	I_I = −18 mA			−1.5			−1.5	V
V_{OH}	V_{CC} = 4.5 V to 5.5 V,	I_{OH} = −0.4 mA	V_{CC}−2		3	V_{CC}−2			V
	V_{CC} = 4.5 V,	I_{OH} = −3 mA	2.4	3.2		2.4	3.2		
	V_{CC} = 4.5 V,	I_{OH} = −12 mA	2						
	V_{CC} = 4.5 V,	I_{OH} = −15 mA				2			
V_{OL}	V_{CC} = 4.5 V,	I_{OL} = 8 mA		0.25	0.4		0.25	0.4	V
	V_{CC} = 4.5 V,	I_{OL} = 16 mA					0.35	0.5	
	(I_{OL} = 24 mA for −1 versions)								
I_{OZH}	V_{CC} = 5.5 V,	V_O = 2.7 V			20			20	µA
I_{OZL}	V_{CC} = 5.5 V,	V_I = 0.4 V			−20			−20	µA
I_I	V_{CC} = 5.5 V,	V_I = 7 V			0.1			0.1	mA
I_{IH}	V_{CC} = 5.5 V,	V_I = 2.7 V			20			20	µA
I_{IL}	V_{CC} = 5.5 V,	V_I = 0.4 V			−0.1			−0.1	mA
I_O §	V_{CC} = 5.5 V,	V_O = 2.25 V	−30		−112	−30		−112	mA
I_{CC}	V_{CC} = 5.5 V	Outputs high		6.5			6.5		mA
		Outputs low		10			10		
		Outputs disabled		12			12		

‡All typical values are at V_{CC} = 5 V, T_A = 25 °C.
§The output conditions have been chosen to produce a current that closely approximates one half of the true short-circuit output current, I_{OS}.

ALS AND AS CIRCUITS

TYPES SN54ALS1240, SN54ALS1241, SN74ALS1240, SN74ALS1241
OCTAL BUFFERS AND LINE DRIVERS WITH 3-STATE OUTPUTS

'ALS1240 switching characteristics (see Note 1)

PARAMETER	FROM (INPUT)	TO (OUTPUT)	V_{CC} = 4.5 V to 5.5 V, C_L = 50 pF, $R1$ = 500 Ω, $R2$ = 500 Ω, T_A = MIN to MAX						UNIT
			SN54ALS1240			SN74ALS1240			
			MIN	TYP†	MAX	MIN	TYP†	MAX	
t_{PLH}	A	Y		9			9		ns
t_{PHL}				9			9		
t_{PZH}	$\overline{G}$	Y		17			17		ns
t_{PZL}				19			19		
t_{PHZ}	$\overline{G}$	Y		7			7		ns
t_{PLZ}				6			6		

'ALS1241 switching characteristics (see Note 1)

PARAMETER	FROM (INPUT)	TO (OUTPUT)	V_{CC} = 4.5 V to 5.5 V, C_L = 50 pF, $R1$ = 500 Ω, $R2$ = 500 Ω, T_A = MIN to MAX						UNIT
			SN54ALS1241			SN74ALS1241			
			MIN	TYP†	MAX	MIN	TYP†	MAX	
t_{PLH}	A	Y		9			9		ns
t_{PHL}				9			9		
t_{PZH}	$\overline{G}$ or G	Y		17			17		ns
t_{PZL}				19			19		
t_{PHZ}	$\overline{G}$ or G	Y		7			7		ns
t_{PLZ}				6			6		

†All typical values are at V_{CC} = 5 V, T_A = 25°C.
NOTE 1: For load circuit and voltage waveforms, see page 1-12.

TYPES SN54ALS1242, SN54ALS1243, SN74ALS1242, SN74ALS1243
QUADRUPLE BUS TRANSCEIVERS WITH 3-STATE OUTPUTS

D2661, DECEMBER 1982 – REVISED DECEMBER 1983

- 2-Way Asynchronous Communication between Data Buses
- P-N-P Inputs Reduce DC Loading
- Low-Power Version of 'ALS242, and 'ALS243
- Three-State Outputs
- Package Options Include Both Plastic and Ceramic Chip Carriers in Addition to Plastic and Ceramic DIPs
- Dependable Texas Instruments Quality and Reliability

SN54ALS1242, SN54ALS1243 . . . J PACKAGE
SN74ALS1242, SN74ALS1243 . . . N PACKAGE
(TOP VIEW)

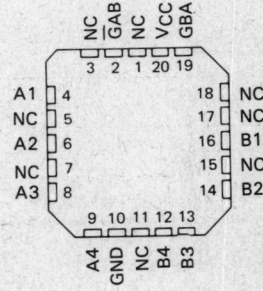

SN54ALS1242, SN54ALS1243 . . . FH PACKAGE
SN74ALS1242, SN74ALS1243 . . . FN PACKAGE
(TOP VIEW)

description

These quadruple bus transceivers are designed for two-way communication between data buses. The control function implementation allows for maximum flexibility in timing.

These devices allow data transmission from the A bus to the B bus or from the B bus to the A bus depending upon the logic levels at the enable inputs (GBA and $\overline{GAB}$).

The enable inputs can be used to disable the device so that the buses are effectively isolated.

The dual-enable configuration gives the 'ALS1242 and 'ALS1243 the capability to store data by simultaneous enabling of $\overline{GAB}$ and GBA. Each output reinforces its input in this transceiver configuration. Thus, when both control inputs are enabled and all other data sources to the two sets of bus lines are at high impedance, both sets of bus lines (8 in all) will remain at their last states. The 4-bit codes appearing on the two sets of buses will be complementary for the 'ALS1242 or identical for the 'ALS1243.

The -1 versions of the SN74ALS' parts are identical to the standard versions except that the recommended maximum I_{OL} is increased to 24 milliamperes. There are no -1 versions of the SN54ALS' parts.

The SN54ALS1242 and SN54ALS1243 are characterized for operation over the full military temperature range of −55°C to 125°C. The SN74ALS1242 and SN74ALS1243 are characterized for operation from 0°C to 70°C.

NC — No internal connection

FUNCTION TABLE

$\overline{GAB}$	GBA	'ALS1242	'ALS1243
L	L	$\overline{A}$ to B	A to B
H	H	$\overline{B}$ to A	B to A
H	L	Isolation	Isolation
L	H	Latch A and B ($A = \overline{B}$)	Latch A and B ($A = B$)

PRODUCT PREVIEW

This document contains information on a product under development. Texas Instruments reserves the right to change or discontinue this product without notice.

Copyright © 1982 by Texas Instruments Incorporated

TEXAS INSTRUMENTS
POST OFFICE BOX 225012 • DALLAS, TEXAS 75265

ALS AND AS CIRCUITS

TYPES SN54ALS1242, SN54ALS1243, SN74ALS1242, SN74ALS1243
QUADRUPLE BUS TRANSCEIVERS WITH 3-STATE OUTPUTS

logic symbols

logic diagrams (positive logic)

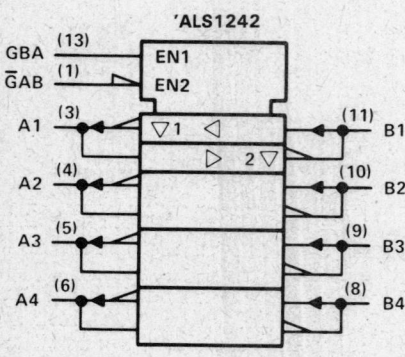
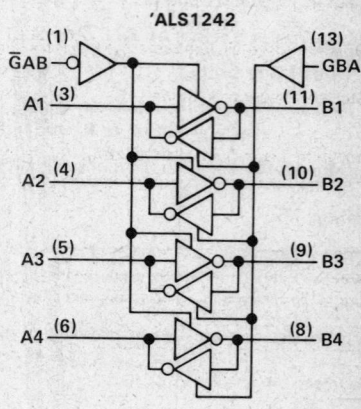

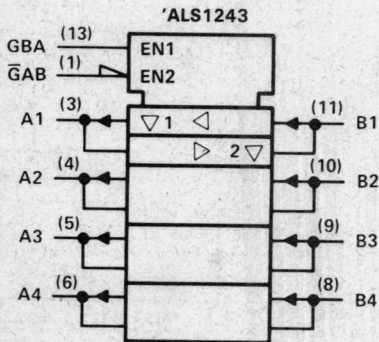
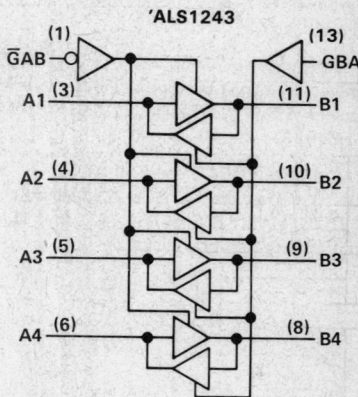

Pin numbers shown are for J and N packages.

TEXAS INSTRUMENTS
POST OFFICE BOX 225012 • DALLAS, TEXAS 75265

TYPES SN54ALS1242, SN54ALS1243, SN74ALS1242, SN74ALS1243
QUADRUPLE BUS TRANSCEIVERS WITH 3-STATE OUTPUTS

absolute maximum ratings over operating free-air temperature range (unless otherwise noted)

Supply voltage, V_{CC} .... 7 V
Input voltage: All inputs .... 7 V
                I/O ports .... 5.5 V
Operating free-air temperature range: SN54ALS1242, SN54ALS1243 .... −55°C to 125°C
                                 SN74ALS1242, SN74ALS1243 .... 0°C to 70°C
Storage temperature range .... −65°C to 150°C

recommended operating conditions

		SN54ALS1242 SN54ALS1243			SN74ALS1242 SN74ALS1243			UNIT
		MIN	NOM	MAX	MIN	NOM	MAX	
V_{CC}	Supply voltage	4.5	5	5.5	4.5	5	5.5	V
V_{IH}	High-level input voltage	2			2			V
V_{IL}	Low-level input voltage			0.8			0.8	V
I_{OH}	High-level output current			−12			−15	mA
I_{OL}	Low-level output current			8			16	mA
							24†	
T_A	Operating free-air temperature	−55		125	0		70	°C

†The extended limits apply only if V_{CC} is maintained between 4.75 V and 5.25 V.
The 24-mA limit applies for the SN74ALS1242-1 and SN74ALS1243-1 only.

electrical characteristics over recommended operating free-air temperature range (unless otherwise noted)

PARAMETER		TEST CONDITIONS		SN54ALS1242 SN54ALS1243			SN74ALS1242 SN74ALS1243			UNIT
				MIN	TYP‡	MAX	MIN	TYP‡	MAX	
V_{IK}		$V_{CC} = 4.5$ V,	$I_I = -18$ mA			−1.5			−1.5	V
V_{OH}		$V_{CC} = 4.5$ V to 5.5 V,	$I_{OH} = -0.4$ mA	$V_{CC}-2$			$V_{CC}-2$			V
		$V_{CC} = 4.5$ V,	$I_{OH} = -3$ mA	2.4	3.2		2.4	3.2		
		$V_{CC} = 4.5$ V,	$I_{OH} = -12$ mA	2						
		$V_{CC} = 4.5$ V,	$I_{OH} = -15$ mA				2			
V_{OL}		$V_{CC} = 4.5$ V,	$I_{OL} = -8$ mA		0.25	0.4		0.25	0.4	V
		$V_{CC} = 4.5$ V,	$I_{OL} = 16$ mA					0.35	0.5	
		($I_{OL} = 24$ mA for -1 versions)								
I_I	Control inputs	$V_{CC} = 5.5$ V,	$V_I = 7$ V			0.1			0.1	mA
	A or B ports	$V_{CC} = 5.5$ V,	$V_I = 5.5$ V			0.1			0.1	
I_{IH}	Control inputs	$V_{CC} = 5.5$ V,	$V_I = 2.7$ V			20			20	μA
	A or B ports§					20			20	
I_{IL}	Control inputs	$V_{CC} = 5.5$ V,	$V_I = 0.4$ V			−0.1			−0.1	mA
	A or B ports§					−0.1			−0.1	
I_O¶		$V_{CC} = 5.5$ V,	$V_O = 2.25$ V	−30		−112	−30		−112	mA
I_{CC}	'ALS1242	$V_{CC} = 5.5$ V	Outputs high		6.5			6.5		mA
			Outputs low		10			10		
			Outputs disabled		12			12		
	'ALS1243		Outputs high		8			8		
			Outputs low		12			12		
			Outputs disabled		14			14		

‡All typical values are at $V_{CC} = 5$ V, $T_A = 25$°C.
§For I/O ports, the parameters I_{IH} and I_{IL} include the off-state output current.
¶The output conditions have been chosen to produce a current that closely approximates one half of the true short-circuit output current, I_{OS}.

Additional information on these products can be obtained from the factory as it becomes available.

ALS AND AS CIRCUITS

TYPES SN54ALS1242, SN54ALS1243, SN74ALS1242, SN74ALS1243
QUADRUPLE BUS TRANSCEIVERS WITH 3-STATE OUTPUTS

'ALS1242 switching characteristics (see Note 1)

PARAMETER	FROM (INPUT)	TO (OUTPUT)	V_{CC} = 4.5 V to 5.5 V, C_L = 50 pF, R1 = 500 Ω, R2 = 500 Ω, T_A = MIN to MAX						UNIT
			SN54ALS1242			SN74ALS1242			
			MIN	TYP‡	MAX	MIN	TYP‡	MAX	
t_{PLH}	A or B	B or A		9			9		ns
t_{PHL}				9			9		
t_{PZH}	$\overline{G}AB$	B		17			17		ns
t_{PZL}				19			19		
t_{PHZ}	$\overline{G}AB$	B		7			7		ns
t_{PLZ}				6			6		
t_{PZH}	GBA	A		17			17		ns
t_{PZL}				19			19		
t_{PHZ}	GBA	A		7			7		ns
t_{PLZ}				6			6		

'ALS1242 switching characteristics (see Note 1)

PARAMETER	FROM (INPUT)	TO (OUTPUT)	V_{CC} = 4.5 V to 5.5 V, C_L = 50 pF, R1 = 500 Ω, R2 = 500 Ω, T_A = MIN to MAX						UNIT
			SN54ALS1243			SN74ALS1243			
			MIN	TYP‡	MAX	MIN	TYP‡	MAX	
t_{PLH}	A or B	B or A		11			11		ns
t_{PHL}				11			11		
t_{PZH}	$\overline{G}AB$	B		19			19		ns
t_{PZL}				21			21		
t_{PHZ}	$\overline{G}AB$	B		9			9		ns
t_{PLZ}				8			8		
t_{PZH}	GBA	A		19			19		ns
t_{PZL}				21			21		
t_{PHZ}	GBA	A		9			9		ns
t_{PLZ}				8			8		

‡All typical values are at V_{CC} = 5 V, T_A = 25°C.
NOTE 2: For load circuit and voltage waveforms, see page 1-12.

Additional information on these products can be obtained from the factory as it becomes available.

PRODUCT PREVIEW

This page contains information on a product under development. Texas Instruments reserves the right to change or discontinue this product without notice.

TEXAS INSTRUMENTS
POST OFFICE BOX 225012 • DALLAS, TEXAS 75265

TYPES SN54ALS1244A, SN74ALS1244A
OCTAL BUFFER AND DRIVER WITH 3-STATE OUTPUTS

D2661, DECEMBER 1982 – REVISED DECEMBER 1983

- Low-Power Version of 'ALS244A
- 3-State Outputs Drive Bus Lines or Buffer Memory Address Registers
- P-N-P Inputs Reduce DC Loading
- Package Options Include Both Plastic and Ceramic Chip Carriers in Addition to Plastic and Ceramic DIPs
- Dependable Texas Instruments Quality and Reliability

SN54ALS1244A . . . J PACKAGE
SN74ALS1244A . . . N PACKAGE
(TOP VIEW)

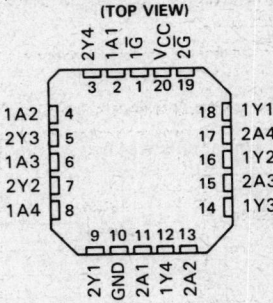

description

This octal buffer and line driver is designed specifically to improve both the performance and density of three-state memory address drivers, clock drivers, and bus-oriented receivers and transmitters. Taken together with the 'ALS1240 and 'ALS1241 this device provides the choice of selected combinations of inverting and noninverting outputs symmetrical $\overline{G}$ (active-low input control) inputs, and complementary G and $\overline{G}$ inputs.

The -1 version of the SN74ALS1244A is identical to the standard version except that the recommended maximum I_{OL} is increased to 24 milliamperes. There is no -1 version of the SN54ALS1244A.

The SN54ALS1244A is characterized for operation over the full military temperature range of $-55°C$ to $125°C$. The SN74ALS1244A is characterized for operation from $0°C$ to $70°C$.

SN54ALS1244A . . . FH PACKAGE
SN74ALS1244A . . . FN PACKAGE
(TOP VIEW)

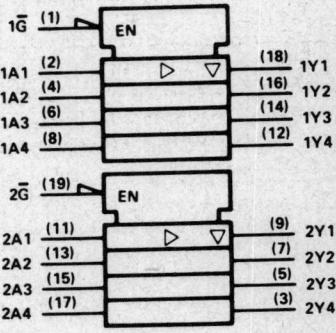

functional block diagram (positive logic)

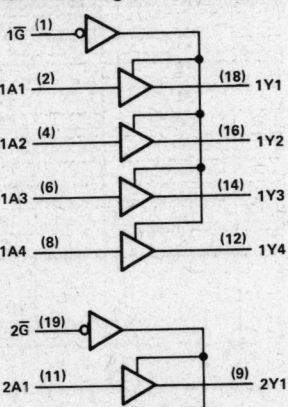

logic symbol

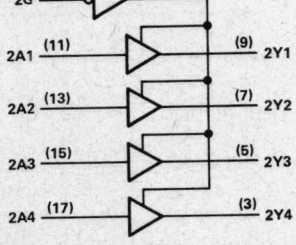

Pin numbers shown are for J and N packages.

Copyright © 1982 by Texas Instruments Incorporated

TYPES SN54ALS1244A, SN74ALS1244A
OCTAL BUFFER AND DRIVER WITH 3-STATE OUTPUTS

absolute maximum ratings over operating free-air temperature range (unless otherwise noted)

Supply voltage, V_{CC} .... 7 V
Input voltage .... 7 V
Voltage applied to a disabled 3-state output .... 5.5 V
Operating free-air temperature range: SN54ALS1244A .... −55°C to 125°C
                                       SN74ALS1244A .... 0°C to 70°C
Storage temperature range .... −65°C to 150°C

recommended operating conditions

		SN54ALS1244A			SN74ALS1244A			UNIT
		MIN	NOM	MAX	MIN	NOM	MAX	
V_{CC}	Supply voltage	4.5	5	5.5	4.5	5	5.5	V
V_{IH}	High-level input voltage	2			2			V
V_{IL}	Low-level input voltage			0.8			0.8	V
I_{OH}	High-level output current			−12			−15	mA
I_{OL}	Low-level output current			8			16	mA
							24†	
T_A	Operating free-air temperature	−55		125	0		70	°C

†The extended limits apply only if V_{CC} is maintained between 4.75 V and 5.25 V.
The 24-mA limit applies for the SN74ALS1244A-1 only.

electrical characteristics over recommended operating free-air temperature range (unless otherwise noted)

PARAMETER	TEST CONDITIONS		SN54ALS1244A			SN74ALS1244A			UNIT
			MIN	TYP‡	MAX	MIN	TYP‡	MAX	
V_{IK}	$V_{CC} = 4.5$ V,	$I_I = -18$ mA			−1.5			−1.5	V
V_{OH}	$V_{CC} = 4.5$ V to 5.5 V,	$I_{OH} = -0.4$ mA	$V_{CC}-2$			$V_{CC}-2$			V
	$V_{CC} = 4.5$ V,	$I_{OH} = -3$ mA	2.4	3.2		2.4	3.2		
	$V_{CC} = 4.5$ V,	$I_{OH} = -12$ mA	2						
	$V_{CC} = 4.5$ V,	$I_{OH} = -15$ mA				2			
V_{OL}	$V_{CC} = 4.5$ V,	$I_{OL} = 8$ mA		0.25	0.4		0.25	0.4	V
	$V_{CC} = 4.5$ V,	$I_{OL} = 16$ mA					0.35	0.5	
	($I_{OL} = 24$ mA for −1 versions)								
I_{OZH}	$V_{CC} = 5.5$ V,	$V_O = 2.7$ V			20			20	μA
I_{OZL}	$V_{CC} = 5.5$ V,	$V_I = 0.4$ V			−20			−20	μA
I_I	$V_{CC} = 5.5$ V,	$V_I = 7$ V			0.1			0.1	mA
I_{IH}	$V_{CC} = 5.5$ V,	$V_I = 2.7$ V			20			20	μA
I_{IL}	$V_{CC} = 5.5$ V,	$V_I = 0.4$ V			−0.1			−0.1	mA
I_O§	$V_{CC} = 5.5$ V,	$V_O = 2.25$ V	−30		−112	−30		−112	mA
I_{CC}	$V_{CC} = 5.5$ V	Outputs high		6	15		6	11	mA
		Outputs low		10	20		10	17	
		Outputs disabled		11	25		11	20	

‡All typical values are at $V_{CC} = 5$ V, $T_A = 25$°C.
§The output conditions have been chosen to produce a current that closely approximates one half of the true short-circuit output current, I_{OS}.

TEXAS INSTRUMENTS
POST OFFICE BOX 225012 • DALLAS, TEXAS 75265

TYPES SN54ALS1244A, SN74ALS1244A
OCTAL BUFFER AND DRIVER WITH 3-STATE OUTPUTS

switching characteristics (see Note 1)

PARAMETER	FROM (INPUT)	TO (OUTPUT)	V_{CC} = 4.5 V to 5.5 V, C_L = 50 pF, R1 = 500 Ω, R2 = 500 Ω, T_A = MIN to MAX				UNIT
			SN54ALS1244A		SN74ALS1244A		
			MIN	MAX	MIN	MAX	
t_{PLH}	A	Y	3	16	3	14	ns
t_{PHL}			3	16	3	14	
t_{PZH}	$\overline{G}$	Y	6	26	6	22	ns
t_{PZL}			6	26	6	22	
t_{PHZ}	$\overline{G}$	Y	2	12	2	10	ns
t_{PLZ}			3	16	3	13	

NOTE 1: For load circuit and voltage waveforms, see page 1-12.

Texas Instruments
POST OFFICE BOX 225012 • DALLAS, TEXAS 75265

2
ALS AND AS CIRCUITS

TYPES SN54ALS1245, SN74ALS1245
OCTAL BUS TRANSCEIVERS WITH 3-STATE OUTPUTS

D2661, DECEMBER 1982 – REVISED DECEMBER 1983

- **'Bidirectional Bus Transceivers in High-Density 20-Pin Packages**
- **Lower-Power Version of 'ALS245**
- **'ALS1245 is Identical to 'ALS1645**
- **Package Options Include Both Plastic and Ceramic Chip Carriers in Addition to Plastic and Ceramic DIPs**
- **Dependable Texas Instruments Quality and Reliability**

SN54ALS1245 . . . J PACKAGE
SN74ALS1245 . . . N PACKAGE
(TOP VIEW)

```
DIR [ 1   20 ] VCC
A1  [ 2   19 ] G
A2  [ 3   18 ] B1
A3  [ 4   17 ] B2
A4  [ 5   16 ] B3
A5  [ 6   15 ] B4
A6  [ 7   14 ] B5
A7  [ 8   13 ] B6
A8  [ 9   12 ] B7
GND [10   11 ] B8
```

description

This octal bus transceiver is designed for asynchronous two-way communication between data buses. The device transmits data from the A bus to the B bus or from the B bus to the A bus depending upon the level at the direction control (DIR) input. The enable input (G) can be used to disable the device so the buses are effectively isolated.

The -1 version of the SN74ALS1245 is identical to the standard versions except that the recommended maximum I_{OL} is increased to 24 milliamperes. There is no -1 version of the SN54ALS1245.

The SN54ALS1245 is characterized for operation over the full military temperature range of $-55\,°C$ to $125\,°C$. The SN74ALS1245 is characterized for operation from $0\,°C$ to $70\,°C$.

SN54ALS1245 . . . FH PACKAGE
SN74ALS1245 . . . FN PACKAGE

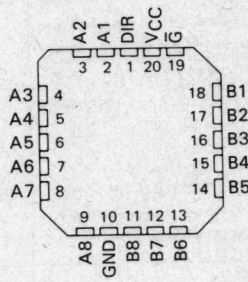

FUNCTION TABLE

CONTROL INPUTS		OPERATION
G	DIR	
L	L	B data to A bus
L	H	A data to B bus
H	X	Isolation

logic diagram (positive logic)

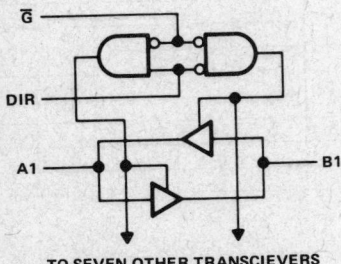

TO SEVEN OTHER TRANSCIEVERS

logic symbol

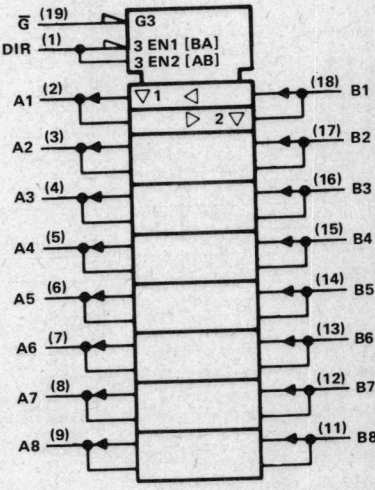

Pin numbers shown are for J and N packages.

Copyright © 1982 by Texas Instruments Incorporated

ALS AND AS CIRCUITS

TEXAS INSTRUMENTS
POST OFFICE BOX 225012 • DALLAS, TEXAS 75265

TYPES SN54ALS1245, SN74ALS1245
OCTAL BUS TRANSCEIVERS WITH 3-STATE OUTPUTS

absolute maximum ratings over operating free-air temperature range (unless otherwise noted)

Supply voltage, V_{CC} .... 7 V
Input voltage: All inputs .... 7 V
                 I/O ports .... 5.5 V
Operating free-air temperature range: SN54ALS1245 .... −55°C to 125°C
                                     SN74ALS1245 .... 0°C to 70°C
Storage temperature range .... −65°C to 150°C

recommended operating conditions

		SN54ALS1245			SN74ALS1245			UNIT
		MIN	NOM	MAX	MIN	NOM	MAX	
V_{CC}	Supply voltage	4.5	5	5.5	4.5	5	5.5	V
V_{IH}	High-level input voltage	2			2			V
V_{IL}	Low-level input voltage			0.8			0.8	V
I_{OH}	High-level output current			−12			−15	mA
I_{OL}	Low-level output current			8			16	mA
							24†	
T_A	Operating free-air temperature	−55		125	0		70	°C

† The extended limit applies only if V_{CC} is maintained between 4.75 V and 5.25 V.
The 24-mA limit applies for the SN74ALS1245-1 only.

electrical characteristics over recommended operating free-air temperature range (unless otherwise noted)

PARAMETER		TEST CONDITIONS		SN54ALS1245			SN74ALS1245			UNIT
				MIN	TYP‡	MAX	MIN	TYP‡	MAX	
V_{IK}		V_{CC} = 4.5 V,	I_I = −18 mA			−1.5			−1.5	V
V_{OH}		V_{CC} = 4.5 V to 5.5 V,	I_{OH} = −0.4 mA	V_{CC}−2			V_{CC}−2			V
		V_{CC} = 4.5 V,	I_{OH} = −3 mA	2.4	3.2		2.4	3.2		
		V_{CC} = 4.5 V,	I_{OH} = −12 mA	2						
		V_{CC} = 4.5 V,	I_{OH} = −15 mA				2			
V_{OL}		V_{CC} = 4.5 V,	I_{OL} = 8 mA		0.25	0.4		0.25	0.4	V
		V_{CC} = 4.5 V,	I_{OL} = 16 mA							
		(I_{OL} = 24 mA for −1 version)						0.35	0.5	
I_I	Control inputs	V_{CC} = 5.5 V,	V_I = 7 V			0.1			0.1	mA
	A, B ports§	V_{CC} = 5.5 V,	V_I = 5.5 V			0.1			0.1	
I_{IH}	Control inputs	V_{CC} = 5.5 V,	V_I = 2.7 V			20			20	µA
	A, B ports§					20			20	
I_{IL}	Control inputs	V_{CC} = 5.5 V,	V_I = 0.4 V			−0.1			−0.1	mA
	A, B ports§					−0.1			−0.1	
I_O¶		V_{CC} = 5.5 V,	V_O = 2.25 V	−30		−112	−30		−112	mA
I_{CC}		V_{CC} = 5.5 V	Output high	20	33		20	32		mA
			Output low	23	39		23	37		
			Output disabled	25	41		25	39		

‡ All typical values are at V_{CC} = 5 V, T_A = 25°C.
§ For I/O ports, the parameters I_{IH} and I_{IL} include the off-state output current.
¶ The output conditions have been chosen to produce a current that closely approximates one half of the true short-circuit output current, I_{OS}.

Texas Instruments
POST OFFICE BOX 225012 • DALLAS, TEXAS 75265

TYPES SN54ALS1245, SN74ALS1245
OCTAL BUS TRANSCEIVERS WITH 3-STATE OUTPUTS

switching characteristics (see Note 1)

PARAMETER	FROM (INPUT)	TO (OUTPUT)	V_{CC} = 4.5 V to 5.5 V, C_L = 50 pF, R1 = 500 Ω, R2 = 500 Ω, T_A = MIN to MAX				UNIT
			SN54ALS1245		SN74ALS1245		
			MIN	MAX	MIN	MAX	
t_{PLH}	A or B	B or A	4	15	4	13	ns
t_{PHL}			4	15	4	13	
t_{PZH}	$\overline{G}$	A or B	10	27	10	25	ns
t_{PZL}			13	32	13	29	
t_{PHZ}	$\overline{G}$	A or B	4	20	4	18	ns
t_{PLZ}			5	23	5	21	

NOTE 1: For load circuit and voltage waveforms, see page 1-12.

TEXAS INSTRUMENTS
POST OFFICE BOX 225012 • DALLAS, TEXAS 75265

ALS AND AS CIRCUITS

2
ALS AND AS CIRCUITS

TYPES SN54ALS1620 THRU SN54ALS1623, SN74ALS1620 THRU SN74ALS1623 OCTAL BUS TRANSCEIVERS

D2661, DECEMBER 1982—REVISED DECEMBER 1983

- Bus Transceivers in High-Density 20-Pin DIPs and the New Plastic and Ceramic Chip Carriers Packages
- Local Bus Latch Capability
- Choice of True or Inverting Logic
- Dependable Texas Instruments Quality and Reliability
- Choice of 3-State or Open-Collector Outputs

DEVICE	OUTPUT	LOGIC
'ALS1620	3-State	Inverting
'ALS1621	Open-Collector	True
'ALS1622	Open-Collector	Inverting
'ALS1623	3-State	True

SN54ALS' . . . J PACKAGE
SN74ALS' . . . N PACKAGE
(TOP VIEW)

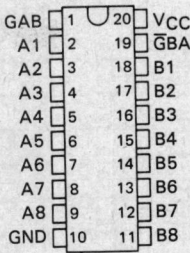

SN54ALS' . . . FH PACKAGE
SN74ALS' . . . FN PACKAGE
(TOP VIEW)

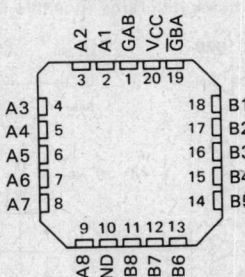

description

These octal bus transceivers are designed for asynchronous two-way communication between data buses. The control function implementation allows for maximum flexibility in timing.

These devices allow data transmission from A bus to the B bus or from the B bus to the A bus depending upon the logic levels at the enable inputs ($\overline{GBA}$ and GAB).

The enable inputs can be used to disable the device so that the buses are effectively isolated.

The dual-enable configuration gives the 'ALS1620 thru 'ALS1623 the capability to store data by simultaneous enabling of $\overline{GBA}$ and GAB. Each output reinforces its input in this transceiver configuration. Thus, when both control inputs are enabled and all other data sources to the two sets of bus lines are at high impedance, both sets of bus lines (16 in all) will remain at their last states. The 8-bit codes appearing on the two sets of buses will be identical for the 'ALS1621 and 'ALS1623 or complementary for the 'ALS1620 and 'ALS1622.

The -1 versions of the SN74ALS' parts are identical to the standard versions except that the recommended maximum I_{OL} is increased to 24 mA. There are no -1 versions of the SN54ALS' parts.

The SN54ALS1620 thru SN54ALS1623 are characterized for operation over the full military temperature range of $-55\,°C$ to $125\,°C$. The SN74ALS1620 thru SN74ALS1623 are characterized for operation from $0\,°C$ to $70\,°C$.

FUNCTION TABLE

ENABLE INPUTS		OPERATION	
$\overline{GBA}$	GAB	'ALS1620, 'ALS1622	'ALS1621, 'ALS1623
L	L	$\overline{B}$ data to A bus	B data to A bus
H	H	$\overline{A}$ data to B bus	A data to B bus
H	L	Isolation	Isolation
L	H	$\overline{B}$ data to A bus, $\overline{A}$ data to B bus	B data to A bus, A data to B bus

PRODUCT PREVIEW

This document contains information on a product under development. Texas Instruments reserves the right to change or discontinue this product without notice.

ALS AND AS CIRCUITS

Copyright © 1982 by Texas Instruments Incorporated

TEXAS INSTRUMENTS
POST OFFICE BOX 225012 • DALLAS, TEXAS 75265

TYPES SN54ALS1620 THRU SN54ALS1623, SN74ALS1620 THRU SN74ALS1623 OCTAL BUS TRANSCEIVERS

logic symbols

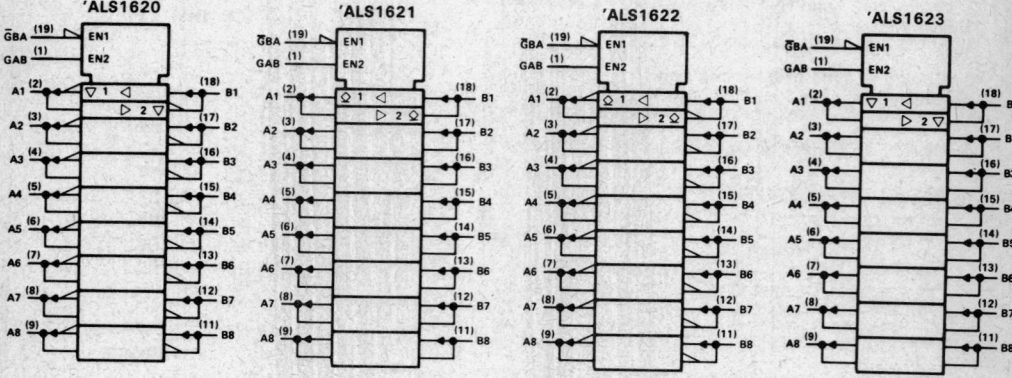

functional block diagrams (positive logic)

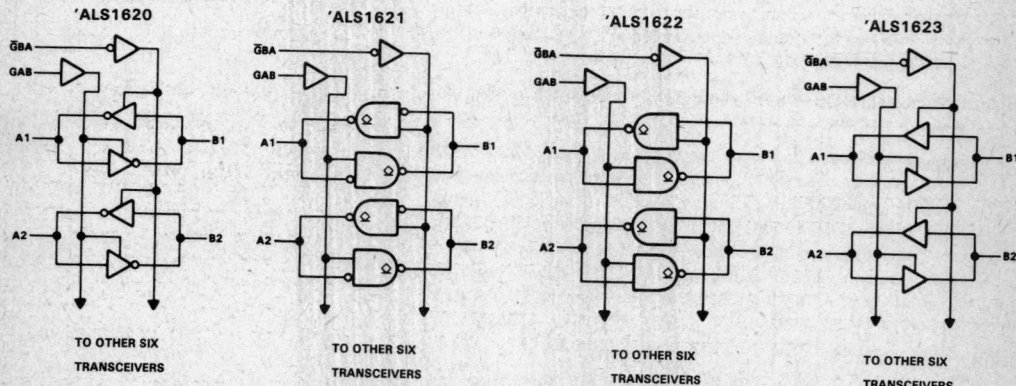

Pin numbers shown are for J and N packages.

TYPES SN54ALS1620 THRU SN54ALS1623
SN74ALS1620 THRU SN74ALS1623
OCTAL BUS TRANSCEIVERS

absolute maximum ratings over operating free-air temperature range (unless otherwise noted)

Supply voltage, V_{CC} .... 7 V
Input voltage: All inputs .... 5.5 V
                I/O ports .... 7 V
Operating free-air temperature range: SN54ALS1620, SN54ALS1623 .... −55°C to 125°C
                      SN74ALS1620, SN74ALS1623 .... 0°C to 70°C
Storage temperature range .... −65°C to 150°C

recommended operating conditions

		SN54ALS1620 SN54ALS1623			SN74ALS1620 SN74ALS1623			UNIT
		MIN	NOM	MAX	MIN	NOM	MAX	
V_{CC}	Supply voltage	4.5	5	5.5	4.5	5	5.5	V
V_{IH}	High-level input voltage	2			2			V
V_{IL}	Low-level input voltage			0.8			0.8	V
I_{OH}	High-level output current			−12			−15	mA
I_{OL}	Low-level output current			8			16	mA
							24[†]	
T_A	Operating free-air temperature	−55		125	0		70	°C

[†] The extended limits apply only if V_{CC} is maintained between 4.75 V and 5.25 V.
The 24-mA limit applies for the SN74ALS1620-1 and SN74ALS1623-1 only.

electrical characteristics over recommended operating free-air temperature range (unless otherwise noted)

PARAMETER		TEST CONDITIONS		SN54ALS1620 SN54ALS1623			SN74ALS1620 SN74ALS1623			UNIT
				MIN	TYP[‡]	MAX	MIN	TYP[‡]	MAX	
V_{IK}		V_{CC} = 4.5 V,	I_I = −18 mA			−1.5			−1.5	V
V_{OH}		V_{CC} = 4.5 V to 5.5 V,	I_{OH} = −0.4 mA	V_{CC}−2			V_{CC}−2			V
		V_{CC} = 4.5 V,	I_{OH} = −3 mA	2.4	3.2		2.4	3.2		
		V_{CC} = 4.5 V,	I_{OH} = −12 mA	2						
		V_{CC} = 4.5 V,	I_{OH} = −15 mA				2			
V_{OL}		V_{CC} = 4.5 V,	I_{OL} = 8 mA		0.25	0.4		0.25	0.4	V
		V_{CC} = 4.5 V,	I_{OL} = 16 mA					0.35	0.5	
		(I_{OL} = 24 mA for −1 versions)								
I_I	Control inputs	V_{CC} = 5.5 V,	V_I = 7 V			0.1			0.1	mA
	A or B ports	V_{CC} = 5.5 V,	V_I = 5.5 V			0.1			0.1	
I_{IH}	Control inputs	V_{CC} = 5.5 V,	V_I = 2.7 V			20			20	µA
	A or B ports§					20			20	
I_{IL}	Control inputs	V_{CC} = 5.5 V,	V_I = 0.4 V			−0.1			−0.1	mA
	A or B ports§					−0.1			−0.1	
I_O¶		V_{CC} = 5.5 V,	V_O = 2.25 V	−30		−112	−30		−112	mA
I_{CC}	'ALS1620	V_{CC} = 5.5 V	Outputs high		14			14		mA
			Outputs low		19			19		
			Outputs disabled		21			21		
	'ALS1623		Outputs high		11			11		
			Outputs low		18			18		
			Outputs disabled		13			13		

[‡] All typical values are at V_{CC} = 5 V, T_A = 25°C.
§ For I/O ports, the parameters I_{IH} and I_{IL} include the off-state output current.
¶ The output conditions have been chosen to produce a current that closely approximates one half of the true short-circuit output current, I_{OS}.

Additional information on these products can be obtained from the factory as it becomes available.

TEXAS INSTRUMENTS
POST OFFICE BOX 225012 • DALLAS, TEXAS 75265

TYPES SN54ALS1620 THRU SN54ALS1623
SN74ALS1620 THRU SN74ALS1623
OCTAL BUS TRANSCEIVERS

'ALS1620 switching characteristics (see Note 1)

PARAMETER	FROM (INPUT)	TO (OUTPUT)	V_{CC} = 4.5 V to 5.5 V, C_L = 50 pF, R1 = 500 Ω, R2 = 500 Ω, T_A = MIN to MAX						UNIT
			SN54ALS1620			SN74ALS1620			
			MIN	TYP†	MAX	MIN	TYP†	MAX	
t_{PLH}	A	B		9			9		ns
t_{PHL}				6			6		
t_{PLH}	B	A		9			9		ns
t_{PHL}				6			6		
t_{PZH}	$\overline{GBA}$	A		14			14		ns
t_{PZL}				17			17		
t_{PHZ}	$\overline{GBA}$	A		7			7		ns
t_{PLZ}				11			11		
t_{PZH}	GAB	B		14			14		ns
t_{PZL}				17			17		
t_{PHZ}	GAB	B		7			7		ns
t_{PLZ}				11			11		

'ALS1623 switching characteristics (see Note 1)

PARAMETER	FROM (INPUT)	TO (OUTPUT)	V_{CC} = 4.5 V to 5.5 V, C_L = 50 pF, R1 = 500 Ω, R2 = 500 Ω, T_A = MIN to MAX						UNIT
			SN54ALS1623			SN74ALS1623			
			MIN	TYP†	MAX	MIN	TYP†	MAX	
t_{PLH}	A	B		8			8		ns
t_{PHL}				8			8		
t_{PLH}	B	A		8			8		ns
t_{PHL}				8			8		
t_{PZH}	$\overline{GBA}$	A		18			18		ns
t_{PZL}				21			21		
t_{PHZ}	$\overline{GBA}$	A		12			12		ns
t_{PLZ}				13			13		
t_{PZH}	GAB	B		18			18		ns
t_{PZL}				21			21		
t_{PHZ}	GAB	B		12			12		ns
t_{PLZ}				13			13		

†All typical values are at V_{CC} = 5 V, T_A = 25°C.
NOTE 1: For load circuit and voltage waveforms, see page 1-12.

Additional information on these products can be obtained from the factory as it becomes available.

TYPES SN54ALS1620 THRU SN54ALS1623, SN74ALS1620 THRU SN74ALS1623 OCTAL BUS TRANSCEIVERS

absolute maximum ratings over operating free-air temperature range (unless otherwise noted)

Supply voltage, V_{CC} .... 7 V
Input voltage: All inputs and I/O ports .... 7 V
Operating free-air temperature range: SN54ALS1621, SN54ALS1622 .... −55 °C to 125 °C
                                SN74ALS1621, SN74ALS1622 .... 0 °C to 70 °C
Storage temperature range .... −65 °C to 150 °C

recommended operating conditions

		SN54ALS1621 SN54ALS1622			SN74ALS1621 SN74ALS1622			UNIT
		MIN	NOM	MAX	MIN	NOM	MAX	
V_{CC}	Supply voltage	4.5	5	5.5	4.5	5	5.5	V
V_{IH}	High-level input voltage	2			2			V
V_{IL}	Low-level input voltage			0.8			0.8	V
V_{OH}	High-level output voltage			5.5			5.5	mV
				8			16	
I_{OL}	Low-level output current						24†	mA
T_A	Operating free-air temperature	−55		125	0		70	°C

†The extended limits apply only if V_{CC} is maintained between 4.75 V and 5.25 V.
The 24-mA limit applies for the SN74ALS1621-1 and SN74ALS1622-1 only.

electrical characteristics over recommended operating free-air temperature range (unless otherwise noted)

PARAMETER		TEST CONDITIONS		SN54ALS1621 SN54ALS1622		SN74ALS1621 SN74ALS1622		UNIT
				MIN TYP‡	MAX	MIN TYP‡	MAX	
V_{IK}		V_{CC} = 4.5 V,	I_I = −18 mA		−1.5		−1.5	V
I_{OH}		V_{CC} = 4.5 V,	V_{OH} = 5.5 V		0.1		0.1	mA
V_{OL}		V_{CC} = 4.5 V,	I_{OL} = 8 mA	0.25	0.4	0.25	0.4	V
		V_{CC} = 4.5 V,	I_{OL} = 16 mA			0.35	0.5	
		(I_{OL} = 24 mA for −1 versions)						
I_I	Control inputs	V_{CC} = 5.5 V,	V_I = 7 V		0.1		0.1	mA
	A or B ports	V_{CC} = 5.5 V,	V_I = 5.5 V		0.1		0.1	
I_{IH}	Control inputs	V_{CC} = 5.5 V,	V_I = 2.7 V		20		20	µA
	A or B ports§				20		20	
I_{IL}	Control inputs	V_{CC} = 5.5 V,	V_I = 0.4 V		−0.1		−0.1	mA
	A or B ports§				−0.1		−0.1	
I_{CC}	'ALS1621	V_{CC} = 5.5 V	Outputs high		11		11	mA
			Outputs low		16		16	
	'ALS1622		Outputs high		13		13	
			Outputs low		18		18	

‡All typical values are at V_{CC} = 5 V, T_A = 25 °C.
§For I/O ports, the parameters I_{IH} and I_{IL} include the off-state output current.

Additional information on these products can be obtained from the factory as it becomes available.

TEXAS INSTRUMENTS
POST OFFICE BOX 225012 • DALLAS, TEXAS 75265

TYPES SN54ALS1620 THRU SN54ALS1623, SN74ALS1620 THRU SN74ALS1623
OCTAL BUS TRANSCEIVERS

'ALS1621 switching characteristics (see Note 1)

PARAMETER	FROM (INPUT)	TO (OUTPUT)	V_{CC} = 4.5 V to 5.5 V, C_L = 50 pF, R_L = 680 Ω, T_A = MIN to MAX						UNIT
			SN54ALS1621			SN74ALS1621			
			MIN	TYP‡	MAX	MIN	TYP‡	MAX	
t_{PLH}	A	B		22			22		ns
t_{PHL}				14			14		
t_{PLH}	B	A		22			22		ns
t_{PHL}				14			14		
t_{PLH}	$\overline{GBA}$	A		33			33		ns
t_{PHL}				24			24		
t_{PLH}	GAB	B		33			33		ns
t_{PHL}				24			24		

'ALS1622 switching characteristics (see Note 1)

PARAMETER	FROM (INPUT)	TO (OUTPUT)	V_{CC} = 4.5 V to 5.5 V, C_L = 50 pF, R_L = 680 Ω, T_A = MIN to MAX						UNIT
			SN54ALS1622			SN74ALS1622			
			MIN	TYP‡	MAX	MIN	TYP‡	MAX	
t_{PLH}	A	B		25			25		ns
t_{PHL}				13			13		
t_{PLH}	B	A		25			25		ns
t_{PHL}				13			13		
t_{PLH}	$\overline{GBA}$	A		31			31		ns
t_{PHL}				28			28		
t_{PLH}	GAB	B		31			31		ns
t_{PHL}				28			28		

‡All typical values are at V_{CC} = 5 V, T_A = 25°C.
NOTE 1: For load circuit and voltage waveforms, see page 1-12.

Additional information on these products can be obtained from the factory as it becomes available.

TEXAS INSTRUMENTS
POST OFFICE BOX 225012 • DALLAS, TEXAS 75265

TYPES SN54ALS1638, SN54ALS1639, SN74ALS1638, SN74ALS1639
OCTAL BUS TRANSCEIVERS

D2661, DECEMBER 1982—REVISED DECEMBER 1983

- Bidirectional Bus Transceivers in High-Density 20-Pin Packages
- Low-Power Version of 'ALS638 and 'ALS639
- Choice of True or Inverting Logic
- A bus Outputs are Open-Collector; B Bus Outputs are 3-State
- Package Options Include Both Plastic and Ceramic Chip Carriers in Addition to Plastic and Ceramic DIPs
- Dependable Texas Instruments Quality and Reliability

SN54ALS1638, SN54ALS1639 . . . J PACKAGE
SN74ALS1638, SN74ALS1639 . . . N PACKAGE
(TOP VIEW)

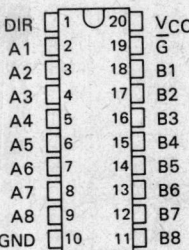

description

These octal bus transceivers are designed for asynchronous two-way communication between open-collector and 3-State buses. The devices transmit data from the A bus (open-collector) to the B bus (3-state) or from the B bus to the A bus depending upon the level at the direction control (DIR) input. The enable input ($\overline{G}$) can be used to enable the device so the buses are effectively isolated.

DEVICE	A OUTPUT	B OUTPUT	LOGIC
'ALS1638	Open-Collector	3-State	Inverting
'ALS1639	Open-Collector	3-State	True

The -1 versions of the SN74ALS' parts are identical to the standard versions except that the recommended maximum I_{OL} is increased to 24 milliamperes. There are no -1 versions of the SN54ALS' parts.

The SN54ALS1638 and SN54ALS1639 are characterized for operation over the full military temperature range of $-55\,°C$ to $125\,°C$. The SN74ALS1638 and SN74ALS1639 are characterized for operation from $0\,°C$ to $70\,°C$.

SN54ALS1638, SN54ALS1639 . . . FH PACKAGE
SN74ALS1638, SN74ALS1639 . . . FN PACKAGE
(TOP VIEW)

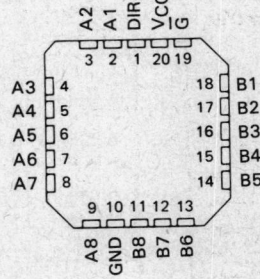

FUNCTION TABLE

CONTROL INPUTS		OPERATION	
$\overline{G}$	DIR	'ALS1638	'ALS1639
L	L	B data to A bus	B data to A bus
L	H	A data to B bus	A data to B bus
H	X	Isolation	Isolation

ALS AND AS CIRCUITS

ADVANCE INFORMATION
This document contains information on a new product. Specifications are subject to change without notice.

Copyright © 1982 by Texas Instruments Incorporated

TEXAS INSTRUMENTS
POST OFFICE BOX 225012 • DALLAS, TEXAS 75265

TYPES SN54ALS1638, SN54ALS1639, SN74ALS1638, SN74ALS1639 OCTAL BUS TRANSCEIVERS

logic symbols

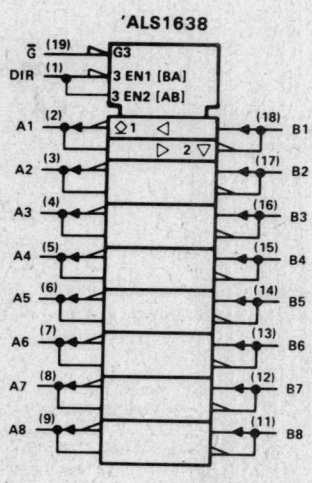

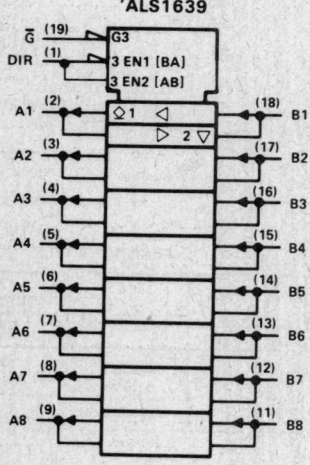

Pin numbers shown are for J and N packages.

functional block diagrams (positive logic)

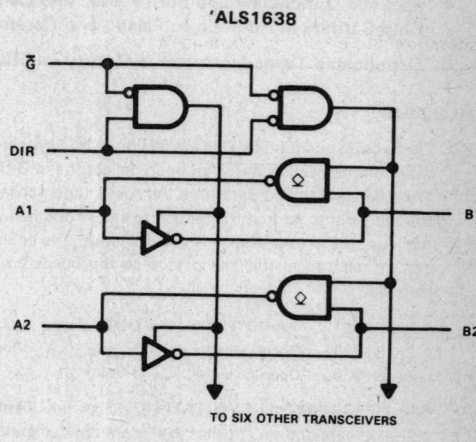

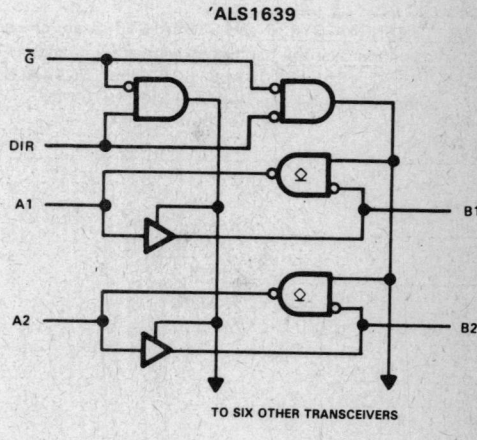

TEXAS INSTRUMENTS
POST OFFICE BOX 225012 • DALLAS, TEXAS 75265

TYPES SN54ALS1638, SN54ALS1639, SN74ALS1638, SN74ALS1639
OCTAL BUS TRANSCEIVERS

absolute maximum ratings over operating free-air temperature range (unless otherwise noted)

Supply voltage, V_{CC} .... 7 V
Input voltage: All inputs .... 7 V
                 A bus I/O ports .... 7 V
                 B bus I/O ports .... 5.5 V
Operating free-air temperature range: SN54ALS1638, SN54ALS1639 .... −55 °C to 125 °C
                                        SN74ALS1638, SN74ALS1639 .... 0 °C to 70 °C
Storage temperature range .... −65 °C to 150 °C

recommended operating conditions

			SN54ALS1638 SN54ALS1639			SN74ALS1638 SN74ALS1639			UNIT
			MIN	NOM	MAX	MIN	NOM	MAX	
V_{CC}	Supply voltage		4.5	5	5.5	4.5	5	5.5	V
V_{IH}	High-level input voltage		2			2			V
V_{IL}	Low-level input voltage				0.8			0.8	V
V_{OH}	High-level voltage	A ports			5.5			5.5	V
I_{OH}	High-level output current	B ports			−12			−15	mA
I_{OL}	Low-level output current	A or B ports			8			16 24†	mA
T_A	Operating free-air temperature		−55		125	0		70	°C

†The extended limits apply only if V_{CC} is maintained between 4.75 V and 5.25 V.
The 24-mA limit applies for the SN74ALS1638-1 and SN74ALS1639-1 only.

electrical characteristics over recommended operating free-air temperature range (unless otherwise noted)

PARAMETER		TEST CONDITIONS		SN54ALS1638 SN54ALS1639			SN74ALS1638 SN74ALS1639			UNIT
				MIN	TYP‡	MAX	MIN	TYP‡	MAX	
V_{IK}		V_{CC} = 4.5 V,	I_I = −18 mA			−1.5			−1.5	V
I_{OH}	A ports	V_{CC} = 4.5 V,	V_O = 5.5 V			0.1			0.1	mA
V_{OH}	B ports	V_{CC} = 4.5 V to 5.5 V,	I_{OH} = −0.4 mA	V_{CC}−2			V_{CC}−2			V
		V_{CC} = 4.5 V,	I_{OH} = −3 mA	2.4	3.2		2.4	3.2		
		V_{CC} = 4.5 V,	I_{OH} = −12 mA	2						
		V_{CC} = 4.5 V,	I_{OH} = −15 mA				2			
V_{OL}	A or B ports	V_{CC} = 4.5 V,	I_{OL} = 8 mA		0.25	0.4		0.25	0.4	V
		V_{CC} = 4.5 V,	I_{OL} = 16 mA					0.35	0.5	
		(I_{OL} = 24 mA for −1 versions)								
I_I	Control inputs	V_{CC} = 5.5 V,	V_I = 7 V			0.1			0.1	mA
	A or B ports	V_{CC} = 5.5 V,	V_I 5.5 V			0.1			0.1	
I_{IH}	Control inputs	V_{CC} = 5.5 V,	V_I = 2.7 V			20			20	μA
	A or B ports§					20			20	
I_{IL}	Control inputs	V_{CC} = 5.5 V,	V_I = 0.4 V			−0.1			−0.1	mA
	A or B ports§					−0.1			−0.1	
I_O¶	B ports	V_{CC} = 5.5 V,	V_O = 2.25 V	−30		−112	−30		−112	mA
I_{CC}		V_{CC} = 5.5 V	Outputs high		21			21		mA
			Outputs low		23			23		
			Outputs disabled		25			25		

‡All typical values are at V_{CC} = 5 V, T_A = 25 °C.
§For I/O ports, the parameters I_{IH} and I_{IL} include the off-state output current.
¶The output conditions have been chosen to produce a current that closely approximates one half of the true short-circuit output current, I_{OS}.

Additional information on these products can be obtained from the factory as it becomes available.

TYPES SN54ALS1638, SN54ALS1639, SN74ALS1638, SN74ALS1639
OCTAL BUS TRANSCEIVERS

'ALS1638 switching characteristics (see Note 1)

PARAMETER	FROM (INPUT)	TO (OUTPUT)	V_{CC} = 4.5 V to 5.5 V, C_L = 50 pF, R_L = 500 Ω (A outputs), R1 = R2 = 500 Ω (B outputs) T_A = MIN to MAX						UNIT
			SN54ALS1638			SN74ALS1638			
			MIN	TYP†	MAX	MIN	TYP†	MAX	
t_{PLH}	A	B		6			6		ns
t_{PHL}				21			21		
t_{PLH}	B	A		6			6		ns
t_{PHL}				8			8		
t_{PLH}	$\overline{G}$, DIR	A		23			23		ns
t_{PHL}				17			17		
t_{PZH}	$\overline{G}$	B		12			12		ns
t_{PZL}				15			15		
t_{PHZ}	$\overline{G}$	B		6			6		ns
t_{PLZ}				7			7		

'ALS1639 switching characteistics (see Note 1)

PARAMETER	FROM (INPUT)	TO (OUTPUT)	V_{CC} = 4.5 V to 5.5 V, C_L = 50 pF, R_L = 500 Ω (A outputs), R1 = R2 = 500 Ω (B outputs), T_A = MIN to MAX						UNIT
			SN54ALS1639			SN74ALS1639			
			MIN	TYP†	MAX	MIN	TYP†	MAX	
t_{PLH}	A	B		7			7		ns
t_{PHL}				21			21		
t_{PLH}	B	A		7			7		ns
t_{PHL}				9			9		
t_{PLH}	$\overline{G}$, DIR	A		23			23		ns
t_{PHL}				19			19		
t_{PZH}	$\overline{G}$	B		14			14		ns
t_{PZL}				17			17		
t_{PHZ}	$\overline{G}$	B		7			7		ns
t_{PLZ}				9			9		

†All typical values at V_{CC} = 5 V, T_A = 25°C.
NOTE 1: For load circuit and voltage waveforms, see page 1-12.

TYPES SN54ALS1640A, SN54ALS1645A, SN54ALS1641 THRU SN54ALS1644
SN74ALS1640A, SN74ALS1645A, SN74ALS1641 THRU SN74ALS1644
OCTAL BUS TRANSCEIVERS

D2661, DECEMBER 1982—REVISED DECEMBER 1983

- Bidirectional Bus Transceivers in High-Density 20-Pin Packages
- Lower-Power Versions of 'ALS640 Series
- Choice of True or Inverting Logic
- Choice of 3-State or Open-Collector Outputs
- Package Options Include Both Plastic and Ceramic Chip Carriers in Addition to Plastic and Ceramic DIPs
- Dependable Texas Instruments Quality and Reliability

DEVICE	OUTPUT	LOGIC
'ALS1640A	3-State	Inverting
'ALS1641	Open-Collector	True
'ALS1642	Open-Collector	Inverting
'ALS1643	3-State	True and Inverting
'ALS1644	Open-Collector	True and Inverting
'ALS1645A	3-State	True

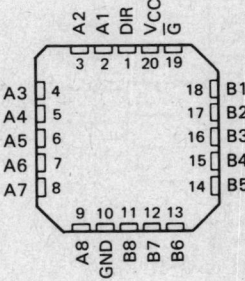

SN54ALS' . . . J PACKAGE
SN74ALS' . . . N PACKAGE
(TOP VIEW)

SN54' . . . FH PACKAGE
SN74' . . . FN PACKAGE
(TOP VIEW)

description

These octal bus transceivers are designed for asynchronous two-way communication between data buses. The devices transmit data from the A bus to the B bus or from the B bus to the A bus depending upon the level at the direction control (DIR) input. The enable input ($\overline{G}$) can be used to disable the device so the buses are effectively isolated.

The -1 versions of the SN74ALS' parts are identical to the standard versions except that the recommended maximum I_{OL} is increased to 24 milliamperes. There are no -1 versions of the SN54ALS' parts.

The SN54ALS' family is characterized for operation over the full military temperature range of —55°C to 125°C. The SN74ALS' family is characterized for operation from 0°C to 70°C.

FUNCTION TABLE

CONTROL INPUTS		OPERATION		
$\overline{G}$	DIR	'ALS1640A 'ALS1642	'ALS1641 'ALS1645A	'ALS1643 'ALS1644
L	L	$\overline{B}$ data to A bus	B data to A bus	B data to A bus
L	H	$\overline{A}$ data to B bus	A data to B bus	$\overline{A}$ data to B bus
H	X	Isolation	Isolation	Isolation

ALS AND AS CIRCUITS

Copyright © 1982 by Texas Instruments Incorporated

TEXAS INSTRUMENTS
POST OFFICE BOX 225012 • DALLAS, TEXAS 75265

TYPES SN54ALS1640A, SN54ALS1645A, SN54ALS1641 THRU SN54ALS1644
SN74ALS1640A, SN74ALS1645A, SN74ALS1641 THRU SN74ALS1644
OCTAL BUS TRANSCEIVERS

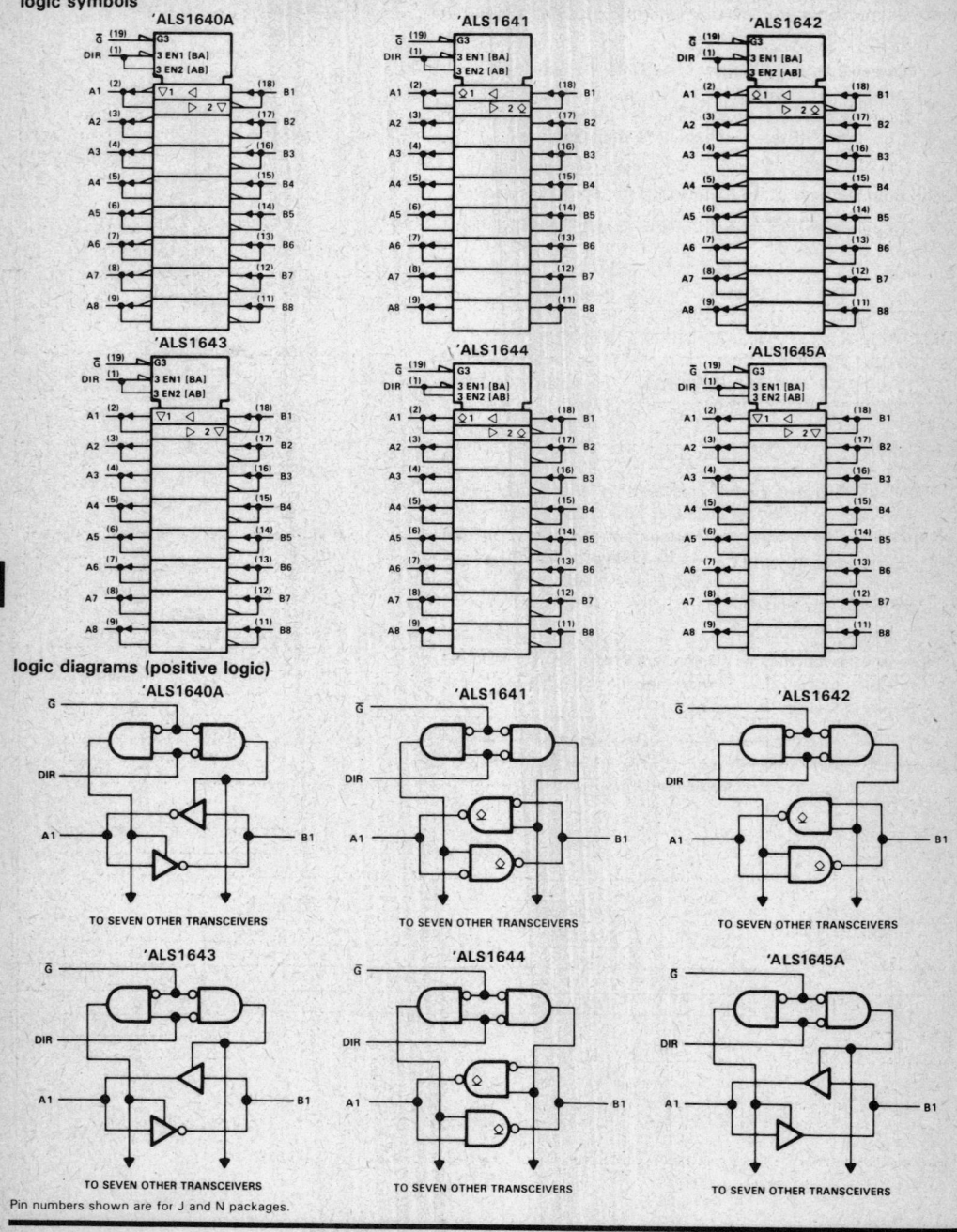

Pin numbers shown are for J and N packages.

TYPES SN54ALS1640A, SN54ALS1643, SN54ALS1645A, SN74ALS1640A, SN74ALS1643, SN74ALS1645A OCTAL BUS TRANSCEIVERS

absolute maximum ratings over operating free-air temperature range (unless otherwise noted)

Supply voltage, V_{CC} .... 7 V
Input voltage: All inputs .... 7 V
                I/O ports .... 5.5 V
Operating free-air temperature range: SN54ALS1640A, SN54ALS1643, SN54ALS1645A ... −55°C to 125°C
                                    SN74ALS1640A, SN74ALS1643, SN74ALS1645A ...... 0°C to 70°C
Storage temperature range .... −65°C to 150°C

recommended operating conditions

		SN54ALS1640A SN54ALS1643 SN54ALS1645A			SN74ALS1640A SN74ALS1643 SN74ALS1645A			UNIT
		MIN	NOM	MAX	MIN	NOM	MAX	
V_{CC}	Supply voltage	4.5	5	5.5	4.5	5	5.5	V
V_{IH}	High-level input voltage	2			2			V
V_{IL}	Low-level input voltage			0.8			0.8	V
I_{OH}	High-level output current			−12			−15	mA
I_{OL}	Low-level output current			8			16 24†	mA
T_A	Operating free-air temperature	−55		125	0		70	°C

†The extended limits apply only if V_{CC} is maintained between 4.75 V and 5.25 V.
The 24-mA limit applies for the SN74ALS1640A-1, SN74ALS1643-1, and SN74ALS1645A-1 only.

electrical characteristics over recommended operating free-air temperature range (unless otherwise noted)

PARAMETER		TEST CONDITIONS		SN54ALS1640A SN54ALS1643 SN54ALS1645A			SN74ALS1640A SN74ALS1643 SN74ALS1645A			UNIT
				MIN	TYP‡	MAX	MIN	TYP‡	MAX	
V_{IK}		$V_{CC} = 4.5$ V,	$I_I = -18$ mA			−1.5			−1.5	V
V_{OH}		$V_{CC} = 4.5$ V to 5.5 V,	$I_{OH} = -0.4$ mA	$V_{CC}-2$			$V_{CC}-2$			V
		$V_{CC} = 4.5$ V,	$I_{OH} = -3$ mA	2.4	3.2		2.4	3.2		
		$V_{CC} = 4.5$ V,	$I_{OH} = -12$ mA	2						
		$V_{CC} = 4.5$ V,	$I_{OH} = -15$ mA				2			
V_{OL}		$V_{CC} = 4.5$ V,	$I_{OL} = 8$ mA		0.25	0.4		0.25	0.4	V
		$V_{CC} = 4.5$ V,	$I_{OL} = 16$ mA					0.35	0.5	
		($I_{OL} = 24$ mA for −1 versions)								
I_I	Control inputs	$V_{CC} = 5.5$ V,	$V_I = 7$ V			0.1			0.1	mA
	A or B ports	$V_{CC} = 5.5$ V,	$V_I = 5.5$ V			0.1			0.1	
I_{IH}	Control inputs	$V_{CC} = 5.5$ V,	$V_I = 2.7$ V			20			20	μA
	A or B ports§					20			20	
I_{IL}	Control inputs	$V_{CC} = 5.5$ V,	$V_I = 0.4$ V			−0.1			−0.1	mA
	A or B ports§					−0.1			−0.1	
I_O¶		$V_{CC} = 5.5$ V,	$V_O = 2.25$ V	−30		−112	−30		−112	mA
I_{CC}	'ALS1640A	$V_{CC} = 5.5$ V			18	35		18	32	mA
	'ALS1643				22			22		
	'ALS1645A				25	40		25	36	

‡All typical values are at $V_{CC} = 5$ V, $T_A = 25$°C.
§For I/O ports, the parameters I_{IH} and I_{IL} include the off-state output current.
¶The output conditions have been chosen to produce a current that closely approximates one half of the true short-circuit output current, I_{OS}.

Additional information on these products can be obtained from the factory as it becomes available.

ADVANCE INFORMATION
This page contains information on a new product.
Specifications are subject to change without notice.

TEXAS INSTRUMENTS
POST OFFICE BOX 225012 • DALLAS, TEXAS 75265

ALS AND AS CIRCUITS

TYPES SN54ALS1640A, SN54ALS1643, SN54ALS1645A SN74ALS1640A, SN74ALS1643, SN74ALS1645A OCTAL BUS TRANSCEIVERS

'ALS1640A switching characteristics (see Note 1)

PARAMETER	FROM (INPUT)	TO (OUTPUT)	V_{CC} = 4.5 V to 5.5 V, C_L = 50 pF, R1 = 500 Ω, R2 = 500 Ω, T_A = MIN to MAX				UNIT
			SN54ALS1640A		SN74ALS1640A		
			MIN	MAX	MIN	MAX	
t_{PLH}	A or B	B or A	5	17	5	15	ns
t_{PHL}			2	13	2	10	
t_{PZH}	$\overline{G}$	A or B	5	23	5	20	ns
t_{PZL}			5	25	5	22	
t_{PHZ}	$\overline{G}$	A or B	2	12	2	10	ns
t_{PLZ}			5	16	5	13	

'ALS1643 switching characteristics (see Note 1)

PARAMETER	FROM (INPUT)	TO (OUTPUT)	V_{CC} = 4.5 V to 5.5 V, C_L = 50 pF, R1 = 500 Ω, R2 = 500 Ω, T_A = MIN to MAX						UNIT
			SN54ALS1643			SN74ALS1643			
			MIN	TYP†	MAX	MIN	TYP†	MAX	
t_{PLH}	A	B		7			7		ns
t_{PHL}				7			7		
t_{PLH}	B	A		8			8		ns
t_{PHL}				8			8		
t_{PZH}	$\overline{G}$	A		18			18		ns
t_{PZL}				21			21		
t_{PHZ}	$\overline{G}$	A		12			12		ns
t_{PLZ}				13			13		
t_{PZH}	$\overline{G}$	B		18			18		ns
t_{PZL}				21			21		
t_{PHZ}	$\overline{G}$	B		12			12		ns
t_{PLZ}				13			13		

'ALS1645A switching characteristics (see Note 1)

PARAMETER	FROM (INPUT)	TO (OUTPUT)	V_{CC} = 4.5 V to 5.5 V, C_L = 50 pF, R1 = 500 Ω, R2 = 500 Ω, T_A = MIN to MAX				UNIT
			SN54ALS1645A		SN74ALS1645A		
			MIN	MAX	MIN	MAX	
t_{PLH}	A or B	B or A	2	15	2	13	ns
t_{PHL}			2	15	2	13	
t_{PZH}	$\overline{G}$	A or B	8	28	8	25	ns
t_{PZL}			8	28	8	25	
t_{PHZ}	$\overline{G}$	A or B	2	14	2	12	ns
t_{PLZ}			3	22	3	18	

†All typical values are at V_{CC} = 5 V, T_A = 25°C.
NOTE 1: For load circuit and voltage waveforms, see page 1-12.

ADVANCE INFORMATION
This page contains information on a new product. Specifications are subject to change without notice.

TEXAS INSTRUMENTS
POST OFFICE BOX 225012 • DALLAS, TEXAS 75265

TYPES SN54ALS1641, SN54ALS1642, SN54ALS1644, SN74ALS1641, SN74ALS1642, SN74ALS1644 OCTAL BUS TRANSCEIVERS

absolute maximum ratings over operating free-air temperature range (unless otherwise noted)

Supply voltage, V_{CC} .... 7 V
Input voltage: All inputs and I/O ports .... 7 V
Operating free-air temperature range: SN54ALS1641, SN54ALS1642, SN54ALS1644 ..... −55°C to 125°C
                                   SN74ALS1641, SN74ALS1642, SN74ALS1644 .... 0°C to 70°C
Storage temperature range .... −65°C to 150°C

recommended operating conditions

		SN54ALS1641 SN54ALS1642 SN54ALS1644			SN74ALS1641 SN74ALS1642 SN74ALS1644			UNIT
		MIN	NOM	MAX	MIN	NOM	MAX	
V_{CC}	Supply voltage	4.5	5	5.5	4.5	5	5.5	V
V_{IH}	High-level input voltage	2			2			V
V_{IL}	Low-level input voltage			0.8			0.8	V
V_{OH}	High-level output voltage			5.5			5.5	V
I_{OL}	Low-level output current			8			16 24[†]	mA
T_A	Operating free-air temperature	−55		125	0		70	°C

[†] The extended limits apply only if V_{CC} is maintained between 4.75 V and 5.25 V.
The 24-mA limit applies for the SN74ALS1641-1, SN74ALS1642-1, and SN74ALS1644-1 only.

electrical characteristics over recommended operating free-air temperature range (unless otherwise noted)

PARAMETER		TEST CONDITIONS		SN54ALS1641 SN54ALS1642 SN54ALS1644			SN74ALS1641 SN74ALS1642 SN74ALS1644			UNIT
				MIN	TYP[‡]	MAX	MIN	TYP[‡]	MAX	
V_{IK}		V_{CC} = 4.5 V,	I_I = −18 mA			−1.5			−1.5	V
I_{OH}		V_{CC} = 4.5 V,	V_{OH} = 5.5 V			0.1			0.1	mA
V_{OL}		V_{CC} = 4.5 V,	I_{OL} = 8 mA		0.25	0.4		0.25	0.4	V
		V_{CC} = 4.5 V, (I_{OL} = 24 mA for -1 versions)	I_{OL} = 16 mA					0.35	0.5	
I_I	Control inputs	V_{CC} = 5.5 V,	V_I = 7 V			0.1			0.1	mA
	A or B ports	V_{CC} = 5.5 V,	V_I = 5.5 V			0.1			0.1	
I_{IH}	Control inputs	V_{CC} = 5.5 V,	V_I = 2.7 V			20			20	µA
	A or B ports[§]					20			20	
I_{IL}	Control inputs	V_{CC} = 5.5 V,	V_I = 0.4 V			−0.1			−0.1	mA
	A or B ports[§]					−0.1			−0.1	
I_{CC}	'ALS1641	V_{CC} = 5.5 V				23			23	mA
	'ALS1642					20			20	
	'ALS1644					22			22	

[‡] All typical values are at V_{CC} = 5 V, T_A = 25°C.
[§] For I/O ports, the parameters I_{IH} and I_{IL} include the off-state output current.

PRODUCT PREVIEW
This page contains information on a product under development. Texas Instruments reserves the right to change or discontinue this product without notice.

TEXAS INSTRUMENTS
POST OFFICE BOX 225012 • DALLAS, TEXAS 75265

ALS AND AS CIRCUITS

TYPES SN54ALS1641, SN54ALS1642, SN54ALS1644, SN74ALS1641, SN74ALS1642, SN74ALS1644
OCTAL BUS TRANSCEIVERS

'ALS1641 switching characteristics (see Note 1)

PARAMETER	FROM (INPUT)	TO (OUTPUT)	V_{CC} = 4.5 V to 5.5 V, C_L = 50 pF, R_L = 500 Ω, T_A = MIN to MAX						UNIT
			SN54ALS1641			SN74ALS1641			
			MIN	TYP†	MAX	MIN	TYP†	MAX	
t_{PLH}	A or B	B or A		22			22		ns
t_{PHL}				14			14		
t_{PLH}	$\overline{G}$ or DIR	A or B		26			26		ns
t_{PHL}				26			26		

'ALS1642 switching characteristics (see Note 1)

PARAMETER	FROM (INPUT)	TO (OUTPUT)	V_{CC} = 4.5 V to 5.5 V, C_L = 50 pF, R_L = 500 Ω, T_A = MIN to MAX						UNIT
			SN54ALS1642			SN74ALS1642			
			MIN	TYP†	MAX	MIN	TYP†	MAX	
t_{PLH}	A or B	B or A		25			25		ns
t_{PHL}				13			13		
t_{PLH}	$\overline{G}$ or DIR	A or B		29			29		ns
t_{PHL}				29			29		

'ALS1644 switching characteristics (see Note 1)

PARAMETER	FROM (INPUT)	TO (OUTPUT)	V_{CC} = 4.5 V to 5.5 V, C_L = 50 pF, R_L = 500 Ω, T_A = MIN to MAX						UNIT
			SN54ALS1644			SN74ALS1644			
			MIN	TYP†	MAX	MIN	TYP†	MAX	
t_{PLH}	A	B		27			27		ns
t_{PHL}				19			19		
t_{PLH}	B	A		24			24		ns
t_{PHL}				17			17		
t_{PLH}	$\overline{G}$ or DIR	A		30			30		ns
t_{PHL}				27			27		
t_{PLH}	$\overline{G}$ or DIR	B		24			24		ns
t_{PHL}				30			30		

‡ All typical values are at V_{CC} = 5 V, T_A = 25°C.
NOTE 1: For load circuit and voltage waveforms, see page 1-12.

Additional information on these products can be obtained from the factory as it becomes available.

PRODUCT PREVIEW
This page contains information on a product under development. Texas Instruments reserves the right to change or discontinue this product without notice.

TEXAS INSTRUMENTS
POST OFFICE BOX 225012 • DALLAS, TEXAS 75265

TYPES SN54AS2620, SN54AS2623, SN74AS2620, SN74AS2623
OCTAL BUS TRANSCEIVERS/MOS DRIVER

DECEMBER 1983

- Bidirectional Octal Bus Transceivers For Driving MOS Devices
- I/O Ports Have 25 Ohm Series Resistors So No External Resistors Are Required
- Local Bus-Latch Capability
- Choice of True or Inverting Logic
- Dependable Texas Instruments Quality and Reliability

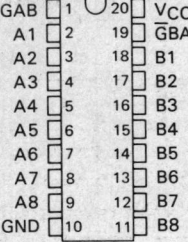

SN54AS'... J PACKAGE
SN74AS'... N PACKAGE
(TOP VIEW)

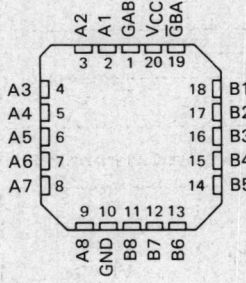

SN54AS'... FH PACKAGE
SN74AS'... FN PACKAGE
(TOP VIEW)

description

These octal bus transceivers are designed to drive the capacitive input characteristics of MOS devices and allow asynchronous two-way communication between data buses. The control function implementation allows for maximum flexibility in timing.

These devices allow data transmission from A bus to the B bus or from the B bus to the A bus depending upon the logic levels at the enable inputs ($\overline{GBA}$ and GAB).

The enable inputs can be used to disable the device so that the buses are effectively isolated.

The dual-enable configuration gives the 'AS2620 or 'AS2623 the capability to store data by simultaneous enabling of $\overline{GBA}$ and GAB. Each output reinforces its input in this transceiver configuration. Thus, when both control inputs are enabled and all other data sources to the two sets of bus lines are at high impedance, both sets of bus lines (16 in all) will remain at their last states. The 8-bit codes appearing on the two sets of buses will be identical for the 'AS2623 or complementary for the 'AS2620.

The SN54AS2620 and SN54AS2623 are characterized for operation over the full military temperature range of −55°C to 125°C. The SN74AS2620 and SN74AS2623 are characterized for operation from 0°C to 70°C.

FUNCTION TABLE

ENABLE INPUTS		OPERATION	
$\overline{GBA}$	GAB	'AS2620	'AS2623
L	L	$\overline{B}$ data to A bus	B data to A bus
H	H	$\overline{A}$ data to B bus	A data to B bus
H	L	Isolation	Isolation
L	H	$\overline{B}$ data to A bus, $\overline{A}$ data to B bus	B data to A bus, A data to B bus

Copyright © 1983 by Texas Instruments Incorporated

TEXAS INSTRUMENTS
POST OFFICE BOX 225012 • DALLAS, TEXAS 75265

TYPES SN54AS2620, SN54AS2623, SN74AS2620, SN74AS2623
OCTAL BUS TRANSCEIVERS/MOS DRIVER

logic symbols

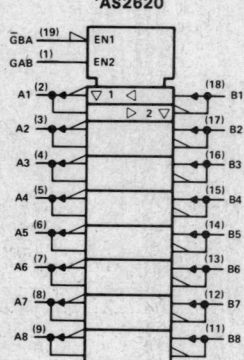
'AS2620

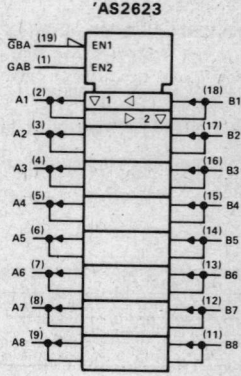
'AS2623

Pin numbers shown are for J and N packages.

logic diagrams (positive logic)

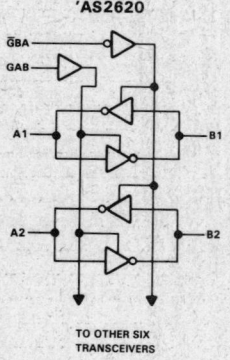
'AS2620

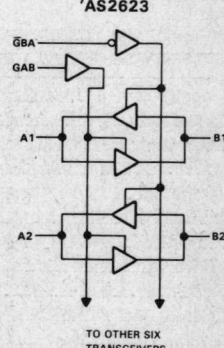
'AS2623

TEXAS INSTRUMENTS
POST OFFICE BOX 225012 • DALLAS, TEXAS 75265

TYPES SN54AS2620, SN54AS2623, SN74AS2620, SN74AS2623
OCTAL BUS TRANSCEIVERS/MOS DRIVER

absolute maximum ratings over operating free-air temperature range (unless otherwise noted)

Supply voltage, V_{CC} .... 7 V
Input voltage: All inputs .... 7 V
                I/O ports .... 5.5 V
Operating free-air temperature range: SN54AS2620, SN54AS2623 .... −55 °C to 125 °C
                                      SN74AS2620, SN74AS2623 .... 0 °C to 70 °C
Storage temperature range .... −65 °C to 150 °C

recommended operating conditions

		SN54AS2620 SN54AS2623			SN74AS2620 SN74AS2623			UNIT
		MIN	NOM	MAX	MIN	NOM	MAX	
V_{CC}	Supply voltage	4.5	5	5.5	4.5	5	5.5	V
V_{IH}	High-level input voltage	2			2			V
V_{IL}	Low-level input voltage			0.8			0.8	V
T_A	Operating free-air temperature	−55		125	0		70	°C

electrical characteristics over recommended operating free-air temperature range (unless otherwise noted)

PARAMETER		TEST CONDITIONS		SN54AS2620 SN54AS2623			SN74AS2620 SN74AS2623			UNIT
				MIN	TYP†	MAX	MIN	TYP†	MAX	
V_{IK}		V_{CC} = 4.5 V,	I_I = −18 mA			−1.2			−1.2	V
V_{OH}		V_{CC} = 4.5 V to 5.5 V,	I_{OH} = −2 mA	V_{CC}−2			V_{CC}−2			V
V_{OL}		V_{CC} = 4.5 V,	I_{OL} = 1 mA		0.15	0.4		0.15	0.4	V
		V_{CC} = 4.5 V,	I_{OL} = 12 mA		0.35	0.7		0.35	0.7	
I_I	Control inputs	V_{CC} = 5.5 V,	V_I = 7 V			0.1			0.1	mA
	A or B ports	V_{CC} = 5.5 V,	V_I = 5.5 V			0.1			0.1	
I_{IH}	Control inputs	V_{CC} = 5.5 V,	V_I = 2.7 V			20			20	µA
	A or B ports‡					50			50	
I_{IL}	Control inputs	V_{CC} = 5.5 V,	V_I = 0.4 V			−0.5			−0.5	mA
	A or B ports‡					−0.5			−0.5	
I_O §		V_{CC} = 5.5 V,	V_O = 2.25 V	−30		−112	−30		−112	mA
I_{OH}		V_{CC} = 4.5 V,	V_O = 2 V	−35			−35			mA
I_{OL}		V_{CC} = 4.5 V,	V_O = 2 V	35			35			mA
I_{CC}	'AS2620	V_{CC} = 5.5 V	Outputs high		62	100		62	100	mA
			Outputs low		74	121		74	121	
			Outputs disabled		48	77		48	77	
	'AS2623	V_{CC} = 5.5 V	Outputs high		57	93		57	93	
			Outputs low		116	189		116	189	
			Outputs disabled		72	116		72	116	

†All typical values are at V_{CC} = 5 V, T_A = 25 °C.
‡For I/O ports, the parameters I_{IH} and I_{IL} include the off-state output current.
§The output conditions have been chosen to produce a current that closely approximates one half of the true short-circuit output current, I_{OS}.

TEXAS
INSTRUMENTS
POST OFFICE BOX 225012 • DALLAS, TEXAS 75265

TYPES SN54AS2620, SN54AS2623, SN74AS2620, SN74AS2623
OCTAL BUS TRANSCEIVERS/MOS DRIVER

'AS2620 switching characteristics (see Note 1)

PARAMETER	FROM (INPUT)	TO (OUTPUT)	V_{CC} = 4.5 V to 5.5 V, C_L = 50 pF, R1 = 500 Ω, R2 = 500 Ω, T_A = MIN to MAX				UNIT
			SN54AS2620		SN74AS2620		
			MIN	MAX	MIN	MAX	
t_{PLH}	A	B	1	9.5	1	8	ns
t_{PHL}			1	7.5	1	6.5	
t_{PLH}	B	A	1	9.5	1	8	ns
t_{PHL}			1	7.5	1	6.5	
t_{PZH}	$\overline{GBA}$	A	1	11	1	10	ns
t_{PZL}			1	12	1	11	
t_{PHZ}	$\overline{GBA}$	A	1	7.5	1	6	ns
t_{PLZ}			1	15	1	12	
t_{PZH}	GAB	B	1	9	1	8	ns
t_{PZL}			1	9	1	8	
t_{PHZ}	GAB	B	1	12	1	11	ns
t_{PLZ}			1	12	1	11	

'AS2623 switching characteristics (see Note 1)

PARAMETER	FROM (INPUT)	TO (OUTPUT)	V_{CC} = 4.5 V to 5.5 V, C_L = 50 pF, R1 = 500 Ω, R2 = 500 Ω, T_A = MIN to MAX				UNIT
			SN54AS2623		SN74AS2623		
			MIN	MAX	MIN	MAX	
t_{PLH}	A	B	1	9.5	1	8.5	ns
t_{PHL}			1	8.5	1	7.5	
t_{PLH}	B	A	1	10	1	9	ns
t_{PHL}			1	9	1	7.5	
t_{PZH}	$\overline{GBA}$	A	1	12.5	1	11	ns
t_{PZL}			1	12	1	11	
t_{PHZ}	$\overline{GBA}$	A	1	8.5	1	7.5	ns
t_{PLZ}			1	13	1	12	
t_{PZH}	GAB	B	1	13	1	12	ns
t_{PZL}			1	13.5	1	12	
t_{PHZ}	GAB	B	1	7.5	1	7	ns
t_{PLZ}			1	14.5	1	12.5	

NOTE 1: For load circuit and voltage waveforms, see page 1-12.

TYPES SN54AS2640, SN54AS2645, SN74AS2640, SN74AS2645
OCTAL BUS TRANSCEIVER/MOS DRIVER
DECEMBER 1983

- Bidirectional Octal Bus Transceivers For Driving MOS Devices
- I/O Ports Have 25 Ohm Series Resistors So No External Resistors Are Required
- Choice of True or Inverting Logic
- Dependable Texas Instruments Quality and Reliability

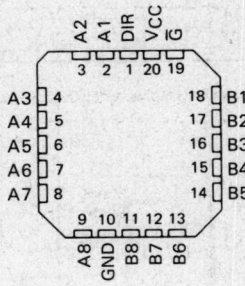

SN54AS' ... J PACKAGE
SN74AS' ... N PACKAGE
(TOP VIEW)

```
DIR  [ 1   20 ] VCC
A1   [ 2   19 ] G
A2   [ 3   18 ] B1
A3   [ 4   17 ] B2
A4   [ 5   16 ] B3
A5   [ 6   15 ] B4
A6   [ 7   14 ] B5
A7   [ 8   13 ] B6
A8   [ 9   12 ] B7
GND  [10   11 ] B8
```

SN54AS' ... FH PACKAGE
SN74AS' ... FN PACKAGE
(TOP VIEW)

description

These octal bus transceivers are designed to drive the capacitive input characteristics of MOS devices and allow asynchronous two-way communication between data buses. The control function implementation allows for maximum flexibility in timing.

The devices transmit data from the A bus to the B bus or from the B bus to the A bus depending upon the level at the direction control (DIR) input. The enable input ($\overline{G}$) can be used to disable the device so the buses are effectively isolated.

The SN54AS' family is characterized for operation over the full military temperature range of −55°C to 125°C. The SN74AS' family is characterized for operation from 0°C to 70°C.

FUNCTION TABLE

CONTROL INPUTS		OPERATION	
$\overline{G}$	DIR	'AS2640	'AS2645
L	L	$\overline{B}$ data to A bus	B data to A bus
L	H	$\overline{A}$ data to B bus	A data to B bus
H	X	Isolation	Isolation

PRODUCT PREVIEW
This document contains information on a product under development. Texas Instruments reserves the right to change or discontinue this product without notice.

TEXAS INSTRUMENTS
POST OFFICE BOX 225012 • DALLAS, TEXAS 75265

TYPES SN54AS2640, SN54AS2645, SN74AS2640, SN74AS2645
OCTAL BUS TRANSCEIVER/MOS DRIVER

logic symbols

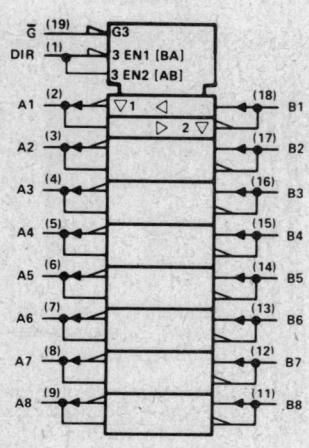

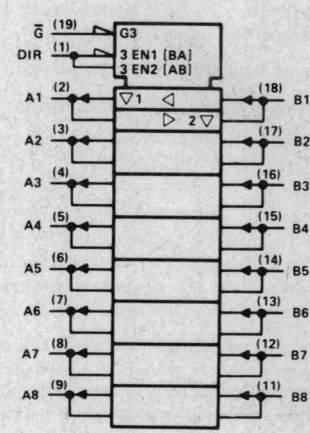

functional block diagrams (positive logic)

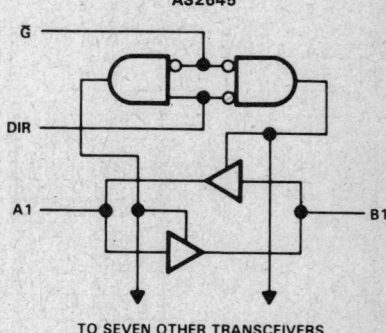

Pin numbers shown are for J and N packages.

TYPES SN54AS2640, SN54AS2645, SN74AS2640, SN74AS2645
OCTAL BUS TRANSCEIVERS

absolute maximum ratings over operating free-air temperature range (unless otherwise noted)

Supply voltage, V_{CC}	7 V
Input voltage: All inputs	7 V
I/O ports	5.5 V
Operating free-air temperature range: SN54AS2640, SN54AS2645	$-55°C$ to $125°C$
SN74AS2640, SN74AS2645	$0°C$ to $70°C$
Storage temperature range	$-65°C$ to $150°C$

recommended operating conditions

		SN54AS2640 SN54AS2645			SN74AS2640 SN74AS2645			UNIT
		MIN	NOM	MAX	MIN	NOM	MAX	
V_{CC}	Supply voltage	4.5	5	5.5	4.5	5	5.5	V
V_{IH}	High-level input voltage	2			2			V
V_{IL}	Low-lvel input voltage			0.8			0.8	V
T_A	Operating free-air temperature	-55		125	0		70	°C

electrical characteristics over recommended operating free-air temperature range (unless otherwise noted)

PARAMETER		TEST CONDITIONS		SN54AS'			SN74AS'			UNIT
				MIN	TYP†	MAX	MIN	TYP†	MAX	
V_{IK}		$V_{CC} = 4.5$ V,	$I_I = -18$ mA			-1.2			-1.2	V
V_{OH}		$V_{CC} = 4.5$ V to 5.5 V,	$I_{OH} = -2$ mA	$V_{CC}-2$			$V_{CC}-2$			V
V_{OL}		$V_{CC} = 4.5$ V,	$I_{OL} = 1$ mA		0.15	0.4		0.15	0.4	V
		$V_{CC} = 4.5$ V,	$I_{OL} = 12$ mA		0.35	0.7		0.35	0.7	
I_I	Control inputs	$V_{CC} = 5.5$ V,	$V_I = 7$ V			0.1			0.1	mA
	A or B ports	$V_{CC} = 5.5$ V,	$V_I = 5.5$ V			0.1			0.1	
I_{IH}	Control inputs	$V_{CC} = 5.5$ V,	$V_I = 2.7$ V			20			20	µA
	A or B ports‡					50			50	
I_{IL}	Control inputs	$V_{CC} = 5.5$ V,	$V_I = 0.4$ V			-0.5			-0.5	mA
	A or B ports‡					-0.5			-0.5	
I_O §		$V_{CC} = 5.5$ V,	$V_O = 2.25$ V	-30		-112	-30		-112	mA
I_{OH}		$V_{CC} = 4.5$ V,	$V_O = 2$ V	-35			-35			mA
I_{OL}		$V_{CC} = 4.5$ V,	$V_{OL} = 2$ V	35			35			mA
I_{CC}	'AS2640	$V_{CC} = 5.5$ V	Outputs high		37	58		37	58	mA
			Outputs low		78	123		78	123	
			Outputs disabled		51	80		51	80	
	'AS2645		Outputs high		58	95		58	95	
			Outputs low		95	155		95	155	
			Outputs disabled		73	119		73	119	

†All typical values are at $V_{CC} = 5$ V, $T_A = 25°C$.
‡For I/O ports, the parameters I_{IH} and I_{IL} include the off-state output current.
§The output conditions have been chosen to produce a current that closely approximates one half of the true short-circuit output current, I_{OS}.

ALS AND AS CIRCUITS

TEXAS INSTRUMENTS
POST OFFICE BOX 225012 • DALLAS, TEXAS 75265

TYPES SN54AS2640, SN54AS2645, SN74AS2640, SN74AS2645
OCTAL BUS TRANSCEIVERS/MOS DRIVERS

'AS2640 switching characteristics (see Note 1)

PARAMETER	FROM (INPUT)	TO (OUTPUT)	V_{CC} = 4.5 V to 5.5 V, C_L = 50 pF, $R1$ = 500 Ω, $R2$ = 500 Ω, T_A = MIN to MAX				UNIT
			SN54AS2640		SN74AS2640		
			MIN	MAX	MIN	MAX	
t_{PLH}	A or B	B or A	1	9.5	1	7.5	ns
t_{PHL}			1	7	1	6.5	
t_{PZH}	$\overline{G}$	A or B	2	11	2	9	ns
t_{PZL}			2	12	2	10	
t_{PHZ}	$\overline{G}$	A or B	1	8	1	7	ns
t_{PLZ}			2	15	2	13	

'AS2645 switching characteristics (see Note 1)

PARAMETER	FROM (INPUT)	TO (OUTPUT)	V_{CC} = 4.5 V to 5.5 V, C_L = 50 pF, $R1$ = 500 Ω, $R2$ = 500 Ω, T_A = MIN to MAX				UNIT
			SN54AS2645		SN74AS2645		
			MIN	MAX	MIN	MAX	
t_{PLH}	A or B	B or A	1	12	1	10	ns
t_{PHL}			1	11	1	9.5	
t_{PZH}	$\overline{G}$	A or B	1	13	1	11.5	ns
t_{PZL}			1	13	1	10.5	
t_{PHZ}	$\overline{G}$	A or B	1	9	1	8	ns
t_{PLZ}			1	13	1	12	

NOTE 1: For load circuit and voltage waveforms, see page 1-12.

TYPES SN54ALS8003, SN74ALS8003
DUAL 2-INPUT POSITIVE-NAND GATES

D2746, JULY 1983—REVISED DECEMBER 1983

- Package Options Include Both Plastic and Ceramic Chip Carriers in Addition to Plastic and Ceramic DIPs.
- Dependable Texas Instruments Quality and Reliability.

SN54ALS8003 . . . JG PACKAGE
SN74ALS8003 . . . P PACKAGE
(TOP VIEW)

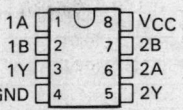

description

These devices contain two independent 2-input NAND gates. They perform the Boolean functions $Y = \overline{A \cdot B}$ or $Y = \overline{A} + \overline{B}$ in positive logic.

The SN54ALS8003 is characterized for operation over the full military temperature range of −55°C to 125°C. The SN74ALS8003 is characterized for operation from 0°C to 70°C.

SN54ALS8003 . . . FH PACKAGE
SN74ALS8003 . . . FN PACKAGE
(TOP VIEW)

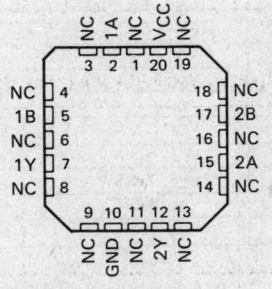

NC—No internal connection

FUNCTION TABLE (each gate)

INPUTS		OUTPUT
A	B	Y
H	H	L
L	X	H
X	L	H

logic symbol

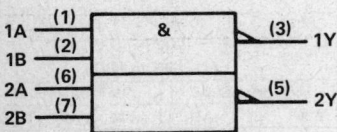

Pin numbers shown are for JG and P packages.

absolute maximum ratings over operating free-air temperature range (unless otherwise noted)

Supply voltage, V_{CC} . . . . 7 V
Input voltage . . . . 7 V
Operating free-air temperature range: SN54ALS8003 . . . . −55°C to 125°C
SN74ALS8003 . . . . 0°C to 70°C
Storage temperature . . . . −65°C to 150°C

Copyright © 1983 by Texas Instruments Incorporated

TEXAS INSTRUMENTS
POST OFFICE BOX 225012 • DALLAS, TEXAS 75265

TYPES SN54ALS8003, SN74ALS8003
DUAL 2-INPUT POSITIVE-NAND GATES

recommended operating conditions

		SN54ALS8003			SN74ALS8003			UNIT
		MIN	NOM	MAX	MIN	NOM	MAX	
V_{CC}	Supply voltage	4.5	5	5.5	4.5	5	5.5	V
V_{IH}	High-level input voltage	2			2			V
V_{IL}	Low-level input voltage			0.8			0.8	V
I_{OH}	High-level output current			−0.4			−0.4	mA
I_{OL}	Low-level output current			4			8	mA
T_A	Operating free-air temperature	−55		125	0		70	°C

electrical characteristics over recommended operating free-air temperature range (unless otherwise noted)

PARAMETER	TEST CONDITIONS		SN54ALS8003			SN74ALS8003			UNIT
			MIN	TYP[†]	MAX	MIN	TYP[†]	MAX	
V_{IK}	V_{CC} = 4.5 V,	I_I = −18 mA			−1.5			−1.5	V
V_{OH}	V_{CC} = 4.5 V to 5.5 V,	I_{OH} = −0.4 mA	$V_{CC}-2$			$V_{CC}-2$			V
V_{OL}	V_{CC} = 4.5 V,	I_{OL} = 4 mA		0.25	0.4		0.25	0.4	V
	V_{CC} = 4.5 V,	I_{OL} = 8 mA					0.35	0.5	
I_I	V_{CC} = 5.5 V,	V_I = 7 V			0.1			0.1	mA
I_{IH}	V_{CC} = 5.5 V,	V_I = 2.7 V			20			20	μA
I_{IL}	V_{CC} = 5.5 V,	V_I = 0.4 V			−0.1			−0.1	mA
I_O[‡]	V_{CC} = 5.5 V,	V_O = 2.25 V	−15		−70	−15		−70	mA
I_{CCH}	V_{CC} = 5.5 V,	V_I = 0 V		0.22	0.43		0.22	0.43	mA
I_{CCL}	V_{CC} = 5.5 V,	V_I = 4.5 V		0.81	1.5		0.81	1.5	mA

[†] All typical values are at V_{CC} = 5 V, T_A = 25°C.
[‡] The output conditions have been chosen to produce a current that closely approximats one half of the true short-circuit output current, I_{OS}.

switching characteristics (see Note 1)

PARAMETER	FROM (INPUT)	TO (OUTPUT)	V_{CC} = 4.5 V to 5.5 V, C_L = 50 pF, R_L = 500 Ω, T_A = MIN to MAX				UNIT
			SN54ALS8003		SN74ALS8003		
			MIN	MAX	MIN	MAX	
t_{PLH}	A or B	Y	3	14	3	11	ns
t_{PHL}			2	10	2	8	

NOTE 1: For load circuit and voltage waveforms, see page 1-12.

The TTL Data Book
Volume 3

General Information	1
ALS and AS Circuits	2
Mechanical Data	3

MECHANICAL DATA

ORDERING INSTRUCTIONS

Electrical characteristics presented in this data book, unless otherwise noted, apply for circuit type(s) listed in the page heading regardless of package. The availability of a circuit function in a particular package is denoted by an alphabetical reference above the pin-connection diagram(s). These alphabetical references refer to mechanical outline drawings shown in this section.

Factory orders for circuits described in this catalog should include a four-part type number as explained in the following example.

EXAMPLE: SN   54ALS01   J   -00

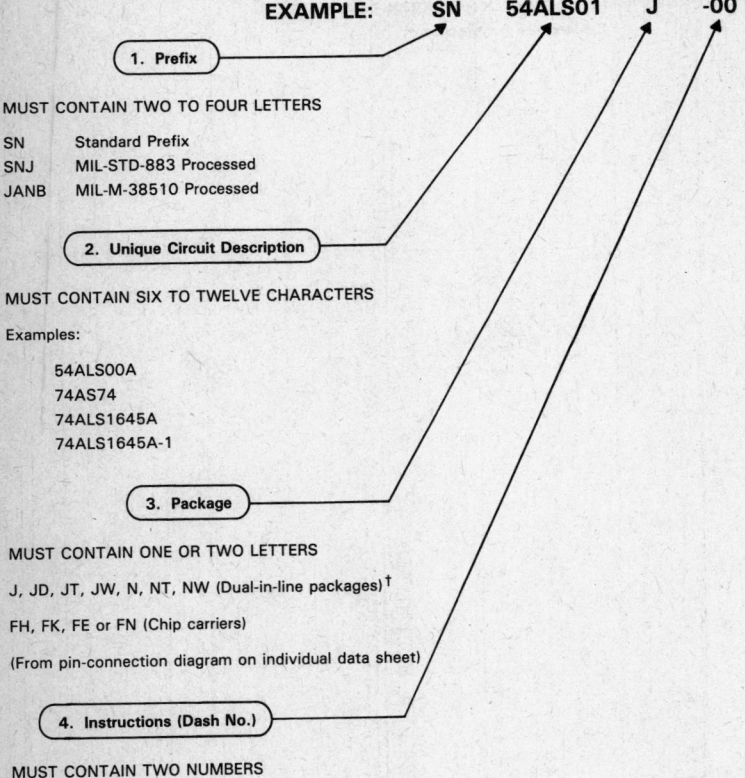

1. Prefix

MUST CONTAIN TWO TO FOUR LETTERS

SN	Standard Prefix
SNJ	MIL-STD-883 Processed
JANB	MIL-M-38510 Processed

2. Unique Circuit Description

MUST CONTAIN SIX TO TWELVE CHARACTERS

Examples:
- 54ALS00A
- 74AS74
- 74ALS1645A
- 74ALS1645A-1

3. Package

MUST CONTAIN ONE OR TWO LETTERS

J, JD, JT, JW, N, NT, NW (Dual-in-line packages)†

FH, FK, FE or FN (Chip carriers)

(From pin-connection diagram on individual data sheet)

4. Instructions (Dash No.)

MUST CONTAIN TWO NUMBERS

— 00 No special instructions
— 10 Solder-dipped leads (N and NT packages only)

† These circuits in dual-in-line packages are shipped in one of the carriers shown below. Unless a specific method of shipment is specified by the customer (with possible additional costs), circuits will be shipped in the most practical carrier. Please contact your TI sales representative for the method that will best suit your particular needs.

Dual-in-line (J, JD, JT, JW, N, NT, NW)

— Slide Magazines
— A-Channel Plastic Tubing
— Barnes Carrier (N only)
— Sectioned Cardboard Box
— Individual Plastic Box

TEXAS INSTRUMENTS
POST OFFICE BOX 225012 • DALLAS, TEXAS 75265

MECHANICAL DATA

FE ceramic chip carrier packages

Each of these hermetically sealed leadless chip carrier packages has a metal cap, a 3-layer ceramic base, and a brazed seal. The packages are intended for surface mounting on solder lands on 1,27-mm (0.050-inch) centers. Terminals require no additional cleaning or processing when used in soldered assembly.

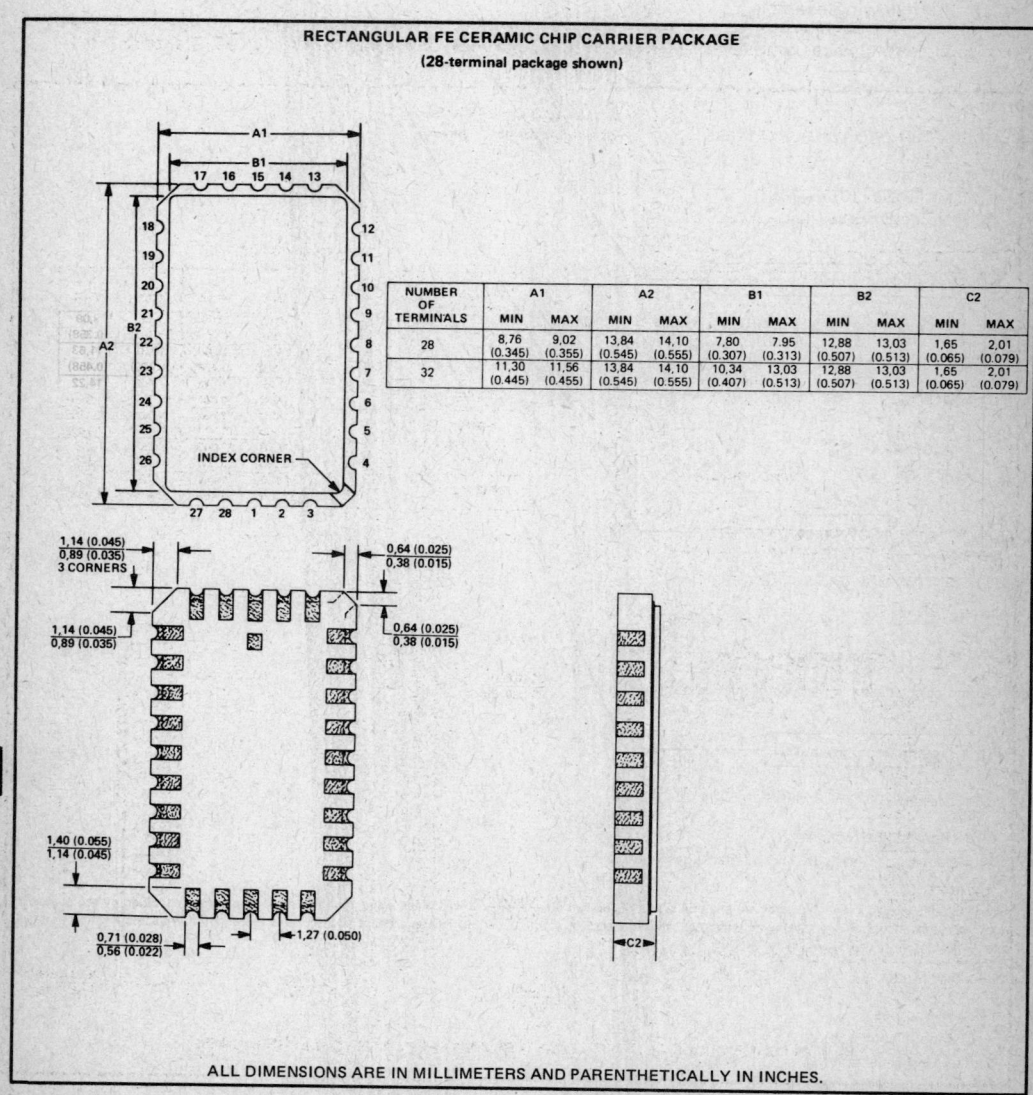

RECTANGULAR FE CERAMIC CHIP CARRIER PACKAGE
(28-terminal package shown)

NUMBER OF TERMINALS	A1		A2		B1		B2		C2	
	MIN	MAX	MIN	MAX	MIN	MAX	MIN	MAX	MIN	MAX
28	8,76 (0.345)	9,02 (0.355)	13,84 (0.545)	14,10 (0.555)	7,80 (0.307)	7,95 (0.313)	12,88 (0.507)	13,03 (0.513)	1,65 (0.065)	2,01 (0.079)
32	11,30 (0.445)	11,56 (0.455)	13,84 (0.545)	14,10 (0.555)	10,34 (0.407)	13,03 (0.513)	12,88 (0.507)	13,03 (0.513)	1,65 (0.065)	2,01 (0.079)

ALL DIMENSIONS ARE IN MILLIMETERS AND PARENTHETICALLY IN INCHES.

TEXAS INSTRUMENTS
POST OFFICE BOX 225012 • DALLAS, TEXAS 75265

MECHANICAL DATA

FH and FK ceramic chip carrier packages

Both versions of these hermetically sealed chip carrier packages have ceramic bases. The FH package has a single-layer base with a ceramic lid and glass seal. The FK package has a three-layer base with a metal lid and braze seal.

The packages are intended for surface mounting on solder lands on 1,27 (0.050-inch) centers. Terminals require no additional cleaning or processing when used in soldered assembly.

FH and FK packages are identical to the FC and FD packages, respectively. The new designations are used to indicate devices whose terminal assignments conform to a forthcoming JEDEC Standard.

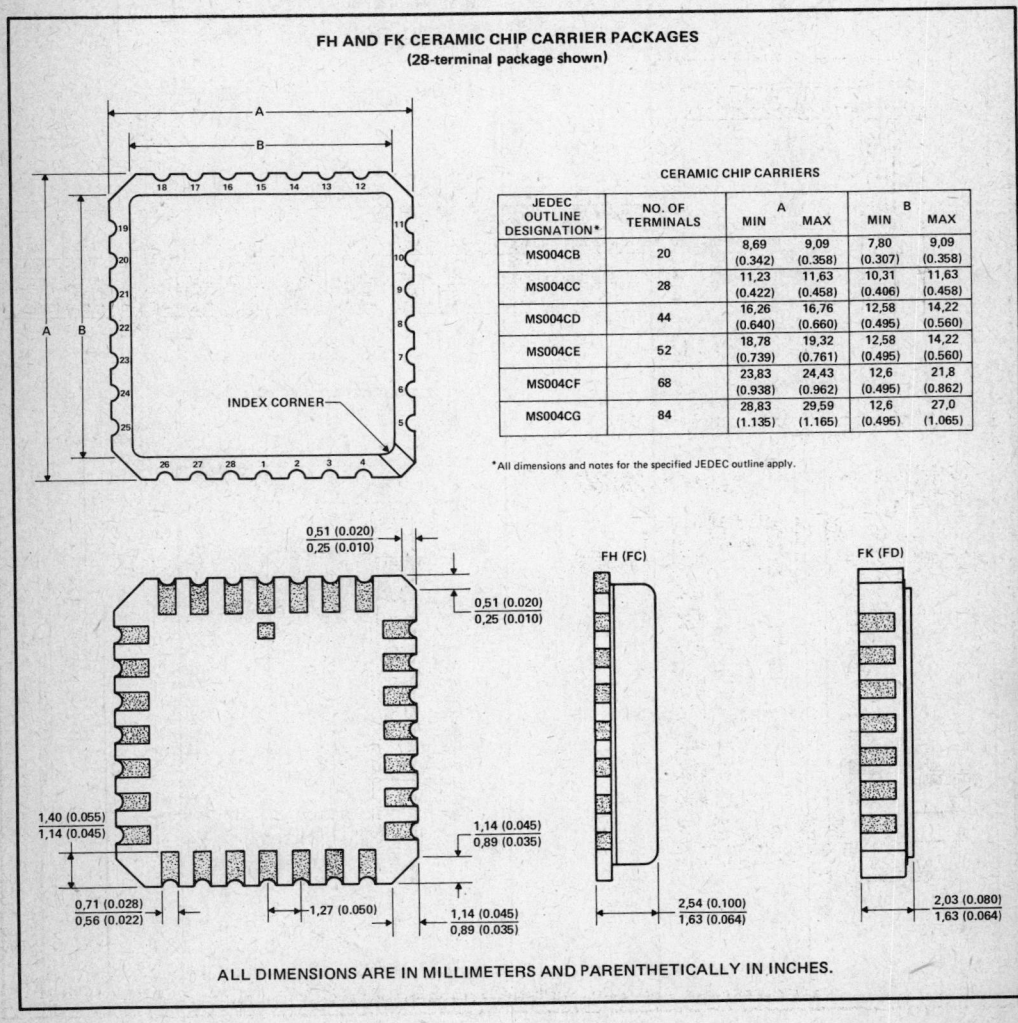

FH AND FK CERAMIC CHIP CARRIER PACKAGES
(28-terminal package shown)

CERAMIC CHIP CARRIERS

JEDEC OUTLINE DESIGNATION*	NO. OF TERMINALS	A MIN	A MAX	B MIN	B MAX
MS004CB	20	8,69 (0.342)	9,09 (0.358)	7,80 (0.307)	9,09 (0.358)
MS004CC	28	11,23 (0.422)	11,63 (0.458)	10,31 (0.406)	11,63 (0.458)
MS004CD	44	16,26 (0.640)	16,76 (0.660)	12,58 (0.495)	14,22 (0.560)
MS004CE	52	18,78 (0.739)	19,32 (0.761)	12,58 (0.495)	14,22 (0.560)
MS004CF	68	23,83 (0.938)	24,43 (0.962)	12,6 (0.495)	21,8 (0.862)
MS004CG	84	28,83 (1.135)	29,59 (1.165)	12,6 (0.495)	27,0 (1.065)

*All dimensions and notes for the specified JEDEC outline apply.

ALL DIMENSIONS ARE IN MILLIMETERS AND PARENTHETICALLY IN INCHES.

TEXAS INSTRUMENTS
POST OFFICE BOX 225012 • DALLAS, TEXAS 75265

MECHANICAL DATA

FN plastic chip carrier package

Each of these chip carrier packages consists of a circuit mounted on a lead frame and encapsulated within an electrically nonconductive plastic compound. The compound withstands soldering temperatures with no deformation, and circuit performance characteristics remain stable when the devices are operated in high-humidity conditions. The packages are intended for surface mounting on solder lands on 1,27-mm (0.050-inch) centers. Leads require no additional cleaning or processing when used in soldered assembly.

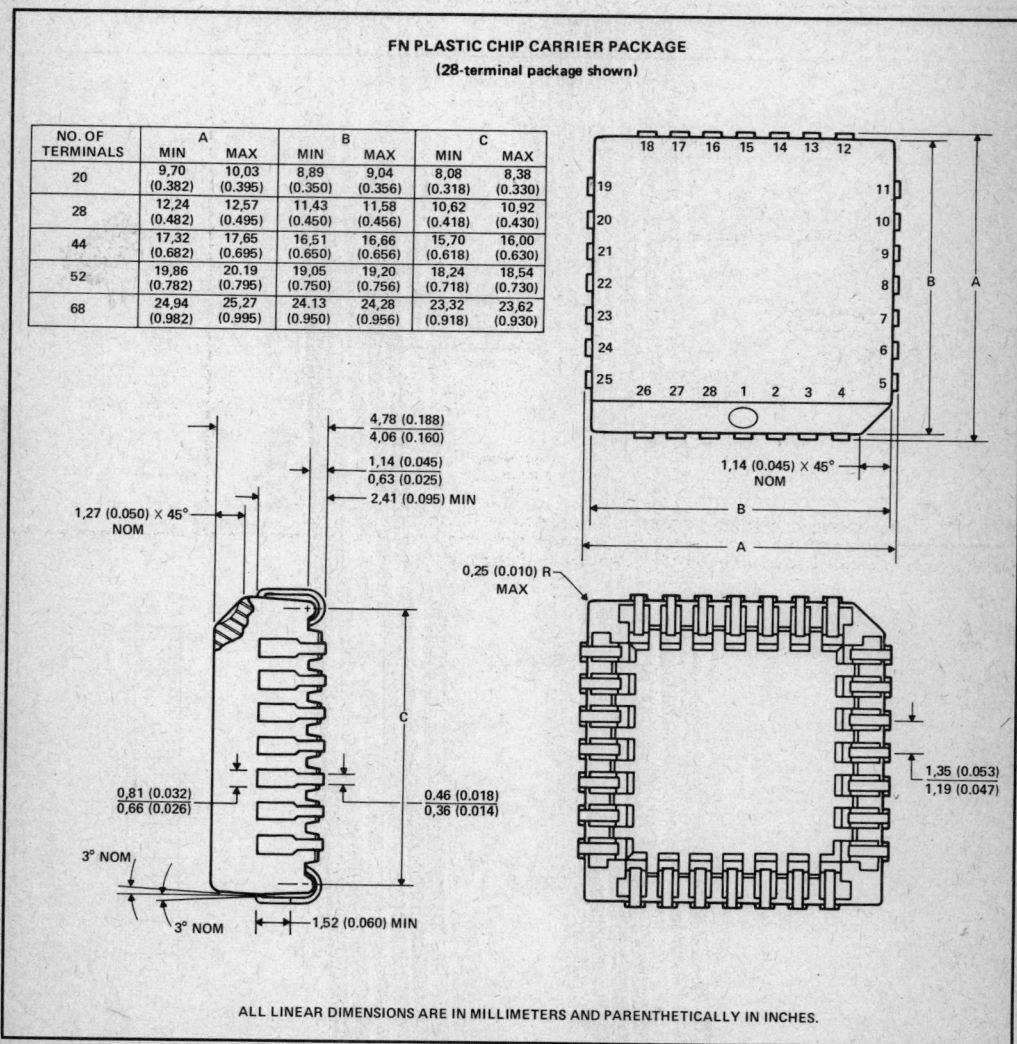

FN PLASTIC CHIP CARRIER PACKAGE
(28-terminal package shown)

NO. OF TERMINALS	A MIN	A MAX	B MIN	B MAX	C MIN	C MAX
20	9,70 (0.382)	10,03 (0.395)	8,89 (0.350)	9,04 (0.356)	8,08 (0.318)	8,38 (0.330)
28	12,24 (0.482)	12,57 (0.495)	11,43 (0.450)	11,58 (0.456)	10,62 (0.418)	10,92 (0.430)
44	17,32 (0.682)	17,65 (0.695)	16,51 (0.650)	16,66 (0.656)	15,70 (0.618)	16,00 (0.630)
52	19,86 (0.782)	20,19 (0.795)	19,05 (0.750)	19,20 (0.756)	18,24 (0.718)	18,54 (0.730)
68	24,94 (0.982)	25,27 (0.995)	24,13 (0.950)	24,28 (0.956)	23,32 (0.918)	23,62 (0.930)

ALL LINEAR DIMENSIONS ARE IN MILLIMETERS AND PARENTHETICALLY IN INCHES.

TEXAS INSTRUMENTS
POST OFFICE BOX 225012 • DALLAS, TEXAS 75265

MECHANICAL DATA

JG ceramic dual-in-line package

This hermetically sealed dual-in-line package consists of a ceramic base, ceramic cap, and 8-lead frame. Hermetic sealing is accomplished with glass. The package is intended for insertion in mounting-hole rows on 7,62 (0.300) centers (see Note a). Once the leads are compressed and inserted, sufficient tension is provided to secure the package in the board during soldering. Non-shiny tin-plated leads require no additional cleaning or processing when used in soldered assembly.

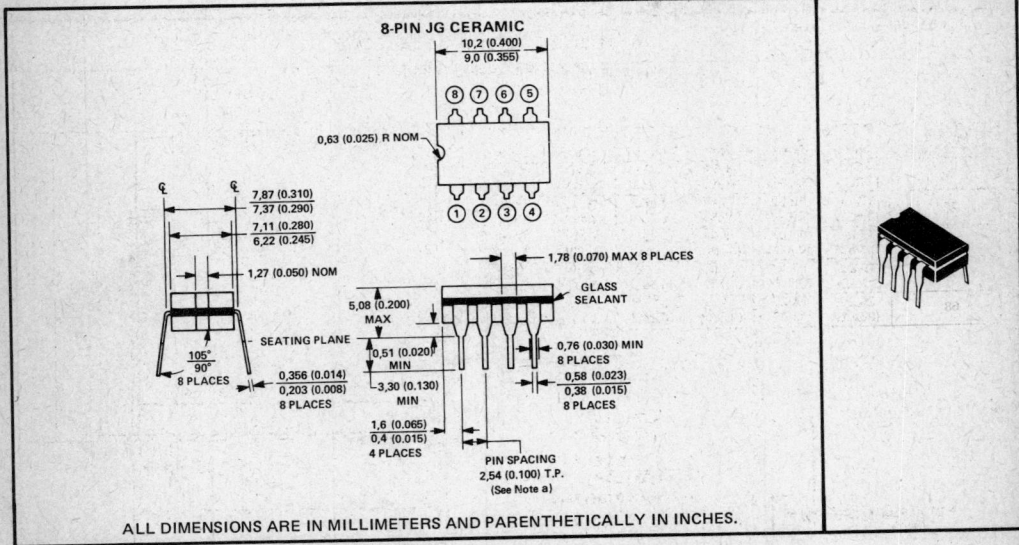

ALL DIMENSIONS ARE IN MILLIMETERS AND PARENTHETICALLY IN INCHES.

NOTE: a. Each pin centerline is located within 0,25 (0.010) of its true longitudinal position.

MECHANICAL DATA

J ceramic packages (including JT and JW dual-in-line and JQ quad-in-line packages)

Each of these hermetically sealed dual-in-line packages consists of a ceramic base, ceramic cap, and a lead frame. Hermetic sealing is accomplished with glass. The JT packages are intended for insertion in mounting-hole rows on 7,62 (0.300) centers, JW packages for mounting-hole rows on 15,24 (0.600) centers, and the JQ quad-in-line package for mounting-hole rows on 15,24 (0.600) and 20,32 (0.800) centers. Once the leads are compressed and inserted sufficient tension is provided to secure the package in the board during soldering. Tin-plated ("bright-dipped") leads require no additional cleaning or processing when used in soldered assembly.

NOTE: For the 14-, 16-, and 20-pin packages, the letter J is used by itself since these packages are available only in the 7,62 (0.300) row spacing. For the 24-pin packages, if no second letter or row spacing is specified, the package is assumed to have 15,24 (0.600) row spacing.

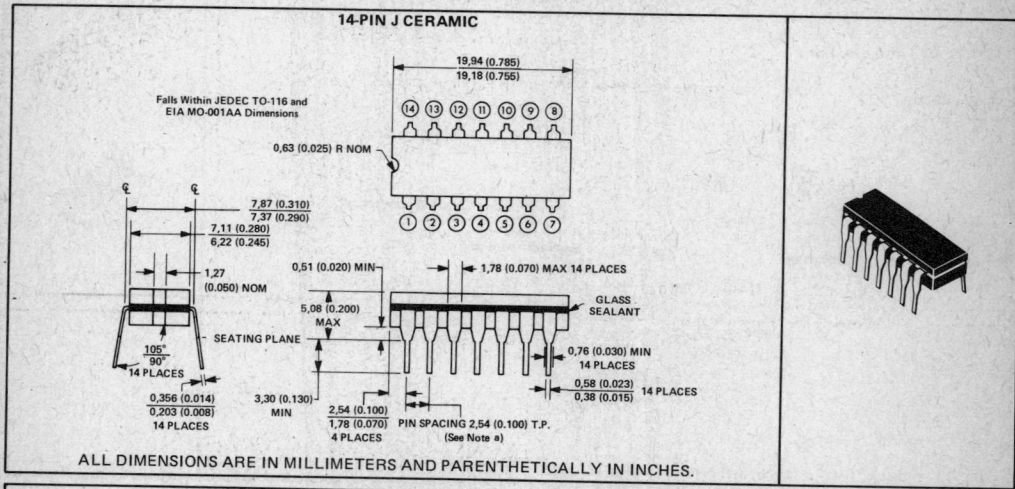

ALL DIMENSIONS ARE IN MILLIMETERS AND PARENTHETICALLY IN INCHES.

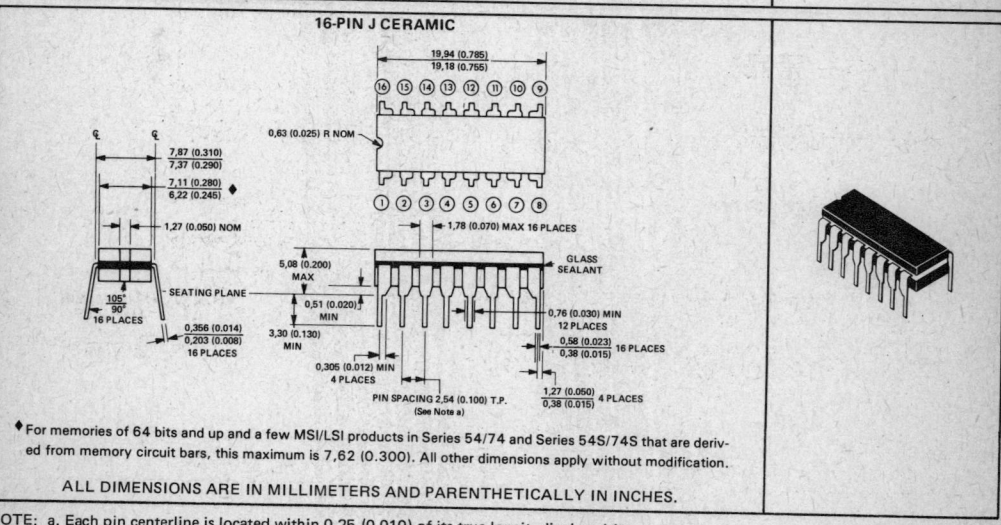

♦ For memories of 64 bits and up and a few MSI/LSI products in Series 54/74 and Series 54S/74S that are derived from memory circuit bars, this maximum is 7,62 (0.300). All other dimensions apply without modification.

ALL DIMENSIONS ARE IN MILLIMETERS AND PARENTHETICALLY IN INCHES.

NOTE: a. Each pin centerline is located within 0,25 (0.010) of its true longitudinal position.

3-8

TEXAS INSTRUMENTS
POST OFFICE BOX 225012 • DALLAS, TEXAS 75265

MECHANICAL DATA

J ceramic dual-in-line packages (continued)

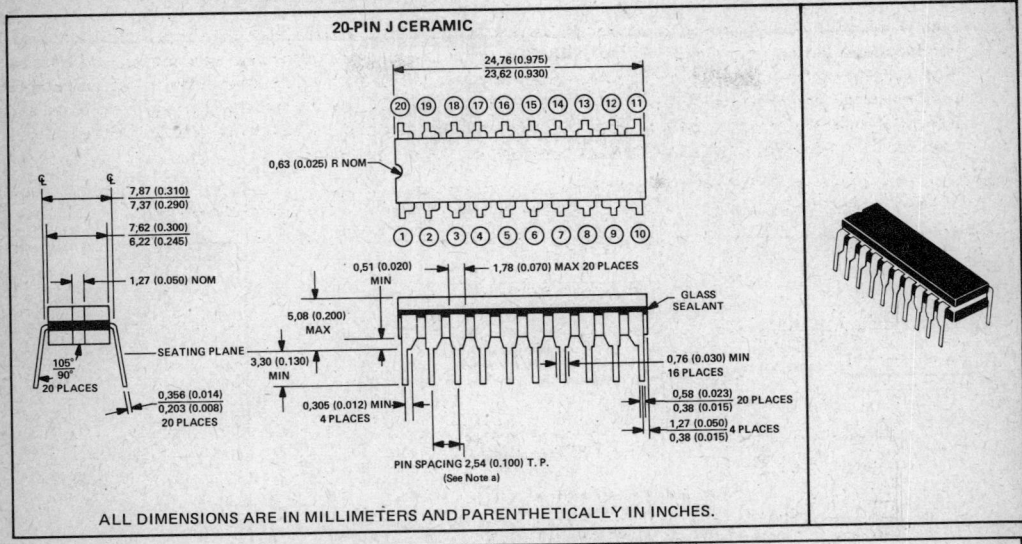

20-PIN J CERAMIC

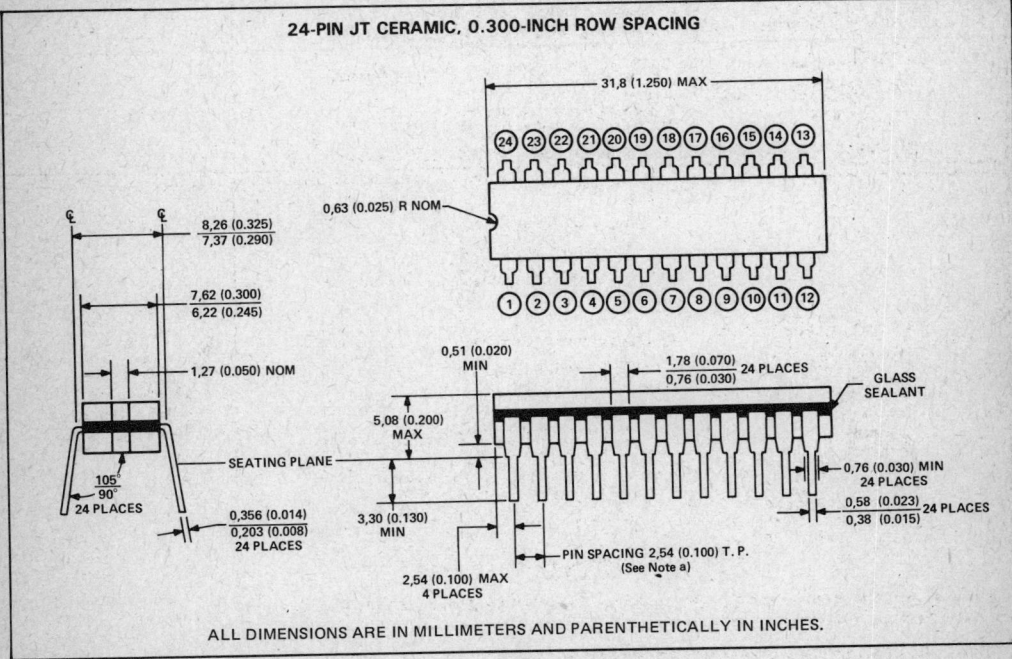

24-PIN JT CERAMIC, 0.300-INCH ROW SPACING

NOTE: a. Each pin centerline is located within 0,25 (0.010) of its true longitudinal position.

MECHANICAL DATA

J ceramic dual-in-line packages (continued)

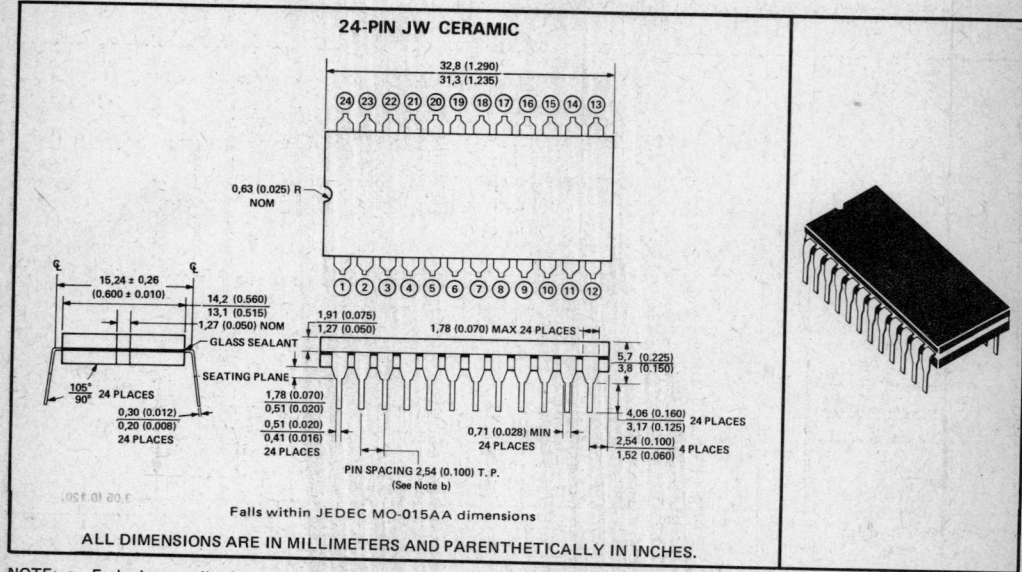

24-PIN JW CERAMIC

Falls within JEDEC MO-015AA dimensions

ALL DIMENSIONS ARE IN MILLIMETERS AND PARENTHETICALLY IN INCHES.

NOTE: a. Each pin centerline is located within 0,25 (0.010) of its true longitudinal position.

MECHANICAL DATA

ceramic packages – side-braze (JD suffix)

This is a hermetically sealed ceramic package with a metal cap and side-brazed tin-plated leads.

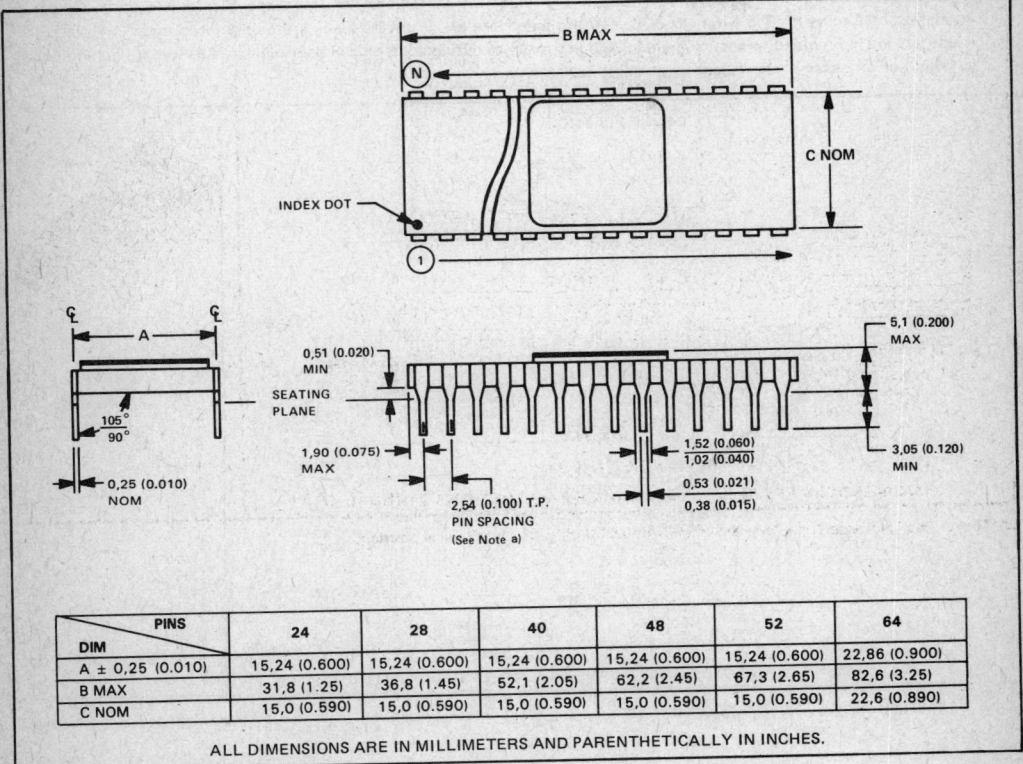

DIM \ PINS	24	28	40	48	52	64
A ± 0,25 (0.010)	15,24 (0.600)	15,24 (0.600)	15,24 (0.600)	15,24 (0.600)	15,24 (0.600)	22,86 (0.900)
B MAX	31,8 (1.25)	36,8 (1.45)	52,1 (2.05)	62,2 (2.45)	67,3 (2.65)	82,6 (3.25)
C NOM	15,0 (0.590)	15,0 (0.590)	15,0 (0.590)	15,0 (0.590)	15,0 (0.590)	22,6 (0.890)

ALL DIMENSIONS ARE IN MILLIMETERS AND PARENTHETICALLY IN INCHES.

NOTE: a. Each pin centerline is located within 0,25 (0.010) of its true longitudinal position.

Texas Instruments
POST OFFICE BOX 225012 • DALLAS, TEXAS 75265

MECHANICAL DATA

P plastic dual-in-line package

This dual-in-line package consists of a circuit mounted on an 8-lead frame and encapsulated within a plastic compound. The compound will withstand soldering temperature with no deformation and circuit performance characteristics remain stable when operated in high-humidity conditions. The package is intended for insertion in mounting-hole rows on 7,62-mm (0.300) centers (see Note a). Once the leads are compressed and inserted, sufficient tension is provided to secure the package in the board during soldering. Solder-plated leads require no additional cleaning or processing when used in soldered assembly.

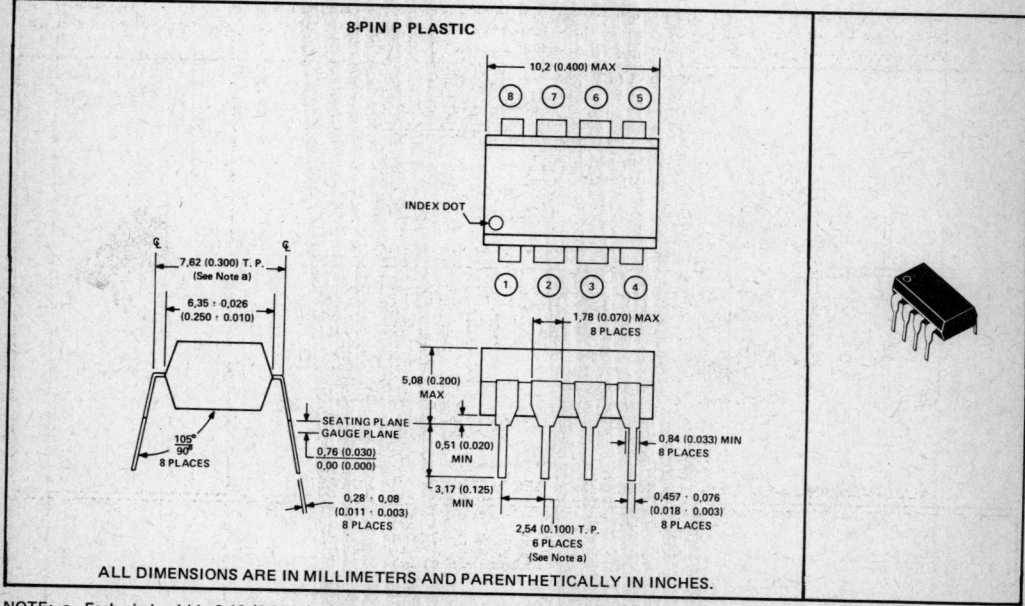

ALL DIMENSIONS ARE IN MILLIMETERS AND PARENTHETICALLY IN INCHES.

NOTE: a. Each pin is within 0,13 (0.0005) radius of true position (TP) at the gauge plane with maximum material condition and unit installed.

MECHANICAL DATA

N plastic packages (including NT and NW dual-in-packages)

Each of these dual-in-line packages consists of a circuit mounted on a lead frame and encapsulated within an electrically nonconductive plastic compound. The compound will withstand soldering temperature with no deformation and circuit performance characteristics remain stable when operated in high-humidity conditions. The packages are intended for insertion in mounting-hole rows on 7,62 (0.300) centers for the NT packages and on 15,24 (0.600) centers for the NW packages. Once the leads are compressed and inserted, sufficient tension is provided to secure the package in the board during soldering. Leads require no additional cleaning or processing when used in soldered assembly.

NOTE: For the 14-, 16-, 20-, and 28-pin packages, the letter N is used by itself since these packages are available in only one row-spacing width — 7,62 (0.300) for the 14-, 16-, 18-, and 20-pin packages and 15,24 (0.600) for the 28-pin package. For the 24-pin package, if no second letter or row spacing is specified, the package is assumed to have 15,24 (0.600) row spacing.

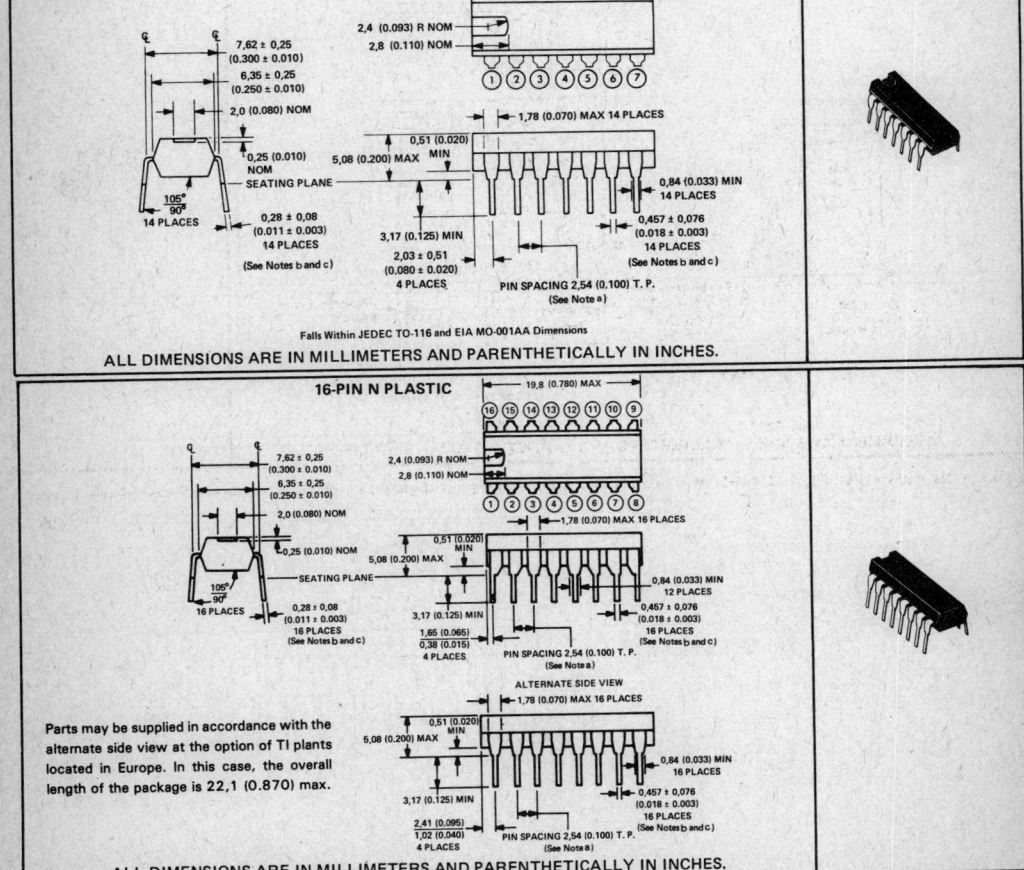

Falls Within JEDEC TO-116 and EIA MO-001AA Dimensions

ALL DIMENSIONS ARE IN MILLIMETERS AND PARENTHETICALLY IN INCHES.

Parts may be supplied in accordance with the alternate side view at the option of TI plants located in Europe. In this case, the overall length of the package is 22,1 (0.870) max.

ALL DIMENSIONS ARE IN MILLIMETERS AND PARENTHETICALLY IN INCHES.

NOTES: a. Each pin centerline is located within 0,25 (0.010) of its true longitudinal position.
b. This dimension does not apply for solder-dipped leads.
c. When solder-dipped leads are specified, dipped area of the lead extends from the lead tip to at least 0,51 (0.020) above seating plane.

Texas Instruments
POST OFFICE BOX 225012 • DALLAS, TEXAS 75265

MECHANICAL DATA

N plastic dual-in-line packages (continued)

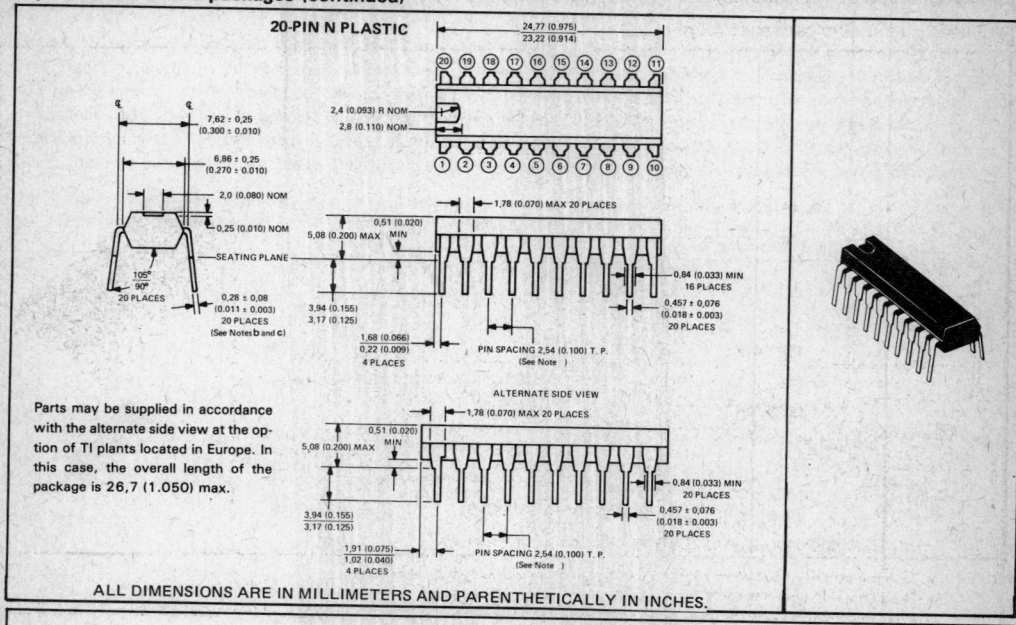

20-PIN N PLASTIC

Parts may be supplied in accordance with the alternate side view at the option of TI plants located in Europe. In this case, the overall length of the package is 26,7 (1.050) max.

ALL DIMENSIONS ARE IN MILLIMETERS AND PARENTHETICALLY IN INCHES.

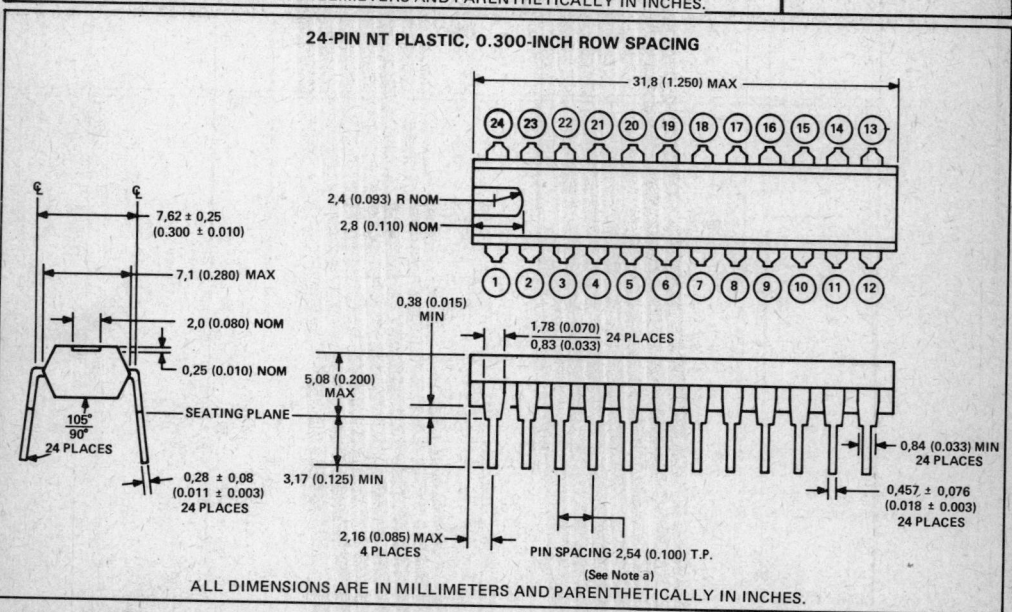

24-PIN NT PLASTIC, 0.300-INCH ROW SPACING

ALL DIMENSIONS ARE IN MILLIMETERS AND PARENTHETICALLY IN INCHES.

NOTES: a. Each pin centerline is located within 0,25 (0.010) of its true longitudinal position.
       b. This dimension does not apply for solder-dipped leads.
       c. When solder-dipped leads are specified, dipped area of the lead extends from the lead tip to at least 0,51 (0.020) above seating plane.

TEXAS INSTRUMENTS
POST OFFICE BOX 225012 • DALLAS, TEXAS 75265

MECHANICAL DATA

N plastic dual-in-line packages (continued)

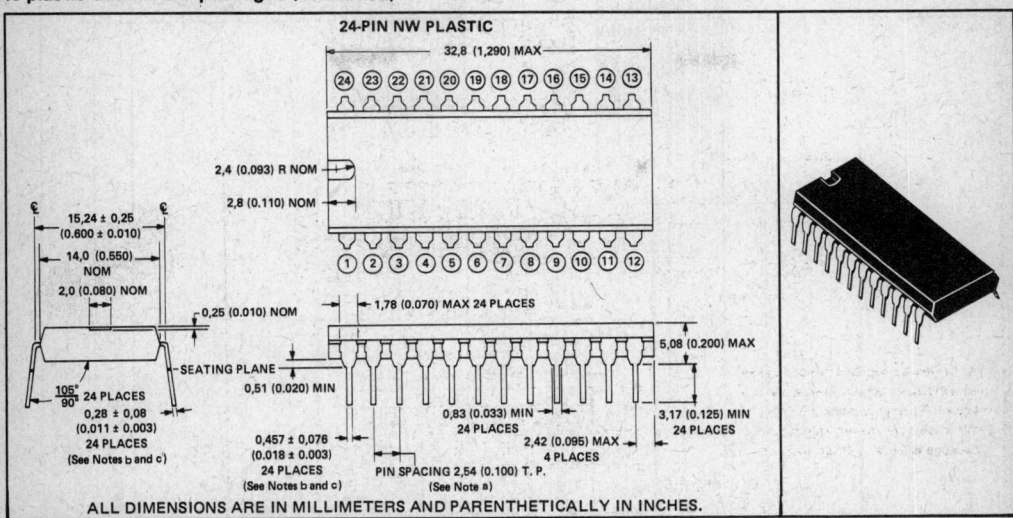

NOTES: a. Each pin centerline is located within 0,25 (0.010) of its true longitudinal position.
b. This dimension does not apply for solder-dipped leads.
c. When solder-dipped leads are specified, dipped area of the lead extends from the lead tip to at least 0,51 (0.020) above seating plane.

MECHANICAL DATA

N plastic packages (continued)

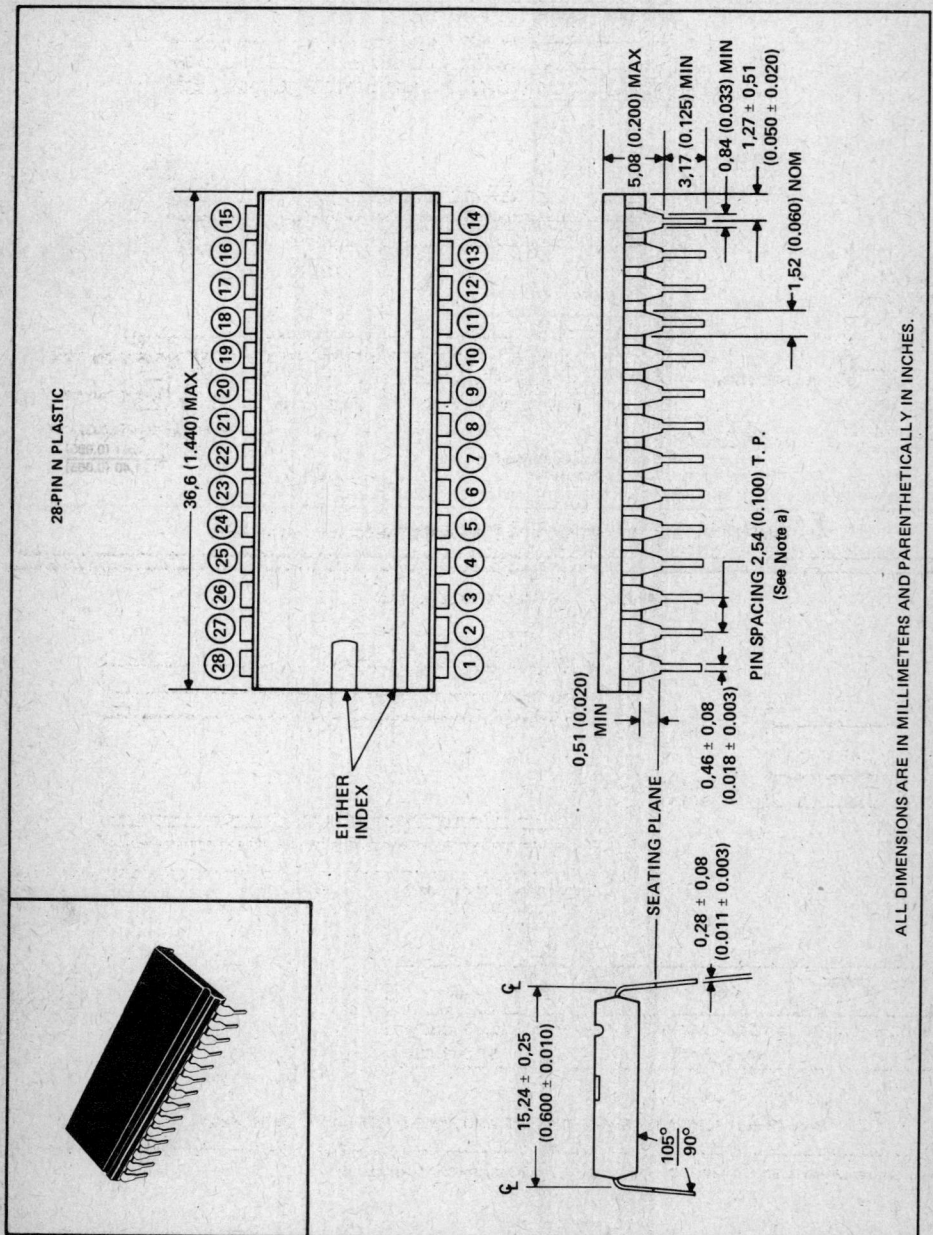

ALL DIMENSIONS ARE IN MILLIMETERS AND PARENTHETICALLY IN INCHES.

NOTE: a. Each pin centerline is located within 0,25 (0.010) of its true longitudinal position.

MECHANICAL DATA

N plastic packages (continued)

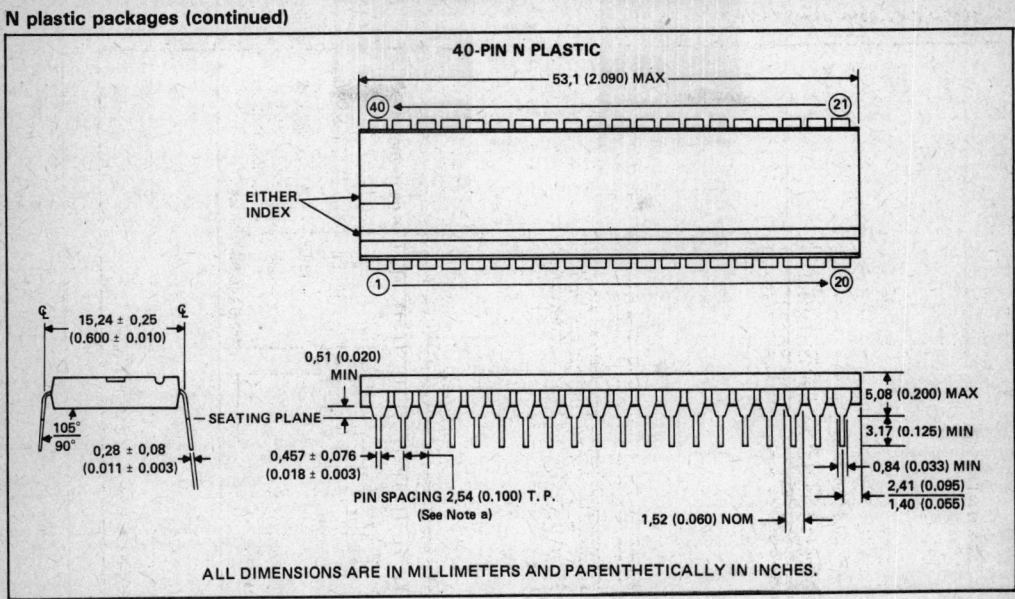

40-PIN N PLASTIC

ALL DIMENSIONS ARE IN MILLIMETERS AND PARENTHETICALLY IN INCHES.

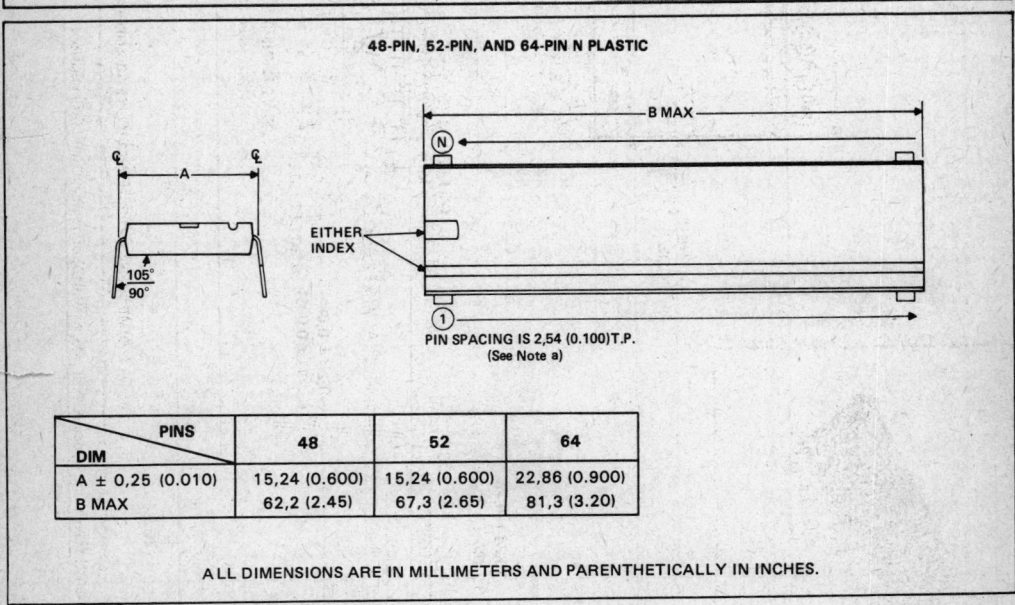

48-PIN, 52-PIN, AND 64-PIN N PLASTIC

PINS DIM	48	52	64
A ± 0,25 (0.010)	15,24 (0.600)	15,24 (0.600)	22,86 (0.900)
B MAX	62,2 (2.45)	67,3 (2.65)	81,3 (3.20)

ALL DIMENSIONS ARE IN MILLIMETERS AND PARENTHETICALLY IN INCHES.

NOTE: a. Each pin centerline is located within 0,25 (0.010) of its true longitudinal position.

Texas Instruments
POST OFFICE BOX 225012 • DALLAS, TEXAS 75265

TI Worldwide Sales Offices

ALABAMA: Huntsville, 500 Wynn Drive, Suite 514, Huntsville, AL 35805, (205) 837-7530.

ARIZONA: Phoenix, P.O. Box 35160, 8102 N. 23rd Ave., Suite A, Phoenix, AZ 85021, (602) 995-1007.

CALIFORNIA: El Segundo, 831 S. Douglas St., El Segundo, CA 90245, (213) 973-2571; Irvine, 17891 Cartwright Rd., Irvine, CA 92714, (714) 660-1200; Sacramento, 1900 Point West Way, Suite 171, Sacramento, CA 95815, (916) 929-1521; San Diego, 4333 View Ridge Ave., Suite B, San Diego, CA 92123, (714) 278-9600; Santa Clara, 5353 Betsy Ross Dr., Santa Clara, CA 95054, (408) 980-9000; Woodland Hills, 21220 Erwin St., Woodland Hills, CA 91367, (213) 704-7759.

COLORADO: Denver, 9725 E. Hampden St., Suite 301, Denver, CO 80231, (303) 695-2800.

CONNECTICUT: Wallingford, 9 Barnes Industrial Park Rd., Barnes Industrial Park, Wallingford, CT 06492, (203) 269-0074.

FLORIDA: Ft. Lauderdale, 2765 N.W. 62nd St., Ft. Lauderdale, FL 33309, (305) 973-8502; Maitland, 2601 Maitland Center Parkway, Maitland, FL 32751, (305) 646-9600; Tampa, 5010 W. Kennedy Blvd., Suite 101, Tampa, FL 33609, (813) 870-6420.

GEORGIA: Atlanta, 3300 Northeast Expy., Building 9, Atlanta, GA 30341, (404) 452-4600.

ILLINOIS: Arlington Heights, 515 W. Algonquin, Arlington Heights, IL 60005, (312) 640-2934.

INDIANA: Ft. Wayne, 2020 Inwood Dr., Ft. Wayne, IN 46815, (219) 424-5174; Indianapolis, 2346 S. Lynhurst, Suite J-400, Indianapolis, IN 46241, (317) 248-8555.

IOWA: Cedar Rapids, 373 Collins Rd. NE, Suite 200, Cedar Rapids, IA 52402, (319) 395-9550.

MARYLAND: Baltimore, 1 Rutherford Pl., 7133 Rutherford Rd., Baltimore, MD 21207, (301) 944-8600.

MASSACHUSETTS: Waltham, 504 Totten Pond Rd., Waltham, MA 02154, (617) 895-9100.

MICHIGAN: Farmington Hills, 33737 W. 12 Mile Rd., Farmington Hills, MI 48018, (313) 553-1500.

MINNESOTA: Edina, 7625 Parklawn, Edina, MN 55435, (612) 830-1600.

MISSOURI: Kansas City, 8080 Ward Pkwy., Kansas City, MO 64114, (816) 523-2500; St. Louis, 11861 Westline Industrial Drive, St. Louis, MO 63141, (314) 569-7600.

NEW JERSEY: Clark, 292 Terminal Ave. West, Clark, NJ 07066, (201) 574-9800.

NEW MEXICO: Albuquerque, 5907 Alice NSE, Suite E, Albuquerque, NM 87110, (505) 265-8491.

NEW YORK: East Syracuse, 6700 Old Collamer Rd., East Syracuse, NY 13057, (315) 463-9291; Endicott, 112 Nanticoke Ave., P.O. Box 618, Endicott, NY 13760, (607) 754-3900; Melville, 1 Huntington Quadrangle, Suite 3C10, P.O. Box 2936, Melville, NY 11747, (516) 454-6600; Poughkeepsie, 385 South Rd., Poughkeepsie, NY 12601, (914) 473-2900; Rochester, 1210 Jefferson Rd., Rochester, NY 14623, (716) 424-5400.

NORTH CAROLINA: Charlotte, 8 Woodlawn Green, Woodlawn Rd., Charlotte, NC 28210, (704) 527-0930; Raleigh, 2809 Highwoods Blvd., Suite 100, Raleigh, NC 27625, (919) 876-2725.

OHIO: Beachwood, 23408 Commerce Park Rd., Beachwood, OH 44122, (216) 464-6100; Dayton, Kingsley Bldg., 4124 Linden Ave., Dayton, OH 45432, (513) 258-3877.

OKLAHOMA: Tulsa, 7615 East 63rd Place, 3 Memorial Place, Tulsa, OK 74133, (918) 250-0633.

OREGON: Beaverton, 6700 SW 105th St., Suite 110, Beaverton, OR 97005, (503) 643-6758.

PENNSYLVANIA: Ft. Washington, 260 New York Dr., Ft. Washington, PA 19034, (215) 643-6450; Coraopolis, 420 Rouser Rd., 3 Airport Office Park, Coraopolis, PA 15108, (412) 771-8550.

TEXAS: Austin, 12501 Research Blvd., P.O. Box 2909, Austin, TX 78723, (512) 250-7655; Dallas, 1001 E. Campbell Rd., Richardson, TX 75080, (214) 680-5082; Houston, 9100 Southwest Frwy., Suite 237, Houston, TX 77036, (713) 778-6592; San Antonio, 1000 Central Parkway South, San Antonio, TX 78232, (512) 496-1779.

UTAH: Murray, 5201 South Green SE, Suite 200, Murray, UT 84107, (801) 266-8972.

VIRGINIA: Fairfax, 3001 Prosperity, Fairfax, VA 22031, (703) 849-1400.

WISCONSIN: Brookfield, 450 N. Sunny Slope, Suite 150, Brookfield, WI 53005, (414) 785-7140.

WASHINGTON: Redmond, 2723 152nd Ave., N.E. Bldg. 6, Redmond, WA 98052, (206) 881-3080.

CANADA: Ottawa, 436 McLaren St., Ottawa, Ontario, Canada, K2POM8, (613) 233-1177; Richmond Hill, 280 Centre St. E., Richmond Hill L4C1B1, Ontario, Canada, (416) 884-9181; St. Laurent, Ville St. Laurent Quebec, 9460 Trans Canada Hwy., St. Laurent, Quebec, Canada H4S1R7, (514) 334-3635.
E

ARGENTINA, Texas Instruments Argentina S.A.I.C.F.: Esmeralda 130, 15th Floor, 1035 Buenos Aires, Argentina, 394-2963.

AUSTRALIA (& NEW ZEALAND), Texas Instruments Australia Ltd.: 6-10 Talavera Rd., North Ryde (Sydney), New South Wales, Australia 2113, 02 + 887-1122; 5th Floor, 418 St. Kilda Road, Melbourne, Victoria, Australia 3004, 03 + 267-4677; 171 Philip Highway, Elizabeth, South Australia 5112, 08 + 255-2066.

AUSTRIA, Texas Instruments Ges.m.b.H.: Industriestrabe B/16, A-2345 Brunn/Gebirge, 2236-846210.

BELGIUM, Texas Instruments N.V. Belgium S.A.: Mercure Centre, Rakerstraat 100, Rue de la Fusee, 1130 Brussels, Belgium, 02/720.80.00.

BRAZIL, Texas Instruments Electronicos do Brasil Ltda.: Av. Faria Lima, 2003, 20 0 Andar—Pinheiros, Cep-01451 Sao Paulo, Brazil, 815-6166.

DENMARK, Texas Instruments A/S, Marielundvej 46E, DK-2730 Herlev, Denmark, 2 - 91 74 00.

FINLAND, Texas Instruments Finland OY: PL 56, 00510 Helsinki 51, Finland, (90) 7013133.

FRANCE, Texas Instruments France: Headquarters and Prod. Plant, BP 05, 06270 Villeneuve-Loubet, (93) 20-01-01; Paris Office, BP 67 8-10 Avenue Morane-Saulnier, 78141 Velizy-Villacoublay, (3) 946-97-12; Lyon Sales Office, L'Oree D'Ecully, Batiment B, Chemin de la Forestiere, 69130 Ecully, (7) 833-04-40; Strasbourg Sales Office, Le Sebastopol 3, Quai Kleber, 67055 Strasbourg Cedex, (88) 22-12-66; Rennes, 23-25 Rue du Puits Mauger, 35100 Rennes, (99) 79-54-81; Toulouse Sales Office, Le Peripole—2, Chemin du Pigeonnier de la Cepiere, 31100 Toulouse, (61) 44-18-19; Marseille Sales Office, Noilly Paradis—146 Rue Paradis, 13006 Marseille, (91) 37-25-30.

TEXAS INSTRUMENTS
Creating useful products and services for you

GERMANY, Texas Instruments Deutschland GmbH: Haggerty-strasse 1, D-8050 Freising, 08161-801; Kurfuerstendamm 195/196, D-1000 Berlin 15, 030-8827365; III, Hagen 43/Kibbelstrasse, D-4300 Essen, 0201-24250; Frankfurter Allee 6-8, D-6236 Eschborn 1, 06196-43074; Hamburger Strasse 11, D-2000 Hamburg 76, 040-2201154, Kirchhorsterstrasse 2, D-3000 Hannover 51, 0511-648021; Arabellastrasse 15, D-8000 Muenchen 81, 089-92341; Maybachstrasse 11, D-7302 Ostfildern 2/Nellingen, 0711-34030.

HONG KONG (+ PEOPLES REPUBLIC OF CHINA), Texas Instruments Asia Ltd.: 8th Floor, World Shipping Ctr., Harbour City, 7 Canton Rd., Kowloon, Hong Kong, 3 + 722-1223.

IRELAND, Texas Instruments (Ireland) Limited: 25 St. Stephens Green, Dublin 2, Eire, 01 609222.

ITALY, Texas Instruments Semiconduttori Italia Spa: Viale Delle Scienze, 1, 02015 Cittaducale (Rieti), Italy, 0746 694.1; Via Salaria KM 24 (Palazzo Cosma), Monterotondo Scalo (Rome), Italy, 06 9004395; Viale Europa, 38-44, 20093 Cologno Monzese (Milano), 02 2532541; Corso Svizzera, 185, 10100 Torino, Italy, 011 774545; Via J. Barozzi, 6, 45100 Bologna, Italy, 051 355851.

JAPAN, Texas Instruments Asia Ltd.: 4F Aoyama Fuji Bldg., 6-12, Kita Aoyama 3-Chome, Minato-ku, Tokyo, Japan 107, 03-498-2111; Osaka Branch, 5F, Nissho Iwai Bldg., 30 Imabashi 3-Chome, Higashi-ku, Osaka, Japan 541, 06-204-1881; Nagoya Branch, 7F Daini Toyota West Bldg., 10-27, Meieki 4-Chome, Nakamura-ku, Nagoya, Japan 450, 052-583-8691.

KOREA, Texas Instruments Supply Co.: Room 201, Kwangpoong Bldg., 24-1, Hwayand-Dong, Sung dong-ku, 133 Seoul, Korea, 02 + 464-6274/5.

MEXICO, Texas Instruments de Mexico S.A.: Poniente 116, No. 489, Colonia Vallejo, Mexico, D.F. 02300, 567-9200.

MIDDLE EAST, Texas Instruments: No. 13, 1st Floor Mannai Bldg., Diplomatic Area, Manama, P.O. Box 26335, Bahrain, Arabian Gulf, 973 - 72 46 81.

NETHERLANDS, Texas Instruments Holland B.V., P.O. Box 12995, (Bullewijk) 1100 AZ Amsterdam, Zuid-Oost, Holland (020) 5602911.

NORWAY, Texas Instruments Norway A/S: Kr. Augustsgt. 13, Oslo 1, Norway, (2) 20 60 40.

PHILIPPINES, Texas Instruments Asia Ltd.: 14th Floor, Ba-Lepanto Bldg., 8747 Paseo de Roxas, Makati, Metro Manila, Philippines, 882465.

PORTUGAL, Texas Instruments Equipamento Electronico (Portugal), Lda.: Rua Eng. Frederico Ulrich, 2650 Moreira Da Maia, 4470 Maia, Portugal, 2-9481003.

SINGAPORE (+ INDIA, INDONESIA, MALAYSIA, THAILAND), Texas Instruments Asia Ltd.: P.O. Box 138, Unit #02-08, Block 6, Kolam Ayer Industrial Est., Kallang Sector, Singapore 1334, Republic of Singapore, 747-2255.

SPAIN, Texas Instruments Espana, S.A.: C/Jose Lazaro Galdiano No. 6, Madrid 16, 1/458.14.58. C/Balmes, 89 Barcelona-8, 253 60 00/253 29 02.

SWEDEN, Texas Instruments International Trade Corporation (Sverigefilialen): Box 39103, 10054 Stockholm, Sweden, 08 - 235480.

SWITZERLAND, Texas Instruments, Inc. Riedstrasse 6, CH-8953 Dietikon (Zuerich) Switzerland, 1-740 2220.

TAIWAN, Texas Instruments Supply Co.: 10th Floor, Fu-Shing Bldg., 71 Sung-Kiang Road, Taipei, Taiwan, Republic of China, 02 + 521-9321.

UNITED KINGDOM, Texas Instruments Limited: Manton Lane, Bedford, MK41 7PA, England, 0234 67466; St. James House, Wellington Road North, Stockport, SK4 2RT, England, 061 442-8448.
BE

TI Sales Offices

ALABAMA: Huntsville, 500 Wynn Drive, Suite 514, Huntsville, AL 35805, (205) 837-7530.

ARIZONA: Phoenix, P.O. Box 35160, 8102 N. 23rd Ave., Suite A, Phoenix, AZ 85021, (602) 995-1007.

CALIFORNIA: El Segundo, 831 S. Douglas St., El Segundo, CA 90245, (213) 973-2571; **Irvine**, 17891 Cartwright Rd., Irvine, CA 92714, (714) 660-1200; **Sacramento**, 1900 Point West Way, Suite 171, Sacramento, CA 95815, (916) 929-1521; **San Diego**, 4333 View Ridge Ave., Suite B., San Diego, CA 92123, (714) 278-9600; **Santa Clara**, 5353 Betsy Ross Dr., Santa Clara, CA 95054, (408) 980-9000; **Woodland Hills**, 21220 Erwin St., Woodland Hills, CA 91367, (213) 704-7759.

COLORADO: Denver, 9725 E. Hampden St., Suite 301, Denver, CO 80231, (303) 695-2800.

CONNECTICUT: Wallingford, 9 Barnes Industrial Park Rd., Barnes Industrial Park, Wallingford, CT 06492, (203) 269-0074.

FLORIDA: Ft. Lauderdale, 2765 N.W. 62nd St., Ft. Lauderdale, FL 33309, (305) 973-8502; **Maitland**, 2601 Maitland Center Parkway, Maitland, FL 32751, (305) 646-9600, **Tampa**, 5010 W. Kennedy Blvd., Suite 101, Tampa, FL 33609, (813) 870-6420.

GEORGIA: Atlanta, 3300 Northeast Expwy., Building 9, Atlanta, GA 30341, (404) 452-4600.

ILLINOIS: Arlington Heights, 515 W. Algonquin, Arlington Heights, IL 60005, (312) 640-2934.

INDIANA: Ft. Wayne, 2020 Inwood Dr., Ft. Wayne, IN 46815, (219) 424-5174; **Indianapolis**, 2346 S. Lynhurst, Suite J-400, Indianapolis, IN 46241, (317) 248-8555.

IOWA: Cedar Rapids, 373 Collins Rd. NE, Suite 200, Cedar Rapids, IA 52402, (319) 395-9550.

MARYLAND: Baltimore, 1 Rutherford Pl., 7133 Rutherford Rd., Baltimore, MD 21207, (301) 944-8600.

MASSACHUSETTS: Waltham, 504 Totten Pond Rd., Waltham, MA 02154, (617) 895-9100.

MICHIGAN: Farmington Hills, 33737 W. 12 Mile Rd., Farmington Hills, MI 48018, (313) 553-1500

MINNESOTA: Edina, 7625 Parklawn, Edina, MN 55435, (612) 830-1600.

MISSOURI: Kansas City, 8080 Ward Pkwy., Kansas City, MO 64114, (816) 523-2500; **St. Louis**, 11861 Westline Industrial Drive, St. Louis, MO 63141, (314) 569-7600.

NEW JERSEY: Clark, 292 Terminal Ave. West, Clark, NJ 07066, (201) 574-9800.

NEW MEXICO: Albuquerque, 5907 Alice NSE, Suite E., Albuquerque, NM 87110, (505) 265-8491.

NEW YORK: East Syracuse, 6700 Old Coilamer Rd., East Syracuse, NY 13057, (315) 463-9291; **Endicott**, 112 Nanticoke Ave., P.O. Box 618, Endicott, NY 13760, (607) 754-3900; **Melville**, 1 Huntington Quadrangle, Suite 3C10, P.O. Box 2936, Melville, NY 11747, (516) 454-6600; **Poughkeepsie**, 385 South Rd., Poughkeepsie, NY 12601, (914) 473-2900; **Rochester**, 1210 Jefferson Rd., Rochester, NY 14623, (716) 424-5400.

NORTH CAROLINA: Charlotte, 8 Woodlawn Green, Woodlawn Rd., Charlotte, NC 28210, (704) 527-0930; **Raleigh**, 2809 Highwoods Blvd., Suite 100, Raleigh, NC 27625, (919) 876-2725.

OHIO: Beachwood, 23408 Commerce Park Rd., Beachwood, OH 44122, (216) 464-6100; **Dayton**, Kingsley Bldg., 4124 Linden Ave., Dayton, OH 45432, (513) 258-3877.

OKLAHOMA: Tulsa, 7615 East 63rd Place, 3 Memorial Place, Tulsa, OK 74133, (918) 250-0633.

OREGON: Beaverton, 6700 SW 105th St., Suite 110, Beaverton, OR 97005, (503) 643-6758.

PENNSYLVANIA: Ft. Washington, 260 New York Dr., Ft. Washington, PA 19034, (215) 643-6450; **Coraopolis**, 420 Rouser Rd., 3 Airport Office Park, Coraopolis, PA 15108, (412) 771-8550.

TEXAS: Austin, 12501 Research Blvd., P.O. Box 2909, Austin, TX 78723, (512) 250-7655; **Dallas**, 1001 E. Campbell Rd., Richardson, TX 75080, (214) 680-5082; **Houston**, 9100 Southwest Frwy., Suite 237, Houston, TX 77036, (713) 778-6592; **San Antonio**, 1000 Central Parkway South, San Antonio, TX 78232, (512) 496-1779.

UTAH: Murray, 5201 South Green SE, Suite 200, Murray, UT 84107, (801) 266-8972.

VIRGINIA: Fairfax, 3001 Prosperity, Fairfax, VA 22031, (703) 849-1400.

WISCONSIN: Brookfield, 450 N. Sunny Slope, Suite 150, Brookfield, WI 53005, (414) 785-7140.

WASHINGTON: Redmond, 2723 152nd Ave., N.E. Bldg. 6, Redmond, WA 98052, (206) 881-3080.

CANADA: Ottawa, 436 McLaren St., Ottawa, Ontario, Canada, K2P0M8,(613) 233-1177; **Richmond Hill**, 280 Centre St. E., Richmond Hill L4C1B1, Ontario, Canada, (416) 884-9181; St. Laurent, Ville St. Laurent Quebec, 9460 Trans Canada Hwy., St. Laurent, Quebec, Canada H4S1R7, (514) 334-3635.

E

TI Regional Technology Centers

CALIFORNIA: Northern, 5353 Betsy Ross Dr., Santa Clara, CA 95054, (408) 748-2220, Hotline: (408) 980-0305. **Southern**, 17891 Cartwright Rd., Irvine, CA 92714, (714) 660-8140, Hotline: (714) 660-8164.

GEORGIA: Atlanta, 3300 Northeast Expressway, Building 8, Atlanta, GA 30341, (404) 452-4682, Hotline: (404) 452-4686.

ILLINOIS: Chicago, 515 W. Algonquin Road, Arlington Heights, IL 60005, (312) 640-2909, Hotline: (312) 228-6008.

MASSACHUSETTS: Boston, 400-2 Totten Pond Road, Waltham, MA 02154, (617) 890-6671, Hotline: (617) 890-4271.

TEXAS: Dallas, 1001 E. Campbell Rd., Richardson, TX 75081, (214) 680-5066, Hotline: (214) 680-5096.

TI Distributors

ALABAMA: Arrow (205) 882-2730; Marshall (205) 881-9235.

ARIZONA: Phoenix, Arrow (602) 968-4800; Kierulff (602) 243-4101; Marshall (602) 968-6181; Wyle (602) 249-2232; **Tucson**, Kierulff (602) 624-9986.

CALIFORNIA: Los Angeles/Orange County, Arrow (213) 701-7500, (714) 838-5422; Kierulff (213) 725-0325, (714) 731-5711; Marshall (213) 999-5001, (213) 442-7204, (714) 556-6400; R.V. Weatherford (714) 634-9600, (213) 849-3451, (714) 623-1261; Wyle (213) 322-8100, (714) 863-9953; **Sacramento**, Arrow (916) 925-7456; Wyle (916) 638-5282; **San Diego**, Arrow (619) 565-4800; Kierulff (619) 278-2112; Marshall (619) 578-9600; R. V. Weatherford (619) 695-1700; Wyle (619) 565-9171; **San Francisco Bay Area**, Arrow (408) 745-6600; Kierulff (415) 968-6292; Marshall (408) 732-1100; Wyle (408) 727-2500; **Santa Barbara**, R. V. Weatherford (805) 965-8551.

COLORADO: Arrow (303) 696-1111; Kierulff (303) 790-4444; Wyle (303) 457-9953.

CONNECTICUT: Arrow (203) 265-7741; Diplomat (203) 797-9674; Kierulff (203) 265-1115; Marshall (203) 265-3822; Milgray (203) 795-0714.

FLORIDA: Ft. Lauderdale, Arrow (305) 776-7790; Diplomat (305) 974-9700; Kierulff (305) 652-6950; **Orlando**, Arrow (305) 725-1480; Milgray (305) 647-5747; **Tampa**, Diplomat (813) 443-4514; Kierulff (813) 576-1966.

GEORGIA: Arrow (404) 449-8252; Kierulff **(404) 447-5252**; Marshall (404) 923-5750.

TEXAS INSTRUMENTS

Creating useful products and services for you.

ILLINOIS: Arrow (312) 397-3440; Diplomat (312) 595-1000; Hall-Mark (312) 860-3800; Kierulff (312) 640-0200; Newark (312) 638-4411.

INDIANA: Indianapolis, Arrow (317) 243-9353; Graham (317) 634-8202; Ft. Wayne, Graham (219) 423-3422.

IOWA: Arrow (319) 395-7230.

KANSAS: Kansas City, Hall-Mark (913) 888-4747; **Wichita**, LCOMP (316) 265-9507.

MARYLAND: Arrow (301) 247-5200; Diplomat (301) 995-1226; Kierulff (301) 247-5020; Milgray (301) 468-6400.

MASSACHUSETTS: Arrow (617) 933-8130; Diplomat (617) 935-6611; Kierulff (617) 667-8331; Marshall (617) 272-8200; Time (617) 935-8080.

MICHIGAN: Detroit, Arrow (313) 971-8200; Marshall (313) 525-5850; Newark (313) 967-0600; **Grand Rapids**, Arrow (616) 243-0912.

MINNESOTA: Arrow (612) 830-1800; Hall-Mark (612) 854-3223; Kierulff (612) 941-7500.

MISSOURI: Kansas City, LCOMP (816) 221-2400; **St. Louis**, Arrow (314) 567-6888; Hall-Mark (314) 291-5350; Kierulff (314) 739-0855.

NEW HAMPSHIRE: Arrow (603) 668-6968.

NEW JERSEY: Arrow (201) 575-5300, (609) 596-8000; Diplomat (201) 785-1830; General Radio (609) 964-8560; Kierulff (201) 575-6750; Marshall (201) 882-0320; Milgray (609) 983-5010.

NEW MEXICO: Arrow (505) 243-4566; International Electronics (505) 345-8127.

NEW YORK: Long Island, Arrow (516) 231-1000; Diplomat (516) 454-6334; JACO (516) 273-5500; Marshall (516) 273-2424; Milgray (516) 420-9800; **Rochester**, Arrow (716) 275-0300; Marshall (716) 235-7620; Rochester Radio Supply (716) 454-7800; **Syracuse**, Arrow (315) 652-1000; Diplomat (315) 652-5000; Marshall (607) 754-1570.

NORTH CAROLINA: Arrow (919) 876-3132, (919) 725-8711; Kierulff (919) 872-8410.

OHIO: Cincinnati, Graham (513) 772-1661; Hall-Mark (513) 563-5980; **Cleveland**, Arrow (216) 248-3990; Hall-Mark (216) 473-2907; Kierulff (216) 587-6558; **Columbus**, Hall-Mark (614) 891-4555, **Dayton**, Arrow (513) 435-5563; ESCO (513) 226-1133; Marshall (513) 236-8088.

OKLAHOMA: Arrow (918) 665-7700; Hall-Mark (918) 665-3200; Kierulff (918) 252-7537.

OREGON: Arrow (503) 684-1690; Kierulff (503) 641-9150; Wyle (503) 640-6000.

PENNSYLVANIA: Arrow (412) 856-7000, (215) 928-1800; General Radio (215) 922-7037.

TEXAS: Austin, Arrow (512) 835-4180; Hall-Mark (512) 258-8848; Kierulff (512) 835-2090; **Dallas**, Arrow (214) 386-7500; Hall-Mark (214) 341-1147; International Electronics (214) 233-9323; Kierulff (214) 343-2400; **El Paso**, International Electronics (915) 778-9761; **Houston**, Arrow (713) 530-4700; Hall-Mark (713) 781-6100; Harrison Equipment (713) 879-2600; Kierulff (713) 530-7030.

UTAH: Diplomat (801) 486-4134; Kierulff (801) 973-6913; Wyle (801) 974-9953.

VIRGINIA: Arrow (804) 282-0413.

WASHINGTON: Arrow (206) 643-4800; Kierulff (206) 575-4420; Wyle (206) 453-8300.

WISCONSIN: Arrow (414) 764-6600; Hall-Mark (414) 761-3000; Kierulff (414) 784-8160.

CANADA: Calgary, Future (403) 259-6408; Varah (403) 230-1235; **Hamilton**, Varah (416) 561-9311; **Montreal**, CESCO (514) 735-5511; Future (514) 694-7710; **Ottawa**, CESCO (613) 226-6905; Future (613) 820-8313; ITT Components (613) 226-7406; **Quebec City**, CESCO (418) 687-4231; ITT Components (514) 735-1177; **Toronto**, CESCO (416) 661-0220; Future (416) 663-5563; ITT Components (416) 630-7971; **Vancouver**, Future (604) 438-5545; Varah (604) 873-3211; ITT Components (604) 270-7805; **Winnipeg**, Varah (204) 633-6190.

BE